Bamboo for Sustainable Development

Bamboo for Sustainable Development

Proceedings of the Vth International Bamboo Congress and the VIth International Bamboo Workshop

San José, Costa Rica, 2–6 November 1998

INBAR Proceedings No. 7

Editors

Arun Kumar, I.V. Ramanuja Rao and Cherla Sastry

///VSP///

Utrecht • Boston • Köln • Tokyo, 2002

INBAR

VSP BV
P.O. Box 346
3700 AH Zeist
The Netherlands

Tel: +31 30 692 5790
Fax: +31 30 693 2081
vsppub@compuserve.com
www.vsppub.com

First published in 2002

ISBN 90-6764-357-2

Printed in The Netherlands by Ridderprint bv, Ridderkerk.

Foreword

Bamboo is a plant that occurs over much of the World. It is probable that billions of the World's population see, eat and touch bamboo every day. Bamboo contributes significantly to the income generation of many poor farmers. Yet it is an "orphan" crop – largely ignored by the main landholding agencies – forestry and agriculture. That is an accident of history. It is part of INBAR's mandate to raise awareness of the contribution bamboo can make to sustainable socio-economic development, so that it is not so ignored in the future.

The joint Vth International Bamboo Congress and the VIth International Bamboo Workshop was held from 2 to 6 November 1998 in San José, Costa Rica. The International Bamboo Association (IBA) joined hands with INBAR in organizing the event, while Funbambu of Costa Rica hosted. As befits such a joint effort, these events were held outside Asia for the first time.

More than 100 papers were presented at the gathering of over 400 scientists, artisans, development practitioners, businesspeople and bamboo enthusiasts. Several plenary and poster sessions were held on a host of scientific, technical and application aspects of bamboo.

This compilation in four parts—Bamboo Resources and Socio-economics, Bamboo Propagation and Management, Bamboo Engineering and Construction, and Bamboo Design and Utilization—includes most of the papers presented at the plenary sessions. The topics covered touch on varied aspects of bamboo: from bamboo resources of Mexico to bamboos of Ethiopia; from bamboo afforestation of a mined area to bamboo as a food and fiber alternative in an island; from bamboo drippers to bamboo wheelchairs; from teaching architecture with bamboo to an international building code for bamboo; from silviculture of Guadua bamboo to flowering of Moso bamboo…

This publication serves to highlight the usefulness of bamboo in aiding developing countries in their sustainable social, economic and environmental development initiatives.

Ian R. Hunter
Director General, INBAR

Contents

Section 4: Bamboo Design and Utilization

Section 1:

Resources and Socio-economics

Bamboo for Sustainable Development, pp. 3–15
Eds. A. Kumar, I.V. Ramanuja Rao and Ch. Sastry

The two bamboos of Ethiopia

WALTER LIESE
Chair for Wood Biology, University Hamburg, Hamburg, Germany

Abstract—Only two bamboo species exist in Ethiopia, one of the largest, densely populated and economically poor countries of Africa. The bamboos occur in two different ecological zones. The monopodial "highland" bamboo *Arundinaria alpina* grows in altitudes ranging from 2 200 to 3 500 m and the sympodial "lowland" bamboo *Oxytenanthera abyssinica* between 700 to 1 800 m. Both are of considerable socio-economic importance for the rural population. The use of the bamboo groves follows traditional customs. The culms are often the only construction material for houses and fences and have also to serve as valuable energy resource. *A. alpina* is processed for simple furniture, made either locally or in Addis Ababa from culms transported over a long distance. Bamboo, an important natural resource of Ethiopia, is endangered by poor management and negligence. A recent consultation, "Study on the bamboo management and utilization in Ethiopia", has elaborated the present situation with recommendations for the improvement of management and utilization of bamboo to benefit the farmers as well as the handicrafts business.

Bamboo can contribute considerably to the sustainable development of the rural areas in Ethiopia. Besides the two bamboo species, two reed species also have to be considered, since they are partly used as a bamboo substitute.

INTRODUCTION

The aim of this paper is to describe the only two bamboo species in Ethiopia, their importance to the rural people and the need for their further promotion. The presentation is based on a study on appropriate bamboo management by Luso Consult (1997) on behalf of the Deutsche Gesellschaft für Technische Zusammenarbeit (GTZ) GmbH, Eschborn, Germany. The team consisted of G. Simon, M. Admassu, E. Boa, G. Hadera, B. Kigomo, W. Liese, G. Melsbach, and W. van Rooij, for fulfilling the various objectives of the study.

Ethiopia has a size three times that of Germany and 56 million inhabitants. The annual consumption of wood for fuel and construction is 1.2 m^3 per capita, amounting to a total of about 67 million m^3. This high demand is mainly met by the about 1.4 million ha of Eucalyptus plantations, with an average yield of 15 m^3/ha, and other plantings. The green belt around the capital of about 45 000 ha Eucalypts produces annually 22 m^3/ha, amounting to about 1 million m^3. Regionally, bamboo plays an important role as a supplement or alternative material. However, in the whole of Ethiopia only two bamboo species occur under specific environmental conditions – the highland bamboo *Arundinaria*

Figure 1. *Arundinaria alpina* at 3 050 m altitude near Hosaina

alpina and the lowland bamboo *Oxytenanthera abyssinica.* This is quite in contrast to neighboring countries, like Kenya, where 14 bamboo species have been planted, partly on an experimental basis (Kigomo 1988; Kigomo and Sigu 1995). Ethiopia has a large bamboo potential, which is hardly recognized by the authorities because bamboo is often considered more as a weed than a valuable forest resource.

The purpose of this study was to evaluate the contribution of a sustainable use of bamboo in selected highland and lowland areas to increase the awareness of bamboo as a valuable natural resource and as a tool for enhancing income, especially for the rural population. Emphasis was also given to the socio-economic factors using a people-bamboo systems approach (not described here). The following presentation concentrates on the characteristics of the two bamboo species and their utilization.

ARUNDINARIA ALPINA

Characteristics

The African alpine bamboo *Arundinaria alpina* (K. Schum.) grows in fertile sites in the highlands of East Africa at an elevation of 2 200 to 3 500 m at temperatures of 10–20°C. In Ethiopia, *A. alpina* exists in the higher and more humid parts of the North-West and South-East highlands which have a rainfall of 1 500 to 2 500 mm and a short dry season. It grows gregariously, not in clumps and has a stout rhizome. It exists in pure stands, and in patches around farmsteads where it is fully integrated into the farmers' economy. *A. alpina* is a monopodial, medium-sized bamboo with straight culms of 12–20 m height and a base diameter of 8–12 (20) cm (Fig. 1). The culms are hollow, with an average wall thickness of about 16 mm at the base and 5 mm at the top. The lanceolate leaves of about 7 cm length do not decompose easily in the cool climate and often form a thick litter cover of 10 cm or more. Occasional flowering, seeding and dying occurs in patches; no extensive flowering has been recorded. (Wimbush 1945; Kigomo and Kamiri 1985).

The mapping work by the study team, mainly from aerial photographs with field verification, showed a total area of highland bamboo of about 130 000 ha. A mature stand has about 6 000 culms/ha, with an average biomass of 51 tons (air-dry). It is noteworthy that owing to neglect in management there are many dead culms left in the stands and a significant number of culms (about 10%) show beetle holes and broken tops. The annual increment of about 1 000 new culms results in a yield of 8.6 tons/ha, which is in the range of other bamboo species. The total annual yield of the *A. alpina* stands would amount to about 160 000 tons. The oven-dry density of the culm wall is about 0.48 g/cm^3. Propagation of the bamboo is done by the villagers, to a certain extent, using one-year-old culm section (cut off at the 5th internode) along with a portion of the rhizome. The survival rate is about 50%.

Besides the typical appearance of *A. alpina,* a few patches of yellow-striped bamboo have been observed which merits further investigation regarding a possible variety of the highland bamboo or another species.

Utilization

The villagers do not customarily eat the young shoots; but as monkeys like the shoots, simple fences are often used to protect the stands. The leaves, which amount to about 5% of the culm's fresh weight, are used as fodder for the cattle. The culms are often cut after one year, resulting in a stand with only medium-sized culms.

The culms are widely used by the rural population, mainly for construction: for house walls and roof, animal sheds and fences for unherded cattle. The typical igloo-shaped bamboo houses of the Ethiopian highlands are either made completely

Figure 2. A typical bamboo farmhouse

Figure 3. Simple furniture, made out of *A. alpina,* displayed at the roadside for sale

out of round and split bamboo culms or partly supported by wooden components (Fig. 2). *A. alpina* is easy to split for making mats, beehives and baskets. Simple furniture – such as chairs, sofas, tables and seats – are made by the people for their own use. They are also sold along the roadside to passers-by and dealers (Fig. 3). The material is often of low quality, as it is taken from young culms or culms discolored by poor storage.

Larger culms are transported to the capital Addis Ababa for further processing in small handicraft workshops.

OXYTENANTHERA ABYSSINICA

Characteristics

The lowland bamboo *Oxytenanthera abyssinica* (A. Rich) Munro grows in the western part of Ethiopia, towards the Savannah Woodlands of Sudan at an altitude

Figure 4. *Oxytenanthera abyssinica* at the lowland near Chagni

of 700 to 1 800 m. It is a hardy species that grows on poor soils in dry vegetation formations. As a bamboo most resistant to drought, it survives in conditions of low rainfall (down to 700 mm) and high temperatures (above 35°C). The sympodial bamboo grows in natural pure stands, covering large areas with considerably dispersed clumps. Their quality is rather low, since there is no management and the cutting is done arbitrarily. The culms are grouped in large dense clumps, erect or leaning, with a length of 6–10 m (Fig. 4). The base diameter varies from 3 to 8 cm. They are solid during shooting and may develop a small central cavity with a thick culm wall later. Hence, they are more difficult to split for weaving. The leaves are about 20 cm long and have deciduous foliage under unfavorable conditions. Scattered individual flowering clumps occur, but not on a large scale (Fig. 5). Propagation is not being done by the farmers, which might be due to the large bamboo area available, its unclear ownership and the low survival rate of the sympodial type in the hot-dry environment (Fanshawe 1972; Kigomo and Kamiri 1985).

O. abyssinica covers 700 000 to 850 000 ha in Ethiopia according to aerial photographs, field verification and further estimations. Kelecha (1980) assumed an area of 1 million ha. Frequent burning is often applied after harvest from December onwards, which leads to a higher mortality of standing culms, but also to a greater number of young shoots. About 8 000 living culms and 4 000 dead culms (caused by regular burning) were counted per ha. New culms, numbering up to 4 000, lead to an annual increment of about 10 tons/ha (oven-dry). The total biomass has been calculated at 19.6 tons/ha. The density of the culm wall amounts to 0.61 g/cm³ (oven-dry), which is more than for the highland bamboo.

Figure 5. *O. abyssinica* flowering (photo – W. van Rooij)

Figure 6. Cattle feeding from a bamboo clump

Some clumps are morphologically different with exceptionally thin culms, while a few others have much taller and thicker culms. The apparent differences could be a result of cross-fertilization during flowering, which leads to a variety of seedlings with growth characteristics different from the individual clumps, quite contrary to the monopodial *A. alpina* in which new culms arise from the expanding rhizome system. A very similar species with united flower-spikelets in dense terminal heads, mentioned briefly by v.-Breitenbach (1963) as *Oxytenanthera bortii* Mattei, warrants further investigation.

Utilization

Fresh bamboo leaves are a valuable fodder, particularly when grass becomes scare during winter, on which cattle can be nourished for a long time. The cattle grazing through bamboo stands during the rainy season eat bamboo leaves, with the result that the clumps become short and dome-shaped (Fig. 6). The young shoots of *O. abyssinica* are eaten by some ethnic groups. In the lowland, which has no Eucalyptus, bamboo culms are the main material for construction, fencing, beehives, mats, baskets, ropes and fuel. The solid culms are used mainly for construction and they are said to be more resistant to termites than the ones with a cavity. The products are made only for the farmers' own use, as there is no local market, transport or trade owing to the lack of tradition, simplicity of the products, longer distances between villages and low processing qualities of *O. abyssinica.*

Of special significance is the use for energy, mainly for cooking of the traditional staple food *injera*. On average, a household needs one bundle (about 20 kg)

Figure 7. Children carrying bundles of bamboo for use as firewood

of dry bamboo every two days for fuel. Whereas the hard work of cutting is normally done by men (and the culms then left standing for drying), the transportation of dried bamboo in bundles to the homes is done by women and children (Fig. 7). Such bundles are also sold on the market. With a yearly consumption of bamboo per household of about 3.6 tons and an average culm weight of 5 kg, about 700 culms (produced from about 0.25 ha) are used per family.

Bamboo does not make a good fuel; although its colorific value of 4.60 cal/g nearly equals the 4.77 cal/g of timber (Zhou 1981), it is not easily cut into pieces, it dries slowly and generates lot of smoke. A bamboo fire must be continuously fed, as it burns quickly. It cannot be converted into charcoal in the usual way, which will yield only light ashes.

BAMBOO PROCESSING AND MARKETING

Bamboo processing on a commercial scale is performed only in Addis Ababa, with culms of the highland bamboo *A. alpina* coming mostly from the Sidamo region, 500–600 km away. The quality of the culms is often low, as they are harvested too young and not sufficiently seasoned, with shrinkage of the products and discoloration caused by bad storage conditions. The production units are small family shops. Most of the workshop owners were trained by a group of Chinese at the Ethiopian Cottage Industries Development Agency about 20 years ago, with an additional input from some Japanese volunteers a few years ago. The processing is mostly manual with simple techniques; surface treatment, like var-

Figure 8. Bamboo furniture at a small shop in Addis Ababa

nish coating, is mostly not done. The products – such as furniture (chairs, sofas, tables, shelves), fruit and flower baskets, and lampshades (Fig. 8) – lack in variety. For example, furniture items produced by various processing shops are mostly similar, with very few exceptions depending on the creativity and workmanship. The tools employed are rather simple, often remakes of original Chinese ones (Fig. 9). Customers are mainly foreigners or wealthy Ethiopians, but not tourists because of export restrictions. The products of standard models and design are mainly made on order with a prepayment. They are also offered at the roadside to travellers. There is no organized marketing for bamboo products; they are not usually seen in department stores or at the huge Mercato market in Addis Ababa. The handicraft work by the Ethiopian Cottage Industries Development Agency shows a wider variety and better quality, but the items are sold only in a public showroom, not on the free market.

Bottlenecks for the production and marketing of bamboo products are:

- Lack of working capital;
- Low quality of culms, tools and techniques;
- Products of low quality and simple design;
- Little space for processing and display of bamboo products;

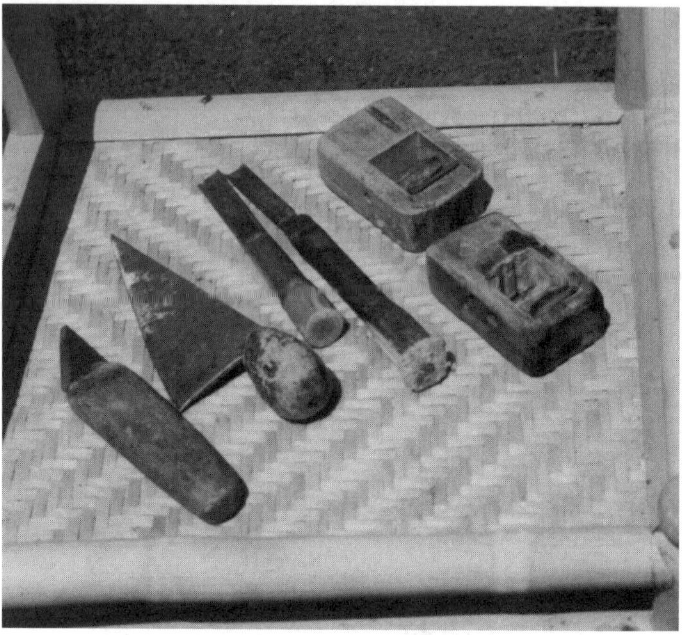

Figure 9. Simple tools used for bamboo processing

- Competition with wooden and plastic commodities; and
- Difficulties in obtaining sufficient culms because of the permit system.

Forest Authorities consider bamboo as part of the forest, and forest products are strictly regulated to control the depletion of the national forest. Transportation of culms for processing to Addis Ababa, even from private stands, requires a trading licence with time and money involved as well as royalties, and there are several checkpoints on the road.

REED "BAMBOO"

Besides the two bamboo species, reeds also should be considered, as they are often used as a bamboo substitute and their products are mixed with those of bamboo origin. Reeds are also the basis of subsistence for a number of families in Addis Ababa. Like the two bamboos, there are two reed species in Ethiopia. As grasses, both belong to the Gramineae family and are structurally similar to the bamboos.

Arundo donax L. grows in the mountains and highland under similar ecological conditions as *A. alpina*. Elsewhere it is known as the "Spanish Reed", native to Mediterranean countries and used there for "bamboo" products. It is a preferred material for reed instruments, like clarinets and oboes. The height of the plant is about 4–6 m and the basal thickness is up to 4 cm. In contrast to *A. alpina,* it has very long green culm sheaths and no branches. It grows widespread in the Addis Ababa region, where some ethnic groups, the Guraghe and the Welayta, make handcraft items using the reed. Young and fresh reeds can be seen as being processed along the roadsides for making chairs, beds, tables, book shelves, beehives, containers, baskets and mats (Figs. 10, 11). Their prices are much lower than those of bamboo articles. The quality is also lower and the service life shorter owing to the softer anatomical structure of the reeds compared with bamboo. These reed products may give the false impression of a wider use of bamboo handcrafting.

Phragmites communis Trin., the East Africa reed, grows in seasonal flooded lowland plains, mid-highlands and riverine areas. Its height is about 4 m and the diameter of the culm may be up to 15 cm. In the mid-highlands, where bamboo is not available, mats and fences are mainly produced from this reed. It is also used in local houses and restaurants as a decorative element.

Figure 10. *Arundo donax* and a mat product (Addis Ababa)

Figure 11. Baskets and furniture made of *A. donax*

The possible impact of reeds is shown in the report "Bamboo and Reeds in Ethiopia" for the Ethiopian Forestry Action Plan (Getahun 1992), where the necessary promotion of reeds is emphasized.

CONCLUSION

The knowledge about bamboo in Ethiopia and its utilization is meager. The bamboo stands are poorly managed or considered as "no man's resources". Bamboo is still available fairly free to the rural population. Nurseries for propagation do not exist. Only few institutions, such as the Forest Research Center and the Wood Utilization and Research Center (which is carrying out tests on lowland bamboo as concrete reinforcement), are interested but have limited working facilities. The Ethiopian Cottage Industries and Development Agency acts as a training center for small handicrafts, including bamboo processing. To promote the development of bamboo, its use and processing, "bamboo awareness" has to be created among the people. Relevant information has to be provided, leaflets prepared, regional cooperation strengthened and international contacts established.

The highland bamboo is widely distributed around homesteads and nearby villages in the respective ecological zones. The care of the stands is according to traditional customs and is governed by the need for utilization and cash. Bamboo cultivation needs to be improved. The lowland bamboo is not managed at all and used as a "no man's resource". Management Guidelines were proposed (Boa in Luso Consult 1997) for consideration by the regional authorities, to secure the most needed continuous supply of bamboo for construction and energy.

Attention should be given to the propagation of the two bamboos at other suitable sites and to the introduction of additional species that have shown promise in Kenya under equivalent ecological conditions. The aim should be to enrich the existing bamboo resource and to fill the gap at the sites and altitudes where so far no bamboo is growing. Millions of rural people could benefit from such an endeavor.

The marketing and utilization of bamboo products are not significant. They are often traditional in design and of low quality. The processing techniques and tools as well as the quality of the raw material require improvement. Bamboo handicrafts and their marketing channels need to be developed.

Several handicaps limit the wider use of bamboo products. These include severe restrictions by the authorities on material supply, lack of capital and restricted facilities for the producers, as well as poor presentation of bamboo products which is inadequate to demonstrate bamboo as a valuable natural, renewable product of Ethiopia.

ACKNOWLEDGEMENT

The author wishes to express his thanks to Luso Consult and GTZ for the permission to refer to the "Study on sustainable bamboo management".

REFERENCES

Fanshawe, D.B. 1972. The bamboo – *Oxytenanthera abyssinica:* its ecology, silviculture and utilization. Kirkia, 8, 157–166.

v. Breitenbach, F. 1963. The indigenous trees of Ethiopia, 2nd Ed. Ethiopian Forestry Association, Addis Ababa, Ethiopia.

Gertuhan, A. 1992. Bamboo and reeds in Ethiopia. Ethiopian Forest Action Plan, Addis Ababa, Ethiopia, 30 pp.

Kalecha, W.M. 1980. The bamboo potential of Ethiopia, Addis Ababa, Ethiopia, 16 pp.

Kigomo, B.N. 1988. Distribution, cultivation and research status of bamboo in East Africa. Kenya Forest Research Institute, Ecolog. Series. 117 pp.

Kigomo, B.N.; Kamiri, J.F. 1985. Observations on the growth and yield of *Oxytenanthera abyssinica* (A. Rich) Munro in plantation. East Africa Agri. For. Journal, 51, 22–29.

Kigomo, B.N.; Sigu, O.G. 1995. Establishment and growth of field trials of exotic and indigenous bamboo species in Kenya. East Africa Agri. For. Journal, 59, 120–129.

Luso Consult. 1997. Study on sustainable bamboo managment. Final Report on behalf of Deutsche Gesellschaft für Technische Zusammenarbeit (GTZ) GmbH, Eschborn, Hamburg, Germany. 85 pp.

Wimbush, S.H. 1945. The African alpine bamboo. Emp. For. Journal, 24, 33–39.

Zhou, F.C. 1981. Measurements of calorific values of 104 species of bamboo and timber. Journal of Nanjing Technical College of Forest Products, 54–69.

Bamboo for Sustainable Development, pp. 17–25
Eds. A. Kumar, I.V. Ramanuja Rao and Ch. Sastry

Indian bamboos in early 21st century

N.S. ADKOLI
Executive Director, Bamboo Society of India, Bangalore, India

Abstract—In India, bamboos occupy 8.96 million hectares of forests and exist as sporadic clumps in agricultural lands and homesteads. About 66% of the growing stock is concentrated in the Northeastern states of the country. Three species make up 78% of the growing stock: *Dendrocalamus strictus* (45%), *Melocanna bambusoides* (20%) and *Bambusa bambos* (13%). There are other species of commercial importance with localized distribution, and some short-sized bamboos in the temperate climate of high hills.

Utilization of bamboos is high in rural and urban areas, particularly by the pulp industry. The estimated annual harvest of 4.5 million tons from recorded sources works out to about 0.5 ton/ha/year. This is far too low in terms of productivity. Privileges granted in the use of forests, permits to village communities and artisans for removal of bamboos by head-load, etc. result in the removal of large quantities from accessible forests. Much of these types of extraction is unaccounted.

Nearly 20% of total forest area receives legal protection in the form of National Parks and Wildlife Sanctuaries. These areas are bound to help in bamboo conservation.

Existing unscientific management of bamboos and their utilization without preservative treatment have to change, if availability of bamboo to the ever-increasing population and usage has to be ensured.

Factors governing the sustenance of bamboo like progression, population growth, management, utilization pattern, etc. have to be analysed and addressed in order to ensure its safety in the 21st century.

This paper analyses the projected future of major bamboo species in India. Provision for progressive yields, germplasm banks, participatory management, research, extension and international cooperation are suggested for maintenance and expansion of bamboo biodiversity in India. International co-operation is suggested for conservation and multiplication of bamboo for centuries to come.

CURRENT STATUS OF BAMBOOS

Although bamboos exist in private land also, the extent of distribution is negligible when compared with their distribution. Sporadic clumps of bamboos are raised in cultivated lands of most eastern, Northeastern states, Goa, Kerala, coastal Maharashtra and Karnataka. Bamboos are widespread in the forests of all Northeastern states and in the deciduous forests of all other states in central, eastern and southern India. Occurrence of bamboo is rare in Rajasthan, Haryana, Punjab and Jammu & Kashmir.

Extent of distribution

According to a status survey of bamboos by the Forest Survey of India, 8.96 million ha of the forest area contains bamboo. This constitutes 11.71% of the 74.96 million ha of forest area of the country. About 66% of the growing stock of bamboos is in the seven Northeastern states. The states having major growing stocks of bamboos are: Assam (16%), Manipur and Mizoram (14% each), Arunachal Pradesh and Madhya Pradesh (12% each), Orissa (7%), Meghalaya (6%), and Nagaland and Maharashtra (5% each) (Chauhan and Rai 1996).

Distribution of species

Clump-forming bamboos make up 67.3% of the total growing stock, with *Dendrocalamus strictus* accounting for 45%, *Bambusa bambos* 13%, *D. hamiltonii* 7%, *B. tulda* 5%, *B. pallida* 4% and all other species together 6%. *Melocanna bambusoides,* a non-clumping bamboo, accounts for 20% of the growing stock and is found mostly in the Northeastern states. A few species of *Phyllostachys* are the other non-clumping bamboos found in India. *Dendrocalamus strictus* is the most widely distributed bamboo, occurring in about eleven states, while *B. bambos* is mostly distributed in moist deciduous forests. *Arundinaria* and *Chimonobambusa* are two genera found in high altitudes in the hills of Western Ghats and outer Himalayas. In total, 135 species are reported in India.

Harvests and yields

The total growing stock in the forest areas is assessed by the Forest Survey of India as 80.428 million tons. The average growing stock works out to be nearly 10 tons/ha. Being tall and larger in diameter, *Bambusa bambos* has higher growing stock per unit area and higher yield. There are moist valleys that contain over 100 clumps of this species per hectare, and the total green weight of each clump is in excess of one ton.

The estimated annual harvest from recorded sources is 4.5 million tons – just about 4.9% of the growing stock. There are extractions from unrecorded sources estimated at half the recorded sources. The current yield of about 0.5 ton/ha is too low and is attributed to improper management.

Bamboo utilization

Bamboo is utilized by farmers for fencing, housing, packaging, rafts, ladders, poles, carts, tools, fishing, fuel, etc. Many articles used in sericulture are made from bamboo. Bamboo shoots are used as food and leaves form fodder for all large herbivores. In woven form, it is used for basketry, roofs, partitions, mat boards, and art and craft items. The core of incense sticks (*agarbathi*) is made of bamboo. The largest single industrial use of bamboo is for pulping – 1.9 million tons annually. Thus, bamboos help provide livelihood and income to millions of villagers and artisans.

Protection

Collection of forest bamboo by head-loads for weaving is a privilege permitted to artisans in most Indian states. Extraction practices are mostly unscientific and this, coupled with the privilege of free and uncontrolled grazing for cattle, hampers the growth of bamboo in forests. Although forest management plans have prescribed rules for bamboo felling, even bulk users such as the pulp mills do not strictly observe these rules. Residues strewn in the forest after bamboo felling often form a cause of fire. Ground fires, which are common in tropical forests, cause damage to bamboo seeds and its regeneration. The lack of proper tending – cleaning of clumps, spacing of culms, soil working and moisture conservation measures – is the main reason for the poor health of Indian bamboos, resulting in poor growth and yield.

Bamboos receive protection only in the National Parks and the core zone of wildlife sanctuaries. Out of the 70 National Parks covering 34 700 km^2, 31 covering about 13 400 km^2 contain bamboos. Bamboos in such National Parks have adequate protection, although some damage by fire and wildlife is inevitable. There are 417 wildlife sanctuaries in India covering about 109 500 km^2 (ICFRE 1995) and, approximately, half of this area is bamboo forests. The core areas of these sanctuaries get legal protection. The total protected area is about 4% of the geographical area and about 20% of the forested area. These protected areas will be of immense value for future conservation of bamboos in terms of biodiversity.

FACTORS GOVERNING THE FUTURE OF BAMBOOS

Progression

As no drastic changes are foreseen in the climatic and edaphic conditions in the bamboo forests of India, adverse conditions for bamboo's future sustenance can be ruled out. Most luxuriant clump-forming bamboos of India are in the deciduous forest zones. Any seral succession to the semi-evergreen condition of these forests may affect the living conditions of bamboos. However, factors such as pressure of population, grazing, ground fires, exercise of privileges and level of bamboo extraction, are bound to check the seral succession and thereby maintain the deciduous condition of bamboo forests. Even in protected areas, which are a small proportion of the total bamboo areas, there are some activities like eco-tourism and maintenance management, and the progression is slow and marginal. Hence, the deciduous forests will continue their current ecological status and harbor bamboos in future.

Population

Growth of both human and cattle populations is the most critical factor affecting the future of bamboos in India. The human population of India, which was 238.40

million in 1901, rose to around 347 million by 1947 when the geographical area reduced owing to separation of areas that are now in Pakistan and Bangladesh. From 361 million in 1961, the population has now reached the staggering figure of 990 million and is showing an annual increase of 2%. The cattle population of India has gone up from 292 million in 1951 to about 450 million in 1995 (ICFRE 1995).

Industrial utilization of bamboo has stabilized at around 1.9 million tons owing to harvest restraints, new conservation policies and the increased demand by a burgeoning population. Increased outputs in the agriculture, horticulture, sericulture and industrial sectors, where bamboo is a major input material, have also resulted in higher utilization. Unless unit area production is increased by scientific management practices and bamboo growing is expanded in both forestry and social forestry plantations, there is definitive danger of acute resource depletion owing to the plant's overexploitation. Overexploitation can also lead to the extermination of species with limited distribution, endangering bamboo's existence in the long term.

Management

Unscientific management practices – such as not strictly observing harvest rules and silvicultural requirements – have already affected yields. Enforceable statutory regulations are required for improving bamboo management. Recent pronouncements of apex courts of India to enforce scientific management of all forest areas should help bamboo conservation in its natural habitats. Investments in scientific management and expansion of bamboo areas by artificial regeneration need to become a feature of the budgetary process. A part of the revenue or a cess on the value of harvested bamboo should be ploughed back for bamboo development. Seeds from gregarious flowering need to be protected to ensure unhindered natural regeneration. Proper tending of both old and new clumps, and restraint on the removal of immature culms are other measures that would help the resource situation. This is possible if harvests are done only by state agencies and not by the user agencies.

Utilization

Over 95% of bamboos used for non-industrial purposes are not given any preservative treatment. This results in total loss of the bamboo by insect attack within a short life span. Simple and cheap preservative treatments have been developed for use in different applications, but are not practiced. It is possible to organize mandatory treatment, if the harvesting and sale are totally controlled by a state agency. The additional cost on treatment is too small when compared with the advantages offered by long service life and therefore justifiable. Wastage in exploitation by user agency in terms of leftovers in bamboo felling, can also be avoided if a responsible agency harvests, grades, gives preservative treatment, before marketing the same to users and consumers. The collection and disposal of felling residues can reduce fire hazard and consequent damage to clumps.

BAMBOOS IN THE 21ST CENTURY

Extent of areas

Forest areas covered by natural bamboos will continue to remain occupied by bamboos in association with deciduous tree species. There will not be any substantial decrease in the extent of bamboo forests, except that the forest areas within walking distance and vicinity of populated villages will continue to decimate and suffer from illegal and over-exploitation. Only a strong political will to enforce strict regulation of illegal felling and put a restraint on privileges can stop decimation. In the present political climate of India, however, such harsh decisions and rigid administrative control cannot be expected.

Distribution of species

The three major species – *Dendrocalamus strictus*, *Bamboosa bambos* and *Melocanna bambusoides* – are dominant in the bamboo forests. *D. strictus* has a 24-year gregarious flowering cycle. Fortunately, this species also excels in sporadic flowering, and the seeds from such flowering are viable. It has already undergone its gregarious flowering in the last 5–6 years and its regeneration has established. Besides, younger clumps of 10–15 years are coexisting with older clumps in several forests. Young rhizome stock from seeds of sporadic flowering is available for vegetative multiplication. Thus, there is no danger of large-scale reduction in *D. strictus* areas during the first 20 years of the 21st century.

The last gregarious flowering of *Bambusa bambos* was during 1961–64 and the species is thus due for the next large-scale flowering around 2005–08. This species does not exhibit sporadic flowering. Hence, the protection of seeded areas and the use of seeds for expanding its natural zone have to be properly planned well in advance. The care required consists of:

- Avoiding harvesting of dead and dying clumps till the seeds are naturally shed;
- Rigid fire protection after the seedfall and establishment of early regeneration for at least 2–3 years;
- Tending of young clumps; and
- Avoiding the removal of tender shoots and immature culms.

The species has a distinct niche in the moist deciduous tropical areas and can be prepared to withstand decimation after the next gregarious flowering. This species has flowered in Thailand during 1990–92 and in Kerala during 1997–98. It would be a big advantage if rhizomes from these places can be introduced elsewhere in India to break the gregarious flowering cycle pattern in India.

Melocanna bambusoides is widespread in the Northeastern states of India in general, and Tripura, Manipur and Meghalaya in particular. It spreads fast both by natural regeneration from seeds and by the creeping rhizome. The slopes of these hill states have ideal climate for its sustenance. The hill slopes are traditionally subjected to shifting cultivation (*zoom*). With the increase in population, the zoom cycle has come down from 7–10 years to 3–5 years in most areas controlled by the tribal councils. The tribal council or its chief allots areas for zoom cultivation. Areas close to habitation are repeatedly hacked and burnt for cultivation in short cycles, resulting in less period for establishment of this species, which is a colonizer of abandoned zoom areas. The state governments are making efforts to wean the local population from this undesirable practice by settling the population on permanent agriculture, horticulture and cash crops (rubber, coffee, etc.). It is thus likely that though the extent of this species may shrink a little, corrective measures as well as the prolific regeneration of the species will help in its sustenance. The species can be a blessing to tribal economy, since weaving of slivers into mats and art objects is a tradition among the villagers. As the mats now form a raw material for plyboard industry, there is scope to manage the *M. bambusoides* areas by rational harvest in 3-year economic cycles.

Other indigenous bamboo species contributing in bulk to the growing stock include *Bambusa tulda, B. pallida, B. balcooa, B. nutans* and *Dendrocalamus hamiltonii.* All are in moist to humid zone, where periodic seeds will not get destroyed by fire and it should be safe to assume that they will continue to thrive in their natural habitat.

Some commercially important species are endemic, and have localized existence and importance. These include *Ochlandra travancorica* in the swamp forests of Kerala, *Oxytenanthera stocksii* in the Western Ghat regions and *B. polymorpha* in North-East India. Though limited in their distribution, they have high commercial importance. With little protection, these species will continue to occupy their natural habitat even in the future because of their ability to regenerate and spread either through natural seeding or prolific vegetative multiplication. Some species of *Arundinaria* and *Chimonobambusa* occupy semi-temperate or temperate regions in higher altitudes of hills like Western Ghats and outer Himalayas. They are commercially less important, but have great genetic and ecological significance. As there are not many adverse factors like fire or heavy grazing in their natural zones, they have survived the repeated flowering cycles in the past and can be hoped to repeat their history.

SUGGESTIONS FOR SUSTENANCE AND DEVELOPMENT OF BAMBOOS

Provision of progressive yields

Increasing bamboo harvest both for industrial and local uses can be expected in view of increase in population and enhanced need for bamboo products for various uses in a developing economy. Since bamboo is the best long-fibered raw material for pulp industries in the tropics, its utility in the industry cannot be avoided despite increasing use of hardwoods and agricultural residues. Although there is scope for expanding bamboo areas, investments are hard to come by in a big way as forestry receives low priority in national planning.

Permitting private investments in bamboo plantations, whether by users themselves or otherwise, is one possibility. There is, however, a wave of public opposition to throwing open even degraded forests to private investments, citing infringement of social benefits to the community. If higher productivity to meet the increasing needs of users is given priority, then the blocking of investments in productivity for any reason is unjustified. Progressive yields can be achieved only by higher investments for better management of natural bamboo forests and plantations. The current average annual yield of about 0.5 ton per hectare is ridiculously low, and can be increased to 2–5 tons/ha by better management.

Germplasm banks

Progress has been made, in both private-owned as well as government-managed research institutes of India, in standardizing technology for rapid macro-propagation and protocols for micro-propagation. It is now possible to create banks for *in vitro* and *in situ* conservation of many bamboo species. *In vitro* germplasm banks have to be created and maintained in a national grid for ensuring supply of propagules for future needs of bamboo plantations. Similarly, a national grid of bambuseta, containing compatible species and provenances for each agro-climatic zone, should be created and maintained. The existing research institutes as well as private investors should be encouraged to take up establishment of such germplasm banks as a sure way to maintain all available bamboo species and provide material for future sustenance and expansion.

Participatory management

Lack of involvement of users and stakeholders in maintenance and development is one reason for the neglect of natural bamboo stands in Indian forests. Whenever lease areas are earmarked for long periods to an industry, it has been possible to involve the industry in investments on scientific management. Similarly, recent experiments in participatory involvement of village community in bamboo management have shown that over-exploitation and neglect of management of bamboo stands can be avoided. For ensuring the sustenance of bamboo stands, it is advisable to provide incentives to users and arrange their involvement in participatory management of bamboo forests on a long-term basis.

Research and extension

Coordinated research in continuous improvement of bamboo management and utilization should become the basic objective of selected institutions. Similarly, selected agencies like the pulp industry, universities and local administration institutions should take up extension services. The services could include supply of propagules, and training of farmers and users in rational inputs for achieving higher yields, better harvesting practices and optimum utilization of such harvests.

International cooperation

Introduction of a species indigenous to one country into a compatible agroclimatic zone of another country has yielded very good results in the past. *Dendrocalamus brandisii,* for example, has done very well in the wet zone of South India. Some species of *Gigantochloa* and *Guadua* have great potential for commercial cultivation and deserve to be multiplied and tried on a mass scale in India.

INTERNATIONAL EFFORTS IN BAMBOO MULTIPLICATION AND CONSERVATION

Conditions in India

When it comes to the introduction of exotic genera and species, the India has the advantage of having a wide range of climate, soil and altitude: hot and humid tropics, desiccated deserts, fresh water and saline water swamps, temperate and alpine locales. All of these can suit one or more bamboo species available in the world. When the issue being considered is the conservation of all bamboo species of the world for the future, India qualifies as much as China, South America or Africa. As a liberal, democratic, developing country with age-old traditions in bamboo science, culture and ethos, the country has the potential to maintain, conserve and multiply almost all bamboo species.

International understanding for bamboo conservation

Bamboos should not become a monopoly of any state, country or region, as they have universal application. It is therefore necessary to treat bamboos as global property and allow free exchange of genetic resources and scientific information on their silviculture, management and utilization. This will assist in its global conservation not only in the 21st century but also in the centuries to follow.

International agencies

In order to achieve the objective of a universal understanding and conservation of bamboo, it is necessary to create responsible and acceptable agencies to coordinate the efforts of all countries. International Network of Bamboo and Rattan (INBAR) and Association of Bamboo Societies (ABS) have emerged as world leaders in coordinating all scientific efforts in bamboo development. These institutions should mobilize resources, create a network of institutions spread in all countries across the world, and take up measures needed for the propagation, conservation and utilization of bamboos for the good of humanity.

FURTHER READING

Chauhan, K.V.S.; Rai, S.N. 1996. Distribution and growing stock of bamboos in India. Proceedings of a National Seminar on Bamboo, Bangalore, India, 28–29 November 1996. Bamboo Society of India, Bangalore, India. pp. 85–87.

ICFRE (Indian Council of Forestry Research and Education). 1995. Forestry Statistics of India, Vol. 1. Assistant Director General, ICFRE, Dehradun, India. pp. 88–92, 115.

Bamboo for Sustainable Development, pp. 27–33
Eds. A. Kumar, I.V. Ramanuja Rao and Ch. Sastry

Bamboo: A natural resource of infinite capabilities

S. SHYAM SUNDER, IFS
Formerly Head of Karnataka State Forest Department, Karnataka, India

Abstract—The paper summarizes the author's interesting experiences with bamboo, during his service of over 37 years as a forester in Karnataka State in South India.
In a bamboo zone, the survival and extension of bamboo depends on the prevention of collection of bamboo seeds (for use as food), elimination of grazing and forest fires.

Broadcast sowing of bamboo seeds, including aerial seeding, can be resorted to only if cattle grazing can be prevented.

With irrigation in the first year, established *Dendrocalamus brandisii* clumps can produce as many as 120 new culms in a year. *Bambusa vulgaris,* if irrigated regularly, produces culms round the year. Bamboo species of the equatorial belt do not stand inundation as well as the species of the tropical zone.

If the forest departments in India have preferred the commonly occurring species of bamboo for plantations, farmers' preferences have been different. For example, it can be seen that *Oxytenanthera stocksii* (a reed) is the preferred species along the West Coast, while it is a thornless strain of *Bambusa bambos* in the Tanjore delta.

These and a few other observations are discussed in the paper.

BAMBOO IN KARNATAKA

This is a record of observations on bamboo of a forester over a period spanning 37 years. Some are just observations, some are reflections on the observations, while a few refer to practices developed based on these observations. The bamboos and reeds discussed are the widely used species that are indigenous or introduced to Karnataka, a state in South India, and, to a lesser extent, found within and outside India.

Karnataka has a land area of 191 770 km^2, from sea level to 1 500 m above sea level. It lies in the tropics, with a rainfall as high as 7 500 mm in the west to as low as 250 mm in parts of the east. The forest area covers 38 600 km^2, with the bamboo zone extending to more than 15 000 km^2 of dry/moist deciduous forests in areas that receive a rainfall of 750 mm to 2000 mm. It does occur in higher rainfall regions, as high as 3 750 mm, but these forests are not climax forests. The bamboo that grows in the dry deciduous forests is mostly *Dendrocalamus strictus,* commonly known as solid bamboo, while in the moist forests it is *Bambusa bambos* (big bamboo). *Oxytenanthera stocksii,* a reed of high economic value, *Bam-*

busa vulgaris (yellow bamboo) and *Dendrocalamus brandisii* (Burma bamboo) are the common exotics.

In early 1950s, bamboo was quite often described as a weed. A weed is an intruder in a disturbed forest or a contender with the introduced species in a plantation; bamboo was described as a weed in the latter sense. The plantations in India at that time were mainly of teak, and the aggressive bamboo, developing from the rootstock in the sites, was often suppressing teak in the initial years of establishment. At this stage, its elimination was a costly task. Interestingly, bamboo, at the same time, was recommended for planting in older teak plantations, after the final thinning. This was prompted by the silvicultural consideration that in the best of teak forests, teak's associate is big bamboo. Unfortunately, this effort was not much of a success, as big bamboo requires full overhead light.

Bamboo is one of the most aggressive groups of plant life in the tropics. Its new shoots are generally toxic; its height and diameter growth are completed within a few months; and its leaves are beyond the reach of most herbivores. Its sporadic seeding is a safeguard against calamities, while its gregarious seeding is a war on other plant life for asserting its domain.

There is a saying in India that bamboo seeding is followed by famine. The gregarious seeding results literally in a mat of seeds around the clumps on the ground. Feeding on this bonanza, the bandicoots and rats multiply to such an extent that when the bamboo seeds have germinated or are exhausted, the armies of these rodents have no alternative but to raid the agriculture fields.

Gregarious flowering of big bamboo had taken place in Karnataka from late 1950s to the end of 1960s, in the form of a wave, from south to the north, taking about 12 years to complete. If the bamboo areas (and forests in general) were lost earlier owing to clearance for extension of agriculture, the post-seeding era saw to a drastic decline in bamboo stocking. If bamboo seeds from sporadic seeding are an exotic culinary item for the rich, the thick mat of seeds after gregarious seeding was a free source of starch for the poor, who traversed long distance to get to the forests for their collection. This should have been prevented but with the general sympathy for the poor, it was explained away that sufficient seeds would still be left amidst the clumps. Once the clumps died and the bamboos were cinder dry, seeds and the tender seedlings over vast areas were destroyed by the inevitable wild fires. The plants that miraculously survived the fires were consumed by the free-ranging cattle. If only some restraint was placed on the removal of the seeds, if forest fires could have been prevented through special efforts and if the cattle were kept out for three years, a full stock of bamboo would have come up at a very low cost. Today, the entire belt has only a meager stock of bamboo.

Bamboo is a grass, the new crop of which is always green and attracts the cattle when the normal grass species, which are annuals, dry up. Bamboo regeneration could cope with the herbivore wildlife for thousands of years, but not with the army of village cattle of the present. In Uttara Kannada District, famed for teak and rosewood, a new program was launched in the early 1980s for providing total protection from cattle to blocks of degraded forests. This was to promote the coppice growth of tree species. In just two years, along with the coppice growth of tree species, the enclosed areas also had a dense growth of whippy shoots of bamboo. This was surprising because bamboo seeding had taken place here nearly two decades earlier! On digging the soil, we found rhizomes the thickness of a thumb. Year after year, these put forth new shoots, but only to be consumed by the cattle. The moment the cattle were kept out, bamboo asserted itself.

In the early 1960s, after bamboo seeding, in the forest areas from which bamboo was conceded to a paper mill, the lessee took up fencing of blocks of forests to provide protection to the new regeneration of bamboo from cattle. There were two views on this among the foresters: one in support of fencing, the other against it. The first view was based on the fact that while bamboo seeding had taken place over a few thousand square kilometers of forest where no protection could be given by the Department to the new regeneration, this was being provided by an outside agency at its cost. It could turn out to be a good measure. The opposite view was that as these were some of the richest teak forests in the country, the dense regeneration of bamboo could smother and eliminate the regeneration of tree species. It was also argued that movement of wildlife would suffer. The situation was settled in favor of the latter argument, not because of the logic of silviculture but because of the strong objection to the fencing by the local population, supported vehemently by their elected representative in the legislature. The fencing was removed and the cattle did the rest.

The best of bamboo forests in Karnataka are in the valleys, along the stream banks, where there is no human inhabitation. The top canopy, at a height of over 30 m, is shared equally between trees (*Lalgerstroemia lanceolata, Pterocarpus marsupium, Tectona grandis, Terminalia tomentosa, T. paniculata,* etc.) with clean boles of over 20 m and the big bamboo. Unfortunately, only a few forests of this class are now left, of which only those included in the National Parks may continue to remain so in the decades to come.

Bamboo forests, in general, have suffered in management owing to the extremely large extent of forests worked annually and the lack of investment on tending. While the felling cycle is 30–45 years in case of trees, it is 3–5 years in case of bamboos. As a result, supervision of extraction gets to be perfunctory. Cutting is often done at heights of 3 m or more to make it easy for the cut piece to be pulled out. This results in congestion of the clumps. As the extraction contract is barely for six months, the blame can easily be passed on to the previous contractor. Different methods of working the congested clumps have been tried, but the best is not to allow congestion to take place. A chain attached to an old 1-hp scooter engine was found to be effective to pull out the severed culms, although it is time-consuming. Heavy compaction of the soil by trampling cattle also affects production. This can be remedied by loosening the soil around the clumps and regulating grazing.

The yield of bamboos from natural forests varies considerably depending on the rainfall, the species and tending work done. The sum of these can make a difference in mean annual increment of between 2 tons and 15 tons per hectare (Stracey 1957; Hussain 1965).

Recognition of the changes that have taken place over decades and adjusting to these are essential for sound forestry. At the time of reservation of forests, areas with sufficient tree cover were set aside for meeting the local needs. While the human population was more or less stable until early 1940s in the forest belt, it increased by 250% in the next four decades (Nadkarni 1989). Simultaneously, the community lands with tree cover – which was 45% of the land area of the country as late as 1928 according to the Report of the Royal Commission on Agriculture – were either brought under the plough or decimated. With this, the source of meeting the requirement of bamboo for agriculture implements, basketry, construction work, incense sticks, etc. shifted to the reserve forests. As long as this was not recognized and accommodated, the removals were by illegal means and the extraction is mostly done in an improper way. During early 1980s, an assessment of the needs of the agriculture sector and of communities hereditarily involved in bamboo handicrafts (Medras and Buruds) was made, and the required quantities were accepted as priority commitment for supply, at the cost of quantities earlier earmarked for paper and pulp mills. This had a social impact in that as long as the bamboo used was illegal, the artisans were at the mercy of the brokers and traders. Incidentally, as a further step, the forest department's, interaction with a handicrafts organization resulted in a simple tool for the production of bamboo incense sticks. This increased the production capacity of the women folk engaged in the task and reduced wastage of bamboo.

SOME POINTERS FOR BAMBOO MANAGEMENT

Instead of assigning the right to extract of bamboo each year based on tenders, the best would be to grant it on a long-term basis to the industry, which has a stake in the bamboo resource. The royalty for the bamboo could be on a sliding scale such that increasing the yield through soil working, moisture conservation measures and scientific harvesting are of advantage to the lessee. As to the harvest, the best pieces should be kept in separate lots and auctioned openly to meet the requirements of trade or sold at prescribed prices to meet genuine needs. The next best quality should be assigned to rural artisans, and the remaining could go to the paper and pulp industry.

When bamboo seeding had commenced in Karnataka, the paper mills were reluctant to use dead bamboo; when green bamboo was denied, however, they used dead bamboo by making some alterations in the chippers. Later, the paper mills were willing to extract dead bamboos from farther distances because the royalty charged was on tonnage basis and dead bamboo is dry and hence weighed less. Similarly, the Medars in Kodagu district were reluctant to use Burma bamboo in lieu of the indigenous big bamboo, to which they were used to, till its non-availability forced them to use the former. Once they got used to it, even when the next crop of big bamboo had come up for harvesting, they preferred to use the Burma bamboo because of its size and thickness of the wall.

As far as the choice of species for raising plantations is concerned, be it in the government sector or in the farm forestry sector, the preferences are many. Some of these are discussed below.

Plantation sites are invariably of a lower ecological status than good natural forests. Thus, if solid bamboo occurs naturally in a rainfall zone of 750 mm to 1 125 mm (received in less than three months), while big bamboo is in the zone of 1 125 mm to 2 000 mm, one should prefer solid bamboo for planting in areas receiving 1 125 mm to 1 500 mm and big bamboo in zones securing 1 500 mm to 3 000 mm rainfall. *Dendrocalamus strictus* is called solid bamboo, but it is hollow at the core. The same species raised in the Mirzapur region (which receives barely 300 mm rainfall), however, is not only solid but also has a diameter of less than 2.5 cm, down from 5–8 cm. This is used for manufacturing *lathis* or batons used by the Indian police. Thus, the environs in which the bamboo is to grow also has a bearing in species selection.

Tanjore delta in Tamilnadu receives less than 500 mm rain and bamboo is not indigenous to the place. The region, however, has subsoil moisture at slightly over 1 m depth for eight months in the year. Here, a thornless variety of big bamboo is cultivated by the farmers. Thorn is the defense mechanism for big bamboo, but in Tanjore belt (the rice granary of India) where the cattle are not allowed free-ranging, the thornless bamboo is safe: a case of bamboo adapting to its environs.

In the coastal lateritic areas of Karnataka and Maharashtra States, the preferred species of farmers is the reed *Oxytenanthera stocksii.* It is non-congesting, clean, tall, straight, solid and while being strong is not brittle. It does seed sporadically but the seeds are infertile. Its propagation is only through rooted cuttings, but as the returns are high (Rs. 41 170 per ha per year against Rs. 22 424 in case of big bamboo and Rs. 35 120 for solid bamboo, Torane et al. 1994), farmers do not mind spending on the planting stock. While tissue culture plantlets have been developed in the case of big bamboo, solid bamboo and yellow bamboo, this has not been successful so far in the case of *O. stocksii.*

If irrigation is feasible, yellow bamboo and Burma bamboo should produce very high yields. The author was much impressed by the yellow bamboo plantations in Ivory Coast in 1965. These were raised for use as props in banana plantations. At that time, yellow bamboo was still a garden variety in Karnataka. After the Ivory Coast experience, yellow bamboo cultivation was taken up on trial basis at a few centers in the higher rainfall zones in the State. In one case, a clump was in the fringe of a nursery bed and was being irrigated round the year. What was surprising was that new shoots were being produced round the year as well. Generally, Indian bamboos put forth new shoots only during the three months when the monsoon rains are received and even in case of clumps irrigated round the year, new shoot formation is seasonal. This was not the case with yellow bamboo. This species would be ideal for the equatorial belt with rainfall round the year. The author has also seen naturally growing big bamboo inundated (in reservoir fringes) for more than 15 days with no visible change. In the case of yellow bamboo, raised on the foreshore of a tank in Bangalore, five days of inundation resulted in loss of all leaves. When inundation continued for more than 15 days, the clumps died.

Karnataka had a few clumps of Burma bamboo (giant bamboo) in Makut in Kodagu District. In 1964, these clumps had seeded and died. The opportunity was used to raise it in several parts of the state. One such site was in Karwar forest division. In 1981, a small portion of this trial plot was converted to a nursery and a water channel happened to irrigate one clump all the time. This clump had about 50 culms aged 2–15 years and 117 culms of the year! This clearly points to the potential of bamboos, if right inputs can be provided.

Another seemingly unusual behavior was that of a clump of Burma bamboo in the research nursery in Mercara. In 1971, the clump was about 6 years old but was already of a magnificent size. Normally, whether it is sporadic or gregarious seeding, the entire clump is involved and after the seeding, the clump dies. With this clump, however, only three culms flowered and seeded (and later died) while the rest of the clump continued to be green.

A FEW LESSONS LEARNED

The nursery practices adopted in Karnataka with solid bamboo and big bamboo, using seeds, are similar to the practices with tree species. Seedlings are grown in raised nursery beds and transplanted after two years one year or less in the case of tree species). In the second year, rhizome formation takes place. Prizing out the seedlings from seedbeds after one year and planting them in transplant beds facilitates better rhizome formation. Transplanting could also be in polythene bags. In case of naked seedlings, clipping the stem at a height of about 25 cm immediately after removal from the nursery bed (by loosening the soil) gives better survival.

In case of yellow bamboo, which is not known to seed, and *O. stocksii,* seeds of which do not germinate, nodal cuttings of second-year culms, raised in nursery beds, have given reasonably good results. In Ivory Coast, the method adopted with yellow bamboo was to use the new shoot (when barely 30 cm tall) with the rhizome. The rhizome is severed from the parent clump along with a portion of the root. This method was initially adopted in Karnataka as well but was given up after the trials with nodal cuttings proved successful.

One interesting trial with Burma bamboo merits a mention. Our experience of Burma bamboo was based on the few clumps at Makut in the evergreen zone receiving over 3 750 mm rain. The clumps and culms there were of a size meriting the species being called as giant bamboo. The seeds of these clumps, which had died after seeding in 1964, were used for planting in the wet zone. The clumps were 3–4 m in diameter and the culms were 20 cm in diameter at the ground level. However, when it was raised at Sirigere, near Shimoga, which receives barely 1 250 mm rainfall, the results were different. The growth was disappointing after three years, with the clumps measuring less than 1 m in diameter and the shoots were mere whips. Obviously, this was not the zone for Burma bamboo. It was then decided that the plantation be used as a source for planting in wet sites. The clumps were dug up during 1967 rains and, except for 2–3 shoots left behind, the rest of the six to eight shoots in each of the clumps were separated with the root system intact to the extent possible. The shoots were next cut at a height of about 50 cm, to reduce transpiration loss. These were used for planting in the wet zone with nearly 100% survival. The shoots left behind in each clump in the Sirigere plantation produced in the next two years another 6–10 shoots. (Shyam Sunder 1970). This trial indicated that seedlings raised in xerophytic conditions could provide planting stock for several years for use in suitable sites for the species. This should apply equally to stock raised from rooted cuttings as well.

Bamboo is a natural resource of infinite capabilities, but we are not even in the threshold of appreciating the potentials. An occasional bamboo of desirable characteristics has been 'cloned'; but have we given any thought to the wide genetic variations for perpetuation?

REFERENCES

Hussain, S. 1965. The preliminary working plan report of Shimoga Division, Government of Mysore.

Nadkarni, M.V. et al. 1989. The political economy of forest use and management, Sage Publications, New Delhi, India.

Shyam Sunder, S. 1970. Extension of bamboo by artificial propagation. Van Vigyan, Sept.–Dec. 1970, Vol. III.

Stracey, P.D. 1957. The rational utilization of India's cellulosic raw materials with particular reference to Assam. Indian Forester, April 1957.

Torane, S.R. et al. 1994. Economics of production of bamboo in Ratnagiri District of Maharashtra State. Bamboo Information Centre – India, Bulletin, June 1994.

Bamboo for Sustainable Development, pp. 35–41
Eds. A. Kumar, I.V. Ramanuja Rao and Ch. Sastry

The geographical distribution of Vietnam bamboo forests

NGUYEN BOI QUYNH

Forest Science Institute of Vietnam, Hanoi, Vietnam

Abstract—Based on the differences in the nature and the distribution of the species, the Vietnam bamboo forests can be divided into 6 main ecological regions.

1. *Dendrocalamus strictus, Sinocalamus hamiltonii, Pseudoxytenanthera albociliata* and *Bambusa tulda* forests are typical in the North-West Region.
2. *Phyllostachys, Sinarundinaria* and *Neohouzeaua* forests are typical in the North-East Region.
3. Economically important bamboo species – e.g. *Bambusa* spp., *Sinocalamus* spp., *Dendrocalamus* spp. – were planted widely in the North Plain Region.
4. Forests with *Neohouzeaua dulloa* and *Sinocalamus patellaris* are typical in the Middle Central Region.
5. *Pseudoxytenanthera* spp. and *Schizostachyum zollingerii* forests are common in the Highland Central Region.
6. The species *Bambusa procea, Dendrocalamus strictus* and *Bambusa* spp. are widely distributed in the South-East Midland Region.

INTRODUCTION

Vietnam is located in the Southeast Asia, the tropical monsoon region. The climate and the soil are very suitable for the growth of bamboo forests. Bamboo grows almost all over the country, from the sea level (like *Bambusa* spp. and *Dendrocalamus strictus* Nees) to the high mountainous areas at about 2 000 m above sea level (like some species of *Sinarundinaria*). However, 1.59 million ha of bamboo forests are at the low altitude (87.5% of which are under 700 m, 11% from 701 to 1 000 m, and 1% above 1 000 m). There are two main types of natural bamboo forests: (1) pure bamboo forests; and (2) mixed Bamboo forests. The area under natural bamboo makes up 16.01% of the total natural forest area, of which 12.8% is homogenous bamboo forests and 4.1% bamboo forests mixed with other timber species.

Bamboo forests are an important source of raw materials for paper mills, for housing construction, for handicrafts, etc. The Forestry Department of Vietnam is tasked with the management of the forests, including natural bamboo forests.

Bamboo is one of the important vegetative resources in Vietnam. Apart from the natural forests that are homogenous or mixed with other broad leaf species, many species are grown by the people. Almost all the villages of Vietnam have cultivated bamboo, which are being managed directly by the forest enterprises, following the techniques and procedures stipulated by the Ministry of Agriculture and Rural Development. These procedures are based on the results of research conducted by the country's Forestry Science Institutes.

The appropriate condition for the growth of bamboo is in those areas having the rainfall from 1 400 mm to 2 200 mm. Some areas, such as Bac Quang of Tuyen Quang province, receives a rainfall of up to 4 465 mm. The temperature ranges from a maximum of about 40°C to a minimum of about 0°C, with an average temperature of 20°C. Bamboo in the country grows on a variety of soils, including the acidic sulfate soil, soil inundated with sea water, sandy soil, etc. In general, however, it prefers the alluvial soil on the banks of waterways, the soil layer varying from average to thick, and from silt to clay. The percentage of mixed rock is average and the water holding capacity is good but not stagnating.

Owing to the variations in the natural conditions among the regions, the number of species is very abundant. Up to the present, 93 species in 17 genera have been investigated, accounting for nearly one-fifth the known bamboo species and one-fourth the genera in the world. This, however, is only partial; when investigations are completed, it is certain that the number of species will significantly increase.

Bamboo in Vietnam finds diversified uses. Since the olden days, bamboo has been a part of the daily life of the Vietnamese. The images of bamboo hedges and bamboo temple roofs have become the symbol of rural Vietnam. Many bamboo species are used as raw materials for various types of well-known national musical instruments like flute, crongput and torung. Bamboo has also entered into poetry and legends. The bamboo culm, with its straight and erect form, is considered the symbol of an honest man. It can be said that a bamboo culture has been formed over the centuries in Vietnam.

Commercial uses of bamboo are also numerous, although a complete study of this topic has not been made so far. In general, the more an industry develops, the more its use of bamboo. For example, today the paper industry uses bamboo to a larger extent than in the past, and in recent times bamboo is also being used for making chopsticks, mats, joss sticks, etc. Nevertheless, compared with countries such as China, the use of bamboo in Vietnam is still very limited.

Studies on bamboo are still very few in Vietnam and there has been no systematic inventory to assess the resource. However, based on the variations in the natural conditions and distribution of species, Vietnam can be divided into six main regions. Some species, owing to their high adaptability, appear in many regions, especially in transition areas. The distribution pattern in these six areas is discussed below.

NORTH-WEST REGION

The North-West of Vietnam is in the left bank of the Red River valley. The western part is adjacent to Laos and the northern part to China. This is an area with high mountain ranges and plateaus. All the highest mountains in the country – such as Phang xi pang (3 142 m), Sa phing (2 870 m) and Puticn vam (2 288 m) – are located in this area.

The North-West region lies deep in the interior and the sea does not affect it. The climate is quite severe: in the summer, it is affected by the hot Laos wind (west wind); in the winter, cold winds keep the temperature lower than in other regions at the same altitude. Every year, there are two or three months with the average temperature of 15–20°C; in the remaining months, the average temperature is about 28°C. The maximum temperature can reach 40–41°C. The temperature difference between day and night is 10–11°C. The rainfall is quite high but is not even in the regions, varying from 1 000 m (Yen Chau, Nam Cum) to 3 164 mm (Muong Te). However, most regions receive an annual rainfall of about 1 500 to 2 000mm, mainly during 2–3 months in the rainy season. The climate in the year is divided into two distinct seasons: the cold and dry winter season, and the hot and humid summer season.

The area under bamboo in this region is quite large, occupying 2.5% of the natural forest area. The species diversity is quite significant and up to now, 40 species have been identified.

Bamboo growing in clusters is typical in this region, and the species found here are those that can survive in the dry and hot climate – such as *Pseudoxytenanthera albociliata* (May lay) and *Bambusa tulda* (May bong). These two species cover large areas, which are homogenous or mixed with broad-leaf species, on the field and along the road. The two most important species are *Sinocalamus hamiltonii* (May hoc) and *Dendrocalamus strictus* Munro (May song), which are large bamboos with a high economic value. These two species also form large natural forests, homogenous or mixed with other timber species, along the Da river.

Other naturally growing sympodial bamboo species in the region are *Neohouzeaua dulloa* (Nua), (*Dendrocalamus* sp. (Giang), and *Dendrocalamus latiflorus* Munro (Mai).

Some monopodial bamboo species, such as *Sinarundinaria* spp. (Vau dang and Vau ngot), can be found growing in low-lying areas. These species also have large culms that have a high economic value. On high mountainous areas, there are some species of *Arundinaria* with small-sized culms growing mixed with other bushes. One often finds a *Chimonobambusa* species (Truc vuong or Tre danh) growing in low-level areas, which have a thin soil layer and bushes as the main vegetation cover.

The species that are cultivated in the villages are mostly large-sized ones like *Sinocalamus latiflorus* (Mai) and *Dendrocalamus membranaceus* (Luong), which and are grown for culms and shoots.

NORTH-EAST REGION

The North-East region lies in high latitude, with the Red River bordering the west, the Eastern Sea to the east and China at the north. This is an area featuring low and medium high mountains. The climate is quite cold (this is the coldest area of the country compared with places at the same latitude) and humid because it is directly affected by the Northeastern wind. The average temperature is 22–23°C and the coldest month is January with an average temperature of 11–16°C. The rainfall is not even in the area, the most common being 1 400 to 3 000 mm (in some areas, like Bac Quang, the rainfall is 4 465 mm). The rain mostly occurs in the months of July-September. The humidity is always high (over 85%).

Bamboos in the region occupy 3.5% of the natural forest area. The species are also quite abundant, especially monopodial species. In areas of over 400–500 m altitude, *Phyllostachys pubescens, Indosasa* sp. and, to a lesser extent, *Chimonobambusa quadrangularis* are distributed in homogenous or mixed forests.

In areas that have a high precipitation (3 000 to 4 000 mm, Bac Quang and Tuyen Quang), *Sinarundinaria* spp. (Vau ngot, Vau dang) grow very well, often forming large homogenous or mixed forests. This area has now become an important raw material source for the paper industry and the construction industry. Another *Sinarundinaria* species (Tre roc), a small-size bamboo found in the low mountainous area near the border, also forms homogenous or mixed forest.

Clumping species – such as *Dendrocalamus* sp. (Giang) and *Neohouzeana dulloa* (Nua) – are also quite abundant in the area, forming large forests that now form an important source of raw material for the paper mills.

In some islands in the Ha Long Bay area, a small-sized *Bambusa* species (Tre tut) also forms homogenous forests, which sometimes occupies the whole land area of the island. Species that can be found in abundance in non-forest setting include *Bambusa sinospinosa* McClure (Tre pheo) and *Lingnaia* sp. (Dung). Cultivated bamboo species are *Phyllostachys pubescens* (Truc can cau, planted in a few thousand hectares in the forest) and *Phyllostachys nigra* Munro (Truc tim, which is planted in the higher areas).

THE PLAINS (NORTH VIETNAM)

The alluvium from the Red River and Thai Binh River makes up the Plains in North Vietnam and the topography of the area is very flat. The average temperature in the area is 23–24°C, and the annual rainfall 1 500 to 2 000 mm. In winter, it is quite cold and humid because of the Northeastern wind; in summer, the region is lashed by the Southwestern wind. The latter is also the season of rains and storms. The soil is fertile and a long time ago, it has been cleared to cultivate the lowland rice. Therefore, there is no natural forest in the region. The planted bamboo species grow mainly in clusters, bracing against strong winds and typhoons.

The hardy species are used for fencing, construction, weaving, and household utensils and containers. Bamboos are also cultivated for their shoots.

The species grown are mostly thorny bamboos, typically the spiny bamboo 'Tre gai', which is planted as fencing around households or around villages to demarcate the border. The bamboo is so popular that the bamboo hedges of the villages and the bamboo roofs of common houses have become the symbol of the countryside in the Plains. Other thorny bamboo species, which are grown widely, are: *Bambusa arundinacea* (Retz.) Willd. (Tre loc ngoc), a large-diameter, tall bamboo species with short internodes, which is used for pillars; and *Bambusa blumeana* J.A. & J.H. Schultes (Tre la nga), another large-diameter, tall bamboo species. In areas adjacent to the mid-mountainous provinces, *Sinocalamus latiflorus* (Mai), *Dendrocalamus* spp. (Dien da) and *Bambusa multiplex* (Lour.) Raeuschel (Hop sao) also appear.

THE MIDDLE REGION

This region stretches from the South of Ma River to the Red river valley, with the border falling between Thua Thien province and Quang Nam-Da Nang province, to the northern part of the Truong Son mountain range. In the West, it is adjacent to Laos and in the East is the Eastern Sea. The climate is affected by the North-eastern wind, which is cold and humid but weak; therefore, the temperature is relatively high. The average temperature increases from 23° to 25°C when moving from the North to the South. Precipitation is high and not evenly distributed among the areas; closer to the South, the rainfall is higher (1 721 to 3 015 mm). The rainy season falls in the months of October-November, and the average humidity is about 80–85%. This region is strongly affected by the dry, hot Western winds (Laos' wind), making the climate quite severe.

This region has a very large bamboo forest area, which occupies 21% of the total natural forest area, with high species diversity. The species that occupy the largest area are *Neohouzeaua dulloa* (Nua) and *Dendrocalamus* sp. (Giang), which form large homogeneous or mixed forests. *Lingnania* sp. (Lung), a large-sized bamboo species that has long internodes of 70–100 cm (some even 150 cm), also grows in homogeneous or mixed forests. This bamboo species is specific to this area. In the south, there is *Schizostachyum zollingeri* (Lo o), which grows naturally or is cultivated. Another important bamboo species is *Dendrocalamus membranaceus* Munro (Luong), which is a large-sized species with high economic value and is extensively planted (60 000 ha) in the forest.

Other widely planted species are: *Bambusa sinospinosa* McClure (Tre gai), *Gigantochloa scribneriana* (Buong), *B. blumeana* J.A. & J.H. Schultes (Tre mo), *B. rigida* Keng. et Keng. f. (Song kim), *Bambusa multiplex* (Lour.) Raeuschel (Hop sao) and *Schizostachyum zollingeri* (Lo o).

HIGH PLATEAU REGION (MIDDLE VIETNAM)

The high plateau region of Middle Vietnam has a topography composed of many high plateaus, lying in the West of Truong Son mountain range, running along the foot of the Ngoc Linh mountain to the South-Eastern region. In the North is the Kontum plateau and Pleiku (600–800 m high). In the middle lies Daklak plateau (300–600 m high), followed by Lam Dong plateau (800–1000 m high). The surface comprising these plateaus is quite flat.

The climate in the west is temperate, while the average temperature in the north is 22–24°C and in the south 18–21°C. The average annual rainfall is 1 500 to 2 000 mm. Pleiku and Boa Loc record the highest rainfall: 2 513 mm/year and 2 763mm/year, respectively. The rainy season is from May to October, during which period 90% of the precipitation occurs, creating two distinct seasons: the rainy season and dry season.

The bamboo forests in this region are quite large, occupied mostly by two species: *Pseudoxytenanthera nigrociliata* and *P. albociliata* (both known by the local name of Le). These form homogeneous forests or mixed forests dipterocarp species shedding leaves in the dry season. In the northern and southern areas of the western plateau, the climate is more humid and the soil more fertile. Therefore, *Schizostachyum zollingeri* (Lo o) forms homogeneous forests or forests mixed with broad-leaf tree species. *Bambusa arundinacea* (Retz.) Willd. (La nga) grows well along the riverbanks and in areas with alluvial soil. This large-sized bamboo species is an important raw material for some industries. Some other bamboo species, which are yet to be identified, also grow in the region. A *Melocanna* species (Tre qua mong) has recently been found in this area.

SOUTH-EAST AND COASTAL REGION (MIDDLE AND SOUTH VIETNAM)

The topography of this region is very flat, except areas adjacent to the mountain and high plateau which have a hilly terrain and terraces. The climate is tropical, with typical seasonal winds. There is no cold winter season. The average temperature is 25–26°C and it seldom changes in the year. The rainfall is not even among the areas, varying from 1 500 mm to 2 200 mm (one small area has a low rainfall of 600–700 mm/year). Rains occur mostly in the rainy season, thus forming a rainy season and a dry season. The dry season lasts 5–6 months and is quite severe. The soil is mainly gray soil, developed on the old alluvial soil, though some areas have mixed 'Bazan' soil.

There is a large area (150 000 ha) under bamboo, with *Bambusa procea* A. Chev et Camus (Lo o) forming vast forests that are homogenous or mixed with broad-leaf tree species, concentrating in the provinces of the South-Eastern region, mainly in Song Be province. This species is quite large, but the culm wall is thin. It is now an important raw material for paper mills. In the coastal provinces of Middle Vietnam, besides *B. procea,* one can find *Schizostachyum zollingeri* (also called Lo o), which has a smaller culm and is very popular. Another much-used species is *B. arundinacea* (Retz.) Willd. (called Tre gai in the region), which has a large reserve and is distributed in the flat area along the riverbanks. This tall large-diameter species is an important raw material for the paper industry, and is used also for making chopsticks.

Bamboos that are widely planted include: *Dendrocalamus strictus* Nees (Tam vong), a solid bamboo used in construction; *D. flagellifer* (Buong), planted for shoots; and *Bambusa nana* (Truc), a small bamboo species used in weaving, and making curtains and fishing rods.

Bamboo for Sustainable Development, pp. 43–48
Eds. A. Kumar, I.V. Ramanuja Rao and Ch. Sastry

Fifteen years of bamboo integrated development in Costa Rica: A personal view

ARCH. ANA CECILIA CHAVES DE SOTO
Executive Director, FUNBAMBU, San José, Costa Rica

THE BAMBOO PROJECT

The Costa Rican National Bamboo Project is the result of a dream inspired by the teachings of Don Rodrigo Chaves, my father. A public official, he used to take my sisters and myself for field trips to where the peasants lived in humble, small, but beautiful wooden houses. He used to tell us that each day the precious wood from the forest was becoming more and more scarce, and that low-income people would soon have to build their houses with zinc sheets and cardboard. I remember he telling me that I would have to seek a solution to this in easily available and easy to produce natural materials, and that this solution would ensure my success in my career as an architect. He suggested bamboo as a possible solution. My dream about bamboo started from that moment; but no dream can become a reality, if not shared with others. One has to communicate dreams and share ideas with others who have similar concerns. Once people accept the general concept, one has to build up a group around it so that the final creation would be a conglomeration of ideas and goals of the individuals. This has been the key to the success of the National Bamboo Project, known today as the National Bamboo Foundation (FUNBAMBU).

The project started with an aid of US$ 26 000 granted by the UN Development Programme (UNDP) and the UN Program for Human Settlements (HABITAT), and a donation of US$ 300,000 from the Royal Government of the Netherlands. At the outset itself, we had the benefit of Dr Jules Janssen from the University of Eindhoven, the Netherlands, evaluating our plan. He gave us full support, and his suggestions motivated our team to develop a clear set of objectives. Since the beginning, we planned to establish bamboo plantations, and to research and develop a technology that takes full advantage of all the physical and mechanical characteristics of bamboo. We established an R&D Department and a Preservation Department, both managed by Ing. Guillermo González, under the guidance of Dr. Walter Liese from the University of Hamburg, Germany.

Designing equipment and tools for the production of bamboo laths was one of the first tasks. We felt that making bamboo a material that could be industrialized would facilitate even more the task of making bamboo products. A system for planting bamboo was established so that several plantation sites could be distributed among the project units. Different propagation systems were tested and optimized before planting was initiated. Special attention was paid to industrialization of production and the prefabrication of housing components. We decided to have several factories in strategic sites all around the country, and to transport the finished products from those sites to their final destination. This greatly helped us to complete 100 houses in 3–4 months instead of the initial program of 9 months.

Several questions have been raised repeatedly about the project. What is the focus of the project? What kind of project is it? Is it a reforestation project? Will it be a social project? Is it a project addressed at women? Is it a development program? The basic answer to all these is that a development project, such as the bamboo project, has to involve all parties involved in the problem. Based on that point, we were able to develop a system that started with the plantation and continued up to the final product – housing. At the same time, we were able to industrialize bamboo and export it.

At FUNBAMBU, the Executive Director is not an expert in all areas that the Foundation works in, but one who respects the knowledge of each professional in his/her area and collaborates with him/her in taking decisions. The Foundation has the services of forestry professionals, civil engineers, structural engineers, architects and social workers. It also has people who know how to handle community groups and motivate them to join forces and work together as a team. This way, the communities involved have been working as a team and was involved in collective decision-making right from the beginning of the project.

For a project to be strong, it has to build bridges with four sectors. First, it has to address a sector with a large demand, one where a lot of need exists for the services offered by the project. The Bamboo Project focused on the poorest communities that have housing needs, and the project became larger day by day. Second, political support is necessary. We approach the local Mayor's office, where we always find the support we need. Third, official institutions in the concerned sector are important. In FUNBAMBU's case, these are the housing and silvicultural sectors. Four, international support may also be required. As FUNBAMBU was looking beyond being a local project, it contacted international institutions such as the UNDP and Habitat for financial support.

IMPACT OF THE PROJECT

The benefits derived from the bamboo project offered are many. These include:

- Development of bamboo for commercial uses;
- Development of housing;

- Development of building technology aimed at the poorest of the community;
- A technology package that can be transferred to places that have similar needs;
- Manufacture of bamboo laths for use in the manufacture of panels for housing;
- Establishment of a panel factory that will lower the cost of construction process;
- Installation of a furniture factory; and
- Establishment of an ecological project for protecting forests from deforestation.

The 3 000 houses built by FUNBAMBU have been financed through the "Sistema financiero de la Vivienda". Their resistance to earthquake and the good quality of construction had to be demonstrated so that the housing units could compete with other traditional construction systems.

Latin America has a high human and technical potential. The problem has been the persistence with the traditional building technologies, which use nonrenewable materials, for over 100 years.

POSSIBILITIES FOR THE FUTURE

FUNBAMBU has to consolidate itself as a center of excellence for the certification of quality in bamboo utilization. Another activity of the Foundation should be the dissemination of its experience to all countries in Central and South America and the Caribbean, especially its know-how in economic self-help house construction processes. FUNBAMBU has to search for sustainable ways to finance research and technology development. Although there was a specific budget for research in the beginning of the Project, the resources disappeared when it became FUNBAMBU and the new Foundation had to self-finance research through the manufacture of construction products. This is a very heavy economic burden; the Foundation has invested in a highly capable and specialized technical team, and the whole region should be able to benefit from it. Any new bamboo development project can benefit from the transfer of the technology developed at FUNBAMBU.

At the national level, the Foundation is working with all governmental institutions in urban and rural areas and is providing them with bamboo construction technology. The main idea is to build schools, community centers, hospitals and health centers, all financed with the funds that the government grants us.

FUNBAMBU hopes to become a regional focal point and a business network, which would facilitate the meeting between supply and demand of bamboo products, at least in Latin America. It aims to establish policies for cooperation in the socio-political sectors. For example, if there is a demand of hotels and tourist projects, FUNBAMBU could participate, or with the revenue by its work, it could finance the construction of low-income houses.

The role of bamboo as a construction material is very unique: as other technologies that use traditional construction materials tend to raise the construction price, bamboo goes the opposite way – price decreases as long as there is a ready availability of the material. For this, however, more area needs to be extensively cultivated with bamboo. The technologies developed in Japan and China demonstrate that we are getting closer to 100% use of bamboo housing components. To ensure access to such technologies, we could develop a mechanism for their transfer through INBAR.

We have to look for a new stage in global development of bamboo. We have to stop thinking at a national or regional level, and cooperate to develop worldwide projects. The Bamboo Bridge has to become the instrument for world development and our strength. Many companies of industrialized countries contribute to environmental pollution. If we succeed in motivating them by asking their governments to contribute to the reforestation of riverbanks and depleted forests in the poorer countries, it would help address many of the undesirable natural climatic phenomena that we are facing today. Reforestation would also provide us with a great quantity of material for housing, one of the basic human necessities, and would stop soil erosion. Why think of only basic needs? The Foundation has had experience in exporting bamboo material – well-preserved, durable and high-quality material – to the markets in developed countries.

The availability of bamboo material could be guaranteed by establishing a bamboo commercialization network. If one country does not have the necessary supply, it could be imported from another country with surplus bamboo.

THE NATIONAL BAMBOO FOUNDATION (FUNBAMBU)

FUNBAMBU, heir to the National Bamboo Project, was born as an organization whose goal is to reproduce systems present in nature. The organization and its working are briefly discussed in the following section.

Some key aspects of the management

The key to the Foundation's effective functioning is a high-level team, created with the aim of imparting to it the ability to find holistic solutions to technical, social and economic problems involved in the project's execution. The most important aspect of the key results obtained by the Foundation has been that all the decisions, from the very beginning of the Foundation, have been taken from the point of view of the whole team under the supervision of a Coordination Committee.

The officials of the International Executive Committee are involved directly with this team, and this has proved to be an important factor in the development of FUNBAMBU.

A specialist from the donor country, the Netherlands, was in charge of the technical supervision of the work – research, development and execution – under various programs. Thus, as the donor was in full knowledge of the results of the work carried out and the investments made towards the work, it was easy to justify the financial and technical support granted by the Netherlands.

The personal involvement of the Resident Representative of the United Nations Development Programme (UNDP) was very important in the progress of FUNBAMBU. This association provided us with better results and considerably reduced the management-related stress that the professionals of the Foundation would otherwise have felt.

A global approach to problem solving

The working style of Foundation is very similar to the same way the human body functions: each organ working independently but, at the same time, coordinating with the functions of all other organs to present a composite whole.

What was once considered a "problem" has become our strength. Our team motto is: "When various factors are part of a problem, they should also be part of the solution". If we have become involved with parts, we have also got involved with the whole. This approach explains how the Project was able to execute 39 of the 40 points in the Agenda 21, even though the organization was not in existence when the Rio de Janeiro Earth Summit took place.

Integration and involvement

Rural base

Development organizations, housing committees and low-income groups are potential allies to the Foundation. Since the beginning, the Foundation has kept close contact with groups that require the services it offers, and has focused attention on rural areas where such groups are located. The Foundation approached its potential clients through a very successful direct advertisement campaign, which proved that the demand for its services surpass its capacity to supply. It then adopted a strategy of making demonstration projects of about 15–20 houses in different zones of the country where the expression of interest was high. The demonstration projects were aimed at convincing people of the options available and demonstrating the feasibility of bamboo housing developed by the Foundation.

Local organizations

These are the conscience of the communities. As democratically elected entities, these organizations are tasked to act for the community's benefit. Without the land, the infrastructure, etc. obtained free of cost under subsidy through them, the cost of housing would be too high for the poor sections of the society to afford.

The national government

The national (and state or provincial) government defines the policies that rule the country (and state/province). It decides the priorities and assigns the resources. In general, it pays attention to the demands of the communities, local organizations and local rulers, and respects the recommendations and suggestions of donor countries and the international organizations.

If the government had not considered the benefits derived from FUNBAMBU as its own, the Foundation might have disappeared by now. This connection with the government has been very important to FUNBAMBU; whenever there was a divergence of views in the Foundation and the government, the Foundation has reverted to the local organizations and the local government in order to work out a mutually acceptable position.

FUNBAMBU has had the active support of many institutions and organizations from its beginning, and it learned a lot from them; now, the Foundation has acquired knowledge on several aspects related to bamboo and housing which it can impart to them. It is opening up this knowledge through technology transfer workshops. I really hope that after this International Bamboo Congress and Workshop, the promotion of extensive use of bamboo shall become the responsibility of all of us.

Bamboo for Sustainable Development, pp. 49–78
Eds. A. Kumar, I.V. Ramanuja Rao and Ch. Sastry

Evaluation of bamboo resources in Latin America

XIMENA LONDOÑO
Instituto Vallecaucano de Investigaciones Cientificas, Cali, Colombia

Abstract—Latin America is the richest region of the Americas in terms of the diversity and number of woody bamboo species. Twenty genera and 429 species of woody bamboos are distributed from approximately 27° North (*Otate acuminata* found in the north-western part of Mexico) to 47° South (*Chusquea culeou* in Chile) (Judziewicz et al. 1999). Of the woody bamboos found in the Americas, only *Arundinaria gigantea* of North America is not found in Latin America. Of the total 1 100 species and 65 genera of woody bamboos known in the world (Judziewicz et al. 1999), Latin America has 39% of the species and 31% of the genera. Brazil has the greatest bamboo diversity (137 species) followed by Colombia (70), Venezuela (60), Ecuador (42) Costa Rica (39), Mexico (37) and Peru (37). A listing of native woody bamboo species by country is provided in this work.

In general, the exploitation of native bamboo in Latin America is limited to the local use of species found close by. It is only in Colombia, Ecuador and Brazil that bamboo plays a more conspicuous role in the local economy.

It is estimated that bamboo in Latin America covers close to 11 million hectares, and that approximately 11% of every square kilometer of Andean forest is occupied by bamboo.

INTRODUCTION

Bamboo is an alternative resource that helps confront some of the problems affecting the majority of the countries. At the same time, it contributes to environmental and habitat protection as well as economic development.

Bamboo's rapid growth, ease of caring and wide distribution make these plants an ideal renewable resource for the development of local economies in Latin America. In this paper, Latin America is defined as the area from Mexico through Central America and South America, including the Caribbean and the West Indies.

Since pre-Columbian times, bamboo has been used in Latin America and recently has been developed successfully in certain countries. However, despite reports that there are uses of bamboo in every country in the region, bamboo in general should be considered a forest resource with little significance in the local economy in majority of these countries. In general, the exploitation of native bamboos in Latin America is limited to local use of species found close by. It is only in Colombia, Ecuador and Brazil that bamboo plays a more conspicuous role in the local economy.

Recently, it has been calculated that the bamboo-dominated areas in the Amazonian region, is close to 180 000 km^2 (Judziewicz et al. 1999), equivalent to 18 million hectares. However, it is mixed with other Amazonian forest species. Keeping in mind that the presence of bamboo in South-western Amazon is not homogeneous or continuous, and that approximately 11% of every square kilometer of Andean forest is covered by bamboo, it may be estimated that bamboo covers close to 11 million hectares in Latin America.

Latin America possesses approximately 47% of the world's tropical forests and 70% of the world's animal and plant species (CESA 1992). The annual deforestation rate of 1.3% (CESA 1992) has been increasing over the last 10 years as a result of the ever more critical socio-economic situation in the region that necessitates its inhabitants and governments to find new sources of income at whatever the costs, including ecological.

Unfortunately, the montane forest of the Andean region, where more of the woody bamboo species are found, is decreasing at a high rate owing to both natural and anthropogenic disturbances (Judziewicz et al. 1999; Clark 1995). If the great diversity of the Andean Region with all its utilitarian species and potentially utilitarian species is to be conserved, then some quick actions are required to find solutions that combine legal protection, sustainable development and reforestation of native species.

Architect Moran once noted that in addition to our historical ties and language, we [the Latin Americans] have other ties that bind us, such as those expressed in our bamboo of which the Guadua, authentically American, is noteworthy as a valuable alternative and instrument in the solution of the serious housing problems that face Latin American countries. In making an evaluation of bamboo resources in Latin America and in analyzing its present situation as well as forecasting its future, the author too believes in this.

This paper is a summary of the final report of an INBAR project on the "Evaluation of Bamboo Resources in Latin America", for which the author was the Project Leader. It indicates for every country the number of woody bamboo species, the economic potential of some native species, geographical distribution of bamboos, and the culm size of important bamboos and their local uses. Special attention is given to Colombia, Ecuador and Venezuela. In this summary, only general comments are furnished and a listing of native woody bamboo species by country is provided. Details may be had from the project report, which INBAR intends to publish in the near future.

BAMBOO RESOURCES IN LATIN AMERICA

Latin America is the richest region of the Americas in terms of the diversity and number of woody bamboo species. Twenty genera and 429 species of woody bamboos are distributed from approximately 27° North (*Otate acuminata* found in the north-western part of Mexico) to 47° South (*Chusquea culeou* in Chile)

(Judziewicz et al. 1999). Of the woody bamboos found in the America, only *Arundinaria gigantea* of North America is not found in Latin America. Of the total 1 100 species and 65 genera of woody bamboos known in the world (Judziewicz et al. 1999), Latin America has 39% of the species and 31% of the genera. Brazil has the greatest bamboo diversity (137 species) followed by Colombia (70), Venezuela (60), Ecuador (42), Costa Rica (39), Mexico (37) and Peru (37).

Woody bamboos are found in almost all Latin American habitats with the exception of desert regions. They are distributed from the humid lowland forest at sea level to the highlands in the Andes up to 4 300 m in the natural formation known as "paramo" (Judziewicz et al. 1999).

Because of the very little economic significance of bamboo resource in the majority of the Latin American countries, there is neither an estimate of bamboo forests (cover area) nor an estimate of bamboo production for any country in this region. However, some survey efforts have been made in a few countries (for example Colombia, Ecuador, Nicaragua and Venezuela).

In general, it may be said that the most useful species in Latin America are found in the genus *Guadua* and in the Asiatic genus *Bambusa,* although the genera of *Apoclada, Aulonemia, Chusquea, Elytrostachys, Otatea* and *Rhipidocladum* also have species that are used in various ways.

The cultivation of bamboos on a commercial scale in Latin America is limited to a few native and introduced species: *Guadua angustifolia, G. amplexifolia, Bambusa vulgaris, B. tuldoides* and *Phyllostachys aurea.*

The following sections give an account of the bamboo resources by region and country.

SOUTHERN SOUTH AMERICA

Argentina (5 genera, 13 species)

In Argentina, bamboo does not play a dominant role in forestry. As in the majority of Latin America countries, bamboo exploitation is generally confined to local use of native species in areas close to the source of supply. There is no control over the removal of native bamboos from riverbanks by people, and this has created a shortage of bamboo resources over many years (Burkart 1969).

The genus *Guadua* is concentrated in the provinces of Corrientes, El Chaco, Misiones and the edge of the Parana River in the province of Santa Fe. In Argentina, the *Guadua* genus includes three species: *G. chacoensis, G. paraguayana* and *G. trinii* (Burkart 1969; Nicora and Rugolo 1987).

Several Asiatic bamboos were introduced into the country approximately a hundred years ago and became widespread, especially in the north-western part of the country. The most widespread and cultivated ones are: *Bambusa tuldoides* (tacuara), *B. vulgaris* var. *vittata* (yellow bamboo), *B. arundinacea* (spiny bamboo), *Dendrocalamus asper* (tacuara gigante), *Phyllostachys aurea* and *P. bambusoides* (cana de indias) (Parodi 1936; Turpe 1975).

Following are the native woody bamboo species of Argentina:

Chusquea andina R.A. Philippi
C. culeou E. Desvaux
C. deficiens L.G. Clark
C. iorentziana Grisebach
C. ramosissima Lindman
C. tenella Nees
C. valdiviensis E. Desvaux
Colanthelia rhizantha (Hackel) McClure
Guadua chacoensis (Rojas) X. Londoño & Peterson
G. paraguayana Doell
G. trinii (Nees) Ruprecht
Merostachys clausenii var. *clausenii* Munro
Rhipidocladum racemiflorum (Steudel) McClure

Chile (1 genus 10 species)

Bamboo in Chile is rarely used as an economic plant, but has some influence on the dynamics and composition of the Chilean Andes forest. The genus *Chusquea* is the only native American bamboo occurring in the country. It is found in the provinces of Cautin, Chiloe, Concepcion, Llanquihue, Malleco, Nuble, Osorno, Santiago, Valdivia and Valparaiso, and is locally known as "culeo" and "colihue"

It is interesting to note that the geographical distribution of the genus *Guadua* includes all southern Latin American countries, except Chile.

The native woody bamboo species of Chile are:

Chusquea andina R.A. Philippi
C. ciliate R.A. Philippi
C. culeou E. Desvaux
C. cumingii Nees
C. fernandeziana R.A. Philippi
C. macrostachya R.A. Philippi
C. montana R.A. Philippi
C. quila Kunth
C. uliginosa R.A. Philippi
C. valdiviensis E. Desvaux

Uruguay (2 genera, 3 species)

Bamboo resources in Uruguay are scarce. There is only one useful species, *G. trinii,* which is sometimes found growing along the forest banks of River Uruguay and its tributaries, in the department of Rio Negro (Rosengurtt et al. 1970). Culms of this bamboo, commonly named "tacuara brava", have been reportedly used in rural construction (Lombardo 1943; Rosengurtt et al. 1970).

Bambusa tuldoides and *Phyllostachys aurea* are cultivated in Montevideo and Soriano Departments (Rosengurtt et al. 1970).

The native woody bamboo species of Uruguay are the following:

Chusquea juergensii Hackel
C. tenella Nees
Guadua trinii (Nees) Ruprecht

Paraguay (3 genera, 5 species)

Bamboo is scarcely used in Paraguay for economic activities and, as in Uruguay, does not form a large resource. There are three *Guadua* species with potential uses: *G. chacoensis* (tacuar), *G. paniculata* and *G. paraguayana* (picanilla).

According to Cabrera, bamboos are an important element of the Paraguayan forest. The forest at the lower levels was cleared of them when they flowered and the culms died, but new bamboo plants had regenerated by the following year and were forming an impenetrable wall.

Following are the native woody bamboo species of Paraguay:

Chusquea ramosissima Lindman
Guadua chacoensis Rojas X. Londoño & L.G. Clark
G. paniculata Munro
G. paraguayana Doell
Merostachys clausenii var. *clausenii* Munro

CENTRAL AND NORTHEASTERN SOUTH AMERICA

Brazil (17 genera, 135 species, 2 subspecies)

Brazil is the country with the greatest diversity and the highest percentage of endemic woody bamboos in Latin America – 137 species (32% of Latin American bamboo species) and 17 genera (85% of Latin American bamboo genera). The states of Sao Paulo, Minas Gerais, Santa Catarina, Bahia and Parana have the greatest diversity of woody bamboos.

Foremost among the world centers of bamboo diversity are the Atlantic forests of Brazil, which extend from the state of Paraiba to Rio Grande do Sul in a mostly narrow coastal strip characterized by abundant rainfall (Moris et al. 1983), and include 22 genera and 62 species (woody and herbaceous) (Judziewicz et al. 1999).

Several exotic bamboo species were introduced to Brazil during the period of colonization by the Portuguese (de Barros and Egashira 1993). These species are widely distributed around the country and, along with some natives ones, have a large number of uses – building material (including raw material for low-cost housing); source of food for people and livestock; raw material for making a wide variety of handicrafts; source of medicine; raw material for paper pulp; for erosion control; as ornamental plants; etc. Bamboo is also under research as possible source of ethanol (Burman and Filgueiras 1993; Azzini and de Barros 1993).

The area of bamboo forests in Brazil is still not clearly determined. However, it is known that the states of Amazonas and Acre have the largest area of bamboo. The extent and distribution of bamboo-dominated forest in south-western Amazon basin are uncertain. Recently, with satellite imagery, aerial photographs and site investigations, it has been calculated that the bamboo-dominant area is close to 180 000 km^2 (Judziewicz et al. 1999), or 18 million hectares.

At least three woody bamboo genera have been reported in this bamboo forest: *Arthrostylidium, Elytrostachys* and *Guadua.* The most dominant genus is *Guadua,* with *G. weberbaueri, G. sarcocarpa, G. superba, G. aff. paraguayana, G. capitata, G. ciliate, G. glomerata* and several unknown species forming an impenetrable tangle of thorny vegetation.

The common view in Brazil is that the bamboo resources of the country are infinite. Even then, the alarming rate of destruction, especially of the ecosystems where the majority of native bamboo species are found, calls for an intensive campaign to protect some bamboo species from indiscriminate felling and extinction (Burman and Filgueiras 1993). It is also recommended that germplasm banks or living collections from different climatic regions of the country be established, and that ecological studies of the species with greatest economic potential be intensified. Some efforts in these areas have already begun; for example, the existing living collections at the Agronomic Institute of the Experimental Center of Campinas in the state of Sao Paulo.

Currently, some Asiatic bamboos, principally the genera of *Bambusa* and *Dendrocalamus,* fill a more important economic role than any native species in Brazil.

Of the 17 genera of woody bamboos native to Brazil, it can be said that only *Actinocladum, Apoclada, Chusquea, Guadua* and *Merostachys* consist of any species that have or could have any potential use.

The native woody bamboo species of Brazil are the following:

Actinocladum verticillatum (Nees) Soderstrom
Alvimia auriculata Soderstrom & X. Londoño
A. gracilis Soderstrom & X. Londoño
A. lancifolia Soderstrom & X. Londoño
Apoclada arenicola McClure
A. cannavieria (Silveira) McClure
A. simplex McClure & L.B. Smith
Arthrostylidium fimbrinodum Judziewicz & L.G. Clark
A. grandifolium Judziewicz & L.G. Clark
A. simpliciusculum (Pilger) McClure
Arthrostachys capitata (Hooker) Benthman
Atractantha amazonica Judziewicz & L.G. Clark

A. aureolanata Judziewicz
A. cardinalis Judziewicz
A. falcata McClure
A. radiata McClure
Aulonemia amplissima (Nees) McClure
A. aristulata (Doell) McClure
A. effusa (Hackel) McClure
A. glaziovii (Hackel) McClure
A. goyazensis (Hackel) McClure
A. radiata (Ruprecht) McClure & L.B. Smith
A. ramosissima (Hackel) McClure
A. setigera (Hackel) McClure
A. ulei (Hackel) McClure & L.B. Smith
Chusquea acuminata Doell
C. anelythra Nees
C. anelytroides Doell
C. attenuata (Doell) L.G. Clark
C. baculifera Silveira
C. bahiana L.G. Clark
C. bambusoides (Raddi) Hackel
C. bradei L.G. Clark
C. caparaoensis L.G. Clark
C. capitata Nees
C. capituliflora var. *capituliflora* Trinius
C. capituliflora var. *pubescens* McClure & L.B. Smith
C. erecta L.G. Clark
C. fasciculata Doell
C. gracilis McClure & L.B. Smith
C. heterophylla Nees
C. ibiramae McClure & L.B. Smith
C. juergensii Hackel
C. leptophylla Nees
C. linearis N.E. Brown
C. meyeriana Doell
C. microphylla (Doell) L.G. Clark
C. mimosa ssp. *australis* L.G. Clark
C. mimosa ssp. *mimosa* McClure L.B. Smith
C. nudiramea L.G. Clark
C. nutans L.G. Clark
C. oligophylla Ruprecht
C. oxylepis (Hackel) E. Ekman
C. pinifolia (Nees) Nees
C. pulchella L.G. Clark
C. ramosissima Lindman

C. riosaltensis L.G. Clark
C. sclerophylla Doell
C. sellowii Ruprecht
C. tenella Nees
C. tenuiglumis Doell
C. tenuis E.G. Camus
C. urelytra Hackel
C. wilkesii Munro
C. windischii L.G. Clark
Colanthelia burchellii (Munro) McClure
C. cingulata (McClure & L.B. Smith) McClure
C. distans (Trinius) McClure
C. intermedia (McClure & L.B. Smith) McClure
C. lanciflora (McClure & L.B. Smith) McClure
C. macrostachya (Nees) McClure
C. rhizantha (Hackel) McClure
Cricuma asymmetrica Soderstrom & X. Londoño
Elytrostachys sp.
Eremocaulon aureofimbriatum Soderstrom & X. Londoño
Glaziophyton mirabile Franchet
Guadua calderoniana X. Londoño & Peterson
G. ciliata X. Londoño & Davidse
G. glomerata Munro
G. latifolia Kunth
G. macrostachya Ruprecht
G. maculosa (Hackel) E.G. Camus
G. paniculata Munro
G. aff. paraguayana Doell
G. sarcocarpa ssp. *sarcocarpa* X. Londoño & Peterson
G. superba Huber
G. tagoara (Nees) Kunth
G. trinii (Nees) Ruprecht
G. virgata (Trinius) Ruprecht
G. weberbaueri Pilger
Merostachys abadiana Sendulsky
M. argentea Sendulsky
M. argyronema Lindman
M. bifurcata Sendulsky
M. bradei Pilger
M. burmanii Sendulsky
M. calderoniana Sendulsky
M. caucaiana Sendulsky
M. ciliata McClure & L.B. Smith
M. clausenii var. *clausenii* Munro

M. clausenii var. *mollior* Doell
M. exserta E.G. Camus
M. filgueirasii Sendulsky
M. fischeriana Doell
M. fistulosa Doell
M. glauca Sendulsky
M. kleinii Sendulsky
M. kunthii Ruprecht
M. lanata Sendulsky
M. leptophylla Sendulsky
M. magellanica Sendulsky
M. medullosa Sendulsky
M. multiramea Hackel
M. neesii Ruprecht
M. petiolata Doell
M. pilifera Sendulsky
M. pluriflora E.G. Camus
M. polyantha McClure
M. procerrima Sendulsky
M. ramosissima Sendulsky
M. riedeliana Doell
M. rodoniensis Sendulsky
M. scandens Sendulsky
M. sellovii Munro
M. skortzovii Sendulsky
M. sparsiflora Ruprecht
M. speciosa Sprengel
M. ternata Needs
M. vestita McClure & L.B. Smith
Myriocladus grandifolius Swallen
M. neblinaensis Swallen
M. paludicolus Swallen
M. virgatus Swallen
Rhipidocladum parviflorum (Trinius) McClure

Bolivia (9 genera, 24 species, 1 subspecies)

Bamboo in Bolivia has been used by Andean communities in traditional production of musical instruments and baskets, and as cattle forage. An estimate of the area of bamboo forest in the highlands or lowland regions of this country does not exist.

The botanical survey conducted during the last decade in the lowland regions of Bolivia has reported two genera: *Actinocladum* and *Elytrostachys,* in addition to three species of *Guadua* to be added to Bolivia's bamboo inventory. Eastern

Bolivia has *G. capitata, G. chacoensis* and *G. paniculata* growing in dry closed-type plant communities; two other species, *G. sarcocarpa* and *G. superba,* occur in north-eastern Bolivia at the border with Brazil and Peru, in the huge bamboo-dominated forest. From the 24 species reported in Bolivia, only three have some economic and/or cultural role: *Aulonemia queko, Rhipidocladum harmonicum* and *Guadua paniculata.*

Guadua paniculata in Bolivia occurs in the departments of Beni and Santa Cruz, growing in dry forests at 200–700 m elevation. It is locally named "guuapa" and is extracted from existing native populations. There is a need for silvicultural plans for the management of the natural stands or for the establishment of new plantations.

Following, the native wood bamboo species of Bolivia:

Actinocladum verticillatum (Nees) Soderstrom
Arthrostylidium canaliculatum Renvoize
Aulonemia boliviana Renvoize
A. herzogiana (Henrard) McClure
A. logipedicellata Renvoize
A. queko Goudot
A. tremula Renvoize
Chusquea delicatula Hitchcock
C. longipendula O. Kuntze
C. lorentziana Grisebach
C. scandens Kunth
C. spicata Munro
Elytrostachys sp.
Guadua capitata (Trinius) Munro
G. chacoensis (Rojas) X. Londoño & Peterson
G. paniculata Munro
G. paraguayana Doell
G. sarcocarpa ssp. *purpuracea* X. Londoño & Peterson
G. sarcocarpa ssp. *sarcocarpa* X. Londoño & Peterson
G. superba Huber
G. weberbaueri Pilger
Merostachys sp.
Neurolepis weberbaueri Pilger
Rhipidocladum harmonicum (Parodi) McClure
R. racemiflorum (Steudel) McClure

Peru (8 genera, 36 species, 1 subspecies)

Bamboo resources in Peru are abundant and could represent a large potential supply. Some efforts have been made by the government to develop bamboo resources, especially in the Amazon region. Some data exist regarding the extent of bamboo area in this region. It is known, for example, that the Ucayali river basin (between the Tambo and Urubamba), has an area of 400 000 ha covered with *Guadua* (Londoño & Peterson, 1991; ONERN, 1976). However, there is no data on the area of each species or genera at the national or regional level.

Peru could be one of the richest Andean countries in terms of bamboo diversity but more fieldwork and taxonomic research need to be done. At present, Peru is known to have 8 genera and 36 species, with the richest bamboo diversity being in the departments of Pasco and Cuzco, and with the largest bamboo area in Madre de Dios and Amazonas.

Guadua weberbaueri and *G. sarcocarpa* are widespread in the lowland Amazonian region of Peru (departments of Amazonas, Cuzco, Huanuco, Junin, Loreto, Madre de Dios, Pasco and San Martin), covering more than 500 000 ha at elevation of 100 to 1 500 m in sites of mature and successional forest on alluvial soil along rivers (Londoño and Peterson 1991). They are locally known by different names such as "paca", "ipa", "Kapiro", "Mame", "Marona" or "chig kan" (McClure 1947; Londoño and Peterson 1991)[1]. These two species are notorious for their well-developed thorns, present on both culms and branches, and for forming dense curtains extremely difficult to penetrate.

The native woody bamboo species of Peru are:

Arthrostylidium simpliciusculum (Pilger) McClure
Aulonemia haenkii (Ruprecht) McClure
A. hirtula (Pilger) McClure
A. humillima (Pilger) McClure
A. parviflora (Presl) McClure
A. queko Goudot
Chusquea aspera L.G. Clark
C. barbata L.G. Clark
C. decolorata Munro
C. delicatula Hitchcock
C. depauperata Pilger
C. dombeyana Kunth
C. exasperata L.G. Clark
C. huantensis Pilger
C. inamoena Pilger
C. neurophylla L.G. Clark
C. peruviana D.G. Camus
C. picta Pilger
C. polyclados Pilger
C. pubispicula Pilger
C. scandens Kunth
C. smithii L.G. Clark
C. spicata Munro
C. straminea Pilger

[1] Also Tovar, A. 1998. Personnal communication on PROBONA Project. Quito, Ecuador.

C. tarmensis Pilger
C. tessellata Munro
Elytrostachys sp.
Guadua glomerata Munro
G. sarcocarpa ssp. *purpuracea* X. Londoño & Peterson
G. sarcocarpa ssp. *sarcocarpa* X. Londoño & Peterson
G. superba Huber
G weberbaueri Pilger
Merostachys brevispica Munro
Neurolepis fimbriligulata ssp. *peruviana* L.G. Clark
N. weberbaueri Pilger
Rhipidocladum harmonicum (Parodi) McClure
R. racemiflorum (Steudel) McClure

The Guyanas (8 genera, 13 species)

Bamboo resources in the Guyanas (French Guyana, Guyana and Surinam) are unimportant to the local economy. Some ethnobotanical uses – such as for making lances, bows, arrowheads, arrow shafts, knives and fishing harpoons – have been reported by the indigenous people.

The most promising species are found in the genus *Guadua,* namely *G. latifolia, G. macrostachya, G. weberbaueri* and *G. glomerata.*

Other uses of bamboo in this region is limited to minor uses such tools and small implements for fishing and hunting, and as a medicinal plant (Grenand et al. 1987).

Following are the native woody bamboo species of the Guyanas:
Arthrostylidium scandens McClure
A. venezuelae (Steudel) McClure
Aulonemia deflexa (N.E. Brown) McClure
Chusquea linearis N.E. Brown
Guadua glomerata Munro
G. latifolia (Humb & Bonpl) Kunth
G. macrostachya Ruprecht
G. weberbaueri Pilger
Merostachys retrorsa McClure
Myriocladus distantiflorus Swallen
Neurolepis angusta Swallen
Rhipidocladum racemiflorum (Steudel) McClure
R. sibilans Davidse, Judziewicz, & L.G. Clark

NORTHWESTERN SOUTH AMERICA

Colombia (70 species, 9 genera)

Colombia has the second highest woody bamboo diversity in Latin America. At present, 9 genera and 70 species are reported, with 24 species being endemic and at least 12 species remaining to be described.

The Andean region has the largest quantity and greatest diversity of woody species (89%) and the Eastern Cordillera is the richest with 55% of all woody bamboos reported until now. About 60% of the woody species are found from 2 000 to 3 500 m altitude (Londoño 1990).

The Colombian departments (states) with the highest woody bamboo diversity are Norte de Santander, Cundinamarca, Cauca, Valle del Cauca, Antiquia, Huila, Narino and Quandio.

The majority of species belong to *Chusquea* genus (30%), with the rest of the species belonging to the genera *Neurolepis, Arthrostylidium, Aulonemia, Elytrostachys, Merostachys, Rhipidocladum, Guadua* and *Otatea.*

Colombia is one of the few countries in Latin America where bamboo plays a notable role in local economy and traditional culture (McClure 1966).

In Colombia, the genera *Aulonemia, Chusquea, Elytrostachys, Guadua* and *Rhipidocladum* include several species that are used by various native and rural mestizo communities to satisfy basic necessities (Londoño 1990). Even though only *G. angustifolia* has any real economic value, there is a history of bamboo culture, processing and utilization in the country. There are population inventories as well as significant research on management, production, marketing and industrial uses for *G. angustifolia*. The impact of the rest of the species on the local economy is minimal. These are utilized only by people close to the source.

Colombia's area of *Guadua* forests is estimated at 51 500 ha, of which 46 261 ha area has natural forests and 5 260 ha cultivated ones. The total area of natural *Guadua* forest in the four central-western departments represents 44% of the national total, with 75% of the *Guadua* stands between 0.4 and 10 ha in size (Bernal 1996; CVC-PAFC 1995). Only 16–20% of this area has proper management (CVC-PAFC 1995).

With an average production of 1 000 culms/ha/year (Castaño 1993), the total *Guadua* culm production for the four central-western departments can be calculated to be 20.3 million of green culms per year, equivalent to 911 745 ton/year in green conditions (45 T-m/ha).

The native woody bamboo species of Colombia are:

Arthrostylidium auriculatum X. Londoño & L.G. Clark
A. chiribiquetensis X. Londoño & L.G. Clark
A. ecuadorense Judziewicz & L.G. Clark
A. pubescens Ruprecht
A. punctulatum X. Londoño & L.G. Clark
A. simpliciusculum (Pilger) McClure

A. venezuelae (Steud) McClure
A. virolinensis X. Londoño & L.G. Clark
A. youngianum L.G. Clark & Judziewicz
Aulonemia bogotensis L.G. Clark, X. Londoño & Kobayashi
A. patula (Pilger) McClure
A. pumila L.G. Clark & Londoño
A. queko Goudot
A. robusta L.G. Clark & X. Londoño
A. trianae (Munro) McClure
Chusquea albilanata L.G. Clark & Londoño
C. angustifolia (Soderstrom & Calderon) L.G. Clark
C. antioquensis L.G. Clark & X Londoño
C. arachniforme L.G. Clark & X. Londoño
C. aff. fendleri Munro
C. grandiflora L.G. Clark
C. latifolia L.G. Clark
C. lehmannii ssp. *lehmannii* Pilger
C. lehmannii ssp. *farinosa* L.G. Clark & X. Londoño
C. ligulata Munro
C. londoniae L.G. Clark
C. longiprophylla L.G. Clark
C. maculata L.G. Clark
C. pallida Munro
C. purdieana Munro
C. scandens Kunth
C. serpens L.G. Clark
C. serrulata Pilger
C. simpliciflora Munro
C. sneidernii Asplund
C. spadicea Pilger
C. spathacea L.G. Clark
C. spencei Ernest
C. spencei X *C. tessellata* (hybrid)
C. subulata L.G. Clark
C. tessellata Munro
C. tuberculosa Swallen
C. uniflora Steudel
Elytrostachys clavigera McClure
E. typical McClure
Guadua amplexifolia Presl
G. angustifolia Kunth
G. angustifolia var. *bicolor* Londoño
G. angustifolia var. *nigra* Londoño
G. glomerata Munro

G. paniculata Munro
G. superba Huber
G. weberbaueri Pilger
Merostachys sp.
Neurolepis acuminatissima (Munro) Pilger
N. aperta (Munro) Pilger
N. aristata (Munro) Hitchcock
N. mollis Swallen
N. nobilis (Munro) Pilger
N. petiolata Davidse & L.G. Clark
N. silverstonei Davidse & L.G. Clark
N. stuebelii (Pilger) Pilger
Otatea fimbriata Soderstrom
Rhipidocladum abregoensis X. Londoño & L.G. Clark
R. angustiflorum (Stapf) McClure
R. geminatum (McClure) McClure
R. harmonicum (Parodi) McClure
R. longispiculatum X. Londoño & L.G. Clark
R. parviflorum (Trinius) McClure
R. racemiflorum (Steudel) McClure

Ecuador (6 genera, 42 species)

Ecuador, for its size, possesses an impressive diversity of woody bamboos (Clark 1997). Up to the present, 6 genera and 42 species have been identified (11 endemics ones), with about 15 species remaining to be described.

It is difficult to evaluate the contribution that the rural communities have made to the country's economy because their productive effort has been statistically underestimated. Unlike other Latin American countries, Ecuador has included some bamboo species of the Andean forest (2 000 m above sea level) in its natural resources management program.

The Ecuadorian woody bamboos are largely montane, with half of the species found at an altitude of 2 500 to 3 500 m (Clark 1997). The greatest diversity in Ecuador is found in the eastern Cordillera, with 74% of the total species, followed by the western Cordillera with 38% (Clark 1997).

The provinces of Loja, Napo, Pichincha, Azuay and Chimborazo have the greatest woody bamboo diversity. Although the Pacific watershed has the least bamboo species diversity, it is where bamboo has primary importance in terms of economy and applications.

Among the 6 woody bamboo genera, *Chusquea* claims the vast majority (41%) of the species. *Arthrostylidium, Aulonemia, Rhipidocladum, Neurolepis* and *Guadua* are also present, but with lower diversity.

G. angustifolia, because of its economic importance, occupies a prime position in the economy of the country. Species of *Chusquea, Aulonemia* and *Rhipidocladum* also have several reported uses among the highlands communities.

The area of *Guadua angustifolia* in the Ecuadorian natural forest is estimated to be 2 000 ha, of which 18 792 ha occurs in the five provinces of the Coastal Region. The total area of natural *Guadua* forest in the Coastal Region represents approximately 75% of the national total, with 93% of the *Guadua* stands below 0.5 ha in size.

With an estimated average production of 1 376 culms/ha/year, the total amount of *Guadua* culm production for this region is 25.86 million of green culms per year, equivalent to 814 500 tons/year in green conditions (43.3 T-m/ha in green condition).

The banana producers have selected *Bambusa vulgaris* as the most recommended bamboo for the production of "cujes" (props) because of its ease of reproduction, rapid growth, greater growth of culms per unit area and smaller culm size (which facilitates harvesting and use).

The native woody bamboo species of Ecuador are the following:

Arthrostylidium ecuadorense Judziewicz & L.G. Clark
A. simpliciusculum (Pilger) McClure
A. youngianum L.G. Clark & Judziewicz
Aulonemia haenkei (Ruprecht) McClure
A. hirtula (Pilger) McClure
A. longiaristata L.G. Clark & X. Londoño
A. patula (Pilger) McClure
A. queko Goudot
Chusquea albilanata L.G. Clark & X. Londoño
C. exasperata L.G. Clark
C. falcata L.G. Clark
C. lehmannii ssp. *lehmannii* Pilger
C. lehmannii ssp. *farinosa* L.G. Clark & X. Londoño
C. leonardiorum L.G. Clark
C. londoniae L.G. Clark
C. loxensis L.G. Clark
C. macclurei L.G. Clark
C. neurophylla L.G. Clark
C. perligulata (Pilger) McClure
C. aff. polyclados Pilger
C. scandens Kunth
C. serpens L.G. Clark
C. simpliciflora Munro
C. subulata L.G. Clark
C. tessellata Munro
C. uniflora Steudel

Guadua angustifolia Kunth
G. superba Huber
G. weberbaueri Pilger
Neurolepis aperta (Munro) Pilger
N. aristata (Munro) A. Hitchcock
N. asymmetrica L.G. Clark
N. elata (Kunth) Pilger
N. fimbriligulata ssp. *fimbriligulata* L.G. Clark
N. nana L.G. Clark
N. nobilis (Munro) Pilger
N. rigida L.G. Clark
N. stuebelii (Pilger) Pilger
N. villosa L.G. Clark
N. weberbaueri Pilger
Rhipidocladum harmonicum (Parodi) McClure
R. racemiflorum (Steudel) McClure

Venezuela (10 genera, 60 species)

Venezuela has the third highest woody bamboo diversity in Latin America. At present 10 genera and 60 species, 18 of them endemics, are reported with several species remaining to be described. A large part of the diversity is in the Guyana region, which has the endemic genus *Myriocladus* as well as endemic species of *Arthrostylidium, Aulonemia, Neurolepis* and *Rhipidocladum.*

The Venezuelan states with the highest woody bamboo diversity are: Bolivar, Amazonas, Tachira, Distrito Federal and Merida. The southern states (Amazonas and Bolivar) hold the greater diversity and number of woody bamboo species (63%), followed by the Andean region (38%).

The majority of the species belong to the genera *Myriocladus* and *Chusquea* (37%). The rest of the species are from the genera *Guadua, Aulonemia, Arthrostylidium, Neurolepis, Rhipidocladum, Elytrostachys, Merostachys* and *Atractantha.*

The genus *Guadua* is well represented in Venezuela and only Brazil has a greater *Guadua* species diversity. Currently in Venezuela, there are seven species described and confirmed with another three or so remaining to be described.

The use of bamboo in the country is not significant and it contributes little to the local economy. However, interest has been expressed by various government agencies in bamboo as a natural resource.

The genera *Guadua* and *Bambusa* are the only ones that might have at least a modest economic value in this country. Bamboo is still used by peasants and indigenous communities for building homes and for making various household utensils (Pittier 1926; Corredor 1966).

Bambusa bambos and *B. vulgaris* are widely cultivated in Venezuela and are mainly used to construct homes, to make racks to dry tobacco and to stabilize riverbanks.

Guadua angustifolia and *B. vulgaris* are the most promising species.

Guadua angustifolia occurs in the states of Merida, Tachira, Barinas and Amazonas, at lower and middle montane forest, mainly at the edge of riverbanks and creeks, at elevations of 100 to 1 000 m. It is locally called "guafa", "guafa real" or "bambu' (Mendoza 1992).

Guadua angustifolia has been studied in the Forestry Reserve of Ticoporo, in the state of Barinas, where several undergraduate theses from the Faculty of Forestry Sciences of the University of Los Andes were carried out (Calderon and Areque 1965; Corredor 1966; Perez 1967; Mejia 1975; Zerpa 1987; Paredes 1991).

The total area of natural *Guadua* forest in Venezuela is difficult to estimate. There is no national or regional inventory of the *Guadua* forest or clumps that could help in calculating an estimated area. The only available inventory is about the 6 000 ha of *Guadua* (2.88%) in the Forestry Reserve of Ticoporo, which extends to 210 000 ha (Calderon and Areque 1965).

According to Corredor (1966), the average green weight of *G. angustifolia* in Venezuela is 59.6 kg, with an average green weight equal to 44.5 T-m/ha, equivalent to an average dry weight of 30.7 T-m/ha. On comparison with Asiatic species, such as *Melocanna bambusoides* (21 T-m/ha), *B. polymorpha* (22 T-m/ha) and *B. arundinacea* (32–36 T-m/ha), *G. angustifolia* represents a bountiful harvest (Corredor 1966).

The native woody bamboo species of Venezuela are:

Arthrostylidium longiflorum Munro
A. pubescens Ruprecht
A. sarmentosum Pilger
A. scandens McClure
A. schomburgkii (Bennett) Munro
A. venezuelae (Steudel) McClure
Atractantha amazonica Judziewicz & L.G. Clark
Aulonemia chimantaensis Judziewicz & Davidse
A. deflexa (N.E. Brown) McClure
A. jauaensis Judziewicz & Davidse
A. purpurata (McClure) McClure
A. robusta L.G. Clark & X. Londoño
A. subpectinata (O. Kuntze) McClure
A. trianae (Munro) McClure
Chusquea angustifolia (Soderstrom & Calderon) L.G. Clark
C. fendleri Munro
C. linearis N.E. Brown
C. maculata L.G. Clark

C. pallida Munro
C. scandens Kunth
C. serpens L.G. Clark
C. simpliciflora Munro
C. spencei Ernst
C. tessellata Munro
Elytrostachys clavigera McClure
E. typica McClure
Guadua amplexifolia Presl
G. angustifolia Kunth
G. ciliata X. Londoño & Davidse
G. glomerata Munro
G. latifolia Kunth
G. paniculata Munro
G. aff. paraguayana Doell
G. weberbaueri Pilger
Merostachys maguireorum McClure
M. retrorsa McClure
Myriocladus cardonae Swallen
M. churunensis Swallen
M. distantiflorus Swallen
M. exsertus Swallen
M. grandifolius Swallen
M. involutus Judziewicz & Davidse
M. longiramosus Swallen
M. neblinaensis Swallen
M. paludicolus Swallen
M. simplex Swallen
M. steyermarkii Swallen
M. virgatus Swallen
Neurolepis angusta Swallen
N. diversiglumis Soderstrom
N. glomerata Swallen
N. mollis Swallen
N. pittieri McClure
N. virgata (Grisebach) Pilger
Rhipidocladum ampliflorum (McClure) McClure
R. angustiflorum (Stapf) McClure
R. geminatum (McClure) McClure
R. parviflorum (Steudel) McClure
R. racemiflorum (Steudel) McClure
R. sibilans Davidse, Judziewicz & Clark

CENTRAL AMERICA

The increase in the use of bamboo in Central America in the last 20 years was driven in large measure by the scarcity of other natural resources owing to deforestation, over-harvesting and overpopulation, and to the limitations of the economies of these countries. This has encouraged people to view this rapidly growing, renewable resource as an alternative material for solving housing problems, for the protection and recuperation of the environment, and for jobs creation through the development of the cottage-level handicrafts industry and the construction sector.

Panama (7 genera, 21 species)

In Panama, bamboo is scarcely used as an economic plant. However, Panama is the third richest country in woody bamboo diversity in Central America: 43% of its species belong to the genus *Chusquea,* and the rest belong to the genera *Arthrostylidium, Aulonemia, Elytrostachys, Guadua, Neurolepis* and *Rhipidocladum.*

G. angustifolia has been cultivated in the Canal Zone Experimental Garden, from where it has been distributed to several Latin American countries, including Peru, Costa Rica, Guatemala and Cuba (US National herbarium). It is doubtful whether this species exists in this country in any natural distribution; it may have spread from the Canal Zone, where it was introduced from Ecuador in 1924 by Mr. Holgar Johansen, according to specimen at the Smithsonian Institution in Washington, D.C.

According to Young and Judd (1992), some of the plants collected as wild have been so suspiciously close to the experiment stations that they might be just escapes. Another unresolved problem in its natural distribution is the role that human cultivation has played in transporting this huge potential species (Young and Judd 1992).

The following woody bamboo species are native to Panama:

Arthrostylidium excelsum Grisebach
A. judziewiczii Davidse
A. pubescens Ruprecht
A. venezuelae (Steudel) McClure
Aulonemia patriae R. Pohl
Chusquea amistadensis L.G. Clark, Davidse & Ellis
C. grandiflora L.G. Clark
C. longifolia Swallen
C. pittieri Hackel
C. pohlii L.G. Clark
C. serpens L.G. Clark
C. simpliciflora Munro
C. subtessellata Hitchcock

C. vulcanalis (Soderstrom & Calderon) L.G. Clark
Elytrostachys clavigera McClure
Guadua angustifolia Kunth
G. macclurei R. Pohl & Davidse
G. paniculata
Neurolepis pittieri McClure
Rhipidocladum panamense R. Pohl
R. racemiflorum (Steudel) McClure

Costa Rica (8 genera, 39 species)

Costa Rica is the country with the largest woody bamboo diversity in Central America. At present, 8 genera and 39 species of woody bamboos are reported (Montiel 1998).

The majority of the Costa Rican native woody bamboos (54%) belong to the genus *Chusquea,* and the rest belong to the genera *Arthrostylidium, Rhipidocladum, Guadua, Aulonemia, Elytrostachys, Merostachys* and *Neurolepis.*

Costa Rica and Mexico are the only countries in Central America where bamboo, especially *G. angustifolia,* has been developed in the last 15 years and where it occupies a significant position in the economy. However, the uses of most native species in Costa Rica have been insignificant (Montiel 1998). It is known that *Chusquea pittieri,* locally called "cana brava" or "caneula", is used in the province of Cartago for Christmas decorations (herbarium specimen: Pohl 13098). Its culms reach 10–20 m in height and up to 5 cm in diameter, are straight at the base and arched at the tip (Davidse et al. 1994). According to Rodriguez (1990), *C. pittieri* is a promising species for light structures and furniture.

Elytrostachys clavigera has been developed in the Valley of Talamanca as a potential material for bamboo weaving (Widmer 1990). It occurs naturally in the provinces of Alajuela, Punctarenas, Limon, San José and Heredia.

The genus *Guadua* in Costa Rica is represented by three species: *G. amplexifolia* (locally called "cana brava") in the Provinces of Guanacaste and Nicoya; *G. macclurei* in the provinces of Alajuela and Puntarenas; and *G. paniculata* in the provinces of Guanacaste and Puntarenas. The presence of *G. aculeata* needs to be confirmed in the northern part of the country. The species *G. angustifolia* is introduced.

Following, the native woody bamboo species of Costa Rica:

Arthrostylidium excelsum Grisebach
A. judziewiczii Davidse
A. merostachyoides R. Pohl
A. pubescens Ruprecht
A. venezuelae (Steudel) McClure
Aulonemia patriae R. Pohl
A. viscosa (Hitchcock) McClure
Chusquea amistadensis L.G. Clark, Davidse & Ellis

C. coronalis Soderstrom & Calderon
C. foliosa L.G. Clark
C. liebmannii Fournier
C. longifolia Swallen
C. longiligulata (Soderstrom & Calderon) L.G. Clark
C. paludicola L.G. Clark
C. patens L.G. Clark
C. pittieri Hackel
C. pohlii L.G. Clark
C. scabra Soderstrom & Calderon
C. serpens L.G. Clark
C. simpliciflora Munro
C. subtessellata Hitchcock
C. subtilis Widmer & L.G. Clark
C. sulcata Swallen
C. talamancensis Widmer & L.G. Clark
C. tomentosa Widmer & L. Clark
C. tonduzii Hackel
C. virgata Hackel
C. vulcanalis (Soderstrom & Calderon) L.G. Clark
Elytrostachys clavigera McClure
Guadua amplexifolia Presl
G. macclurei Pohl & Davidse
G. paniculata Munro
Merostachys latifolia R. Pohl
Neurolepis pittieri McClure
Rhipidocladum clarkiae R. Pohl
R. maxonii (Hitchcock) McClure
R. pacuarense R. Pohl
R. pittieri (Hackel) McClure
R. racemiflorum (Steudel) McClure

Nicaragua (4 genera, 9 species)

Nicaragua is a country with relatively few species of native bamboos but with widespread *Guadua* species covering the eastern alluvial plains. The genus *Guadua* includes four species: *G. amplexifolia, G. aculeata, G. macclurei* and *G. paniculata.*

G. amplexifolia occurs in greatest abundance in eastern Nicaragua, in a narrow but more or less continuous band along the rivers emptying into the Caribbean between Puerto Cabezas and Bluefields. Occasionally, it occurs in openings in the forest either as pure stands or mixed with trees and shrubs. It is known to the Mesquite community as "cauro" and is used in local construction (McClure and Stenvenson 1946).

McClure and Stevenson (1946) estimated a *G. amplexifolia* cover of 12 000 ha; the largest *Guadua* forests were found along the rivers Prinzapolca, Wawa, Coco and Rio Grande. It is probable that the most preserved bamboo area is in the north-eastern part of the country.

In 1945, *Guadua angustifolia* was introduced from the Canal Zone, Panama, into the Agricultural Experimental Station E1 Recreo in Bluefields, where other exotic bamboos were also introduced (Widmer 1990).

It can be said that Nicaragua is a country where government and private institutions have made intensive efforts to explore and develop its bamboo resources.

The native woody bamboo species of Nicaragua are:

Arthrostylidium excelsum Grisebach
A. venezuelae (Steudel) McClure
C. simpliciflora Munro
Guadua amplexifolia Presl
G. aculeata Fournier
G. macclurei R. Pohl & Davidse
G. paniculata Munro
Rhipidocladum pacuarense R. Pohl
R. pittieri (Hackel) McClure
R. racemiflorum (Steudel) McClure

Honduras (6 genera; 11 species)

Of the 11 native woody bamboo species occurring in Honduras, only 4, besides several Asiatic ones, have economic potential.

Otatea fimbriata, Guadua amplexifolia, G. aculeata and *G. paniculata* are the native bamboos that can be promoted as useful in environmental protection and in economic activities.

At present, *G. aculeata* grows wild at the western end of Cordillera de Nombre de Dios at 700–900 m elevation, and probably also in the Lancetilla valley. *G. amplexifolia* occurs in the departments of Cortez and Comayagua at 200–700 m elevation. *G. paniculata* is the most widespread *Guadua* species in the country; it is found along riverbanks between Guatemala and Honduras border, and in the departments of El Paraiso, Olancho and Cortez at elevations of 300–600 m.

During the height of banana production, the United Fruit Company made in Lancetilla and extensive experimental planting of native and introduced bamboos for trial as banana props (McClure 1951). About 61% of the exotic bamboos species occurring in Central America are growing in this Experimental Station. *Guadua* was an important resource for the company (McClure 1996). Natural populations of the genus have now diminished by over-harvesting.

Following are the native woody bamboo species of Honduras:

Arthrostylidium excelsum Grisebach
Aulonemia clarkiae Davidse & R. Pohl

C. deflexa L.G. Clark
C. lanceolata Hitchcock
Guadua amplexifolia Presl
G. aculeata Fournier
G. macclurei R. Pohl & Davidse
G. paniculata Munro
Otatea fimbriata Soderstrom
R. pittieri (Hackel) McClure
R. racemiflorum (Steudel) McClure

El Salvador (6 genera, 11 species)

El Salvador is the smallest country in Central America, and it has the highest deforestation and overpopulation. Probably, some of the 11 species reported in the country do not exist any more.

There are several reports on the use of native bamboos – belonging to the genera *Guadua, Otatea* and *Elytrostachys* – that occur below 1 000 m elevation.

G. amplexifolia occurs in areas adjacent to Tecoluca (San Vicentre Department), Zacatecoluca (La Paz Department) and Canton Soledad (Cabañas Department) at elevations of 400 to 800 m. *G. aculeata* occurs in the vicinity of Ahuachapan (Ahuachapan Department) at 900 to 1 050 m elevations. *G. paniculata* (locally known as "cana brava") is found in the Departments of Cuscatlan and Cabañas growing wild on dry ridges at elevations of 540–870 m. More recently, *G. paniculata* has been reported in Chalatenango Department, at 300 m elevation.

McClure brought around 20 species of Asiatic bamboos from Puerto Rico, the Canal Zone and El Paraiso (Honduras) and planted them at San Andres. At present, the National Center of Agricultural Technology (CENTA) has the largest bamboo collection in the country, with 17 species of native and exotic bamboos, mostly from McClure's introductions (Widmer 1990).

The native woody bamboo species in El Salvador are the following:

Chusquea coronalis Soderstrom & Calderon
C. deflexa L.G. Clark
C. liebmannii Fournier
Guadua amplexifolia Presl
G. aculeata Fournier
G. paniculata Munro
Elytrostachys clavigera McClure
Merostachys latifolia R. Pohl
Otatea fimbriata Soderstrom
Rhipidocladum pittieri (Hackel) McClure
R. racemiflorum (Steudel) McClure

Guatemala (5 genera, 17 species)

Guatemala is a country with many species of native woody bamboos, majority of them belonging to the genus *Chusquea* and the rest to the genera *Arthrostylidium, Guadua, Otatea* and *Rhipidocladum.*

Guatemala is one of the first Central American countries to focus its attention on bamboo as a potential renewable resource. Native woody bamboos are widespread in the country. Although 53% of the species belong to the genus *Chusquea*, only two – *C. longifolia* and *C. nelsonii* – are reported as useful.

Guadua is represented by *G. aculeata, G. longifolia* and *G. paniculata.* The largest in size is *G. aculeata,* which occurs at 50–500 m elevations in the departments of Izabal, Quirigua, Retalhuleau and Suchitepequez. The species is locally called "taro" and used for house construction.

G. angustifolia was introduced into Guatemala from Panama in 1946 at the Experimental Gardens of the Instituto Agropecuario Nacional (Hda. Chocola). In the same year, *G. aculeata* from Honduras was also brought and cultivated here. At present, 15 species of Asiatic bamboos are cultivated in the farms at Chocola, Colombia, La Maquina and Pantaleon (Widmer 1990).

The native woody bamboo species of Guatemala are:

Arthrostylidium excelsum Grisebach
Chusquea coronalis Soderstrom & Calderon
C. deflexa L.G. Clark
C. lanceolata Hitchcock
C. liebmannii Fournier
C. longifolia Swallen
C. nelsonii Scribner & J.G. Smith
C. pittieri Hackel
C. simpliciflora Munro
C. sulcata Swallen
Guadua aculeata Fournier
Guadua longifolia (Fournier) R. Pohl
G. paniculata Munro
Otatea fimbriata Soderstrom
Rhipidocladum bartlettii (McClure) McClure
R. pittieri (Hackel) McClure
R. racemiflorum (Steudel) McClure

Belize (4 genera, 4 species)

This small flat country is the only one in Central America without any *Chusquea* species. Belize, however, is well known for the *Guadua longifolia* populations along the Belize river and its tributaries on the Yucatan Peninsula.

Following are the native woody bamboo species of Belize:

Guadua longifolia (Fournier) R. Pohl

Merostachys pauciflora Swallen
Otatea fimbriata Soderstrom
Rhipidocladum bartlettii (McClure) McClure

Mexico (8 genera, 35 species, 2 subspecies)

Mexico is the second most diverse in Central America in terms of woody bamboos. At present, 8 genera and 37 species are reported with one genus, *Olmeca,* and 14 species being endemic.

The majority (47%) of its species belong to the genus *Chusquea,* and the remaining to *Arthrostylidium, Aulonemia, Guadua, Olmeca, Otatea* and *Rhipidocladum.*

In Mexico, some *Guadua* and *Otatea* species are used by local communities. There are several common names for the *Chusquea* species – such as "Carrizo", "otatillo", "camalote" and "chajiche" – and there are reports that the culms of these are used to make baskets.

Guadua has 5 species, almost all of them are utilized for various purposes: *G. aculeata, G. amplexifolia, G. longifolia, G. paniculata* and *G. velutina.*

In the last 5 years *G. angustifolia* has been cultivated in Chiapas and Tabasco. The initial material was brought from Quindio, Colombia, and more than 4 000 ha are cultivated in southern Mexico for commercial purposes. Other commercial plantations are being raised with *G. amplexifolia.*

The most common introduced Asiatic bamboos in Mexico are *Bambusa vulgaris* and *B. vulgaris* var. *vittata.* Probably, there are more Asiatic bamboos but they are not documented in herbarium collections (Davidse et al. 1994).

The woody bamboo species native to Mexico are:

Arthrostylidium excelsum Grisebach
Aulonemia clarkiae Davidse & R. Pohl
A. fulgor Soderstrom
A. laxa (F. Maekino) McClure
Chusquea aperta L.G. Clark
C. bilimekii Fournier
C. circinata Soderstrom & Calderon
C. coronalis Soderstrom & Calderon
C. foliosa L.G. Clark
C. galeottiana Munro
C. glauca L.G. Clark
C. lanceolata Hitchcock
C. liebmannii Fournier
C. longifolia Swallen
C. muelleri Munro
C. nelsonii Scribner & Smith
C. repens ssp. *oaxacacensis* L.G. Clark & X. Londoño
C. repens ssp. *repens* L.G. Clark & X. Londoño

C. perotensis L.G. Clark, G. Cortes & Chazaro
C. pittieri Hackel
C. simpliciflora Munro
C. sulcata Swallen
Guadua amplexifolia Presl
G. aculeata Fournier
G. longifolia (Fournier) R. Pohl
G. paniculata Munro
G. velutina X. Londoño & L.G. Clark
Merostachys sp.
Olmeca recta Soderstrom
O. reflexa Soderstrom
Otatea acuminata ssp. *acuminata* Calderon & Soderstrom
O. acuminata ssp. *aztecorum* (McClure & E. Smith) Guzman et al.
Otatea fimbriata Soderstrom
Rhipidocladum bartlettii (McClure) McClure
R. martinezii Davidse & R. Pohl
R. pittieri (Hackel) McClure
R. racemiflorum (Steudel) McClure

THE CARIBBEAN AND WEST INDIES

The native bamboo species in the Caribbean and West Indies are confined to small-sized bamboos less than 1 cm in diameter. Because of this, they do not have any economic value. However, Asiatic species have entered conspicuously into the local economy on some islands.

At present, the Caribbean and West Indies are known to have 4 genera and 36 species of native woody bamboos, with the richest bamboo diversity being in Cuba. Like Chile, the Caribbean the West Indies islands also do not have any natural distribution of the genus *Guadua.*

Bambusa vulgaris is perhaps the most widely cultivated exotic species in the Caribbean region. It is most abundant in the islands of Cuba, Jamaica, Dominican Republic and Puerto Rico, where it has become an important wood source for multiple uses.

The Caribbean not only initiated the industrialization of bamboo in the region, with a small mill located in near St. Augustine in Trinidad, but also initiated the introduction and cultivation of exotic and useful bamboos from the Old World.

The Tropical Agricultural Research Station (TARS) in Mayaguez, Puerto Rico, began the introduction and cultivation of bamboo at the time of its foundation in 1901, and continues to the present day (Edelman et al. 1985). The 1930s and 1940s were perhaps the most active period at TARS with respect to the development of bamboo. During this time, bamboo species from the collection were widely distributed in Puerto Rico and other parts of the Caribbean as well as in several countries of Latin America.

The species *Guadua angustifolia* has been introduced to several Caribbean islands. Besides Puerto Rico, it is cultivated also in the Botanical Garden of Cienfuegos, Cuba, and it is probable that the species has been introduced to the Dominic Republic and Jamaica.

The native woody bamboo species of the Caribbean and West Indies are:

Arthrostylidium angustifolium Nash
A. banaoense Catasus Guerra
A. cubense Ruprecht
A. distichum Pilger
A. ekmanii Hitchcock
A. excelsum Grisebach
A. farctum (Aublet) Soderstrom & Lourteig
A. fimbriatum Grisebach
A. haitiense (Pilger) Hitchcock & Chase
A. multispicatum Pilger
A. obtusatum Pilger
A. pinicolium Catasus Guerra
A. reflexum Hitchcock & E. Ekman
A. sarmentosum Pilger
A. urbanii Pilger
A. venezuelae (Steudel) McClure
Chusquea abietifolia Grisebach
Neurolepis virgata (Grisebach) Pilger – Trinidad
Rhipidocladum prestoei (Munro) McClure – Trinidad

ACKNOWLEDGEMENTS

The author is grateful to the International Network for Bamboo and Rattan, and Dr Cherla Sastry, Dr Ramanuja Rao and Mr. William Reidhead for making it possible to present this work at the International Bamboo Congress and Workshop. Research on the evaluation of bamboo resources in Latin America was supported by INBAR Grant No. 96-8300-01-4. Without the help and support of large number of people, the completion of this study would not have been possible. I extend my gratitude to all of them, especially Lynn Clark, Trevor Dagilis, Bernie Lemus and Hugo Cota who provided valuable insights and advice, particularly in translating the final report into English. I am grateful to the many curators and staff of the various herbaria in Colombia, Ecuador, Venezuela and the United States who provided specimens and extended hospitality throughout this work.

REFERENCES

Azzini, A.; de Barros, A.L. 1993. Posibilidades del bambu como materia prima industrial. *In* Memorias del Primer Congreso Mundial Bambu/Guadua, Pereira, Colombia, 8–15 August 1992. Servicio Nacional de Aprendizaje, Bogota, Colombia. pp. 139–144.

Bernal L.E. 1996. Actualizacion del inventario departmental de Guadua. Final Report, Pereira, Colombia. (Unpublished). 7 pp.

Burkart, A. 1969. Bambusoideae. *In* Flora ilustrada de Entre Rios, Vol. 2, Coleccion Cientifica del I.N.R.A., Buenos Aires, Argentina. pp. 32–36.

Burman, A.G.; Filgueiras, T.S. 1993. A review of the woody bamboo genera of Brazil (Gramineae: Bambusoideae: Bambuseae). Thaiszia, 3, 53–88.

Calderon, J.; Areque, 1965. Estimacion de las existencias de bambu en la Reserva Forestal de Ticoporo, Estado Barinas. Thesis, Universidad de los Andes, Facultad de Ciencias Forestales, Merida, Venezuela.

Castano, F. 1993. La silvicultura de la guadua en Colombia. *In* Memorias del Primer Congreso Mundial Bambu/Guadua, Pereira, Colombia, 8–15 August 1992. Servicio Nacional de Aprendizaje, Bogota, Colombia. pp. 69–72.

CESA (Central Ecuatoriana de Servicions Agricolas). 1992. El deterioro de los bosques naturales del callenjon interandino del Ecuador. Central Ecuatoriana de Servicions Agricolas, Quito, Ecuador.

Clark, L.G. 1995. Diversity and distribution of the Andean woody bamboos (Poaceae: Bambuseae). *In* Churchill, S.P.; Baslev, H.; Forero, E.; Luteyn, J., ed., Biodiversity and Conservation of Neotropical Montane Forests. New York Botanical Garden, Bronx, New York, USA. pp. 501–512.

Clark, L.G. 1997. Diversity and biogeography of Ecuadorian bamboos (Poaceae: Bambusoideae) and their allies. *In* Valencia, R.; Balslev, H., ed., Estudios sobre Diversidad y Ecologia de plantas. Memorias del II Congreso Ecuatoriano de Botanica, 16–20 Octobrr 1995. Pontificia Universidad Catolica del Ecuador, Quito, Ecuador. pp. 51–63.

Corredor, J.R. 1966. Algunos aspectos relacionados con la medicion del bambu para las negociaciones de compre-venta. Thesis, Universidad de los Andes, Facultad de Ciencias Forestales, Merida, Venezuela.

CVC-PAFC. 1995. La Guadua: Alternativa Economica para el Desarrollo Sostenible. Corporacion Autonoma Regional del Valle del Cauca and PAFC, Cali, Colombia.

Davidse, G.; Sousa, M.; Chater, A.O. 1994. Flora Mesoamericana Vol. 6, Alismataceae a Cyperaceae. Bambusoideae. Universidad Nacional Autonoma de Mexico, Mexico City, Mexico.

de Barros, A.S.; Egashira, Y. 1993. Cultivo commercial del bambu para approvechamiento industrial en celulosa, papel y etanol. *In* Memorias del Primer Congreso Mundial Bambu/Guadua, Pereira, Colombia, 8–15 August 1992. Servicio Nacional de Aprendizaje, Bogota, Colombia. pp. 73–75.

Edelman, D.; Soderstrom, T.R.; Dietzer, G.F. 1985. Bamboo introduction and research in Puerto Rico. Journal of the American Bamboo Society, 6(1–4), 43–57.

Grenand, P.; Moretti, C.; Jacquemin, H. 1987. Pharmacopee traditionnelles en Guyane. Creoles, Palikur, Wayapi. Collection Memories 108, Editions de Rostom, Paris, France.

Judziewicz, E.J.; Clark, L.G.; Londoño, X.; Stern, M.J. 1999. American bamboos. Smithsonian Institution Press, Washington, D.C., USA. (In press).

Lombardo, A. 1943. Noticias de la vegetacion de la costa oriental del Rio Uruguay en los departamentos de Paysandu, Salto y Artigas, Comun. Bot. Mus. Hist. Nat., Montevideo, 1, 1–9.

Londoño, X. 1990. Estudio botanico, ecologico, silvicultural y economico-industrial de las Bambusoideae de Colombia. Cespedesia, 16/17, 51–78.

Londoño, X.; Peterson, P.P. 1991. *Guadua sarcocarpa* (Poaceae: Bambuseae), a new species of Amazonian bamboo with fleshy fruits. Systematic Botany, 16, 630–638.

McClure, F.A. 1947. Introduccion, propagacion y utilizacion del bambu en Tingo Maria. Subproyecto PE–12–6. Departamento de Agricultura Tropical, Seccion Silvicultura. Report. (Unpublished).

McClure, F.A. 1951. Letter to Mr. Pierre Trever. Personal Files. Smithsonian Institution, Washington, D.C., USA. (Unpublished).

McClure, F.A. 1966. The Bamboos: A fresh perspective. Harvard University Press, Cambridge. Reprint 1993. The Bamboos, Smithsonian Institution Press, Washington D.C.

McClure, F.A.; Stevenson, D.D. 1946. The occurrence and exploitation possibilities of bamboo in Nicaragua. Report. (Unpublished).

Mejia, J. 1975. Caracteristicas tecnologicas de *Guadua angustifolia* para la produccion de pulpa de papel en la Reserva Forestal de Ticoporo, Estado Barinas. Thesis, Universidad de los Andes, Facultad de Ciencias Forestales, Merida, Venezuela.

Mendoza, S. 1992. Estudios sobre el bambu. Recursos, 41, 14–14.

Montiel, M. 1998. Planting and using bamboo in the Neotropics. Editorial Universidad de Costa Rica, San Jose, Costa Rica.

Moris, S.A.; Boom, B.; de Carvalho, A.M.; dos Santos, T.S. 1983. Southern Bahian moist forest. Botanical Review, 49, 155–232.

Nicora, E.G.; Rugolo, Z.E. 1987. Los generos de Gramineas de America Austral. Editorial Hemisferio Sur Sociedad Anonima, Buenos Aires, Argentina.

ONERN (Oficina Nacional de Evaluacion de Recursos Naturales). 1976. Mapa ecologico del Peru. Oficina Nacional de Evaluacion de Recursos Naturales, Lima, Peru.

Paredes, E. 1991. Estudio preliminar floristico del bambu en los Estados Barinas, Portuguesa, Cojedes y Yaracuy. Thesis, Universidad de los Andes, Facultad de Ciencias Forestales, Merida, Venezuela.

Perez, T.L. 1967. Algunas observaciones sobre la morfologia y desarrollo del bambu en la Reserva Forestal de Ticoporo, Estado Barinas. Thesis, Universidad de los Andes, Facultad de Ciencias Forestales, Merida, Venezuela.

Parodi, L.R. 1936. Las Bambuseas indigenas en la Mesopotania Argentina. Revista Argent. Agron.,10, 89–110.

Pittier, H. 1926. Manual de las plantas usuales de Venezuela. Litografia del Comercio, Caracas, Venezuela.

Rodriguez, F. 1990. Propiedades fisico-mecanicas del bambu *Chusquea pittieri* Hackel, de tres sitios en Costa Rica. Departamento de Ingenieria Forestal, Instituto Tecnologico de Costa Rica, Cartago, Costa Rica.

Rosengurtt, B.B.; Arrillaga, R.; Isaguirre, P. 1970. *Guadua. In* Gramineas Uruguayas. Universidad de la Republica, Departmento de Publicaciones, Montevideo, Uruguay.

Turpe, A.M. 1975. Los generos de Gramineas de la Provincia de Tucuman (Argentina). Opera Lilloana, 24, 14–16.

Widmer, Y. 1990. Situacion del bambu en America Latina, con enfasis en America Central y Costa Rica. El Chasqui, 24, 14–23.

Young, S.M.; Judd, W.S. 1992. Systematics of the *Guadua angustifolia* complex (Poaceae: Bambusoideae). Annals of the Missouri Botanical Garden, 79, 737–769.

Zerpa, G.B. 1987. Establecimiento de bases para desarrollo pormocional del bambu nacional. Thesis, Universidad de los Andes, Facultad de Ciencias Forestales, Merida, Venezuela.

Bamboo for Sustainable Development, pp. 79–87
Eds. A. Kumar, I.V. Ramanuja Rao and Ch. Sastry

Bamboo in latin America: Present situation and projection for the 21st century

JORGE MORAN UBIDIA
Universidad Catolica y Estatal de Guayaquil, Guayaquil, Ecuador

Abstract—The situation in Latin America, which is confronting the challenges thrown up by high environmental damage and poverty, requires urgent reflection. The processes of globalization also have made the future of million of families uncertain, and demand stronger efforts from the Third World countries in general and Latin American countries in particular.

Latin America has resources of extraordinary value, and bamboo is one of them. Bamboo and its appropriate utilization will depend, to a large extent, on a rapid overall development, especially human development.

This paper examines the present situation of bamboo in Latin America and makes projections for the 21st century. The current situation is examined by means of a generalized study of bamboo in terms of: natural resource status; taxonomy; cultivation, reforestation and management; scientific research; the people involved; congresses, workshops and seminars; and the current utilization. The study also groups the existing problems and gives a detailed account of the positive experiences of some countries, such as Costa Rica, Ecuador and Colombia.

The projections made allow a glimpse of the strengths and weaknesses that could be generated around bamboo in the next century. The analysis also allows the presentation of a hypothesis to counteract the weaknesses and augment the strengths, as well as to specify new focuses for bamboo research and development.

INTRODUCTION

Out of the 90 genera and 1 100 species of bamboo that exist in the world (McClure 1966), 42 genera and 547 species are in Latin America. In other words, approximately 50% of the bamboo diversity is in the Americas.

Bamboo is found in every country in the American continent, from 40° latitude North to 47° latitude South, from the sea level to 4 000 m above it in the Andes (Londoño 1992). However, it should be mentioned that within the great variety and number of bamboos in the Americas, the subtribe Guaduinae stands out because of its spread and importance. It has approximately 30 species that cover the greater part of Mexico, the whole of Central America and 70% of South America.

No less than 50 million people in the continent use bamboo in their daily life, in traditional or modern uses that range from simple craft items to complex structural constructions.

Bamboo has been always known as an exotic Asian product until about the latter half of the 20th century, when the existence of the American bamboo from pre-Colombian times was brought to by archeological discoveries (Donnan 1964; Stothert 1988)

Bamboo has been, is and will be a valuable natural resource of the Americas. Because of this, it is necessary to study its current situation and its future in the years to come.

PRESENT SITUATION

The resource status

While the taxonomic investigations have identified most genera and species, the quantification of resources by countries, areas and species is far from satisfactory, as only general estimates exist. Only a few inventories have been executed, like the one in Ecuador (Universidad Laica Vicente Rocafuerte de Guayaquil, 1985), which also has to be discounted because it is devoted to one region of the country and to one species. Colombia and Ecuador have partial estimates exclusively on *Guadua angustifolia.*

It is essential to start inventorying to bring the resource data up to date, and to repeat the exercise periodically. This would help make a reliable assessment on the extent of bamboo depletion that is visible in most of the countries, with the exception of Costa Rica and part of Colombia.

It is roughly estimated that there are 50 000 acres of *Guadua* bamboo area in Colombia and 15 000 acres in Ecuador. Although the estimates for Peru and Brazil are too general, there is no doubt that the natural bamboo cover in these countries are of great magnitude. For example, in Brazil, the RADAM project (on remote sensing) found in the state of Acre 8 500 acres of natural forests of pure bamboo (Azzini 1993).

Taxonomy

The complex botanical structure of bamboo, combined with its exclusion from college studies, has made bamboo taxonomy a difficult subject. The works of McClure, Calderón, Soderstrom, Ellis and others are considered pioneering efforts in the botanic classification of bamboo in the New World. The contributions of Lynn Clark (USA) and Ximena Londoño (Colombia) in the last few years have been notable and have helped in the discovery of new species and in the betterment of knowledge of their classification.

The specialized work of Clark and Londoño at the Smithsonian Institution (Washington) and Iowa University has generated important data on bamboo in general and on *Chusquea* genera in particular.

Despite these efforts and the information generated in different workshops, seminars, and publications – such as the one brought out by INBAR, "Diversity and Distribution of New World Bamboo, with Special Emphasis on the Bambuseae" (Londoño 1996) – there are still many unexplored regions in the amazon area of Colombia, Ecuador, Peru and Brazil that require exhaustive studies.

Reforestation and management

Costa Rica, Colombia, Brazil and, to a lesser extent, Venezuela and Ecuador are starting programs of cultivation and reforestation of some species of bamboo, especially *Guadua.*

It is gladdening to note that Colombian institutions – such as CRQ, CRAMSA, CARDER, CVC, CDMB and UNIVALLE – have developed multi-purpose regional programs with support from INCIVA, National Federation of Coffee Dealers, town councils and, especially, rural communities. At this point, it is important to mention the National Center for the Study of Guadua Bamboo, near the Colombian city of Armenia, and the Regional Corporation of Quindio (CRQ) in Cordoba, Colombia. Both institutions demonstrate different types of bamboo dwellings, uses of bamboo, bamboo nurseries, and endemic and exotic bamboo species to interested visitors, albeit with special emphasis on *Guadua.*

The El Niño phenomenon of 1998 highlighted in Ecuador the necessity of reforesting the banks of the rivers with bamboo to protect the riverbanks and control the erosion of river basins. Bamboo, it became evident, would also supply much-needed material for the immediate rehabilitation of dwellings, construction of temporary bridges and erection of electricity posts.

In this context, the project "Diffusion and Promotion of Guadua Cultivation for Multiple Purposes" needs a mention. The project is being executed in Colombia under the sponsorship of Interamerican Development Bank, the independent Corporation of the Cauca Valley (CVC) and the Colombian Environment Ministry. It aims to reforest 750 acres of the Cauca Valley with *Guadua angustifolia,* disseminating forestry management techniques of the species for the betterment of the region's economy, while taking advantage of the characteristics of the species. The project has the participation of the local communities.

Some countries, like Ecuador, lack policies on natural *Guadua* forest management. This is causing a sustained process of deforestation and an obvious erosion of the genetic variety of the plant.

The experience of CRAMSA-CORPOCALDAS of Colombia deserves a detailed study, as it involves the application of bioengineering for controlling erosion. The experiences of the National Bamboo Project of Costa Rica are also worth studying, as they cover the cultivation, management, preservation and use of bamboo, particularly *G. angustifolia,* and present an example for the rest of Latin America to emulate.

Scientific research

Universities

Universities and specialized institutions are, without a doubt, the bases of the scientific research. From the 1980s, the departments of architecture, civil engineering, agronomy and forestry in many universities, as well as polytechnic schools Latin America have promoted study on bamboo. Most of such studies have taken the form of thesis, at post-graduate level, on different aspects of bamboo. Unfortunately, the majority of them did not have any continuation and very few have resulted in specific projects that allow the verification of the outlined hypothesis.

The higher education institutes in Colombia, Ecuador and Venezuela are known for the number of thesis-level investigations carried out.

Institutions

The first institution to promote the study and classification of bamboo of the New World was the Smithsonian Institution of Washington D.C. The investigations of McClure, Soderstrom, Calderón and Ellis, and those of Londoño and Clark later, have contributed much to dispel the notion of the New World as a region of less consequence as far as bamboo wealth is concerned.

The regional corporations of Colombia – such as CRQ and CVC – and other institutions of Ecuador and Costa Rica have pursued theoretical and practical research on different aspects of bamboo with excellent results.

The Institute for Housing Studies of the Netherlands has facilitated the participation of some Latin American sponsors, whose monographic works were oriented toward topics related to bamboo.

It would be very difficult to list out all the research efforts in the last 20 years that have promoted the scientific research of bamboo; however, two more organizations need a mention. One is the American Bamboo Society (ABS), which has promoted bamboo studies through its periodical publications. The other is the International Network for Bamboo and Rattan (INBAR), which has taken major initiatives to launch studies on American bamboo species from different points of view.

The people

It will be apt to mention, at this point, the men and women who have been pioneers in the study of bamboo in Latin America.

Oscar Hidalgo Lopez, the architect from Colombia, is definitely the initiator of actions aimed at rescuing the bamboos in Latin America. His publications and initiative for the organization of the first two Latin American symposiums on bamboo, several meetings, workshops, and national and regional seminars in Colombia, Ecuador, Costa Rica and other countries, make him a pioneer of bamboo studies and the reason for the attention paid to bamboo today.

Other names that merit mention include: Francisco Castaño, Ximena Londoño, Dicken Castro, Herbert Soto, Hayro Acero, Jorge Arcila and Luis Fernando Botero (Colombia); Anicio Azzini and Antonio Luis de Barros de Salgado (Brazil); Jorge Gutiérrez, Arturo Vanegas and Cecilia Chavez (Costa Rica); Misael Acosta Solis and Olaf Holm (Ecuador); and Horacio Saleme (Argentina).

It is also necessary to highlight the scientific work of Dr Walter Liese, Director of the Department of the Tree Biology at the University of Hamburg, in light of his research on bamboos in Latin America.

Congresses, workshops and seminars

Beginning 1981, a number of congresses, workshops and seminars were organized in Latin America. The World Congress in Pereira (Colombia) in 1992 and the two Latin American Symposiums (Manizales 1981 and Guayaquil 1982) need to be mentioned for the number participants and the quality of the presentations.

Since the mid-1980s, several important seminars have taken place in different countries: Medellin (1989), Tulua (1989) and Ocana (1993) in Colombia; Portoviejo (1991), La Libertad (1992) and Quito (1993) in Ecuador; Caracas (1992) in Venezuela; Mayaguez (1985) in Puerto Rico; San José (1998) in Costa Rica; and La Habana (1996) in Cuba. Brazil, Peru, Mexico and the Central American countries have also organized seminars and workshops in different years.

All these events have been favorable forums to share experiences, exchange opinions, and formulate conclusions and recommendations. The repetitiveness of the recommendations in these meetings is a reminder that there is a lot that remains to be done within wide topic of the American bamboo.

Current Utilization

Crafts

The principal use of Latin American bamboos, in general, and *Guadua,* in particular, is in decorative craft items and utilitarian objects. Colombia and Ecuador lead in this area and it could be said that there is no house in the Ecuadorian coast that does not have a variety of utilitarian articles made of *Guadua.*

Construction

Informal housing. Hundreds of thousands of families from different Latin American countries have found in bamboo the most suitable material for their houses, owing to its low cost and ease of transport. However, the high levels of poverty and lack of basic services have turned this asset into sign of poverty and misery.

Social housing. Among various social housing programs, the one in operation in Guayaquil, Ecuador, stands out. The Viviendas Hogar de Cristo, an organization founded and run by Jesuit priests, has been constructing low-cost prefabricated dwellings for the poor for the past 25 years. The number of houses constructed has reached 30 000 units and during the El Niño phenomenon, it produced 30 houses per day to make it possible to give shelter to 3 000 families in the Ecuadorian coast.

Some Costa Rican and Colombian institutions have also created programs for rural housing. The National Project of Bamboo of Costa Rica, which won in 1995 the World Habitat prize instituted by the English Foundation for Dwellings of Social Interest, merits a special mention.

Housing projects. Renowned architects and structural engineers such as Simón Velez, Jorge Arcila, Jaime Mogóllon and Rafael Rojas (Colombia), and Eliseo Guzman (Peru) have carried out residential projects and eco-tourism projects at different sites in Colombia, Peru and Ecuador. Their designs have demonstrated the feasibility of *Guadua* as a suitable material for construction.

Other uses in construction. The platforms for carrying out masonry work and painting in most of the conventional constructions of Ecuador and part of Colombia are made of *Guadua.* Even the slabs of concrete used in such work have *Guadua* reinforcement.

In Colombia, coffins of *Guadua* are much used in place of reinforced concrete slabs. The bamboo – particularly the opened out *Guadua,* also named esterilla – is also used as preformed patterns to give texture to concrete finishing.

Furniture

The use of *Phyllostachys aurea* initiated in Colombia and Costa Rica has been popularized in other countries. The exquisite furniture designed by Marcelo Villegos (Colombia) and Jose Miguel Coto (Bolivia) may be mentioned to highlight the use of *Guadua* and other species of bamboo in furniture.

Paper, food, cellulose, ethanol and others

Research on the industrialization of bamboo for the extraction of paper pulp, as well as for the production of food, starch, ethanol and charcoal, has reached significant levels, especially in Brazil, thanks to the works of Azzini and Barros de Salgado. However, scales of economy have to be worked out, as large-scale applications might require a great deal of resources, investment and time.

Current problems

The main problems that confront the bamboo sector in Latin America are the following.

Communication

There is a lack of communication among the countries. This should be addressed through a network that allows access to general and specific information. The publications about bamboo are scarce and are considered as specialized literature despite the efforts of various institutions, especially the regional corporations of Colombia.

Diseases

Bamboos in the region, in general, are resistant to diseases. There are, however, some specific causes of concern, such as the *Ralstonia* bacterial strain that is affecting Colombian bamboos. CVC is already investigating this bacterial strain, which destroys the plant within a period of 20–30 days. It is evident that a lot of more research needs to be done in order to identify potential dangers and find the right remedies.

Deforestation
The extent of deforestation in America is huge. Though not specifically directed at bamboo, large-scale cutting down of trees is contributing to the destruction of *Guadua* bamboos.

Ecologically unsuitable commercialization practices, such as the one in Ecuador that allows the farmers to clear-cut natural *Guadua* stands, are also aiding bamboo deforestation. The lack of sustainable forestry techniques is evident in the region.

Lack of technicians
The lack of specialists in taxonomy, silviculture, management and industrialization, poses a major problem in some Latin American countries.

Lack of awareness
In the absence of state measures to aid national programs that promote the cultivation, rational management and optimized use of bamboo, there is a general lack of awareness about bamboo and its advantages. Schools and agricultural colleges must be induced to start academic programs that study *Guadua* and other bamboos.

Farmers should be approached with integrated programs that promote bamboo cultivation, management and use. They should be made aware of the problems that may arise and how to tackle such problems, and should be taught how to commercialize the cultivation of bamboo.

PROJECTIONS FOR THE 21ST CENTURY

Observing the overexploitation of the natural resources since the discovery of the continent (particularly in Chuquisaca, Rosario and Guayaquil), Simon Bolivar, the Liberator of Latin America, exhorted the people to preserve all natural resources, public and private, in the Americas. The efforts, initiated during the independence of Latin America following Bolivar's message, did not have any impact on the people and the overexploitation of the resources have continued to this day.

Almost two centuries later, in 1983, the presidents of some of the countries met in Caracas to issue the "Manifesto of the American Countries", which presents the thoughts of Bolivar for a unified America, asking as a basic condition, the preservation of its natural resources. There have been several other manifestos and agreements; however, in the final balance, hundreds of forests have been, and are continuing to be, destroyed.

The climatic changes, which began in 1982 with the advent of El Niño and its return in 1998, have had far-reaching and disastrous ecological, social and economic consequences in the region. Unbridled industrialization and the consequent pollution increase have also had negative effects.

The hope

Bamboo reforestation, rational management of natural forests, sustainable exploitation, and the optimum utilization are the hope for the Americas for the 21st century.

It is, by now, clear that bamboo is the hope as:

- It is the plant with the fastest growth.
- It is the plant closely linked to the culture of the people, from the pre-Colombian era to the present time.
- It is the ideal material because of its versatility and range of application.
- It is the material that has, for centuries, met the housing needs of farmers and other rural folk.
- It is the material that makes possible the fast rehabilitation of the environment.

The means

For bamboo realize its fill potential in the 21st century, some concrete actions need to be initiated:

- Implementing in each country laws that facilitate the execution of national bamboo programs, under the aegis of Agriculture Ministries, for detailed inventories, plantation development, reforestation, creation of germplasm banks, rational management and of natural stands, and optimum use of bamboo resources in housing construction, handicrafts, furniture and industrial processes.
- Developing human resources through universities to build up capabilities in bamboo taxonomy, cultivation, forestation, construction and industrialization.
- Exchanging information through a continental network that would facilitate access to information on the progresses in the studies on different aspects of bamboo.
- Establishing an academic plan for the dissemination of *Guadua* bamboo in every single country in the region (as G. Pedraza recommended in the World Congress of Pereira in 1992).
- Promoting publications on diverse aspects of bamboo with the purpose of reaching every strata of the society, particularly in the rural areas, to induce community participation.
- Establishing Popular Didactic Centers for *Guadua* bamboo, similar to CRQ in Cordoba (Colombia), to promote the setting up of recreational places like "El parque de la vida" in Armenia.

- Encouraging the use of bamboo in housing construction and eco-tourism projects, both under public and private ownership, that would allow the tapping of the extraordinary characteristics of bamboo, especially *Guadua*.

The above-mentioned measures are necessary to convert the hope of today into a reality of tomorrow, to ensure that the eyes of our children's children will also behold this wonderful plant species.

REFERENCES

Azzini, A. 1993. Bambusa Guadua en Brasil. *In* Memorias del Primer Congreso Mundial Bambu/Guadua, Pereira, Colombia, 8–15 August 1992. Servicio Nacional de Aprendizaje, Bogota, Colombia.

Donnan, C. 1964. An early house of Chilca, Peru. American Antiquity, 30, 137–144.

Londoño, X. 1992. Distribucion, morfologia, taxonomia, anatomia, silvicultura y uso de los bambues del Nuevo Mundo. Cespedecia, 19(62–63), 87–137.

Londoño, X. 1996. Diversity and Distribution of New World Bamboo, with Special Emphasis on the Bambuseae. INBAR Working Paper No. 8. International Network for Bamboo and Rattan, New Delhi, India. 25 pp.

McClure, F.A. 1966. The Bamboos: A fresh perspective. Harvard University Press, Cambridge. Reprint 1993. The Bamboos, Smithsonian Institution Press, Washington D.C.

Stothert, K.E. 1998. Cultura Las Vegas. La prehistoria temprana de la Peninsula de Santa Elena-Ecuador. Museo del Banco Central del Ecuador, Guayaquil, Ecuador. 94 pp.

Bamboo for Sustainable Development, pp. 89–95
Eds. A. Kumar, I.V. Ramanuja Rao and Ch. Sastry

Native species of bamboo in Mexico

GILBERTO R. CORTÉS and ROSA INÉS AGUILAR
Instituto Tecnológico de Chetumal, Chetumal, Quintana Roo, México

Abstract—The following study presents an updated list of Mexico's native bamboo species; also mentioned are their geographical distribution and the endemic species. Eight genera and 35 woody bamboos species, and 3 genera and 4 herbaceous bamboos species have been recognized as being from Mexico. The 39 native Mexican species represent 10% of the present species in the Americas. Although they inhabit principally the tropical region of the country, 13 endemic species are known to be growing in diverse sites in Mexico.

INTRODUCTION

In 1981, T.R. Soderstrom (Soderstrom 1981) published the discovery of a woody bamboo, distinguished by its fleshy fruits, inhabiting the southern side of the states of Veracruz and Oaxaca. The existence of two more new species of the *Olmeca* genus motivated the Mexican botanists to make a taxonomic review of the bamboo groups that, until then, had remained relatively unknown.

Upon checking the principal herbaria in Mexico for bamboo collections, it was observed that few specimens were properly collected and with information adequate for the preparation of the list of species. The literature search on the bamboos present in Mexico, conducted at the beginning of the 1980s, turned up only a few species, mostly from quite old publications such as the studies by Fournier (1881) and Munro (1868).

Rzedowski (1991) estimates that Mexico's floristic riches are more than that of 22 000 vascular plant species. Valdés and Dávila (1995) mentioned the presence of 206 genera and a little more than 1 000 species of grasses (Poaceae), a family that occupies the third place in terms of the number of plant species in the country.

Soderstrom et al. (1988) mentioned that the diversity of bamboos in Mexico is considered as "moderately rich" compared with other Neotropical regions. Studies on bamboos in the states of Jalisco (Anaya 1989) and Veracruz (Cortés 1993) show the existing interest to know with certainty the species present in Mexico and their distribution.

Beginning 1990, Dr Lynn G. Clark of Iowa State University initiated a series of field collections in Mexico that increased, to a great deal, the number of species registered earlier (Clark 1990). The studies also provided new information on the known species and descriptions of several new species.

This study is an attempt to update the list of Mexico's native bamboos, both woody and herbaceous, and identify their geographical distribution based principally on the revision of the herbaria field collections and recent bibliographical information.

METHODOLOGY

For updating the list of *Bambusoideae* species of Mexico, a revision of the collection in the herbariums XAL, MEXU, CHAPA in Mexico and a visit to the herbarium in Washington D.C., United States, on behalf of the first author were conducted (numerous field collections have been carried out since 1981 and the specimens deposited at the herbariums XAL and the United States). Literature search yielded information valuable for the preparation of the list. Identification of species was based on selected literature, using identification keys developed by several authors. Dr Clark contributed greatly in completing the taxonomic knowledge of Mexican bamboos.

RESULTS

According to Soderstrom et al. (1988), bamboos have been divided into two large groups, herbaceous bamboos, *Olyreae,* and woody bamboos, *Bambuseae.* Of the 17 herbaceous genera reported for the Americas, 3 are found in Mexico. Woody bamboo genera present in Mexico number 8.[1] The genera are:

Herbaceous genera

- *Cryptochloa* Swallen
- *Lithachne* P. Beauvois
- *Olyra* L.

Woody genera

- *Arthrostylidium* Ruprecht
- *Aulonemia* Goudot
- *Chusquea* Kunth
- *Guadua* Kunth
- *Merostachys* Sprengel
- *Rhipidocladum* McClure
- *Olmeca* Soderstrom
- *Otatea* (McClure & E.W. Smith) C. Calderón & Soderstrom

[1] Clark 1998, personal communication.

Figure 1. The natural distribution of the native bamboo species in Mexico (inside the thick line)

Distribution

The states situated in Mexico's southeast region adjacent to the Gulf of Mexico and Guatemala are those that present the highest number of bamboo species (Fig. 1). The eastern states and the areas situated towards the Pacific Ocean contain some species that are known to be endemic. The distribution towards the north almost ends at the Tropic of Cancer.

Chiapas, at state at the south of the country's border with Guatemala, has the greatest number of bamboo species in the country and is the richest in terms of plant diversity. According to one estimate, the species in Chiapas flora total 7 018.

Genera and species of herbaceous bamboos and their distribution

1. Cryptochloa

One species, *Cryptochloa strictiflora* Swallen, is present principally in the evergreen forest of Veracruz, Oaxaca and Chiapas (Davidse et al. 1994).

2. Lithachne
Of the 4 species that are included in this genus, only one is found in Mexico – *Lithachne pauciflora* (Swallen) P. Beauv. The species is abundant in various states of the Republic (Soderstrom 1980).

3. Olyra
Widely distributed in the Americas, *Olyra latifolia* L. is the most common species in the tropical region of the country, growing in conserved forest regions. In Veracruz and the Yucatan Peninsula, *O. glaberrima* Raddi is common in some specific locations (Soderstrom and Zuloaga 1989).

Genera and species of woody bamboos and their distribution

1. Arthrostylidium
A. excelsum Griseb. is the only species found in Mexico. It is found in specific locations in the state of Chiapas (Davidse et al. 1994).

2. Aulonemia
A. laxa (Maekawa) McClure, is the least abundant species in the country. Its habitat is restricted to some specific locations in the humid forest of Veracruz, Chiapas and Guerrero. *A. clarkiae* Davidse & R. Pohl grows only in Chiapas, while *A. fulgor* Soderstrom can be found growing in certain mountainous areas in Oaxaca and Veracruz (Soderstrom 1988; Davidse and Pohl 1992; Davidse et al. 1994).

3. Chusquea
In Mexico, L.G. Clark identified 17 species of this most diverse woody bamboo genus (more than 130 species).[2] A recent addition to the list of *Chusquea* species in the Americas is *C. perotensis* L. by Clark et al. (1997) in Mexico. It is important to mention this, as an example, to emphasize the point that there are several bamboo species in the country yet to be assigned a taxonomic status. The following *Chusquea* species grow in the mountains in various states in Mexico (Clark 1986, 1989; Clark & Londoño 1991, 1991b):

1. *C. aperta* L.G. Clark
2. *C. bilimekii* Fournier
3. *C. circinata* Soderstrom & C. Calderón
4. *C. coronalis* Soderstrom & C. Calderón
5. *C. foliosa* L.G. Clark
6. *C. galeottiana* Ruprecht ex Munro
7. *C. glauca* L.G. Clark
8. *C. lanceolata* A. Hitchcock
9. *C. liebmannii* Fournier
10. *C. longifolia* Swallen

[2] Clark 1998, personal communication.

11. *C. muelleri* Munro
12. *C. nelsonii* Scribner & J.G. Smith
13. *C. repens* L.G. Clark & Londoño
    C. repens ssp. *repens*
    C. repens ssp. *oaxacacensis* L.G. Clark & Londoño
14. *C. perotensis* L.G. Clark, Cortés & Cházaro
15. *C. pittieri* Hackel
16. *C. simpliciflora* Munro
17. *C. sulcata* Swallen

4. Guadua

Five species of this genus are found in the lower regions, mainly in the eastern and south-eastern states in Mexico. *G. aculeata* Rupr. ex Fourn. Found in Veracruz is the largest bamboo in Mexico. *G. amplexifolia* J.S. Presl is distributed mainly in Veracruz and Chiapas, as is *G. longifolia* (Fournier) R. Pohl. *G. paniculata* Munro is distributed extensively in the country, while *G. velutina* Londoño & L.G. Clark is found in Tamaulipas, San Luis Potosí and Veracruz (Young 1992; Davidse et al. 1994).[3]

5. Merostachys

Only one *Merostachys* species is found in Mexico, that too in a very restricted area in the state of Chiapas. However, because of the absence of flowering in this species, it has been impossible to establish its taxonomic status. By studying Belice's *Merostachys* specimen, the authors noted a similarity to the species called *M. pauciflora.* The species present in Mexico is now identified as *M.* aff. *pauciflora* Swallen (McClure 1955).

6. Olmeca

The only endemic bamboo genus in Mexico, this is distinguished by its fleshy fruits. Its two described species – *O. recta* Soderstrom and *O. reflexa* Soderstrom – grow in the rainforest in the states of Veracruz, Oaxaca and Chiapas (Soderstrom 1981).

7. Otatea

Two *Otatea* species are known: *O. acuminata* (Munro) C. Calderón & Soderstrom and *Otatea fimbriata* Soderstrom. The former has 2 subspecies, *Otatea acuminata* (Munro) Calderón & Soderstrom ssp. *acuminata* and *Otatea acuminata* (Munro) Calderón & Soderstrom ssp. *aztecorum* Guzmán, Anaya & Santana. Growing mainly in the dry regions of the country, *O. acuminata* is surely one of the most abundant bamboo species in Mexico, occupying large tracts in the central and western states. *O. fimbriata* is found in the states of Chiapas and Jalisco (Calderón and Soderstrom 1980; Soderstrom 1983; Guzmán et al. 1984).

---

[3] Also Londoño 1998, personal communication.

8. Rhipidocladum
Of the 15 species described in this genus, 4 have been reported as being from Mexico. *R. racemiflorum* (Steudl) McClure is found widely distributed in many states in the country. On the other hand, *R. pittieri* (Hackel) McClure and *R. martinezii* Davidse & R. Pohl are restricted to the state of Chiapas. *R. bartlettii* (McClure) McClure is present in Chiapas, Campeche and Quintana Roo states (McClure 1966; Clark and Londoño 1991).

Endemic species of *Bambusoideae* in Mexico

1. *Olmeca recta* Soderstrom
2. *Olmeca reflexa* Soderstrom
3. *Guadua velutina* Londoño & L.G. Clark
4. *Chusquea aperta* L.G. Clark
5. *Chusquea bilimekii* Fournier
6. *Chusquea galeottiana* Ruprecht ex Munro
7. *Chusquea glauca* L.G. Clark
8. *Chusquea muelleri* Munro
9. *Chusquea repens* L.G. Clark & Londoño (2 subspecies)
10. *Chusquea perotensis* L.G. Clark, Cortés & Cházaro
11. *Rhipidocladum martinezii* Davidse & R. Pohl
12. *Aulonemia fulgor* Soderstrom
13. *Aulonemia laxa* (Maekawa) McClure

CONCLUSION

According to the results obtained after several years of study, it was established that in Mexico there are a total of 4 herbaceous bamboo species and 35 woody bamboo species. However, although some species of the *Chusquea* genus have been detected, their taxonomic status has not been established because of the absence of flowers. The distribution of bamboos in the country is genuinely tropical. The greater number of bamboo species is in the state of Chiapas, which has 25 woody species and four herbaceous species, followed by Veracruz, which has 17 woody species and four herbaceous species.

The 13 endemic woody bamboo species occupy different habitats from the "selvas altas perennifolias" to mountain forests.

ACKNOWLEDGEMENTS

The authors extend their thanks to Dr. Lynn Clark, whose written work and field collections have increased to a large extent the knowledge on Mexico's bamboos, and to Dr. Ximena Londoño for her contribution in clarifying the status of *Guadua* species.

REFERENCES

Anaya, M.C. 1989. Estudio de la subfamilia Bambusoideae (Poaceae) con revisión taxonómica para el estado de Jalisco. Tesis de Licenciatura, Universidad de Guadalajara, Mexico.

Caledrón, C.E.; Soderstrom T.R. 1980. The genera of Bambusoideae (Poaceae) of the American continent: keys and comments. Smithsonian Contr. Bot., 44, 1–27.

Clark, L.G. 1986. Eight new species of *Chusquea* (Poaceae: Bambusoideae). Iowa State Journal of Research, Vol. 61, No. 1, 99–122.

Clark, L.G. 1989. Systematics of *Chusquea* sect. *Swallenochloa*, sect. *Verticillatae*, sect. *Serpens* and sect. *Longifoliae* (Poaceae: Bambusoideae). Syst. Bot. Monogr., 27, 1–127.

Clark, L.G. 1990. Diversity and biogeography of Neotropical bamboos (Poaceae: Bambusoideae). Acta Bot. Brasil, 4(1), 125–132.

Clark, L.G.; Londoño, X. 1991. Miscellaneous new taxa of bamboo (Poaceae:Bambusoideae) from Colombia, Ecuador and Mexico. Nord. J. Bot., 11, 323–331.

Clark, L.G.; Londoño, X. 1991b. A new species and new sections of *Rhipidocladum* (Poaceae: Bambusoideae). Amer. J. Bot., 78 (9), 1260–1297.

Londoño, X.; Cortés, R.G.; Cházaro, M. 1997. An unusual new species of *Chusquea* (Poaceae: Bambusoideae) from Mexico. Systematic Botany, 22(2), 219–228.

Cortés, R.G. 1983. Revisión taxonómica de los bambúes leñosos de Veracruz (Poaceae: Bambusoideae). Tesis de Licenciatura, Universidad Veracruzanna, Veracruz, Mexico.

Davidse, G.; Pohl, R.W. 1992. New taxa and nomenclatural combinations of Mesoamerican grasses (Poaceae). Novon, 2, 81–110.

Davidse, G.; Sousa, S.; Chater, A.O. 1994. ed., Flora Mesoamericana, 6, 1-xvi, 1–543.

Fournier, E. 1881. Mexicanas plantas. Pars secunda, Gramineae. Imprimerie National, Paris, France. 160 pp.

Guzmán, M.R.; Anaya, M.; Santana, F. 1984. El género *Otatea* (Bambusoideae) en México y Centroamérica. Bol. Inst. Bot., Universidad Guadalajara, 5(10), 2–20.

McClure, F.A. 1955. Bamboos. *In* Grasses of Guatemala. Flora of Guatemala, Fieldiana: Botany 24, Part II.

McClure, F.A. 1966. Genera of bamboos native to the New World (Gramineae: Bambusoideae). Smithsonian Contr. Bot., 9, 1–148.

Munro, W. 1868. A monograph of the Bambusaceae, including descriptions of all the species. Transactions of the Linnean Society of London, 26, 1–157.

Rzedowski, J. 1991. Diversidad y orígenes de la flora fanerogámica de México. Acta Botánica de Mexicana, 14, 3–21.

Soderstrom, T.R. 1980. A new species of *Lithachne* (Poaceae: Bambusoidea) and remarks on its sleep movements. Brittonia, 32(4), 495–501.

Soderstrom, T.R. 1981. *Olmeca*, a new genus of Mexican bamboos with fleshy fruits. Amer. J. Bot., 68, 1361–1373.

Soderstrom, T.R. 1983. *Otatea. In* Flora Novo-Galiciana, 14, 278–282.

Soderstrom, T.R. 1988. *Aulonemia fulgor* (Poaceae: Bambusoideae), a new species from Mexico. Brittonia, 40, 22–31.

Soderstrom, T.R.; Calderón, C.E. 1978. The species of *Chusquea* (Poaceae: Bambusoideae) with verticillate buds. Brittonia, 30, 154–164.

Soderstrom, T.R.; Judziewicz, E.J.; Clark, L.G. 1988. Distribution patterns of Neotropical bamboos. *In* Vanzoini, P.E.; Heyer, R.E., ed., Proceedings of a workshop on Neotropical distribution patterns. Academia Brasileira de Ciencias, Río de Janeiro, Brazil. pp. 121–157.

Soderstrom, T.R.; Zuloaga, F.O. 1985. New species of grasses in *Arberella, Cryptochloa* and *Raddia* (Poaceae: Bambusoideae: Olyreae). Brittonia, 37(1), 22–35.

Soderstrom, T.R.; Zuloaga, F.O. 1989. A Revision of the genus *Olyra* and the new segregate genus *Parodiolyra* (Poaceae: Bambusoideae: Olyreae). Smithsonian Contr. Bot., 69, 1–79.

Valdés, J.; Dávila, P.D. 1995. Clasificación de los géneros de gramíneas (Poaceae) Mexicanas. Acta Botánica Mexicana, 33, 37–50.

Young, S.M. 1992. Systematics of the *Guadua angustifolia* complex (Poaceae: Bambusoideae). Ann. Missouri Bot. Gard., 79, 737–769.

Bamboo for Sustainable Development, pp. 97–105
Eds. A. Kumar, I.V. Ramanuja Rao and Ch. Sastry

Native and exotic bamboos of Cuba

RAMONA OVIEDO
Instituto de Ecologia y Sistematica, La Habana, Cuba

XIMENA LONDOÑO
Instituto Vallecaucano de Investigaciones Cientificas, Cali, Colombia

Abstract—Cuba is the island in the Americas with the richest bamboo diversity. A total of 7 genera and 19 species of native bamboos, and 7 genera and 22 species of exotic bamboos are found here. This study examines the diversity, distribution and habitat of the Cuban native bamboos, and mentions the exotic species introduced here since the last century.

INTRODUCTION

In the Americas, there are about 41 genera and 530 species of bamboos (Bambusoideae sub-family) growing in habitats ranging from lower to upper mountain forests, from "paramos" to dry regions (such as "Cerrado" in Brazil), and from sea level up to 4 300 m in the Andean highlands (Soderstrom and Calderon 1979; Judziewicz et al. 1999).

Bamboos are either woody or herbaceous. Woody bamboos represent a monophyletic group and are recognized as a single tribe, Bambuseae, based on:

1. Complex vegetative branching;
2. Dimorphic shoots and culm leaves;
3. External ligules; and
4. Gregarious and monocarpic flowering.

The herbaceous bamboos do not represent a monophyletic group and are allocated to a number of separate tribes, such as: Olyreae, Streptogyneae, Buergersochloeae, Guadualleae and Puelieae (Clark et al. 1995; Londoño 1996).

In Cuba, the Bambusoideae is represented by 7 genera and 19 species of native bamboos (Annex 1), and 7 genera and 22 species (and 4 varieties) of exotic bamboos (Annex 2).

NATIVE BAMBOOS

In Cuba, of the 7 genera and 19 species of native bamboos, 2 genera and 12 species are woody bamboos, while 5 genera and 7 species are herbaceous bamboos. In view of this, Cuba could be considered as the island in the Caribbean and West Indies, and probably in the whole of America, with the largest Bambusoideae diversity: (1) 86% of all the species existing in the region; and (b) the highest bamboo endemism – 3 genera and 11 endemic species (Catasus 1977a, 1977b; Lopez et al. 1989).

The largest generic and specific bamboo diversity in Cuba is found in:

1. The northern-central part of the eastern side of the island (Nipe-Moa-Baracoa, Holguin and Guantanamo);
2. The western part (Los Organos Mountain Range and Sierra del Rosario); and
3. The central-southern section (Guamuhaya, Cienfuegos, Sancti Spiritus).

The Nipe-Moa-Baracoa region has 15 of the 18 species found in Cuba, making it a priority center for the study of the subfamily as well as the evaluation of diversity, conservation and resource management.

Native woody bamboos

There are two native genera in Cuba – *Arthrostylidium* and *Chusquea* – and 12 species of woody bamboos. Six of the *Arthrostylidium* are endemic to the country. These woody bamboos are known by the common names of "tibisi", "large tibisi", "solid tibisi" and "parrilla". Roig Mesa (1988) states that cattle feed on the leaves of woody bamboos and that the canes are used to make fishnets and birdcages.

In Cuba, the genera *Chusquea* and *Arthrostulidium* are distributed from sea level up to 1 974 m above sea level right on top of Pico Turquino, accounting for the largest number of species located between 0 and 1000 m above sea level (Annex 3). They are found in different habitats such as: thorny xeromorphic thickets ("cuabal"); subthorny xeromorphic thickets ("charrascal"); sensu latum savannah; pine forests; evergreen forests; semi-deciduous forests; gallery forests; mountain and plains pluvial forests; cloudy forests and sub-Alpine thickets (Capote and Berazain 1984).

The level of conservation of woody bamboos in Cuba is acceptable. Out of the 12 woody species, three are considered as vulnerable (VU) (Annex 3), and one with less risk (Lrca). The rest of the species maintain a large or relatively high presence with a great capacity to grow even in disturbed habitats.

The *Arthrostylidium* genus from the Arthrostylidiinae sub-tribe reaches the highest representation and diversity in the Caribbean and West Indies with 15 from the 31 species known in the Americas (Judziewicz et al. 1999). Cuba is the main center of diversity in this region with a total of 11 species, six of them being endemic (Annexes 1 and 3).

Arthrostylidium species are most abundant and diversified at altitudes from 200 to 1 100 m above sea level, occupying different habitats such as mogote forests (Borhidi 1987), xeromorphic thickets, semi-deciduous forests, gallery forests and pluvial forests, growing in dense colonies (locally known as "tibisales") that are almost impenetrable to humans. No studies are known in Cuba regarding the relationship existing among these colonies and native fauna, although we may suppose that they are hosts to a large number of insects, birds and rodents.

Some species of *Arthrostylidium* have been associated with other woody Bambusoideae, as is the case of *A. farctum* with *A. cubense* and *A. farctum* with *A. fimbriatum* in thorny and sub-thorny xeromorphic thicket vegetation in the serpentine areas, and near rivers, or in gallery forests. Another examples are that of *A. farctum* with *A. urbanii* in semi-denuded forests, and *A. multispicatum* and *Chusquea abietifolia* in mogote forests and mountain pluvial forests.

Native herbaceous bamboos

There are 5 genera and 7 species of herbaceous bamboos in Cuba, pertaining to the Olyreae tribe (Annex 1). Two species of *Ekmanochloa*, one of *Lithachne*, one *Mniochloa* and one *Piresiella* are endemic to the country (Annex 3). They grow at altitudes ranging from 50 to 900 m above sea level.

Olyra latifolia is the species reaching the largest altitudinal range, from sea level up to 1 000 m. All other species grow more restricted ranges; for instance, *Lithachne pineti* is found from 700 to 800 m and *Ekmanochloa subaphylla* from 400 to 500 m (Annex 3).

The level of conservation of herbaceous bamboo in Cuba is rather critical. Of the seven known species, three are in critical risk and two are vulnerable. The five endemic species are found in restricted areas; in certain places there is only a small population, numbering less than 100 individuals. These endemic species are found in specific habitats and there is no association among them. Each grows in a different place with specific characteristics. For instance, *Ekmanochloa subaphylla* is found in mogote forests in the Sierra de Nipe (Holguin) and Florida Blanca (Santiago de Cuba); *Ekmanochloa aristata* grows in pine forests; *Mniochloa pulchella* grows in mogote forests in Yunque de Baracoa, Guantanamo and Moa (Holguin). On the other hand, *Olyra latifolia* and *Lithachne pauciflora* are found growing together in the same habitat or are associated with some of the endemic species. Sometimes they even grow in disturbed areas.

According to Oviedo,[1] and taking into consideration the categories established by IUCN regarding the conservation of species (Metodologia de Categorias de la IUCN 1994), *Ekmanochloa aristata*, *E. subaphylla* and *Lithachne pineti* are considered as species under critical risk (CR), while *Piresiella strephiodes* and *Mniochloa pulchella* are considered as vulnerable species (VU). Therefore, we could say that 71% of the Cuban herbaceous bamboos are already in critical con-

[1] Personal communication to X. Londoño, based on fieldwork, 1998.

dition and close to extinction. Immediate action is necessary concerning the study, propagation, habitat protection and conservation of these species.

Herbaceous bamboos are mainly ornamental plants (Calderon and Soderstrom 1973); they also help stabilize the soil and control land erosion. A clear example of this was observed in the Pinar del Rio Province where *Piresiella strephiodes* was found protecting the banks of a stream.

Ekmanochloa consists of two species – *E. aristata* and *E. subaphylla* (Zuloaga et al. 1933) – both of them endemic to rocky hills and pinelands in ultamafic (serpentine) areas in lowlands of Oriente Provinces in Eastern Cuba (Judziewicz et al. 1999). The generic name honors Erik L. Ekman (1883–1931), prolific Danish collector of the West Indian flora (Judziewicz et al. 1999). Little is known of the natural history of these interesting plants. The last-known collection and observation of *E. subaphylla* was made by R. Oviedo in 1998 in Sierra de Nipe.

Lithachne is a genus with a wide distribution range, from Mexico and Cuba to Paraguay and northern Argentina. The generic name means "stone chaff" and alludes to the hard, stony female floret (Judziewicz et al. 1999). In Cuba, there are two of the four species recognized in the genus: *L. pauciflora* and *L. pineti*. *L. pauciflora* is woody and is often found growing in several ecosystems in the western, central and eastern parts of Cuba or along trails through secondary forests. *L. pineti* is endemic to the pinelands in eastern Cuba.

Mniochloa is a monotypic and endemic genus of Cuba. It grows in shaded limestone cliffs at elevations less than 900 m, near the northern coast of the eastern tip of Cuba. The generic name means "moss grass" and alludes to its delicate appearance (Judziewicz et al. 1999). The one species recognized, *M. pulchella,* is classified as vulnerable (VU) and the last observation was made in 1960 (Judziewics et al. 1999).[2]

Olyra is the most widespread herbaceous bamboo genus in the Americas. In Cuba, only *O. latifolia* has been reported and it is the most frequent herbaceous bamboo growing in the understory of gallery forests, mountain rain forest and mogote forests, or in disturbed forests at elevations of 0 to 1 000 m above sea level.

Piresiella is one of the most delicate and smallest of world's Bambusoideae endemic to Cuba. The one species recognized – *P. strephiodes* (Zuloaga et al. 1993) – is found in western Cuba, at an altitude of less than 200 m above sea level, in the Provinces of Pinar del Rio and La Habana. The generic name for this bamboo alludes to the plant's resemblance to a diminutive species of *Piresia*, another herbaceous bamboo genus.

An individual of *P. strephiodes* collected by Londoño and Oviedo in 1994 was planted outdoors in La Habana, and has thrived for over three years in good condition and with minimum of care. According to Judziewicz et al. (1999), if this type of bamboo could be cultivated, it would make a very interesting species.

---

[2] Also, L. Catasus, in personal communication to T.R. Soderstrom, 17 May 1982.

Exotic woody bamboos

There are 7 genera, 22 species and 4 varieties of exotic bamboos in Cuba, introduced into the country by the end of the last century.

Bambusa vulgaris was the first species introduced in 1889 (Ricardo et al. 1995) and as of today, it remains as the most widely distributed and used bamboo in the island. In Cuba, the common name given to this type of bamboo is "bambu". However, in the western parts, it is usually known as "caña brava" and "Pito" and in the Eastern side as "canambu". This species has been well integrated to the Cuban wilderness, and is usually seen in large numbers mainly in the central-western regions of Cuba, as well as near rivers and streams.

All other species arrived in Cuba from 1900 to 1950 and were introduced by the Harvard University to the Botanic Garden La Soledad in Cienfuegos, popularly known as Cienfuegos Botanic Garden (Anonymous 1993). The American species known as *Guadua angustifolia* was introduced into Cuba in 1940, and has adapted satisfactorily to the conditions found at the Cienfuegos Botanic Garden, where a few individuals are now thriving.

The Cienfuegos Botanical Garden, Cuba, as well as the Tropical Agricultural Research Station in Mayaguez, Puerto Rico, are considered important genetic bamboo banks not only for the Caribbean Region and the West Indies but also for Central and South America.

Among the bamboo collection at the Cienfuegos Botanical Garden, there are some species that are seen in greater abundance and frequency: *Bambusa bambos* (38 clumps), *B. vulgaris* var. *vulgaris* (15 clumps), *B. longispiculata* (15 clumps), *B. polymorpha* (12 clumps), *B. tuldoides* (12 clumps), *B. oldhamii* (10 clumps) and *Dendrocalamus strictus* (20 clumps). Some other species found less abundant and with less frequency include *Bambusa vulgaris* var. *vittata*, *B. malingensis*, *B. balcooa, B. ventricosa*, *Dendrocalamus asper*, *D. membranaceus*, *Cephalostachyum pergracile*, *Gigantochloa apus*, *G. verticillata, Guadua angustifolia, Phyllostachys aurea, P. flexuosa* and *Pseudosasa japonica*.

Among these exotic species, the most widespread in the island after *Bambusa vulgaris* are: *B. multiplex, B. bambos, Dendrocalamus strictus* and *Phyllostachys aurea*. The second largest bamboo collection in Cuba is found at La Habana in the National Botanical Garden (8 spp.). In addition, there are some Recreation and Tourist Centers and State Study Farms that may have 1–3 species, most of them coming from the Cienfuegos Botanical Garden.

The exotic bamboo cultivation in Cuba has not had any stability through the years. Although there have been some sound actions taken, these have not been followed up. For instance, though some valuable species were introduced, there has been no further effort to propagate them. During the period 1930–50, some large-scale projects were established, most of them to grow *Bambusa vulgaris*. Later on, during 1990–95, there was a great interest in propagating and developing these species and some measures were carried out under a national project by some national and international bamboo specialists. This was followed by some activities intended to spread information regarding the ecology and some botanical aspects of bamboo, and to promote the plant's propagation. An example of this effort is now found in the experimental farms of Instituto de Investigaciones Forestales in Topes de Collantes and at Loma Ventana in Sierra de Escambray, in the Provinces of Cienfuegos and Sanctis Spiritus. The propagation by seed of *Dendrocalamus asper* and *D. strictus* found in the premises of the Empresa Nacional was part of efforts to protect the Bayamo Province Flora and Fauna.

Bamboo resources in Cuba are not considered very important as far as the country's economy is concerned; however, they are used for different types of crafts, ornamentation, furniture, rural building construction, as well as for the protection of water basins, erosion control and as fuel.

Seven different use categories for bamboos and the most appropriate species for these are presented below.

Construction: *Guadua angustifolia, Dendrocalamus asper, Bambusa polymorpha* and *B. bambos.*

Furniture: *Dendrocalamus strictus, Phyllostachys aurea, Bambusa tuldoides, B. oldhamii* and *B. polymorpha.*

Handicrafts: *Bambusa tuldoides, B. multiplex, B. polymorpha* and *Phyllostachys aurea.*

Protection of water basins and soil: *Dendrocalamus strictus, D. membranaceus, Guadua angustifolia, Bambusa longispiculata, B. bambos, B. multiplex, B. oldhamii, B. polymorpha, B. vulgaris* and *Phyllostachys aurea.*

Agricultural use: *Dendrocalamus strictus, Guadua angustifolia, Bambusa longispiculata, B. polymorpha* and *B. vulgaris.*

Food: *Dendrocalamus asper* and *D. strictus.*

Ornamentation: All species, but the most attractive are: *Bambusa vulgaris* var. *vittata*, *B. multiplex* var. *silver stripe*, *B. polymorpha*, *Cephalostachyum pergracile* and *Pseudosasa japonica.*

REFERENCES

Anonymous 1993. Catalogo de plantas. Jardin Botanico de Cienfuegos. Edit. Academia, La Habana, Cuba. 188–190.

Borhidi, A. 1985. Phytogeographic survey of Cuba. Acta Botanica Hungarica, 31(1–4), 3–34.

Borhidi, A. 1987. The main vegetation units of Cuba. Acta Botanica Hungarica 33(3–4), 151–185.

Calderon, C.E.; Soderstrom, T.R. 1973. Morphological and anatomical considerations of the grass subfamily Bambusoideae based on the new genus *Maclurolyra*. Smithsonian Contr. Bot., 11, 1–55.

Capote, R.P.; Berazain, R. 1984. Clasificacion de las formaciones vegetales de Cuba. Rev. Jar. Bot. Nac. 5(2), 27–76.

Catasus, L. 1997a. Las gramineas (Poaceae) de Cuba. I. Fontequeria XLVI, Madrid, Spain. 259 pp.

Catasus, L. 1997b. Manual de agrostologia. Edit. Academia, La Habana, Cuba. 98 pp.

Clark, L.G.; Weiping Zhang; Wendel, J.K. 1995. A phylogeny of the grass family Poaceae based on *ndh*F sequence data. Systematic Botany, 20(4), 436–460.

Judziewicz, E.J.; Clark, L.G.; Londoño, X.; Stern, M.J. 1999. American bamboos. Smithsonian Institution Press, Washington, D.C., USA. 400 pp. (In Press).

Londoño, X. 1990. Aspectos sobre la distribucion y ecologia de los bambues de Colombia (Poaceae: Bambusoideae). Caldasia, 16(77), 139–153.

Londoño, X. 1996. Diversity and Distribution of New World Bamboo, with Special Emphasis on the Bambuseae. INBAR Working Paper No. 8. International Network for Bamboo and Rattan, New Delhi, India. 25 pp.

Lopez, A.; Pouyu, A.E.; Catasus, L. 1989. El endemismo de la familia Poaceae en Cuba. Acta Botanica Cubana, 82, 1–11.

Ricardo, N.E.; Pouyu, R.; Herrera, P.P. 1995. The Synantropic flora of Cuba. Fontqueria, 42, 367–429.

Roig Mesa, J.T. 1988. Diccionario botanico de nombres vulgares cubanos. Edit. Cientifico-Tecnica, La Habana, Cuba. 1142 pp.

Soderstrom, T.R.; Calderon, C.E. 1979. Ecology and phyto-sociology of bamboo vegetation. *In* Numata, ed., Ecology of grasslands and bamboo lands in the World. pp. 223–226.

Zuloaga, F.; Morrones, O.; Judziewicz, E.J. 1993. Endemic herbaceous bamboo genera of Cuba (Poaceae: Bambusoideae: Olyreae). Ann. Missouri Bot. Gard., 80, 846–856.

Annex 1.
Native Bamboo Species of Cuba and Phyto–geographic distribution

Bamboo species	Phytogeographic regions		
	Western	Central	Eastern
Woody bamboos			
Arthrostylidium angustifolium Nash	–	–	X
A. banaoense Catasus G.	–	–	X
A. cubense Ruprecht	X	X	–
A. distichum Pilger	X	X	X
A. farctum (Aublet) Soderstrom and Lourteig	X	X	X
A. fimbriatum Grisebach	–	–	X
A. multispicatum Pilger	–	X	X
A. pinifolium Catasus G.	–	–	X
A. reflexum Hitchcock and E. Ekman	X	–	–
A. sarmentosum Pilger	–	X	X
A. urbanii Pilger	–	X	X
Chusquea abietifolia Grisebach	–	X	X
Herbaceous bamboos			
Ekmanochloa aristata A. Hitchcock	–	–	X
E. subaphylla A. Hitchcock	–	–	X
Lithachne pauciflora (Swartz) P. Beauvois	X	X	X
L. pineti (Grisebach) Chase	–	–	X
Mniochloa pulchella (Grisebach) Chase	–	–	X
Olyra latifolia L.	X	X	X
Piresiella strephiodes (Grisebach) Judziewicz, Zuloaga and Morrone	X	–	–

Annex 2.
Exotic Bamboo Species of Cuba

Bambusa bambos (L.) Voss in Vilmorin (syn. *Bambusa arundinacea*)
B. balcooa Roxburg
B. longispiculata Gamble ex Brandis
B. aff. malingensis McClure
B. multiplex (Loureiro) Raeuschel (syn. B. glaucescens)
B. multiplex cv. Alphonse-Karr
B. multiplex cv. Fern leaf
B. multiplex cv. Silver stripe
B. oldhamii Munro
B. polymorpha Munro
B. tuldoides Munro
B. ventricosa McClure
B. vulgaris Schrader ex Wendland var. vulgaris
B. vulgaris var. vittata A. and C. Riviere
Cephalostachyum pergracile Munro
Cephalostachyum sp.
Dendrocalamus asper (Schultes f.) Backer in Heyne
D. membranaceus Munro
D. strictus (Roxburgh) Nees
Gigantochloa apus (Schultes) Kurz
G. verticillata (Wildenow) Munro
Guadua angustifolia Kunth
Phyllostachys aurea A. and C. Riviere
P. flexuosa A. and C. Riviere
P. meyeri McClure
Pseudosasa japonica (Siebold and Zuccarini) Makino

Annex 3.
Native Bamboos: Altitudinal Range, Endemism & Conservation Category

Bamboos	Altitude	Endemic	Conservation
Woody bamboos			
Arthrostylidium angustifolium	500–600	X	Vulnerable
A. banaoense	50–300	X	Infrequent
A. cubense	50–300	X	Infrequent
A. distichum	200–900	–	Less risk
A. farctum	30–1000	–	Frequent
A. fimbriatum	50–1200	–	Frequent
A. multispicatum	500–1200	–	Frequent
A. pinifolium	100–1000	X	Vulnerable
A. reflexum	600–800	X	Vulnerable
A. sarmentosum	450–1200	–	Infrequent
A. urbanii	80–1100	X	Frequent
Chusquea abietifolia	500–1974	–	Frequent
Herbaceous bamboos			
Ekmanochloa aristata	50–500	X	Critical
E. subaphylla	400–500	X	Critical
Lithachne pauciflora	0–900	–	Frequent
L. pineti	700–800	X	Critical
Mniochloa pulchella	50–900	X	Vulnerable
Olyra latifolia	0–1000	–	Frequent
Piresiella strephiodes	50–200	X	Vulnerable

Bamboo for Sustainable Development, pp. 107–113
Eds. A. Kumar, I.V. Ramanuja Rao and Ch. Sastry

Prospects and challenges of bamboo industries in Malaysia in the 21st century

ABD. LATIF MOHMOD
Forest Research Institute Malaysia (FRIM), Kuala Lumpur, Malaysia

Abstract—Bamboo continues to be an important source of livelihood for rural folks, especially those who are at the bottom of the economic ladder. This is particularly true in the sense that it has been put for many uses ranging from food and hunting devices to utensil items and structural materials, and as remedies in treating diseases such as jaundice and kidney problems. In addition to stimulating development in rural areas in the 21st century, bamboo-based industries have the potential to become one of the major sources of income for the rural populace. The challenges related to the future development of the industry include the steady supply of good quality raw material, transfer of suitable technology for more value-added products, upgrading skills of workers, enhancing management capabilities and removing institutional constraints. This paper discusses these issues and highlights some of the key strategies for the future development of bamboo-based industries.

INTRODUCTION

Bamboos were once considered weeds in forestry practice and their growth prevented or controlled (Salleh and Wong 1987). Now bamboo has become the most important non-timber forest produce in Malaysia after rattan, owing to a relatively rapid expansion of bamboo-based industries in the last 15 years (Abd. Latif and Abd. Razak 1994).

Bamboo in Malaysia, particularly in the Peninsula, is viewed as a multipurpose plant with a wide range of uses, including in construction, for paper pulp, utensils such as skewers and chopsticks, hunting devices, musical instruments, and as food. In certain parts of the country, bamboo shoot, roots and stem are used as traditional remedies for bone fracture and treating several diseases such as gonorrhoea, jaundice and kidney problems (Abd. Latif and Liese 1995).

Lately, there has been an increased utilization of bamboo by industries. This could probably be owing to: the increasing difficulty in obtaining commonly used timber from the forest; hygienic concerns, especially the demand for cheap disposable items used in contact with food; and the social awareness about the need to sustain the surrounding green environment. The popularization of bamboo as an environment-friendly material (fast-growing, high-yielding, renewable resource and a secondary forest species) has also resulted in the rejuvenation of bamboo industries – especially the mechanized ones – after its seven stagnant years from 1988 to 1995. Domestic demand for machine-made products such as skewers and chopsticks, for example, increased from the average USD 3 million per annum during 1988–95 to about USD 5 million in 1996 (Abd. Latif 1989; Abd. Latif and Abd. Razak 1994; Abd. Latif 1996). Although the increasing local demand for disposable bamboo products signifies a bright future for the industry, several problems encountered by it need to be adequately addressed to sustain and speed up its growth. This paper examines these issues.

UTILIZATION ASPECTS OF BAMBOO

At present, the bamboo-based industry is confined mainly to the manufacturing of limited number of end-products. It employs a few workers, and sometimes operates on a part-time basis (Abd. Latif and Abd. Razak 1994). A small number of industrial units may have about 15–75 employees. The products are the following:

1. Poultry cages and vegetable baskets
2. Joss sticks and joss paper
3. Handicraft items
4. Utensils
5. Panel products

Of the total 300 bamboo mills operating in the country, 75% are engaged solely in making vegetable baskets, 15% in producing handicrafts, and 9% in making utensils (such as skewers), bamboo strips for incense sticks, and panel products. The remaining 1% makes joss paper for ritual ceremonies. The manufacturers of joss paper, panel products, skewers and bamboo strips fall within the classification of medium to large, and the others are grouped as cottage and small-scale enterprises (Abd. Latif 1996). The units engaged mainly in manufacturing products for the domestic market are mostly located in the west coast, where a large market and basic facilities are readily available (Abd. Latif and Abd. Razak 1994).

THE BAMBOO INDUSTRY

The major problems in fostering the development of bamboo-based industries in Malaysia and some solutions to these are given below.

Steady raw material supply

Malaysia has an ample number of natural bamboo stands. Still, the lack of steady supply of good-quality culms in large quantities is frequently reported to hinder the manufacture of quality products. Indiscriminate harvesting of wild bamboo, without due consideration to the maturity of culms for their intended use, is the prime factor for higher wastage in mechanized production (Hartley Tham 1997).

In the manufacture of skewers, bamboo strips of about one internode length (bought by weight) are preferred to 3–4 m poles as the former entails less transportation and processing costs, and at the same time helps maintain the quality of final products.

Natural characteristics of bamboo

Malaysian bamboos grow in clumps (sympodial), often thick enough to cause the culms to be bent and thus be under constant stress. The straight and older culms are found at the less accessible clump center (Abd. Latif 1993; Abd. Razak and Abd. Latif 1995). Furthermore, these bamboos grow wild and scattered and are practically unmanageable. Lack of control over harvest has resulted in uncontrolled price of bamboo culms (Abd. Latif and Abd. Razak 1994).

Systematic harvest of bamboo is still lacking. High processing wastage owing to the use of culms of unsuitable age and inappropriate species, relatively low natural durability of the culms to agents of decay, splitting and cracking that occur easily owing to high initial moisture content, and lower processing performance by imported machinery are among the indicators pointing to the need for intensive studies on the properties of this resource (Abd. Latif and Liese 1995).

Technology transfer

While technology transfer is essential for the development of the industry, it is equally important to ensure that the technology being transferred is appropriate to local conditions. About a decade ago, the local industry received a shot in the arm when Taiwan experienced a serious bamboo shortage. The local manufacturers of disposable bamboo products managed to get a foothold in markets previously monopolized by Taiwan. Depleting bamboo resources in Taiwan also prompted many of its entrepreneurs to relocate to Malaysia or seek joint ventures. Malaysia's strong Taiwan connection stems from the fact that majority of local bamboo producers use Taiwanese machines, which are cheaper and more easily procured than those from Japan. As the Taiwanese offered a buy-back arrangement (buying up the entire quantity produced), newcomers to the business who were yet to find markets for their products got attracted (Lee 1989). The poor quality products produced by Taiwanese machines, which are more suited to Taiwanese bamboo that have different processing characteristics, and the stringent quality standards of consumer countries like Japan and Korea, however, resulted in minuscule export earnings. Consequently, majority of the factories closed down or shifted to other activities.

Management and market

A majority of the bamboo-based industries is either one-man-show, family-run or part-time. The entrepreneurs have minimal academic background and their activities are often hampered by lack of capital, which prevents them from expanding technological capabilities and acquiring permanent business sites. They rarely conduct market surveys to keep abreast of business and product trends. In the export markets, the low prices offered (though for a large volume) for relatively high quality products deter them from entering these markets.

Skilled labor

The problem in getting skilled local workers is rather critical in almost half the industries. As the general wages for local workers have kept increasing by about 75% over the last decade, the current trend is towards foreign sources of labor. The relatively high levy imposed on foreign workers (USD 800 per worker per year) and the poaching (weaning away skilled workers from another company by offering higher salary) prevalent in the industry are part and parcel of this issue.

Work procedures and product design

A majority of bamboo-based industries thrives on handed-down, traditional work procedures. The treatment of bamboos is an example. Because of the limited financial resources and the operators' low level of education, the procedures are relatively labor-intensive and employ traditional devices that have low technology level and production capacity. Usage of jigs and efficient mechanization are confined mainly to exporting manufacturers, who boast of high quality and prompt delivery. Product designs often fail to receive sufficient attention. The local designers do try to meet local tastes but without much success. Most local manufacturers cannot afford to engage qualified designers and have to rely on existing ethnic designs or copy ideas from other sources for new product designs.

Institutional constraints

The waiting period for applying for a processing license, which involves an approval of land conversion from agriculture to industrial land, is reported to be a prime business constraint. As there is no guarantee for the extension of bamboo extraction license, the industrial units usually start operation on an illegal basis after the expiry of a year's permit. The problem becomes more serious when it involves a local authority dealing with foreign workers.

The forest policy does not yet provide for investors to manage natural bamboo stands within a concession area and since no guarantee is given on the extension of harvest license, the resource is usually overexploited. The clear-felling system also contributes to the area becoming unproductive and unprofitable for the following bamboo licensee.

Lack of a proper pricing policy for bamboo and bamboo products continues to be a major drawback. The criteria adopted by the government in determining the price or royalty of bamboo materials is yet to be spelt out clearly.

MEASURES FOR INDUSTRY DEVELOPMENT

The bamboo industry offers a wide variety end-products of different designs and has a high potential for development (Abd. Latif and Abd. Razak 1994). Harmonized efforts, however, must first be geared towards overcoming the immediate problems faced by the industry. At the outset, the program should be rural-based since that is where the raw material is. Moreover, it is within the government's ambit to stimulate production through a rural-based program, advocating self-help for its populace to prosper.

An early step to be taken is to revise the current institutional systems and procedures pertaining to bamboo resources. This also means that the current policy of issuing permit for bamboo extraction should be revised to include a provision for the licensees to manage the natural bamboo stands within a given area. On a larger context, setting up bamboo plantations should be encouraged. An institutional proposal submitted by the author in 1992 urged policy-makers to designate (1) the large areas of bamboo forests (about 187 000 ha) within the north-western part of the Peninsula and (2) the large bamboo areas resulting from the shifted cultivation practices of the aborigines, as bamboo estates and lease them out to the private sector to manage on a sustained basis for the production of shoots or culms. With intensive and careful management, these natural bamboo resources can be gradually converted to equivalent plantation stands. With proper silvicultural treatment and systematic harvesting practices, sustainability of quality bamboo supply can be ensured.

A thorough and reliable inventory of total bamboo areas and production – especially the extent of useful bamboo species available within the country (especially within the states of Sabah and Sarawak of East Malaysia) – should be carried out in order to facilitate planning for development of the bamboo-based industries. Only with a continuous and regular supply of the raw material would the industry be able to develop rationally.

Establishment of centralized and strategically located information, service, processing and manufacturing centers close to the resource base should also be considered. It would be practical and sensible to link up such a program to the aborigines (amounting to 105,000 people), as bamboos are abundantly found in the areas they inhabit. These centers would help utilize the raw material efficiently and act as marketing outlets and training venues for teaching basic and advanced production techniques. A proper training program, especially on the production of popular products and maintenance of related machinery, could also be included. An appropriate incentive scheme could be introduced for accomplished trainees to induce them utilize all the equipment and facilities provided at the centers, and special attention may be given to newcomers until they are able to work independently.

Amalgamation, conversion or corporatization of either the entrepreneurship training programs or the facilities available at related governmental research institutions is another practical alternative towards realizing a training-cum-production center. Apart from ensuring that all government-funded facilities remain in good condition, the maintenance of machinery used on a rental basis could be handled easily by selected industrial units.

The levy imposed on the introduction of foreign skilled workers could be lowered as an incentive to the industry. This is particularly important in view of the current serious shortage of local workers in plantations and industrial sectors. Apart from obtaining new knowledge on skilled and artistic work to the benefit of the local workforce, introduction of foreign skilled labor could also be expected to effect an integration of more diversified value-added products from different cultures.

Information on the technological properties of Malaysian bamboos, particularly in relation to their industrial applications, should be expanded through studies. Since many bamboo species remain unutilized, research is needed to determine the properties of such species and to develop their appropriate utilization technologies. Emphasis should also be given to encourage locally fabricated machines designed to process local species, not only to achieve less wastage but also to bring down the cost of machinery imports. There is also a need to form a team of industrial designers to produce new and attractive designs for products solely of bamboo or bamboo in combination with other materials such as plastic, metal or less popular wood species. This would help maintain and upgrade the existing products without sacrificing their original identity, as well as help sustain the interest of buyers in bamboo-based products. Furthermore, to expand the utilization of bamboo, consideration should be given to promote or upgrade the existing research and development institutions associated with small-scale bamboo industries.

CONCLUSION

Bamboo has the potential to be developed as an important industrial resource, especially for the rural populace. Several constraints encountered by the bamboo-based industries need serious attention and practical solutions to further improve production in both quantity and quality. With proper planning and collaboration among the industrialists and government agencies, the bamboo-based enterprises should be able to develop and even compete globally with other bamboo-producing countries.

REFERENCES

Abd. Latif, M. 1989. Current status of machine intensive bamboo industries in Peninsular Malaysia. Paper presented at the International Bamboo Workshop, Nanjing, China, 24–27 July 1989. 9 pp.

Abd. Latif, M. 1993. Effects of age and height on machining properties of Malaysian bamboos. Journal Tropical Forest Science, 3(1), 66–71.

Abd. Latif, M. 1996. Non-wood resources and products of Peninsular Malaysia: Management and research considerations. Proceedings of the Malaysia-Indonesia Bilateral Meeting of Group D: Research, Training and Extension, Palembang, Sumatera, Indonesia, 26–27 August 1996. Forestry Research Support Programme for Asia and the Pacific, Bangkok, Thailand; Forest Research Institute Malaysia, Kuala Lumpur, Malaysia. pp. 121–132.

Abd. Latif, M.; Abd. Razak, O. 1994. Availability, distribution of bamboo resources and its industrial status in Peninsular Malaysia. *In* Bamboo in Asia and the Pacific. Proceedings of the 4th International Bamboo Workshop, Chiangmai, Thailand, 27–30 November 1991. International Development Research Centre, Ottawa, Canada; Forestry Research Support Programme for Asia and the Pacific, Bangkok, Thailand. pp. 60–67.

Abd. Latif, M.; Liese, W. 1995. Utilization of bamboos. *In* Abd. Razak, O.; Abd. Latif, M.; Liese, W.; Norini, H., ed., Planting and utilization of bamboo in Peninsular Malaysia. FRIM Research Pamphlet 118, 50–102.

Abd. Razak, O.; Abd. Latif, M. 1995. Bamboos in Peninsular Malaysia". *In* Abd. Razak, O.; Abd. Latif, M.; Liese, W.; Norini, H., ed., Planting and utilization of bamboo in Peninsular Malaysia. FRIM Research Pamphlet 118, 1–17.

Lee, E.E. 1989. Chopsticks makers in a fix. Malaysian Business, 16–31 August 1989, 37–39.

Hartley Tham, T.S. 1997. Personal communication with the owner of Bertham Bamboo Industrial Enterprise, Semenyih, 43500 Selangor, Malaysia.

Salleh, M.N.; Wong, K.M. 1987. The bamboo resources in Malaysia: strategies for development. *In* Rao, A.N.; Dhanarajan, G.; Sastry, C.B. ed., Recent Research on Bamboo. Proceedings of the International Bamboo Workshop, Hangzhou, China, 6–14 October 1985. Chinese Academy of Forestry, Beijing, China; International Development Research Centre, Ottawa, Canada. pp. 45–49.

Bamboo for Sustainable Development, pp. 115–123
Eds. A. Kumar, I.V. Ramanuja Rao and Ch. Sastry

The bamboo industry of China: Looking back and ahead

WANG SHUDONG
China National Bamboo Research Center, Hangzhou, China

Abstract—At the turn of the century, a retrospection is being made on the development of China's bamboo-based industry. The industry went through three distinct development phases: (1) the stage in which the industry underwent a natural development and bamboo usage was at primary level; (2) the next phase when the industry took on an orderly development and bamboo went into wider usage; and (3) the period in which the industry developed rapidly and bamboo usage became extensive. The paper also looks at the achievements of bamboo sector in resources, exploitation and research. It examines the current problems the sector faces, and takes a look at the prospects in the 21st century in terms of industry development guidelines, targets, tasks and effective measures to be taken.

INTRODUCTION

Bamboo is an important component of the forest resources in China, which has one of the richest bamboo resources in the world. The country would ranks first in the number of bamboo species, bamboo forest area and bamboo stock. The deep-rooted bamboo culture that developed through the history of China since ancient times has given rise to a unique "bamboo civilization". Since the 1990s, China's bamboo-based industry has registered a boom, and has developed into a resurgent and vigorous bamboo sector in which the industry has integrated with cultivation, processing, utilization and export.

CHINA'S BAMBOO INDUSTRY: A RETROSPECTION

The development of the industry

Stage 1: Natural development and early usage
China has the earliest history of bamboo usage in the world. Some bamboo weavings, unearthed at the Banpo village, Xian City, date back to over 5 000 years. Various articles belonging to the Neolithic Stone Age have been discovered also in Zhejiang Province. Bamboo was used for making arrows, books and wares during the Shang Dynasty (1766–1122 BC), for building palaces in the Han Dynasty

period (202 BC – 220 AD), and for pulping and papermaking by the (Western) Jin Dynasty (265–316 AD). A rather comprehensive description of bamboo cultivation and usage was found among the records of the later Wei Dynasty. During the Ming and Qing Dynasties, some "bamboo hostels" were specially set up in Boai, Hunan, for accommodating bamboo business people, which accelerated the production and trade of bamboo there. In the long history of Chinese culture, bamboo can be found everywhere, be it for clothing, food, housing, transport, agriculture, fine arts or for any other day-to-day usage.

Stage 2: Orderly development and general usage

Since the founding of the People's Republic of China in 1949, the Chinese government has paid great attention to pushing bamboo forward as a means to accelerate prosperity of mountainous economies and enriching people's material comforts and livelihood. Bamboo was listed as one of main planting tree species in the first National Notice on "Instructions on Planting in Spring" issued by the Ministry of Forestry and Reclamation in March 1952. In the Notice on the Protection and Development of Bamboo Forest, promulgated by the State Council in June 1956, governments at all the levels were requested to take effective measures to protect and develop bamboo forests. A national conference on bamboo was held in Jiangsu Province in 1962. A meeting on bamboo forest management in 12 southern bamboo-growing provinces/autonomous regions was convened by the Ministry of Forestry in Beijing in July 1982.

During this stage, significant success has been scored in such areas as introduction and cultivation of crop bamboo, planting of southern bamboo in the north, control of harmful bamboo pests, bamboo weaving, and arts and handicrafts. There was also an effort made to orient bamboo cultivation and management for shoots. In short, bamboo industry development was put in a regular and well-oiled track.

Stage 3: High-speed development and extensive use

Since the 1990s, China's bamboo industry growth has entered a high-speed phase, thanks to the implementation of a series of policies and measures by the Chinese government. This is particularly notable since the National Bamboo Conference held in Hangzhou in 1991, in which the guidelines for bamboo industry development in China was put forward covering:

- comprehensive planning;
- reasonable layout;
- strengthening of administrative management;
- optimum exploitation bamboo resources;
- tapping all potential for bamboo processing and utilization;
- opening up market for raising economic benefits; and
- invigorating bamboo industry through science and technology.

Guided by this policy, and after many years of efforts, the bamboo industry has shown some encouraging trends in terms of changes:

- From the sole pursuit of area expansion to a combination of new planting and low-yield bamboo stand improvement;
- From an extensive management approach to an intensive one;
- From the management of single bamboo species to that of multiple ones;
- From forests managed for timber alone to those for timber and shoots; and
- From low added value to high added value.

Thus, a bamboo cultivation system with distinct characteristics has formed, with a bamboo processing industry functioning at the appropriate scale and economic efficiency. Ten bamboo-growing counties/cities have been nominated as "China Bamboo Town", ten demonstration bases for forest bamboo harvesting have been established, and a new bamboo development plan tested. The founding of China National Bamboo Research Center (CBRC) in Hangzhou in 1989 and China Bamboo Industry Association (CBIA) in Beijing in 1993, and the implementation of some key national bamboo projects (under eighth five-year plan and ninth five-year plan) have all helped to a large extent in further pushing forward China's bamboo industry development.

Achievements of China's bamboo industry

Bamboo resources

During the eighth five-year plan period, a bamboo forest area of over 0.33 million ha has been earmarked for new planting and 1 million ha for productivity improvement. So far, bamboo forest area has reached 4.4 million ha, with 3.1 million ha (1.1 million ha is low-yield sites) exclusively for Phyllostachys pubescens (Moso): an increase of 0.86 million and 0.53 million hectares, respectively, over the area at the end of the seventh five-year plan. Standing culms of Moso is up to 2 070 culms/ha. The annual Moso culm yield is about 400 million pieces, annual culm yield of other bamboos is 11.8 million tons, and annual Moso shoot yield is 0.16 million tons. Annual national bamboo output value is estimated at USD 1 450 million.

Bamboo exploitation

1. Bamboo timber. Since the 1990s, a bamboo timber processing industry of appropriate economic scale and efficiency has developed in the country. Production of bamboo timber products in China is as follows:

Table 1.
China's bamboo timber processing and production (in 1998)

Product	Annual production capacity (m^3)	No. of factories
Plywood	20,000	over20
Floorings (including decorative items)	30,000–100,000	over 100
Building moldboard	100,000	20–30
Laminated board	20,000–30,000	over 10
Particleboard	–	5

2. Bamboo art and handicraft, bamboo articles of daily usage. Bamboo arts and handicrafts, one of the traditional and important products of China, enjoy a good reputation in international market. Bamboo carving, weaving, etc. has a long history. Since the 1990s, owing to extensive use of bamboo processing machinery and new technologies, machine-woven bamboo mats have received a quantum jump in terms of quality production. They are sold all over China and even exported to Southeast Asian countries.

3. Bamboo shoots. Fresh shoot yield in 1997 was 1.91 million tons, of which 1.2 million tons were used for making dried shoot and 0.2 million tons for canned shoot. Export value of shoot and its derivatives was USD 155 million – 38.8% of the total bamboo exports valued at USD 400 million. Besides fresh and canned Moso shoots, preserved shoots and other products from species such as *Phyllostachys praecox* have also appeared in the market.

4. Bamboo pulping and papermaking. There are more than 70 bamboo paper mills in China. Bamboo pulp yield in 1990 was only 0.21 million tons, or 1.6% of the total pulp yield in the same period. Shortly, several large-scale bamboo paper mills will go into production, and the pulp yield is expected to reach 0.4–0.5 million tons.

5. Bamboo chemical outlets. Bamboo juice contains rich nutrients and unique flavor. Bamboo juice beverage, wine, beer, even rice (seeds) have been successfully exploited. Bamboo leaves are used for extracting useful chemical compounds such as flavonoids.

Bamboo research

1. Bamboo classification and flora. A classification system has been drawn up, and accordingly there are 39 genera and about 500 species of bamboo in China. Division and planning of bamboo forest regions have been completed, and some bamboo genetic gardens for *ex situ* conservation have been established.

2. Fundamental research. Since the 1980s, research has been started in the area of bamboo ecology and physiology. Notable results have been achieved in areas such as: bamboo individual growth and differentiation; mechanism of bamboo flowering; photosynthesis and carbon distribution; phenomena of on-year and off-year; mineral nutrition; hormones; sympodial bamboo breeding; bamboo stand biomass; dynamics and degradation of leaf litter; and underground water and nutrient cycling. Significant progress has been made also in the area of physical and mechanical properties of bamboo.

3. Bamboo cultivation. A full set of propagation techniques has been developed for mother bamboo transplanting, seedling, rhizome-, node- and culm-offsets, culm layering, and tissue culture. Notable successes have been registered in harvest cultivation techniques covering aspects such as rational remaining density of bamboo plant, thinning of shoots, soil loosening, weeding, composite fertility, and improvement of on-year and off-year Moso stands. Important strides have been made also in orientation cultivation of bamboo forest – such as shoot forest, timber forest, shoot and timber forest, pulping forest, landscaping forest. Another

success was is in the development of counter-season cultivation of *Phyllostachys praecox* forest for shoots.
4. Control of bamboo pests. There are over 600 species of bamboo pests in 280 genera. Among these, research has been conducted on about 200 species, covering aspects such as their occurrence patterns and control measures. In general, bamboo pests have been kept under control.
5. Bamboo shoot and processing techniques. Nutrient components in over 40 main bamboo species shoots have been analysed. Research has been done also on preservation of shoots by boiling, and on flavored canned shoots.
6. Bamboo timber processing technologies. Different types of bamboo plywood have been developed with success, and put into a large-scale industrial production. These include bamboo mat plywood, building concrete moldboard, bamboo laminated plywood, bamboo particleboard, bamboo flooring, bamboo wood composite board and reinforced bamboo board.
7. Bamboo chemical utilization. In-depth research has been carried out on bamboo xylose, bamboo leaf extracts and other bioactive substances.

Current problems
Despite significant achievements in the bamboo-based industry, one needs to consciously acknowledge that, from a macroscopic view, the bamboo development level is still low and confronted with many problems.

1. Less financial input for bamboo science and technology, and unsuitability of newly developed products in terms of market needs. For many years, the financial input for bamboo science and technology has been very low, and lagged behind the needs of the bamboo industry to a large extent. This financial constraint has made it very difficulty for the industry to exploit some key products that have very promising market advantage. Nearly no progress has been made on the application of high technology for bamboo products and, furthermore, the development of some products that have major market prospects have ceased because of funding shortage.
2. Small industrial scale, poor management system and low economic benefits. Most bamboo processing enterprises are small in scale, employ outdated technologies and have low profits. The combination of these detrimental factors has even forced some of them to close.
3. Weak administration and delayed production. Majority of bamboo processing enterprises are scattered in remote mountainous areas. Lack of access to information and technology makes it difficult to respond in time to market changes and to effective administer the industry in general.

CHINA'S BAMBOO INDUSTRY: PROSPECTS

Bamboo is a vital and unique component of forestry, and plays a very important role in ecological and environmental improvement, sustainable development, and poverty alleviation. China, with a very fragile ecology, is scarce in forest resources including timber trees. Bamboo, which has a very strong root system and is a renewable resource once planted, not only effectively conserves water and soil, but also provides a great deal of timber for replacing wood. This enables us to protect "wood with bamboo".

At present, China's bamboo industry is in a bad condition. However, it is faced with a rare opportunity, a challenge, to break free of the present stagnation. The issue is what attitude needs to be adopted to meet that challenge. The opportunity is provided by the convergence of a series of individual events: the global focus on environmental protection and sustainable development; the Chinese government decision lowering national timber cutting quota; the setting up of INBAR Headquarters in Beijing; and the accelerated establishment of China National Bamboo Research Center (CBRC). The convergence of these events has provided the bamboo sector with a very favorable climate and good opportunity to cover lost ground and to move ahead, fully understanding the current problems and the difficulty they pose.

Guidelines for bamboo industry development

On the precondition of ensuring sustainability of bamboo, following socialist market system and taking bamboo cultivation as the base, the industry needs to:

- take low-yield bamboo stand improvement as a breakthrough;
- look to science and technology for guidance;
- consider market needs for orientation; view bamboo product exploitation as a "dragon-head";
- keep comprehensive economic benefit as the target; and
- try opening up domestic and overseas markets.

Such an approach would help gradually set up a modern bamboo industry that integrates bamboo cultivation, processing, production and trade with the national economy and social development needs.

Targets for bamboo industry development

The targets for the development of the industry are:

- to complete, by the year 2010, improvement of low-yield bamboo stand in an area of 3.1 million ha and planting in an area of 1.2 million ha;
- to raise the annual Moso culm yield to 800 million culms and the yield of other bamboos to 24 million tons per year;
- to increase the annual fresh shoot yield to 3.2 million tons; and
- to boost the annual bamboo output value to USD 290 million, annual export value to USD 800 million (double the index at present).

Main measures for bamboo industry development

1. Step up macroscopic guidelines and streamline management

The administration and management of all the stages in the bamboo production-to-consumption chain – such as cultivation, conservation, utilization and regeneration – should be streamlined. Policies and regulations related to bamboo harvest and transportation by license, and market administration of bamboo products and shoots need to be implemented. Special funds need to be set up for increasing the bamboo resource. In a nutshell, all measures required to put the bamboo industry back on track must be adopted.

2. Strengthen support to bamboo R&D Institutions

CBRC was established, with the approval of the Ministry of Science and Technology and the Forestry Administration, tasked with three missions:

a. Undertake, organize and coordinate key bamboo projects in China and abroad;
b. Engage extensively in international economic and technical exchange and co-operation, including personnel training; and
c. Form China Bamboo Industry Group gradually to act as the backbone and engine for the development of the country's bamboo industry.

A multifunction China Bamboo Plaza, which integrates research, education and training, is to be completed by the end of 1998. It is being built on an area of 15 000 m^2 at an investment of USD 6 million. It is hoped that the Plaza would play a major role in the development of the bamboo industry in China and elsewhere.

The headquarters of the International Network for Bamboo and Rattan (INBAR) was established in Beijing in November 1997 through the concerted efforts of the Chinese and Canadian governments, and some other bamboo-growing countries and international organizations. The Chinese government has always attached importance and offered support to such international initiatives, under the follow-up activities envisaged by the Earth Summit and as a backing to global sustainable development plan of action. It is believed that INBAR will play a lead role in global bamboo research and development.

3. Appropriate management regimes and markets initiatives

Appropriate management regimes are required to combine bamboo processing with bamboo stand silviculture. Different kinds of bamboo stands – such as timber stand, shoot stand, landscaping stand and pulp stand – have different technical parameters oriented to their purposes. Management systems that take into consideration these differences are needed to optimize the benefits. Strengthening bamboo processing capabilities, enhancing production level and product variety, and realizing multiple value addition are the other measures necessary. Domestic and overseas markets need to be exploited, and production and management systems should take into account the nuances of market economy.

4. Use science and technology for overall bamboo development

Bamboo research should be deepened and promoted, particularly in the following areas: resource assessment and conservation, including bio-diversity and genetic improvement; sustainable management techniques; bamboo processing and integrated utilization techniques; and optimal management models for bamboo processing enterprises.

Active dissemination of bamboo science and technology results and speeding up technology transfer have to be taken up to enhance the contribution of science and technology to the bamboo industry.

There is a need to actively probe for a new mechanism that would incorporate science and technology into production in the context of a socialist economy, encouraging technical personnel to disseminate technical knowledge and provide consultation to production enterprises.

5. Enhance financial input and accelerate bamboo industry growth

A set of financial input mechanisms suitable to bamboo industry development should be drawn up. An appropriate proportion of funds should be set aside from the Agricultural Comprehensive Development Funding and Special Interest-deduced Forest Loan for bamboo industry development. Besides, some levy should be imposed on the sale of bamboo timber and products and this revenue should be ploughed back into bamboo cultivation to ensure a reasonable balance between cultivation and processing.

6. Tap research institutes and forestry universities for technical and policy consultation

CBIA and its provincial branches, CBRC, and other relevant forest research institutes and universities should be tapped for using their full potential in offering policy consultation to governments, and technical consulting service to bamboo enterprises.

BAMBOO INDUSTRY DEVELOPMENT AND THE ENVIRONMENT

Worldwide, governments and people have attached a great importance to global environmental protection because of the threat to survival and development of human beings posed by phenomena such as global warming, environmental pollution, floods, desertification and soil erosion. What stance must we take, stepping into the 21st century, and what future should we leave to the posterity? These thought are now the concern of people from all over the world, and the most common of these reached the Global Environment Conference.

Recently, China has focused on environment protection and pollution prevention, and has taken a series of measures. However, despite some achievements, there is much room for improvement. There are various causes for the recent devastating floods in the reaches of the rivers Yangzhi and Songhua, which have brought direct economic loss of about USD 20 billion. However, two important causes are the severe destruction of forest vegetation along the banks of, and particularly the source of, Yangzhi River as well as the lack of appropriate management of forestation there. The Chinese Government has decided to earmark 30 billion yuan from the central fiscal budget for 5 years for Natural Forest Protection Engineering, as a fundamental measure to prevent the recurrence of such incidents. Water conservation projects would be, at best, a measure of control, and will not have much effect on water and soil erosions, riverbed rising, and worsening climate caused by vegetation destruction.

The development of bamboo industry is listed in the ninth five-year plan among the pillar engineering projects aimed at environment protection. It is not difficult to conclude, considering the unique features of bamboo, that including bamboo industry among the pillar engineering projects for environmental protection is a multi-purpose and justified measure that would be beneficial to posterity. This should be an important pointer to other bamboo-growing countries.

Bamboo has the advantage of being a rapid-growth, short-cycle, sustainable plant that provides fast returns on a small investment. These important features make it unique as compared with other woody species. Considering the present capacity of China, it should be easy for the country to use bamboo to afforest the areas south of the Yellow River. However, it would be much better if done in combination with the natural forest protection engineering projects, and more so with pillar engineering projects for environment protection. The problem is how to reach a common understanding among the bamboo circles for making a concerted publicity effort, especially to raise people's awareness of the issues involved, and to focus on realizing a revolutionary change in the bamboo-based industry to fuel its development.

FURTHER READING

Li Shidong. 1998. China's bamboo development processing and development strategies towards 21 century. Journal of Bamboo Research, Vol. 17, No. 1, 1–5.

Bamboo for Sustainable Development, pp. 125–135
Eds. A. Kumar, I.V. Ramanuja Rao and Ch. Sastry

Bamboo as an alternative material to timber in Malaysia: Yield, production and utilization

ABD. LATIF MOHMOD, ABD. RAZAK OTHMAN and NORINI HARON
Forest Research Institute Malaysia (FRIM), Kuala Lumpur, Malaysia

Abstract—Bamboo, once considered a weed in forestry practice, has is now becoming an important forest resource with industrial uses. In view of the current pressure for Forest Certification under Sustainable Forest Management System and the possible consequences of the timber shortage foreseen in the near future, bamboo has been put as one of the secondary forest species to be planted on a commercial basis in Malaysia. This paper discusses the potential of bamboo, with emphasis on commonly utilized species (*Bambusa blumeana*, *B. vulgaris, Dendrocalamus asper, Gigantochloa levis* and *G. scortechinii*) as an alternative planted and industrial material to popular timber species (such as *Acacia mangium, Azadirachda excelsa, Hevea braziliensis* and *Tectona grandis*). The potential of bamboo is examined based on its high productivity, short maturity cycle, economic viability, and its promising physical and mechanical characteristics. With developments in technological know-how, supported by sustainable resources and management, the bamboo industry is expected to see further growth in Malaysia in the near future.

INTRODUCTION

Bamboo, which was once considered a weed in forestry practice (Salleh and Wong 1987), has now become an important forest resource with several industrial uses. This could probably be associated to the increasing difficulty in obtaining commonly used forest timbers, growing hygienic concerns among the population (especially for cheap disposable items used in contact with food), and the heightened environmental awareness of the people on the need to sustain the surroundings healthy and green.

With the International Tropical Timber Organization (ITTO) guidelines and the pressure for Forest Certification procedure in Sustainable Forest Management Systems, harvesting (and hence the utilization) of natural stand timbers from the tropical forest is expected to reduce towards the year 2000. To address the consequences of timber shortages in the near future, integrated approaches to forest production and management need to be looked into. If backed by adequate research and development, bamboo could be a potential alternative to meet the demand for timber-based products. Bamboo has traditionally played an important role in the lives of people, particularly those in the rural areas, owing to its easy availability, renewable nature and multiple uses.

BAMBOO RESOURCES

A comprehensive survey by Abd. Latif and Abd. Razak (1994) on the availability and distribution of bamboo and its industrial profile disclosed that bamboo is mainly distributed in the north-western part of Peninsular Malaysia. Because of this pattern of distribution, the units that manufacture bamboo products for the local market is predominantly located in the west coast region so that the transport of bamboo from the forest to the factories is easy. The survey also revealed that the most common species extracted from the forest, amounting to about 2 million poles of 10 feet length or about 70% of the total harvest, was *Gigantochloa scortechinii* (Abd. Latif and Abd. Razak 1994). This was followed by *Dendrocalamus pendulus* (about 20%), and *G. wrayi*, *G. ligulata, Schizostachyum grande* and *S. zollingerii* (total 10%).

Kamaruzaman (1992) indicated that about 587 million bamboo culms were available from both undisturbed and disturbed forests of Peninsular Malaysia. Lockman et al. (1994) estimated that more than 110 million culms were available from about 420 000 ha of forest reserve areas in Peninsular Malaysia. The data obtained from the Third National Forest Inventory (1991–93) revealed that the total number of bamboo clumps estimated within the peninsula amounted to about 82.2 million, with about 1.4 billion standing culms (6 m/culm).

There was an increase in the number of bamboo poles harvested from the forest reserve in 1991, 1992 and 1993 – amounting to 1.6, 2.1 and 2.4 million culms, respectively – in order to meet the increased demand from the industry. Most of these came from Pahang, Perak, Kelantan and Selangor (Poh et al. 1994).

A NEED FOR BAMBOO PLANTATION

Peninsular Malaysia has at least 60 bamboo species of which 25 are indigenous, and the rest are known only in cultivation (Abd. Latif and Abd Razak 1994). Most of the raw material for the bamboo-based industries comes from natural stands, especially the stands in disturbed areas such as logged-over forests, degraded land, marginal localities at forest fringes, and riverbanks.

Although bamboo resources are quite substantial, some factors associated with the distribution and growth of natural bamboo stands make large bamboo-based industries unviable. The lack of steady supply of good quality culms in large quantities is frequently reported to hinder the production of quality products. A study showed that only 4.7% of the total standing culms of *G. scortechinii* in the natural stands were tagged as suitable for harvest (Abd. Razak 1994). Also, indiscriminate harvesting of wild bamboos, without due consideration to their suitability in terms of age for the intended use, causes high wastage during machining.

Considering the economic importance of bamboo and the difficulty in assuring steady supplies of high-quality bamboo culms for industrial uses, it is felt that there is a need to establish bamboo plantations on a large scale.

Choice of species

Research has identified a few thick-walled bamboo species as suitable for the manufacture of various products. These are *Gigantochloa levis*, *G. scortechinii*, *Gigantochloa* sp. (brang type) and *Dendrocalamus asper*. The lower to middle portions of the culms can be used to produce high-value bamboo products – such as bamboo board, panel, parquet and laminates – while the upper portion can be used for the production of items such as toothpicks and skewers. The distribution and culm characteristics of the species are given in Table 1.

Table 1.
Distribution and culm characteristics of selected bamboo species

Species	Distribution	Culm height (m)	Internode length (cm)	Diameter at DBH (cm)	Culm-wall thickness (mm)
Dendrocalamus asper	Cultivated	18–23	35–60	9–13	10–14
Gigantochloa levis	Cultivated	18–23	30–60	11–13	11–15
G. scortechinii	Wild	17–20	35–59	9–11	7–12
Gigantochloa sp. (brang)	Wild	15–20	30–50	6–14	10–12

Bamboo growth and production

The productivity of any bamboo plantation depends on the annual number of shoots that emerge from a bamboo clump and then successfully develop into culms. In a bamboo plantation, the harvesting of bamboo shoots and culms starts 3 and 4 years, respectively, after planting in the field. The production will continue for a period of 20 years or more if properly maintained. Bamboo culms that are 3 years old are considered mature enough to be harvested for use.

Culm/shoot growth and biomass production

Bamboo consists of woody stems or culms that emerge from underground rhizome buds. These buds initially grow into shoots before developing into culms. In Peninsular Malaysia, shoots commonly emerge from May to September, which is also called the shoot production season (Abd. Razak and Hashim 1993).

Results obtained from trial plots of selectively planted bamboos species (3–6 years of age) show that the average number of shoots that emerge annually ranged from 9 to 15, depending on species. Only 7 to 11 of these were productive – that is, grew and developed into bamboo culms. *G. levis* produced the highest number of productive shoots annually (11), followed by *D. asper* (10), *Bambusa vulgaris* (9), *G. scortechinii* (8) and *B. blumeana* (7) (Table 2). The data further showed that the percentage of non-productive bamboo shoots – consisting of shoots that are dormant, injured by animals or dried during the dry season – ranged from

19.1% to 28.2%. In general, a bamboo plantation with a planting density of 400 clumps/ha produces between 2 800 to 4 400 culms/ha/year, depending on the bamboo species planted.

Table 2.
Average number of annual shoot/culm production of selected bamboo species

Species	Number of shoots			Culm productivity
	Productive	Non-productive	Total	per ha per year
Bambusa blumeana	7	2	9	2 800
B. vulgaris	9	3	12	3 600
Dendrocalamus asper	10	4	14	4 000
Gigantochloa levis	11	4	15	4 400
G. scortechinii	8	2	10	3 200

Bamboo, considered as one of the fastest growing woody plant, takes only 5–6 months to reach its maximum height (Liese 1985; Abd. Razak 1992). Although the growth decreases at this early age, its chemical components and physical characteristics keeps on changing until the bamboo culm reaches maturity.

Data on the total above-ground biomass of mature bamboo culms (3 to 4 years old) gathered from 30 culms sampled from each bamboo species selected randomly in 6-year-old bamboo trial plots are shown in Table 3. The total above-ground biomass of *B. vulgaris, G. levis* and *G. scortechinii* was 6.4, 13.4 and 7.2 kg/culm, respectively. In a bamboo plantation aged 5 years and above, the estimated annual culm biomass production was estimated to be 13.75, 16.32 and 29.04 t/ha for *B. vulgaris, G. scortechinii* and *G. levis,* respectively. From the total annual culm biomass production, 60–70% is usually harvested for consumption.

PHYSICAL AND MECHANICAL PROPERTIES OF BAMBOO

For any material to replace timber, it should at least have physical and mechanical properties comparable to those of timber. For example, consider the oven-dry density of bamboo, which varies from about 0.5 to 0.9 g/cm^3 (Abd. Latif and Liese 1995). It falls within the range of light to heavy hardwoods such as *Shorea teysmanniana* (Meranti bunga, 0.42 g/cm^3) and *S. ochrophloia* (Balau, 0.88 g/cm^3), the commercial timbers used for furniture and heavy construction, respectively.

Table 3.
Biomass production of mature bamboo culms (3–4 years old) in 6-year-old trial plots

Parameter	*B. vulgaris*	*G. levis*	*G. scortechinii*
Average DBH (cm)	10.5	11	8.5
Average height (m)	12	15	13.5
Average dry weight (kg)			
Culm	3.82	6.6	5.1
Branches	2.14	5.2	1.0
Leaves	0.44	1.6	1.1
Total dry weight	6.4	13.4	7.2
No. of culms/ha/y	3 600	4 400	3 200
Culm dry weight (t/ha/y)	13.75	29.04	16.32
No. of culms/t	261	151	196

Abd. Rahman (1972) has compared the properties of Malaysian timbers in terms of shear strength, compression strength, MOR, MOE and stress at proportional limit. Similar data for bamboo, except MOR and MOE, place it under the strength group A, in the same basic grade as *Madhuca utilis* (Bitis), *Shorea laevis* (Balau kumus) and *Intsia palembanica* (Merbau), the heavy hardwoods used for construction, piling, posts, bridges and heavy-duty flooring. The modulii of rupture and elasticity, however, classify bamboo under a standard to basic grade, which is a mixture of strength groups C and D. Combined with its relatively high stress at proportional limit, bamboo offers a satisfactory material for general construction purposes if properly treated with preservatives.

Abang Abdullah (1983) demonstrated that the compressive, tensile and bending strengths of bamboo are equivalent to those of wood or steel, but much weaker in shear. This low shear strength is a disadvantage for structural use, particularly in bending and joints, because shear plays an important role in the loss of cohesion between fiber bundles. This property, however, is useful in basketry, handicraft and in the manufacture of parquet strips wherein split bamboo is used.

Wear resistance is another important factor in considering the utilization of a material for flooring purposes. The results obtained from a study conducted in the Forest Research Institute Malaysia (FRIM) showed that the abrasive resistance of two bamboo species – *B. vulgaris* and *G. scortechinii* – was about 30% superior to that of *Koompasia malaccensis* and about five times better than rubber wood (Abd. Latif et al. 1990). Hence, bamboo could be considered suitable for flooring purposes under medium to heavy traffic conditions. Other advantages of bamboo are that it can be cut, polished, varnished and painted easily.

Table 4.
Comparative strengths of bamboo and timber

Species	Shear parallel to grain (MPa)	Compression parallel to grain (MPa)	Static bending		
			Modulus of rupture (1000 MPa)	Modulus of elasticity limit (Mpa)	Stress at proportional
Bambusa blumeana	4.0–4.9	19.5–28.5	43.1–156.4	2.6–5.6	21.2–38.9
B. vulgaris	3.3–4.8	20.5–30.0	46.1–78.4	4.1–8.1	28.7–42.6
Dendrocalamus asper	4.4–6.3	28.3–34.6	48.9–22.4	3.8–8.8	32.2–46.8
Gigantochloa scortechinii	3.6–4.9	21.6–32.3	35.9–68.9	3.7–5.9	31.1–42.2
G. levis	3.7–5.8	37.3–42.8	37.6–119.4	3.7–6.5	35.7–48.7
Balanocarpus hemii (cengal)*	13.7	69	122	1.8	–
Koompasia malaccensis (kempas)*	10.0	54.7	100	1.7	–
Madhuca utilis (bitis)*	15.4	90.3	171	23.8	12.48
Shorea laevis (balau kumus)*	15.0	76.0	142	20.1	9.8
Intsia palembanica (merbau)*	12.5	58.2	116	15.4	9.2

Note: * Data collected from Engku Abd. Rahman (1972)

ECONOMIC ANALYSIS OF BAMBOO CULTIVATION

Economic analysis is a key consideration for investment. In view of the expected timber shortage, measures have been initiated to increase timber production from plantations. Three most commonly used measures of a project's viability – 1) Net Present Value (NPV); 2) Internal Rate of Return (IRR); and 3) Benefit-Cost Ratio (B/C ratio) – are used in the analysis.[1]

The management of a commercial bamboo plantation could be broadly classified into three main stages: 1) site preparation, which includes planting; 2) silviculture treatments; and 3) harvesting. Detailed breakdown of cost and revenue of establishing and managing a 500 ha of commercial bamboo plantation is presented in Appendix 1.

[1] All costs incurred and revenue received throughout the project cycle are discounted to present value. For the purpose of this analysis, a discount rate of 10% is assumed as the rate recommended by Economic Planning Unit of the Prime Minister's Department and/or the rate normally charged on the capital borrowing by commercial banks in Malaysia. All costs incurred are based on the establishment and maintenance of *G. levis* planted in FRIM and FELDA Mempaga 3, Bentong, Pahang.

To test the financial feasibility of a large-scale planting, an analysis of a 500-ha bamboo plantation based on a rotation of 15-year cycle was conducted. The total expenditure for the whole project duration was estimated to be MYR 7.6 million, while the total revenue collected was almost MYR 14.9 million.[2]

Most of the expenditure – 78% of the total costs incurred – is for fertilizer, maintenance and harvesting. The annual culm production in 4, 5–6, 7–9 and 10–15 years after field planting were estimated to be 400 000, 600 000, 800 000 and 1 million culms, respectively. The harvesting of shoots, with revenue worth MYR 4.9 million, is considered supplementary as the silvicultural management of the bamboo clumps is aimed mainly at the production of high-quality culms. The net revenue for the whole project is about MYR 7.3 million. The IRR of the project is 22.59%. The NPV discounted at 10% is MYR 1.99 million and the B/C ratio is >1. Thus, the project can be considered financially feasible and attractive for investment as compared with plantation timber species with a rotation cycle of 15 years (Table 5).

Sensitivity analysis

A sensitivity analysis, with respect to increase in price of culm and operational costs, is summarized in Table 6. Sensitivity analysis carried out for the culm production indicated that the project is still viable even when the operational costs increased by 10%. The IRR of 19.56% recorded is much higher than the discounted rate and within the acceptable level of investment.

The project would be more attractive if the establishment of bamboo plantation were to be accompanied by downstream activities and product development. Economic analysis of bamboo flooring production, from the establishment of a 500-ha plantation to the finished product, gives an IRR of 48%, NPV of MYR 39.8 million and B/C ratio of 2.2 (Abd. Razak et al. 1997).

Table 5.
Financial analysis of selected forest plantation species in comparison to bamboo

Species	Area (ha) rate (%)	Discount (Ringgit)	NPV	IRR (%)	B/C ratio
Bamboo	500	10	1 988 032	22.6	>1
[1]*Acacia mangium*	2000	10	5 077 001	14.5	1.4
[2]*Tectona grandis* (teak)	2000	10	23 759 552	17.3	>1
[2]*Azadirachda excelsa* (sentang)	2000	10	17 906 868	16.4	>1
[3]*Hevea brasiliensis* (rubber wood)	2000	10	1 420 000	11.1	1.1

Source: [1] Thai et al. 1997; [2] Krishnapillay et al. 1997; [3] Najib et al. 1997

[2] MYR (Malaysian Ringgit) 1 = approx USD 0.263.

Table 6.
Sensitivity analysis of commercial bamboo plantation with respect to increased price and operational costs

Variable simulation	NPV @10% (MYR)	IRR (%)
Price increased annually by:		
2%	2 110 284	23.22
5%	2 293 663	24.15
10%	2 599 294	25.65
Operational cost increased annually by:		
2%	1 905 540	21.96
5%	1 781 802	21.04
10%	1 575 572	19.56

The main advantage that a bamboo plantation has over a timber plantation is that the return on investment is fast, starting from the production of shoots in the third year and culms in the fourth year. Other advantages of bamboo plantation are that the plant regenerates fast, it has a short harvesting cycle and the production continues for 20 years or more. This is financially a very attractive proposition: repayments on loan taken can begin early, and return on investment continues for a fair term.

Products

The bamboo industry, in general, offers a wide variety of end products in different designs and has a high potential for development. Current activities, which are confined to traditional products, serve mainly the limited local market (Abd. Latif and Ashaari 1991). The machine-intensive bamboo industry, on the other hand, has been introduced to fulfill the increasing demand for disposable products such as chopsticks, skewers and vegetable baskets.

Based on its high productivity and physical and mechanical properties, bamboo can be used for high-value products such as boards, panels, parquet and laminates, similar to the ones produced from timber. Furthermore, bamboo boards and laminates can be used to produce other high-quality end products such as furniture and furniture components.

The concept of zero wastage can be adopted in bamboo production and utilization. As said earlier, the lower portion of bamboo culms (up to 8 m) can be used for bamboo boards and panels, while the middle and upper portions can be utilized for lower-end products such as toothpicks and skewers. In addition, bamboo wastes and branches can be used for making cement-bonded particleboards, plastic-bamboo or activated carbon.

The bamboo-based industry is a profitable business. Average comparative ratio of prices for bamboo and its products are summarized in Table 7.

Table 7.
Comparative ratio of selected bamboo products

Product	Price ratio
Raw bamboo culm	1.0
Raw shoot	1.0–3.0
Canned shoot	2.5–7.0
Bamboo split	3.5–4.0
Chopstick	5.0–6.5
Skewers/toothpicks	6.0–9.0
Vegetable baskets	3.0–6.0
Blinds	30.0–50.0
Parquet and panels	15.0–65.0
Plastic bamboo (waste, culm and branches)	150.0–300.0

Source: Abd. Latif 1997

CONCLUSION

Based on productivity, economic viability and utility, bamboo is a promising option to replace timber in high-quality products. With proper planning in assuring the availability of quality and continuous supply of bamboo, and accompanied with technology development, there is no reason why the industry cannot grow further. The success of the industry will not only help reduce the high consumption rate of timber but also conserve and preserve precious forests.

REFERENCES

Abang Abdullah, A.A. 1983. Utilization of bamboo as a low cost structural material. *In* Proceedings of a Symposium on Appropriate Building Material for Low-cost Housing. Nairobi, Kenya, 7–14 November 1983. pp. 117–182.

Abd. Latif, M. 1997. Prospect of non-timber forest produce as an industrial resources. Paper presented at the Malaysian Indigenous People Entrepreneurs Seminar, Langkawai, Malaysia, 26–29 August 1997. 22 pp.

Abd Latif, M.; Abd. Razak, O. 1994. Availability, distribution of bamboo and its industrial status in Peninsular Malaysia. *In* Bamboo in Asia and the Pacific. Proceedings of the 4th International Bamboo Workshop, Chiangmai, Thailand, 27–30 November 1991. International Development Research Centre, Ottawa, Canada; Forestry Research Support Programme for Asia and the Pacific, Bangkok, Thailand. pp. 60–67.

Abd Latif, M.; Ashaari, M.A. 1991. Non-wood forest products as main industrial resources for rural development. *In* Proceedings of the Third Asian Conference on Rural Technology, Kuala Lumpur, Malaysia, 12–14 August 1991. pp. 286–292.

Abd. Latif, M.; Liese, W. 1995. Utilization of bamboo. *In* Abd. Razak O.; Abd. Latif, M.; W. Liese; Norini, H. ed., Planting and utilization of bamboo in Peninsular Malaysia. FRIM Research Pamphlet No. 118. Forest Research Institute Malaysia, Kuala Lumpur. pp. 50–102.

Abd. Latif, M.; Mohd. Tamizi, M.; Mohd. Rashid, S.; Mohd. Shukari, M. 1990. Wear resistance of two commercial bamboo species in Peninsular Malaysia and their suitability as a flooring material. *In* Ramanuja Rao, I.V.; Gnanaharan, R; Sastry, C.B., ed., Bamboos: current research. Proceedings of the International Bamboo Workshop, Cochin, India, 14–18 November 1988. Kerala Forest Research Institute, Kerala, India; International Development Research Centre, Ottawa, Canada. pp. 223–230.

Abd. Rahman, E.C. 1972. Basic and grade stresses for strength group of Malaysian timber. Malaysian Forester, 35(2), 131–134.

Abd. Razak, O. 1992. Shoot production and culm growth of two commercial bamboo species in Peninsular Malaysia. Bamboo Information Centre (India) Bulletin, 2(1), 5–7.

Abd. Razak, O. 1994. Culm composition and above-ground biomass of *Gigantochloa scortechinii* in a natural stand and a three-year-old plantation. Journal of Tropical Forest Science, 7(2), 280–285.

Abd. Razak, O.; Hashim, M.N. 1993. Shoot production at initial stage of planting. Paper presented at the Symposium on Tropical Resources, Kucing, Sarawak, 6–11 December 1993.

Abd. Razak, O.; Shamsudin, I.; Chong, P.F.; Wan Mohd. Shukri, W.A.; Abd. Latif, M.; Norini, H.. 1997. Manual of bamboo plantations for high-value products. Forest Research Institute Malaysia. (Unpublished).

Kamaruzaman, A.B. 1992. Bamboo resources in Peninsular Malaysia. Buletin Buluh (Bamboo), 1(1), 8–9.

Krishnapillay, B.; Mohd Noor, M.; Norini, H.; Zuhaidi, A.Y.; Ab. Rasip, A.G.; Mahmud, A.W. 1997. Viability of planting teak and sentang in Malaysia. Paper presented at the Seminar on Commercial Cultivation of Teak, Sentang, Acacia and Hevea for Timber, Kuala Lumpur, Malaysia, 9 January 1997.

Liese, W. 1985. Bamboos – biology, silvics, properties, utilization. Deutsche Gelleschaft für Technische Zusammenarbeit (GTZ), Nr. 180, TZ Verlagsges., Roßdorf 1, 132 pp.

Lockman, M.S.; Mohd. Shahwahid, O.; Yong, P.L.; Saroni, J. 1994. Distribution of bamboo and the potential development of the bamboo industry. *In* Wan Razali W.M.; Aminuddin, M., ed., Towards the management, conservation, marketing and utilization of bamboos. Proceedings of the National Bamboo Seminar 1, Kuala Lumpur, Malaysia, 2–4 November 1992. Forest Research Institute Malaysia, Kuala Lumpur, Malaysia. pp. 6–11.

Najib, L.A.; Mohamad Johari, M.H.; Abdul Ghani, I.; Mahdan, B.; Dan, Y.M.; Mahmud, A.W. 1997. Viability of Hevea plantation for wood production. Paper presented at the Seminar on Commercial Cultivation of Teak, Sentang, Acacia and Hevea for Timber, Kuala Lumpur, Malaysia, 9 January 1997.

Salleh, M.N.; Wong, K.M. 1987. The bamboo resources in Malaysia: strategies for development. *In* Rao, A.N.; Dhanarajan, G.; Sastry, C.B. ed., Recent Research on Bamboo. Proceedings of the International Bamboo Workshop, Hangzhou, China, 6–14 October 1985. Chinese Academy of Forestry, Beijing, China; International Development Research Centre, Ottawa, Canada. pp. 45–49.

Poh, L.Y.; Mohd. Shahwahid, O.; Saroni, J. 1994. Kajian Industri Buluh Di Semenanjung Malaysia. Ibu Pejabat Perhutanan Semenanjung Malaysia, Jalan Sultan Salahuddin, Kuala Lumpur, 176 pp.

Thai, S.K.; Mahdan, B.; Ridzuan, S. 1997. Planting of *Acacia mangium* for general utility timber and reconstituted wood products. Paper presented at the Seminar on Commercial Cultivation of Teak, Sentang, Acacia and Hevea for Timber, Kuala Lumpur, Malaysia, 9 January 1997.

Appendix 1.

Estimated cost and revenue per annum of commercial bamboo plantation (500 ha) for 15 years period

Operation	Total	1	2	3	4	5	6	7	8	9	10	11	12	13	14	15
1. Site preparation (including land clearing)	259 000	259 000														
2. Planting material @RM2.00/stem	400 000	400 000														
3. Planting (RM0.30/ planting point)	60 000	60 000														
4. Road	20 000	20 000														
5. Boundary	10 000	10 000														
6. Fertilizer: Organic–MYR50/t & NPK Green–MYR700/t	908 000	82 000	96 000	70 000	110 000		110 000		110 000		110 000		110 000		110 000	
7. Replanting:																
Planting material	40 000	40 000														
Replanting cost	6 000	6 000														
8. Pest/disease protection	5 000	5 000														
9.Maintenance	1 400 000	400 000	200 000	200 000	100 000		100 000		100 000		100 000		100 000		100 000	
10. Land rent@ $50/ ha/y	375 000	25 000	25 000	25 000	25 000	25 000	25 000	25 000	25 000	25 000	25 000	25 000	25 000	25 000	25 000	25 000
11. Harvesting cost (MYR 0.20/culm or shoot)	3 480 000	–	–	40 000	200 000	240 000	240 000	280 000	280 000	280 000	320 000	320 000	320 000	320 000	320 000	320 000
Sub-total	6 963 000	1 307 000	321 000	335 000	435 000	265 000	475 000	305 000	515 000	305 000	555 000	345 000	555 000	345 000	555 000	345 000
Contingency (5%)	348 150	65 350	16 050	16 750	21 750	13 250	23 750	15 250	25 750	15 250	27 750	17 250	27 750	17 250	27 750	17 250
Total Cost	7 311 150	1 372 350	337 050	351 750	456 750	278 250	498 750	320 250	540 750	320 250	582 750	362 250	582 750	362 250	582 750	362 250
1. Bamboo shoots (400 g/shoot)	4 864 000	–	–	256 000	384 000	384 000	384 000	384 000	384 000	384 000	384 000	384 000	384 000	384 000	384 000	384 000
2. Bamboo culms (10 m/culm) (MYR 1.00/culm MYR 1.60/kg shoot)	10 000 000	–	–	–	400 000	600 000	600 000	800 000	800 000	800 000	1 000 000	1 000 000	1 000 000	1 000 000	1 000 000	1 000 000
Total Revenue	14 864 000	–	–	256 000	784 000	984 000	984 000	1 184 000	1 184 000	1 184 000	1 384 000	1 384 000	1 384 000	1 384 000	1 384 000	1 384 000
Net Cash Flow	7 552 850	(-1 372 350)	(-337 050)	(-95 750)	327 250	705 750	485 250	863 750	643 250	863 750	801 250	1 021 750	801 250	1 021 750	801 250	1 021 750

Bamboo for Sustainable Development, pp. 137–153
Eds. A. Kumar, I.V. Ramanuja Rao and Ch. Sastry

Bamboo industrialization in the context of sustainable development

TREVOR D. DAGILIS and DAVID J. TURCKE
Department of Civil Engineering, Queen's University, Ontario, Canada

Abstract—Modern society is in many ways unsustainable. The significant reliance, for instance, on plastics, steel and concrete as base materials for products is harmful, as long-term durability is gained through excessive energy consumption, harmful raw material extraction, and inefficient and detrimental manufacturing processes. Fortunately, natural fiber products, such as those made from bamboo, offer sustainable alternatives.

This paper shows that bamboo is a natural fiber raw material that can be used to produce a number of products that can compete directly with, and can be substituted for, products made of plastic, steel, concrete and clear-cut timber, thereby helping to reduce the harmful impact of our consumption-based societies.

Bamboo industrialization projects offer significant benefits to rural regions in developing countries. Bamboo plantations can be established and operated with little capital investment in equipment and infrastructure, allowing for the involvement of small-scale landholders. Associated manufacturing plants can provide a range of products that can be used locally at costs below those of comparable imported goods. In addition, higher production capacities can ensure product for export. Therefore, the controlled and sensible production of composites, laminates, flooring, bamboo mat-board, pulp and paper must be supported and encouraged as the global population grapples with its future.

INTRODUCTION

With increasing global population and wealth, and growing demand on raw materials for development, there is concern about the sustainability of human economic activities. This paper reviews some of the primary definitions and concerns of "sustainable development", and shows that bamboo has a tremendous opportunity to provide needed products within a framework of environmentally sustainable practices.

SUSTAINABLE DEVELOPMENT

"Sustainable development" is a term that many attribute to The World Commission on Environment and Development that was chaired by Gro Harlem Bruntland (therefore, often called the Bruntland Commission). The Commission's findings were published in the book *Our Common Future* (WCED 1987), which includes much of the basis for the discussion and efforts towards reinterpreting development on a global scale.

The most often quoted passage from the book includes the definition of sustainable development as meeting "the needs of the present without compromising the ability of future generations to meet their own needs." This concept is sometimes taken out of context to suggest that human beings must return to entirely benign activities as the planet already shows signs of being close to its ultimate limits of natural resource survival. The Commission, however, continues on to say that sustainable development "does imply limits – not absolute limits but limitations imposed by the present state of technology and social organization on environmental resources and by the ability of the biosphere to absorb the effects of human activities." It goes on, and correctly so, to point out that "technology and social organization can be both managed and improved to make way for a new era of economic growth."

The imperative of "economic growth for nations in which the majority are poor" is linked with "an assurance that those poor get their fair share of the resources required to sustain that growth." Many fail to accept this critical concept, that economic growth is not simply for the continued benefit of already rich countries and companies. It is a mechanism that will help alleviate poverty, distribute wealth and fulfill the goal of sustainable development in "meeting the basic needs of all and extending to all the opportunity to fulfill their aspirations for a better life." Industrialization that leads to equitable economic growth is therefore a positive aspect of development.

Of course, many will cite past examples to demonstrate that industrialization specifically avoids the "equitable" distribution of wealth. Owing to the economic reality of the cost of capital, required return on investment, ownership and distribution of profits, manufacturing entities, for example, are not always established with the benefit of local communities as a first priority. The failure to do so in the past, however, should not be taken as an imperative that changes in social, market and political systems will not endorse such aspirations and the actions needed for these goals to become a reality.

Proponents of "deep ecology" would suggest that the needs of the environment are paramount, and the need for increased standards of living for the world's population should not be of consideration, especially as it leads to further environmental disruption. This argument is a moral one. The United Nations Development Program (UNDP), for example, has poverty eradication as a main objective. This is clearly, and appropriately, anthropocentric. Yet, it does not imply, as deep ecologists would contend, an abuse of the planet's environment. The common global development perspective is that the needs of people in poverty are paramount. "The Habitat Agenda" (Chapter I – Preamble) states that "human beings are at the centre of concerns for sustainable development, including adequate shelter for all and sustainable human settlements, and they are entitled to a healthy and productive life in harmony with nature."

Considering all these points, it can be seen that sustainable development is a desirable concept. It means, in a practical sense, that long-term awareness of our natural resources is essential. At the same time, it also promotes the idea that economic growth will help alleviate poverty and improve the lives of the world's citizens.

SMALL IS BEAUTIFUL

Many supporters of bamboo feel that the plant should primarily be used in its most natural forms. By growing small stands of bamboo in their yards, rural people can take advantage of its numerous beneficial uses. This is usually accomplished with sustainable harvests (as the people are stewards of their own resource) and simple conversion techniques to transform the natural resource into useful forms. These products are usually chemical-free and biodegradable after the useful lifetime.

There are many publications that identify the many traditional uses of bamboo (Farrelly 1984), and the important role of bamboo in the lives and economies of much of the world's population (INBAR 1997). Through small-scale growth, management and utilization, bamboo benefits billions of people throughout the world. There is no doubt that bamboo is a wonderful material of nature.

CHOICES

The small-scale use of bamboo, however, does not happen by chance but out of necessity. People need housing, implements, tools, baskets, food, etc. and bamboo fulfils many of these needs. If bamboo were not available, people would have searched for other materials to meet their needs. As people gain wealth, they often change to other materials out of practical or social concerns. It is no secret that bamboo, still considered "the poor man's timber", is not a desired building material. However, as it is available, inexpensive and the best choice (in terms of price, ease of use, durability, aesthetics, etc.), it is widely used.

When other materials (usually processed materials such as concrete, steel, plastic) are cheaper in cost, easier to use, more durable or improve performance, people have a choice in the materials that they use. The sight of plastic litter in many developing countries is a symptom of these changes. Where banana leaves were once used as fresh food containers, plastic is now often used. The pattern of use of the container remains the same – it is discarded when not needed. The non-biodegradable plastics are now seen where once natural fiber materials (such as banana leaves) were left to decay.

Changes of materials are made often for increased durability, ease of use and other such worthy improvements. If people have an option of other competitively priced products that out-perform traditional materials, changes in consumption patterns will take place. The effects of these choices are multiplied when considered on a global scale. How does bamboo compare with other common materials?

HOUSING MATERIALS

Instead of discussing a broad range of materials and their uses throughout the world, we will focus on materials used for housing. The United Nations recognizes "the right of everyone to an adequate standard of living for himself and his family, including adequate food, clothing and housing, and to the continuous improvement of living conditions".

Housing is an essential aspect in development – home provides shelter, a social base for the family, geographic stability, a possible location for commercial activity, etc. Yet, large portions of the world's population live in sub-standard housing (not counting the people who have no house) and there is a shortage of housing throughout the world. Housing, therefore, is a priority of the United Nations, and most developing countries around the world.

The construction of this number of basic shelters will have tremendous ecological impacts. Nevertheless, considering the important role of housing in development, the consumption of materials for this application is a necessity. The negative impact of this construction can be decreased through careful selection of the building materials and methods.

Social interest housing projects exist across the world. They are generally government-supported housing projects for low-income families. Priorities of quality, size, social services, infrastructure, etc. vary greatly but in general, these projects are aimed at the economically poor population.

Concrete, wood and steel

Concrete, wood and steel are most often used in housing construction projects throughout the world as they lend themselves to engineered design. Through knowledge of material performance, and established practices of design and construction, engineers can design and build structures with a high degree of certainty regarding performance and safety. In addition, in much of the world, concrete houses increases social prestige and are therefore preferred over local traditional materials, even if they are unsuitable to local conditions (such as high temperatures and solar radiation).

Concrete and steel have tremendous impacts on the environment in terms of the acquisition, processing and possibility of recycling of raw materials, and transformation of the structure at the end of its life. To discuss the environmental sustainability of these materials in construction, it is important to consider some of the associated environmental impacts.

Highly processed construction materials

Concrete and steel are manufactured materials. This allows for careful control of physical and mechanical properties, further enhancing the reliability of engineered designs. As design reliability is improved, costly over-design is reduced. The safety and ease of design (combined with durability and other benefits) make concrete and steel the choice materials of designers and engineers in large-scale housing projects.

Unfortunately, extraction of raw materials and the processing of steel and cement have large negative environmental impacts. Binkley (1993) noted that cement production is the second largest source of CO_2 emissions in Canada after fossil fuel consumption. Koch (1992) noted that replacing wood studs by steel in construction requires nine times more energy, and replacing wooden floor with concrete requires 21 times more energy.

Meil (1993) provided estimates of the impacts of two construction systems (2×4 inch wood stud construction, and 20 gauge non-structural steel). His group modelled the materials needed to construct a non-structural wall, arbitrarily designed to be 3 m high and 30 m long. Figs. 1–5 show Meil's estimation of manufacturing effluents and environmental impact for the steel and wood wall.

If criteria such as energy use, water demand and a variety of air emissions are taken, wood construction showed significantly less environmental impact than steel construction. The only category in which wood construction performed poorly was the comparative mass of solid wastes.

This study, although conducted by an economist within the wood industry, was intended to "encourage the wider use of all materials in any particular building

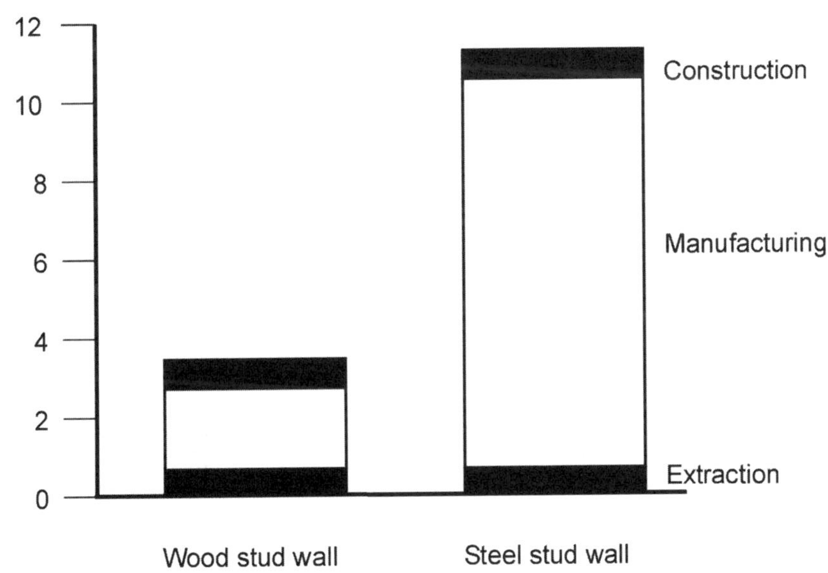

Figure 1. Comparative energy use

design such that together they minimize the building's environmental impact". The author correctly recognized that steel and concrete do offer benefits in terms of performance and applicability in construction. Construction based on steel and concrete, however, is difficult to recycle. This is in contrast to the marketing information put out by modern steel organizations to entice public purchases, with advertisements mentioning, "Steel is the most recycled material on earth". In addition, the impact of mining, quarrying and refinement of raw materials has huge negative implications, and the finite amount of minerals that exists makes the activity unsustainable by definition.

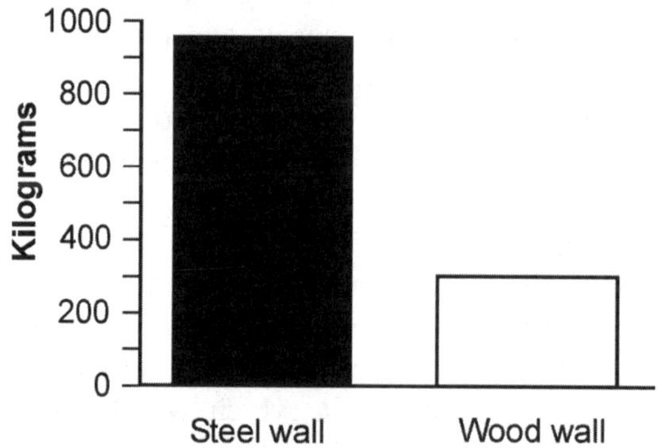

Figure 2. Comparative carbon dioxide emissions

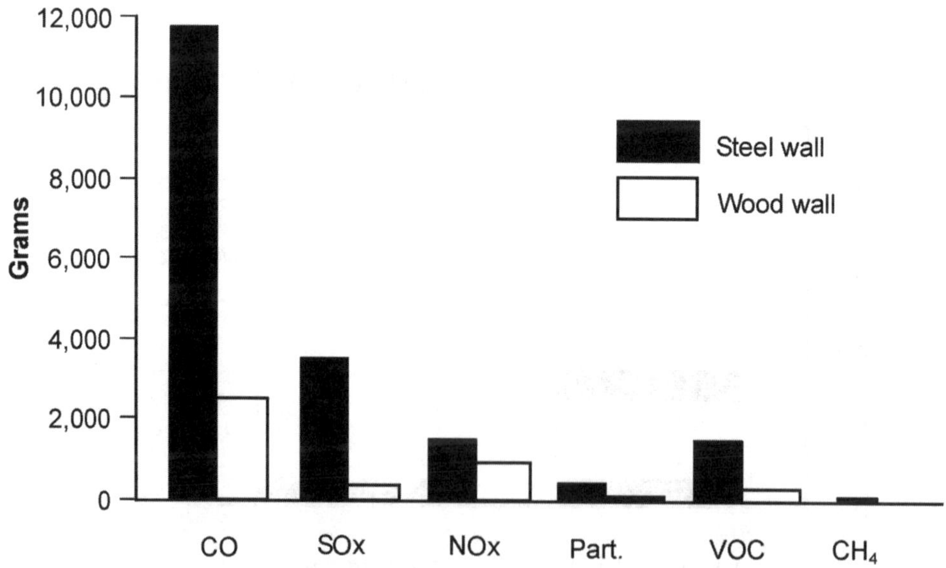

Figure 3. Other air emissions

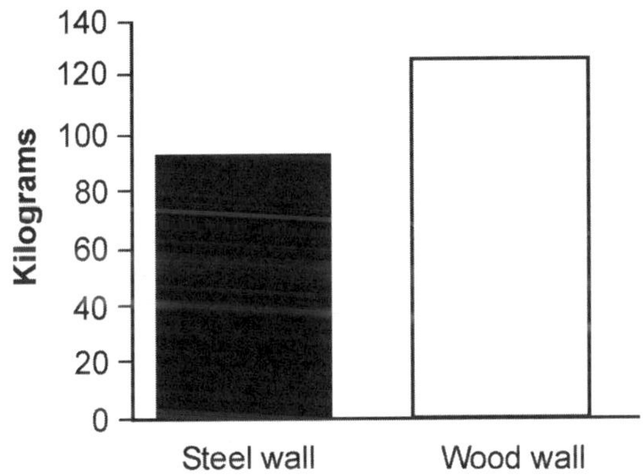

Figure 4. Comparative solid wastes

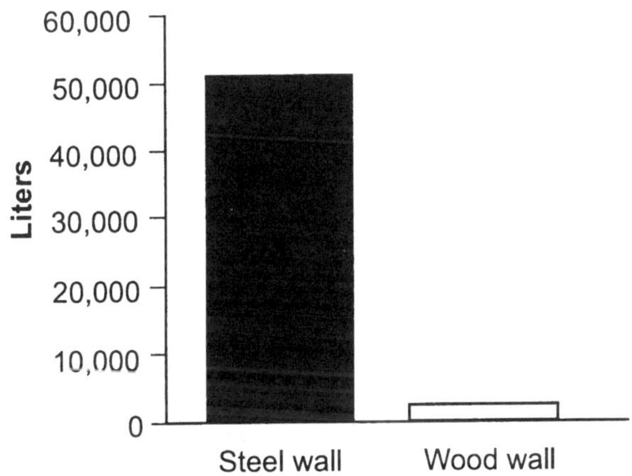

Figure 5. Water demand

WOOD

Utilization

On the other extreme is wood use in construction. Traditional uses of wood include timber members that require only small amounts of processing. Despite rigorous testing of material properties, the natural variability within wood samples (even within the same species) leads to larger factors of safety in construction design. This uncertainty and requirement of safety factors leads to non-optimal wood utilization, thereby increasing costs.

Recent advancements in the utilization of wood have largely been based on manufactured products. Through a controlled manufacturing process, the producers can largely engineer the properties of the final product (hence the name "engineered composite products"). The control of product properties is considered the most important advantage of composites. In addition to dictating the size and thickness of products, the actual physical and mechanical properties themselves can be largely controlled. Plywood, particleboard and oriented-strand board are examples of engineered composite materials.

Engineered composites also allow for the use of a wider range of raw materials. Wood species that suffer from too much variation in physical and mechanical properties might produce suitable composite products, given proper controls in the manufacturing process.

In the production of composite products, the refined raw material can be quite small (e.g. particles, fibers, etc.), which leads to less waste of the original roundwood tree. This minimization of waste in the wood products industry is now at exceptional levels, and almost all parts of harvested trees are used in some way (see Table 1).

Table 1.
Roundwood yield efficiency for wood products

Product	Yield efficiency from roundwood
Timber	30%–70%
Veneer	approx. 50%
Particleboard	75% – 90%

The controlled manufacturing of timber and wood products has allowed for the introduction of standards for product performance and for specific product applications. This is an essential component for the wide use of wood in construction, especially in technologically advanced countries aiming to protect public safety through safe engineering design practices.

In summary, "engineered wood products represent one of the best options to extend our fiber basket, providing for the efficient use of a wide variety of wood fiber and species sources" (Bykhovsky 1993).

Demand

The Food and Agriculture Organization (FAO) of the United Nations (FAO 1993) stated that the global production for particleboard was 40.2 million cubic meters in 1980, which increased by 23% to 49.5 million cubic meters in 1991. In the North and Central American regions, production was 7.0 million cubic meters and 9.9 million cubic meters for 1980 and 1991, respectively (an increase of 42%). South American production of particleboard rose by 27% during the same period, from a volume of approximately 1.0 million cubic meters in 1980 to approximately 1.3 million cubic meters.

FAO (1997) noted that the production of all wood-based panels in the world increased from 69.9 million cubic meters in 1970 to 126.9 million cubic meters in

1994, with similar consumption volumes in different regions. In South America, for example, wood-based panel production grew from 1.4 million cubic meters in 1970 to 5.1 million cubic meters in 1994.

The United Nations Economic Commission for Europe noted that the apparent consumption of particleboard (including non-wood particleboard and waferboard) in North America was expected to rise from 20.4 million cubic meters in 1995 to 22.7 million cubic meters in 1997 (United Nations 1996a). The Commission estimated that in 1997 about 5.0 million cubic meters of particleboard would be imported into the United States alone.

The market data also demonstrate the trend of North American users to proportionately increase their use of composite panels as mature trees for veneer production decrease in availability (Table 2).

Table 2.
Apparent consumption of wood-based panels (excluding veneer sheets) in North America during 1991–1995

	Volume (million m^3)					Percent of total	
	1991	1992	1993	1994	1995	1991	1995
Particleboard*	14.37	16.82	18.1	19.05	20.31	37.7	44.9
Plywood	18.38	18.81	18.86	19.19	18.89	48.3	41.7
Fibreboard	5.32	5.50	5.65	5.91	6.06	14.0	13.4
Total	38.07	41.13	42.61	44.15	45.26	100.0	100.0

Source: United Nations 1996b

Note: * Particleboard refers to particleboard and oriented-strand board

The trend of increased composite panel use is expected to continue. RISI (1996) has projected that the particleboard consumption will grow from 9.0 million m^3 in 1995 to 13.0 million m^3 in 2010 (Table 3) in the United States alone. Market prices are expected to increase throughout that period, after a lull in the late 1990s owing to increases in medium density fiberboard production from new plants coming on-line.

In summary, wood composite panels are well known throughout the world, and as large trees decrease in availability, composite panels will continue to take larger market shares. With ever-increasing demands for wood in all forms, the wholesale prices for panel products are expected to increase.

Table 3.
Particleboard market projections in the United States

Year	Industrial particleboard prices (USD/m^3, ¾ inch)		Particleboard demand (million m^3)
	South (East) region	West (Coast) region	United States
1995	163.00	158.00	9.0
2000	161.00	151.00	9.7
2005	188.00	180.00	11.1
2010	246.00	234.00	13.0

Source: RISI 1996

NATURAL FIBER

Sources

Natural fiber-based products are clearly in common use throughout the world in many forms. As growing population and wealth boost the demand for construction materials, energy and paper, natural fiber sources can offer a sustainable source of raw material.

The supply of wood is in question. Sutton (1993) noted that "there appears to be a limited scope globally to increase wood harvest levels from traditional sources." Wood harvest levels from natural forests might even decline significantly if other forest values (like biodiversity, protection of endangered species, etc.) are considered by society to be more important and to be incompatible with wood harvesting. RISI (1993) stated that "the question is not whether timber supply conditions will get worse but how fast and to what degree." Plantations of fast-growing species are thought to be one solution.

With increased wealth, there is generally a greater demand for wood products (Sutton 1993). Given the current population growth pattern and the increasing levels of development, pressures on existing material supplies, including timber supplies, are bound to increase.

Utilization

Hart noted (1997) that there are three ways of stabilizing or reducing the global population's "environmental burden" – decreasing population, lowering levels of affluence (and therefore consumption) or "changing fundamentally the technology used to create wealth".

The last point involves transformations (or "advances") of technology in many sectors of society. Most of our current technological advances are due to commercial effort, or through research and development that have commercial application of technologies as its goal. As Hart notes, "technology is the business of business." Perhaps the large-scale industrialization that takes advantage of economies of scale is the bridge for commercial interests to move away from unsustainable industrial activity, and to concentrate on sustainable practices and products using natural fiber sources.

Alternative fiber sources – such as bamboo, hemp and agricultural wastes – can offer an alternative supply for the production of construction material composites, bio-energy and pulp.

Demand for other uses

Bio-energy: Given the international efforts to restrict fossil fuel combustion owing to its impact on climate change, bio-based energy sources offer alternative energy supplies. Through fermentation or burning, the energy of the process of photosynthesis is released. By-products of the burning process (such as carbon dioxide) are subsequently taken up by the growth of new woody materials (such as bamboo). This creates a closed cycle of raw material use with a long-term sustainable supply of energy.

FAO (1997) estimated that 2 billion people in the world use wood or charcoal as their main or sole source of energy for cooking and heating. It cited estimates by unnamed sources that predict 3 billion people by the year 2000 would rely on wood fuel as their main domestic energy source. Bamboo is known to have caloric values up to 4,600 to 5,400 calories/g (CAF 1993).

Pulp and paper: FAO (1991) estimated the world capacity for pulp, paper and paperboard for the years 1990–1995. In 1995, for example, the total global capacity for wood pulp paper and paperboard was estimated to be approximately 186 million tons (air dry) per year.

The total capacity of bamboo pulp for paper and paperboard worldwide was estimated to be 1.7 million tons (air dry) per year. Of this, 42 000 tons were in Bangladesh, 20 000 tons in Myanmar, 200 000 tons in China, 11 000 tons in Indonesia, 1 000 tons in Cambodia, 64 000 tons in Vietnam and 1.3 million tons in India.

BAMBOO

Why bamboo?

Bamboos are generally found between 40° northern and southern latitudes. They are most commonly found in tropical and sub-tropical environments, with a mean annual temperature of 20–30°C (Liese 1985), although some bamboos can survive –20°C. Bamboos flourish at low altitudes (0–800 m), and are found in well-drained soils in regions with an annual rainfall of 1 000 to 2 000 mm.

There are many benefits of bamboo, as recognized by a growing number of publications and conferences. Because of its prevalence around the world, and the astonishing number of traditional uses of bamboo, many see bamboo as a miracle material that is well suited to environmental protection while providing a useful service to its users.

The global interest in bamboo is increasing in the North, owing largely to the increased environmental awareness of recent years. As customers boycott clear-cut timber products, the environmental impact of materials has now become an issue of frequent discussion. With the recent awareness of bamboo, new demands are made on bamboo products (such as flooring) that are not traditional uses of bamboo, but adaptations of traditional uses. This provides a turning point for the support of bamboo as a useful material with worldwide applications.

The positive aspects that lend conventional construction materials (concrete, steel and timber) to engineered design are some of the basic factors that limit the use of adobe, natural bamboo and other materials in larger scale construction projects.

As far as construction applications using natural bamboo culm are concerned, it has been found that bamboo has very high strength, measured macroscopically (Montero 1991) and microscopically (Liese 1985), and a unique geometrical cross-section that makes it favourable for lightweight post and beam constructions. Because of the nodes and variances in wall thickness and cross-section, however, the use of culms pose a large element of uncertainty that ultimately leads to overdesign in order to ensure safe construction. This leads to an over-use of bamboo in housing construction. Recent investigations into the improvement of joints in traditional bamboo construction are in part intended to remove some of the engineering uncertainty, and thus reduce over-design of bamboo structures and increase bamboo utilization and efficiency.

The use of bamboo as a raw material can be improved in much the same way that the efficiencies of global wood utilization have been improved through the development and use of engineered wood products. Using bamboo as the raw material for composite panels will remove the variability in culms, and the final product performance can be controlled through engineered manufacturing processes to meet targeted product standards (Turcke and Dagilis 1996; Dagilis and Turcke 1996, 1997). Engineered bamboo products offer one of the greatest opportunities to use bamboo fiber as a large-scale raw material to replace wood, steel and concrete. Considering the numerous benefits of natural fiber, the use of bamboo in place of concrete and steel (as mentioned above, using wood as an example) would have positive environmental impacts on a global scale.

Large-scale benefits

From a broader perspective, sustainable bamboo cultivation promotes soil erosion control, carbon sequestration, stability in hydrological cycles, oxygen production and many other benefits of natural plant growth. Apart from these, bamboo cultivation is beneficial also to the ecosystem as a whole.

As a resource, bamboo yields are high per unit area (Liese 1985; Shi et al. 1987). It can be asexually propagated, allowing for quick reforestation of denuded areas and the rapid development of bamboo plantations. Owing to the rapid growth of bamboo, plantations and forests absorb large amounts of CO_2 and are therefore useful carbon sinks that can offset global pollution caused by the burning of fossil fuels.

Economically, bamboo plantations have excellent potential. The annual yields could be up to 17 tons/ha for temperate bamboos (Shi et al. 1987) and 40 tons/ha for tropical bamboos (de Barros Salgado and Egashira 1992).

Bamboo harvesting costs are also practical. In Ecuador, according to official sources (DFO 1998), a representative bamboo plantation of 1 ha with an initial plant spacing of 8×8 m will have 156 clumps of bamboo. At four years of age, there will be 7 to 10 harvestable (mature) culms per clump. This provides a total harvest of 1 092 to 1 560 culms per year.

Each full culm can produce one or two bamboo pieces (6 m in length) for construction and one top piece used for supporting growing fruit bunches on banana plantations. According to DFO, current market values are approximately USD 0.33 per top piece and USD 0.44 per bottom/middle piece. This provides total revenue per hectare of about USD 840 to USD 1 201 (conservatively, assuming only one 6-m bottom/middle piece per culm).

During informal site visits in Ecuador, similar prices were noted in June 1998. Market prices varied, but averages prices at the road access point adjacent to the harvesting site were approximately USD 0.30 per top piece and USD 0.40 for bottom/middle piece (including loading onto trucks for subsequent transportation).

Harvesting costs (not mentioned by DFO, but obtained by surveying bamboo workers) averaged USD 0.08 per section (i.e. USD 0.24 for an entire culm). For each marketable piece, the worker had to fell, trim and transport the piece to the agreed-upon roadside pick-up location. It was also noted that most mature culms provided two full bottom/middle pieces and a top piece (in contrast to the single bottom/middle piece used in the DFO example). Each worker harvests approximately 30 full culms per day (80–100 marketable pieces). The difference between harvest costs and market costs are attributed to payment to landowners and payment of government harvest taxes, which are applicable to all wood products harvested from non-plantation sources.

Given an average external diameter of 10 cm, a wall thickness of 1 cm and a dry density of approximately 600 kg/m^3, the dry mass of the average bottom/middle piece is approximately 10 kg. This gives a harvested cost of USD 8.00 per dry ton, with the bamboo being carefully extracted in full 6-m lengths.

For a small composites factory (annual production of 25 000 m^3 per year), a plantation would employ approximately 150 people full-time to manually harvest the bamboo.

It is interesting to note that STCP (1993) conducted a survey of the wood industry in Ecuador, and found that the total harvest costs (not including administration) was USD 16.39/m^3, or assuming a density of 600 kg/m^3 (possibly a high estimate), USD 27.20/ dry ton. Wood raw material costs – including administration, transportation, government harvest taxes and payment to the landowner – were approximately USD 97.00/m^3 for small sawmills, USD 78.00/m^3 for large sawmills and USD 134.00/m^3 for plywood production. Owing to past deforestation, the transportation distances between standing wood and the mills is now substantial.

Even though there is no large-scale global trade in bamboo fiber, the Ecuadorian example provides an interesting benchmark. Over 30% of houses in Guayaquil, Ecuador's largest city, were made of bamboo according to a 1982 census (INEC 1983). Garcia-Rios (1992) noted that 52% of houses in squatter settlements have walls made from bamboo (split open and used as wall sheathing) and 40% made from bamboo and wood.

Accurate data on the quantity of new squatter construction using bamboo are not available, although the squatter settlements form a significant part of the city. However, since the bamboo is sold through privately owned construction material deposits, the free market prices cited for bamboo from private lands are likely to reflect the true bamboo harvesting costs. These costs are very competitive with large-scale, industrial tree-wood harvesting costs.

Is bamboo industrialization sustainable?

Bamboo industrialization projects can create hundreds of new jobs. These jobs include conventional agricultural jobs common in developing regions, and higher skilled jobs required to process bamboo culms into value-added products.

As noted by Fundación Ecuador (1996), there is a close relationship between the percentage of manufactured goods exports of a country and the quality of life there. Currently less than 5% of exports in Ecuador are manufactured or are value-added products made from the existing natural resources. Equitable global development depends on countries transforming natural resources, not simply supplying raw materials for further processing abroad.

Bamboo plantations, because of their high yield per unit area, allow companies to economically grow their own fiber for processing. Because these companies are stewards of their own resource, they are more likely to pursue sustainable management of the bamboo. The unique nature of bamboo makes selective harvesting programs rewarding with increased yields.

Wood substitution technologies

Let us consider what the potential global role of bamboo could be. If bamboo is considered as a fiber source, it can be directly compared with the global supply and demand of wood and wood products; but is that a valid comparison?

There has been recent work in the development of viable bamboo products throughout Asia. The Chinese Academy of Forestry (CAF 1994) provides a comprehensive view of bamboo substitution for wood in China. Unfortunately, cost issues are not discussed.

While bamboo products often have properties superior to the properties of similar wood products, manufacturing with bamboo usually requires significantly more labor, and different and more costly bonding cycles owing to the presence of silica in relatively large amounts. This leads to bamboo products that, for the most part, cannot compete directly with wood products, unless there are unique economic considerations such as those in China. For that reason, work that overcomes the economic barriers of bamboo composite manufacturing is critical (Dagilis and Turcke 1996).

Private commercial development of bamboo products is presently unlikely, as the limited scientific knowledge of bamboos is not widely known by commercial operations, which are familiar with wood-based manufacturing. In addition, the international wood industry conventionally takes advantage of existing supplies of well-known species. Bamboo industrialization requires investment and the acceptance of alternative fiber sources, something that many corporate leaders are unwilling to acknowledge because alternative fibers may be considered too risky for conservative CEOs of the wood industry attempting only to "stay on course".

Given the expected rate of increase of wood costs, alternative fiber sources will become more important in the future. Those companies and public organizations that begin now will have a commanding lead in the supply of low-cost fiber for a wide variety of applications, and the know-how required to install and manage bamboo plantations efficiently. It is likely that a combination of private initiatives combined with public support for environmental protection and rural development will be required to diversify risk and provide access to capital. This consideration is important if it is expected that bamboo industrialization begin in the near future in non-traditional bamboo areas.

CONCLUSION

Bamboo is a material that is currently useful to many small-scale users worldwide. Given the many attributes of this natural fiber material, bamboo should be supported and encouraged in continued traditional uses. However, if we are to take bold steps forward in global economic development and wish the development to be environmentally and socially sustainable, then large-scale bamboo growth and utilization need to be encouraged. Through modern processing techniques, many of which are recently developed, bamboo raw material can be transformed into many products that can compete directly with wood products in price and performance.

Environmental protection in a larger context will gain from increased use of bamboo in large-scale, sustainably managed projects, largely because natural fiber products are far more sustainable in the long-term than products of plastic, steel or cement. Bamboos are efficient users of land, and produce more fiber per unit area than most other tree species. Conventional manual harvesting of bamboo is economical, maximizes annual yields, and is a positive factor in mitigating erosion and in watershed management.

These factors, and others, make bamboo an ideal material for future sustainable growth. The bamboo supporters and international funding agencies should support large-scale bamboo industrialization in efforts to meet the existing goals of environmental protection, rural development, job creation and low-cost housing.

REFERENCES

Binkley, C.S. 1993. Adapting to global supply constraints. *In* Proceedings of The Globalization of Wood: Supply, Processes, Products and Markets, Portland, Oregon, USA, 1–3 November 1993.

Bykhovsky, A. 1993. Managing change: the revolution has begun. Keynote address at the International Particleboard/Composite Materials Symposium, Pullman WA, USA, 30 March–1 April 1993.

CAF (Chinese Academy of Forestry). 1993. Substitute bamboo for wood in China, interim report. Chinese Academy of Forestry and the International Tropical Timber Organization, Beijing China.

CAF (Chinese Academy of Forestry). 1994. Substitute bamboo for wood in China, final report. Chinese Academy of Forestry and the International Tropical Timber Organization, Beijing China.

Dagilis, T.D.; Turcke, D.J. 1996. An economical bamboo particleboard. *In* Ganapathy, P.M.; Janssen, J.A.; Sastry, C.B. ed., Bamboo, people and the environment, Vol. 3, Engineering and utilization. Proceedings of the Vth International Bamboo Workshop, Ubud, Bali, Indonesia, 19–22 June 1995. International Network for Bamboo and Rattan, New Delhi, India. pp. 165–173.

Dagilis, T.D.; Turcke, D.J. 1997. Bamboo industrialization in North America. *In* Proceedings of the Pacific North-West Bamboo Agroforestry Workshop, Port Townsend, USA, June 1997. (In print).

de Barros Salgado, A.L.; Egashira, Y. 1992. Cultivo comercial del bambú para aprovechamiento industrial en celulosa, papel, y etonol. I Congreso Mundial de Bambú/Guadua, Pereira, Colombia, 8–15 August 1992.

DFO (Desarrollo Forestal en el Occidente de Pinchincha). 1998. Caña guadua, reproducción y manejo. DFO, Honorable Consejo Provincial de Pinchincha, Ecuador.

FAO (Food and Agriculture Organization of the United Nations). 1991. Pulp and paper capacities survey: 1990–1995. FAO, Rome, Italy.

FAO (Food and Agriculture Organization of the United Nations). 1993. FAO yearbook: forest products 1991. FAO Forestry Series No. 26, FAO Statistics Series No. 110, FAO, Rome, Italy.

FAO (Food and Agriculture Organization of the United Nations). 1997. State of the world's forests: 1997. FAO, Rome, Italy.

Farrelly, D. 1984. The book of bamboo. Sierra Club Books, San Francisco, USA. 332 pp.

Fundación Ecuador. 1996. Ecuador: sus ventajas competitivas en el mercado internacional. Guayaquil, Ecuador.

Garcia-Rios, L.E. 1992. Low income housing in squatter settlements in Guayaquil, Ecuador. Master of Architecture Thesis, Faculty of the Graduate School, State University of New York, Buffalo, USA.

Hart, S.L. 1997. Beyond greening: strategies for a sustainable world. Harvard Business Review, January–February 1997.

INBAR (International Network for Bamboo and Rattan). 1997. INBAR Newsmagazine, Special Edition, 5(3). INBAR, Beijing, China.

INEC (Instituto Nacional de Estadistica y Censos). 1983. IV Censo de población y III de vivienda 1982, resultados definitivos, Provincia del Guayas, Republica de Ecuador. Quito, Ecuador, 1983.

Koch, P. 1992. Wood versus non-wood materials in US residential construction: some energy related global implications. Forest Products Journal, 42(5), 1992.

Liese, W. 1985. Bamboos – biology, silvics, properties, utilization. Deutsche Gelleschaft für Technische Zusammenarbeit (GTZ), Nr. 180, TZ Verlagsges., Roßdorf 1, 132 pp.

Meil, J.K. 1993. Environmental measures as substitution criteria for wood and non-wood building products. *In* Proceedings of The Globalization of Wood: Supply, Processes, Products and Markets, Portland, Oregon, USA, 1–3 November 1993.

Montero, J.S. 1991. Propiedades fisico-mecanicas de la especie "guadua" en Costa Rica. Proyecto Nacional del Bambú, Costa Rica.

RISI (Resource Information Systems, Inc.), 1993. Wood Products Review: Softwood Lumber, Panels and Board, 24(1), March 1993. RISIS, Bedford, Massachusetts, USA.

RISI (Resource Information Systems, Inc.). 1996. Wood Products Review, 27(2), July 1996. RISI, Bedford, Massachusetts, USA.

Shi, Q-T.; Bian, Y-R; Wang, Y-X. 1987. Study on the application of chemical fertilizer to the timber and paper-pulp stands of *Phyllostachys pubescens*. *In* Rao, A.N.; Dhanarajan, G.; Sastry, C.B. ed., Recent Research on Bamboo. Proceedings of the International Bamboo Workshop, Hangzhou, China, 6–14 October 1985. Chinese Academy of Forestry, Beijing, China; International Development Research Centre, Ottawa, Canada. pp. 87–90.

STCP (STCP Engenharia de Projetos Ltda.) 1993. Estrategias para la industria sostenida de la madera en el Ecuador- informe final. Instituto Ecuatoriano Forestal y de Areas Naturales y Vida Silvestre (INEFAN) Proyecto PD 137/91.

Sutton, W.R.J. 1993. The world's need for wood. *In* Proceedings of The Globalization of Wood: Supply, Processes, Products and Markets, Portland, Oregon, USA, 1–3 November 1993.

Turcke, D.J.; Dagilis, T.D. 1996. The development of engineered bamboo construction products. *In* Proceedings of the CSCE 1st Structural Speciality Conference, Edmonton, Alberta, USA, June 1996.

United Nations Economic Commission for Europe. 1996a. Forest products markets in 1996 and prospects for 1997. Timber Bulletin, XLIX (6), 1996. United Nations, New York, USA.

United Nations Economic Commission for Europe. 1996b. Forest products annual market review 1995–1996. Timber Bulletin, XLIX (3), 1996. United Nations, New York, USA.

WCED (World Commission on Environment and Development). 1987. Our common future. Oxford University Press, Oxford, USA.

Bamboo for Sustainable Development, pp. 155–171
Eds. A. Kumar, I.V. Ramanuja Rao and Ch. Sastry

Economic reforms, bamboo production and local economy: A case study of Linan County, Zhejiang Province, China

SHASHI KANT and MARIAN CHIU
Faculty of Forestry, University of Toronto, Ontario, Canada

Abstract—China, with over 4 million hectares under bamboo and an approximate annual output of one million culms, is a top bamboo-producing country. The policy reforms initiated in the country in 1979 put production, investment and marketing decisions in the hands of local governments and local enterprises, and allowed foreign capital and technology to enter into China and helped promote the establishment of foreign enterprises. Land tenure and management reforms (1982), abolishment of price controls (1985), relinquishing state control of foreign trade (1988) all helped to open up the domestic market and economy.

Several studies have attempted to evaluate the impact of these reforms on economic efficiency of the agricultural sector. No serious initiative, however, has been taken to quantify the impact of these reforms on the forestry sector, especially on bamboo PCS. In this paper, which is an outcome of a project funded by INBAR, the impacts of three policy reforms – land tenure, industrial production and market – on bamboo production are examined, and the role of returns from bamboo on income distribution is evaluated. The study is based on the data collected from Linan County, one of the ten counties known as bamboo counties, in the Zhejiang Province of the People's Republic of China.

INTRODUCTION

China is home to more than one-fourth of the world's total bamboo area, with an approximate annual output of one million culms. Bamboo, sometimes referred to as 'the poor man's timber' because of the ability of the resource-poor to enter into bamboo production-to-consumption system (PCS), plays an influential role in the economy of China. There are millions of Chinese farmers who grow bamboo as a component in integrated farming systems, bringing over 4 million hectares in China under bamboo and thus further emphasizing its economic importance. In its search for economic efficiency, China has undergone considerable changes in policies related to land tenure, industrial production and competitive markets in the last two decades. Policy reforms were initiated in 1979 with the transformation of the centrally planned and centrally controlled economy of the country to one where production, investment and marketing decisions were placed in the hands of local governments and local enterprises. An open door policy was adopted in the Chinese economy, which was earlier closed to foreign investment.

This allowed foreign capital and technology to enter into China and helped promote the establishment of foreign enterprises. Differences in the quota and above-quota prices were abolished in 1985, and this precipitated the establishment of an open domestic market, in which farmers were able to directly sell their products on a local or provincial level. In 1988, state control of foreign trade was relinquished, further promoting foreign investment and the establishment of joint ventures.

Land tenure also underwent reforms in the early to mid-1980s. In the past, most of the 4 million hectares of bamboo land in China belonged to the state and was subject to collective management by villages. Individual farmers who participated in the production system received their reward through the unified income allocation system. However, through changes in the structure of land tenure, management of the land became the responsibility of each household through the implementation of the Farmers' Household Responsibility System, whereby each household was allowed to independently organize and control production. Although the land still belonged to the collective, by 1983, the majority of households were effectively utilizing the Household Responsibility System.

Many attempts have been made to evaluate the impact of these reforms on economic efficiency of agricultural sector. No serious attempt, however, has been made to quantify the impact of these reforms on the forestry sector or, specifically, on the production-to-consumption system of bamboo. The International Network for Bamboo and Rattan (INBAR) is attempting to fulfill this gap in the literature by funding projects on related topics. The first study, funded by INBAR, attempted to evaluate the economic efficiency of bamboo production and bamboo processing units. However, this study did not specifically evaluate the impact of separate policy reforms on bamboo production and resulting income distribution.

This paper, which is an outcome of a project funded by INBAR, examines the impacts of three policy reforms – land tenure, industrial production and market – on bamboo production and evaluates the role of returns from bamboo on income distribution. The study is based on the data collected from Linan County, which is one of the ten counties known as bamboo counties, in Zhejiang Province of the People's Republic of China. Zhejiang is one of the most prosperous provinces in China, and this is partly owing to the predominance of bamboo and ideal conditions for agriculture. Zhejiang is located in the south-eastern corner of China, bordered by Shanghai, Jiangsu, Anhui, Nanchang and Fuzhou Provinces and the East China Sea. This study was conducted during June-August 1998, and data were collected using Participatory Rural Appraisal (PRA) methods and questionnaires.

In this paper, first, literature is reviewed. Second, specifics about the area of study, Linan County, and policy changes in this county are summarized. Third, the impact of the policy changes on bamboo production and the role of returns from bamboo on income distribution are discussed. Finally, paper is concluded with policy suggestions.

LITERATURE REVIEW

China's economic reforms have been the focus of economists in the last decade as have the land tenure and land-use systems. The latter two issues have always been controversial in both theoretical and practical discussions. One focus of discussion has been the possible changes in ownership of farmland (Liu 1989), while another has concentrated on how to obtain economies of scale by expanding farm size (Dong and Chen 1989). The comparison between the collective farm system and family farm system is often related to these discussions (Zhou 1989).

Other studies have examined recent developments in the market for rural land use. Zhang and Makeham (1992) examined developments in land transfer processes in detail in a case study of Nanhai County, Guangdong Province. A two-field system was established in Nanhai in which land was contracted by individual households as a grain ration field and by groups of villagers as responsibility fields. This was to decrease the farming of small, scattered plots, which was unfavorable to farmers, and to increase the ability to contract land use by households with large enough labor forces. Several trends can be seen in the Wengfang and Makeham study. One is that the use and transfer of land with compensation are the two most common types of market activity in many parts of China, and another is that the farmers are now not forced by necessity to enter the land use market, but rather enter out of a desire to increase production and income.

The majority of the above discussions, however, are focused in the agricultural context. There is minimal literature on the impact of reforms on the bamboo sector. Only the studies by Zhong et al. (1998) and Ruiz-Perez et al. (1998) on the bamboo sector in Anji County, Zhejiang Province, come to attention. Zhong et al. (1998) report that management patterns have shifted from extensive to intensive plantations, increasing forest productivity and economic benefits; however, it was found that a lack of standard plantation measures led to lower output. The study addresses many economic issues such as composition of farmers income at the aggregate level, aggregate economic benefits from bamboo plantations, cost-benefits analysis of bamboo products and the impact of the bamboo industry in rural economy. However, it is difficult to draw an inference about the comparison of pre- and post-reform economic efficiency from this study; nor does it address issues such as the impact of the bamboo PCS on income distribution. Ruiz-Perez et al. (1998) examined the role of bamboo plantations in rural development, using a multiple regression analysis to study the factors that influenced farmers' incomes. It was found that bamboo has the highest relative importance for middle-class farmers as opposed to poor farmers as previously hypothesized. Conclusions were drawn concerning the effectiveness of development projects aimed at improving the situations of the poor in rural communities. The study has also suggested that wealthy farmers have diverted their income resources owing to limitations on farmland and increasingly available alternate sources of income. Thus, while wealthy farmers still receive the largest income from bamboo, their total income is complemented by other income-generating activities, both on and off the farm. Both these two studies have not evaluated, in specific terms, the impact of reforms on bamboo production and the change in income composition over a period.

THE STUDY AREA: LINAN COUNTY

Linan County is located 50 km west (118°51' latitude, 119°52' longitude) of Hangzhou, the capital of Zhejiang Province, and falls under Hangzhou's administrative jurisdiction, which includes 38 other townships. The county has a total area of 313 478 ha. Forested land covers 71% of this area with 32 000 ha of economic or commercial forests and 47 000 ha of bamboo. Standing timber volume is an estimated 50 million m^3. The county has a population of 504 254, 87% of which are farmers. Linan has a sub-tropical monsoon climate with an average annual rainfall and temperature of 1 426.4 mm and 15.8°C, respectively. The county has two important water systems for the province, the Tai Lake, China's largest freshwater lake, and the Qiantang River.

Linan is considered a middle class county in Zhejiang Province, although it was recognized as one of the top 100 strongest counties in China. In 1996, the GDP per capita in Linan was 11 534 yuan. The county is well known for its agriculture and forestry resources, both of which have been the focus of reforms since 1983. Prior to 1983, agriculture was the top priority; however, the recognition of the vast and rich forest resources has shifted this focus. Improved management and strong support to conserving its natural resources have helped Linan to be selected as one of eighty science and technology demonstration counties in China. The county is also home to two national reserves, one of which, the West Tianmu Mountain Reserve, has been designated as a United Nations Biosphere Reserve.

Much of Linan's economic prosperity, however, can be linked to the dominance of bamboo. It is one of the only ten bamboo-growing counties in a country of 2 000 counties. As mentioned earlier, Linan has 47 000 ha under bamboo production, divided approximately equally into areas for fresh shoots, dried shoots and wild bamboo. Shoot production in Linan in 1997 was 86 000 tons, valued at 262 million yuan. Over 60 000 tons of timber from wild bamboo, valued at 33 million yuan, was harvested and bamboo-based processing grossed 175 million yuan. There is a wide range of bamboo processing in Linan County – dried shoot processing, plywood, mats, scaffolding, clothing, handicrafts and a variety of miscellaneous products – carried out in 32 factories. There are also a variety of entry points for the farmers to enter the bamboo PCS, ranging from growing and distribution of bamboo shoots, to processing and export. For example, 90% of canned shoots are exported to Japan and 60% of flooring material is exported to the EU and Southeast Asia. Many joint ventures with Taiwan-based companies have also been established in the bamboo flooring industry.

Linan further compounded its bamboo industry with corresponding policy reforms that helped it to utilize another of the resources: its people. In conjunction with the new policies, the organization of its farmers helped to promote responsible forestry. Linan County experienced the same policy reforms the rest of China experienced. Pre-1979, all land and factories in Linan were owned either collectively or by the state. After the dissolution of this type of land tenure in 1982, land was distributed to individual households in three different categories: private use land, family responsibility system (FRS) land and contract land. The private use land is mainly to meet basic subsistence needs of the household, and distribution of land is based on the total population of the household. The area of land distributed under FRS is proportional to the number of laborers in a household (family). Hence, allotment under these two categories is dependent upon the demographic composition of the household, and equity overrides efficiency. The third category of land – contract land – is distributed based on management capacity of a household, and hence efficiency overrides equity. The bamboo area is distributed under the FRS and contract land. Many farmers, however, do not have any idea about the difference between these three categories of land, and all land under these categories is known as Land Managed by Farmers. In addition to these three tenures of farmers-managed land, there are state-owned farms and collective farms.

All land is still owned by the state, but private-use land is inheritable by future generations and is increasingly well managed because of the promise of continued tenure. Land under FRS and contract land was initially given for 15 years tenure. The contract land is distributed using three methods: auction, rent and share holding. Auctioning of land was through the village government and forestry bureau, both of which decide on the quality of land and the lowest acceptable bid. Private auction sales are forbidden and there is a limit on individual purchase, which must be less than double the average area per household when the total area of the village is divided amongst the households. For example, a household in a village occupying 100 ha with 50 households is allowed to buy through the auction system a maximum of 4 ha ($100 \div 50 = 2, \times 2$). Lower quality land is auctioned first, with successful bids paid as a one-time payment. To aid poorer households in purchasing land, only 10% of land is made available for auction with the rest available through the rental system. This system allows poor households to enter the land market, as rent payment is not due until the 15th year of rental ownership. This allows households a chance to work the land and generate income for payment. To compensate for the scattering of small land plots, households have been allowed to transfer land to others for a fee. The shareholding system is based on the use of pooled resources cooperative management. Thus far, shareholding has been done on 3 500 ha of land in Linan. This system can take one of four main forms: state-run enterprise and village shareholding (14%), village and farmer shareholding (65%), forestry bureau and township shareholding (19%), and forestry college and village shareholding (1–2%).

Since the 1982 reforms, however, many problems surfaced concerning farmers-managed as well as collective land. Farmers-managed land was further degraded despite the new changes, as farmers, fearful of the short term of tenure and continuously changing policies that might relieve them of their current holdings, quickly harvested all available timber from these lands in return for quick profit. The plots of land were small and scattered, thus incentives to re-plant and maintain responsible stewardship was lessened. Collective land was degraded because they were difficult to manage by the collective owing to their large area and thus, only minimum income was generated from these lands for re-planting efforts. The minimal species variety on these lands also made it difficult to diversify away from timber, thus the valuable old growth was quickly harvested with little re-planting efforts or future planning.

With the recognition of these new issues arising out of the initial policy reforms, the Linan County government started an experiment on land tenures in Hongchow township. Period of tenure was extended from 15 years to 50 years, and this effectively decreased the deforestation and land degradation that resulted from farmers being motivated by only a short-term planning horizon. Furthermore, during the period of contract, the inheritance or transfer of land was made possible. At contract expiry, the owner was allowed a chance to re-contract the land or, if the land was contracted to a new owner, to be entitled to 50% of the next three years' profit. These conditions had a three-fold purpose: firstly, to facilitate the transfer of land to maximize land use; secondly, to increase and promote proper land management through profit sharing; and thirdly, to increase individual household profits and thus increasing farmers' purchasing power.

Based on the experience of this experiment, a new land tenure policy has been developed over the last few years addressing the continuing need to maintain the stability of land and forest ownership. Details of these new policies were finalized in the summer of 1998, with projected implementation in September 1998. The new policies include:

- Extension of management periods from 15 years to 30–50 years;
- Security deposit of a 20–50 yuan/mu to ensure forestry development on newly acquired land within 2 years of acquisition (if done, this deposit is returned);
- Priority to local farmers over outsiders with respect to land tenures;
- The encouragement of managing larger tracts of land (over 5 mu); and
- The requirement of a permit to harvest wild bamboo culms and timber.

The new polices will be implemented on barren and/or deserted land, forested land and burnt or clear-felled land. The desired outcomes include increased farmer-initiated tree planting, increased attention to growing marketable species such as bamboo, and increased development of new technology. Reforestation in Linan is continually increasing with 50 000 mu land newly planted each year. This increase is due to: (1) the utilization of previously unused, mountainous land by farmers to increase their income; (2) increased attention to forestry owing to demand for green products; and (3) the increased confidence in policy reforms that promote good management.

In addition to policy reforms, bamboo production is encouraged also through financial incentives and demonstrations. Many households, chosen as demonstration households, receive subsidies for allowing and managing bamboo-growing experiments on their land. Other households receive a 300 yuan subsidy from the local forestry bureau for each hectare of land converted to bamboo. Linan has also established several associations for bamboo and related industries: in 1985, the Association for Bamboo Industry was formed; and in 1998, the Association for Bamboo Product Assessment was established.

Policy reforms did not only manifest themselves in the form of changes in land tenure but also as market and industrial transitions. In 1987, the bamboo-processing sector was opened for private enterprises. Consequently, private entrepreneurs purchased many state-owned enterprises, characterized by large debts and declining production. The large debts carried by state enterprises is due to: the large number of retired state workers, who are entitled to 35–60% of their salaries as pension; complex marketing and purchasing channels, which had varying stages of surcharges and fees; and inefficient administration. Conditions of the above sales generally included the shouldering of the debt by the buyer; however, factories – including the machinery, buildings and land – have been purchased at a fraction of the cost. An example of this is the Linan Silk Factory. Previously state-owned and operated, the factory was purchased by a private group for 38 million yuan. This transaction included the stipulation that the group takes responsibility for the factory's debts, which totalled several million yuan. Although the amount of debt constituted a substantial share of the buying price, the price also included the right to management, buildings, machinery and land. The group re-hired most of the workers, which significantly decreased their labor training and start-up costs. Thus, the overall purchase of the silk factory was at a fraction of its value.

The loosening of the previously strict market and purchasing regulations also allowed the purchasing of state enterprises to take place under differing conditions. Private buyers were able to purchase the right to manage the factories and market the products, but the land and buildings remained state or collectively owned, and were rented by the buyer or used free in exchange for shouldering a percentage of the debt. This allowed the state to relieve itself of its debts while keeping some of the assets under its control. There are many examples of these differing types of transactions in Linan County.

In 1990, state removed their control from bamboo prices, resulting in the establishment of a fresh bamboo shoot (FBS) market in Gaohong township. Gaohong township has a population of approximately 10 000 people, the majority of whom derive some part of their income from growing and selling fresh bamboo shoots. Before the establishment of the market, households had minimal channels through which to market their shoots and FBS collectors lacked a singular place from which to collect bamboo shoots from, resulting in inefficiencies in the grower-collector relationship. The FBS market in Gaohong was established in 1994 and within the first 5 months, 10 000 tons of fresh shoots were traded, valued at 45 million yuan. Shoot price also increased by 1.5 yuan/kg, making it possible for FBS growing households to maximize the potential of their product, as there was also no cost to them for using the Gaohong market. The market was funded by the townships forestry department and by the user fee charged to collectors. The collectors also benefited, as there was now a single place of trading in the region, thus minimizing their transactions costs. The FBS market in Gaohong is one of its kind in the county and has invited outside business interests and ventures into Linan.

IMPACTS OF REFORMS ON BAMBOO PRODUCTION

The land, industry and market policy reforms and other initiatives, such as establishment of bamboo association, made a great impact on the bamboo production to consumption system in Linan county. The GDP per capita rose from 1 721 yuan in 1985 to 2 205 yuan in 1991 and 11 534 yuan in 1996. The average contribution of bamboo to farmers' income was 28% in 1995. Ten townships had more than 10 million yuan income from bamboo during 1995–97. In 1996 and 1997, 46 farmer families had more than 50 000 yuan/year from bamboo. Four families in 1996 and two families in 1997 earned more than 100 000 yuan from bamboo. Within the period 1983–87, the value of fresh bamboo shoot production increased from 824 000 yuan to over 21 million yuan. Minimal change in production value was seen between 1987 and 1992, averaging around 29 million yuan. However, in 1993, production values more than doubled to 71 million yuan. This significant increase continued in 1994, 1995 and 1996, with values of over 100 million yuan, 150 million yuan and 240 million yuan, respectively. Dried bamboo shoot production values also experienced significant changes between 1983 and 1987, with an increase in value of 328 600 yuan over this 5-year period. It was only after 1991 that values began to rise significantly, finally peaking at a maximum of 1.9 million yuan. Increases in DBS values, however, are more of a result of significantly increasing annual yields as opposed to the increases in price as experienced by FBS.

The policy reforms made an enormous impact on area under bamboo, bamboo prices and bamboo production, specifically fresh bamboo shoots, and these impacts resulted in increased income from bamboo to households. The three policy reforms, covering land tenure, industry and market, were introduced in 1982, 1987 and 1990, respectively. All the three reforms had pronounced impact on land area under bamboo, bamboo production and bamboo prices. The impacts on fresh bamboo are markedly different from impacts on dry bamboo and wild bamboo. With the introduction of Farmer's Responsibility System, farmers were given the freedom to select a crop of their choice to grow on their land. Hence, with the introduction of reforms under the new land tenure system caused an increase in the area under fresh bamboo shoots production. Later on, as the new policy reforms related to industrial production units and market increased the comparative advantage of bamboo compared with other main agriculture crops (such as rice, tea and silk), land was diverted from these agriculture crops to grow bamboo shoots (Fig. 1).

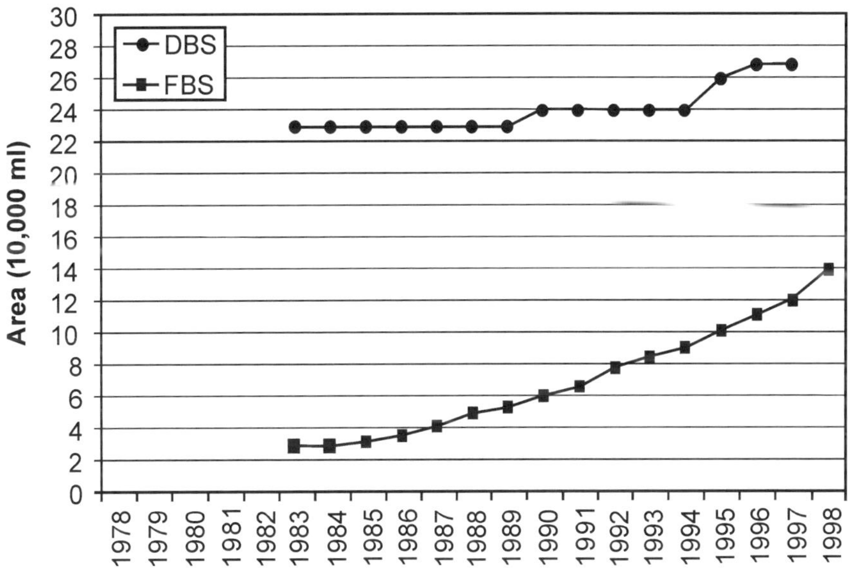

Figure 1. Bamboo growing area in Linan from 1978 to 1998

On average, annual increase in area under bamboo averaged 3 000 mu from 1983 to 1987, 4 000 mu from 1988 to 1991, and 9 000 mu from 1992 to 1998. Hence, in terms of policy reforms, these increases – 3 000, 4 000 and 9 000 mu/year – can be attributed to land, industrial and market reforms, respectively. However, it is possible that the impact of land and industrial reforms continued even after market reforms were started. In such a case, 3 000, 1 000 and 4 000 mu/year should be attributed to land, industrial and market reforms, respectively. However, in 1990, a new technology of fresh bamboo shoot production (covering technology) was introduced. Hence, the increase since 1990 is a combined effect of market reforms and technological development.

In the case of dry bamboo shoots, area remained constant at 23 000 mu from 1983 to 1989, increased to 24 000 mu in 1990 and remained stagnant up to 1994. In 1995, it changed to 26 000 mu and in 1997 to 27 000 mu. Thus, the area under dry bamboo shoots did not change continuously as in the case of fresh bamboo shoots, but changed only during some years. Time series data of area under wild bamboo were not available.

A trend similar to area is observed in prices of fresh and dry bamboo shoots (Fig. 2). As mentioned earlier, up to 1987, bamboo shoots could be sold to the government agencies at controlled prices. In 1988, private enterprises were allowed market entry and in 1990, government withdrew price control on bamboo products. The impacts of these policy changes can be clearly discerned in the prices of fresh bamboo shoots (FBS). During 1983–87, increases in FBS administered prices were marginal – 1.90 yuan/kg. From 1988 to 1990, although prices

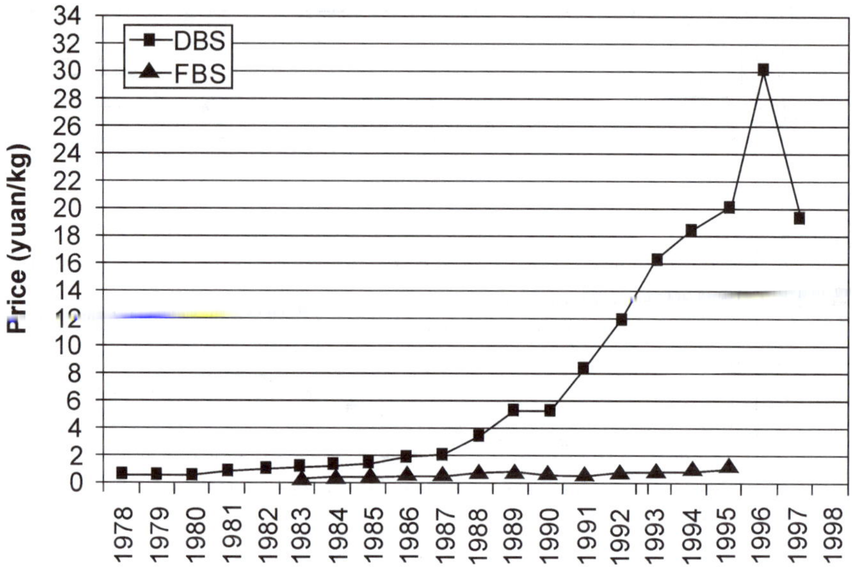

Figure 2. Bamboo shoot production during 1978–98

were controlled by the government, the entry of private enterprises caused an increase in FBS prices to 3.50 yuan/kg in 1988 and to 5.40 yuan/kg in 1989. After the 1990 market reforms, FBS prices increased to 8.36, 12.00 and 16.70 yuan/kg in 1991, 1992 and 1993, respectively. During 1994–95, prices increased at a slower rate of about 1.10 yuan/kg/year. In 1996, they suddenly jumped to 30.5 yuan/kg, but dropped in 1997 to slightly below the 1995 level.

In addition to market reforms, the use of the covering technique also contributed to these increased prices. With this technique, fresh shoots were made available during the high demand season, the Spring Festival in January, and farmers were able to charge maximum selling prices. Although the coverage technique did increase the supply of shoots to the market, supply did not exceed the demand during the peak season, thus sustaining the increased selling prices.

In the case of dry bamboo shoots (DBS), changes in prices were very marginal, even though trends were almost similar to the fresh bamboo shoots, reflecting the preference of farmers to go in for fresh shoot production owing to the higher profit margin (Fig. 2). In government controlled price period (1983–87), price of DBS increased from 0.34 yuan/kg to 0.56 yuan/kg, but in the next two years, prices jumped to 0.90 yuan/kg. Price increase before 1987 is generally attributable to the improvement of shoot quality. After the opening of agricultural product marketing in 1987, demand for DBS increased dramatically, which resulted in decreased attention to shoot quality as producers scrambled to fill orders. However, by 1987, with improvements in the drying process of shoots, which decreased the salt content from 30–40% down to 3%, and the demand for increased quality from

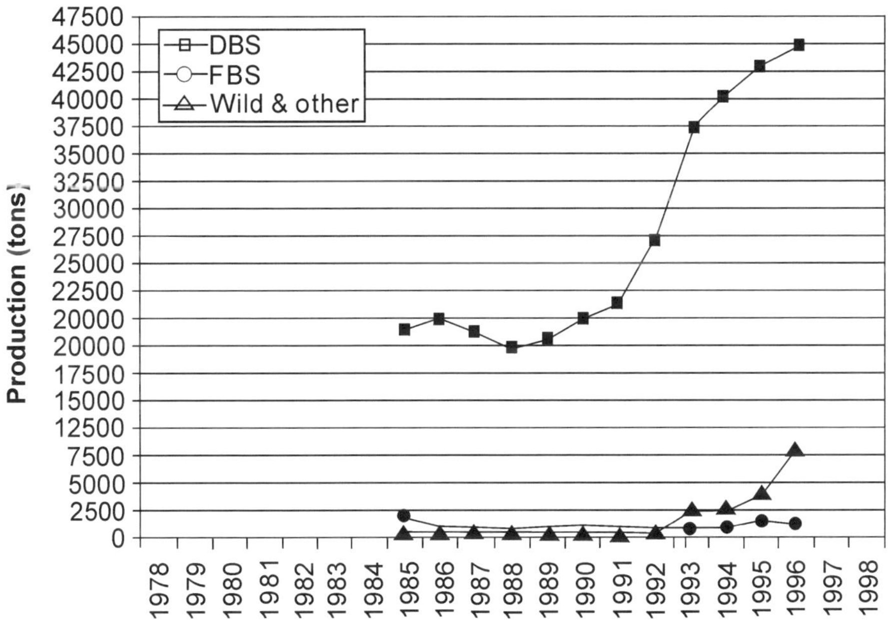

Figure 3. Bamboo production during 1978–98

consumers, DBS quality improved. This subsequently increased shoot prices. They decreased again in 1989 and this is hypothesized to be a result of the Tiananmen Square incident, after which DBS factories and enterprises lost money as the result of a large fall in sales. It is not known whether this was a result of disrupted marketing channels or fears from the customers of political instability spilling over into trade relations. DBS prices began picking up again in 1992 and have shown steady increase since then; but still the price of DBS was only 1 yuan/kg in 1997.

Production of all three types of bamboo products (culms, fresh shoots and dry shoots) also follows this positive trend (Fig. 3). Production will depend mainly upon the area under bamboo, technology available and the intensity of management. Crop management intensity will depend upon the trend of prices. Higher prices will provide incentives to individuals to use intensive management. Hence, a statistical analysis was undertaken to find out the impact of different factors on the production of fresh and dry bamboo shoots. The results of this analysis provided interesting results. In the case of FBS, the only factor that is responsible for increases in production was the area under production; all other factors, such as increased prices and new technology, encouraged farmers to bring more area under fresh bamboo shoots. The correlation coefficients of FBS production with area, price, technology change and market reforms were 0.92, 0.91, 0.80 and 0.73, respectively. Correlation coefficients of area with price, technology and market were 0.96, 0.86 and 0.83, respectively. Similarly, the correlation coefficients of price with technology and market reforms were 0.80 and 0.73, respectively. All these values were significant at 5% significance level. These correlation coefficients indicated a high level of multi-co-linearity between these variables, and a regression analysis proved detrimental. This problem was resolved by breaking the impact analysis into two stages. Firstly, area was regressed over prices and secondly, production was regressed over area. Results of these two equations are given below:

Area = 3.81 + 0.2719 × Price, where area is in 10 000 mu, and price in yuan/kg. (11.60)*

Adjusted $R^2 = 0.92$, F value = 134.58, and significance of F = 0.00.

Production = -5068.64 + 4633.02 × Area, where area is in 10 000 mu, and production in tons. (7.67)*

Adjusted $R^2 = 0.84$, F value = 58.79, and significance of F = 0.00.

These equations indicate that change in price explains 92% of change in area under FBS, and area explains 84% change in production. The area equation gives that, on an average, an increase of 1 yuan/kg in FBS prices caused an increase of 2 719 mu land area under FBS, and one additional mu of area resulted in 0.46 tons of production of FBS.

In the case of DBS, the relationships between production, area, prices and market reforms were not as strong as in the case of FBS. The correlation coefficients of production with area, price and market reforms were 0.61, 0.43 and 0.26, respectively, and none of them was significant at a 10% significance level. The correlation coefficients of area with price and market reforms were 0.77 and 0.52, respectively, but only correlation with price was significant at 10% significance level. The correlation of price with market reforms was 0.52 but not significant at 10% significance level. Change in production was determined in two stages, similar to FBS. The two regression equations are:

Area = 21.08 + 3.68 × Price, where area is in 10 000 mu, and price in yuan/kg. (3.85)*

Adjusted $R^2 = 0.56$, F value = 14.81, and significance of F = 0.00.

Production = –2959.50 + 163.16 × Area, where area is in 10 000 mu, and production in tons. (2.43)*

Adjusted $R^2 = 0.31$, F value = 5.93, and significance of F = 0.03.

Hence, in the case of DBS, price explains only 56% of the variation in area, and area explains only 31% variation in production. Inclusion of any other variable, such market reforms or industrial reforms, did not increase the explanatory power of the regression equations. In fact, it created the problem of multi-co-linearity. One constraint of this exercise is limitation of the data points.

* T-statistics

The area equation indicates that an increase in prices by 0.10 yuan will lead to an increase of 3 680 mu land area under DBS, and an increase 1 mu in area will lead to an increase of 0.016 tons of production of DBS. However, it should be remembered that these results are based on the data from 1984 to 1996. Hence, there are only 12 data points, and degree of freedom is equal to 11.

In summary, the changes in FBS prices were more sensitive to policy changes, and increase in prices brought more area under FBS production resulting in increased production of FBS. The elasticity of area to prices, however, was more in the case of DBS.

IMPACT OF BAMBOO RETURNS ON INCOME DISTRIBUTION AND POLICY IMPLICATIONS

In order to examine the impact of bamboo returns on income distribution, one village was chosen in which 50 households were surveyed on a comprehensive list of economic and management variables. This study examined the data on household income, land distribution and income from land in detail. Similar data from 1987 from the same village was also available from local researchers. Thus, analysis of village-level data over an 11-year period (1987–98) was undertaken. The village chosen was Hong Qiao village in the township of Gaohong. Hong Qiao village is constituted of approximately 90 households, which were divided into ten village groups during the 1950s, based on the proximity of each household to another: all households occupying the same area were placed in the same village group. Each group had 5–14 households in it, but the administrative duties remained with the village and township governments.

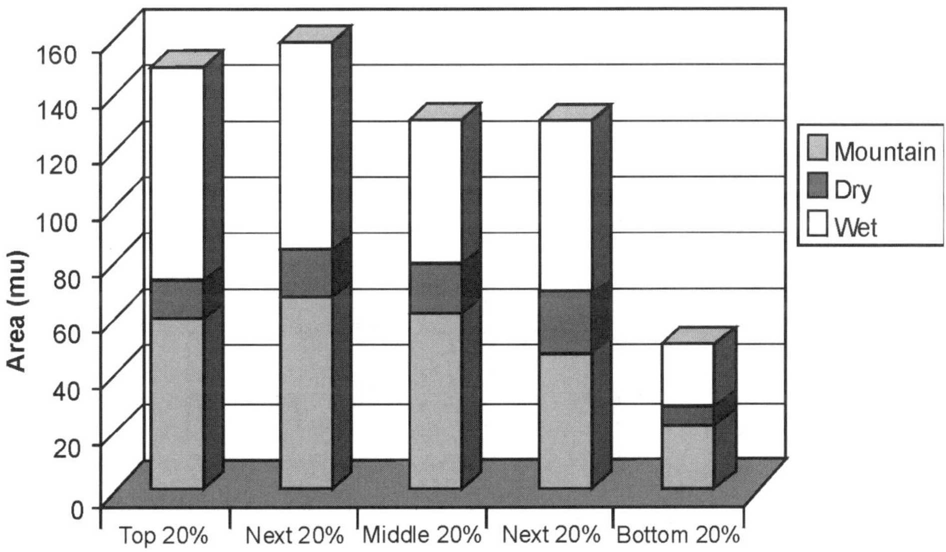

Figure 4. Distribution of land types based on 1987 income ranking

The land distributions in 1987 and 1998, based on the total household income ranking, are given in Figs. 4 and 5. In 1987, the second highest income group has the highest land area while the highest income group has the second highest land area. The middle group and the second lowest group have almost the same land area, and lowest income group has the lowest land area. The total land of the lowest group is about one-third of the land of the two highest groups and about 40% of the middle and the second lowest group. The land distribution in 1998, however, shows a markedly different picture. The middle 20% group of households has the maximum amount of land, but the lowest 20% income group has the lowest amount of land – almost half that of the middle income group. The second lowest and the second highest income groups have almost the same land area. Land holdings of the top income group are approximately 88% of the land holdings of the middle group. Hence, it could be said that in 1998, the overall land inequality reduced. It was also confirmed by Gini coefficients of land distribution that were 0.167 in 1987, and 0.113 in 1998. The reduction in total land holdings in the top three income groups also indicates that with the policy reforms, higher income group people diversified their income resources, and after reduction in the land area, they could manage incomes from other sources such as industrial production, marketing enterprises etc.

At this point, it is important to note that the segregation of land types into dry and wet land refer to land used for the growing of fresh bamboo shoot species and rice fields, respectively. Percentage of dry land increased between 1987 and 1998 in all groups, with significant increases in the higher income groups – upwards of 10% from 1987. While the actual amount of land decreased in all groups, the significant increase of dry land area in all but the lower two income groups reflects the income-earning potential of bamboo shoots. For example, the amount of dry land in the top 5% income group increased from 5% to 25% during the 11-year

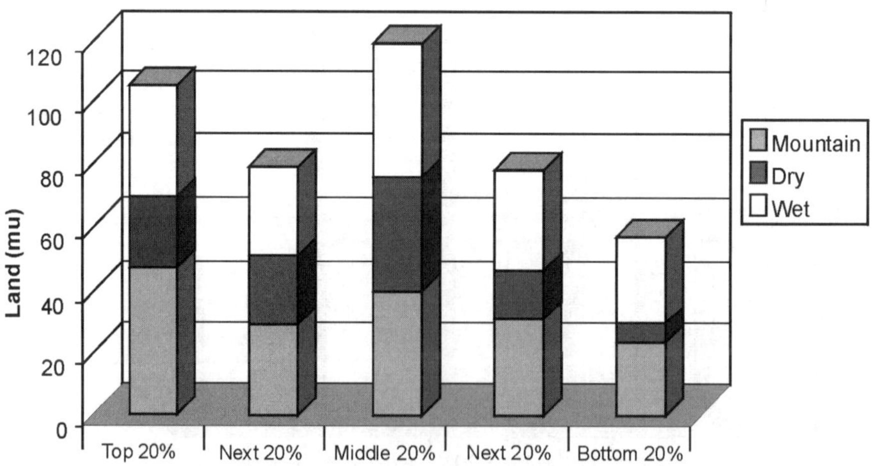

Figure 5. Distribution of land types based on 1998 income ranking

period. Dry land area increased only minimally for the bottom 40% income groups and this may be a reflection of the inability of the rural poor to fully realize the potential of bamboo shoot production, as suggested by Ruiz-Perez et al. (1998)

The land distribution based on the land income ranking for 1987 and 1998 are given in Figs. 6 and 7, respectively. In 1987, the land distribution is not in accordance to the income from land; but in 1998, land distribution follows the same trend as of the land income. This change in land distribution indicates the possibility of almost similar returns from land to land income groups and hence, the possibility of distribution of returns from economic reforms to all land income groups.

The contribution of bamboo income to the total household income for five income groups based on the total income is given in Fig. 8. The percentage of bamboo income to the total household income changed from 3.9 to 18, 17 to 28, 18 to 36, 8 to 22 and 12 to 38, from the top income group to the lowest income group, respectively. This indicates that the relative increase in bamboo income is highest for the highest income group, but the second highest is the lowest income group, and the third highest is the second lowest group. Hence, because of industrial reforms and market reforms related to bamboo during this period of 11 years, the richest people as well as poorest people are benefited to the greatest extent. However, the middle class people did not get a similar benefit. During this period, the Gini coefficient of the total income changed from 0.21 to 0.25, but the Gini coefficient of the income from bamboo decreased from 0.31 to 0.25. Hence, the overall economic reforms increased the income inequality marginally, but bamboo reforms improved the income distribution.

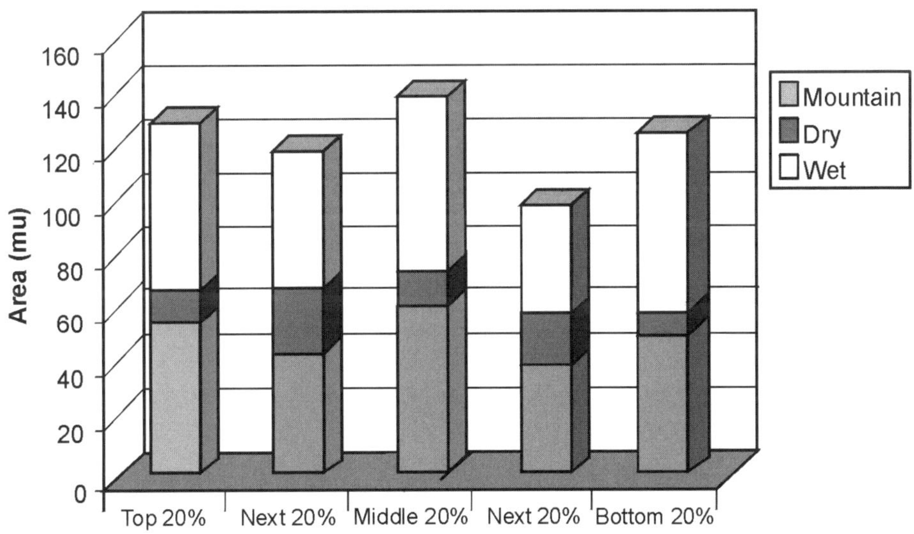

Figure 6. Distribution of land types based on 1987 land income ranking

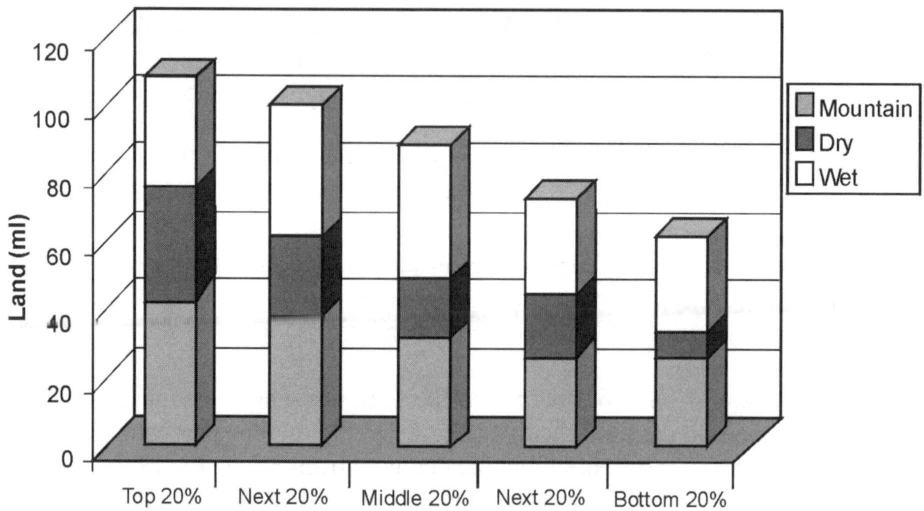

Figure 7. Distribution of land types based on 1998 land income ranking

Figure 8. Distribution of income based on total income ranking (1987 & 1998)

CONCLUSION

All policy reforms related to land, industrial production and market, and other initiatives such as technological developments and establishment of associations for the promotion of bamboo production, processing and marketing have contributed to enhance the role of bamboo sector in the economy of Linan County. However, owing to limitation of data points, it was not possible to separate the impact of each policy separately. One possibility to find the effect of each policy and innovation separately is to collect the cross-section data from all the ten bamboo counties in China.

The impact of bamboo policy reforms on income distribution indicates that the richest income group derived the maximum benefit, but the poorest group was the next in terms of proportional increase in the benefits (during the period of 1987–98). The middle-income groups did not get proportional benefits. These results are contradictory to the results of Ruiz-Perez et al. (1998), who found that bamboo has the highest relative importance for middle-class farmers as opposed to poor farmers. Hence, it seems that the hypothesis of bamboo improving the economic condition of poor people cannot be rejected based one or two case studies. To find the role of bamboo in rural poor economies, a number of studies have to be taken up in diverse socio-economic and ecological environments.

This paper contains only a small part of a much larger study conducted by the authors. Other aspects that were studied in this INBAR-funded project are economic efficiency of enterprises producing different bamboo products (such as bamboo flooring, furniture, handicrafts, etc.), economic efficiency of production of some bamboo products (fresh bamboo shoots, dry bamboo shoots, bamboo timber), and comparative economic study of bamboo, rice, tea and silk production. These results will be available shortly.

ACKNOWLEDGEMENTS

The authors would like to acknowledge and extend thanks to: the staff and faculty of Zhejiang Forestry College, in particular Professor Zhou Guomo, Zeng Yenru, Shen, Q.; the Linan County Forestry Bureau, in particular Wong Anguo; and the Goahong Forestry Department, in particular Mrs. Ma.

REFERENCES

Dong, J.; Chen, X. 1989. On economy of farm scales. Bai, 1989, China.

Ling, Z.; Jiang, Z. 1993. From brigade to village community: the land tenure system and rural development in China. Cambridge Journal of Economics, 17, 441–461.

Liu, F. 1989. Inquiry into privatization of farm land, Bai, 1989, China.

Ruiz-Perez, M.; Zhong, M.G.; Belcher, B.; Xie C.; Fu Maoyi; Xie, J.Z. 1998. The role of bamboo plantations in rural development: the case of Anji county, Zhejiang, China. World Development, 27(1), 101–114.

Zhang, W.F.; Makeham, J. 1992. Recent developments in the market for rural land use in China. Land Economics, 68, 139–162.

Zhong, M.G.; Xie, C.; Zheng, W.; Fu Maoyi; Xie, J.Z. 1998. Bamboo in Anji, China: a case study of an intensive production-to-consumption system. INBAR Working Paper No. 15. International Network for Bamboo and Rattan, Beijing, China. 47 pp.

Zhou, X. 1989. Investigation and analysis of the family farm system. Bai, 1989, China.

Bamboo for Sustainable Development, pp. 173–186
Eds. A. Kumar, I.V. Ramanuja Rao and Ch. Sastry

Bamboo-dependent livelihood patterns of fishing communities: A case study of the West Coast of India

SANJAY V. DESHMUKH
PRASAD C. KARNIK
BALAKRISHNA BHATKHANDE
AMOL B. DESHMANE
Manav Sadhan Vikas Sanstha (MSVS), Mumbai, India
ANIL K. PATIL
Department of Botany, J.S.M. College, Alibag, Maharashtra, India
MANOJ R. BORKAR
Department of Zoology, Carmel College for Women, Nuvem, Goa, India
KEDAR KULKARNI
Abhikalpan (Architects & Planners), Naupada, Thane, India

Abstract—A detailed study of 44 coastal fishing villages, covering about one-fourth of Indian coastline, revealed the ecological and environmental benefits of bamboo resources, and their use in sustaining traditional occupations of fishing communities. A categorization of livelihood patterns of fishing communities that contribute to measurement economic value of bamboo resources was also made.

The study helped to identify gaps in availability of bamboo, opportunities (demand for bamboo, new products and new technologies) as well as constraints (supply of raw material and need for better products) for improving the livelihood patterns of Indian fishing folk. Though short-term economic gains from exploiting bamboo appeared more attractive than the long-term benefits of conservation, our study revealed that this was due to the scarcity of applied economic analysis of bamboo-dependent livelihood patterns of coastal fishing communities.

INTRODUCTION

Bamboo is an invariable constituent of most Indian rural houses in one form or the other. Coastal stretches of India are home to a great variety of ecosystems as well as cultural diversity. Nearly 30% of Indians live within 80 km of the shoreline, which extends to over 7 500 km. Bamboo, with a utility of over 60% in fishing and other occupations, is an integral part of the livelihoods of coastal fishing communities. It plays an important role in the traditional beliefs and legends, and has contributed to the conservation of biological resources of coastal regions.

Sustenance of the livelihoods of the coastal communities can be attributed to the diverse utilization pattern of natural resources, in general, and of bamboo, in particular, which continues to serve the needs of people and their environments since. The age-old knowledge and wisdom of these rural people about the properties of bamboo has been seldom documented.

COASTAL GEOGRAPHY OF MAHARASHTRA AND GOA

Maharashtra

The coastline of Maharashtra State can be categorized into three prominent types – hills with steep cliffs entering the sea, sandy beaches and mudflats. The coastal region of Maharashtra is traditionally known as the "Konkan", which is a narrow terrain with a width of up to 60 km between the coast and the ridge of the Sahyadri Mountains. The coastline of Maharashtra measures about 720 km, sharing its northern boundary with the State of Gujarat in the North and the southern boundary with Goa in the South. This region has five districts – Thane, the urban area of Greater Mumbai, Raigad, Ratnagiri and Sindhudurg.

Goa

The State of Goa is located below Maharashtra, along the Central-West coast of India. The coastal strip of Goa is highly indented with sea cliffs, notches and promontories, alternating with rivers and estuaries. Major rivers such as Mandovi and Zuari with Cumbarjua Canal, which form the largest estuarine complex, drain the coastal regions of Goa. Most fishing villages of Goa are located in the upstream regions of these estuaries and not directly on the coast facing the Arabian Sea. This is attributed to the fact that the State Tourism Department commercially exploits the entire coastline of Goa.

OBJECTIVES OF THE STUDY

A study on bamboo-related livelihood patterns of coastal fishing communities of Maharashtra and Goa States, along the West Coast of India, was initiated with the specific objectives of:

- Documenting present use of bamboo resources by coastal fishing communities in sustaining their traditional occupations;
- Identification of gaps, opportunities (demand for bamboo, new products and new technologies) as well as constraints (supply of raw material and need for better products) for improvement of livelihood patterns of fishing communities; and
- Recommending, based on the above studies, a few short and medium-term technical and socio-economic interventions aimed at increasing the availability,

utility and value of bamboos to the coastal fishing communities, on a sustainable basis.

METHODOLOGY

Studies were conducted in 44 coastal fishing villages (Table 1) along the coastal States of Maharashtra (30) and Goa (14).

The following methodology was adopted for conducting detailed studies:

- Explorations and surveys were conducted in the coastal regions of Maharashtra and Goa for identification of fishing villages for detailed studies. A list of various bamboo species most commonly utilized by the coastal people in the region of study was prepared during field visits.
- A questionnaire (see Annex) was developed for documenting the demographic profile of the villages under study. Using the same questionnaire, bamboo resource use by the fishing communities was also studied with respect to: (a) traditional beliefs, (b) bamboo in fishing, (c) bamboo in agriculture, (d) bamboo in housing, and (e) bamboo in household items.

Table 1.
Fishing villages selected along the West Coast of India for the study

State	District	Name of village
Maharashtra	Thane	Vasai – Pachubunder, Satpati, Motha Dahanu, Dhakti Dahanu, Kalyan, Diva, Mumbra
	Mumbai	Cuffe Paradc, Mahul, Versova, Khardanda, Worli
	Raigad	Kachali, Akshi Sakhartar, Akshi Raiwadi, Agrav, Korlai, Navedar Navgaon, Bodhni Koproli, Mandwa, Alibag, Koliwada
	Ratnagiri	Dabhol, Burondi, Paj Pandhari, Harne, Ratnagiri Kasarweli, Ratnagiri Mirkarwada
	Sindhudurg	Varavade, Deogad Aanandwadi, Deogad Taramumbri
Goa	North Goa	Neura, Carambolim, Tivrem, Marcel, Cumbarjua, Sant Estev, Gaundale, Brittona, Salvador de Munde, Pomburpa
	South Goa	Quellosim, Lotulim, Borim, Kavlem

Note: Total number of villages – 44.

- In addition, an assessment of bamboo availability (accessibility to the resource) to the fishing communities was also carried out. This helped in understanding the economic potential of the coastal belt, with respect to availability and accessibility of bamboo.
- Based on the information gathered during the study, an assessment of current livelihood patterns of the coastal communities, in relation to the accessibility

and utilization of bamboo resources, was done and recommendations for their management were drafted.

OBSERVATIONS

Bamboo – the resource

While bamboo is widespread, botanists seldom collect it because the plants flower rarely and are very cumbersome to collect. If one has to undertake systematic sampling, then taxonomically critical vegetative parts, such as new culm leaves and branch complements, also need to be sampled. Hence, the taxonomy of bamboo remains relatively poorly understood. Following six species of bamboo are commonly observed in the western Indian region, which include the States of Maharashtra and Goa.

- *Bambusa arundinacea* Wild.
- *Bambusa vulgaris* Schrad.
- *Dendrocalamus giganteus* Munro
- *Dendrocalamus strictus* Nees
- *Oxytenanthera monostigma* Beddome
- *Oxytenanthera stocksii* Munro

Figure 1. Bamboo, as a sign of auspicious beginning, at the entrance of homes in coastal villages

Of the six species mentioned above, *B. arundinacea, B. vulgaris, D. strictus, O. monostigma* and *O. stocksii* were the most commonly used by the fishing communities in the region of study.

People and bamboo

Over 200 million Indians, whose livelihoods are largely dependent on bamboo, are fairly distributed all over the country, following the trail of the resource itself.

Figure 2. Bamboo in agriculture: a storage device

Numerous tribal and rural communities work with this resource (Fig. 1). Although their products play a crucial role in the rural ago-industrial web (Fig. 2), these groups do not enjoy any clout in the society or any control over the resource. Almost without exception, these groups of bamboo workers fall in the scheduled caste/scheduled tribe categories. Coastal regions of India, however, appear to be an exception; fishing communities here enjoy utilizing bamboo for almost every segment of their livelihood options including fishing, which is the primary source of income.

Bamboo-based livelihood patterns of fishing communities in the 44 fishing villages in coastal Maharashtra and Goa are summarized in Table 2.

Table 2.
Fishing communities of western India: bamboo-based livelihood patterns

Livelihood types	Involvement in activity (%)		Fish landing (%)
	Male	Female	
Active fishing population (78.75% of the total village population with a standard deviation of 16.56)			
Traditional (58.14%)	65.18	34.82	67.00
	(15.80)	(09.58)	(26.65)
Commercial (41.86%)	64.68	35.32	33.00
	(21.40)	(21.40)	(22.91)
Active (seasonal) farming population (about 60% of the total active fishing population with a standard deviation of 21.73)	NS	NS	NA

Notes: NS = not significant; NA = not applicable; Figures in parentheses indicate standard deviation.

It may be noted here that the average population of the fishing villages studied (total 44 villages) was 4 034, the maximum being 38 500 and the minimum 55, with a standard deviation of 7 890.88.

Indigenous bamboo knowledge systems

While examining the "man-bamboo interaction" in the coastal region of western India, it was essential to identify and list various bamboo-dependent livelihood activities and the respective user groups, the type of output/products, and the degree and nature of stress on the utilization of bamboo resources. The relevance of bamboo in the life of coastal fishing communities in Maharashtra and Goa is discussed below.

Many traditional societies all over the world value a large number of plant species from the wild for a variety of purposes – food, fiber, shelter or medicine (Ramakrishnan et al. 1998). Usually these are nurtured in what are known as "sa-

Figure 3. Bamboo used in house construction

cred groves" and defined as "traditionally protected small patches of vegetation types and managed by the local communities through a wide range of management practices". Till recently, this biological heritage was conserved largely by a cultural and religious tradition of dedicating patches of forests to various deities (Gadgil and Vartak 1976).

One of the deities ("Kalkai") associated with sacred groves found in and around the coastal villages in Maharashtra and Goa is named after a bamboo species – *Bambusa vulgaris* Schrad. Bamboo was not only observed as abundant in the region of study but was also seen used for making flags (indicating existence of the deity) that are placed at the entrance of sacred groves. Various kinds of bamboo baskets have significance in life and death of the coastal communities. For the welfare of the newly born child, small bamboo baskets are hooked to a village temple tree to cast away the evil. Bamboo also features in the construction of houses (Fig. 3).

Bamboo resource accessibility

Accessibility of fishing communities to bamboo resources was categorized as: (a) availability within the village, (b) availability within 20-km range of the village, (c) availability within the State, and (d) availability outside the State. Table 3 highlights the percentage availability documented during the current study.

Table 3.
Indian fishing communities' accessibility (% availability) to bamboo resources

Location of bamboo	Availability of bamboo (%)	
	Maharashtra	Goa
Within the village	25.77 (19.66)	28.93 (12.74)
Within 20 km of the village	48.80 (23.68)	46.78 (12.80)
Within the State	25.13 (31.89)	24.29 (18.38)
Outside the State	00.30 (00.88)	00.00 (00.00)

Note: Figures given in parentheses indicate standard deviation.

It can be seen from the Table 3 that at least 25% of the annual requirement of the fishing communities was met with locally grown bamboo, whereas about 50% of the resource still came from locations 20 km away from the village.

Bamboo craftspersons and products

Unlike many other craft traditions, bamboo skills are not restricted to one particular tribe or community. In a village, most people are familiar with bamboo work, though the levels of skills vary greatly from person to person. This was confirmed during field visits, when it was observed that traditional fishermen from most of the villages had acquired the skills of making bamboo-based items useful primar-

Figure 4. A bamboo tray for fishes

ily for livelihood support, rather than income generation (Fig. 4). Various products made out of bamboos are grouped based on their uses and are highlighted in Table 4.

In all the fishing villages, it was observed that agriculture, as an income generating activity, was practiced as a stopgap arrangement and that too in only 12 out of total 44 fishing villages. During non-fishing season, most of the remaining villages did not make any effort to enhance livelihood activities by exploring alternative measures of income.

Fig. 5 highlights estimates of bamboo use in livelihood activities (in percent) and the annual bamboo requirement (in number of culms) in each in the fishing villages studied. It can be seen that the maximum utilization of bamboo resources is in making various fishing devices, followed by housing and household-related items, agricultural implements, boat building accessories, and other devices.

Table 4.
Bamboo based utilization patterns observed in fishing communities along West coast of India

Livelihood activity	Bamboo utilization
Fishing	Devices for fishing nets, fish traps, grading & measuring, and transporting and storing fish
Boat building	Scaffolding for boat building, parking slots for canoes
Agriculture	Trellises, props, fish catching devices specially designed for use during farming activity, walkways
Housing & household uses	Construction, fencing material, household items such as small baskets, mats, water storing/drinking devices, pot-holders

Note: Agriculture is seasonal activity for most of the fishing villages.

BAMBOO FOR LIVELIHOOD SUSTENANCE

Bamboo has an important role to play in development. It is a natural tool with which to encourage sustainable, integrated farming systems and an excellent resource on which to build a variety of income and employment-generation opportunities (Fig. 6). With its multiple uses and high value in a range of products aimed at national and international markets, there is great deal of potential for value-adding operations and many different entry points for development interventions that are accessible to and appropriate for resource-poor people. In order to do this efficiently, an improved understanding is required on the bamboo sector, the people involved and the main problems they face.

It was observed during the studies that, despite the differences in cultural and geographical backgrounds, the coastal fishing communities from both states have been utilizing bamboo resources in a variety of similar ways. It also became clear that bamboo could be used to strengthen cultures of rural, indigenous as well as migratory communities, as well as provide them with a means to deal with various stresses of environment and development in a rapidly changing world. The

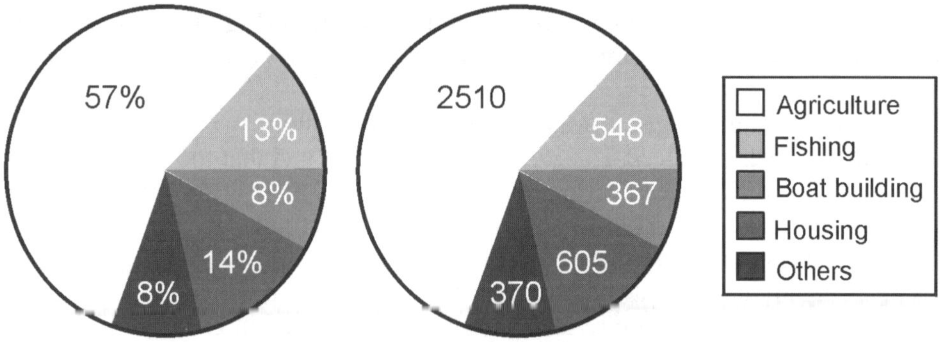

Figure 5. Estimates of bamboo use pattern (left) and bamboo culm requirement (right) in livelihood activities of the fishing communities

Figure 6. Bamboo growers of Goa State

knowledge on (a) conservation, (b) restoration, (c) classification, (d) evaluation and (e) utilization of bamboo among people most often exists only as oral tradition, which is passed on from one generation to another.

WHERE DO WE GO FROM HERE?

The socio-economic aspects of bamboo are significant because of the commodity's role as a major non-wood forest product and they warrant detailed research.

Although a large number of studies have been done in the past on various aspects of bamboo, only limited attention has been paid to socio-economic and policy aspects of bamboos in the coastal regions, including production-to-utilization systems and sustainable resource development. Appropriate economic interventions are proposed as an additional measure for achieving the following short-term and medium-term goals.

- To generate employment and income in rural areas as also to meet the subsistence need of the poor people.
- To minimize the pressure on forests by popularizing bamboo as an alternative to wood in constructional activities, and also as an income source through participatory management.
- To increase the productivity of bamboo resources with high-yielding planting material through scientific management practices.
- To motivate villagers to undertake bamboo cultivation in their common property resources for enhancing income.
- To minimize the dependency on forest by the local/indigenous population with respect to timber and fuel wood.
- To control/eliminate underprivileged nature of livelihoods through alternative means of income generation.

Participatory management & incentives

It would be desirable to promote a participatory management approach for bamboo development with a particular aim to generating employment in rural areas, and consequently alleviate poverty. In the North-East Indian region, Joint Forest Management schemes of the Forest Departments provide various incentives to the JFM participants involved in bamboo development. They are:

- Wage earning through collection and planting of bamboo rhizomes/seedlings and protection of planted area including gap filling.
- Rights to returns from the produce, after the final harvest of bamboos, at 50:50 profit sharing basis.
- Wage earnings from the manufacture of value-added products in the JFM village.
- Entitlement to bamboo produce to meet the minimum domestic need of the beneficiaries.
- Income through local sale of handicrafts, mats, etc. and export of value-added products.

It may be noted here that in a country such as India, Forest Departments plays a very important role in the process of JFM, such as (a) selection of beneficiary village, (b) identification of beneficiaries under participatory management and (c) facilitate plantation activity.

CONCLUSION

Knowledge of indigenous people on use of bamboo, as observed during the studies, exists only in the oral tradition, being passed on from one generation to another. It is necessary to document myths and cultural practices (rites, rituals and taboos) associated with bamboo since they form part of the living tradition. Similarly, data related to the use of bamboo by the fishing communities in house building is also of great value. Any variation in product owing to cultural difference could indicate the level of skill, innovation and productivity of a bamboo worker. Finally, data on bamboo as a commercial resource would prove to be most useful in understanding the resource flow and cash flow from the growing site to the point of utilization. Further studies with regard to (a) who gains or loses, and (b) who may buy bamboo and sell it at distant markets, will help understand better the resource flow, the cash flow and the magnitude of transactions involving bamboo.

In the last few years, India has initiated several measures to develop its enormous bamboo (and rattan) resources to their full potential. India's prospects in bamboo sector are high, as there is no dearth of talent and opportunities for entrepreneurs. However, current utilization practices are wasteful, and resources are either sold to the pulp and paper industry for less than their value or made into cheap and poor quality handicraft items for local markets and use.

This country now has a unique opportunity to consolidate the diverse activities of this sector and focus on accelerating, where possible, the pace of strategic, adaptive and developmental work, thereby enhancing the micro-enterprise development and increased employment.

ACKNOWLEDGEMENTS

The work on which this paper is based was part of a consultancy project financially supported by the International Network for Bamboo and Rattan (INBAR). The authors would like to thank INBAR for the support provided, particularly Dr Cherla Sastry, Director General, and Dr I.V. Ramanuja Rao, Senior Manager (Programs). The first author gratefully acknowledges the cooperation and research facilities provided by Dr Asad R. Rahmani, Director, Bombay Natural History Society (BNHS), during the course of this work.

REFERENCES

Gadgil, M.; Vartak, V.D. 1976. Sacred groves of India: a plea for continued conservation. Journal of Bombay Natural History Society, Vol. 72(2), 313–320.

Ramkrishnan P.S.; Saxena, K.G.; Chandrashekara, U.M., ed. 1998. Conserving the sacred groves for biodiversity management. Oxford and IBH Publishing Co. Pvt. Ltd., New Delhi, India. 480 pp.

ANNEX: QUESTIONNAIRE FOR SURVEY

I. General information on the fishing village

1.	Name of the fishing village & hamlet if any	:	
2.	Type of fishing village	:	
3.	Location	:	
4.	State	:	
5.	District/Taluka	:	
6.	Accessibility	:	

II. Demographic details

1.	Population		
	Total population	:	
	Population of fishing community	:	
2.	No. of Households	:	
3.	Literacy rate [Male : Female ratio]	:	

III. Cultural background

1.	Religion	:	
2.	Caste	:	
3.	Sub-caste	:	
4.	Tribal / Non tribal.	:	
5.	Occupation		
	Primary	:	
	Secondary	:	
	Other (seasonal)	:	
6.	Livelihood pattern / occupation	:	

IV. Bamboo resources

1.	Forest	:	
2.	Cultivated	:	
	Ownership	:	
	Common property	:	
	Area	:	
3.	No. of bamboo species occurring in the village	:	
4.	Conservation techniques practised, if any -	:	
	a) For (natural) forest area	:	
	b) For cultivated (Individual/common property)	:	
5.	People's knowledge on bamboos.	:	
	a) Folk lore	:	
	b) Song	:	
	c) Dances	:	
	d) Rituals	:	
	e) Festivals	:	

V. Utilization of bamboo resources

1.	Access : by outsiders, if any	:	
2.	Uses (Traditional):		
	a) Fishing & related equipment	:	
	b) Agriculture	:	
	b) Other uses		
	Livelihood/ Housing	:	
	Household	:	
3.	Community-based Utilization (Annual need):	:	

VI. Market

1.	Is bamboo sold?
2.	Are bamboo products sold?

Bamboo for Sustainable Development, pp. 187–191
Eds. A. Kumar, I.V. Ramanuja Rao and Ch. Sastry

Earthquakes, poverty and bamboo

MARIANTONIA PINEDA and HENY DOING
The Flores Bamboo Development Foundation, Flores, Indonesia

Abstract—The Flores Bamboo Development Foundation emerged from a natural disaster – an earthquake and tidal wave – that struck Flores Island, Indonesia. Apart from the work involved in promoting bamboo as a safe building material, it would seem that the Foundation's role should be focused on the research and development of a kind of "rubber joint system", that would make buildings behave as completely flexible structures. In reality, the main issue has been about changing people's perceptions of using bamboo. In spite of bamboo being a traditional building material that has been used for centuries, it has a "poor" image among Florinese people and Indonesians in general. As it is associated with the poorer sections of the society, living in crude and temporary dwellings, it was inappropriate devising a high-tech bamboo system, as no one wanted to use bamboo as a building material. This paper deals with the social factors, as well as the aesthetic and technical reasons, that support the perception and status of bamboo houses. It also explains the strategies to be used to change the image in promoting bamboo, not just as a safe building material, but a dignified one through attractive, innovative and resistant solutions.

INTRODUCTION

Flores Island is part of Nusa Tenggara Timur, the poorest province of Indonesia. An earthquake and a tidal wave struck Maumere in December 1992. The Southeast Asia Association of Seismology and Earthquake Engineering now classifies Flores Island as a "very high risk" seismic zone. The disaster caused horrendous damage to the local community housing, schools and hospital. It was then that bamboo came to the fore as a safe and economic building material, and the obvious choice to prevent the recurrence of such large-scale damage.

The Flores Bamboo Development Foundation (FBDF), a non-governmental organization, was established to find solutions to the problems facing the Flores communities using bamboo. Apart from promoting the sowing and use of this material, the first approach was the development of new techniques to make bamboo buildings respond more efficiently to earth movements. Time has proved that when trying to promote bamboo, the focus in the first stage has to be on social factors. The work on technical issues will not be valuable if the people in the communities are not willing to accept the material.

Even though there was an increase in the use of bamboo after the earthquake, it was because the material was cheap and easily available and not because of a social acceptance. Two years after the disaster, it was clear that the shift towards bamboo buildings was only a temporary measure (Edwards and Doing 1996). Six years have passed since then, and the situation tends to be the same as before the earthquake. If there is no real understanding of the social factors that produced this phenomenon, the situation will not change.

The people of Flores already recognize some advantages of building with bamboo. It seems obvious then that the work of FBDF should be concentrating on solving problems related to the disadvantages of using bamboo, from a Florinese perspective. The technological improvements should be left for the second stage, as they are based on qualities inherent to the material.

With the fascinating possibilities that building with bamboo offer, however, it is easy to veer towards technical matters, forgetting the real focus: People, their wishes and their needs. The strategy to be used has to be based on the result of a research in the villages. By collecting and organizing opinions about real problems, one will be able to find solutions related to the needs, aspirations, identity and building traditions of the Florinese.

ADVANTAGES RECOGNIZED BY FLORINESE PEOPLE

1. Availability

The availability of bamboo is apparent to most of the Florinese.

2. Economics

Bamboo can be obtained by the people themselves; but even if purchased, it is much more cheaper than any other building material.

3. Flexibility

During the earthquake, and the constant minor tremors that still haunt Flores, bamboo structure demonstrated its ability to move safely without collapsing. Even when it collapsed, it proved to be less dangerous than other structures.

4. Easy and quick to transport and build with

People in Flores carry bamboo on their shoulders or in handmade carts (wheel barrows). They build tents and other improvised buildings in bamboo in a matter of hours.

5. Cool spaces

Even those who can afford another building material, have a bamboo hut in the back of the house. This is used for cooking and other activities during the hot peak hours of the day.

DISADVANTAGES FROM A FLORINESE PERSPECTIVE AND MAIN PROBLEMS

1. Vulnerability

a) Weakness and damage of bamboo pieces because of pest attacks: beetles, termites and fungus infestation.
b) Deterioration of walls, ceiling, floor, structure, etc., resulting from a direct exposure to climate: sun, rain, humidity.
c) Building loss owing to fire: poor electric installations, neglected kerosene lamps, no precautions for the cooking place.

2. Durability

a) All the above contribute to shorten the life span of a bamboo house.
b) Deformation of the building caused by the absence of important structural parts, like trusses and triangulation systems.
c) Permanent jointing of the main structure with other parts of the building, making it impossible to change individual pieces without demolishing the entire building. Usually bamboo structures last longer than the mats used for walls, which can be easily changed if the structure is a separable entity.
d) Difficulties in solving problems of details in joints and in bamboo used in conjunction with other building material.
e) Splitting of bamboo pieces caused by the use of nails, and other inappropriate construction practices.
f) Lack of maintenance of the building.
g) Lack of a proper foundation for the building.
h) Selection of bamboo culms that are inappropriate for specific purposes in the building's structure.

3. Image

a) Poor aesthetic image because of the reasons mentioned above, plus the fact that no one invests time, effort and money in making a good design, as a bamboo building is not supposed to last.

STRATEGY PROPOSAL

All buildings for the headquarters of the Flores Bamboo Development Foundation will be constructed in bamboo, using a different structural concepts and construction systems in each of them. FBDF will become a center for permanent open exposition. Apart from models, drawings and the written information available in this center, the buildings will be a practical example of what can be done with bamboo and how to do it. The headquarters will consist of an office, library, showroom, training center and four middle class prototype houses for the staff.

Living and working in bamboo buildings will heighten FBDF's reliability in the community. These four houses will be prototypes of bamboo in combination with other building materials: concrete, brick, wood and iron. All these materials, including corrugated iron sheets for roofing, reflect the real situation in Flores. Each of these offers an advantage for a specific purpose. The right approach will be, not to avoid other materials but to find solutions for them to work in combination with bamboo.

Earthquakes affect the whole of the community, irrespective of their socio-economic status. In fact, the middle class was more adversely affected as the poorer people were already living in bamboo houses. Part of the program will aim to increase the interest of people in a better economical position. The increase of middle class houses and official buildings constructed with bamboo will automatically raise the status of bamboo, making it easier to promote the material among the poorer people. This will begin to separate the image of bamboo from the poverty stereotype.

The construction of FBDF buildings will provide the staff the opportunity to have the control over all the processes involved in working with bamboo: the quality of the bamboo to be used, harvesting, transport, storage, curing and preservation, design, and the maintenance of the building. The result of this process will be a long-term strategy, but possibly a very effective one. This will show how all stages of the process play an important role when using bamboo, and how it can make a difference in terms of vulnerability and durability of bamboo.

On the other hand, the buildings will show how concrete might be used for the most vulnerable parts of the building like foundations, bathrooms, kitchens, and in the lower half of the walls in contact with the earth. In case of an earthquake, concrete will be for spaces infrequently used, taking into account how the material would respond to ground movement. Bamboo will then be used for structures, for the upper half of walls and in spaces frequently used, such as the living room and the bedroom. These buildings will show how to separate bamboo structures from concrete in order to maintain bamboo's flexibility and the possibility of replacement of deteriorated parts. These buildings will also show how to ventilate a space efficiently when using tin sheet for roofing and how to assemble it with a bamboo structure. All the above points would show how to take maximum advantage of the diverse qualities of different building materials and the importance of a good design.

Different structural systems like hyperbolic paraboloid, bridge-style structure, column and truss systems, Simón Vélez's principle of "pares y nudillos and flying roofs" (Vélez 1989), and other attractive and innovative ideas will be used to attract people's attention about the versatility of bamboo.

It is not necessary to promote the use of bamboo among community members from the poorest sector. They use it, as they have no alternative. The objective here will be the improvement in the quality of these buildings so that they can consider them as permanent structures. Even when their economic situation improves, the people would continue to feel comfortable about using bamboo rather than, or in combination with, other building materials. In that sense, FBDF has to become a 'tool' to interpret their needs, wishes, identity and tradition for transferring these concepts into practice with the right technical approach. The Foundation must work with the people in the communities as practical trainers in construction technologies, which are based on simple construction systems that do not require sophisticated tools or high technology. Theories mean little to them if they cannot be demonstrated practically.

The Foundation will continue the research and development of simple preservation technologies, relying on what is available locally (e.g. lime and tobacco). It will organize housing design competitions at the local level to promote the people's interest in spending time and effort in the design of bamboo dwellings.

Finally, FBDF will develop prototypes that offer new advantages of bamboo over other building materials. For example, it will develop a prototype kit for a low-cost, all-bamboo, modular house. This prototype will provide the possibility of moving the house from one community to another, to replace parts, to enlarge or to improve the house in the future.

In the second stage, the system will be industrialized. At the workshop, FBDF will aim to produce the kit house in large quantities in order to reduce cost and offer better quality. This would be the right time to develop the "rubber joint system" so that people can buy the house and assemble it on site.

CONCLUSION

The solution for the problems related to bamboo building depends on different factors, each of them playing an important part in the process that working with bamboo would involve. A successful result in terms of vulnerability, durability and image would be possible, as all these factors consume no money or technology but effort and time, and they are the only resources most of the Florinese people have here.

REFERENCES

Edwards, K.; Doing, H. 1996. The importance of bamboo and housing construction: a case study in Flores. *In* Ganapathy, P.M.; Janssen, J.A.; Sastry, C.B. ed., Bamboo, people and the environment, Vol. 3, Engineering and utilization. Proceedings of the Vth International Bamboo Workshop, Ubud, Bali, Indonesia, 19–22 June 1995. International Network for Bamboo and Rattan, New Delhi, India. pp. 33–37.

Vélez, S. 1989. La arquitectura, futuro de la guadua. *In* Villegas, M., ed., La cultura del cafe, bambusa *Guadua*. pp. 40–45, 86, 170.

Bamboo for Sustainable Development, pp. 193–204
Eds. A. Kumar, I.V. Ramanuja Rao and Ch. Sastry

Bamboo craft development: A tool for the development of tribal and rural communities

RITU VARUNI
Architect & designer, Arunachal Pradesh, India

Abstract—Both the North-East Indian hill states of Arunachal Pradesh and Nagaland abound in bamboo and together account for about 40 of the 136 species of bamboo found in India. Most of these species happen to be some of the best in the world. Even better, this resource is owned, managed and grown by the people. Not owning a bamboo grove is unthinkable in this part of India. Every male has a working knowledge of bamboo craft, though these skills are on the decline. In totally forest- and land-based economies such as these, it is obvious that development of any sort must be based on the incredibly rich natural resource base of the area in a way that does not exploit this wealth unwisely. It is in this context that bamboo assumes an extremely crucial role, being the single largest replenishable bio-resource in the region. With proper inputs of innovative research, design, appropriate technology, training, marketing and production management geared towards its proper and improved utilization, bamboo has an enviable potential as a tool of change that is based on real needs.

A wide range of issues surround bamboo craft development. There is skill that comes from traditional knowledge lying at the very core, of which craft and livelihood are direct resultants. There is bamboo itself, a natural resource, which translates into environment, where bamboo plays the role of a soil retainer. In many cases, bamboo forests and plantations lie at the base of the entire economy. There is the processing or transformation of the bamboo that constitutes technology – bamboo technology becomes small-scale industry, which again feeds the economic process. Thus, the development of bamboo craft becomes synonymous with the development of communities that depend upon it.

INTRODUCTION

Bamboo, when managed and used properly, can transform the lives of people who have come to depend on it since generations. In most tribal rural communities in the states of Arunachal Pradesh and Nagaland in India, bamboo has been synonymous with daily life. These states are the home of a total of 42 major tribes, which together form a colorful mosaic of diverse and distinct cultures, having developed independently in comparative geographical isolation. The last 40 years have seen a gradual opening out of this area to external influences, effecting some rapid changes in the otherwise traditional societies. The link between these tribes is their bamboo culture, reflecting a deep understanding and dependence on the material. From bridges and houses to basket and a ladle, the wide range of traditional applications serves to underline the importance of the material and people's familiarity with it. Though the traditional use of bamboo does not lack in any way, the present trend is to replace it almost entirely with non-local, energy-intensive and expensive materials like plastic, cement and metal.

Development in this region, as in other parts of the country, has tended towards superimposing external systems without a recognition or sensitivity towards the strengths of existing knowledge systems. This has brought a break in the intrinsic continuity that assures the survival and growth of these traditional systems and consequently a gradual loss of skills. It has stripped a community of its self-sufficient character, while creating demands that seem like aspirations. In both Nagaland and Arunachal Pradesh, an unhealthy dependence on the government machinery for employment has developed, culminating in the aspiration for a government job and consequently hampering the development of any kind of private enterprise.

With changing lifestyles and formal education hastening the gradual breakdown of old social orders, the lack of alternative livelihoods is sorely felt in an area bereft of industry and local markets. The low densities of population make local markets too small to fully sustain livelihoods. The lack of communication facilities resulting from difficult terrains has increased isolation and prevented proper connections to centers of trade and commerce in the country. This has led to high transportation costs for any goods coming in or going out of the area. In addition, there has been an unequal exchange of goods – a steady inflow of goods into the region with little or no outflow of goods. This situation has added to the loss of basic self-reliance of the local communities, which depend on outside supply even for essentials like rice and cotton yarn. It is odd that communities that have managed for centuries on their own suddenly do not seem to have any resources of their own, nothing to offer to the other parts of the country. Nothing could be more farther from the truth. This area is rich in natural resources with a biodiversity that is unequalled in the country and traditional systems of land and forest management that are an example anywhere. All these resources either lie dormant or are brutally exploited like in the case of timber trees, the frenzy of which prompted a ban by the Supreme Court on the export of any raw forest produce from the area.

In this scenario, where growing unemployment and alternative means of livelihood are primary concerns, where forest and land lie at the base of sustenance, development needs to be redefined. Economic self-reliance must lie at the core of the development process. The self-sufficiency of any rural community is possible only through the proper development of locally available resources like bamboo. It is then strange that the government has entirely ignored the role of bamboo while designing development programs for a community that has looked at bamboo for centuries for fulfilling its essential needs. There are few plentiful natural materials like bamboo, which can meet such a range of requirements so comfortably. If development is having decent livelihood opportunities for all and creating avenues for income generation, then what can be a better tool than bamboo craft in a region where practically every man is a bamboo artisan? If development includes proper housing and sanitation for all, then bamboo is the traditionally used and key material for housing in the area, as it has a tremendous potential as a building material and is accessible to all in the region. If development means the promotion of non-polluting industries that provide employment without adversely affecting the environment, then small-scale industries based on bamboo craft – such as bamboo ply, bamboo furniture and home based industry like dried and fermented bamboo shoots – are the answer. If development means improved communication facilities through road and bridge construction and infrastructure development, then too bamboo-reinforced structures and, at another level, bamboo craft have a role to play.

An added advantage is that people maintain their own bamboo stands, while having access to community-owned natural bamboo groves. Bamboo has always been viewed as a poor second to rattan in the region owing to its comparatively low durability. However, the availability and rapid growth of bamboo tips the scales in its favor, especially when durability can be greatly increased by preservative treatments. There is very little that bamboo cannot be used for. It is this versatile quality of bamboo, when developed, that makes it such a wonderful instrument for change. The economy of a region like Arunachal Pradesh or Nagaland can be largely sustained by bamboo craft development, which involves the development of both livelihood and housing-sanitation aspects in rural settlements.

STATUS OF BAMBOO CRAFT

Bamboo craft is not merely a creation of beautiful objects of art in the tribal rural economies. It is an intrinsic part of daily life and an activity that produces certain essential items of daily use, like baskets to carry firewood or for food and grain storage, vessels, containers, spoons, fishing and hunting traps, etc. Another aspect of bamboo craft finds expression in traditional housing and public engineering, including bridges, fencing and pathways where bamboo is a major ingredient along with timber. Here, bamboo craft can be extended to include housing and rural engineering, which traditionally employ basic craft skills. It is actually the weaving of bamboo splits or flattened bamboo culms, that finds application in various ways, causing this unique extension of the craft form into unorthodox areas.

Traditional bamboo craft products and skills upgrading

Fine detailing and intricate weaves are the hallmarks of traditional bamboo and rattan products of the region that comprises Arunachal Pradesh and Nagaland. There are more than a hundred types of baskets to be found here, each designed for a specific purpose: large conical forms with straps for carrying firewood or water (in bamboo tubes) or for marketing; lidded ones with legs for storage of personal articles; and cylindrical, rounded forms for carrying livestock (chickens and pigs). Open trays for drying grain, back-packs for hunting expeditions (Adi and Nishi tribes of Arunachal Pradesh), salt and bamboo shoot containers, winnowing trays and fans of varying forms, raised woven tables to support leaf plates (Konyak Naga), and knife covers are some of the numerous uses that bamboo is put to in the traditional tribal context. The Chang of Nagaland and the Monpa, Apatani and the Nocte tribes of Arunachal Pradesh make elegant containers and dishes from heat-flattened green bamboo using an ingenious traditional technique that could be put to great use in the development of new products. There are a variety of hats and caps, with helmet-like hardness, said to withstand strokes from sharp swords. There are woven containers that boast of not letting water pass through them. There are mats that reflect the coolness of the earth for both sleeping and drying grains.

The use of full bamboo is less developed in terms of finish though not application. It is fashioned into mugs, rice beer jugs, cooking and serving spoons, kitchen tongs and spoon holders. Bamboo splits form the basis of all hunting equipment from bow-and-arrow and crossbows to animal and fish traps. Bamboo is everywhere in the traditional home. All these skills and knowledge lay hidden, not shared with the rest of the country and disappearing a little each day. There is a dearth of proper and extensive information available on traditional crafts of the region. It is therefore necessary to study and document existing craft techniques and products to provide a base from where to begin the craft development process. There is also the need to identify the many existing traditional products that can be sold, mainly through active promotion, in markets outside the region. It is also important that new products developed to have a distinct flavor of the local culture: for example, the use of the Indian bison head as a decorative motif for mugs, jugs and boxes, which is a traditional symbol of wealth and prosperity among most tribes.

Since all craft products are always made for home use, not much attention is paid to treatment or finish. Smoking over the fireplace is one extra treatment given to the product, which acquires an attractive gloss through constant use and exposure to smoke, besides protecting it from fungus and insect attack. Other methods of treatment include immersion of bamboo poles in river water for up to 4 weeks to leach out sugars. These processes though effective, are extremely time-consuming. Proper seasoning and preservative treatments using Boucherie pressure treatment equipment and chemicals have tangible benefits and are much faster processes. There is a great need to set up such facilities and provide related training to local people at the village level to ensure the protection of bamboo products against splitting, fungus and borer attack. Moreover, craft, unlike its common perception in most other areas of India, has never been a means of livelihood in these communities and is still not commonly perceived as being able to provide one. Therefore, bringing craft out of the home and into the market economy, which would provide alternative livelihoods, implies a change in the way bamboo craft is perceived and produced by communities in the region. Local artisans need some basic training in managerial skills and to develop an entrepreneurial attitude towards craft production.

At present, the tools used for bamboo processing are the *dao* (a broad blade machete), a small knife and sometimes a hacksaw. There is no organized production system visible anywhere. Transforming a home-use product into a commercial high-value product catering to urban markets requires a revamping of the entire production process. This implies some basic improvements in craft tool technology and equipment, to the extent of a small degree of mechanization at the splitting, cutting and finishing stages. The use of splitting machines, sanders, simple wood turning lathes, electric drills and band saws can revolutionize craft production. Training in the use of this equipment and wood tools (like chisels, hammers, files, hand drills and blow torches) that can be applied to bamboo work will have to be a simultaneous process. Production technology is the starting point for all research and training inputs in the development of bamboo craft to increase production speed, reduce drudgery and improve the quality of the finished product. A great deal of practical experimentation with bamboo craft products is required in this process, since bamboo is a severely neglected material as far as appropriate tools and equipment for its processing are concerned.

Traditional designs need modification to suit urban tastes and lifestyles, while new design development is a prerequisite for maintaining a hold on the metro markets. New products must take into consideration local skills and their upgrading. Design and related training inputs, through interactions between designers and craft workers, become crucial at the next stage of craft development process. This is possible through a series of practical workshops that begin the process of sample product development and explore product potential through finishes and market testing. Proper sizing and standardization need to be addressed through the development of jigs and fixtures. Superior finishes in terms of surface coatings – like melamine, lacquer, spirit polish and varnish – and decorative techniques – like carving, engraving, inlay, branding, painting and vegetable dyeing – can put a high value on a craft product, bringing high returns and an alternative livelihood possibility for young people in the area. Marketing inputs, such as exposure through exhibitions and identification of sale outlets, will complete the process of craft product development. This fine-tuning will prepare local craft workers for urban and metro markets, since local markets are too small to completely support livelihoods. An understanding and exposure to market needs and trends, and proper costing of craft products through training workshops, are essential for the local artisans.

The combination of bamboo with other natural materials has a special relevance in the context of both these states. In bamboo product development, the weaving of bamboo splits in combination with cotton yarn for blinds, purses, lampshades, mats and with clay (ceramic) for other products are the two major areas of concentration. Both these materials and skills(weaving and hand made pottery) are traditional, complement bamboo and further enhance the quality and value of the craft product. The use of metals like steel, copper, brass, silver in conjunction with bamboo are other avenues for exploration, but require outside expertise and special training, besides being expensive investment.

The inputs in bamboo craft product development may be summarized as:

1. Seasoning and preservation treatment facilities for bamboo;
2. Documentation of traditional bamboo craft techniques and products;
3. Development and improvement of bamboo craft technology (tools and equipment);
4. Design inputs through interactions between designers and artisans;
5. Improvements and development in the production process of bamboo craft products with emphasis on quality control and finish;
6. Sample bamboo product development with standardization of design and size through the development of jigs and fixtures;
7. Training of bamboo artisans through workshops at every stage of the development process; and
8. Marketing study and identification of proper sale outlets.

Upgrading traditional bamboo housing techniques

All traditional houses and structures of the region have a large bamboo component and employ related construction techniques that need proper documentation before they disappear completely, a trend that is apparent in most parts of Nagaland. Little documentation exists on the tremendous variety of house types and interesting forms that are found in the different tribal communities of the region. The bamboo tile roofing with *Phyllostachys bambusoides*, locally called *bije*, employed by the Apatani tribe of Arunachal Pradesh lasts 10–12 years without treatment and is preferred as it is less susceptible to fire than a thatch roof. All Apatani people maintain bamboo groves of this species for housing purposes, as about 80% of the traditional Apatani house is bamboo-based. It is not unreasonable to assume that chemically treated tiles of the same bamboo could last at least 25 years. This theory must be put to field test. The Konyak, Yimchunger, Chang and Sema tribes of Nagaland use a variety of thick bamboo weaves and full bamboo for walling, which smoked continuously by the hearth fire, can last up to 20 years. The Ao Nagas make a beautiful floor of large woven bamboo culm planks, which form a durable, springy surface. Bamboo twine is used by the Adis of Arunachal Pradesh to tie palm thatch leaves for roofing. Other tribes of the area make beautifully crafted pivoted doors. Bamboo post and beam joinery, usually secured with cane twine, is common to all the tribes of the area. The compilation of all these techniques into a single documentation could provide an encyclopedia of information on bamboo architecture. Experiments on the structural uses of bamboo and information regarding the comparative strength and durability of the different species useful for construction, is crucial to the development of bamboo craft in architecture. The basis of housing programs in the region should be centered on furthering and strengthening traditional construction techniques based on bamboo by research and experimentation.

The very indispensability of bamboo naturally elevated it to an object and part of local faith and ritual worship. It remains the main element of tribal shrines and dedications to various spirit gods. Gates and arches for various local festivals are predominantly made of bamboo. These unusual forms and designs can be modified to suit the needs of interior decor in urban areas, providing yet another avenue for income generation.

The practice of community building, wherein all the members of a clan or village come together to build or renovate any house or structure in the settlement, is common to all the tribes of the region. When a house is to be repaired or built, the community is informed through an established pattern. Each house in the village knows that it must participate in this annual ritual, for its turn will come sooner or later. A system of fines controls errant members of the community. Forest and home gardens provide the necessary building materials – bamboo, rattan, timber and thatch. Materials are dried and prepared well in advance by the family. The processes have been perfected over time – through observing, learning, practicing and understanding – so that when the current formal school-based education takes youngsters away, the links begin to weaken. A house is taken down and rebuilt in just three days. At the end of the work, pigs are butchered and the village is feasted, an opportunity to bring community and clan together in dance and song. There seems to be no effort at all in the smooth functioning of this system – no apparent leader, no instruction. Its very perfection, so rooted in its context and environs, is what puts it in peril, as it cannot be replaced without destroying its very spirit.

The potential of traditional bamboo craft techniques in housing, interior decoration and public engineering works (like bridges, farm pathways and steps) is considerable. However, though this form of bamboo craft addresses an important aspect of developmental activity, neither the central nor the state government considers bamboo as a material for rural or urban housing development. This is surprising, considering its traditionally important role in this sector and its ready availability to people in the region. Instead of developing the use of bamboo, all efforts have been to replace it with energy-intensive, expensive and imported materials like brick, steel and concrete. High transport costs make the use of these materials impractical and inaccessible to the large majority of people. If and when these imported materials replace bamboo and rattan, new skills would be required for which old systems are ill prepared. As a result, traditional building practices would start disintegrating. It would also mean the loss of self-sufficiency and the beginning of dependence for a very basic need that could easily be met within the community and area.

A balance is needed whereby the use of imported materials like cement is limited to uses like foundations, water tanks, sanitation units and the parts of the house that are in direct contact with the ground or water. In this context, the training of local people in basic masonry, carpentry, electricity, plumbing and steel work is also required to cater to changing needs and to retain the local building practices, while providing added opportunities for local employment. The use of bamboo can be optimized and continued, but in a more effective manner, through measures such as the introduction of large-span bamboo trussing to eliminate the traditionally used central post. Heavy bamboo species like the *Bambusa balcooa*, which is commonly grown by most communities, could replace timber posts almost entirely in house construction. Seasoning and preservation facilities and related training programs would greatly enhance the durability of bamboo and change its perception as a building material of temporary quality. Rural housing programs need to focus on these issues, rather than on building small, culturally inappropriate houses of brick and concrete, since such houses are irrelevant in an area where homelessness is a non-issue.

Bamboo craft in traditional engineering

Bamboo suspension bridges are built by the Monpas of Arunachal Pradesh (Ranjan et al. 1986). These bridges have suspension cable ropes of twisted bamboo splits joined together to form the required bridge span. The one-meter wide walkway is made of flattened bamboo culms, while the sides are completed with similar bamboo planks woven together in an open weave that reduces wind loads on the structure. The cane (rattan) and bamboo suspension bridges built by the Adi Padam tribe of Arunachal Pradesh can span lengths of up to 380 m, as observed in Damro village on the Yamne River in Upper Siang District. The walkway is made of fine bamboo split pieces, while the main cables are primarily of twisted cane. This bridge is yet another example of traditional technology that has tremendous potential for development. The substitution of the traditionally used 21-cane cables by six steel cables, while retaining the basic bamboo-cane structure, was an intervention that saved cane and increased the durability of the bridge without making it expensive. This change was made 50 years ago, after the previous bridge was irreparably damaged in the 1950 earthquake. It has also managed to retain the community building tradition, which involves the annual repair of the bridge. These kinds of bridges were earlier used primarily for accessing fields, forest and connecting neighboring villages.

The government agencies, which have now taken over the job of bridge building from the community, usually build steel suspension bridges. Although these bridges have their own value in places, they are very expensive, take a long time to construct and require the import of a great deal of labor, materials and skills. Improvised traditional technology could still be used at a smaller scale for connecting villages and forest areas at a much lower cost and in much less time. A larger number of areas may be connected at a lower cost by this technology, which also has some potential in places of tourist interest.

Adi community of Arunachal Pradesh, and the Sema, Ao and Konyak tribes of Nagaland have traditionally used bamboo piping to transport water from springs to the center of their settlement. Halved bamboo culms, tilted slightly to ensure free flow, are supported on crossed bamboo or timber posts. The design of this crudely constructed system can be further improved by better joinery and some technology that can use whole bamboo (with nodes removed) and steel joints. With proper research inputs and innovation, promising local technologies can be identified and further developed for better use by the communities of the region. Bamboo in a variety of design and detail is extensively used as primary fencing material by all the tribes of the area. Treatment of bamboo would prevent the need for frequent replacement of fencing.

Although a great deal of research has been done in some parts of the country in the field of bamboo as a reinforcement material, there is little information or practice of these techniques in the North-East region. It is necessary that the results of research are disseminated to the grassroots level and duplication of effort avoided. This field has wide application in tanks for rainwater storage and sanitation units.

The inputs required in housing and engineering can be summed up as follows:

1. Seasoning and preservative treatment facilities with related training for local communities;
2. Documentation of current and traditional building practices;
3. Research, design and experimentation of construction and joinery techniques related to the structural use of bamboo through the building of experimental structures and forms;
4. Training in improved bamboo construction techniques;
5. Technological inputs to develop the use of improved tools and equipment for working with bamboo; and
6. Classification of local bamboo species in terms of structural and mechanical properties relating to their strength and durability.

Bamboo resource management

Raw material management lies at the core of the entire craft development process. Fortunately, bamboo is a part of every home garden and community forestland in most communities of the region. The requirement is to maintain and enlarge this plentiful resource by setting up bamboo seed banks and bamboo and rattan nurseries, and introducing some new bamboo species at the village level. Raw material depots can emphasize the importance of bamboo as a cash crop and ensure its regular supply for craftwork. Such depots for collection, distribution and preservative treatment of bamboo, along with nurseries and seed bank units, could be easily set up in each hamlet or settlement under the different government schemes for community service. If the settlement is large, there could be more than one depot of this kind.

THE BAMBOO CRAFT DEVELOPMENT MODEL

All the necessary inputs required for the development of bamboo craft can be catered to by a typically multidisciplinary center that takes care of all the aspects of the development process, from preservative treatment to marketing. Any kind of system that does not tackle all these aspects together, will fail to make significant impact. It is thus important to have a single system that caters to all these needs to ensure proper linkages at every stage.

Role and functioning of the development center

Such a center is visualized as a model for the development of the typically small hill tribe communities of Arunachal Pradesh and Nagaland. A series of such small independent centers at the village or district level, run with community participation, can provide opportunities to fully explore bamboo in an innovative frame. They will provide the inputs required in the entire development process through research, documentation, technology, design, training, production and marketing, and can become the prime agencies for people's development. Looking at the basic problems of communication and unequal exchange with other areas, the center will serve as a forum for interaction between people working in related fields across the country. Setting up teams of the local youth for various development programs can be taken up by the center. This will involve capacity building, skills upgrading and exposure and exchange workshops. The character of such a center must be flexible and constantly evolving to take care of changing needs and specific situations of a given area.

Non-government groups or grassroots development workers are best suited to take up the running of such centers. Tribal hill populations are typically small with single-tribe populations ranging from 1 500 to 100 000 people, and small grassroots-based catalytic agencies could work well in such areas. This concept is being applied to a center of this kind planned for Ziro area (Apatani community) in the Lower Subansiri District of Arunachal Pradesh. The center is to be run by a small group of professional grassroots-level workers.

Self-financing nature

This kind of center will require financial support for the initial period to set up basic infrastructure and to get the major programs off the mark. Thereafter, it should become an independent and primarily self-supported organization, providing training and research-design services to the community and building up independent small-scale, craft-based enterprises at the village level. It will generate funds, on a no-profit no-loss basis, for its running and maintenance from the services it provides.

Work areas

A summary of the work areas and components of the center is given below.

Research inputs based on practical experimentation

1. Exploration and upgrading of bamboo utilization in traditional housing;
2. Use of bamboo in rural engineering works such as bridges, culverts and pathways;
3. Study of tools, jigs and fixtures for bamboo craftwork; and
4. Combination of bamboo with other materials that optimize the use of bamboo.

Documentation

1. Existing craft techniques and products; and
2. Traditional building practices, construction techniques and shelter types.

Technological inputs in bamboo craft

1. Developing appropriate tools, fixtures and equipment for bamboo;
2. Providing information on existing technologies;
3. Providing information and service on the preservation and treatment methods for bamboo; and
4. Developing the use of bamboo in small-scale industries that involves large craft inputs.

Training of bamboo artisans for skill upgrading

1. Improved production processes related to tools and equipment;
2. Design, jointing and finishing techniques and quality control;
3. Managerial skills and marketing strategy, including the understanding of urban and world markets and costing of craft products; and
4. Seasoning and preservative treatment of bamboo.

Marketing inputs for bamboo products

1. Marketing links and identifying proper sale outlets for bamboo artisans; and
2. A central marketing agency at the village level for different craft groups.

Bamboo product design development

1. Designer-craft worker interactions;
2. Sample production; and
3. Market testing.

Bamboo resource management

1. Introduction of new priority species suitable for craft and housing purposes;
2. Creating bamboo seed banks at the village level;

3. Setting up bamboo nurseries and raw material depots in villages; and
4. Classification of different local species according to their characteristics and properties related to their use for a given purpose, and determining their susceptibility to insects and fungal attack, splitting tendency and strength.

The different components of the center would include:

1. Building craft research and design workshop;
2. Training center;
3. Bamboo and rattan seasoning and treatment section;
4. Low-cost sanitation section;
5. Natural resource management section, including a bamboo and rattan nursery and seed bank;
6. Craft product development section; and
7. Sales and marketing section.

CONCLUSION

This model of development is not exclusively based on bamboo, but has coordinated research and development of bamboo craft at its core as a means of self-reliance of local communities. It has been illustrated how this can be achieved through a series of small interventions or efforts that build on existing strengths and do not require huge investments and imported materials. The term 'sustainable development', which has now become jargon but is still operative, thus becomes synonymous with the development of bamboo craft in this region. The development of bamboo craft is an attitude to development that bases it on earth and not in the sky.

REFERENCES

Ranjan, M.P.; Iyer, N.; Pandya, G. 1986. Bamboo and cane crafts of North-East India. Development Commissioner of Handicrafts, New Delhi, India. pp. 40–41.

Bamboo for Sustainable Development, pp. 205–219
Eds. A. Kumar, I.V. Ramanuja Rao and Ch. Sastry

Industrialized handicraft (In-Hand): The Abra experience

CARMELITA BERSALONA
In-Hand Abra Foundation, Abra, The Philippines

Abstract—Industrialized handicraft (In-Hand) is a development process based on the reality that people needs to actualize themselves in their lifetime by bringing out their highest potentials through work.

There are basically two kinds of labor in the Philippines: (1) Handicraft – handcrafted production based on intuition, imagination, and creative skills; and (2) Industrial – machine-oriented, mass production based on repetitive tasks and undergoing technical training. In-Hand is a synthesis of the two.

The object of In-Hand is self-sufficiency through production networks with the capacity to efficiently produce the basic requirements of food, shelter, clothing, health and education. This is achieved through the continuous development of raw materials and transfer of the appropriate technology and mechanized processes, all geared to the development of communities and responsive to their needs. Furthermore, in the process of earning adequate wages, participating in productivity to increase the purchasing power of the community members to buy the products they produce ("prosumerism"), with parallel efforts undertaken to research and develop raw materials, new products and markets to optimize labor, machinery, technology and material resources through maximized value addition.

The product of In-Hand in the Philippines is creative people, who are able to manipulate their tools towards achieving their wants.

In the process of industrialization, it is important that people control the process and actualize their dreams and their ability to control machines instead of becoming another machine in the chain of industrialization. This process would take people from the subsistence level of handicraft work to being average wage earners through increased productivity, owners of the means to production, able managers who translate technology into production, teachers for the transfer of technology, and designers of appropriate technology and products for increased benefits.

In-Hand uses three strategies to industrialization: (1) Networking of resources; (2) Adherence to a productivity formula; and (3) Instituting continuous innovations into the systems. The effect of these enables the community to invest in appropriate tools, purchase appropriate production technology, utilize the appropriate expertise, and build up the required capital.

In the Philippines, the In-Hand Abra Foundation's experience of setting up a plyboo (plybamboo) factory demonstrates that the concept of In-Hand hastens the development of the Filipino rural worker and the community when it is founded on a material such as bamboo, which is a part of their way of life. Parallel developments of the raw material, technology, support services, product range and market happen simultaneously.

INTRODUCTION

This paper outlines the experience, beginning 1986, of setting up a Plyboo (bamboo-ply board, or plybamboo) factory in the depressed and landlocked province of Abra situated on Luzon Island in the Philippines.

This factory was set up after a process of consultation, community organization and training based on an approach dubbed "Industrialized-Handicraft (In-Hand)". Medilen Singh developed this approach in the 1980s based on her experiences in the setting up and running of a wooden toy factory.

The following sections discuss the In-Hand philosophy and its strategy for development. Detailed descriptions of In-Hand's operations and the Plyboo Project are included as Annexes.

THE IN-HAND PHILOSOPHY

Industrialized handicraft (In-Hand) is a development process based on the reality that people needs to actualize themselves in their lifetime through work.

The Philippines has essentially two types of labor: Handicraft – handcrafted production based on intuition, imagination, creative skills, a feel for materials and learning on-the-job; and Industrial – machine-oriented mass production based on routine, repetitive tasks and learning through a mix of formal subject learning and vocational/technical training. In-Hand is a synthesis of the two.

In-Hand is a concept to uplift the urban and rural poor into being a strong, educated middle class through the development of communities into viable small and medium-scale production units. These communities would then become self-managed, networked industries with the capacity to produce basic requirements – food, clothing, shelter, health, and education – thus developing self-reliance and, at the same time, owning, as individuals and community organizations, the accumulated profits as seed capital for other productive investments.

While other industrialization processes are focused on industry, In-Hand is focused on the workers to draw out their highest potential through work. The process takes them from subsistence level (knowing the materials of the craft), to average wage earners through increased productivity, to able middle managers who ensure systems implementation, and to designers who reconfigure existing patterns and systems for increased benefits. Thus, the product of In-Hand is creative people able to manipulate tools towards achieving their wants.

The In-Hand concept involves a strategy to increase and upgrade the production capability of communities to benefit the participating individuals/families of the community through the introduction of effective systems and procedures, processes and appropriate technology. This strategy introduces the production/assembly line procedure, wherein manufacturing resources are organized in a logical and efficient manner to promote smooth flow of operations and increased productivity. It calls for the introduction of innovative, appropriate and cost-effective tools and jigs, machinery and equipment, and the building up of raw material resources through research and development.

In-Hand encourages its communities to link up with raw material suppliers/farmers, as well as with other communities producing similar products to form a self-sustaining network. Part of this network could also be a community specializing in R&D on either raw materials or fabrication of appropriate machinery, or some function integral to the whole. In this manner, market forces of demand and supply could be harnessed, such that the networked communities, in their endeavor to cooperate with and maximize productivity through each other, would create steady demand for raw materials. This could result in price stability, which in turn would promote competitive costing of finished products while maintaining quality levels and reasonable earnings.

In-Hand uses three approaches in order to help communities make their development plans without compromising their existing resources (see Annex A). These are:

1. Networking resources and expertise;
2. Adherence to a productivity formula; and
3. Instituting continuous innovation in the systems.

In time, these three strategies will foster sustainable development as they enable the individual and community to:

1. Invest in appropriate tools to increase productivity;
2. Purchase the appropriate production technology required to convert raw materials into value-added finished goods, while sustaining raw material replenishment;
3. Tap the right expertise required to design a product range from its raw material, allowing the feasibility of a long-range plan; and
4. Build the required capital necessary for investment in growth opportunities – made possible by research, development, design and marketing activities – thus expanding opportunities for the community on a trade-not-aid policy.

THE ABRA EXPERIENCE

The setting

Abra is a landlocked mountainous province located in the northern part of the largest island of the Philippines – Luzon. Though the province is financially poor, it is rich in raw materials that could support rural industry development, as forestlands constitute roughly three-fourths of the total land area of 397 800 ha.

Although several bamboo species grow in the province, *Schizostachyum lumampao* (buho) and *Gigantochloa levis* (bolo) are the most abundant. Other raw materials include tiger grass, gravel, *nito* and rattan. There had been attempts at mining, livestock and agricultural farms in the province, but they failed, ostensibly because of the presence of an insurgency movement (New People's Army) and poor accessibility to road and communication facilities.

Average annual income in 1985 was 23 175 pesos per household. There were 208 registered cottage industries, of which about 59 per cent were engaged in bamboo working. Other popular industries are piggery, poultry, loom weaving and woodcraft.

In 1986, there were no small, medium or large industries in the Province. A pulp mill was erected in 1973 to produce long-fiber Kraft pulp out of pinewood from the uplands. The cutting down of the trees on ancestral lands without prior consultation and remuneration caused the upland dwellers (the Tinggians) to revolt, forcing the factory to cease operations and finally close in 1986 when its license was cancelled. The lack of alternative employment opportunities in the province led to the migration of mill's managerial and technical experts.

There are two main ethno-linguistic groups in Abra. Town dwellers in the valley are mostly Ilocanos, who speak the language of the lowland coastal areas. The other group is the Tinggians, who inhabit the upland areas that form part of the great backbone of Northern Luzon: the Cordillera Mountains.

The beginning: organizing for development

The Philippine People Power Revolution of 1986 was one of the impetuses of the Abra Experience. Using the local chapter of Namfrel (a national election watch group) as a core group, the author and some like-minded people formed the Concerned Citizens of Abra. After a series of meetings, a decision to pursue a peaceful economic revolution was arrived at.

Casting around for alternative development models, we came in contact with Medilen Singh of Integrated Development Systems, a proponent of In-Hand approach. We invited Ms. Singh Abra Province in December 1987 for a seminar on Industrialized-Handicraft. Representatives of government and non-government organizations involved in development attended this 2-day orientation seminar. The seminar was an eye-opener as to the immense possibilities of human and natural resources, and how the quality of lives could be improved by pursuing development through industrialization.

There were several conditions unique to Abra, arising from characteristics of the cultures of its peoples, which needed to be addressed before an initiative could be launched. For example, with regards the upland Tinggians, there were the following aspects to consider:

1. The *lapat* system of forest conservation;
2. The Council of Elders governance system;
3. The pulp mill experience; and
4. The clamor for people's participation in decision-making processes.

A series of meetings and consultations involving the various groups in Abra were held, the highlight of these orientation seminars being the Abra Highland

Elders and Leaders Conference held in February 1988. There were 120 participants representing the Church, the Government, the Council of Elders, People Voluntary Organizations, the business sector, and the leaders from the highland municipalities of Abra. The group unanimously expressed their acceptance and support of the "Investment in Man Program", using In-Hand as the strategy to achieve real autonomy in Abra.

An 8-week Training Workshop for the Core Group, selected by the different community leaders, was held in Abra from 3 April to 27 May 1988, with 64 participants representing 9 Highland and 6 Lowland municipalities. It was the first time that leaders and elders from the highlands and lowlands came together to talk about their differences and dreams (some participants never before had seen the sea or been to the mountains). The success of the seminar cannot be overemphasized, as it brought hope, mutual understanding and a bonding that did not exist before.

In keeping with the perspectives of the In-Hand philosophy, the first four weeks of the workshop were focused on awareness of:

1. Self – through drawing, the participants were made aware of how the left and right sides of the brain worked;
2. Community – the two different cultures of Tinggians and Ilocanos;
3. Past and Present Eras of the World – in terms of agriculture, industrial, information; and
4. Industrialized Handicraft strategies of development.

The fifth week was dedicated to developing a vision, starting with the production of a scale model of each municipality represented. The Vision for Abra put together is as follows:

> "A self-actualized community of individuals – fully aware of their self-worth, dedicated to the sincere service for the preservation of all life, actively participating in the process of self-governance, productively sharing in the fruits and responsibility of self-sufficiency through the effective control and management of resources for its economic benefit through the understanding, knowledge and use of appropriate and indigenous technology while maintaining ecological balance."

The last three weeks focused on mapping out a three-year development plan, and learning about baseline surveys and environmental scanning.

At the end of the Workshop, elections were held to form the network organization in Abra, touching the provincial, district and municipal levels to ensure communication and information dissemination. Thus, In-Hand Abra came into being as an organization. It ventured to put together the required environmental scan, baseline study and community orientation in preparation for selecting a seed project. From the results, Intech Dev. Systems, the support organization of In-Hand, was asked to look into a possible product made of *Schizostachyum lumampao* (buho), abundant in the upland (in three municipalities alone, 15 000 ha of dense stands). The study of Intech Dev. Systems on the production of Plyboo – a bamboo-ply board – was presented to In-Hand Abra on March 1989.

The Plyboo Project

Plyboo is a trade name of a three-layer panel board measuring 3/8 inch × 4 ft × 8 ft, made of crushed *S. lumampao* planks, comparable to the strength of a 3/4-inch plywood board. It is water-resistant and can be used for interior and exterior walls and ceiling. Its purpose was to substitute ordinary plywood for the low-cost housing.

The project was a joint effort with Intech Dev. Systems with soft loans from government. It hoped to link In-Hand Abra to the housing and construction industry of the Philippines, which was booming at that time and shortage of construction materials was causing a rapid increase in housing costs, making it increasingly difficult to construct low-cost housing units at affordable levels for 60% of the Filipinos.

The Project Plan had three major components:

1. The raw material management, harvest and provision of access roads;
2. The processing of raw materials and its waste into finished products;
3. A workforce capability development program and infrastructure support.

The workforce development program addressed three levels of project participants:

1. Those who planted and harvested the raw material (60 project participants, 5 of whom were women and 90% were either elementary graduates or with no schooling at all);
2. Those who run the machines (70 project participants, 60% were technical and vocational graduates, 40% were high-school graduates); and
3. Those who translated technology into production (15 project participants, 40% were professionals and 60% technical graduates).

Of the above only 5% had work experience. Not one had the experience of ever working in a factory. There was no sense of time in terms of seconds and minutes, particularly in the harvest areas.

The training

Owing to the size of its domestic market alone, the Plyboo Project was seen as a long-term project that could be used to train future middle managers and potential entrepreneurs of Abra. All successful graduates of the hands-on training could in the future spin-off to handle downstream enterprises, such as cabinet and furniture manufacturing. At the same time, the replanting and management of the bamboo plantations would eventually lead the communities involved into learning the skills of seedling care and management of their forest reserves, and lead to opening of roads in the uplands.

The training of workers by the community managers, supervisors and facilitators consisted of three modules:

1. The first module comprised formal lectures encompassing the In-Hand vision – material units, time units, standards, time and motion, process and cost sheets, the productivity sharing formula and a budget control system. Serving as the actual training ground were:

a. The road construction, totaling 17 km, in the three harvest areas;
b. Development of 1 ha of land for the factory site; and
c. The construction of 1 200 m^2 factory building.

2. The second was characterized by hands-on training on basic machine operation conducted at a trade school. Machine shop work and machine/tool maintenance and set-up was further taught when the machines were being fabricated and improved. The trainees were able to take apart the machines and rebuild them to improve performance during testing.

It was interesting to note that while testing the performance of the machines, the participants were amazed at how the crushing of an 8-ft pole would take only about 10 seconds while by hand it took them 10 minutes. This drummed up interest in the design and manipulation of tools, jigs and equipment.

3. The third module comprised training the workers on each production process as per product specifications in the harvest areas and factory while conducting trial production. This phase saw the workers actualizing each production process themselves, with the supervisors, maintenance and machine set-up crew carefully monitoring execution time and motion, giving on-the-spot instructions in the event of errors.

As most of the training was on the job, the participants were given the equivalent apprenticeship allowance of 50 pesos per day.

Manufacturing issues

1. Electricity supply was a major issue. Aside from its supply being unreliable, and with much fluctuation, the cost of electricity in Abra was double that of Metro Manila. Technical expertise to service 3-phase transformers was not available in the province and had to be imported from neighboring provinces. A generator was finally purchased to avert the breakdown of motors.

2. Communication system within the province was non-existent. The telephone system was confined to the capital, Bangued, and other means of communication was through the province's radio station or by hand.

3. Roads to the upland municipalities were opened only during the dry season. After the earthquake, if the communities did not take it upon themselves to open the roads themselves, it took government several years to restore them to their former state. River transport and hiking became the norm.

4. Industry services, such as machine shop services and electric motor rewinding, were unavailable or unreliable. These had to be sent out to other provinces for service.

5. Functional literacy was low, requiring the training modules to be altered to more basic training.

6. Because of the density of the bamboo forests, a large percentage of the poles were crooked. This was later solved with the development of another product by stripping the poles into slivers first before pressing them together.

Major hindrances

1. The plant was operational in June 1990 doing trial production when the earthquake struck, destroying the little infrastructure there was in the province and cutting-off all raw material supply to the factory.

2. The earthquake was accompanied by another phenomenon – the flowering of the bamboo, which normally occurs every 50 years. Though the salvaged poles were excellent material for bamboo ply-boards if cut within a year from flowering, there was no way of transporting them down to the factory area without breakage owing to the steep river-ways.

Efforts were put to open roads in areas where the flowering had not started. However, strong winds of the super typhoon Trining in October that year also destroyed the main factory building and some of the machinery.

Consequently, because the factory participants had no means of livelihood while the project was being re-planned, participation in the project diminished substantially as those who were given technical training started their own businesses utilizing the skills they had learned or went abroad.

In 1992, a local architect designed a low-cost housing unit, with walls made entirely of Plyboo. Though it did drum up interest causing President Aquino to send the Secretary to look into the needs of the project's infrastructure, the flowering had become extensive causing government and creditors to proceed with caution. Efforts were then put into the creation of plantations of other species in the lowlands.

The project took a turn in character from being a development project to a private endeavor of applied research by In-Hand Abra Foundation and Intech Dev. Systems. Beginning 1993, the project's efforts were focused on processing new product lines utilizing lowland bamboo accessible by roads. This was geared to high-end products for the local and international markets. However, owing to the temporary roof over the factory structure and poor transport conditions during the rainy season, the project operated only during the dry season in a small way.

This brought up three marketing issues:

1. Orders had to be served within 30 days of order. To do this, bamboo had to be stockpiled during the dry season, creating problems of space and inventory costs.

2. Being a new material for offices and residences, clients wanted a 10-year warranty on Plyboo. Extra expenses for waterproofing cement floors had to be made, as moisture would seep in during the rainy season and cause mold to grow on the bamboo.
3. Black marks would sometimes occur on the floor owing to heavy metal chairs used. It was found that the lowland bamboo species was not consistent in quality as it was not cultivated in plantations.

Because of these problems, the participants were eventually set up to do on their own house-ware items for the local market, and the project concentrated on propagating giant bamboo in lowland plantations for future production.

Still, two major concerns affect the future of the Plyboo Project in Abra:

1. The lack of knowledge of government agencies on the flowering of bamboo caused the project creditors to stop supporting the project. Something needs to be done to avert this in the future.
2. Earthquakes, typhoons and other natural calamities in the region can wipe out projects completely even before it gets a chance to prove itself. Hence, ways have to be found whereby a project can insure itself against these calamities in the future.

GAINS OF THE PROJECT

1. A low-cost housing unit, utilizing Plyboo for its wall panels and lahar (volcanic ash) for its roof and floor, was designed by the renowned Filipino architect, Francisco Mañosa, who presented the model to President Aquino in 1992. The house was called the Pinatubo House. The cost of the 36 m^2 house was estimated at 150 000 pesos (about US$ 6 000 then) and seemed a possible answer to the country's housing and rural industrialization problems.

2. *S. lumampao*, a low-value bamboo species thought of as a weed, was given a value. How this simple grass, with no value before, could be used as a vehicle for development has made people conscious of giving value to the vast resources they possess. The technology of harvesting and curing *S. lumampao* has been demonstrated and disseminated. Since then, it has been used for high-quality products. The importance of plantation management and responsibility of maintaining roads has been emphasized. The roads built to access the bamboo are now being used by the people for marketing mangoes.

3. The project's general impact on the beneficiaries were:

a. Awareness of time, units, standards, processes, production systems, and costs and its control;
b. The use of tools, jigs and equipment for increased productivity; and
c. A better quality of life resulting from increased knowledge, allowing the workers to take better jobs elsewhere or start their own enterprises.

4. The product evolved from an ordinary crushed bamboo panel board to a high-quality interior panel (rift grain). Plyboo led to further technology development with zero waste, showing the backward and foreword linkages of industry, as high-quality bamboo paper and house-ware were developed, involving the bamboo weavers. These are sold to high-end local shops and traders for the local and export market.

Also developed were furniture components (such as table tops and cabinet doors), furniture items (such as bookcases and consul tables), floorboards and interior wall boards. These were supplied to local architects who introduced them in resorts and residences.

5. Plyboo led to the acceptability of bamboo as a material to be used in the homes of the rich, changing its reputation from that of a poor man's timber to a high-quality material. This led to other provinces taking interest in their bamboo resource. There is now an organization for bamboo growers and manufacturers.

6. Having developed uses for the different bamboo species found in the province and neighboring provinces, In-Hand Abra is now more aware of which species to propagate for the benefit of the growers and manufacturers.

CONCLUSION

The Abra Plyboo Project was encumbered by many problems. These problems should not detract from the fact that the goals of the project, however modest, were achieved. People were able to realize what they can do by themselves, for themselves and among themselves through a cycle of forward and backward linkages. Its effect in the community can be perceived as a microcosm of what meaningful development can be.

If industrialization is founded on something familiar to the rural worker, such as a craft or skill, then the process of industrialization is without resistance, as its benefits are immediately felt.

In the Philippines, a rural worker's skill is usually related to bamboo, as the plant is found all over the country. It is a material that is part of Philippine life: in food, shelter, lantern, furniture, transport, utensils and livelihood.

While the objective of In-Hand Abra is self-reliance, with production to provide income for a greater number of people, it does not believe in perpetuating labor-intensive processes. Its mission and ideology call for the continuous development and transfer of the appropriate technology and mechanized processes geared to constantly improve product quality and to increase productivity so they can produce the products they need. Further, in the process, earn adequate wages and share in the productivity to increase the people's purchasing power to buy the products they produce: "prosumerism". Likewise, parallel efforts will be undertaken to research and develop raw materials, new products and markets to optimize labor, machinery, technology and material resources through maximized value addition.

Thus in the Philippines, the development of the rural worker and the development of bamboo – plantations, technology and markets – go hand in hand.

ANNEX A: OPERATIONAL DETAILS OF IN-HAND ABRA

The strategy

In-Hand's strategy is based on the reality today that in most communities in the country, especially in the rural areas, there are no specific advantages left-mostly due to its past mismanagement of resources – save the abundance of unskilled labor.

The In-Hand concept uses three approaches in order that these communities are able to make their development plans without compromise to their existing resources/capabilities. These are:

1. *Networking Resources* among communities, organizations, corporations and individuals with specific advantage in each area of concern to work together on a market-led project benefiting each as well as the "whole", while providing the individual in the community with the opportunity to build his/her capability, work for a living, and accumulate experience and profits to eventually manage and control the same.

The following *9 Key Functions* of a straight-line technology are the areas of resource networks.

1. Raw material	– supply
2. Production facilities	– economies of scale
3. Product design	– function and form
4. Production technology	– intermediate, appropriate
5. Product range and variety	– spread for R&D expenses
6. Application engineering	– lab applied (VD& VE)
7. Sales force	– quality and quantity
8. Distribution	– market share
9. Services	– industry support

2. *Adherence to a Productivity Formula*, which assures that each function profits commensurate to its inputs, thus motivating each to work as a team rather than individually, balancing the distribution of critical resources needed in the industry.

The norm	In-Hand's distribution of critical resources
40% labor and material	40% skilled and un-skilled labor and raw materials
10% overhead	10% factory overhead
30% administrative	5% administrative cost
	5% tools, jigs, and equipment
	10% design, marketing, R&D, technical assistance
20% profits	10% intracapital (capital build-up for new ventures)
	6% labor share
	6% management share
	6% capital share
	2% interest cost (until loan is paid) & community development fund
100% Sales	100% Sales

3. *Instituting a Continuous Innovation System*

3.1 Use of a simple Budget Control System that brings detailed information to the level of the worker, enabling to see the whole and parts of the cost and processes of production.

3.2 With actual experience on the floor and guided by cost allocations in the budget, the worker is able to introduce innovation in the production systems and processes to increase output.

3.3 A Production Reporting Management Tool taht cuts across production, maintenance, accounting, materials control and personnel. In seeing the whole, the workers are able to coordinate with each other and manage themselves. Consequently, administrative costs are reduced from 30% to 5% of sales, allowing the remainder to be allocated to:

- Tools, equipment and machinery
- Research and Development
- Design and marketing
- Intracapital fund (capital build-up)

In time, these three strategies will foster the following results, enabling sustainable development as they enable the individual and community to:

- Invest in appropriate tools thus releasing them from hard labor. Energies instead are channelled to acquiring capabilities in problem solving, thinking, planning, designing and managing while upgrading technical awareness and skills necessary to industry.
- Purchase the appropriate production technology required to manufacture their indigenous raw material into value-added finished goods, while sustaining raw material replenishment.
- Tap the right expertise required to design a product range from its raw material allowing the feasibility of a long-range plan.

- Build the required capital necessary for investment in growth opportunities that R&D, design and marketing make possible, thus expanding opportunities for the community, on a trade-not-aid policy.

The formula's strength is in the ratio of fixed cost to variable costs, which allows profits to be divided equally in three among capital, labor and management. This allows management and labor's inputs and results to be turned into investments in the venture, giving the laborers the opportunity for added income through investment without touching their wages and, at the same time, become owners of the means of production.

The phases

Phase 1 – Orientation seminar in the concept and strategy of In-Hand
Phase 2 – An Intensive Core Group Training to map out a 3-year development plan
Phase 3 – Formation of a Network Organization in the province
Phase 4 – Environmental scan and baseline surveys
Phase 5 – Preparation for seed project
  a. Development of a marketable product using indigenous materials that can sustain the program.
  b. Feasibility study
  c. Financing
Phase 6 – Organization and training for the seed project
Phase 7 – The seed project
Phase 8 – Higher training for the graduates of the seed project to prepare them in managing subsequent ventures.
Phase 9 – Set up of basic industries
Phase 10 – Set up of secondary industries
Phase 11 – Spin-off tertiary industries (small entrepreneurial and livelihood projects)
Phase 12 – Self-sufficiency

The object is to make people realize what they can do by themselves; for themselves and among themselves through a cycle of forward and backward linkages.

ANNEX B: THE SEED PROJECT – PLYBOO PANEL BOARDS

The purpose of the seed project was to actualize the "Investment in Man Program" utilizing the In-Hand strategies. Further, it would provide employment and technical know-how to some 500 unemployed and under-employed inhabitants of Abra – that is, lift them out of the poverty level – thus forming a nucleus to spearhead development in the province. The projections given below would illustrate how this seed project would achieve its goals.

Phase A – Bamboo harvesting

For a loan of 1.1 million pesos loan from the National Livelihood Support Fund, the following would be generated in three years:

1. Equipment worth 619 000 pesos to be owned by the group involved in this phase;
2. The full loan of 1.1million pesos plus interests to be fully paid back after 16 months;
3. Wages and salaries amounting to 3 548 736 pesos to be paid out over a period of three years.
4. An Intracapital Fund of 1 254 240 pesos is generated to be placed as capital in the next venture, which in turn will generate more jobs, create wealth and be owned by the people themselves.
5. The Labor Profit Share Fund has accumulated 752 544 pesos to be put into whatever community improvements or projects the people want, generating more jobs and creating wealth.
6. The Capital Share Fund has accumulated 752 544 that shall be put back into replanting and management of bamboo plantations, generating additional jobs and helping in the tree planting scheme of the Nation.

Phase B – Plyboo factory

For the consideration of 3.96 million pesos loan from the NLSF, the following is generated after three years:

1. Machines worth 2 215 000 pesos plus a second-hand truck (budgeted at 179 700 pesos) and a temporary warehouse (budgeted at 759 000 pesos) are owned by the group involved in this phase, thereby owning the means of production.
2. The full loan of 3 960 000 pesos plus interests at 6% are fully paid back after 18 months.
3. Wages and salaries amounting to 5 834 400 pesos are paid out over a period of three years.
4. An Intracapital Fund of 2 276 352 pesos is generated from the 17th month (after all loans are paid back) to be placed as capital for the next venture, which in turn will generate more jobs and create more wealth and be owned by the people.
5. The Labor Profit Share Fund has accumulated 1 365 811 pesos to be put into community improvements or other projects, which will generate more jobs, creating more wealth.
6. The Capital Share Fund (which is owned, after the 17th month, by all the persons involved) has accumulated 1 365 811 pesos, which can be used either to purchase land and facilities or to invest into some other projects to begin the cycle again.

Project Development Process

1. Registration papers, originally submitted with the SEC, are amended, turning In-Hand Abra Foundation into In-Hand Abra Provincial Cooperative Federation;
2. Loan outstanding with the NLSF is repaid. Copies of loan repayment documents are provided to project participants after each payment is done. Loan repayment represents the actual and gradual acquisition of ownership of the project by all its participants/beneficiaries.
3. Within one week after the NLSF loan is fully repaid, the Equity Share Certificate is issued to each project participant/beneficiary with the percentage of ownership of the project in ratio and proportion according to the Equity Share Formula.
4. The project is registered anew as a distinct cooperative and a member of the In Hand Abra Provincial Cooperative Federation. The original participant /beneficiaries, who now individually and collectively own the enterprise, become its incorporators, with the percentage of ownership per individual stipulated in the Equity Share Certificate.

Section 2:
Bamboo Propagation and Management

Bamboo for Sustainable Development, pp. 223–233
Eds. A. Kumar, I.V. Ramanuja Rao and Ch. Sastry

Culm characteristics of two bamboos in relation to age, height and site

ABD. LATIF MOHMOD
Forest Research Institute Malaysia (FRIM), Kuala Lumpur, Malaysia
WALTER LIESE
Chair for Wood Biology, Hamburg University, Hamburg, Germany

Abstract—Culms of 1-, 2-, 3- and 4-year-old *Bambusa vulgaris* and *Gigantochloa scortechinii* grown in four different localities in Peninsular Malaysia were assessed. The above-ground biomass (culm, branches and leaves weight), culm length, number of internode, internodal length, internodal diameter and culm wall thickness were correlated with culm age, height and site. All the culm characteristics (except culm wall thickness) varied significantly with site. Both the bamboo species from Sik area, with the lowest annual rainfall, exhibited the lowest above-ground biomass and internodal number, but possessed the longest internode. Bamboos from Sungai Lembing, the area with highest rainfall in Malaysia, had the highest branch/leaf weight but the shortest internode. The culm wall thickness of bamboo was not significantly affected by topographical factors, implying that site specificity has no role when selecting bamboo for a product based on culm thickness. With regards to age, the general trend was that the above-ground biomass and culm wall thickness tend to increase with maturity. With the exception of internodal length, the other culm characteristics displayed significant negative correlation to height increment from the base to the top.

INTRODUCTION

The importance of knowing and assessing the culm characteristics of bamboo is related to the end-use of the material. For instance, culm wall thickness, which is correlated to the amount of vascular bundles, influences the mechanical properties that characterize its final intended use (Abd Latif and Mohd. Zin 1992). The above-ground biomass, on the other hand, is related to the area where the bamboo grows and is thus associated with its adaptability to survive stress conditions. It also helps to define the rationale of any silvicultural or plant improvement practice in planting or managing bamboo at different localities. In view of this, this study was aimed at examining the trends of variation in culm characteristics – such as culm height, culm wall thickness, and internodal length and diameter – of the two most commonly found bamboo species in Peninsular Malaysia – *Bambusa vulgaris* Schrad. ex. Wendl and *Gigantochloa scortechinii* Gamble – at four different ages and from four localities with different climatic conditions. The interrelationships of age and height on culm characteristics were also evaluated to serve as a guide to their potential application.

MATERIALS AND METHODS

Source of materials

Samples of the two bamboo species were harvested from wild clumps growing in four research plots established in Kedah, Selangor, Johor and Pahang (Fig. 1). The growth of the one- to four-year-old samples was monitored from the sprouting stage. Brief descriptions of the study sites are given in Table 1.

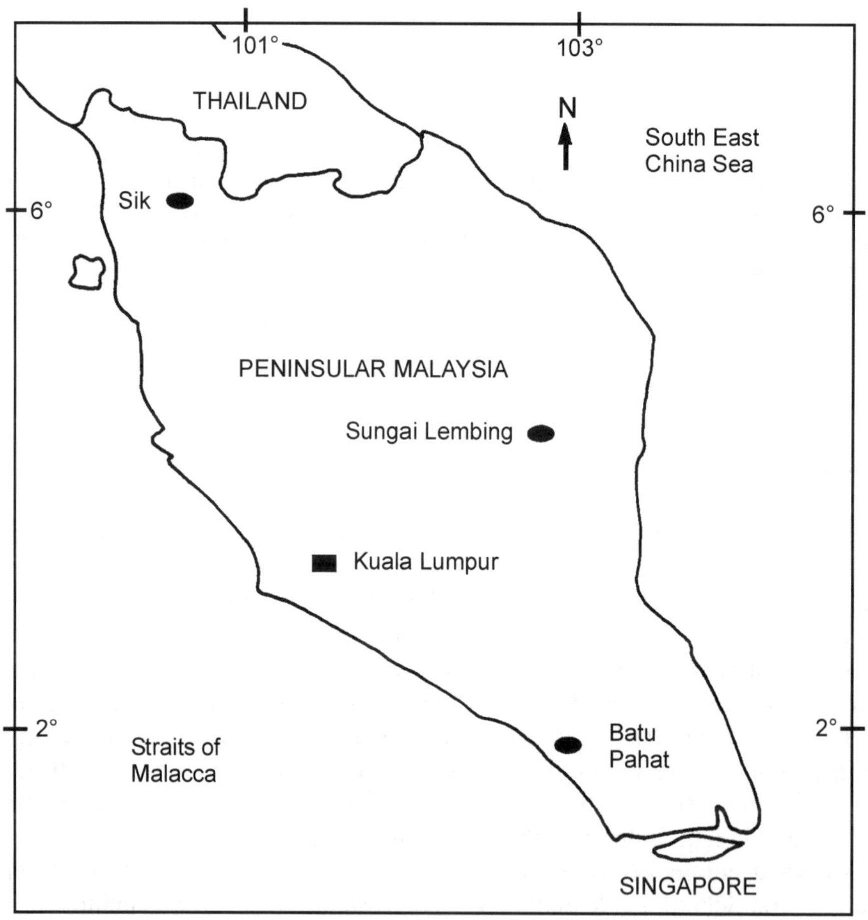

Figure 1. Location of study sites

Table 1.
Description of study sites

Location	Altitude (m)	Soil type	Climate [5]		
			Rainfall (mm)	Temperature (°C)	Rel. humidity (%)
Sungai Namek, Sik, Kedah	90–120	Riverine [1]	Annual – 2 540 Driest – 5, Jan. Wettest – 473, May	19–35	70–85
Sungai Lembing, Kuantan, Pahang	85–150	Riverine [2] alluvium	Annual – 3 048 Driest – 78.2, June Wettest – 950, Dec.	22–34	76–85
FRIM, Kepong, Kuala Lumpur	80–90	Granitic [3]	Annual – 2 970 Driest – 49, June Wettest – 386, Apr.	22–34	76–87
Parit Broleh, Batu Pahat, Johor	18–30	Marine [4] clay	Annual – 2 400 Driest – 65, Aug. Wettest – 452, Dec.	22–33	78–87

Sources:

1 – Debaveye et al. 1983; 2 – Libby and Chan 1965; 3 – Amir Husni et al. 1995;

4 – Panton et al. 1965; 5 – Malaysian Meteorological Service 1995.

Field procedure

Bamboo culms – 1-, 2-, 3- and 4-year-old – were cut at about 30 cm above ground level. The diameter, thickness, girth and standing height were measured from the cut base towards the tip. The above-ground biomass (the fresh weight of culm and branches) was recorded for each. Each culm was then cut at about 4.0 m intervals into three portions: basal, middle and top. The number of specimens taken was 60 poles for each age, species and study site.

RESULTS AND DISCUSSION

The physical characteristics of the two bamboo species from the four sites are presented as mean values. The above-ground biomass, culm length, number of internodes, internodal length, internodal diameter and culm wall thickness are discussed separately.

The significant difference between means of each characteristic of the two bamboos at different levels of age, height and site was tested by Duncan's Multiple Range Test (DMRT). Linear regression analysis was used to determine the correlation coefficients of physical characteristics with age and height of bamboo at the different study sites. Details were given by Abd. Latif (1996).

Above-ground biomass

Table 2 presents the average results of the above-ground biomass of both species at different sites. The above-ground biomass varied from 15 kg to 55 kg, regardless of species, age and site. The lowest mean above-ground biomass was observed in the 1-year-old *G. scortechinii* (15.4 kg) and *B. vulgaris* (20.6 kg) from Sik, while the highest was found in the 4-year-old *G. scortechinii* of the Sungai Lembing (55.0 kg) and the 3-year-old *B. vulgaris* from FRIM area (43.7 kg).

Table 2.
Average above-ground biomass of the two bamboo species at different ages and sites

Site and part of plant	Species & age (years)							
	G. scortechinii				*B. vulgaris*			
	1	2	3	4	1	2	3	4
Batu Pahat								
Culm	18.0	18.5	30.0	39.0	23.8	20.7	19.7	22.4
Branches	1.7	2.6	7.3	7.9	2.8	3.6	5.2	4.4
Leaves	1.5	4.2	9.6	8.2	1.3	1.0	1.4	1.3
Total	21.2	25.4	46.9	55.1	27.8	25.3	26.3	28.0
FRIM								
Culm	27.2	36.9	28.8	31.8	25.6	27.8	37.3	31.2
Branches	3.0	5.2	3.9	3.9	1.9	6.2	5.7	2.0
Leaves	2.0	2.8	1.8	2.1	1.1	0.8	0.6	1.8
Total	32.2	44.9	34.4	37.9	28.6	34.8	43.7	35.0
Sik								
Culm	13.5	15.7	18.6	15.9	16.3	17.2	19.4	14.3
Branches	0.8	0.9	1.3	1.0	3.7	4.0	4.1	1.6
Leaves	1.1	0.8	1.1	1.0	0.6	1.6	1.8	0.6
Total	15.4	16.4	21.0	17.9	20.6	22.8	25.7	16.4
Sungai Lembing								
Culm	17.9	18.2	28.1	38.9	23.1	19.6	18.5	21.8
Branches	1.7	2.6	7.2	7.9	7.5	3.7	5.1	4.3
Leaves	1.5	4.2	9.8	8.3	1.2	0.7	1.5	1.3
Total	21.2	25.1	45.1	55.0	31.8	24.0	25.0	27.3

The fresh culm weight decreased from about 31 kg (FRIM) followed by 24 kg (Batu Pahat), 23 kg (Sungai Lembing) and 16 kg in Sik. The highest and lowest mean branch weights were observed at Sungai Lembing (5.0 kg) and Sik (2.2 kg), the leaf weight was found to be higher at Batu Pahat (3.6 kg) and Sik (1.1 kg). The lowest biomass at Sik could be because the area is relatively drier than the other study sites.

With regard to the age of culms, the general trend showed an increase of the biomass with maturity. This might be related to the development of more mature tissue. The partial decrease of branch weight with age could probably be due to broken tips and drooping culms in older ones. A similar decrease of branch weight with age was also reported by Abd. Latif et al. (1993a, b) for *B. blumeana* and *B. heterostachya*.

Culm length

The culm length of *G. scortechinii* and *B. vulgaris* varied from 15 to 22 m and 12 to 19 m, respectively (Table 3). The lowest mean culm length for *G. scortechinii* was observed in 1-year-old culms (15.5 m at Sik) and the highest in 4-year-old culms (22.0 m at Batu Pahat and Sungai Lembing). The shortest (12.3 m) and longest (18.9 m) culm lengths for *B. vulgaris* were found in the 3-year-old culms at Sik and FRIM, respectively. As revealed by DMRT (Abd. Latif 1996), this implies that bamboo almost reaches its maximum height within a year, and as it grows older, incidences of broken tips or drooping increase, leading to shorter culm length (Abd. Latif et al. 1993b). The lowest mean above-ground biomass and the lowest mean culm length observed for both species at Sik could probably be due to the drier condition there.

Table 3.
Average culm length (m) of the bamboo species at different sites

Site	Species & age (years)							
	G. scortechinii				*B. vulgaris*			
	1	2	3	4	1	2	3	4
Batu Pahat	16.1	19.0	18.4	22.0	12.9	16.4	16.4	16.1
FRIM	20.3	19.4	17.5	20.6	15.5	14.1	18.9	15.8
Sik	15.5	15.5	17.7	17.2	13.9	13.2	12.3	12.9
Sungai Lembing	16.1	19.0	18.4	22.0	12.9	16.4	16.5	16.1

Internodal number

The mean internodal number of *B. vulgaris* was observed to be higher than that of *G. scortechinii* (about 51 and 49, respectively), regardless of site and age. The internodal number was observed to be highest at FRIM (about 52.3) followed by Sungai Lembing (51.4), Batu Pahat (51.0) and Sik (43.8) (Table 4). Again, this could be due to the drier environment at Sik. Gnanaharan (1994) also reported lower internodal number for *Dendrocalamus strictus* sampled from drier places than those from the moist areas.

Table 4.
Average number of internodes per culm of the bamboo species at different site

Site	Species & age (years)							
	G. scortechinii				*B. vulgaris*			
	1	2	3	4	1	2	3	4
Batu Pahat	47.2	43.5	46.2	56.0	41.0	62.2	51.1	61.3
FRIM	64.8	55.8	50.6	51.0	52.4	39.2	55.0	50.0
Sik	46.0	42.5	44.5	45.1	43.5	40.0	49.1	40.0
Sungai Lembing	48.0	40.0	40.0	58.0	40.0	62.0	61.4	62.0

Internodal length

Generally, the internodal length is longer in *G. scortechinii* than in *B. vulgaris* (average of about 44 cm and 34 cm, respectively). The highest and lowest mean internodal lengths of *G. scortechinii* were observed in the middle portion of the 4-

Table 5.
Average internodal length (cm) of the two bamboos at different sites

Site & portion	Species & age (years)							
	G. scortechinii				*B. vulgaris*			
	1	2	3	4	1	2	3	4
Batu Pahat								
Basal	39.0	44.0	40.9	43.2	32.3	30.5	30.0	34.3
Middle	50.4	50.5	56.3	58.5	41.9	42.5	37.4	41.0
Top	35.4	34.7	43.6	38.5	36.9	31.6	32.7	30.0
Average	**41.6**	**43.1**	**46.9**	**46.7**	**37.1**	**34.9**	**33.4**	**35.1**
FRIM								
Basal	35.4	36.2	35.3	39.1	38.8	35.3	36.1	30.3
Middle	50.4	44.7	52.5	51.1	41.9	38.4	44.0	37.2
Top	37.0	46.5	38.6	41.6	39.7	36.3	41.0	36.5
Average	**40.9**	**42.5**	**42.1**	**43.9**	**40.1**	**36.7**	**40.4**	**34.5**
Sik								
Basal	39.1	44.1	41.1	43.5	32.5	33.7	30.3	34.6
Middle	50.5	50.8	56.6	58.6	42.3	43.1	38.1	41.6
Top	36.2	34.9	43.7	38.6	37.2	32.0	33.1	30.5
Average	**41.9**	**43.2**	**47.2**	**46.9**	**37.4**	**36.2**	**33.8**	**35.6**
Sungai Lembing								
Basal	25.9	36.5	37.5	40.6	25.3	20.5	22.8	22.9
Middle	45.5	48.4	55.9	59.1	31.6	29.1	28.8	29.4
Top	43.9	39.0	52.3	51.2	28.6	33.6	33.4	32.0
Average	**38.5**	**41.3**	**48.6**	**50.3**	**28.5**	**27.8**	**28.3**	**28.1**

year-old culm (59 cm) and basal portion of the 1-year-old culm (about 26 cm), respectively, sampled from Sungai Lembing (Table 5). The respective values for *B. vulgaris*, on the other hand, were found to be greater in the middle portion of the 3-year-old culm sampled from FRIM (about 44 cm) and the basal portion of the 2-year-old culm of Sungai Lembing (about 20 cm).

Duncan's analysis (Abd. Latif 1996) indicated that the internodal length for both bamboo species increased from the basal region to the middle portion of the culm (from about 35 cm to 45 cm) and decreased towards the top (about 37 cm). In the linear regression perspective, the internodal length increases insignificantly with age and culm height ($r = 0.11$ at $p<0.05$). This reflects the internodal elongation during the height growth of the shoot which begins at the basal portion towards the middle and decreases towards the top (Ueda 1960; Hsiung et al. 1980). With regard to age, the internodal length remains almost constant ($r = 0.30$ at $p<0.05$), from about 38 cm in the 1- and 2-year-old culms to about 40 cm in 3- and 4-year-old culms (about 40 cm). Since the length of an internode cannot increase with age, the different values indicate the considerable variation within one age class, even with a sampling of 60 culms.

Internodal diameter

The internodal diameter of both bamboo species is presented in Table 6. The mean diameter of *G. scortechinii* was observed as ranging from 3.0 to 7.5 cm (Batu Pahat), 5.0 to 8.9 cm (FRIM), 3.1 to 7.5 cm (Sik) and 4.8 to 8.8 cm (Sungai Lembing). The mean diameter of *B. vulgaris* varied from 2.9 to 6.5 cm (Batu Pahat), 5.2 to 8.2 cm (FRIM), 3.2 to 6.7 cm (Sik) and 6.5 to 8.3 cm (Sungai Lembing). The highest and lowest mean diameters of *G. scortechinii* were observed in the middle portion of the 2-year-old culm from FRIM (8.9 cm) and the top portion of the 1-year-old culm of the Batu Pahat area (3.0 cm), respectively. The highest and lowest mean diameters of *B. vulgaris*, on the other hand, were found in the basal portion of 1-year-old culm from Sungai Lembing (8.3 cm) and the top portion of the 4-year-old culm of the Sik area (3.2 cm), respectively.

In general, bamboo tapers from the base towards the tip with a gradual decrease in diameter (for *B. vulgaris,* $r = -0.45$ at $p<0.05$ and for *G. scortechinii,* $r = -0.71$ at $p<0.01$). As the bamboo ages, the diameter does not increase significantly ($r = 0.10$ and $r = 0.19$ at $p<0.05$ in *B. vulgaris* and *G. scortechinii*, respectively) (Abd. Latif 1996). This signifies that the new bamboo sprout emerges with almost full diameter and does not increase further with age (Simao 1961; McClure 1967; Ueda 1981; Chaturvedi 1988). Bamboo, unlike the dicotyledons, does not show any secondary thickening owing to lack of a secondary meristem and thus attains its final diameter during the sprouting stage (Liese 1985).

Table 6.
Average internodal diameter (cm) of the two bamboos at different sites

Site & portion	Species & age (years)							
	G. scortechinii				*B. vulgaris*			
	1	2	3	4	1	2	3	4
Batu Pahat								
Basal	6.9	6.9	7.5	7.5	5.4	6.0	6.5	5.6
Middle	5.8	5.7	6.6	6.7	5.9	6.1	6.4	5.4
Top	2.9	2.8	3.7	3.8	4.3	4.3	4.7	2.9
Average	**5.2**	**5.1**	**5.9**	**6.0**	**5.2**	**5.5**	**5.9**	**4.6**
FRIM								
Basal	8.6	8.7	8.9	8.3	7.4	7.1	7.5	7.9
Middle	8.3	8.9	8.6	7.7	7.3	7.2	8.1	8.2
Top	6.1	6.9	6.5	5.0	5.2	5.7	6.4	7.0
Average	**7.6**	**8.2**	**8.0**	**7.0**	**6.6**	**6.7**	**7.4**	**7.7**
Sik								
Basal	7.1	7.0	7.5	7.5	5.6	6.3	6.9	5.8
Middle	6.0	5.9	6.8	6.8	6.2	6.3	6.7	5.7
Top	3.2	3.1	4.0	4.2	4.6	4.4	4.9	3.2
Average	**5.4**	**5.3**	**6.1**	**6.2**	**5.5**	**5.6**	**6.2**	**4.9**
Sungai Lembing								
Basal	7.1	7.3	7.8	7.9	8.1	7.4	6.0	7.5
Middle	6.2	7.2	8.2	8.9	8.3	8.0	7.0	7.8
Top	3.9	4.8	6.0	7.2	6.5	6.8	6.5	6.9
Average	**5.7**	**6.4**	**7.3**	**8.0**	**7.6**	**7.4**	**6.5**	**7.4**

Culm wall thickness

The average results (Table 7) indicate that the culm wall thickness of *G. scortechinii* and *B. vulgaris* varies respectively from 0.4 to 1.6 cm and 0.5 to 1.6 cm at Batu Pahat, 0.5 to 1.8 cm and 0.4 to 2.2 cm at FRIM, 0.4 to 1.7 cm and 0.5 to 1.7 cm at Sik, and 0.5 to 2.0 cm and 0.6 to 2.0 cm at Sungai lembing. Further, Duncan's analysis (Abd. Latif 1996) suggested that the culm wall thickness was not significantly higher at Sungai Lembing and FRIM (1.0 cm) or at Sik and Batu Pahat (0.9 cm). This implies that the culm wall thickness of bamboo is not significantly affected by topographical factors. This is an advantage as it renders the use of a given bamboo species site-independent for a product requiring a specific culm wall thickness.

The culm wall thickness was also observed to not correlate significantly with age ($r = 0.01$ and 0.02 at $p<0.05$ for *B. vulgaris* and *G. scortechinii,* respectively) but decreased significantly with height ($r = -0.89$ at $p<0.01$ in both species) (Abd. Latif 1996). As in the case of internodal diameter, this is associated with the natural characteristics of bamboo, which shows no further increment of culm wall thickness with age.

CONCLUSION

Till now, there have been only a few observations on the natural variation of main culm characteristics – such as culm biomass and length, number, length and diameter of internodes, and culm wall thickness – in bamboos of different age classes and on the possible influence of site factors. Such basic information, however, is important to assess the quality of bamboo culms for their efficient processing and optimum use. Ueda (1960) reported that, generally, the quality of bamboo is influenced by environmental factors, especially soil conditions. The Sundanese of West Java prefer the culms of *Gigantochloa pseudo-arundinaceae* harvested from the groves on the hill slopes to those growing in the valley. A detailed analysis by Soeprayitno et al. (1990) revealed higher specific gravity and

Table 7.
Average culm wall thickness (cm) of the two bamboos at different sites

Site & portion	Species & age (years)							
	G. scortechinii				*B. vulgaris*			
	1	2	3	4	1	2	3	4
Batu Pahat								
Basal	1.3	1.3	1.6	1.6	1.4	1.5	1.6	1.4
Middle	0.5	0.5	0.5	0.6	0.7	0.8	1.0	0.8
Top	0.4	0.4	0.4	0.4	0.6	0.5	0.6	0.5
Average	**0.7**	**0.7**	**0.8**	**0.9**	**0.9**	**0.9**	**1.1**	**0.9**
FRIM								
Basal	1.6	1.6	1.8	1.7	2.1	1.9	1.4	2.0
Middle	0.6	0.6	0.6	0.6	0.9	0.7	0.8	0.8
Top	0.5	0.5	0.5	0.5	0.6	0.5	0.4	0.6
Average	**0.9**	**0.9**	**1.0**	**0.9**	**1.2**	**1.0**	**0.9**	**1.1**
Sik								
Basal	1.5	1.5	1.7	1.7	1.5	1.6	1.7	1.5
Middle	0.6	0.5	0.5	0.6	0.7	0.8	1.1	0.9
Top	0.4	0.4	0.4	0.4	0.6	0.6	0.6	0.5
Average	**0.8**	**0.8**	**0.9**	**0.9**	**0.9**	**1.0**	**1.1**	**1.0**
Sungai Lembing								
Basal	2.0	1.4	1.4	1.4	2.0	1.3	1.6	1.9
Middle	0.9	0.6	0.8	0.8	1.0	0.7	0.8	1.0
Top	0.6	0.5	0.6	0.5	0.6	0.6	0.6	0.7
Average	**1.2**	**0.8**	**0.9**	**0.9**	**1.2**	**0.8**	**1.0**	**1.2**

bending and tensile strengths for the former, although no significant differences were noted in the anatomical parameters such as like fiber length, wall thickness and frequency of fibrovascular bundles. According to Gnanaharan (1994), *Dendrocalamus strictus* growing in a moist area within Kerala State, India, has culms with longer internodes and larger diameter than the same species grown in a dry place. The latter, however, was found to be stronger although its culm length, diameter and internode length were lower. These indications were confirmed by the present study conducted on about 1 000 culms for each of the two species.

All the morphological characteristics of the bamboos differed significantly with site, species, age (except for culm wall thickness) and culm height. Sik area, which is relatively drier than the other three study sites, exhibited a lower total culm weight, shorter culms, and fewer but slightly longer internodes. The total above-ground biomass (except for branch weight), culm length, internodal length and diameter of *G. scortechinii* were significantly greater than those of *B. vulgaris*. The variation in branch and leaf weight of both bamboo species at all sites was found to be greater in 3-year-old culms, while the other morphological characteristics statistically decreased somewhat with increasing age and height of the culm (except for culm wall thickness and internodal length).

REFERENCES

Abd. Latif, M. 1996. Some selected properties of two Malaysian bamboo species in relation to age, height, site and seasonal variations. Ph.D. thesis, Universiti Pertanian Malaysia, Serdang, Malaysia.

Abd. Latif, M.; Mohd. Zin, J. 1992. Culm characteristics of *Bambusa blumeana* and *Gigantochloa scortechnii* and its effects on physical and mechanical properties. *In* Zhu, S.; Li, W.; Zhang, X.; Wang, Z. ed., Bamboo and its use. Proceedings of the International Symposium on Industrial Use of Bamboo, Beijing, China, 7–11 December 1992. International Tropical Timber Organization; Chinese Academy of Forestry, Beijing, China. pp. 118–128.

Abd. Latif, M.; Abd. Razak, O.; Jamaludin, K.; Abd Jalil, A. 1993a. Variability in physical properties of *Bambusa heterostachya.* Bamboo Journal, 11, 20–28.

Abd. Latif, M.; Ashaari, A.; Jamaludin, K.; Mohd. Zin, J. 1993b. Effects of anatomical characteristics on the physical and mechanical properties of *Bambusa blumeana.* Journal Tropical Forest Science, 6(2), 159–171.

Amir Husni, M.S.; Ahmad Zuhaidi, Y.; Mohd. Nor, M.; Ghazali, H.; Suhaimi, W.C.; Lok, C.H. 1995. Growth and current stocking of plantation species at Bukit Hari and Bukit Lagong Forest Reserve, Selangor in relation to soils distribution pattern. FRIM Report 65, Forest Research Institute Malaysia, Kuala Lumpur, Malaysia. pp. 1–18.

Chaturvedi, A.N. 1988. Management of bamboo forests. Indian Forester, 114 (9), 489–495.

Debaveye, J.; Meirvenne, M.V.; Abd. Rahman, A.B.; Juzhar, J. 1983. Reconnaissance soil survey of Sik District, Kedah". Ministry of Agriculture, Kuala Lumpur, Malaysia. 38 pp.

Gnanaharan, R. 1994. Physical and strength properties of *Dendrocalamus strictus* grown in Kerala. *In* Bamboo in Asia and the Pacific. Proceedings of the 4th International Bamboo Workshop, Chiangmai, Thailand, 27–30 November 1991. International Development Research Centre, Ottawa, Canada; Forestry Research Support Programme for Asia and the Pacific, Bangkok, Thailand. pp. 188–192.

Hsiung, W.Y.; Ding, Z.; Li, Y. 1980. Intercallary meristem and internodal elongation of bamboo plant. Scientia Silvae Sinicae, 16, 81–89.

Libby, D.A.; Chan, Y.K. 1965. Schematic reconnaissance soil survey of Northeast Pahang. Malayan Soil Survey Report 4, 59 pp.
Liese, W. 1985. Bamboos – biology, silvics, properties, utilization. Deutsche Gelleschaft für Technische Zusammenarbeit (GTZ), Nr. 180, TZ Verlagsges., Roßdorf 1, 132 pp.
Malaysian Meteorological Service. 1995. Annual summary of meteorological observations in 1991–1995. Ministry of Science, Technology and Environment, Kuala Lumpur, Malaysia.
McClure, F.A. 1966. The bamboos: a fresh perspective. Harvard University Press, Cambridge, Mass., USA. 347 pp.
Panton, W.P.; Soo, S.M.; Nuruddin, M. 1965. Reconnaissance soil survey of the Southwest Johor coconut area. Malayan Soil Survey Report 3, Ministry of Agriculture, Kuala Lumpur, Malaysia, 40 pp.
Simao, S. 1961. Measurement of bamboo. Rev. Agric Piracicaba, 36(4), 233–236.
Soeprayitno, T.; Tobing, T.L.; Widjaja, E. 1990. Why the Sundanese of West Java prefer slope-inhabiting *Gigantochloa pseudoarundinaceae* to those growing in the valley. *In* Ramanuja Rao, I.V.; Gnanaharan, R; Sastry, C.B., ed., Bamboos: current research. Proceedings of the International Bamboo Workshop, Cochin, India, 14–18 November 1988. Kerala Forest Research Institute, Kerala, India; International Development Research Centre, Ottawa, Canada. pp. 215–217.
Ueda, K. 1960. Studies on the physiology of bamboo. Bulletin of Kyoto University of Forestry, 30, 167.
Ueda, K. 1981. Bamboo industry in Japan – present and future. *In* Higuchi, T. ed., Bamboo production and utilization. Proceedings of the XVII IUFRO World Congress, Kyoto, Japan, 6–17 September 1981. Kyoto University, Kyoto, Japan. pp. 2–7.

Bamboo for Sustainable Development, pp. 235–246
Eds. A. Kumar, I.V. Ramanuja Rao and Ch. Sastry

Dynamical analysis on endogenous phytohormones of *Phyllostachys pubescens* during its growth and development from shoot to young culm

YANG TON
Chemistry & Biology Department, Zhejiang Normal College, Hangzhou, China
DING XINGCUI
China National Bamboo Research Center, Hangzhou, China

Abstract—The dynamic changes in endogenous phytohormone (IAA, GA_3 and ABA) contents and bioactivity of *Phyllostachys pubescens* shoots in different representative stages of growth and differentiation were analysed by High-Pressure Liquid Chromatography and bioassay. It was found that the changes in phytohormone content and bioactivity were parallel to the growth and development of the shoot. Traces of ABA were detected only in the growth-inactive stages – "winter shoot" (0.28 μg/G.F.W.) and "height-growth stagnation" (0.092 μg/G.F.W). IAA and GA_3 contents increased correspondingly with the shoot growth rate and reached their peak 3.6 μg/G.F.W. and 11.4 μg/G.F.W., respectively, on the date when the fastest growth was recorded. When the growth rate declined, IAA and GA_3 contents decreased sharply and were only 0.85 μg/G.F.W. and 4.2 μg/G.F.W., respectively. The spatial distribution of IAA and GA_3 contents decreased from top to bottom, but it was the reverse in the case of ABA content. The IAA and GA_3 capacities of shoot were very strong in synthesis power, and contents were about 10 times higher than the general high. There were several GA_3 in shoot; perhaps, the mutual transforming of each GA_3 played an important role. The supposed causes for the rapid growth of bamboo are: (1) very high contents of IAA and GA_3, but traces of ABA; (2) high content of IAA produced and maintained to supply the nutrients required for its rapid growth; (3) harmonious coordination of IAA and GA_3 in spatial and time distribution.

INTRODUCTION

The rhizome bud of *Phyllostachys pubescens* is formed during late summer and early winter, and is in the state of dormancy in winter. It resumes growth in the following spring, with an average growth rate of 30–50 cm per day, even 100 cm at peak. Later, the growth rate gradually decreases until the stage when height growth ceases and it becomes a young culm.

Such distinctive changes in bamboo growth – dormancy, stagnation, fast growth and growth ceasing – must be dependent on its internal physiological causes. So far, no systematic research on these has been made in the light of physiology and biochemistry, but some scholars attribute them to the action of certain enzymes, such as transferase (Li Liren et al. 1979), DNase (Lu Xianhui et al. 1986), cytopigment oxidase and ascorbic acid oxidase (Cheng Shaoyu et al. 1986). Despite some studies on endogenous phytohormones, they have been constrained to hormone qualitative analysis without relation to bamboo growth. For example, Koshimizu discovered IAA in 1956 in *Phyllostachys pubescens* and an unknown GA, which is similar to GA_9 in bioactivity, in 1963 (Koshimizu 1956, 1963). Jiro Kato found a GA in a mutant of *Bambusa oldhamii* in 1963. Wu Guanming and Mcdoskey (1983) proved the existence of zealin-9-D-ribonucleoside, Zealin, GA_3 and IAA.

The main purposes of this research study were:

1. To determine the relationship between endogenous phytohormones, and bamboo growth and development; and
2. To probe the causes for the rapid growth of bamboo from the angle of physiology and biochemistry through a dynamic analysis of content and bioactivity of IAA, GA_3 and ABA in different growth and differentiation phases, by means of High-Pressure Liquid Chromatography (HPLC) and bioassay.

MATERIALS AND METHODS

Materials

Shoots of *Phyllostachys pubescens*: winter shoot (rhizome bud) and late shoot (young culm).

Methods

Sampling
Healthy and vigorous hoots were sampled from 8 March to 14 May at the arboretum of Nanjing Forestry University with the criteria that the early shoots should be uniform and late shoots must be of the same height and diameter. Only the top part of shoots was cut for hormone analysis, with 50 g for each sample in duplicate.

Extraction and purification of endogenous IAA, GA_3 and ABA
Carried out in accordance with the method of Ding Qin et al. (1973).

Bioassay of growth promoter and inhibitor by paper chromatography
The crude extraction was dropped on a 30×3.5 cm chromatography paper which, after drying, was chromatographed in a chromatography solution (isopropanol: ammonia:water – 8:1:1 v/v) in dark and 25°C, until reaching 25 cm high. Then the paper was cut into 10 pieces, and each piece put into a small bottle containing 2 ml of 2% sugar phosphoric acid buffer, along with a 0.5 cm long wheat shoot,

and kept for inoculation in an oscillator (speed of 20 cycles per minute) for 20 hours in dark and at 25°C. Then, the increased length of each shoot was measured.

IAA and ABA purification on silica glue GF254 chromatography plate
A 24×10 cm GF254 plate was made of GF254 silica glue, using 0.6% CMC, an adhesive, and activated at 110°C for half an hour. Chromatography solution consisted of isopropanol, ammonia and water at a ratio of 2:8:1 (v/v). The crude extraction was dropped on the GF254 plate and chromatographed. The spots were stripped off in the same locations with sample IAA and ABA, and extracted 3 times in 90% alcohol.

GA purification on silica H thin plate
H thin plate was made of silica glue, with 6% CMC adhesive, and was activated at 110°C for half an hour. Chromatography solution used consisted of chloroform, acetic ester, and acetic acid (10:8:2 v/v).

Bioassay
Wheat shoot method of Ding Qin et al. (1979) was followed. Wheat variety used was Yang Wheat 5. The barley endosperm bioassay method, developed by Paleg L.G. et al. (1964) for GA_3, was also followed. H cation exchange resin used was 734.

Analysis on IAA, GA_3 and ABA by HPLC
The instrument used was Japanese Waters HPLC, with stainless steel column of 0.39×30 cm, uBondapack-Cl8. Chromotagraphy conditions were:

Column temperature: 35°C
Eluant: 40% (v/v) methyl alcohol
Sensitivity: 0.05 AuFs

Measurement of recovery rate of IAA, GA_3 and ABA
For each standard IAA, GA_3 and ABA, 10 g was added into 500 ml of 80% methyl alcohol following the same procedures as with sample analysis, such as solvent partition chromatography and thin plate chromatography. Finally, the content of each residue was determined by HPLC, and the recovery rates of IAA, GA_3 and ABA were found to be 47%, 65% and 72%, respectively.

OBSERVATIONS AND RESULTS

Biological activity of IAA, ABA and GA_3

IAA biological activity
After Thin Layer Chromatography (TLC), a noticeable purplish strip of absorption at UV254 nm was observed at Rf 0.4 on the thin plate, which was the same as standard IAA Rf (0.39–0.42). IAA biological activity change pattern was the

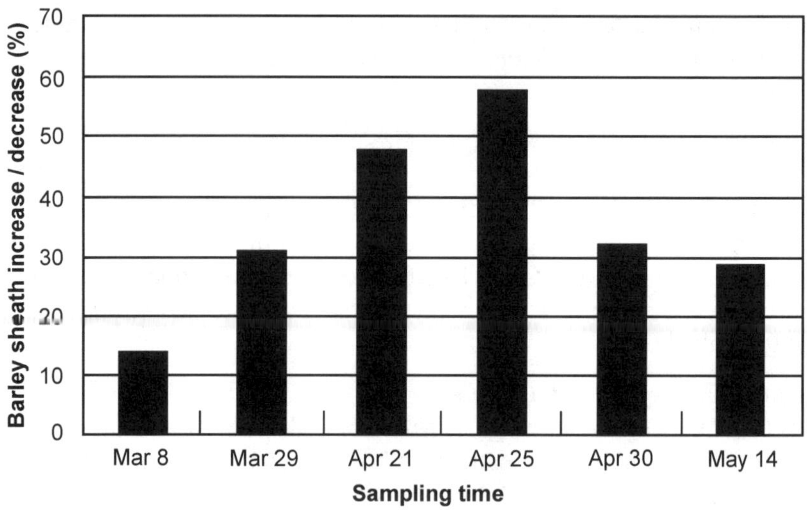

Figure 1. IAA activity change of the bamboo shoot acid extract during different growth periods

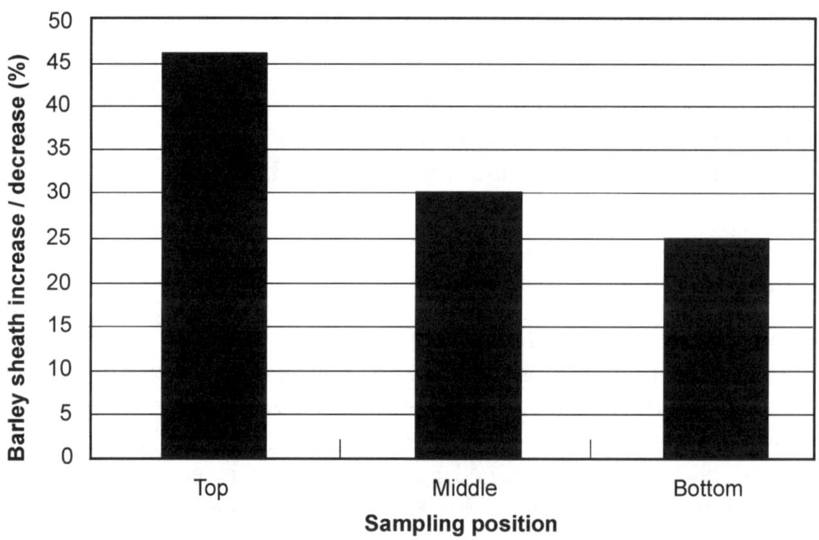

Figure 2. IAA activity change of the bamboo shoot acid extract at different positions

weakest on 8 March, and then increased gradually to reach its peak on 25 April (Fig. 1). Later, the activity change pattern dropped to the lowest level on 14 May, when the height growth almost ceased. The change in different positions was the highest at the top, middle at the medium part and the lowest at the bottom (Fig. 2).

ABA biological activity

The bioactivity of ABA was found only on 8 March and 14 May. On the former date, the shoot was still in dormant stage and showed higher activity, but on the latter date it was in height-growth ceasing stage and showed lower activity (Fig. 3). Nothing was detected at other stages. It was worth noticing that growth inhibition that occurred from 29 March to 30 May 30 was caused by inhibitors other than ABA. This was proved by HPLC, which determined that ABA did not exist during that period. ABA bioactivity was the highest at the bottom, medium at the medium part, and lowest at the top (Fig. 4).

GA biological activity

After TLC of the crude extraction and spraying with sulfuric acid, 4–5 fluorescent specks were found at UV 254 nm (Table 1). Specks 1, 2, 4 and 5 could always be observed for various growth stages, but Speck 3 was seen only on the May 21 sample with weaker fluorescent strength. It was proved, by comparison with standard GA_3 and $GA_{4/7}$, that Speck 2 and 5 might well be GA_3 and GA_4 because their Rf values and colors were the same as the corresponding ones.

Fluorescent strength of Speck 5 (GA_4) was stronger in the sample of 29 March, and weaker on 21 May. Adler et al. (1963) concluded that it was feasible and reliable to estimate GA content by comparing fluorescent strength. Hence, the content of GA_4 was assumed as decreased from 29 March to 21 May. The bioactivity change pattern in different stages was the lowest on 8 March, and then increased to a maximum on 25 May. Later, the activity decreased until 14 May when height growth nearly stopped (Fig. 5). The bioactivity of GA_3 in spatial distribution was highest at the top, medium at the middle and weakest at the bottom (Fig. 6).

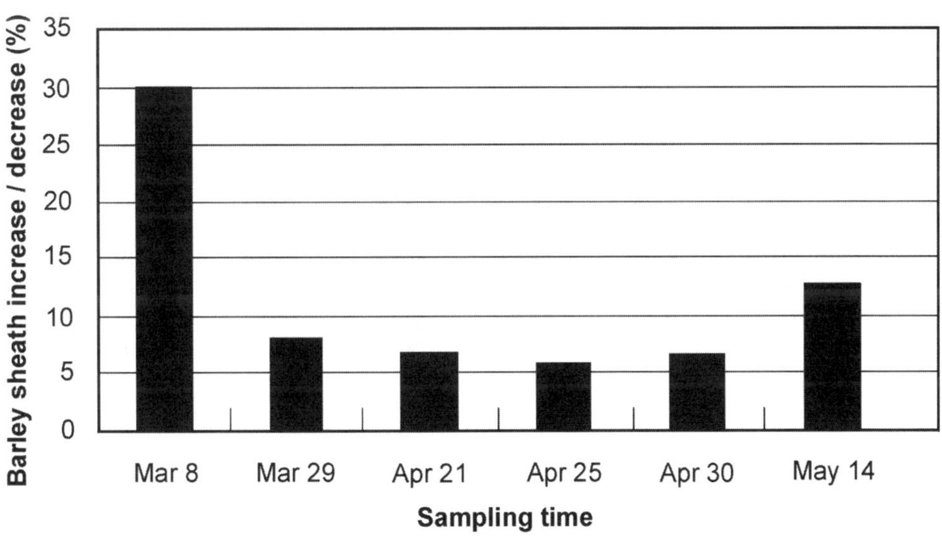

Figure 3. ABA activity change of the bamboo shoot acid extract during different growth periods

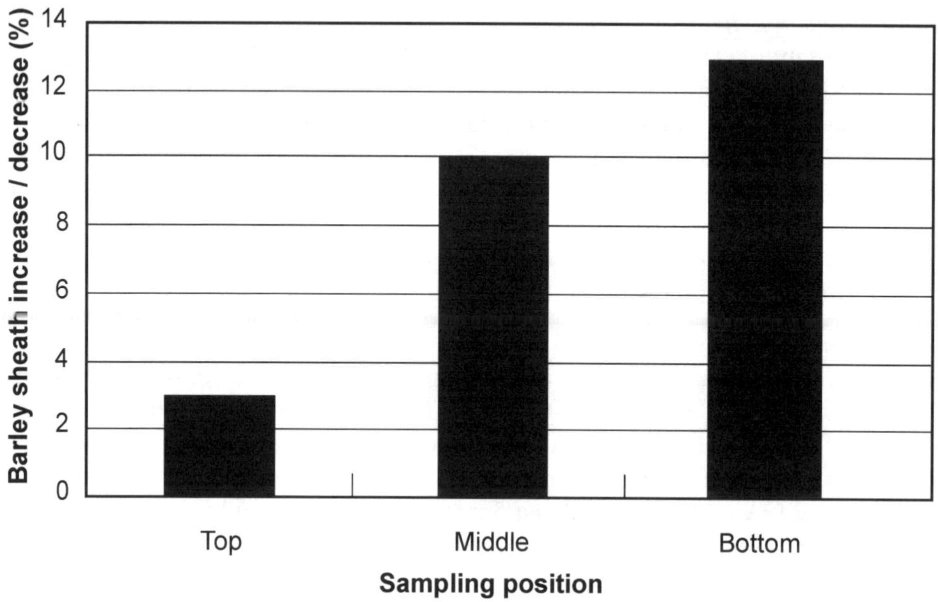

Figure 4. ABA activity change of the bamboo shoot acid extract at different positions

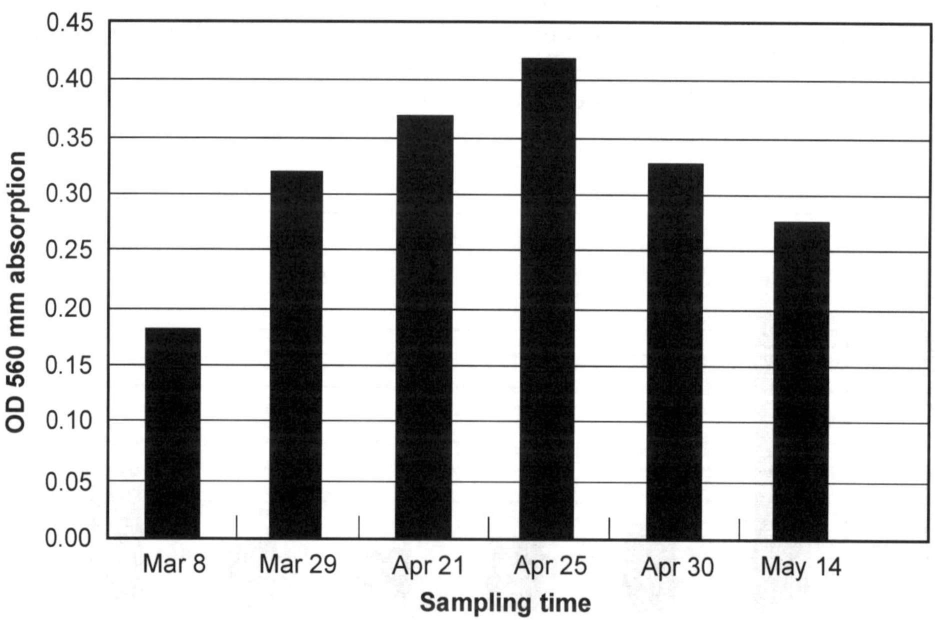

Figure 5. GA_3 activity change of the bamboo shoot acid extract during different growth periods

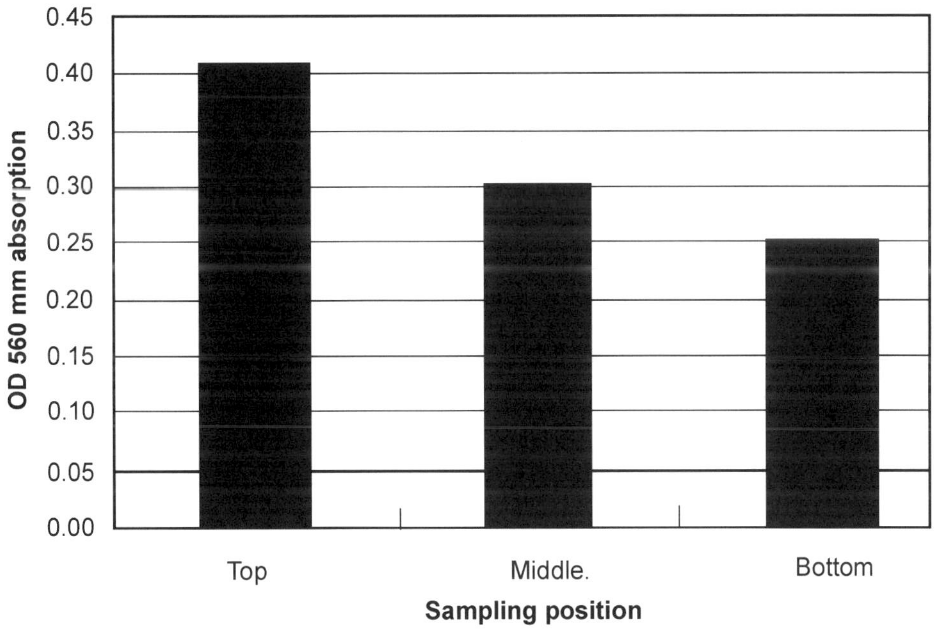

Figure 6. GA_3 activity change of the bamboo shoot acid extract at different positions

Table 1.
GA_3 fluorescent behavior of bamboo shoot acid extract on thin layer chromatography at UV 254 nm after sprayed with 70% H_2SO_4

No.	Sample Rf value	Color	Intensity	Standard GA	
				Gas	Rf value
1	0–0.05	Purplish	Weaker	?	–
2	0.32	Greenish	Stronger	GA_3	0.31
3	0.47	Purple blue	Weak	?	–
4	0.55	Dark blue	Weak	?	–
5	0.70	Dark blue	Strong	GA_4	0.72

Qualitative and quantitative analysis on IAA, GA_3 and ABA by HPLC

IAA analysis by HPLC

As shown by the HPLC spectrum of IAA detected at UV221 nm, the retention time (RT) of Peak 1 was 1.86 min, which was same as the RT of standard IAA. The Peak Addition Way (Fig. 7) also testified indirectly that it was IAA. Ultra violet scanning between 190–290 nm on Peak 1 demonstrated that its absorptive curve was same as the standard IAA in shape, and the maximum absorptive wave-

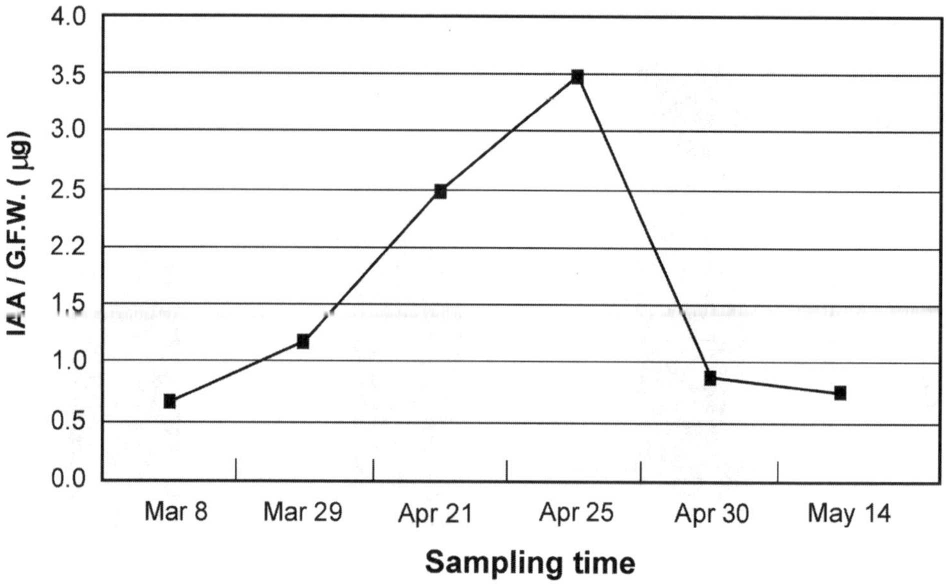

Figure 7. IAA content change of the bamboo shoot acid extract during different growth stages

length was 221 nm and other two absorptive sub-peak wavelengths were 196 nm and 275 nm. The eluant of HPLC was proved by wheat shoot assay as having bioactivity. IAA content change in different phases (Fig. 7) behaved in the same way as proved by bioassay – IAA content was lowest on 8 March (0.5 μg/G.F.W.), increasing to a maximum on 25 May (3.6 μg/G.F.W.) and then decreasing to a low (0.85 μg/G.F.W.) on 14 May when height growth was stagnating. IAA content distribution at different positions was the highest at the top (2.8 μg/G.F.W.), medium at the middle (1.2 μg/G.F.W.) and lowest at the bottom (0.8 μg/G.F.W.) (Fig. 8). The rate of change was $2.8 \div 1.2 = 2.3$ from top to medium, and $1.2 \div 0.8 = 1.5$ from medium to bottom.

GA_3 analysis by HPLC

It has been proved that bamboo shoot has extremely high GA_3 content – 11.4 μg/G.F.W., which is far higher than the normal level for tall plants (0.001–1 μg/G.F.W.). GA_3 content change at various stages is shown in Fig 9. It was minimum on 8 March (0.5 μg/G.F.W.), but increased rapidly to reach its maximum of 11.4 μg/G.F.W. on 25 April. Later it decreased till 14 May, falling to 4.2 μg/G.F.W. The GA_3 spatial distribution is shown in Fig. 10: the top is 9.4 μg/G.F.W., the medium 4.3 μg/G.F.W. and the bottom 1.6 μg/G.F.W. The change gradient is nearly same from top to medium and from medium to bottom.

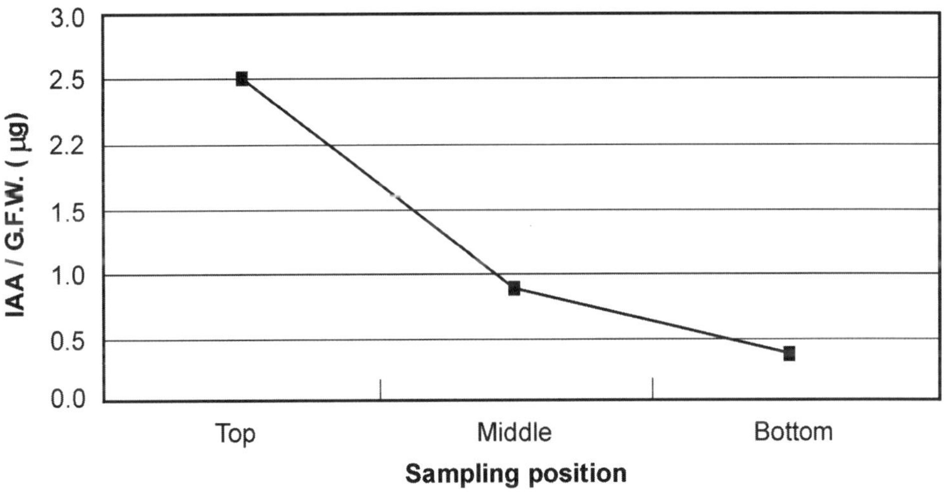

Figure 8. IAA content distribution at different positions

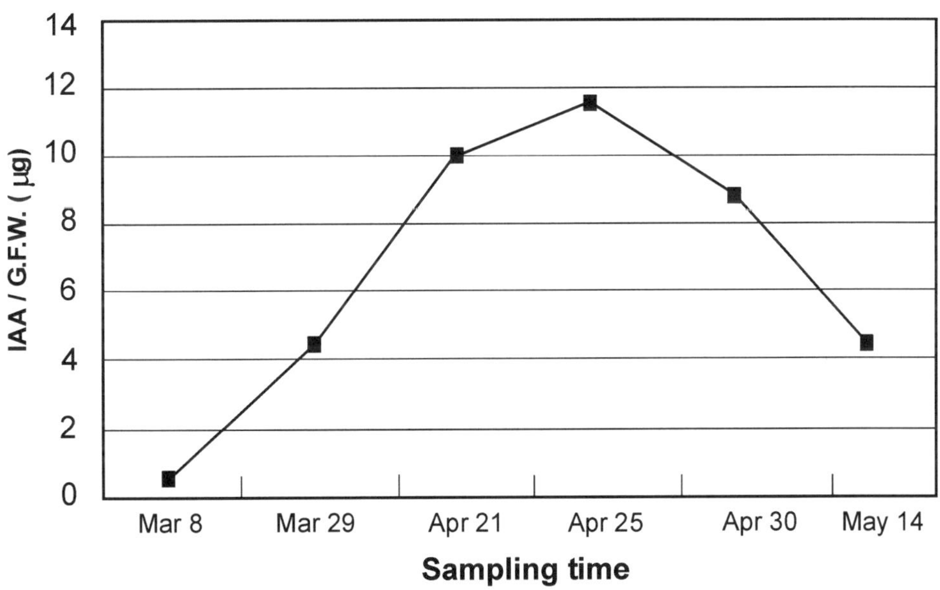

Figure 9. GA_3 content change of the bamboo shoot acid extract during different growth stages

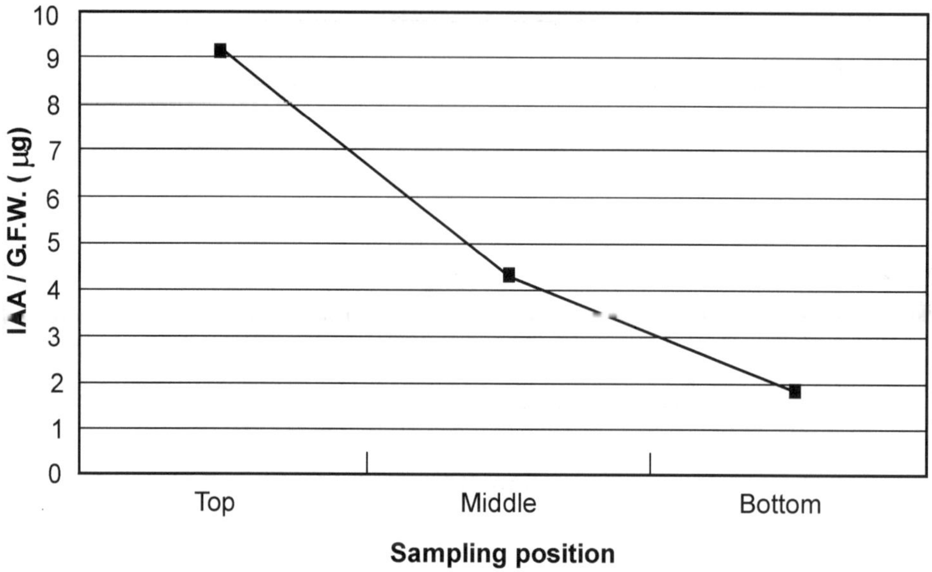

Figure 10. GA_3 content distribution at different positions

ABA analysis by HPLC

ABA contents were detected only in the two samples of 8 March and 14 May, which were 0.28 μg/G.F.W. and 0.092 μg/G.F.W., respectively. As for ABA spatial distribution, the sample of 21 April may be taken as an example. Nothing was detected at the top, but the medium and the bottom ones were 0.061 μg/G.F.W. and 0.16 μg/G.F.W., respectively. The distribution pattern of ABA, therefore, was the reverse of IAA and GA_3.

DISCUSSION

The relationship of IAA, GA_3 and ABA with bamboo shoot growth and differentiation

The lateral bud on underground rhizome is differentiated into shoot during late summer and early winter. It is in dormancy in winter, and then re-starts its growth into a sprout /culm in the following spring. Thus, it should have the same two processes as the buds of other woody plants: bud dormancy, and breaking of dormancy for further growth. In most woody plants, endogenous phytohormones have been proved to be playing an important role in these two processes; but no such evidence has been reported so far in the case of bamboo. This study found that in the underground "winter shoot" stage, ABA was predominant but GA_3 and IAA were less. As soon as a bamboo shoot sprouted out of ground, on 29 March, no ABA was detected by HPLC, possibly because of a sudden large drop in ABA content. On the other hand, GA_3 and IAA contents showed a sudden peak rise, particularly GA_3 that had a 8-fold increase. Thus, it may well be concluded that the dynamic changes of ABA, GA_3 and IAA contents play an important role in the two processes of bamboo bud dormancy and breaking of dormancy, as in the case of most woody plants. Eagles and Wareing (1964) established bud dormancy breaking of *Acer pseudoplatnus* as related to the balance between growth promoters (GA_3) and growth inhibitors (ABA). As far as the role of IAA content rise from 0.5 µg/G.F.W. on 8 March to 1.7 µg/G.F.W. on 29 March is concerned, the direct role is not so clear as compared with the GA_3 and ABA contents; but indirectly, one can surmise that it provides a condition suitable for the continued growth of bamboo shoot once the dormancy breaks down and a conducive environment is available.

The change pattern of GA_3, IAA and ABA contents was in conformity with bamboo shoot height growth curve (the ‘S’ curve). As observed in field, bamboo shoot growth pattern was very slow on 29 March, peaks on 25 April when it reached 80 cm height, and then decreased gradually till the height-growth ceasing stage on 14 May, which was synchronous with IAA and GA_3 in positive relation but with ABA in negative relation. The role of GA_3 in promoting plant stalk growth is mainly by elongation of each stalk internode, and not by increasing the number of internodes. The rapid growth of bamboo is by elongation of each predifferentiated internode. Hence, it is very obvious that high GA_3 content is synchronous with high-speed growth. This conclusion is supported by the fact that bamboo grows very fast into a thin and tall culm following an external injection of ‘920’, a growth regulator that has GA_3 as its main component.

IAA synthesis capacity is extremely strong, with the average content reaching up to 0.1 µg/G.F.W., about twice as high as in most tall woody plants. During growth when the distance between shoot apex (which consumes nutrients) and underground rhizomes (which store nutrients and ensure their timely supply to consumption areas) increases, this kind of high concentration and apex domination is maintained. It is critical to sustain a high concentration of IAA gradient along shoot apex, as IAA has a vital role in promoting the transportation of assimilates along the phloem and in the timely unloading of nutrients.

Height growth appeared to slow down during the later period of growth (14 May), when IAA, GA_3 content was decreasing and ABA content was increasing. Hence, it can be concluded that the process of growth is regulated by IAA, GA_3 and ABA, as demonstrated by the research on sugarcane by Fewkes in 1979.

Apart from the levels of endogenous phytohormone levels, suitable surroundings are also an important factor in the growth of bamboo shoot. For example, IAA and GA_3 contents showed marked increase from 8 to 29 March, but the growth speed did not increase much. Temperature and moisture during that time was low, and this might have restrained the shoot's growth.

IAA, GA_3 and ABA spatial distribution

The distribution of IAA, GA_3 and ABA in various positions of shoot depends on age. Bamboo growth is effected by cell division in meristem located above each node which causes internode to elongate bottom to top. While the spatial distribution of IAA and GA_3 decreases from top to bottom, it is the opposite in the case of ABA, thus ensuring meristem activity.

Suppositions on the rapid growth of bamboo

Bamboo is among the fastest growing plants. There are meristems at the bottom of each internode – that is, along the vertical direction – which is an anatomical advantage. From the physiological point, the reasons for the rapid growth may be summarized as follows.

1. Very high contents of IAA, GA_3 and traces of ABA;
2. A high concentration of IAA is produced and maintained to ensure constant supply of nutrients from the rhizome up to the shoot tip; and
3. Harmonious coordination of IAA and GA_3 in spatial and time distribution.

ACKNOWLEDGEMENTS

The authors record here their appreciation for the services rendered in sample collection and chemical analysis by colleagues at Zhejiang Normal College, Nanjing Forestry University, China National Bamboo Research Center.

REFERENCES

Cheng Shaoyu, et al. 1986. A research on oxidase activity and aspiration action of *Phyllostachys pubescens* shoot growth and degradation. Fujian Forestry College Journal, 6(2), 11–18.

Ding Qin, et al. 1979. Endogenous phytohormones extraction, purification and identification. Plant Physiology Communications, 2, 27–39.

Eagles, C.F.; Wareing. 1964. The role of growth substance in the regulation of bud dormancy. Plant Physiology, 17, 697–709.

Koshimizu, K. 1956. Growth substance and free amino acids in bamboo shoots (*Phyllostachys edulis*). Journal of Agricultural Chemistry Society of Japan, 30, 63–66.

Koshimizu, K. 1963. Gibberellin-like substance in bamboo shoot. Nature, 196, 1306–1307.

Li Liren, et al. 1979. The relationship between bamboo shoot transferase and shoot growth rate. Plant Physiology, 5(1), 65–69.

Lu Xianhui, et al. 1986. DNase activity change of *Phyllostachys pubescens* shoot sprouting. Plant Physiology, 8(2), 187–192.

Adler, N. et al. 1963. Evidence for the widespread occurrence of a gibberellic acid-like substance in higher plants. Advances in Chemistry Series No. 287, 18, 473–486.

Paleg, S.C., et al. 1964. A barley endosperm bioassay for gibberellins. Nature, 199, 823–834.

Wu Guanming; Mcdoskey. 1983. Endogenous phytohormones analysis by gas chromatography. *In* Proceedings of National Tree Physiology and Bio-chemistry Symposium. pp. 32–33.

Bamboo for Sustainable Development, pp. 247–255
Eds. A. Kumar, I.V. Ramanuja Rao and Ch. Sastry

Pattern of mass-loss and nutrient accumulation and release in leaf litter in *Phyllostachys pubescens* (moso bamboo) stands

CAO QUNGEN, FU MAOYI and LI ZHENGCAI
Subtropical Forestry Research Institute, Chinese Academy of Forestry, Fuyang, Zhejiang, China

Abstract—Decomposition of leaf litter originating from moso bamboo stands was measured at two trial sites over 104 weeks during 1993 to 1995. The stands were receiving five different treatments and had different stand densities and stand components. The mass-loss rate can be expressed as the exponential function of decomposing time. The trial treatments showed marked influence on the mass-loss rate, which was higher in bamboo-oak mixed stands and in pure bamboo stands with a density of 3 015 culms per hectare than in other bamboo stands. Patterns of N concentration change in decomposing leaf litter were found to be very similar to those of P; both increased during the whole decomposing process. In contrast, the concentration of K, Ca and Mg decreased. No obvious influence by the different treatments was found on the dynamics of N, K and Mg concentrations. N was initially accumulated and later released, while K, Ca and Mg were released from the decomposing leaf litter during the whole observation period. The release of N and P during the late stage was caused mainly by mass-loss. As for K and Ca release, leaching was more important than mass-loss, especially in the early stage of the decomposition.

INTRODUCTION

Phyllostachys pubescens (moso bamboo) is the most important bamboo species in China, the distribution area of which is about 280 ha. During the late 1980s, some basic research was conducted on leaf litter biomass and its decomposition in pure moso stands in different site types. However, no such research was done on how the stand structure affects the decomposition of leaf litter. It is very important to have an optimum stand structure to ensure high bamboo production. The aim of this paper is to discuss the pattern of leaf litter mass-loss, and nutrient accumulation and release in moso bamboo stands. It also aimed to analyse the effect of stand density and stand component on leaf litter decomposition, and to provide data to determine the optimum structures of moso bamboo stand in relation to the nutrient cycling in bamboo forest ecosystems.

MATERIALS AND METHOD

Research site

Two research sites were selected. Site 1 was in the forestry station of Zhejiang Anji Forestry Research Institute (31°59' N, 119°45' E), where annual rainfall was 1485.4 mm, soil type was yellow-red loam and elevation 34 meters. Site 2, at Xiashu Forestry Station of Nanjing Forestry University (31°59' N, 119°14' E), had an annual mean temperature of 15.2°C, yellow-brown loam soil and an elevation of 200 meters.

Trial plots

Nine plots were laid out in Site 1 and six plots in Site 2. Of the five stand structure treatments (Table 1), two treatments were for bamboo culm standing density and three treatments for stand component. Each plot area was 400 m^2.

Table 1.
Stands structure treatments

Site	Treatment	Stand component	Culm standing density (culm/ha)	Diameter at breast height (cm)	Canopy close degree
1	1	10 bamboo	1 740	9.38	0.75
	2	10 bamboo	3 015	9.23	0.90
	3	10 bamboo	3 360	8.90	0.95
2	4	10 bamboo	3 300	8.84	0.90
	5	6 bamboo + 4 oak	1 320 (bamboo)+ 420 (oak)	9.20	0.95

Sample preparation

Leaves that were ready to fall down during the bamboo leaf falling season were collected as samples. The sample leaves were air-dried and put into 300 Nylon bags with 1×1 mm mesh size. Bag sizes were 30×40 cm and 25×30 cm. Each bag contained 31–51g air-dried leaf samples. Twenty sample bags were put on the ground around the stands in the trial plots. One sample bag from each plot was taken for analysis after 4, 8, 14, 22, 32, 44, 56, 68, 80, 92 and 104 weeks.

Litter mass-loss and nutrient content analysis

Sample bags were randomly taken and the leaf litter sample purified. Leaf litter samples from the same treatment were mixed together and weighed to determine mass-loss rate of leaf litter. The samples were analysed for their content of nitrogen (N) by a modified Kjelddahl producer, potassium (K) by emission, and calcium (Ca) and magnesium (Mg) by an atomic absorption spectrophotometer. P was determined calorimetrically on a Technicon II Auto-analyser.

Measurement of air temperature, humidity, soil temperature and soil water content

A soil thermometer was placed in a randomly selected trial plot to measure soil surface temperature. A sample was taken to measure soil water content at 0–10 cm depths, and a dry-wet ball thermometer used to measure air temperature and relative humidity at 50 cm above the ground inside the stand.

Soil microbe analysis

Microbiological method (Collins and Lyne 1976) was used.

RESULTS AND DISCUSSION

Patterns of mass-loss

Mass-loss rates were all rapid for all treatments in the early stage of decomposition but slowed down later (Fig. 1). The mean mass-loss rate during the first 44 weeks was 46.36%, which dropped to 22.68% after 60 weeks.

The mass-loss rate can be expressed as an exponential function of decomposition time: $M_t/M_o = e^{-kt}$, which was in accordance with leaf litter mass-loss pattern of many other tree species. The result (Table 2) by regression analysis showed that the decomposition parameter sequence of different treatments was: $K_{Treatment5} > K_{Treatment4} > K_{Treatment2} > K_{Treatment1} > K_{Treatment3}$. The decomposition rate of bamboo-oak leaf litter was the fastest, of which the half-life parameter and 95% loss time were 1.06 years and 4.61 years, respectively. For the slowest treatment 3,

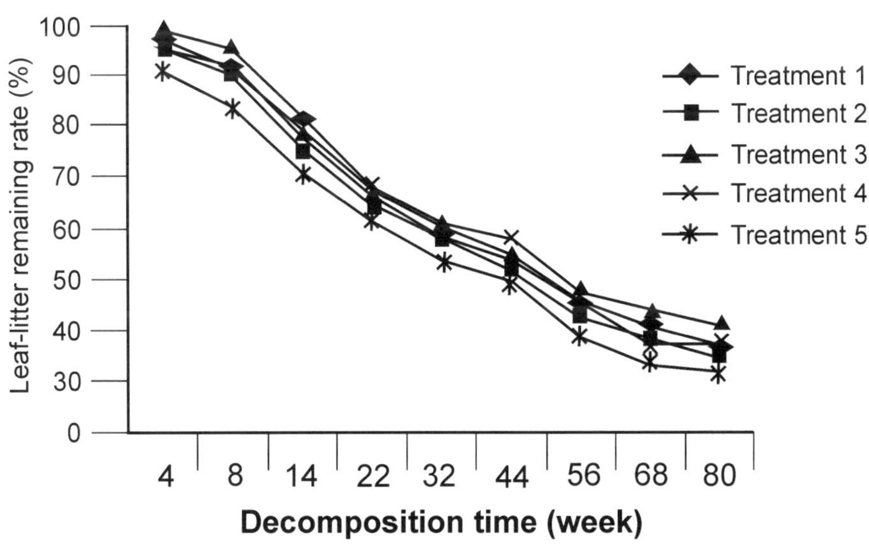

Figure 1. Dynamics of leaf litter decomposition under different treatments

these were 1.35 years and 5.48 years, respectively. The remaining rate of litter among different treatments varied significantly (Table 3). At Site 1, the variation of litter decomposition was mainly because of the difference in ecological factors and the mass of decomposing microbe (Table 4). At Site 2, however, the composition of the litter was another factor that greatly affected the decomposition. Lignin content, especially, had significant influence on decomposition rate, as verified research results in China and abroad.

Table 2.
Decomposition equation and decomposition period

Treatment	Decomposition equation	Decaying parameters (K)	Half-life of decomposition (0.693/k, year)	95% decomposition life (3/k, year)
1	$Mt=Mo/e^{0.5574t}$	0.5574	1.25	5.38
2	$Mt=Mo/e^{0.5736t}$	0.5736	1.21	5.23
3	$Mt=Mo/e^{0.5139t}$	0.5139	1.35	5.84
4	$Mt=Mo/e^{0.5761t}$	0.5761	1.20	5.21
5	$Mt=Mo/e^{0.6510t}$	0.6510	1.06	4.61

Table 3.
Variance analysis of leaf litter remaining rate after decomposition

Resource	Degree of freedom	Sum of squares of deviation	F	Fa
Between class	4	7,142.778	1,785.694	4.69
Between treatment	50	19,027.771	380.555	
Sum	54			

Table 4.
Comparison of factors affecting leaf litter decomposition of different treatments

Factors	Treatment 1	Treatment 2	Treatment 3	Treatment 4	Treatment 5
Surface soil temp. (°C)	15.35	15.35	15.04	14.11	14.75
Air temperature at 50 cm above ground	15.54	15.63	15.46	14.89	15.92
Relative air moisture at 50 cm above ground (%)	88.44	88.08	88.07	84.16	84.84
Soil water content at 0–10 cm underground (%)	24.36	24.01	23.48	23.78	26.64
Soil microbe mass (× 10^7pcs/g dry soil)	79.62	92.02	27.98	22.16	61.73

Nutrient concentration variation within decomposition

Each nutrient showed distinct characteristics of concentration variation with decomposition process. Concentration of N and P increased, while K, Ca and Mg decreased (Table 5). The mean concentrations of N, P, K, Ca and Mg of different treatments during the 80th week were 166.5%, 202.7%, 21.4%, 25.3% and 24.8%, respectively, of their initial content. The trend of N and P concentration variation was similar, but very different among K, Ca and Mg. The concentration of K decreased greatly till 4 weeks and then slowly; the concentration of Ca and Mg decreased rapidly till 8 weeks and 44 weeks, respectively, and then slowed down.

Table 5
Nutrient concentration variation during decomposition of leaf litter

Treatment	Nutrient	Decomposition time (weeks)									
		0	4	8	14	22	32	44	56	68	80
1	N	1.023	1.332	1.084	0.927	1.020	1.350	1.470	1.490	1.824	1.720
	P	0.033	0.054	0.039	0.033	0.075	0.060	0.053	0.046	0.054	0.067
	K	0.377	0.173	0.174	0.216	0.240	0.269	0.138	0.156	0.101	0.135
	Ca	0.569	0.472	0.259	0.203	0.183	0.131	0.202	0.172	0.193	0.182
	Mg	0.218	0.269	0.334	0.104	0.125	0.131	0.075	0.061	0.062	0.063
2	N	1.023	1.395	1.237	0.885	1.390	1.810	1.040	1.431	1.732	1.863
	P	0.033	0.054	0.038	0.035	0.070	0.077	0.055	0.055	0.053	0.070
	K	0.377	0.218	0.218	0.213	0.216	0.226	0.150	0.137	0.113	0.132
	Ca	0.569	0.471	0.227	0.190	0.106	0.124	0.198	0.183	0.202	0.175
	Mg	0.218	0.272	0.264	0.253	0.144	0.158	0.079	0.058	0.060	0.050
3	N	1.023	1.431	1.356	1.514	1.190	1.760	1.190	1.432	1.646	1.905
	P	0.033	0.053	0.049	0.041	0.078	0.076	0.066	0.051	0.068	0.070
	K	0.377	0.163	0.201	0.173	0.249	0.205	0.110	0.130	0.105	0.116
	Ca	0.569	0.483	0.219	0.380	0.202	0.118	0.192	0.165	0.193	0.162
	Mg	0.218	0.218	0.330	0.126	0.142	0.126	0.087	0.062	0.071	0.056
4	N	1.483	1.564	0.973	1.700	1.510	2.700	2.070	1.275	1.676	1.750
	P	0.050	0.067	0.055	0.071	0.090	0.047	0.090	0.071	0.081	0.083
	K	0.813	0.199	0.181	0.221	0.185	0.126	0.087	0.160	0.110	0.133
	Ca	0.813	0.776	0.392	0.401	0.282	0.233	0.472	0.363	0.283	0.273
	Mg	0.245	0.243	0.193	0.258	0.191	0.196	0.091	0.083	0.063	0.054
5	N	1.153	1.321	0.906	1.040	1.430	2.280	1.520	1.220	1.665	2.240
	P	0.036	0.047	0.045	0.043	0.093	0.071	0.068	0.060	0.073	0.083
	K	0.925	0.185	0.192	0.181	0.139	0.161	0.107	0.090	0.078	0.099
	Ca	0.925	0.865	0.459	0.427	0.425	0.337	0.533	0.463	0.402	0.352
	Mg	0.150	0.177	0.172	0.125	0.230	0.157	0.063	0.052	0.048	0.047
Mean	N	1.141	1.409	1.111	1.213	1.308	1.980	1.460	1.370	1.709	1.900
	P	0.037	0.053	0.045	0.045	0.081	0.066	0.066	0.057	0.066	0.075
	K	0.574	0.188	0.194	0.201	0.206	0.197	0.118	0.135	0.079	0.123
	Ca	0.689	0.613	0.311	0.320	0.240	0.189	0.319	0.269	0.253	0.174
	Mg	0.210	0.236	0.298	0.173	0.166	0.154	0.095	0.063	0.061	0.052

Table 6.
Correlation-regression among nutrient concentrations (%) and remaining rate of decomposing leaf litter

Nutrients	Regression equation	Coefficient	F	Between treatments+
N	$Y_N = 740.72–54.31X_N$	0.70	7.67*	1.21
P	$Y_P = 138.56–1226.90X_P$	0.73	9.15**	2.21
K	$Y_K = 42.35+118.07X_K$	0.69	7.10*	0.06
Ca	$Y_{Ca} = 29.79+107.70X_{Ca}$	0.79	12.93**	5.01
Mg	$Y_{Mg} = 25.03+273.38X_{Mg}$	0.94	61.93**	0.32

Notes: * Significantly correlated; ** Very significantly correlated – $F_{0.05(1,8)} = 5.32$ and $F_{0.01(1,8)} = 8.65$; + $F_{0.1\ (4,45)} = 2.07$ and $F_{0.01(4,45)} = 3.77$

The nutrient concentration variation among the treatments was significant for Ca at 0.01 level and for P at 0.1 level, but not significant for other nutrients (Table 6). There was a significant linear functional relationship between decomposition remaining rate and nutrient concentration: highest with (0.94) and lowest with K (0.69). This indicated that the variation of nutrient concentration during decomposition was caused mainly by the mass-loss and that, with regard to the variation of K concentration, leaching was also an important cause.

Gray-system analysis showed some relationship among the nutrients that remained in the decaying leaf litter. When E was at 0.5 level, the gray co-relationships of $R_{N.P}$, $R_{K.Ca}$, $R_{K.Mg}$ and $R_{Ca.Mg}$ were 0.68, 0.81, 0.68 and 0.72, respectively.

Nutrient return by litter decomposition

N accumulated in the decomposing leaf litter during 4–14 weeks and then released, although some increment or reduction in N concentration occurred (Figs. 2–6). The variation pattern of P was similar to that of N but release occurred later and slower. N and P were found releasing, though their concentrations in the decomposing litter increased, showing that it was mainly the mass-loss rate that determined whether the nutrient accumulate or release. Except for Mg accumulation in the early decomposition stage of 4 weeks, K, Ca and Mg were released throughout the decomposition process, rapidly in the initial 14 weeks and then slowing down. Comparing the mass-loss curve, the releases of K and Ca were faster than leaf litter mass-loss, which reflected the importance of leaching on the movement of K and Ca. When leaf litter decomposed for 80 weeks, the mean moving rate sequence of the nutrient was: K (91.28%) > Ca (89.2%) > Mg (88.2%) > N (40.08%) > P (29.46%). The nutrient moving rate was also related to the trial site condition and treatments. Moving rates of N and P were slower in bamboo-oak mixed leaf litter than in pure bamboo leaf litter. In contrast, Ca and Mg showed faster moving rates in bamboo-oak mixed leaf litter than in pure bamboo leaf litter.

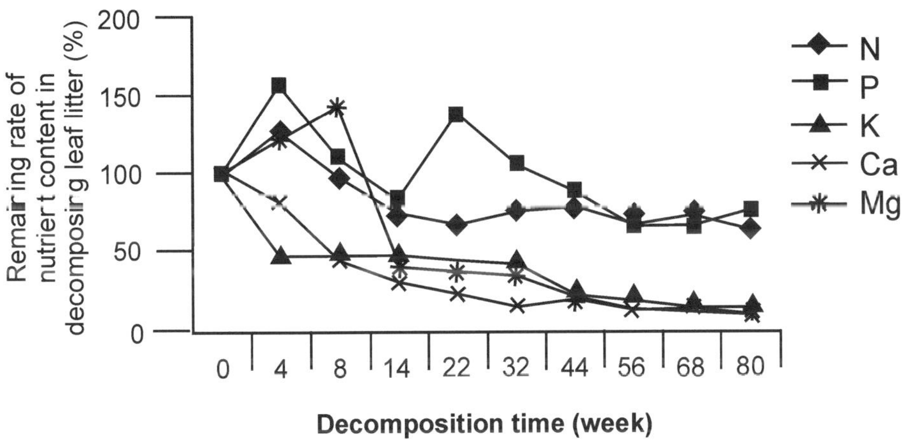

Figure 2. Variation of nutrient content in decomposing leaf litter of Treatment 1

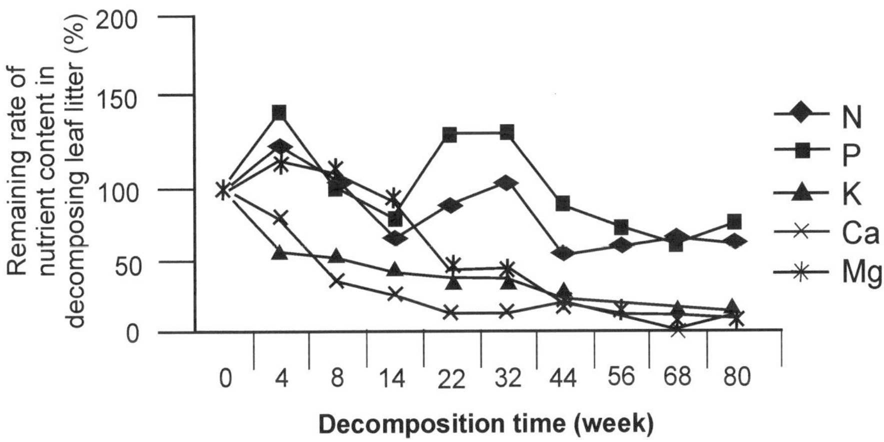

Figure 3. Variation of nutrient content in decomposing leaf litter of Treatment 2

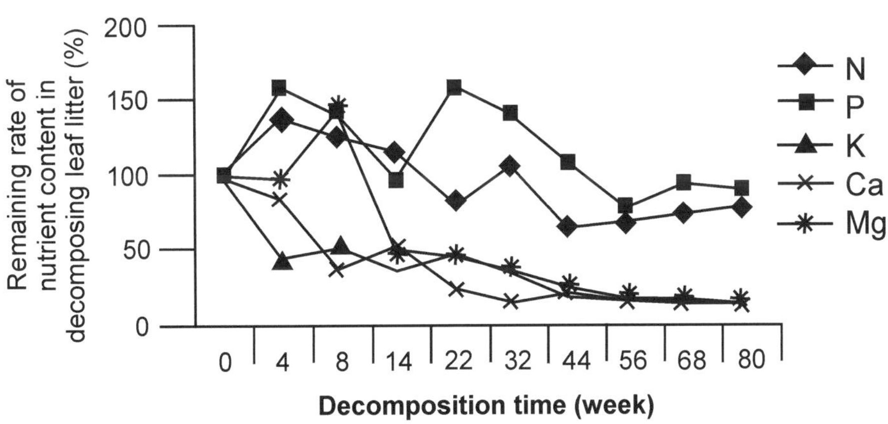

Figure 4. Variation of nutrient content in decomposing leaf litter of Treatment 3

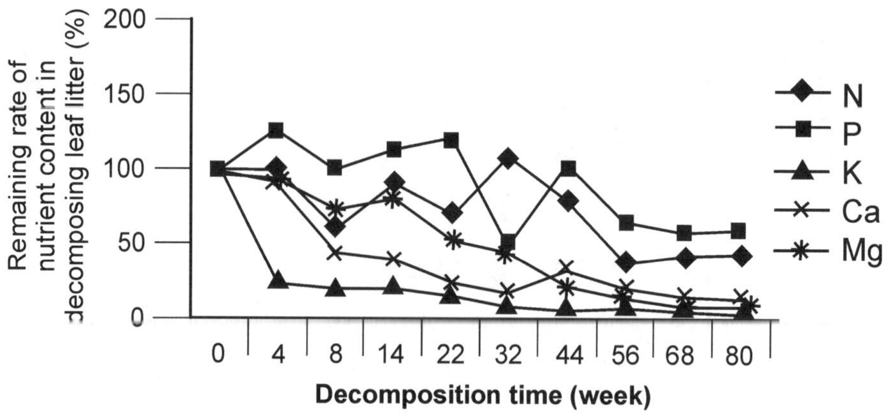

Figure 5. Variation of nutrient content in decomposing leaf litter of Treatment 4

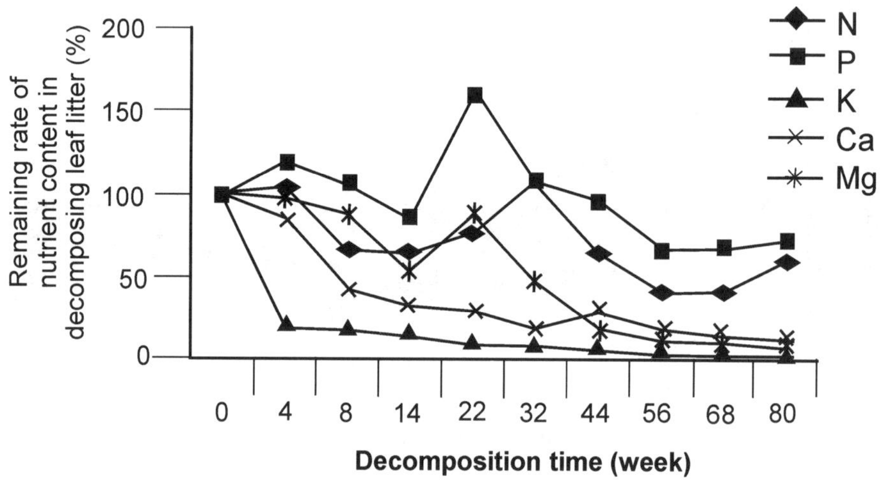

Figure 6. Variation of nutrient content in decomposing leaf litter of Treatment 5

CONCLUSION

Relationship of an exponential function was noticed between leaf litter remaining rate and decomposition time. The stand density and component had obvious effect on decomposition rate: bamboo leaf litter under Treatment 2 had shorter decomposition period than the other two treatments; and bamboo-oak mixed leaf litter decomposed faster than pure bamboo leaf litter. To enhance bamboo production, it would be prudent to adopt a reasonable stand density and suitable stand component to accelerate nutrient cycling of the bamboo forest ecosystem and thus sustain soil fertility.

Each nutrient displayed distinct concentration variation characteristics during decomposition. The concentrations of N and P increased with the process of decomposition, while K, Ca and Mg decreased. Mass-loss by decomposition had a significant effect on nutrient concentration variation. Leaching showed a great effect on the concentration of K, which has a very active and fast-moving ion. Treatments did not show significant effect on the variation of N, K and Mg concentrations in the remained leaf litter.

There was some relationship among the nutrients in the decomposing leaf litter. It would be very beneficial for the sustainable production of bamboo stands if full use of these relationships were to be made to adjust nutrient balance of bamboo forest ecosystems.

N and P were accumulated in the initial stage of decomposition and then released, while K, Ca, Mg were released throughout the decomposition period. Mass-loss determined the release of N and P. Leaching was also an important factor that caused the release and return of K and Ca. The sequence of release rate was: K > Ca > Mg > N > P. Stand density and stand component had great influence on the nutrient accumulation and release during bamboo leaf litter decomposition.

ACKNOWLEDGEMENTS

This paper is based on part results of the research program "Study on the mechanism of leaf litter decomposition of moso bamboo stands", which was supported by the Science Foundation of Zhejiang Province.

REFERENCES

Collins, C.H.; Lyne, P.M. 1976. Microbiological methods. Butterworths, London, UK.

FURTHER READING

Fu Maoyi; Fang Mingyu; Xie Jingzhong. 1989. Leaf litter and its decomposition in bamboo timber stands. Forestry Research, 2(1), 25–32.

Olson J.S. 1963. Energy storage and the balance of producers and decomposers in ecological systems. Ecology, 44, 322–331.

Rustad, L.E. 1994. Element dynamics along a decay continuum in a red spruce ecosystem in Maine, USA. Ecology, 75(4), 867–878.

Upadhyay, V.P.; Singh, J.S.; Meentemeyer, V. 1989. Dynamics and weight loss of leaf litter in central Himalayan forests: abiotic versus litter quality influence. Journal of Ecology, 77, 147–161.

Berg, B.; Staaf, H.; Wessen, B. 1987. Decomposition and nutrient release in needle litter from nitrogen-fertilized Scots pine stands. Scandinavian Journal of Forest Research, 2, 399–415.

Liu Shifeng; Guo Tianbang. 1991. The Gary System Theory and its application. Henan University Press House, Kaifeng, China.

Bamboo for Sustainable Development, pp. 257–268
Eds. A. Kumar, I.V. Ramanuja Rao and Ch. Sastry

Moisture content of two Malaysian bamboos in relation to age, culm height, site and harvesting month

ABD. LATIF MOHMOD
Forest Research Institute Malaysia (FRIM), Kuala Lumpur, Malaysia

WALTER LIESE
Chair for Wood Biology, Hamburg University, Hamburg, Germany

Abstract—This paper studies the variation in the initial moisture content of 1-, 2-, 3- and 4-year-old *Bambusa vulgaris* and *Gigantochloa scortechinii* collected from four experimental sites in Peninsular Malaysia over 12 consecutive harvesting months. The initial moisture content differs significantly with species, site, age, culm height and harvesting month. The highest and lowest moisture contents in both species occurred at the basal rather than at the top portion, and in the younger than in the older culm. With regards to seasonal changes, the initial moisture content in each bamboo species increased with increase in monthly rainfall but decreased with the increasing temperature at each site. The difference in moisture content levels in both species between the driest and the wettest months is relatively higher at the basal than at the top portion, and in the younger than in the older bamboo.

INTRODUCTION

The yield of a bamboo plantation is an important criterion for judging its production capacity. Such data are needed when comparing the productivity of bamboo plantations with plantations of fast-growing tree species, and also for estimating the quantum of material that would be available for processing and utilization. While assessing the moisture content of bamboo, often the fresh/green weight is taken for practical convenience and a reduction factor for the water content applied to obtain the air-dry weight. Such measures are applied during the delivery of culms to paper mills, where the price is calculated on the basis of truckloads of fresh culms from the forest. Investigations on the properties of bamboo often refer to the moisture content in relation to seasoning procedures and shrinkage behavior also (Rahman and Ishaq 1947). A variation of the moisture content values in both young and old culms as well as a distinct fluctuation through the year was stated for Indian bamboos by Liese and Grover (1961). An investigation the possible influence of site conditions by comparing the same species in different environments has not been conducted so far.

For the naturally occurring Malaysian bamboos, the variation of moisture content in the standing culms is not known. This information, however, is important as moisture content is of consequence for estimating yield and identifying end-uses. Therefore, the general objective of this study was to observe the moisture content and its variation in two commonly utilized Malaysian bamboo species – *Bambusa vulgaris* Schrad. and *Gigantochloa scortechinii* Gamble – from four areas over 12 harvesting months and to examine the data in relation to the age and height of culm, and to investigate the possible influence of external factors – such as site, rainfall and surrounding temperature – on moisture content. Understanding the relationship of species, age, culm height and felling month to the related technological properties of bamboo will facilitate proper species selection for industrial applications.

MATERIALS AND METHODS

Source of materials

The bamboo samples evaluated in this study are part of a larger investigation on some selected properties of these two species by Abd. Latif (1996). They were collected over 12 successive months (January-December 1991) from four research plots: Sik in Kedah (established in 1988); Sungai Lembing in Kuantan, Pahang (1988); FRIM in Kepong, Selangor (1986); and Broleh in Batu Pahat, Johor (1987). The study sites are described in detail by Abd. Latif (1996), and in brief by Abd. Latif and Liese (this volume). Laboratory studies were conducted at FRIM.

From each culm, two diametrical samples of one cm length were taken from the basal, middle and top portions. Five culms from each age group, month and site – a total of 5 760 samples from 960 culms for each species – were used in the study.

Field procedures

On every first week of the month, selected 1-, 2-, 3- and 4-year-old bamboo culms were cut at about 30 cm above ground level. Each cut culm was then marked and sectioned at about 4 m intervals into three portions: basal, middle and top. A coat of paraffin wax was applied to the cut surfaces, to minimize evaporation and prevent fungal and insect attacks, before the samples were taken to FRIM laboratory for further studies. Despite the precaution, some loss of moisture was envisaged during storage and transportation. Once in the samples were in FRIM, their both ends were sealed with epoxy resin and wrapped with polythene. The samples were then refrigerated at 5°C until test to ensure their freshness.

Determination of moisture content

The method used for the determination of moisture content of bamboo was based on Indian Standard of Testing Round Bamboo (IS 6874: 1973). Fresh duplicate samples (1 cm × 1 cm × thickness of bamboo) from the basal, middle and top portions were cut, weighed and dried in an oven at a temperature of 103 ±2°C until constant weight. The samples were then cooled for 30 minutes in a dessicator prior to weighing. The moisture content of the test sample was expressed as a percentage loss in weight of the final oven-dry weight to that of the green one.

Statistical analysis

Correlations were traced between the moisture content, age and culm height in order to relate the importance of the harvestable age and height within and between sites. In addition, the moisture content was correlated with harvesting month (the parameters being rainfall and temperature) to determine its sensitivity to change with different environmental conditions at each site. The data on moisture content were statistically analysed using multi-factor analysis of variance (ANOVA), the factors being age, height, site and harvesting month. Duncan's New Multiple Range Test (DMRT) was used to compare mean values of moisture content in both bamboo species (Abd. Latif 1996).

RESULTS AND DISCUSSION

The average results on moisture content of both bamboo species at different ages and culm heights enumerated from samples collected monthly at Batu Pahat, FRIM, Sik and Sungai Lembing were tabulated (Appendixes 1–4). The DMRT is given in Table 1, and regression analysis on the effects of age, height and seasonal factors (such as rainfall and temperature) at each study site is shown in Table 2.

As shown in Table 1 and the Appendixes, the two bamboo species exhibit different moisture values, regardless of age, height, site and harvesting month. The mean values vary between 50% to 221% in *G. scortechinii* and 47% to 157% in *B. vulgaris*.

The analysis of variance (Abd. Latif 1996) indicates that the initial moisture content differs significantly with species, site, age, culm height and harvesting month. The initial moisture contents were observed to be:

G. scortechinii – 55–134% (Batu Pahat); 50–221% (FRIM); 55–134% (Sik); and 54–132% (Sungai Lembing).

B. vulgaris – 51–123% (Batu Pahat); 49–128% (FRIM); 52–166% (Sik); and 57–149% (Sungai Lembing).

Table 1.
Mean values of moisture content (%) of bamboos at different age, height and harvesting month at the study locations

Parameters	Batu Pahat		FRIM		Sik		Sungai Lembing	
	A	B	A	B	A	B	A	B
Culm age								
1 year	98a	104a	104a	100a	97a	116a	97a	108a
2 years	93b	95b	78b	99b	82b	90b	84b	88b
3 years	82c	85c	72b	75c	72d	79c	65c	80c
4 years	79d	81c	76b	61d	77c	68d	75d	62d
Culm portion								
Basal	101a	99a	85a	91a	94a	95a	89a	91a
Middle	86b	92b	74b	85b	80b	89b	80b	83b
Top	77d	83c	66c	75c	72c	81c	70c	73c
Harvest month								
January	84c	94c	68d	82b	72d	75e	76c	90b
February	82c	98b	77c	83b	74c	79d	73d	87c
March	87c	97b	78c	82b	88b	86c	71d	78d
April	95b	97b	88b	91a	77c	88c	73d	80d
May	85c	96b	82b	83b	90a	98b	76c	80d
June	92b	71d	68d	75c	77c	90c	69e	74e
July	86c	74d	75c	83b	83b	96b	75c	78d
August	77d	99a	79c	89b	84b	101a	72d	77e
September	88c	94b	88b	92a	85b	89c	77b	75e
October	86c	89c	101a	94a	97a	100a	79b	94b
November	93b	93c	90b	85b	87b	89c	78b	92b
December	102a	103a	98a	85b	76c	82d	82a	104a

Notes: Values with the same alphabet down the column are not significantly different at $p<0.05$; A: *G. scortechinii*; B: *B. vulgaris*

The variation in the initial moisture contents of the two bamboo species at the four different sites was almost similar. Moisture content was more at the basal portion than the top, and more in the younger culms than in the older. When the moisture content was correlated with age and culm height in each species at different sites (Table 2), it was seen to significantly decrease with culm height and age (r = –0.34 and –0.66 in *G. scortechinii* and r = –0.27 and –0.71 in *B. vulgaris*, respectively). These observations were supported also by the DMRT (Table 1). While the initial moisture content was about 104% and 100%, respectively, for 1-year-old culms of *G. scortechinii* and *B. vulgaris* at FRIM, lower moisture values were recorded for 4-year-old culms (76% and 61%). Specimens from FRIM also showed higher moisture values at the basal portion (85% in *G. scortechini* and 91% in *B. vulgaris*) of the culm than at the top (66% and 75%, respectively).

Table 2
Regression analysis between moisture content of bamboo against rainfall, temperature, age and culm height at the study locations

Regression	Bamboo	Site				
		Batu Pahat	FRIM	Sik	Sungai Lembing	Average
Moisture content/ Rainfall	A	0.28*	0.16*	0.35**	0.16*	0.22*
	B	0.14*	0.12ns	0.17*	0.15*	0.14*
Moisture content/ Temperature	A	–0.13ns	–0.03ns	–0.09ns	–0.04ns	–0.08ns
	B	–0.64**	–0.16*	–0.26*	–0.24*	–0.45**
Moisture content/ Culm age	A	–0.71**	–0.50**	–0.70**	–0.84**	–0.66**
	B	–0.71**	–0.76**	–0.78**	–0.68**	–0.71**
Moisture content/ Culm height	A	–0.43**	–0.25*	–0.43**	–0.41**	–0.34**
	B	–0.28*	–0.30*	–0.24*	–0.31**	–0.27*

Notes: A: *G. scortechinii;* B: *B. vulgaris*
ns: not significant at $p<0.05$; *: significant at $p<0.05$; **: highly significant at $p<0.01$

This confirmed the general observation that older culms have lower moisture content: the decrease of moisture with height is related to the lower percentage of parenchyma cells as the water storage site in the upper portion of the culm (Liese and Grover 1961).

Seasonal changes in the mean initial moisture content in *G. scortechinii* and *B. vulgaris* according to the study sites were given in Table 1. Correlating the moisture content with mean monthly rainfall and temperature (Table 2) provides a satisfactory explanation. In general, the moisture content increases with increase in monthly rainfall but decreases with the increasing surrounding temperature. Figs. 1 and 2, for example, clearly show the relation between initial moisture content with rainfall and temperature over the entire study period for both the bamboo species at Sungai Lembing.

Regardless of age and culm height, the highest and lowest mean initial moisture contents in *G. scortechinii* at Batu Pahat were observed in December (102%) and August (77%), respectively, and in *B. vulgaris* in December (103%) and June (71%), respectively. At FRIM, the highest moisture content occurred in October (101% and 94% for *G. scortechinii* and B. *vulgaris,* respectively) and the lowest in January (68% for *G. scortechinii*) and June (75% for *B. vulgaris*). The lowest reading for the mean moisture content at Sik was in January (72% in *G. scortechinii* and 75% in *B. vulgaris*), while the highest was in October (97% for *G. scortechinii* and 100% for *B. vulgaris*). At Sungai Lembing, the highest and lowest mean initial moisture contents of *G. scortechinii* and *B. vulgaris* were in December (82% and 104%, respectively) and June (69% and 74%, respectively). Liese (1959) reported for 3- to 4-year old *Dendrocalamus strictus* a distinct trend of high moisture content during the rainy season when compared with the dry season and related differences between the bottom and top parts (Fig. 3). The rela-

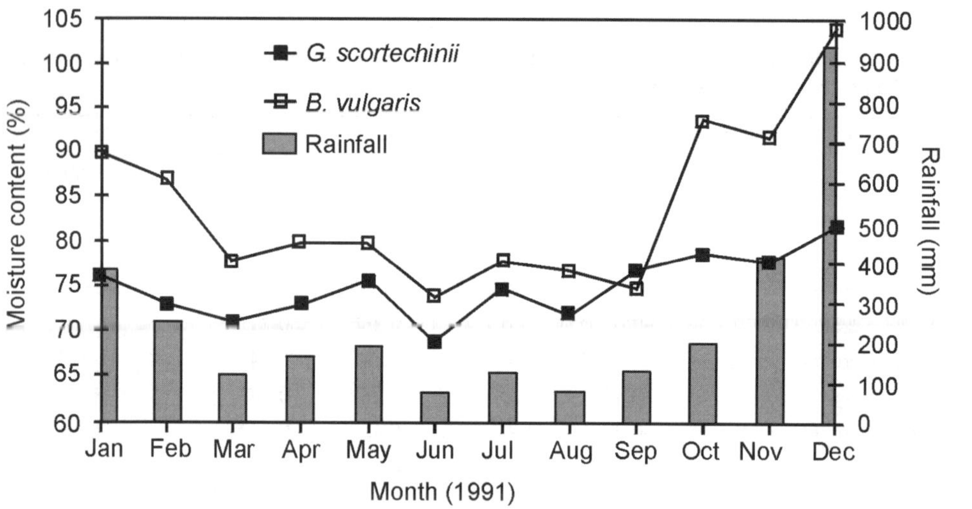

Figure 1. Correlation of monthly moisture contents of the two bamboo species with monthly rainfall at Sungai Lembing

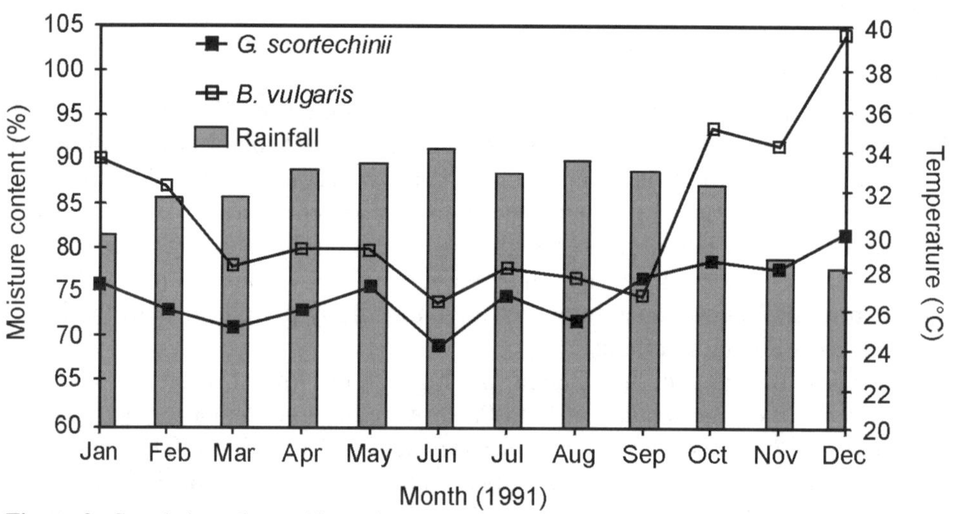

Figure 2. Correlation of monthly moisture contents of the two bamboo species with monthly temperature at Sungai Lembing

tively higher soil moisture, accompanied by a considerably low surrounding temperature but high humidity, could be the reason for the high moisture content in the wettest months.

The difference in moisture content levels in both bamboo species between the driest and wettest months was observed to be relatively higher at the basal than the top portion. This is because more parenchyma cells are located within the thicker culm wall of the basal portion than the top. The results (Appendixes 1–4) further show that differences in moisture content are considerably lower in the younger culms than in the older. This may be attributed to the comparatively equal distribution of water in young culms (Liese and Grover 1961). The largest difference in moisture content between the driest and wettest months may amount to at least 100% (Appendixes 1–4). For example, the difference in moisture content of *G. scortechinii* was observed to be about 200% (at the basal part of a 3-year-old culm). A similar finding was reported also for *Dendrocalamus strictus* (Fig. 3). Besides the difference in the water-holding capacity of parenchyma cells (Liese and Grover 1961), the variation in moisture content between and within the species, culm height and age at different harvesting month could also be related to the variation in the dissolution of cell sap substances (such as mineral salts, sugar and organic acid), which could influence the water vapor pressure in the plant (Noggle and Fritz 1979).

The results are in agreement with observations on the influence of culm height and age for *Phyllostachys pubescens* by Jiang et al. (1995). Lu et al. (1985) reported a decrease in moisture content for *Phyllostachys glauca* culms of up to 7 years old from four sites.

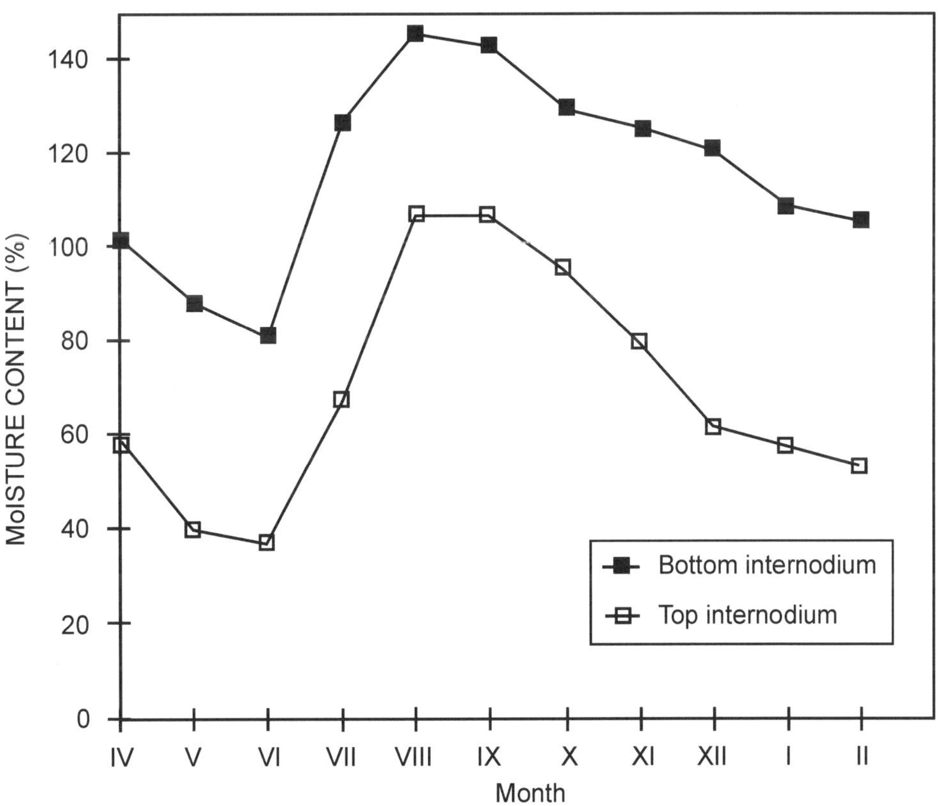

Figure 3. Seasonal changes of moisture content in *D. strictus*

CONCLUSION

The variation in the moisture content of the two bamboo species at four different sites was almost similar – it decreased with age and height of the culm. The initial moisture content of both species was observed to be increase with the amount of rainfall but tended to decrease with the increasing temperature at each site. These variations suggest an importance of topographical factors and harvesting months that is equal to the intrinsic factors of age and culm height during the cutting of bamboo. The average moisture content of the two bamboo species over the entire study period ranged from 50% to 221% in *G. scortechinii* and 47% to 157% in *B. vulgaris*. These considerable differences have to be considered when determining the yield of material from fresh bamboo culms and for various aspects of bamboo processing (Abd. Latif 1993).

REFERENCES

Abd. Latif, M. 1996. Some selected properties of two Malaysian bamboo species in relation to age, height, site and seasonal variations. Ph.D. thesis, Universiti Pertanian Malaysia, Serdang, Malaysia.

Abd. Latif, M. 1993. Effects of age and heights on machining properties of bamboos. Journal Tropical Forest Science, 5(3), 528–535.

Jiang, Z.; Meng, Z.; Zhang, Y. 1995. Moisture content of *Phyllostachys pubescens*. Journal of Bamboo Research, 14 (2), 58–67.

Liese, W. 1959. Bamboo preservation and soft rot. F.A.O. Report No. 1106 to the Government of India. 37 pp.

Liese, W.; Grover, P.N. 1961. Untersuchungen über den Wassergehalt von indischen Bambushalmen. Ber. Deut. Bot. Gesellschaft, 74: 105–117.

Lu, X.; Wang, K.; Yi, X.; Liou, J.; He, J. 1985. A study on the physico-mechanical properties of culmwood of *Phyllostachys glauca* of Shandong. Journal of Bamboo Research, 4 (2), 98–106.

Noggle, R.G.; Fritz, J.G. 1979. Introductory plant physiology. Prentice Hall Biological Science Series, 687 pp.

Rahman, M.A.; Ishaq, S.M. 1947. Seasoning and shrinkage of bamboo. Indian Forest Records, 4 (2), 1–22.

APPENDIXES

Appendix 1.
Mean monthly moisture content (%) of the bamboos at Batu Pahat, Johor

Month	Age (years)/Portion											
	1			2			3			4		
	Basal	Middle	Top	Basal	Middle	Top	Basal	Middle	Top	Basal	Middle	Top
G. scortechinii												
January	106	96	86	101	92	82	87	77	63	83	73	63
February	106	96	81	91	82	82	87	78	58	83	73	64
March	108	99	89	104	94	85	90	80	65	85	75	66
April	109	100	90	195	95	86	91	81	66	86	77	67
May	107	97	88	103	93	83	88	79	64	84	74	65
June	112	102	92	107	98	88	103	93	69	89	79	70
July	107	97	87	102	93	83	98	88	64	84	74	64
August	98	88	79	84	84	69	89	80	55	75	65	56
September	111	101	91	96	97	87	92	83	68	88	78	69
October	111	101	91	96	97	82	92	82	68	88	78	48
November	118	71	81	109	74	80	109	69	68	136	94	108
December	134	101	98	94	80	93	137	107	91	103	88	99
Average	111	96	88	107	90	83	97	83	67	90	77	70
B. vulgaris												
January	114	107	99	100	102	94	96	88	75	91	84	76
February	120	112	104	105	98	90	101	93	81	97	89	81
March	117	110	102	103	95	88	109	91	78	94	87	79
April	119	110	103	104	97	89	100	92	79	96	88	80
May	114	106	98	100	102	94	105	98	75	91	83	76
June	91	83	75	76	78	71	72	64	51	68	60	52
July	94	86	78	80	82	74	75	68	55	71	63	55
August	119	111	103	104	107	99	100	92	79	96	88	80
September	112	104	96	98	100	92	103	96	73	89	81	73
October	107	99	91	92	95	87	98	90	68	84	76	68
November	116	108	100	101	94	96	87	79	77	93	85	77
December	123	115	107	108	111	103	104	96	85	100	92	84
Average	112	104	96	98	97	90	96	87	73	89	81	73

Appendix 2.
Mean monthly moisture content (%) of the bamboos at FRIM, Selangor

Month	Age (years)/Portion											
	1			2			3			4		
	Basal	Middle	Top	Basal	Middle	Top	Basal	Middle	Top	Basal	Middle	Top
G. scortechinii												
January	73	74	65	69	66	65	71	64	52	86	77	64
February	97	80	81	92	77	45	88	69	58	81	79	74
March	102	104	111	102	86	64	62	51	41	78	74	61
April	108	103	134	79	67	62	98	95	73	102	82	50
May	90	91	82	77	65	71	115	85	78	90	73	63
June	81	74	86	74	80	78	59	54	45	62	63	54
July	89	63	76	77	89	67	84	63	57	93	81	62
August	75	78	69	81	68	73	99	105	42	101	81	81
September	151	120	111	83	83	69	67	64	61	90	89	63
October	107	96	89	126	101	95	146	95	93	93	91	69
November	120	105	113	105	94	88	88	77	50	87	80	62
December	221	220	194	63	59	50	59	45	40	77	63	61
Average	110	101	101	86	78	69	86	74	58	87	78	64
B. vulgaris												
January	112	96	91	106	97	92	66	61	61	64	62	60
February	102	98	88	101	100	89	89	89	52	60	59	57
March	121	96	94	106	99	83	63	61	58	61	64	59
April	116	104	91	108	102	97	98	105	49	65	69	67
May	104	91	81	94	93	92	87	84	78	62	60	51
June	89	70	76	130	134	65	50	54	65	52	51	49
July	104	91	81	94	94	92	86	84	78	62	60	50
August	119	99	92	106	101	89	87	89	89	59	61	58
September	118	104	98	121	111	98	95	88	52	72	69	57
October	128	121	96	105	98	98	100	98	75	67	63	61
November	121	122	95	101	98	86	75	71	54	62	63	55
December	112	99	92	105	98	92	83	64	63	65	68	63
Average	112	99	89	106	102	89	82	79	64	63	62	57

Appendix 3.
Mean monthly moisture content (%) of the bamboos at Sik, Kedah

Month	Age (years)/Portion											
	1			2			3			4		
	Basal	Middle	Top	Basal	Middle	Top	Basal	Middle	Top	Basal	Middle	Top
G. scortechinii												
January	97	87	78	83	73	63	68	59	59	74	64	55
February	99	90	80	85	76	66	71	61	52	77	67	57
March	114	104	95	100	90	80	85	76	66	91	81	72
April	102	93	83	88	79	69	74	64	55	80	70	60
May	116	106	96	102	92	82	87	78	68	93	83	74
June	102	93	83	88	78	69	74	64	54	79	70	60
July	108	99	89	94	84	75	80	70	60	85	86	66
August	110	100	91	96	86	76	81	72	62	87	77	58
September	118	71	81	109	64	70	99	59	62	116	94	88
October	134	101	98	94	70	83	127	97	86	103	88	79
November	113	103	93	98	89	79	84	74	65	90	80	70
December	102	92	82	87	78	68	73	63	54	79	69	59
Average	110	95	87	94	80	73	84	70	62	88	77	67
B. vulgaris												
January	111	103	95	66	49	52	69	86	77	73	60	52
February	104	97	89	90	82	75	76	68	60	62	74	66
March	115	107	99	101	93	85	86	79	71	62	64	57
April	116	108	101	102	94	86	88	80	72	73	66	58
May	126	118	111	112	104	96	97	90	82	83	75	68
June	118	110	103	104	96	88	89	82	74	75	67	60
July	124	116	108	110	102	94	95	88	80	81	73	65
August	157	159	166	97	89	81	82	74	67	88	77	62
September	117	109	102	103	95	87	89	81	73	74	67	59
October	128	120	112	114	106	98	99	92	84	85	77	70
November	136	152	145	88	78	75	77	63	51	62	62	60
December	108	100	92	94	86	78	79	71	64	75	67	59

Appendix 4.
Mean monthly moisture content (%) of the bamboos at Sungai Lembing, Pahang

Month	Age (years)/Portion											
	1			2			3			4		
	Basal	Middle	Top	Basal	Middle	Top	Basal	Middle	Top	Basal	Middle	Top
G. scortechinii												
January	109	99	89	94	85	75	80	70	61	86	76	66
February	106	96	86	91	82	72	77	67	58	83	73	63
March	103	94	84	89	79	70	75	65	55	80	71	61
April	105	95	86	91	81	71	77	67	57	82	73	63
May	109	99	89	94	85	75	80	70	61	86	76	66
June	102	92	83	88	78	68	73	64	54	79	69	60
July	108	99	89	94	84	75	80	70	60	85	76	66
August	105	95	86	91	81	71	74	67	57	82	72	63
September	101	112	99	99	65	54	64	60	55	85	78	68
October	98	89	79	84	75	65	70	60	51	86	76	76
November	106	96	86	92	82	72	77	68	58	83	77	75
December	132	118	97	103	94	78	72	71	56	90	87	71
Average	107	99	88	92	81	70	75	67	57	84	75	67
B. vulgaris												
January	114	118	117	95	87	90	96	98	85	74	61	48
February	124	117	109	77	68	69	96	80	60	102	83	62
March	107	100	92	93	85	78	79	71	63	64	57	49
April	109	101	93	95	87	79	80	73	65	66	58	51
May	109	101	94	95	87	79	81	73	65	66	58	51
June	104	96	88	89	82	74	73	67	60	61	53	45
July	108	100	92	93	86	78	79	71	63	65	57	49
August	106	98	90	92	84	76	77	69	62	63	55	47
September	105	97	89	90	83	75	76	68	61	62	54	47
October	124	116	108	109	102	94	95	87	80	81	73	65
November	121	113	105	107	99	91	92	85	77	78	70	63
December	143	149	123	111	103	95	107	113	81	82	74	66
Average	110	105	95	92	85	77	84	79	68	72	63	54

Bamboo for Sustainable Development, pp. 269–278
Eds. A. Kumar, I.V. Ramanuja Rao and Ch. Sastry

A study on the effects of inoculating associated nitrogen-fixing bacteria on moso bamboo seedlings

WU XIAOLI and GU XIAOPING
Research Institute of Subtropical Forestry, CAF, Fuyang, Zhejiang, China

Abstract—Four strains – *Bacillus polymyxa* ($2^{\#}$) and *Bacillus licheniformis* ($12^{\#}$ and $14^{\#}$), isolated from *Phyllostachys heterocycla* var. *pubescens* rhizosphere, another bacterial species ($7^{\#}$, the species name has not been determined) isolated from *Phyllostachys meyeri* rhizosphere – were cultivated, prepared into liquid inoculums and inoculated on to *Phyllostachys heterocycla* var. *pubescens* (moso bamboo) seedlings. Potted culture (sand culture) under artificial and sterile conditions was then conducted. Statistical analysis showed that all the four strains were able to raise the survival rate and promote seedlings' growth in terms of several indexes – plant fresh weight, root fresh weight, dry weight of seedling's above-ground part, seedling height, seedling's leaf number and plant nitrogen content.

INTRODUCTION

The use of chemical fertilizers in agriculture today has given rise to a host of problems such as environment pollution and energy resources shortage. Therefore, research on biological nitrogen fixation has started receiving more attention.

Dobereiner and Day (1976) had reported on the habit of associated nitrogen fixation exhibited by maize. Since then, several associated nitrogen-fixing bacteria have been isolated from grain crops such as rice, maize, wheat and some pasture grass. Gramineae plants are also considered to have the character of associated nitrogen fixation (Boddey and Dobereiner 1988). In the field of research on biological nitrogen fixation, studies on associated nitrogen-fixing bacteria have become an important sub-field, hopeful of resolving the problem of supplying adequate nitrogen to main grain crops through the biological route. Currently, several scientists are engaged in studying its many aspects such as: ecology, physiology, biochemistry, molecular biology and genetic engineering. Some nitrogen-fixing bacteria, which have been used in fertilizer, are reported to have achieved notable results (Chalk 1991).

As Gramineae plant, do bamboos possess the habit of associated nitrogen fixation? In 1994, this question was answered, with Gu and Wu (1994) confirming associated nitrogen fixation in bamboo plants. It was the first time this line of research, which was carried out only in agriculture ecosystem till then, was being applied to forestry ecosystem. Since then, several *Azotobacter* strains have been isolated from several bamboo species. Some of them have very high nitrogenase activity. The strains isolated from *Phyllostachys heterocycla* var. *pubescens* (moso bamboo) and *P. meyeri* rhizospheres have been identified. To study the interactive mechanism between associated nitrogen-fixing bacteria and bamboo plants further, it is necessary to select suitable other Azotobacter strains that have effective associated nitrogen-fixing system with bamboo. Some representative bacterial strains with high nitrogenase activity were inoculated on to moso bamboo seedlings. This paper discusses the results of this research.

MATERIALS AND METHODS

Tested strains

The strains tested were $2^{\#}$ (*Bacillus polymyxa*) and $12^{\#}$ and $14^{\#}$ (*B. licheniformis*), which are isolated from roots of moso bamboo, and $7^{\#}$ (the species name is yet to be confirmed) isolated from roots of *P. meyeri*.

Tested plants

Tested plants were moso bamboo seedlings. After the seeds of moso bamboo were sterilized with 0.15% silver chloride ($HgCl_2$) solution and washed with sterilized water, they were put in sterilized culture dishes containing water. The dishes were kept at 25°C to induce the seeds germinate.

Medium

Sand from Fuchun River was used as the medium for the trial. It had been sterilized at 170°C for 4 hours, and put into plastic cups (300 g/cup) after cooling to room temperature. Sterilized water (100 ml) was poured into the cup and the water was allowed to permeate into the sand naturally for two days. The physical and chemical characteristics of the sand can be found in Table 1:

Table 1.
Physical and chemical characteristics of the soil (medium)

Medium	pH	Organic matter (%)	Whole N (%)	Whole P (%)	Whole K (%)	Hydrolyzable nitrogen (PPM)	Quick-acting P (PPM)	Quick-acting K (PPM)
Sand	7.13	0.054	0.024	0.015	4.21	30	7.74	49.5

Experiment design

Two blocks were used in the trial. Each block had four treated groups with one control (CK), and each group had 15 repetitions. In same block, the seed germination and inoculation times were identical.

Inoculation

Azotobacteria, which have been identified and purified, were inoculated into Dobereiner non-nitrogen liquid medium and cultivated for 24 hours. The bacteria cells were collected by centrifugal method. Then 45 ml sterilized water and 5 ml Dobereiner non-nitrogen medium was mixed to become 50 ml azotobacteria suspension (about 10^8 cells/ml) for inoculation. Same azotobacteria suspension was used for CK group, but sterilized by high temperature and pressure before inoculation. Two different inoculation methods were employed for the different blocks. The method employed in Block I comprised selecting just-germinating seeds from the culture dish and putting them into the plastic cup that contains wet sand. Then 2 ml above-mentioned suspension was dropped on the seeds and the adjacent sand, and the seeds were covered by sand to about 1cm thickness. For Block II, just-germinating seeds from the culture dish were put on the wet sand in the plastic cup and covered by sand about 1cm thick. After the seedlings became 2 cm high with 2–3 leaves, 2 ml of azotobacteria suspension was dropped around the root of the seedlings.

Culture of seedlings

The plastic cups for cultivating seedlings were transparent and colorless, with a wall of 1 mm thickness. The diameters of the top end and the bottom of the cup were 80 mm and 52 mm, respectively. For draining surplus water from the sand, 100 small holes (diameter 2 mm) were provided at the bottom of the cup. Each cup was covered and overlapped at its top and bottom ends by two other cups, each of which had on its side a hole covered by ventilation film. This ensured that while the plants can ventilate, water and bacteria could not penetrate into the cup. The overlapping cup at the bottom was smaller, just enough to reach up to the middle of the cup containing the sand. Its purpose was to hold the water draining from the sand in the culture cup. Ventilation film was used to cover the parts where these three cups join. All these operations were carried out in sterile environment so that the seedlings not contaminated with any other microbe. The seedlings were then allowed to grow under artificial illumination for 14 hours/day. In this trial, the preparation for germinating the seed was carried out in early March, the inoculation of strains in late March, and the measurement of growth indexes in early November.

Testing

Whole nitrogen was tested by Kjeldahl method. Nitrogen-fixing activity was tested with excised roots by means of a gas chromatograph, using acetylene reduction method. Other items were measured by regular methods.

RESULTS AND ANALYSIS

Effect of inoculating associated nitrogen-fixing bacteria on the seedling survival rate

After seven months, the survival rate of seedlings was observed. Seedlings that had withered or turned yellow completely were recorded as dead. Seedlings that had one or more green leaves were recorded as alive. It can be seen from Fig. 1 that, on comparing with CK, the survival rates of the four treatment groups are higher. The survival rate increased by 14.8–18.5%. The growth of the seedlings in the four treatment groups in growth was also better than CK. The seedlings in the treatment groups were green and vigorous, while the seedlings of CK were weak with most of the leaves turning yellow, and some were even dead. Investigation on seedling growth at different times shows that the seedlings of CK group were vigorous in the early stage but became weak later. It is supposed that in the early stage, the seeds supplied the nutrients needed by the CK GROUP seedlings. However, in the later stage, when the nutrients in the seeds were used up and the nutrient content in the sand was poor, the growth of seedlings slowed down or even stopped. Comparatively, the seedlings subjected to inoculation treatment showed the same growth as CK seedlings in early stage, but were still green and sprouting new leaves in the later stage. It is supposed that the nitrogen-fixing bacteria were supplying some nitrogen element to the seedlings.

Effect of inoculating associated nitrogen-fixing bacteria on seedling growth

Several indexes of growth have shown that inoculating associated nitrogen-fixing bacteria on to the moso bamboo seedlings can promote seedling growth in various levels (see Table 2 for average value of different treatments).

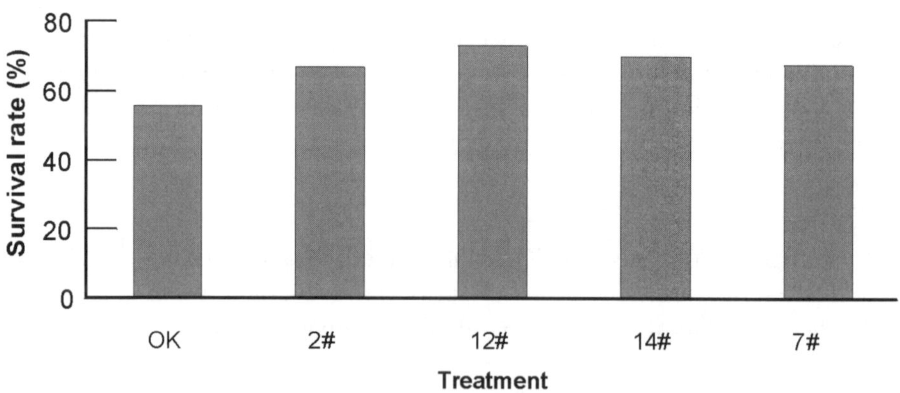

Figure 1. Effect of inoculating azotobacteria on the survival rate of seedlings

Table 2.
Effect of inoculation on seedling growth

Treat-ment	Plant fresh weight increment percentage		Root fresh weight increment percentage		Dry weight of above-ground part percentage		Seedling height increment percentage		Leaf number increment percentage	
	mg	%	mg	%	mg	%	mg	%	mg	%
2#	375.1 aA	32.0	177.2 aA	66.7	73.5 aA	8.4	14.39 aA	30.2	14.9 aA	29.2
12#	388.9 aA	36.9	188.0 aA	76.9	75.5 aA	11.4	13.36 bB	20.9	14.1 aA	3.0
14#	360.7 bA	29.9	153.2 aB	44.1	69.5 aA	2.5	12.79 cB	15.8	13.5 aB	17.4
7#	366.1 bA	28.9	161.1 aA	51.6	71.5 aA	5.4	11.93 dC	8.0	13.4 aB	16.8
CK	284.1 bA	0.0	106.3 bB	0.0	67.8 aA	0.0	11.05 dC	0.0	11.5 bB	0.0

Note: The data are average values for Block I and Block II. In SSR test data, the differentiation in lowercase letters (a, b) indicate that the differences are significant at $t_{0.05}$ level, and the differentiation in uppercase letters (A, B) indicate that the differences are remarkably significant at $t_{0.01}$ level.

Plant fresh weight
It was seen that all inoculation treatment groups had increments of 28.9–36.9% when compared with CK group. In SSR test (Duncan's New Multiple Range Test), the differences between CK and 2# and 12# strain treatment groups reached significant levels.

Root fresh weight
SSR test showed that the differences of root fresh weight between CK and the four treatment groups were significant. Among them, the differences in 2#, 12# and 7# treatment groups were remarkably significant. Therefore, it may be inferred that inoculating associated nitrogen-fixing bacteria can promote the growth of roots. A comparison with CK showed increases of 44.1–76.9%. These results were in line with some reports (Li and An 1997; Watanebe and Lin 1984). It is, therefore, inferred that inoculating associated nitrogen-fixing bacteria can change the form of roots, stimulate the growth of lateral roots and increase the number of root hair.

Dry weight of seedling's above-ground part
The dry weight increments in inoculated treatment groups were 2.5–11.4%. SSR test showed that the differences between CK group and the treatment groups were not significant. From this result, it is supposed that the effect of inoculating associated nitrogen-fixing bacteria in seedling's early stage is the stimulation of root growth to increase the absorption of mineral salt and water and then to promote the overall growth of seedlings.

Height growth of seedling

Seedlings of the inoculation treatment groups were taller than the seedlings of CK group by about 8–30%. Statistical test revealed that the differences between CK group and 2#, 12# and 14# treatment groups were remarkably significant.

Number of leaves

Statistics test showed the effects of inoculation to be significant in terms of the number of leaves; the increments were about 16.8–29.2%.

Effect of inoculating associated nitrogen-fixing bacteria on plant's nitrogen content

The experiment results showed (Fig. 2) that after inoculating nitrogen-fixing bacteria, plant nitrogen content increased significantly. Comparing with CK, the increment percentages were about 22.0–38.1%. To understand the relationship between the nitrogen content of plant and the nitrogen-fixing activity of the inoculated bacteria further, the nitrogen fixing activities of roots were determined. The analysis of the correlation between plant nitrogen content and nitrogen fixing activities of roots showed the correlation coefficient to be 0.7493. Hypothesis testing, $t=2.771>t_{0.05,6}=2.447$, revealed a significant positive correlation between the nitrogen content (N%) of plant and the nitrogen fixing activity of the roots: the higher the nitrogen-fixing activity of the root, the more the nitrogen content of the plant.

Effect of inoculation time on seedling growth

To prevent competition of other bacteria in soil and to ensure that the inoculated bacteria are in a dominant position on host plant roots, the bacteria inoculation re-

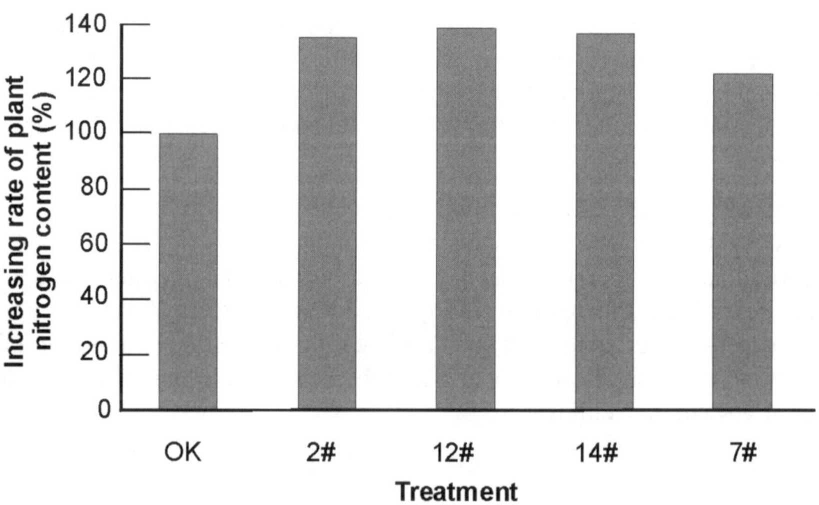

Figure 2. Effect of inoculation on plant's nitrogen content

search on Gramineae crops is now conducted usually by soaking the seeds with azotobacteria suspension or wrapping the seeds with bacteria-manure mix.

The results of this experiment show (Table 3 and Fig. 3) that it is better to inoculate associated nitrogen-fixing bacteria in seedling stage (Block II) than in seed germinating stage (Block I):

1. Statistical test, which compared experiment results with CK for all growth indexes, proved that when inoculation was carried out in seed geminating stage, the number of treatments that reached significant level was less than when inoculation was done in seedling stage.
2. Statistical test, which compared experiment results with CK for all treatments, proved that when inoculation was carried out in seed geminating stage, the increments in all growth indexes were lower than when inoculation was done in seedling stage.

All growth indexes of all treatments in both blocks were subjected to statistical analysis for differences, and the result was $t=3.637>t_{0.01}=2.861$. All treatments in both blocks were analysed statistically for differences in plant nitrogen content, and the result showed $t=3.815>t_{0.05}=3.182$. Inoculating nitrogen-fixing bacteria in seedling stage has more significant effect on plant growth than when it is done in seed germinating stage. Further, the increasing rate of plant nitrogen content is more significant when inoculating in seedling stage than in seed geminating stage. It is supposed that the reason for this maybe the competition for energy resources between seedlings and bacteria. The bacteria consume the nutrients of seed for the energy required for reproduction; but as the germinating seeds cannot carry out photosynthesis, their germination is prevented.

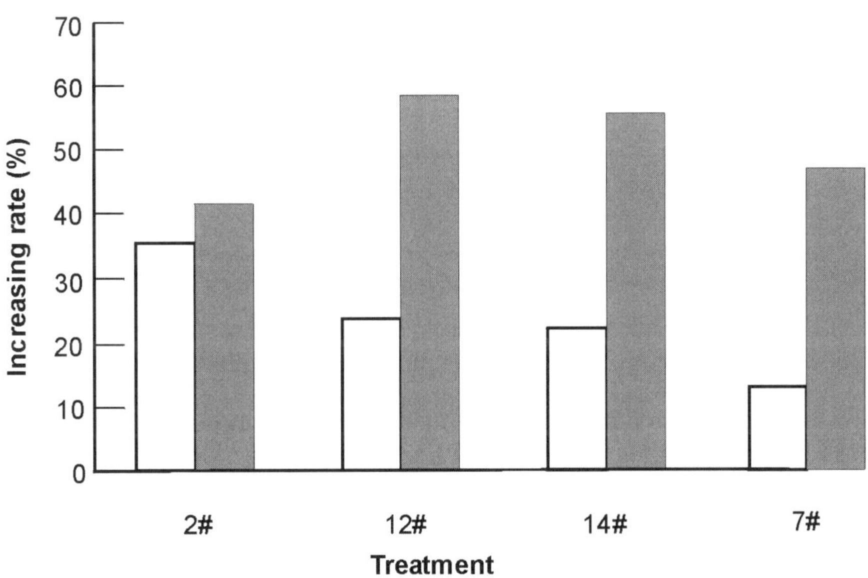

Figure 3. Effect of inoculation time on increase rate of plant nitrogen content

Table 3.
The effect of inoculation time on seedling growth

Block & Treatment	Plant fresh weight increment percentage		Root fresh weight increment percentage		Dry weight of above-ground part percentage		Seedling height increment percentage		Leaf number increment percentage	
	mg	%	mg	%	mg	%	mg	%	mg	%
I, 2#	384 bA	5.2	143 bA	0.1	74 aA	4.4	14.58 aA	19.0	15.3 aA	16.4
I, 12#	462 aA	26.6	213 aA	33.9	73 aA	3.7	13.08 bA	6.8	14.2 bB	7.6
I, 14#	379 bA	3.8	153 bA	–3.8	63 aA	–9.6	12.83 bA	4.7	14.0 bB	6.3
I, 7#	465 bA	11.0	195 bA	22.6	71 aA	0.1	12.10 bA	–1.2	12.0 bB	–8.9
I CK	365 bA	0.0	159 bA	0.0	71 aA	0.0	12.25 bA	0.0	13.2 bB	0.0
II, 2#	456 aA	51.5	189 aA	122.4	73 aA	14.9	14.17 aA	32.6	14.3 aA	31.6
II, 12#	432 aA	43.5	171 aA	101.2	78 aA	22.8	13.28 bB	24.2	13.9 aA	27.6
II, 14#	454 bA	50.8	158 aA	85.9	76 aA	19.6	12.61 bC	18.0	14.2 aB	18.4
II, 7#	418 bA	38.9	141 aB	65.9	72 aA	13.6	11.83 cC	10.7	14.2 aA	30.6
II CK	301 bA	0.0	85 bB	0.0	63 aA	0.0	10.69 cC	0.0	10.9 bB	0.0

CONCLUSION

Recent research has shown that inoculation of nitrogen-fixing bacteria benefit the growth of herbage and grain crops in various environments and soils. However, up to now, the effective mechanism of inoculation has not been fully recognized. This inoculation experiment of bamboo seedlings has resulted in the following observations and conclusions:

1. The associated nitrogen-fixing strains, which have been used to inoculate in this experiment, can significantly promote the growth of moso bamboo seedlings, especially in the fresh weight of root. Therefore, these strains have high potential for further studies and utilization. Rao et al. (1983) reported that inoculating *Azospirillum* bacteria could promote the growth of plant, especially in young seedling stage. This maybe mainly because the hormone produced by *Azospirillum* (Watanabe and Lin 1984) can affect parasitic plants' respiratory rate, metabolism, increase of root hair and secondary root, and then affect the absorption of mineral salts and water (Li and An 1997). The good growth of roots is the base for accumulating organic material for tillers. It is perhaps logical using this theory to explain the results of the moso bamboo experiment. However, if the azotobacteria used in this experiment stimulate the root of moso bamboo seedling by producing hormone, then the kind of hormone they produce still need to be studied.
2. Inoculating associated nitrogen-fixing bacteria can increase the nitrogen content of plant significantly and there is a positive correlation between the nitrogen-fixing activity of roots and the nitrogen content of plant. The results have

shown that inoculating nitrogen-fixing bacteria is beneficial in supplying nitrogen to young seedlings. They concur with other experiments involving the inoculation of associated nitrogen-fixing bacteria in some grain crops (Song 1990; Lin and You 1992). However, to evaluate the correlation of associated nitrogen fixation to the nitrogen supply to moso bamboo seedlings, further studies using ^{15}N dilution method are required.
3. Inoculating associated nitrogen-fixing bacteria in seedling stage gave better results than inoculating in seed germinating stage, providing some clear indications as to the best time and method of inoculation. It could be assumed that after inoculating nitrogen-fixing bacteria, the azotobacteria can reproduce in the rhizosphere of moso bamboo seedlings. The nutrients and energy for reproduction and nitrogen fixing come from the seed and roots. By selecting the appropriate inoculation method and time, mutually beneficial and reciprocal relationship between the bacteria and bamboo can be established.
4. Among different treatment strains, some differences in promoting seedling growth have been observed. Selecting adaptive strain is the key to establishing an effective nitrogen fixing system between bamboo and associated nitrogen-fixing bacteria.
5. This experiment was carried out under artificial conditions. Hence, more research studies – such as the competition between inoculating bacteria and native ones, the effects of environment condition on the growth of nitrogen-fixing bacteria, etc. – are needed before the azotobacteria can be used in practice.

ACKNOWLEDGEMENTS

This paper is an outcome of a research supported by Zhejiang Provincial Natural Science Foundation, to which the authors would like to express their thanks.

REFERENCES

Boddey, R.M.; Dobereiner, J. 1988. Nitrogen fixation associated with grasses and cereals: recent results and perspectives for future research. Plant and Soil, 108, 53–65.

Chalk, P.M. 1991. The contribution of associative and symbiotic nitrogen fixation to the nitrogen nutrition of non-legumes. Plant and Soil, 132, 29–39.

Dobereiner, J.; Day, J.M. 1976. *In* Proceedings of the 1st International Symposium on Nitrogen Fixation. University of Washington Press, Pullnam, USA. pp. 512–538.

Gu, X.; Wu, X. 1994. Study on associated nitrogen fixation of bamboo rhizosphere. Forest Research (China), 7(6), 618–623.

Li, J.; An, Q. 1997. Advances in associated nitrogen fixation research. Chinese Bulletin of Botany, 14(3), 14–21.

Lin, M.; You, C. 1992. Advance in the study on mechanism of associated nitrogen fixation in rhizosphere. Plant Physiology Communication, 28(5), 323–329.

Rao, V.R.; Nayak, D.N.; Charyulu, P.B.B.N. 1983. Yield response of rice to root inoculation with *Azospirillum.* Journal of Agricultural Science (Camb.), 100, 689–691.

Song, W. 1990. Associated nitrogen fixation study of rice rhizosphere. You, C.B. ed. Beijing Agriculture Press, Beijing, China.

Watanabe, I.; Lin, C. 1984, Response of wetland rice to inoculation with *Azospirillum ipoferum* and *Pseudomonas* Sp. Soil Science & Plant Nutrition, 30, 117.

Bamboo for Sustainable Development, pp. 279–285
Eds. A. Kumar, I.V. Ramanuja Rao and Ch. Sastry

Photoperiodic and hormonal control of growth and proliferation in seedlings of some bamboos

MOHINDER PAL
Botany Division, Forest Research Institute, Dehra Dun, India

Abstract—Investigations were carried out to evaluate the effect of photoperiod and growth regulatory substances on the growth and proliferation behavior of seedlings of *Dendrocalamus strictus*, *Bambusa Bambos* and *Bambusa tulda.* The aim was to accelerate the rate of proliferation and thereby, the rates of propagation of bamboo seedlings using macroproliferation technique. The seedlings were treated with 100 ppm each of gibberellic acid (GA_3), Ethrel, triiodobenzoic acid (TIBA) and a combination of GA_3 and Ethrel. These were maintained either under short day (8 h) normal day (12 h) or long day (24 h) photoperiod. Short day conditions and treatments of Ethrel and TIBA suppressed height growth, while long day conditions and treatment with GA_3 increased it. Higher rates of proliferation were achieved under the normal day conditions and by the treatment of seedlings with Ethrel and TIBA. The species differed in their quantitative response to the treatments. This paper discusses the importance of the findings in relation to the improvement in multiplication rates for mass production of bamboo propagules using macroproliferation method.

INTRODUCTION

Bamboos are a cultural feature of tropical India. A large segment of rural poor and tribal people in bamboo-growing areas depend on it for livelihood and shelter. Although pulp and paper industries in India depend mainly on bamboo as raw material, about 80% of the 20 million tons of bamboo harvested annually is used locally (Tewari 1992). Earlier, bamboo formed a perpetual resource because of its abundance in forests and vigorous vegetative growth. At present, however, overexploitation associated with growing populations, destruction of tropical forests and increasing demands on the resources for industrial uses have together contributed to the severe loss of stock from the once-sprawling bamboo forests of Asia. The greatest losses are borne by rural poor, as the abundant and cheap resource is now becoming scarce and expensive (Banik 1997). There is an urgent need to improve the bamboo resource by raising plantations. However, the limited availability of planting stock is a major hurdle to implement such a program. As is well known, most of the economically important bamboos produce seeds after very long intervals – only 2–3 times in a century – and seed viability period is very short. Hence, it has become essential to develop simple vegetative propagation methods for mass-producing bamboo planting stock.

The recently developed macroproliferation method (Banik 1987; Adarsh Kumar 1991) is becoming very popular with bamboo nurseries in India because of its simplicity and cost-effectiveness. However, the multiplication rates obtained using this method need further improvement to enhance the efficiency and increase the acceptability of this method. Therefore, an investigation was undertaken to assess if the growth and proliferation rate in bamboo seedlings can be increased by manipulating photoperiodic and chemical treatments. Three species from among the economically most important bamboos in India were selected for experimentation.

MATERIAL AND METHODS

Six-month old seedlings of *Bambusa bambos*, *Dendrocalamus strictus* and *Bambusa tulda* were selected, ensuring uniformity of culm number and size, from the nursery stock grown at the Forest Research Institute, Dehra Dun. These were separated into culm units, each having a rhizome and root system, in the month of March as described by Adarsh Kumar (1991). Each unit was replanted in polythene bags filled with 2 kg mixture of forest soil, river sand and farmyard manure in 1:1:1 ratio. The propagules were allowed to stabilize for 15 days. Thereafter, the seedlings of each species were divided into three groups to be exposed to 8-hour (short day, SD), 24-hour (long day, LD) and normal (normal day, ND) photoperiods. Under each photoperiodic regime, the seedlings were treated with 100 mg/l solutions of gibberellic acid (GA_3), Ethrel (ET), triiodobenzoic acid (TIBA), or a combination of Ethrel and GA_3, through a small trench made in the soil in each bag. The treatments were administered at bi-weekly intervals, and each time 50 ml of the chemical solution was applied. The control plants did not receive any chemical treatment.

The seedlings were maintained under natural environmental conditions, expect those receiving SD were kept in the dark from 5.00 PM to 9.00 AM of the next day, each day, and those maintained under LD received supplementary white light (120 lux at plant level) during the night. The average maximum and minimum temperatures varied from 28.0°C to 20.4°C and the photoperiod from 13.30 hours in June and July to about 10.40 hours in March and October. The plants were irrigated frequently to avoid development of water stress. The observations on height growth and the number of culms were recorded in the month of October, 6 months after the treatments were initiated, 20 replicates were kept for treatment and the results were analyzed statistically using students 't' test.

RESULTS AND DISCUSSION

The results on the height and culm number are presented in Figs. 1–6. The average height and culm number of the seedlings varied with the species, the photoperiod to which these were exposed and the chemical treatment. While maximum height was attained by the seedlings of *Dendrocalamus strictus* (Fig. 3), the seedlings of *Bambusa bambos* (Fig. 5) had the maximum height. The average number of culms was maximum in *B. bambos* (Fig. 6) and minimum in *D. strictus* (Fig. 4), which indicates that there was an inverse correlation between the height and culm number among these three species. Khullar (1998) has reported that within species, there was an inverse, but statistically insignificant, relationship between culm height and culm number in seedlings of *D. strictus* and *B. bambos*. However, more research is needed to establish the relationship between the culm height and culm number, so that appropriate treatment can be selected to increase the culm number, and thereby increase the multiplication rates in bamboo seedlings. The results clearly show that photoperiod has a strong effect on the height growth and proliferation rate in bamboo seedlings. In general, short day conditions decreased height growth and long day conditions increased it (Figs. 1, 3 and 5), but the response varied quantitatively with the species.

Photoperiodic conditions also affected the proliferation rates in seedlings of bamboos. Short day conditions decreased the number of culms in the seedlings of all the three species of bamboos. While normal days caused maximum culm production in *B. tulda* and *D. strictus*, long days were the most effective in this regard in *B. bambos* seedlings. Although the effect of photoperiod on growth and development on a number of plant species were known, there is hardly any report available in literature on the effect of photoperiod on the growth of bamboos.

GA_3 treatment increased height growth in seedlings of all three species under all photoperiodic conditions. In general, height growth of GA_3-treated seedlings increased with the length of photoperiod irrespective of the species. Treatments with Ethrel and TIBA suppressed height growth under all photoperiods and in all three species. GA_3 treatment could overcome Ethrel-caused suppression in height growth of the seedlings. However, the quantitative effects of these growth regulatory substances varied with species and photoperiodic conditions under which the seedlings were maintained (Figs. 1, 3 and 5). GA_3 treatment decreased the culm number in seedlings under all photoperiods and in all species tested. Ethrel and TIBA treatments increased the culm number; their effect was significantly higher under normal day and long day conditions than under short day conditions. Maximum culm number was observed with Ethrel treatment under normal days in *B. tulda*, with Ethrel treatment under long days in *D. strictus* and with TIBA treatment and under long days in *B. bambos* (Figs. 2, 4 and 6, respectively). While the effect of GA_3, Ethrel and TIBA on growth and development of a large number of plant species is well known (Krishnamurthy 1981), very little information is available on the effect of these chemicals on the growth and proliferation behavior of bamboo seedlings. The findings of this investigation clearly show that photoperiod and treatment with growth regulatory substances modify the growth and proliferation behavior of bamboo seedlings and these treatments can be effectively used to enhance the proliferation rates, and thereby vegetative multiplication rates, in macroproliferation method, of seedlings of these species of bamboos.

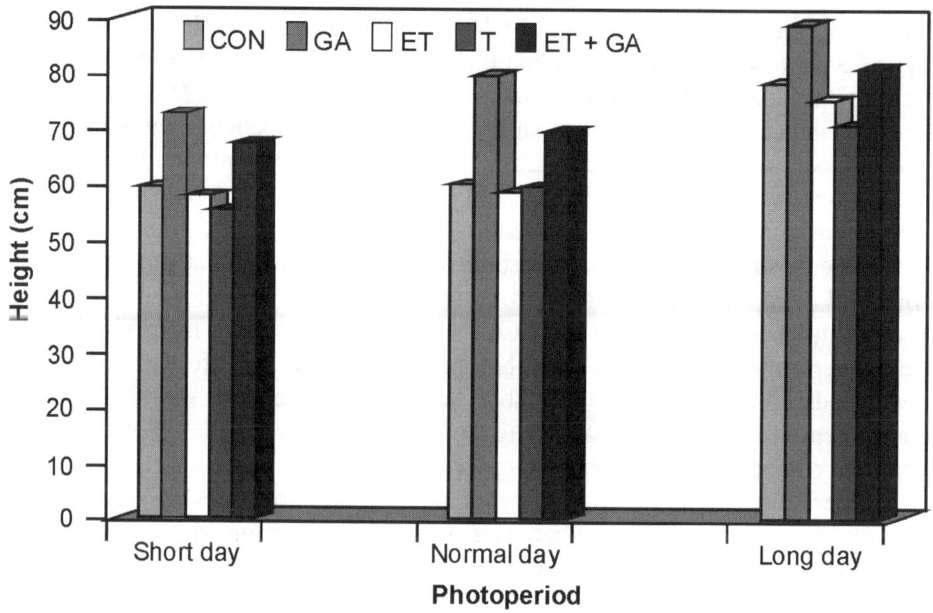

Figure 1. Average height of *B. tulda* seedlings treated with regulatory substances and grown under varying photoperiods.

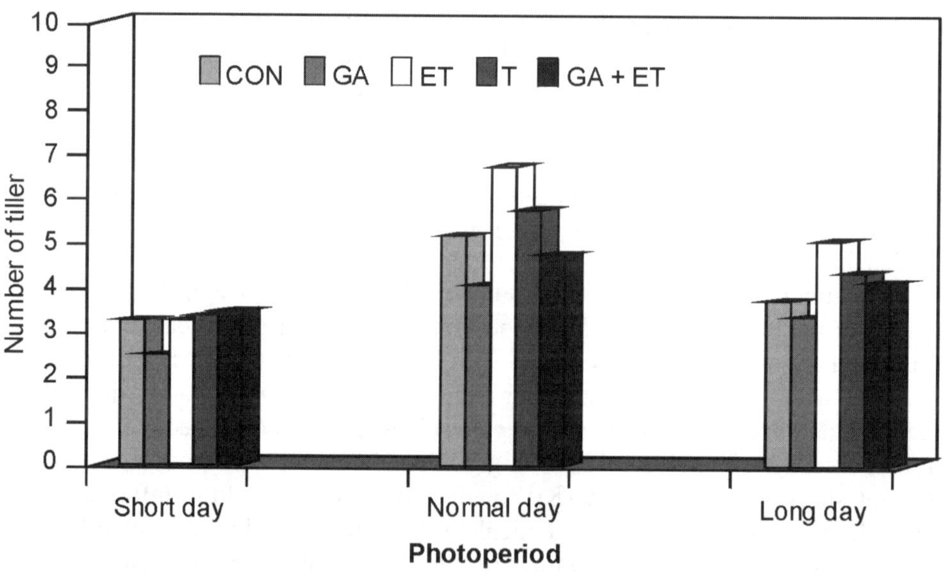

Figure 2. Average number of culms in *B. tulda* seedlings treated with growth regulatory substances and grown under varying photoperiods.

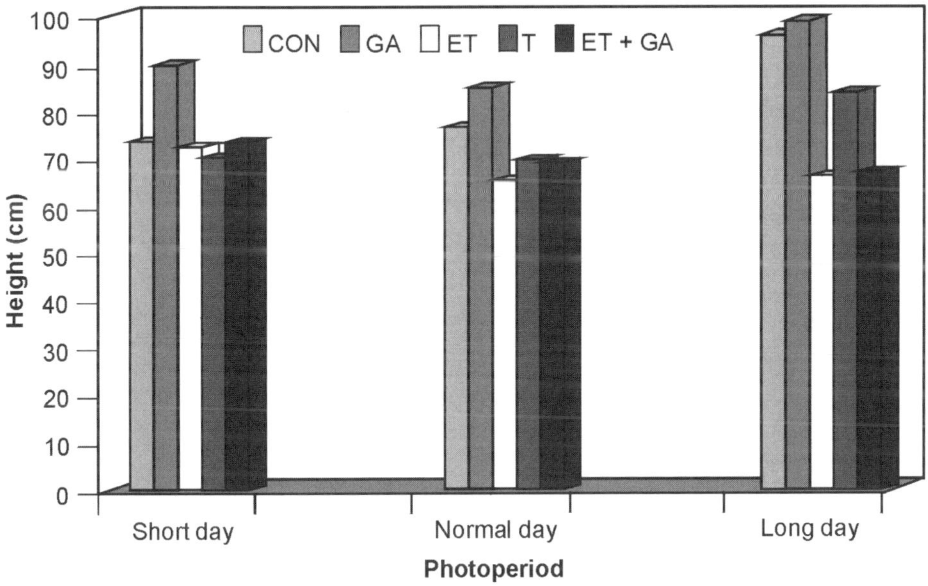

Figure 3. Average height of *D. strictus* seedlings treated with growth regulatory substances and grown under varying photoperiods.

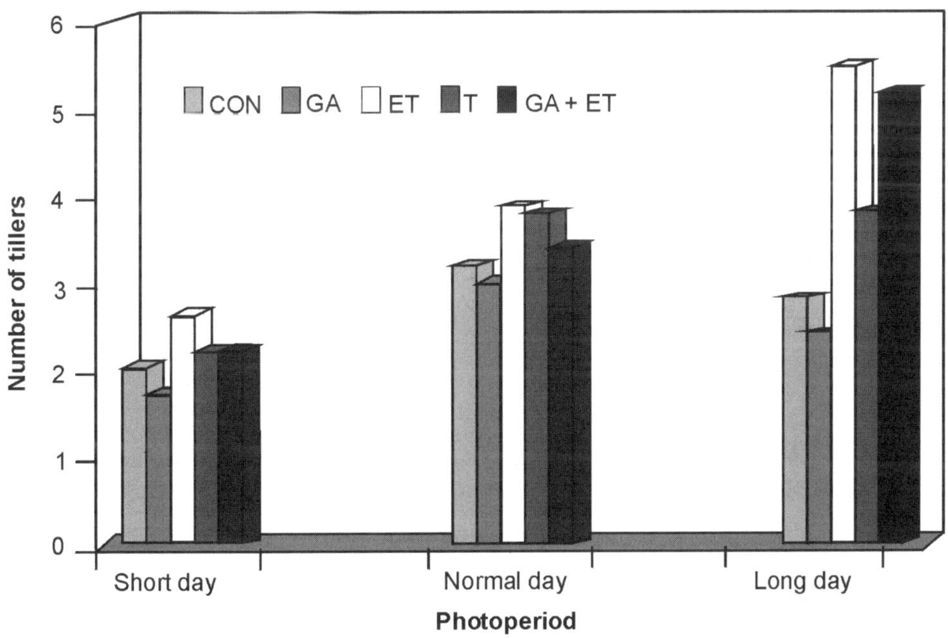

Figure 4. Average number of culms in *D. strictus* seedlings treated with growth regulatory substances and grown under varying photoperiods.

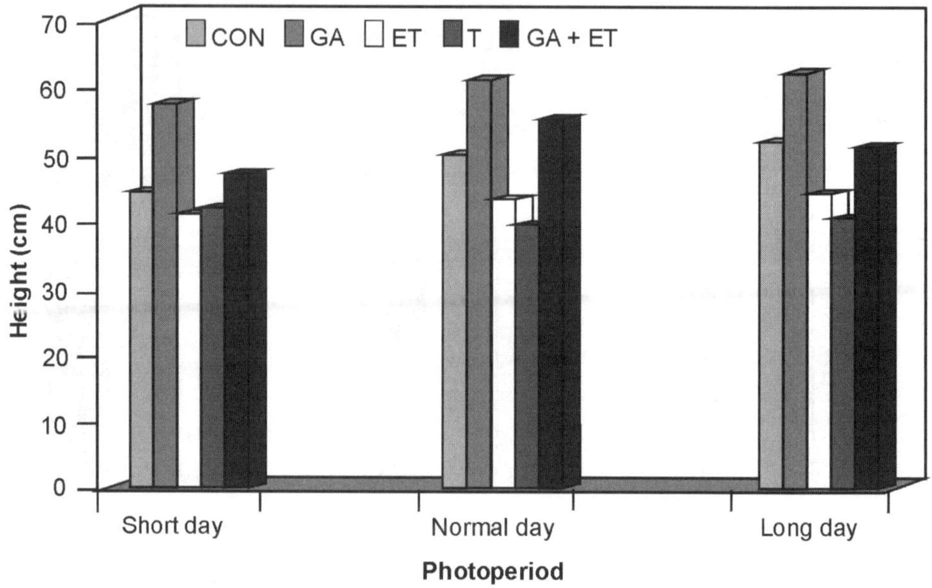

Figure 5. Average height of *B. bambos* seedlings treated with growth regulatory substances and grown under varying photoperiods.

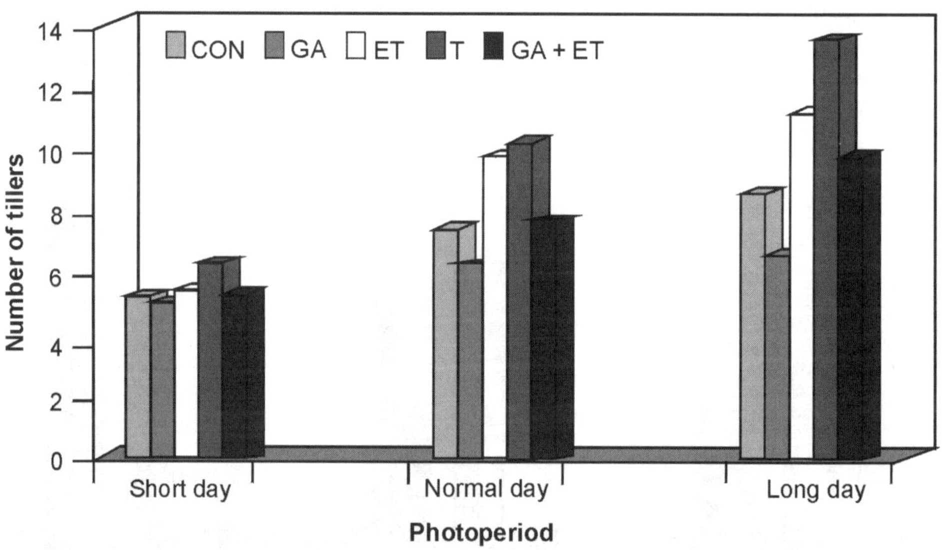

Figure 6. Average number of culms in *B. bambos* seedlings treated with growth regulatory substances and grown under varying photoperiods.

REFERENCES

Adarsh Kumar. 1991. Mass production of field planting stock of *Dendrocalamus strictus* through macroproliferation: a technology. Indian Forester, 117 (12), 1046–4052.

Banik, R.L. 1987. Techniques of bamboo propagation with special reference to pre-rooted and pre-rhizomed branch cuttings and tissue culture. *In* Rao, A.N.; Dhanarajan, G.; Sastry, C.B. ed., Recent Research on Bamboo. Proceedings of the International Bamboo Workshop, Hangzhou, China, 6–14 October 1985. Chinese Academy of Forestry, Beijing, China; International Development Research Centre, Ottawa, Canada. pp. 160–169.

Banik, R.L. 1997. Domestication and improvement of bamboos. INBAR Working Paper No. 10. International Network for Bamboo and Rattan, New Delhi, India. 53 pp.

Khullar, P. 1998. Studies on growth attributes and proliferation behavior of seedlings of some sympodial bamboos as affected by silvicultural manipulations. Ph.D. thesis submitted to Deemed University, FRI, Dehra Dun, India.

Krishnamurty, H.N. 1981. Plant growth substances in applications in agriculture. Tata McGraw-Hill Publishing Co. Ltd., New Delhi, India.

Tewari, D.N. 1992. A monograph on bamboo. International Book Distributors, Dehra Dun, India. 498 pp.

Bamboo for Sustainable Development, pp. 287–297
Eds. A. Kumar, I.V. Ramanuja Rao and Ch. Sastry

Characteristics of three giant bamboo species in Brazil

A.L. BERALDO
Agricultural Engineering, Universidade Estadual de Campinas, São Paulo, Brazil

A. AZZINI
Instituto Agronômico de Campinas, São Paulo, Brazil

Abstract—Pachymorph bamboo species were introduced into Brazil during the Portuguese colonization. Today, there is a wide distribution of several bamboo species in the country. Among these, the giant bamboos are of importance for handicrafts and construction purposes. In this study, three culm portions (basal, middle and apical) of three giant bamboo species (*Dendrocalamus giganteus, Gigantochloa verticillata and G. angustifolia*) were studied. The sample culms were cut into three portions, and each portion was measured along its axis and weighed in three moisture conditions (wet, air-dry and over-dry). Three replications were carried out in each case. The results showed that the weight of the bamboo samples changes quickly with moisture conditions. Radial shrinkage of the samples was sometimes greater than tangential shrinkage; longitudinal shrinkage was negligible. Analysis using imaging technique proved very useful to evaluate the distribution of the anatomic components of bamboo culms. Culm chips were shredded to extract starch, and to separate the fibrous and residual fractions. Shredding was conducted in a domestic mixer with inverted cup. The composition ranged from 2% to 6% for starch and from 50.2% to 61.5% for cellulosic fiber.

INTRODUCTION

Several bamboo species were introduced into Brazil during the Portuguese colonization. The most important species among these is *Dendrocalamus giganteus* or the "giant bamboo". Another giant bamboo species that can be found in Brazil, although in lesser quantities, is *Gigantochloa verticillata*. During the 1970s, remote sensing performed in the Western part of Amazonia revealed a major concentration of the *Guadua* species, another giant bamboo. The species also occurs in the Pantanal region (central-west), where this bamboo is known as the *taquarucu* (giant bamboo in native language).

Despite bamboo presenting several possible uses, its utilization in Brazil has been much restricted when compared with many countries in Asia and South America. In these regions, bamboo has come to a level where it can compete against other more sophisticated materials. One can cite poor technical knowledge about this renewable, easy-to-process and widely distributed material as among the major reasons that prevent a higher utilization of bamboo in Brazil.

Several parameters must be considered in the determination of the physico-mechanical properties of bamboo. Research on *Guadua angustifolia* by Lopez (1974) proved that the culm portion from which the sample is taken has an effect on the compression strength. Other important parameters are culm age, harvest season and ecological conditions.

The starch content in bamboo culms varies with the species. For *Bambusa vulgaris*, which is the most utilized species in Brazil for producing paper pulp, the granular starch content varies from 4.6% to 9.6%, according Azzini and Gondim-Tomaz (1996).

In general, the anatomy of bamboo culms indicates 40% fiber bundles, 50% parenchyma cells and 10% vessels (Liese 1987). Nevertheless, this anatomical distribution changes with the radial position in the culm. Ghavami and Hombeck (1981) reported for *D. giganteus* 40–70% of fiber bundles in the peripheral layers and 15–30% in the internal layers. Beraldo (1994) evaluated the anatomical distribution of *Phyllostachys viridis* using imaging analysis techniques. Average values for fiber bundles were 63%, 37% and 25% for peripheral, middle and internal layers, respectively. Parenchyma cell distribution was 27%, 50% and 63%, respectively, for the same regions. Vessel diameters, which were larger in the internal layers, became smaller in the peripheral layers. Surface vessel occurrence, however, was almost the same for all regions of the culm (8–12%).

This study was aimed at contributing to the determination of certain physico-mechanical, chemical and anatomical properties of three giant bamboo species in Brazil, mainly for rural construction purposes.

MATERIALS AND METHODS

In November 1997, in the Agronomic Institute of Campinas, Brazil, three seasoned culms of three giant bamboo species – *Dendrocalamus giganteus, Gigantochloa verticillata* and *Guadua angustifolia* – were collected. For each culm, two internodes each were selected from three portions: basal (bottom) middle and apical (top).

Anatomical analysis

Samples having a length of 40 mm each were polished using the Struers mechanical polisher and Norton abrasives (number 200 to 1 000). After this preliminary preparation, digital images of the samples were obtained using HP Scanjet scanner. The quality of image scanned was enhanced with Khoros software.

Physical analysis

Green bamboo samples were stamped with waterproof ink in three directions (longitudinal, radial and tangential). The samples were measured under three moisture conditions: green, natural air-dried and oven-dry. Shrinkage of the samples was calculated based on the Brazilian wood recommendation test.

Mechanical analysis

The objective of this analysis was to trace the correlation between the compression strength of the samples and the speed of ultrasonic waves. The propagation time was determined using a Steinkamp BP-5 apparatus with a sensibility of 0.1 μs. Solotest equipment was used to measure compression strength.

Chemical analysis

The starch and cellulosic fiber contents were determined using laminated chips (0.5 mm thickness) obtained from the basal, middle and apical regions of the bamboo culm. The shredding of chips for starch extraction and fibrous fraction separation was conducted in a domestic mixer with inverted cup. The shredding time was 30 seconds, and the chips and water were in a ratio 1:33. The fibrous fractions were converted into cellulosic fibers using an oxidizing solution composed of concentrated acetic acid (50%), hydrogen peroxide (40%) and distilled water (10%).

RESULTS

Internode dimensions

Fig. 1 shows the average measurements of the samples (9 replications) from the three giant bamboo species. The values show the predominance in height of the middle internodes (M) for the bamboos *G. verticillata* (GM) and *D. giganteus* (DM). This characteristic – combined with a larger diameter – suggests their potential use for irrigation purpose, as the head loss will be small. *G. angustifolia* seems more appropriate for structural utilization because its culm is homogeneous, and it has a low height and, hence, a greater stiffness.

Dry-air culm

Figs. 2–4 show the drying curves of the air-dry specimens from the three portions of the culms. The higher the portion from where the specimen comes, the lesser the moisture content. After 20 days of natural drying conditions, all the samples had 12% average moisture content. The drying rate of culms follows an exponential law. For the basal region of *G. angustifolia*, the following equation was derived:

$$h = 11.20 + 111.76\,[\exp(-0.18211 \times t^{1.1027})]$$

where h = percentage of moisture content and t = time in days

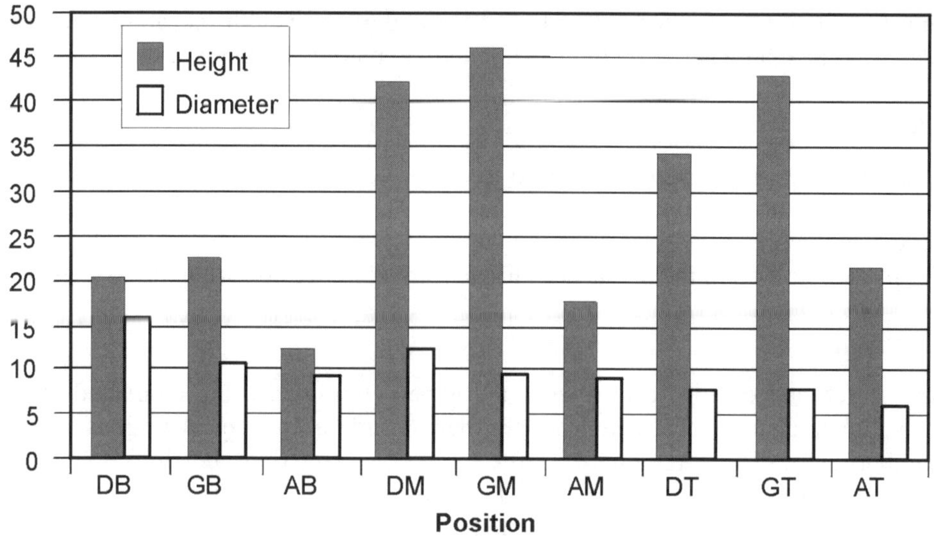

Figure 1. Measurements of bamboo samples

Note: G = *G. verticillata;* D = *D. giganteus;* A = *G. angustifolia;* B, M, P = basal, middle, apical

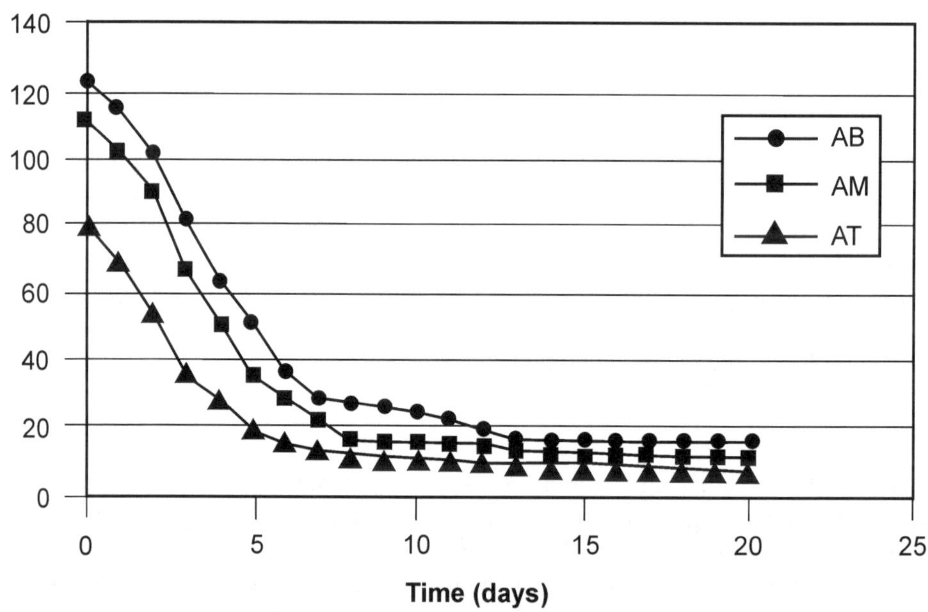

Figure 2. Drying curves for *G. angustifolia*

Note: B, M, P = basal, middle, apical

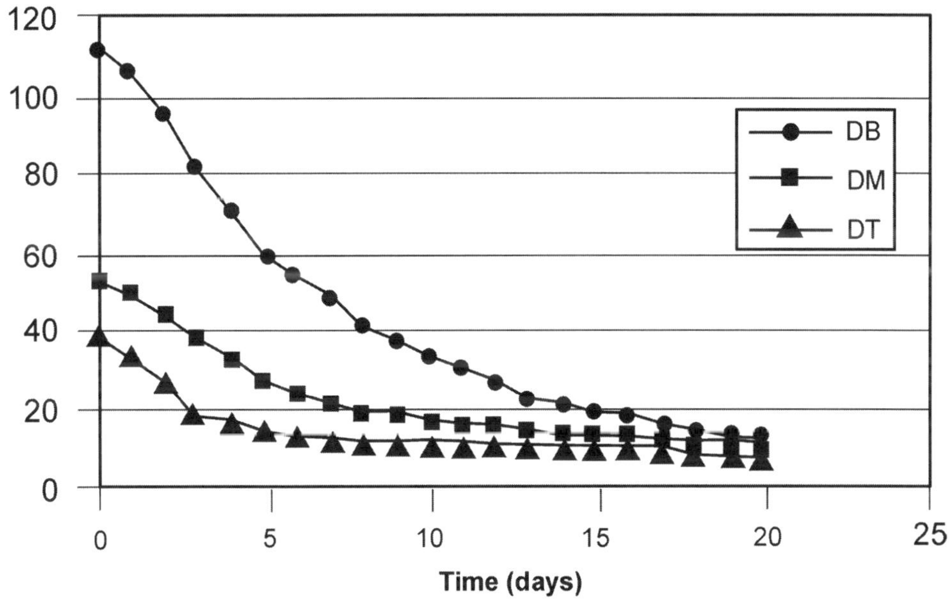

Figure 3. Drying curves for *D. giganteus*
Note: B, M, P = basal, middle, apical

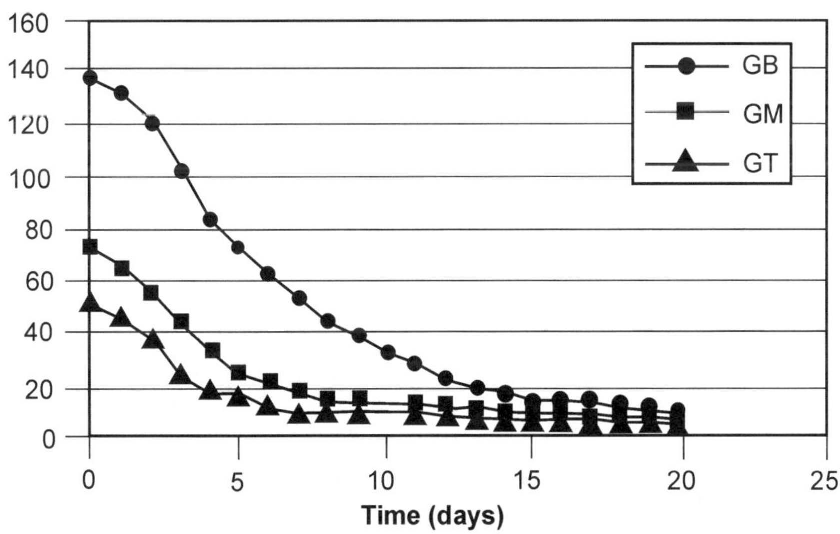

Figure 4. Drying curves for *G. verticillata*
Note: B, M, P = basal, middle, apical

Anatomical analysis

Images of polished bamboo samples were separated into three portions of the same height. Figs. 5, 6 show the digitized images of culm wall cross-sections of *D. giganteus* and *G. angustifolia*, respectively. Fiber bundles (dark regions) are more concentrated towards the bottom, the peripheral (outer) culm region. Table 1 presents the distribution of fiber bundles across the culm (height and diameter position) for *D. giganteus* and *G. angustifolia* species. There was no significant statistical difference among the culm positions of the samples.

Table 1.
Fiber bundle distribution (%)

Bamboo species & specimen position	Culm layer		
	Outer	Middle	Inner
D. giganteus			
bottom	54	40	26
middle	59	37	28
top	57	38	23
G. angustifolia			
bottom	44	30	17
middle	46	37	14
top	38	36	12

Physical analysis

Results of the shrinkage of bamboo culms are presented in Figs. 7, 8. Natural shrinkage of culms was greater than shrinkage of rehydrated dry culms. Dimensional variations are most significant when the green culms are cut without nodes.

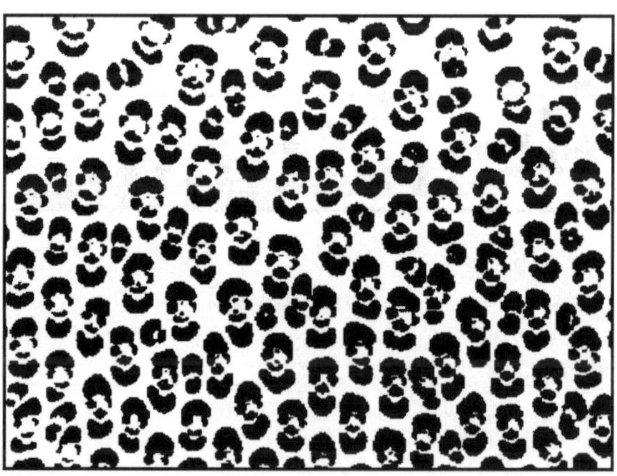

Figure 5. Digitized image of culm wall of *D. giganteus*

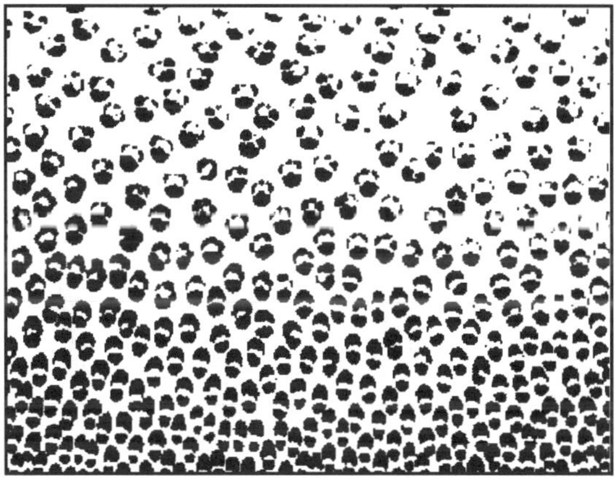

Figure 6. Digitized image of culm wall of *G. angustifolia*

Shrinkage analysis for the utilization of bamboo culms is a more interesting approach than weight stabilization of the samples. Axial shrinkage of the bamboo culm samples was negligible (compared with wood species), but radial (R) and tangential (T) shrinkage did not present the same trend as observed in wood species (tangential shrinkage was greater than radial shrinkage).

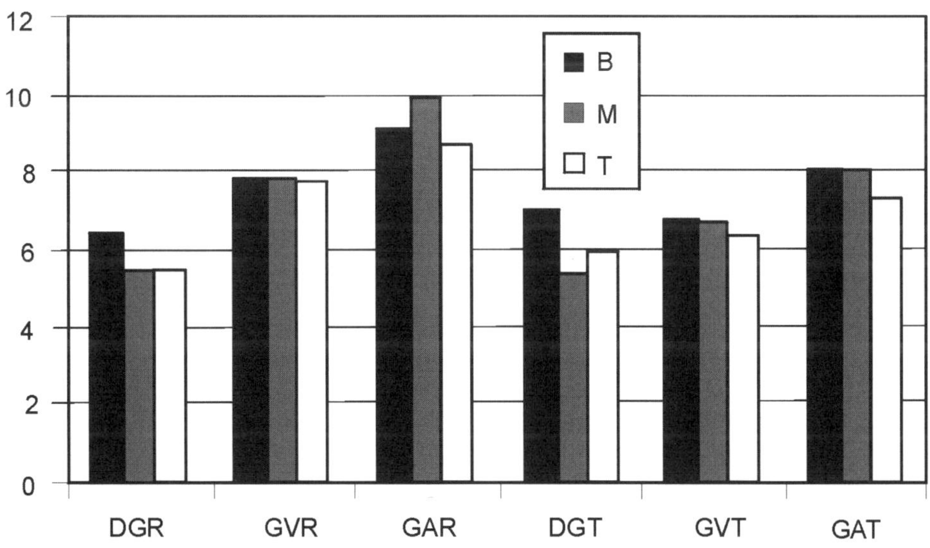

Figure 7. Shrinkage after rehydration

Note: GL = *G. verticillata;* DG = *D. giganteus;* GA = *G. angustifolia;* R, T = radial, tangential

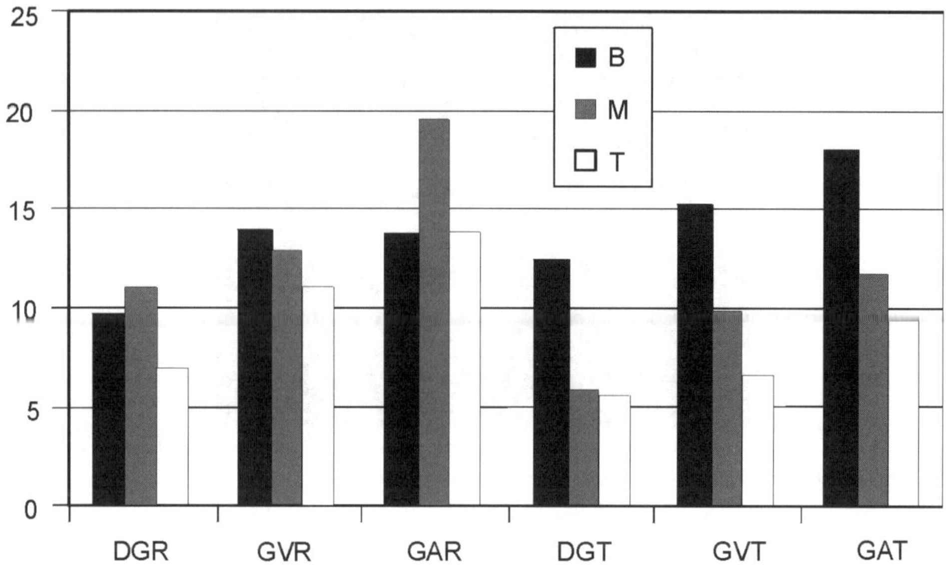

Figure 8. Natural shrinkage of bamboo samples

Note: GL = *G. verticillata;* DG = *D. giganteus;* GA = *G. angustifolia;* R, T = radial, tangential

Mechanical analysis

Results of compression strength tests on bamboo samples are presented in Figs. 9, 10. In most cases, the sample position on the culm does not influence the results. However, like a woody material, natural air-dried bamboos were found to be stronger than green culms. The density of natural air-dried samples ranged from 0.66 (GB) to 0.97 (DP), while the density of green samples was between 0.83 (GB) and 1.15 (DP). The density presented only small variations. Samples from the top portion (P) had a greater density when compared with the bottom portion (B). A significant correlation between density and compression strength was not observed.

Non-destructive evaluation tests using ultrasonic waves did not show any notable correlation with density. This might have been because of small wave speed variations, which ranged from 4 (AP) to 5.6 km/s (GM) for green samples, and from 5 (AP) to 6.1 km/s (GM) for natural air-dried samples.

Chemical analysis

The amount of extracted starch (1.7–3.8%) and cellulosic fiber (50.2–61.5%) did not change with species, but were a function of the culm portions (basal, middle or apical) in *D. giganteus* and *G. verticillata* species. For these species, the highest starch contents were observed in the middle culm regions (Figs. 11, 12). The starch quantity obtained from these species was smaller compared with *B.vulgaris* (4.7–9.6%). The cellulose fiber contents were similar in all the analysed species.

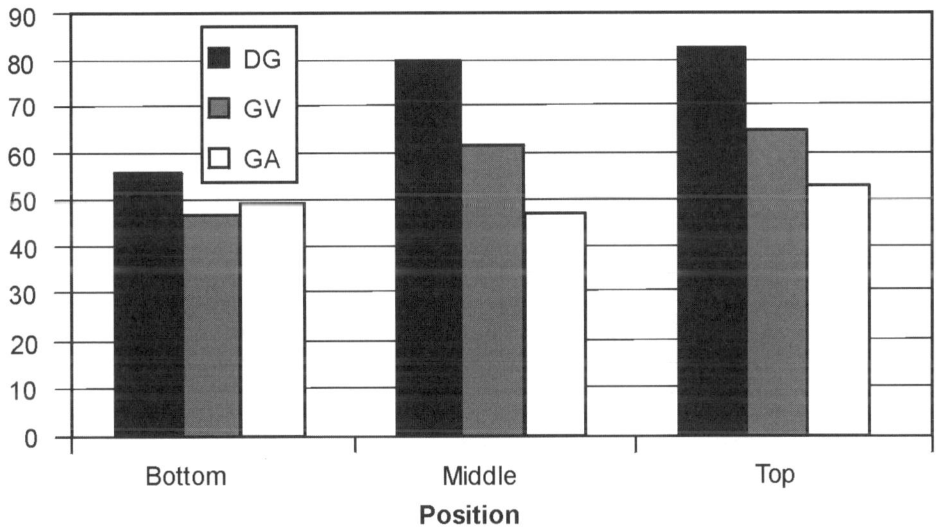

Figure 9. Compression strength of air-dried culms

Note: A = *G. angustifolia;* G = *G. verticillata;* D = *D. giganteus;* B, M, P = basal, middle, apical

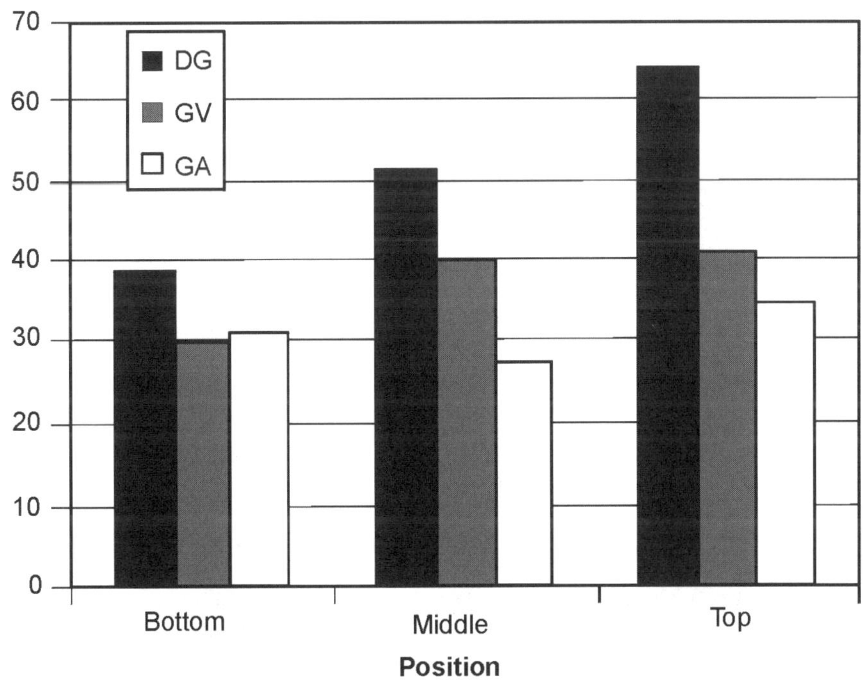

Figure 10. Compression strength of green culms

Note: A = *G. angustifolia;* G = *G. verticillata;* D = *D. giganteus;* B, M, P = basal, middle, apical

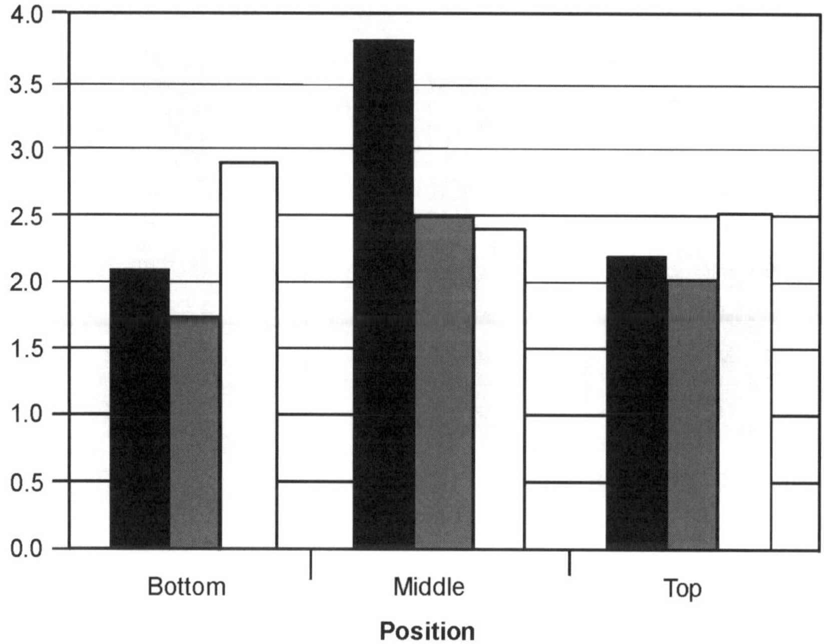

Figure 11. Extracted starch contents

Note: A = *G. angustifolia;* G = *G. verticillata;* D = *D. giganteus;* B, M, P = basal, middle, apical

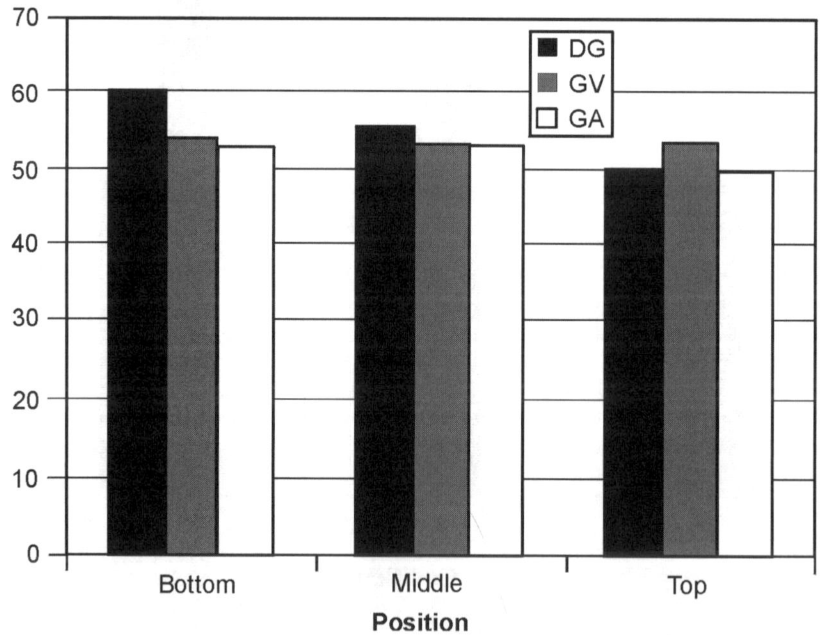

Figure 12. Cellulose fibers content

Note: A = *G. angustifolia;* G = *G. verticillata;* D = *D. giganteus;* B, M, P = basal, middle, apical

CONCLUSION

The three giant bamboo species did not show any marked variation among them in their properties. However, samples collected from three culm positions (basal, intermediate and apical) presented differences. Fiber bundles are more concentrated in the external 1/3rd of culm layers. Starch contents ranged from 1.7% to 3.8%, while cellulosic fiber contents varied from 50.2% to 61.5%. There was no correlation between density and compression strength of the samples or ultrasonic wave speed.

REFERENCES

Lopez, O.H. 1974. Bambu-su cultivo y aplicaciones. *In* Fabricacion de papel, construccion, arquitectura, ingenieria, artesania, estudios tecnicos Colombianos, Cali, Colombia. 318 pp.

Azzini, A.; Gondim-Tomaz. 1996. Extracao de amido em cavacos de bambu (*B. vulgaris*) tratados com solucao diluida de hidroxido de sodio. Bragantia, Caminas, Brazil, 55 (1), 137–140.

Liese, W. 1987. Reserch on bamboo. Wood and Science Technology, 2(3), 189–209.

Ghavami, K.; Hombeck, R.V. 1981. Application of bamboo as a construction material. *In* Proceedings of the Latin American Symposium on Applied Low-Cost Housing, IPT, Brazil. pp. 49–66.

Beraldo, A.L. 1994. Generalisation et optimisation de la fabrication d'un composite biomasse vegetale-ciment a variations dimentionsnelles limitees vis-a-vis des variations de I'humidite. Ph.D. Thesis, Universite Henri Poincare, Nancy, France. 222 pp.

Bamboo for Sustainable Development, pp. 299–307
Eds. A. Kumar, I.V. Ramanuja Rao and Ch. Sastry

Studies on microbial diversity and their activity in soil under bamboo plantation

K.S. MAHESH KUMAR, A.R. ALAGAWADI AND V.C. PATIL
University of Agricultural Sciences, Dharwad, Karnataka, India

Abstract—Investigations were carried out on the population and activities of rhizosphere microflora under bamboo plantation during different seasons. The enumeration of microflora in the rhizosphere of different species of bamboo grown at different locations indicated that *Dendrocalamus strictus* harbored the maximum number of total bacteria and N_2-fixers, while *Bambusa bambos* recorded the highest number of P-solubilizers and *Bambusa vulgaris* recorded the highest populations of fungi and actinomycetes. Bamboo plants grown in medium black soil showed higher population of different groups of microorganisms than those grown in red soils.

Out of the 32 representative major bacterial isolates obtained from bamboo rhizosphere, 16 were found to be N_2-fixers, six were P-solubilizers and 10 were predominant general bacterial isolates. The N_2-fixers mainly belonged to the genera *Azotobacter* and *Azospirillum*. The P-solubilizers belonged to *Pseudomonas, Xanthomonas* and *Erwinia*. Predominant bacterial isolates belonged to *Bacillus, Erwinia, Pseudomonas* and *Xanthomonas*. Among N_2-fixers, *Azospirillum* isolates were found to fix higher amount of nitrogen than *Azotobacter* isolates. In case of P-solubilizers, strains PSER–1 and PSER–2 showed maximum solubilization of tricalcium phosphate with maximum reduction in pH of the broth. All N_2-fixers and P–solubilizers were found to produce IAA and GA.

Inoculation of bamboo seeds with selected efficient nitrogen fixers or P–solubilizers, either individually or in combination, showed increased growth of bamboo seedlings. In general, single inoculation treatments gave better results than combined inoculations. All the inoculation treatments enhanced the seed germination, root and shoot length, as well as dry matter accumulation of bamboo seedlings.

The microbial load and enzyme activities in soil under bamboo leaf litter were significantly higher than those of control soil (not planted). Among the bamboo plant densities, the soil under highest plant density showed maximum population of bacteria, fungi, actinomycetes, N_2-fixers, P–solubilizers and cellulose degraders, as well as dehydrogenase, urease and phosphatase activities.

INTRODUCTION

Bamboos are well known for their fast growth and are important in agro-forestry and in supplementing afforestation programs. The demand for bamboos is much more than their availability. Bio-inoculants as source of nutrients may be advocated as an eco-friendly approach to improve the productivity of bamboo plants. This investigation was undertaken to study the population density and functional diversity of microflora associated with bamboo species.

MATERIALS AND METHODS

For enumeration and isolation of microflora from rhizosphere, rhizoplane and endorhizosphere of bamboo, two different locations of an established bamboo plantation were selected: Site 1 – Prabhunagar forest area in Dharwad District; and Site 2 – a 20-year-old plantation of *Dendrocalamus strictus* (Roxb.) Nees at the main campus of the University of Agricultural Sciences (UAS), Dharwad. For studies on microflora and enzymatic activities in soil under bamboo leaf litter, bamboo planted at different plant population densities was selected and soil sampled in different seasons.

Enumeration and isolation of microflora from bamboo rhizosphere, rhizoplane and endorhizosphere were carried out following standard serial dilution plate count method, using soil extract agar (Bunt and Rovira 1955) for total bacteria, rose bengal agar (Martin 1950) for fungi, Kuster's agar (Kuster and Williams 1964) for actinomycetes, Norris N-free agar (Norris 1959) for *Azotobacter*, Pikovskaya's agar (Pikovskaya 1948) for P–solubilizing bacteria and semi-solid N-free sodium malate medium (Okan et al. 1979) *for Azospirillum.* VAM spore count was taken by following the procedure of wet sieving and decanting (Gardemann and Nicolson 1963).

Identification of bacterial isolates

The representative general bacteria, nitrogen-fixing bacteria and phosphate-solubilizing bacteria isolated were identified up to generic level, based on morphological and biochemical properties (Holt et al. 1994). Efficiency of isolates to fix atmospheric nitrogen, to solubilize tricalcium phosphate and to release growth promoting substances – such as IAA and GA – were carried out by following Microkjeldahl method (Jackson 1973), phosphomolybdicate blue color method (Jackson 1973) and paper chromatography method (Vancura and Macura 1960). The organic acid profile for P–solubilizers was determined using paper chromatography method (Nordmann and Nordmann 1960). Activities of dehydrogenase, urease and phosphatase enzymes were determined by following the procedures described by Casida et al. (1964), Pancholy and Rice (1973) and Eivazi and Tabatabai (1979), respectively.

RESULTS AND DISCUSSION

Bamboo rhizosphere, in general, harbored the maximum population of total bacteria than all other groups of microorganisms. Among the species of bamboo, *Dendrocalamus strictus* harbored the maximum number of total bacteria and nitrogen fixers, whereas *Bambusa bambos* (syn. *B. arundinacea*) recorded the highest population of P–solubilizers. *Bambusa vulgaris* showed the highest populations of actinomycetes and fungi (Table 1). These results indicate preferential stimulation of different groups of microorganisms by each species. It is well established that different plant species release different organic compounds into their rhizosphere. Variation in exudation patterns from different plant species was studied by Rovira (1959) and Vancura (1964). Bamboo plants grown in medium black soil at Prabhunagar and UAS campus showed higher populations of different microflora, whereas red soil of UAS campus and eroded soils at both the locations showed lower populations. The soil type showed marked influence on the rhizosphere microflora of bamboo. Monrozier et al. (1991) had observed higher microbial popu-

Table 1.
Population of bacteria, fungi, actinomycetes, nitrogen fixers and phosphate solubilizers in the rhizosphere of bamboo species from different locations.

Bamboo species	Location and soil type	Rhizosphere region	Bacteria ($cfu \times 10^6/g$)	Fungi ($cfu \times 10^4/g$)	Actinomycetes ($cfu \times 10^2/g$)	N-fixers ($cfu \times 10^4/g$)	P-solubilizers ($cfu \times 10^2/g$)
Bambusa vulgaris	Prabhunagar	RS	8.80	10.70	9.20	3.20	2.40
	(medium black	RP	5.70	5.80	4.80	2.80	2.00
		ER	2.20	0.00	0.00	0.00	0.00
		Total	16.70	16.50	14.60	6.00	4.40
	UAS, Dharwad	RS	7.30	11.00	3.20	3.40	2.40
	campus (red soil)	RP	5.40	5.70	2.80	3.20	2.30
		ER	0.00	0.00	0.00	0.00	0.00
		Total	12.70	16.70	6.00	6.60	4.70
	Mean		14.70	16.60	10.30	6.30	4.55
Dendrocalamus	Prabhunagar	RS	9.80	6.30	3.70	7.30	2.70
strictus	(eroded soil)	RP	5.90	3.90	2.90	5.40	2.40
		ER	0.00	0.00	0.00	0.00	0.00
		Total	15.70	10.20	6.60	12.70	5.10
	UAS, Dharwad	RS	19.66	8.74	8.47	8.63	2.57
	campus (medium	RP	11.00	4.88	4.00	4.87	2.68
	black soil)	ER	3.85	0.00	0.00	0.00	1.80
		Total	34.51	13.59	12.47	13.50	7.09
	Mean		25.10	11.90	9.54	13.10	6.10
Bambusa	UAS, Dharwad	RS	7.90	9.50	3.20	4.70	3.20
bambos	campus (eroded	RP	5.80	2.40	2.80	4.40	3.10
	soil)	ER	0.00	0.00	0.00	0.00	0.00
		Total	13.70	11.90	6.00	9.10	6.30

Notes: RS = Rhizosphere soil; RP = Rhizoplane; ER = Endorhizosphere

lation in vertisols than in alfisols and attributed it to higher organic matter content of vertisols. Eroded soils are known to have low fertility (Harold 1985) and hence, cannot support higher population of microorganisms.

The predominant rhizosphere bacterial isolates – identified as belonging to the genera *Pseudomonas, Xanthomonas, Erwinia* and *Bacillus* – were subjected to morphological and biochemical tests. *Pseudomonas* was found to be predominant (64%), followed by *Bacillus* (22%); *Xanthomonas* (13%) and *Erwinia* (8%) were less predominant indicating the species diversity. Morphological characters of five predominant fungal isolates identified as belonging to the genera *Fusarium, Penicillium, Aspergillus, Rhizopus* and *Cladosporium* were studied. Actinomycete isolates belonged to the genera *Nocardia* and *Streptomyces*. The nitrogen-fixing isolates mainly belonged to the genera *Azotobacter* and *Azospirillum*.

All the six phosphate-solubilizing bacteria isolated were Gram-negative, rod-shaped, non-spore formers and belonged to the genera *Pseudomonas, Xanthomonas* and *Erwinia*. Isolate *Pseudomonas* sp. PSER–1 produced the characteristic pink pigment in Pikovskaya's broth after four days of incubation, and produced attractive dark orange pigment on nutrient agar. Isolate *Pseudomonas* sp. PSRS–2 produced dark green pigment after three days of incubation which subsequently turned blackish.

In general, *Azospirillum* isolates fixed higher amount of nitrogen compared with *Azotobacter* isolates. Among the *Azotobacter* isolates, AZDR–1, an isolate from *Dendrocalamus strictus* rhizome, fixed the maximum amount of nitrogen. ADSR–1, an *Azospirillum* isolate from the roots of *D. strictus,* fixed the highest amount of nitrogen among the *Azospirillum* isolates Table 2).

Table 2.
Nitrogen-fixing efficiency of *Azotobacter* and *Azospirillum* isolates of bamboo rhizosphere

Azotobacter isolate	N (mg) fixed per gram of carbon source	*Azospirillum* isolate	N (mg) fixed per gram of carbon source
AzRS–1	7.00	ABVR–1	16.83
AzRS–3	6.72	ABVR–2	19.60
AzRS–4	5.04	ABVR–3	15.68
AzRP–1	4.76	ADSR–1	22.40
AzRP–2	7.00	ADSR–2	15.68
AzRP–3	9.24	ADSR–3	17.92
AzRP–4	4.23		
AzBR–1	9.24		
AzDR–1	10.36		
AzS–1	7.00		

Isolates PSER–1 and PSER–2 showed maximum solubilization of tricalcium phosphate and also showed the maximum drop in pH of the medium at all the intervals (Table 3). Phosphate-solubilizing isolates were also analysed for organic acid production. PSRS–1, PSRS–2 and PSER–4b produced oxalic acid; PSER–1, PSER–2 produced citric acid and PSER–4a produced both oxalic acid and citric acid (Table 4). Almost all the nitrogen-fixing *Azotobacter* and *Azospirillium* isolates are known to release both IAA and GA. Phosphate-solubilizing isolates also produced both IAA and GA. Besides fixing atmospheric nitrogen/solubilization of insoluble phosphates, several such beneficial organisms are known to produce plant growth promoting substances (Gaur 1990; Omay et al. 1992).

Table 3.
Change in the pH of the medium and percentage of P released from TCP by the P-solubilizing bacteria of bamboo rhizosphere

Isolate no.	Incubation time					
	5 days		10 days		15 days	
	Solubilization (%)	pH drop	Solubilization (%)	pH drop	Solubilization (%)	pH drop
PSRS–1	8.55	6.00	10.14	5.50	12.85	5.50
PSRS–2	9.39	6.00	10.79	5.50	13.58	5.50
PSER–1	16.09	5.50	19.53	5.00	19.87	5.00
PSER–2	14.35	5.50	17.03	5.00	18.54	5.00
PSER–4a	11.98	6.00	18.02	5.50	19.02	5.00
PSER–4b	9.32	6.00	15.44	5.50	18.15	5.00
P. striata	19.56	5.50	21.34	5.50	26.74	5.00

Note: Initial pH of the medium - 6.80

Table 4.
Organic acid production by the phosphate solubilizing isolates from bamboo rhizosphere

Isolate No.		Organic acids				
		Oxalic acid Succinic acid	Gluconic acid	Tartaric acid	Citric acid	Malic acid
PSRS–1	++	–	–	–	–	–
PSRS–2	++	–	–	–	–	–
PSER–1	–	–	–	+	–	–
PSER–2	–	–	–	+	–	–
PSER–4a	++	–	–	+	–	–
PSER–4b	++	–	–	–	–	–
P. striata	++	+	+	–	–	+

Notes: – : not produced; + : weak production; ++ : strong production

Soil microflora and enzyme activities in soil under bamboo leaf litter

Microbial population

1. Bacteria: The plant density and sampling seasons had significant influence on the population of total bacteria. Among the plant density treatments, bamboo planted at highest density (10 000 plants/ha) recorded the highest bacterial population (Table 5). Highest bacterial population among the sampling seasons was recorded in early monsoon. Interaction effect between plant density and sampling season on total bacterial population was significant. However, all the plant density treatments were superior over the control (no plants).

Table 5.
Population of total bacteria, fungi and actinomycetes in soil under bamboo leaf litter of in different seasons

Plant density (plants/ha)	Populations in different seasons														
	Total bacteria (cfu × 10^6/g soil)					Fungi (cfu × 10^4/g soil)					Actinomycetes (cfu × 10^4/g soil)				
	A	B	C	D	Mean	A	B	C	D	Mean	A	B	C	D	Mean
1 111	6.06	7.06	8.42	5.72	6.81	3.48	3.77	4.02	3.24	3.63	5.28	4.51	6.26	4.26	5.08
3 086	6.29	7.07	7.34	6.37	6.77	4.11	4.06	4.26	3.43	3.97	4.63	4.53	5.25	4.08	4.62
10 000	16.20	14.38	15.69	14.16	15.10	4.19	5.11	5.04	4.16	4.63	4.45	5.86	5.15	4.02	4.87
Control (no plants)	4.28	5.44	4.63	5.21	4.89	2.91	3.05	3.10	2.87	2.98	3.24	3.50	4.20	3.24	3.54
Mean	8.21	8.49	9.02	7.86	–	3.68	4.00	4.11	3.42	–	4.40	4.60	4.90	5.21	–
				C.D. at 1%				C.D. at 1%					C.D. at 1%		
Plant density (E)				0.255				0.156					0.156		
Season (F)				0.255				0.156					0.156		
Interaction (E × F)				0.532				0.280					0.320		

Notes: A = Mid-summer; B = Late summer; C = Early monsoon; D = Mid-monsoon.

2. Fungi: The population of fungi in soil under leaf litter of bamboo was significantly higher than in control soil where there was no plantation. Fungal population was also higher in high density planting (Table 5). Soil samples collected in early monsoon showed highest fungal population. Interaction effects of plant density and sampling seasons were also significant.
3. Actinomycetes: Actinomycete population in soil under leaf litter was highest with low density planting and all plant density treatments were superior to the control. Soil samples collected in early monsoon recorded highest population (Table 5). Among the treatment combinations, samples taken in early monsoon from lowest density planting recorded maximum population.

The population of beneficial micro-organisms like N_2-fixers P-solubilizers and cellulose degraders was highest in high density planting and during early monsoon sampling. Mycorrhizal spore count was also significantly influenced by plant density and sampling season (Table 6). Soil under the density of 10,000 plants/ha recorded the highest spore count. All density treatments were significantly superior to control. Early monsoon sampling showed highest spore count among sampling seasons.

Table 6.
Mycorrhizal spore count in soil under bamboo leaf litter at different seasons

Plant density	Mycorrhizal spore count/50 g of soil				
(plants/ha)	Mid-summer	Late summer	Early monsoon	Mid-monsoon	Mean
1,111	252	268	344	145	252
3,086	298	302	493	312	351
10,000	245	325	621	338	382
Control (no plants)	109	92	142	126	117
Mean	226	247	400	230	–
		C.D. at 1%			
Spacing (A)		27.70			
Season (B)		27.70			
Interaction (A × B)		55.40			

Enzyme activities

Plant density, sampling season and their interactions significantly influenced dehydrogenase, urease and phosphatase activities in soil under bamboo leaf litter (Table 7). Dehydrogenase, urease and phosphatase activities were highest in high-density planting (10,000 plants/ha). All spacing treatments were superior over control soil. Among sampling seasons, samples collected in early monsoon showed the highest activity of these enzymes. Interaction between plant density and sampling season revealed that combination of 10 000 plants/ha and early monsoon sampling recorded significantly higher activities of enzymes than all other combinations. Bamboo is deciduous in nature. Hence, there will be a high accumulation of leaf litter. In case of high-density planting, there will be large amounts of diverse organic compounds in the form of root exudates, such as sugars, amino compounds, organic acids, fatty acids, growth factors and other components as reported by Bolton et al. (1993).

Among the seasons, highest population of total bacteria, fungi, actinomycetes, free-living nitrogen-fixers, phosphate-solubilizers and cellulose degraders observed during early monsoon could be attributed to higher soil moisture and congenial temperature. The activities of the enzymes dehydrogenase, urease and phosphatase were also higher in soil under high-density planting and during early monsoon sampling which could be due to higher population of organisms. Significant correlation has been found between enzyme activities and organic matter content. Jha et al. (1992) observed highest microbial population and enzyme activity during July, and suggested soil enzymes to be the indication of microbial activity.

Table 7.
Dehydrogenase, urease and phosphatase activities in soil under bamboo leaf litter

Plant density (plants/ha)	Enzymatic activity in different seasons														
	Dehydrogenase (µg TPF/g soil/day)					Urease (µg NH_4–N/g soil/day)					Phosphatase (µg PNP/g soil/hour)				
	A	B	C	D	Mean	A	B	C	D	Mean	A	B	C	D	Mean
1 111	4.80	4.79	4.96	4.39	4.73	4.06	5.64	6.55	3.97	5.05	27.75	27.71	31.09	25.73	28.04
3 086	5.70	5.85	6.12	4.91	5.64	6.16	8.25	9.84	5.84	7.52	27.86	30.45	38.98	24.89	30.55
10 000	5.78	6.73	7.36	5.15	6.25	6.51	8.78	10.52	6.29	8.03	38.36	36.90	45.13	27.69	36.27
Control (no plants)	3.38	3.42	3.43	3.76	3.50	3.45	4.21	6.29	3.40	3.90	25.39	26.55	26.15	24.67	25.69
Mean	4.91	5.19	5.47	4.55	–	5.04	6.72	8.03	4.88	–	29.07	30.40	35.34	25.74	–
				C.D. at 1%				C.D. at 1%					C.D. at 1%		
Plant density (E)				0.156				0.182					0.983		
Season (F)				0.156				0.182					0.983		
Interaction (E × F)				0.295				0.331					1.938		

Notes: A = Mid-summer; B = Late summer; C = Early monsoon; D = Mid-monsoon.

CONCLUSION

The investigations clearly indicated the diversity of population and activity of microorganisms in the bamboo rhizosphere. The studies also indicated the possibility of exploiting the diverse functional groups of organisms in improving the growth and biomass production of bamboo grown on poor and marginal lands.

REFERENCES

Bolton, H.J.; Fredrickson, J.K; Elliot, L.F. 1993. Microbial ecology of the rhizosphere. *In* Metting, Jr. F.B., ed., Soil microbial ecology. pp.27–65.

Bunt, J.S.; Rovira, A.D. 1955. Microbiological studies of sub-Antartic soils. Journal of Soil Science, 6, 119.

Casida, L.E.; Klein, D.A.; Santoro, T. 1964. Soil dehydrogenase activity. Soil Science, 98, 371–376.

Eivazi, Z.; Tabatabai, M.A. 1979. Phosphatase in soils. Soil Biology and Biochemistry, 9, 167–172.

Gardemann, J.W.; Nicolson, J.H. 1963. Spores of mycorrhizal endogene species extracted from soil by wet sieving and decanting. Transactions of British Mycological Society, 42, 235–244.

Gaur, A.C. 1990. Phosphate-solubilizing microorganisms as biofertilizers. Omega Scientific Publishers, New Delhi, India.

Harold, G. 1985. Erosion:its causes and care. International Book and Periodicals Supply Service, New Delhi, India.

Holt, J.G.; Kreig, N.R.; Sneath, P.H.A.; Staley, J.T.; William, S.T. 1994. Bergey's manual of determinative bacteriology, Ninth Edition. Williams and Wilkinsons, Waverly Company, UK.

Jackson, M.L. 1973. Soil chemical analysis. Prentice-Hall of India Pvt. Ltd., New Delhi, India.

Jha, D.K.; Sharma, G.D.; Mishra, R.R. 1992. Soil microbial population numbers and enzyme activities in relation to altitude and forest degradation. Soil Biology and Biochemistry, 24, 761–767.

Kuster, E.; Williams, S.T. 1964. Selection of media for isolation of streptomyces. Nature, 202, 926–929.

Martin, J.P. 1950. Use of acid rose bengal and streptomycin in plate method for estimating soil fungi. Soil Science, 69, 215–232.

Monrozier, J.L.; Ladd, J.N.; Fijzpatrick, R.W.; Foster, R.J.; Raupach, M. 1991. Components and microbial biomass content of size fraction in soils of contrasting aggregation. Geoderma, 50, 37–62.

Nordmann, J.; Nordmann, R. 1960. Organic acids in chromatographic and Electrophoretic Techniques. Intersciences Publishers Inc., New York, USA.

Norris, J.R. 1959. Isolation and identification of *Azotobacter*. Lab Practice, 8, 239–240.

Okan, Y.; Albrecht, S. Z.; Burris, R. H. 1977. Methods for growing *Spirillum lipoferum* and for counting it in pure culture and in association with plants. Applied and Environmental Microbiology, 33, 85–87.

Omay, S.H.; Schmidt, W.A.; Martin, P. 1992. Indole acetic acid production by the rhizosphere bacterium *Azospirillum brasilense* cd under in-vitro conditions. Canadian Journal of Microbiology, 39, 187–192.

Pancholy, S.K.; Rice, E.L. 1973. Soil enzymes in relation to old field succession: amylase, invertase, cellulase, dehydrogenase and urease. Soil Science Society of America, 34, 47–50.

Pikovskaya, R.I. 1948. Mobilization of phosphorus in soil in connection with vital activity of some microbial species. Mikrobiologiya, 17, 362–370.

Rovira, A.D. 1959. Root excretions in relation to the rhizosphere effect: IV Influence of plant species, age of plant, light, temperature and calcium nutrition on exudation. Plant and Soil, 9, 53–69.

Vancura, V. 1964. Root exudates of plants: analysis of root exudates of barley and wheat in their initial phases of growth. Plant and Soil, 21, 231–248.

Vancura, V.; Macura, J. 1960. Indole production in *Azotobacter* culture. Folia Microbiologia, 5, 293–298.

Bamboo for Sustainable Development, pp. 309–321
Eds. A. Kumar, I.V. Ramanuja Rao and Ch. Sastry

Diseases and disorders of bamboos in Asia

C. MOHANAN
Kerala Forest Research Institute, Peechi, Kerala, India

Abstract—Bamboos form the backbone of the rural economy in many Asian countries. Bamboo resource base in this region is dwindling at a very fast rate owing to various biotic and abiotic interferences, including diseases and pests. Recently, an exhaustive study has been undertaken and information on diseases and disorders of bamboos from different bamboo growing countries in Asia collated. The study revealed that about 170 species of bamboos belonging to 26 genera in this region are affected by various diseases and disorders. A total of 440 fungi, three bacteria, two viruses, one phytoplasma and one bacteria-like organism have been reported as associated with these diseases and disorders. Only a few among these diseases are recognized as potentially serious ones, affecting the bamboo industry as well as the rural economy as a whole. Potentially serious diseases of bamboos in Asia include: culm blight caused by *Sarocladium oryzae* in the village groves in Bangladesh and in the coastal areas of Orissa state, India; rot of emerging and growing culms of industrially important bamboo species in India caused by *Fusarium* spp.; witches'-broom incited by *Balansia* spp. in different species of bamboos in Japan, China, Taiwan and in reed bamboos in India; little leaf of *Dendrocalamus strictus* caused by phytoplasma (MLO) in the dry tracts of southern India; culm mosaic caused by bamboo mosaic virus (BoMV) in Taiwan; culm rust caused by *Stereostratum corticioides* and top blight of *Phyllostachys* spp. caused by *Ceratosphaeria phyllostachydis* in China.

INTRODUCTION

Bamboos are fast-growing versatile plants with multiple end uses. For centuries, they have been closely related to agriculture, cottage industries, arts, culture and day-to-day life of more than half of the world population. Because of their versatility, adaptability to different eco-climatic conditions and myriad end uses, bamboos form the backbone of the rural economy in many Asian countries.

Bamboos are estimated to comprise about 1 200 species and varieties belonging to 75 genera. They have a wide range of distribution from the tropic to temperate zones and from sea level to the alpine elevations of 3 000 to 4 000 meters. However, most bamboos occur in the warm, humid areas in the tropical and subtropical Asia, South America and Africa. Of these, Asia has the largest resource with about 18 million hectares of bamboo-growing lands and the highest diversity with more than 800 species and varieties of bamboo in 45 genera (Table 1).

The production potential of bamboo stands is greatly affected by various biotic and abiotic factors. Bamboos, like many other forestry crops, suffer from numerous diseases and disorders that affect the availability of healthy planting stock, overall productivity of the stands, as well as quality and quantity of post-harvest culms. Available information on disease and disorders of bamboos in Asia is widely scattered, and it is not possible to make an accurate assessment on their present status or their impact upon bamboo cultivation and culm production in a particular country in this region. A detailed study was carried out in bamboo-growing countries in Asia, under an Internship awarded by INBAR (Mohanan 1997). This highlights the status of potentially serious diseases of bamboos, economic losses caused by diseases and disorders, and available control measures.

RESEARCH METHODOLOGY

Bamboo-growing areas in different parts of India, China, Thailand, the Philippines and Malaysia were visited and information on bamboo species grown, occurrence of diseases and disorders, economic losses caused, etc. collected. Disease specimens from bamboo nurseries, plantations and natural stands, and decay specimens from storage yards, mill sites and natural stands in different parts of India were collected, and casual organisms isolated and identified. For authentic identification/confirmation, the fungal and bacterial cultures were referred to the Commonwealth Agricultural Bureau (CAB), U.K. Light and Fluorescent Microscopy, Transmission Electron Microscopy and pathogenicity trials were carried out to confirm the fungal, bacterial, viral and phytoplasma diseases of bamboos. Literature searches were carried out in libraries and institutions in different bamboo-growing countries in Asia and available information collated.

BAMBOO DISEASES

Nursery diseases

Bamboo seedlings are raised in conventional seedbed nurseries in this region. Planting stock is also raised through different vegetative propagation methods. Usually planting stock has to be maintained in the nursery for 6 to 19 months before planting out. Diseases affect the nursery stock right from the radicle stage to the out-planting phase and cause considerable damage, depending on the prevailing microclimate in the nursery and virulence of the associated pathogens. Although bamboo seedlings are being raised in different countries in Asia, diseases affecting the nursery stock have been reported only from India, Thailand and the Philippines. Altogether 14 nursery diseases have been recorded and most of these were from India, where bamboo seedlings are raised in conventional seedbed nurseries (Mohanan 1994a, 1994b; Mohanan and Liese 1990). Damping-off (caused by *Fusarium moniliforme, F. oxysporum, Rhizoctonia solani*), wilt (*R. solani*), web blight (*R. solani*), leaf rust (*Dasturella divina*) leaf spot (*Exserohilum rostratum, E. holmii, E. halodes, Dactylaria bambusina, Colletotrichum gloeo*

Table 1.
Area under bamboos, number of genera and species, and economically important species in different Asian countries

Country	Area (10^4 ha)	Genera	Species	Important bamboos
Bangladesh	60.0	13	33	*Bambusa balcooa, B. vulgaris, B. nutans, B. tulda, Dendrocalamus strictus, Melocanna baccifera*
Bhutan	–	8	11	*B. Nutans, D. hamiltonii, Cephalostachyum* spp.
Cambodia	28.7	–	–	*D. giganteus, Cephalostachyum* spp.
China	400.0	33	400	*B. textilis, B. pervariabilis, B. spinosa, B. blumeana, D. latiflorus, D. strictus, D. giganteus, Phyllostachys pubescens, P. flexuosa, P. glauca, P. nuda, P. bambusoides, P. tanaka*
India	1003.00	23	128	*B. bambos, B. nutans, B. tulda, B. vulgaris, D. hamiltonii, D. strictus, Ochlandra ebracteata, O. scriptoria, O. travancorica, M. baccifera*
Indonesia	5.00	14	125	*B. bambos, Gigantochloa apus, G. atter, G. verticillata*
Japan	14.10	13	165	*Phyllostachys bambusoides, P. edulis, P. nigra, P. reticulata, P. pubescens, Pleoblastus simonii, P. hindsii*
Korea	0.80	10	13	*Phyllostachys reticulata, Pleoblastus* sp.
Laos	9.80	–	–	*G. albociliata, Cephalostachyum* sp. *Thyrsostachys* sp.
Malaysia	3.29	10	50	*Gigantochloa levis, G. ligulata, G. scortechinii, G. wrayi, Thyrsostachys* sp.
Myanmar	217.00	18	90	*B. polymorpha, B. tulda, B. vulgaris, D. calostachys, D. giganteus, D. longispathus, D. scriptoria, Cephalostachyum pergracile, Melocanna bambusoides, Thyrsostachys siamensis*
Nepal	–	5	30	*Bambusa* sp., *D. strictus, D. hamiltonii*
Pakistan	–	–	–	*B. tulda, D. Strictus, D. hamiltonii*
The Philippines	0.80	12	55	*B. blumeana, B. vulgaris, D. merillianus, D. latiflorus, G. aspera, G. levis, Schizostachyum lima, S. lumampo*
Sri Lanka	8.00	7	14	*B. bambos, B. vulgaris, O. stridula, D. giganteus, D. membranaceus, D. asper, D. strictus*
Taiwan	17.56	18	56	*B. oldhamii, D. latiflorus, P. edulis, P. makinoi*
Thailand	100.00	12	50	*B. bambos, B. blumeana, B. longispiculata, B. nana, B. tulda, B. polymorpha, D. hamiltonii, D. Membranaceus, D. brandisii, D. strictus, C. pergracile, G. albociliata, G. nigrociliata, G. haskarliana, G. macrostachys, T. oliveri, T. siamensis*
Vietnam	13.0	15	20	*B. bambos, B. tulda, P. bambusoides*

sporioides, Curuvularia pallescens, Alternaria alternata), leaf blight (*Bipolaris maydis, B. urochloae, B. bambusae*), rhizome rot (*Rhizostilbella hibisci*), and leaf striping and stunting caused by a virus are the diseases recorded on bamboo seedlings.

Among these, web blight caused by *Rhizoctonia solani* is the most widespread disease. Web blight of bamboo seedlings has been reported in India, and in the Philippines as 'sheath blight'. In India, the disease is widespread in bamboo nurseries, especially in South India. The disease affects *Bambusa bambos, Dendrocalamus strictus, D. brandisii* and *Thyrsostachys siamensis* seedlings that are 20–30 days old. The disease usually occurs as small patches comprising 5–10 seedlings in the seedbed, and then increasing in size as more and more seedlings become affected gradually under favorable microclimatic conditions. The affected seedlings are killed outright in about 10–20 days of infection, leaving large circular to irregular patches of dried-up seedlings in the seedbed.

Disease management measures suggested for the web blight include sanitation, modification of nursery cultural practices and use of fungicide (Mohanan 1993, 1994b). Solarization of seedbed and treatment of seed and seedbed with antagonistic fungi, *Trichoderma harzianum* and *T. viride* are also recommended for reducing the disease incidence (Mohanan 1994a, 1994b, 1995). Regulation of the shade over the nursery beds (no shade) and watering quantity (120 liters per standard bed per day) and low seeding density (500 g seeds per standard seedbed) are suggested (Mohanan 1995). Application of fungicide (carboxin 0.2% a.i.) after 7 and 21 days of seedling emergence is also recommended for controlling the diseases.

Diseases in bamboo plantations and natural stands

Bamboos in stands are vulnerable to various diseases at different stages of their growth. The woody stems arising from woody-root stock or rhizomes are called culms. The culms that are newly emerging or growing supple are generally susceptible to diseases. Diseases affecting the bamboo culms include: rot of emerging and growing culms, culm blight, die-back, witches'- broom, little leaf, thread blight, mosaic, wilt, culm rust, smut, top blight, base rot and necrosis of culm internodes. Apart from this, a large number of diseases affecting roots, rhizomes, branches and twigs, foliage, culm sheath, inflorescence, etc. have also been reported from various countries in this region. More than 400 fungi, 3 bacteria, 1 bacteria-like organism, 1 virus and 1 phytoplasma (MLO) have been reported as associated with these diseases. Among the diseases mentioned above, only a few are considered as potentially serious (Table 2), which may affect the culm production and stand productivity considerably.

Table 2.
Potentially serious diseases of bamboos in Asia.

Potential Disease	Causal organism	Bamboo species	Country
Rot of emerging culm	*Fusarium moniliforme* var. *intermedium*	*B. balcooa*	India
		B. bambos	
		B. polymorpha	Bangladesh
		B. vulgaris	
		D. longispathus	
		D. strictus	
Rot of growing culm	*F. equiseti*	*B. balcooa*	India
		B. bambos	
		B. polymorpha	
		D. brandisii	
		D. longispathus	
		D. strictus	
Blight	*Sarocladium oryzae*	*B. bambos*	Bangladesh
		B. balcooa	
		B. tulda	
		B. vulgaris	
		B. nutans	India
Top blight	*Ceratosphaeria phyllostachydis*	*P. edulis*	China
		P. heterocycla	
		P. pubescens	
Witches'-broom	*Balansia take*	*Phyllostachys* spp.	China,
	B. linearis	*Ochlandra* spp.	Taiwan,
			Japan,
			India
Culm rust	*Stereostratum corticioides*	*Phyllostachys* spp.	China
		Bambusa spp.	
Little leaf	*Phytoplasma*	*D. strictus*	India
Bamboo mosaic	*Bamboo mosaic virus*	*B. oldhamii,*	Taiwan
		D. latiflorus	

POTENT BAMBOO DISEASES

Bamboo blight

Bamboo blight has been reported to affect five species of bamboos in village groves – *Bambusa bambos, B. balcooa, B. tulda* and *B. vulgaris* in Bangladesh, and *B. nutans* in the coastal areas of Orissa, India. The disease was first recorded in 1970 by Rahman and Zethner (1971) as a potentially serious problem of village bamboos in Bangladesh. Death of the entire affected clump within 3–4 years of infection has been reported. The disease was studied in detail and about 3%–66% infection of clumps in different areas (Chittagong, Dhaka, Comilla, Rahshahi) has been reported (Boa and Rahman 1987). In India, large-scale mortality of culms in Balashwar, Puri and Ganjam districts of Orissa state has been recorded for the past 8 years. At present, the disease seems to be spreading at a very fast pace to the disease-free neighboring groves in the coastal areas. *Sarocladium oryzae* is considered as the principal fungus associated with the disease, which causes a sequential die-back of culms in their first season of growth. Silvicultural measures recommended for controlling the disease include: cutting and removing the blighted culms, burning the debris of clumps *in situ* and addition of new soil around the clumps. Light surface fire before the onset of monsoon is suggested for reducing the inoculum potential of the pathogen, which exists in the debris or in the top few centimeters of the soil. Application of fungicides (carbendazim (0.15% a.i.) combined with mancozeb (0.3% a.i) is also recommended for controlling the disease.

Rot of emerging culms

This disease affects *B. bambos, B. tulda, B. vulgaris* and *B. balcooa* in village groves in Bangladesh. In Pakistan, large-scale mortality of emerging shoots in *D. strictus* has been reported. In India, the disease is widespread in the southern states (mainly Kerala and Karnataka) affecting the bamboos in plantations as well as natural stands. Bamboos affected seriously include: *B. balcooa, B. bambos, B. polymorpha, B. vulgaris, D. longispathus, D. strictus, O. travancorica, T. oliveri*, etc. Severe infection and mortality of emerging culms occur in bamboo stands situated in low-lying, high-rainfall areas of Kerala. Among the bamboo species, the highest mortality (34%) was recorded in *B. bambos* natural stands in the northern parts of Kerala state (Mohanan 1994a). The disease is caused by *Fusarium moniliforme* var. *intermedium*. Cultural control measures – such as removal of debris from around the clumps before the onset of monsoon and light burning of the debris over the ground, loosening soil around the clump before the culm emergence, etc. – are suggested.

Rot of growing culm

The disease affects bamboos in plantations and natural stands in India (Mohanan 1990, 1992). The disease attained an epidemic status during 1987–91 in high-rainfall areas in Kerala, and affected most bamboo species grown in plantations and natural stands. In plantations, *B. bambos, B. polymorpha* and *D. strictus* were the severely affected species. Though the disease occurs in China on *Phyllostachys pubescens* (Anonymous 1982), it has not attained the status of a serious disease. The disease is caused by *Fusarium equiseti* and the infection is predisposed by injury of culm sheath and internode caused by a sap-sucking insect, *Purohitha cervina*. Since the incidence and severity of the disease can be well correlated with the heavy build-up of the insect population in bamboo stands during the culm elongation phase, control measures suggested include checking the insect population build-up of the sap-sucking insect by pesticide application as well as the prophylactic fungicide treatment on the emerging and growing culms.

Top blight

The disease – also named as 'withered tip disease' and 'die-back' – occurs in *Phyllostachys edulis, P. pubescens* and *P. heterocycla* in China. The disease is widespread in Jiangsu, Zhejiang, Anhui, Jiangxi, Fujian and Shanghai provinces, causing heavy damage in bamboo groves. The disease mainly affects new culms of the season. Infection causes browning and necrosis of the culm and internodes, and later withering of branches, twigs and whole culms. Die-back of the affected culms occurs owing to infection in successive years. The causal organism is *Ceratosphaeria phyllostachydis*. Peak incidence of disease occurs during July-September. The disease has been studied in detail and a disease prediction model generated for forecasting the infection in *P. heterocycla* stands (Ou and Zhang 1993).

Silvicultural measures suggested for controlling the disease are cutting and removing the infected branches and culms to reduce the inoculum, clearing up of miscellaneous trees and tending the young bamboos year-around, and eradicating the alternative hosts of the pathogen. Quarantine measures to check the transport of diseased bamboos out of the infected areas are recommended. Application of fungicides (carbendazin 0.05% a.i., thiophanate methyl 0.02% a.i.) is also suggested to manage the spread of the disease.

Witches'-broom

The disease affects different species of *Phyllostachys* in China, Taiwan, Japan, Vietnam, *Gigantochloa* spp. in Indonesia, and *Ochlandra* spp. in India. In China, the disease is widespread in Hunan province causing 95% to 100% infection (Lin and Wu 1987). In India, the disease is widespread in the reed bamboos *Ochlandra travancorica, O. scriptoria* and *O. ebracteata* in Kerala state. The disease causes development of numerous highly shortened shoots at the nodes of mature culms and from the rhizomes. Successive development of a large number of thin wiry shoots in tuft from the nodes of the infected culm/rhizome gives rise to the appearance of witches'-broom. *Balansia take* and *B. linearis* are the important fungi associated with the disease. To control the disease, silvicultural measures, such as pruning and removing infected twigs, have been suggested. Fungicides and antibiotics such as tetracycline and oxytetracycline have been recommended in China to contain the disease. Cutting the old and weak mother culms, removing the diseased clumps and burning them outside the forest are recommended to manage the disease in *Phyllostachys* spp. in Japan and in *Ochlandra* spp. in India.

Culm rust

In China, culm rust poses a major problem in bamboo stands. The rust fungus *Stereostratum corticioides* affects more than 16 species of bamboos, including *Phyllostachys glauca, P. conjesta, P. propinqua, P. meyeri* and *Pleoblastus vaginatus*. Since 1978, the rust has spread rapidly in Jiangsu, Hunan, Zhejiang and Anhui provinces of China. Incidence of rust infection in *P. glauca* ranges from 30% to 90% (Zhu and Zhang 1988). A serious outbreak of culm rust, which destroyed about 200 ha of *P. glauca* stands, has been reported recently. The rust infection has reduced bamboo culm and edible shoot production and affected development of the bamboo industry. Silvicultural measures and chemical application against rust infection have been developed in China. Felling and removal of severely affected bamboo culms and chemical treatment in groves for 3–4 years successively are necessary to minimize the disease incidence.

Little leaf

The disease, characterized by the development of numerous highly reduced and abnormal bushy shoots from the nodes of newly emerged culms and from culm branches, occurs in *Dendrocalamus strictus* stands in southern India (Kerala, Karnataka, Tamil Nadu and Andhra Pradesh). In Kerala, the disease incidence in dry tracts and sandalwood reserves is very high (Mohanan 1994a). Association of a Phytoplasma (MLO) with the disease was confirmed (Mohanan 1994c). In severely affected stands, all culms produced from rhizomes in a growing season become infected. Although clump-to-clump spread of the disease is rather slow, it is posing a serious threat to the *D. strictus* stands raised in the dry tracts of the state.

Bamboo mosaic

Mosaic disease is reported to affect the two major cultivated species of bamboos in Taiwan – *D. latiflorus* and *B. oldhamii*. Infection causes mosaic on leaves, and brown internal streaking in emerging and growing culms. Infected shoots exhibit shortened internodes, poor growth and development. They have a hard texture and are of low quality for eating and canning. The disease is caused by a bamboo mosaic virus (BoMV). Owing to large-scale exploitation of vegetative propagules of bamboos, the disease has spread all over the bamboo growing areas of the country. Strict quarantine measures are recommended to control the disease.

DECAY AND BIODETERIORATION OF BAMBOOS IN STORAGE

Bamboo is a major raw material for the pulp and paper industry in the Asian region. Large quantities of bamboo culms are consumed also by traditional industries. Bamboo culms extracted for utilizing in this industry have to be stored for a considerable period. The natural durability of bamboo is low and varies from 1 to

36 months, depending on the species and climatic conditions. In humid tropical areas, enormous quantities of culms under storage are decayed and deteriorated by fungal organisms. About 20–25% damage during 12-month storage has been reported in India. Most bamboos used for structural purposes in housing deteriorate within a couple of years, putting heavy pressure on the resources caused by increased demands for frequent replacements.

Decay and biodeterioration of culms are caused mainly by fungi, including soft rot, brown rot and white rot. Bacteria also deteriorate culms in post-harvest facilities. One or more of these organisms may attack the culms in succession. The extent of colonization by the microorganisms as well as the severity of attack depend on moisture and starch contents in the culm, ambient temperature, humidity, etc. Bamboo consists of 50–70% hemicellulose, 30% pentosans and 20–25% lignin. Most of the hemicellulose (90%) is xylan, with a structure between hardwood and softwood xylans. Bamboo is known to be rich in silica (0.5–4%), but the entire silica is located in the epidermal layers, with hardly any in the rest of the cell wall. Bamboo also has minor amounts of resins, waxes and tannins. However, none of these compounds has any effect on the natural durability. High starch content makes bamboo very susceptible to attacks by the staining and decay fungi.

A large number of fungi causing decay and deterioration of culms in storage have been reported from various countries (Mohanan 1997). Decay and deterioration have been considered as a serious problem in bamboo storage yards, particularly in pulp mills. Decay fungi seriously affect the pulp yield (up to 25% loss in one year of storage) and pulp strength is reduced by 15–40%. Fungal staining also affects the brightness of the pulp and consequently, the consumption of bleach increases by about 10% to bring stained pulp to acceptable brightness. Thus, in general, fungal infection increases pulping costs, as well as operational costs in terms of additional costs incurred for the treatment of wastewater with high alkali content.

Decay and biodeterioration of bamboo culms during storage outdoors can be checked to a large extent by adopting good storage yard management practices, such as systematic storage on a first-come-first-off basis. Horizontal stacking of culms over raised walls is recommended to facilitate drainage and air circulation. Chemical preservation is generally more effective for the protection of bamboo under storage than non-chemical methods, though they may not be economical or applicable in all cases. Chemical protection ensures a longer life for bamboos. Treatments can be given using a variety of chemicals depending upon the culm conditions – green or dry – and the end use. Various chemical treatments recommended for fresh bamboo (green) to increase the service life include: steeping, sap displacement, diffusion process, Boucherie process, etc. Dry bamboo culms are treated by soaking in preservative solution, hot-cold process, pressure treatment, etc. Chemical treatment on a large scale has been carried out in India, Taiwan and Japan. Spraying of borax-boric acid (2% boric acid equivalent) at the rate of 0.3 l/m^2 at monthly intervals on the exposed surface of the stack has been suggested (Gnanaharan et al. 1993; Kumar et al. 1994; Liese 1993).

IMPORTANCE OF BAMBOO DISEASES AND ECONOMIC LOSSES

The relative importance of different diseases affecting bamboo in the region is difficult to assess because of the general lack of information accompanying each disease record. Similarly, lack of quantitative data on the extent of damage caused to the bamboo stands makes it difficult to evaluate the economic loss caused by these diseases. The overall impression, however, is that many of the bamboo diseases have only limited economic importance. In general, although a large number of foliage diseases – leaf spots, leaf blight, leaf rust, etc. – have been recorded, they cause comparatively minor damage to bamboo stands than culm diseases. Diseases affecting the emerging and growing culms are much more serious and cause extensive damage to bamboo stands.

Bamboo blight, the most dreaded of diseases, affects the village groves of four species of bamboos throughout Bangladesh, and one species in the coastal belt of Orissa state in India. Losses from diseases have been recorded highest in *B. vulgaris* and the jawa/kata jalli variety of *B. tulda*. Although the disease severely affects bamboos in village groves in Bangladesh and individual farmers in these areas are experiencing considerable economic hardship as a result of damage to the culms, the available data are not sufficient to make an overall assessment of the economic losses caused by the disease or impact of the disease on culm production. In India, large-scale mortality of *B. nutans* owing to blight disease occurred in 1988 in the coastal areas of Ganjam, Puri and Cuttack (all in Orissa), considerably affecting the economy of the village. The villagers in these areas earn Rs. 2 000 to Rs. 5 000 annually from sale of culms from their groves. The infection was so widespread and severe that it threatened to damage all the groves, each consisting of 50–100 culms, within a period of 2–3 years. During 1988, the disease reached an epidemic proportion and about 55–67% of the clumps in Cuttack and Puri districts were affected. At present, the disease is spreading at a very fast pace affecting village groves, and farmers are so desperate that they have started abandoning the cultivation of bamboo in their groves.

Rot of emerging culm reported in various bamboo species from Bangladesh, India, Pakistan, China, Thailand and the Philippines appears to be one of the factors for the large-scale reduction in culm production. In India, the disease is widespread in natural bamboo stands and plantations in the southern states. Severe infection and large-scale mortality of emerging culms occur in bamboo stands in high-rainfall areas of Kerala state. About 34% mortality of emerging culms was recorded in *B. bambos* natural stands in different locations in Wynad Forest Division of Kerala during 1988–91. In bamboo plantations, mortality of emerging culms ranged from 5.5% to 26% during this period. Usually very high disease incidence and mortality occur in emerging culms of 15–30 cm height. It has been reported that economic losses owing to the disease is high in unmanaged natural stands than in plantations. Rot of growing culms reported from India and China is another potentially serious disease affecting culm production. In India, the disease has been recorded in different species of bamboo grown in Kerala state and the disease incidence varied from 3% to 25% during 1997–91 in different localities. Young clumps (2–4 years old) of *B. bambos, D. longispathus* and *D. strictus* were severely affected. The disease affects the quality as well as quantity of the culms.

Large-scale mortality of bamboos owing to top blight occurred in 1983 in China. Economic loss caused the disease has been reported to be very high in bamboo stands growing in drought-affected areas and in poor soils. Witches'-broom, which causes malformation of the culms and affects culm production, is posing a serious threat to the bamboo industry. However, data on economic losses are not available from most countries. In India, the disease is widespread among reed bamboos in Kerala state, and about 6–15% disease incidence was recorded in different localities during 1988–92. Severe infection has not been assessed; but in view of the widespread incidence and systemic nature of the infection, it is suspected that culm production and stand productivity among reed bamboos may be adversely affected in the near future.

Bamboo mosaic disease affecting the two major cultivated species of bamboos in Taiwan, *D. latifolia* and *B. oldhamii,* is widespread and about 70–82% disease incidence has been reported from different localities. Economic loss from the disease is very high as the disease affects the export of edible shoots as well as the shoots canning industry. A survey on little leaf disease during 1988–92 recorded 6–13% increase of disease incidence in *D. strictus* in different localities in Kerala state, India. Spread of the disease over the last four years is very high in the dry tracts of the state, with the farmers hardly getting any culms from their groves.

Bamboo seeds are invaded by microorganisms during their different developmental stages as well as after the seed fall. Seeds are also infected during storage and subsequent handling prior to sowing. Poor quality of seeds in terms of viability and vigor has been the cause for a high degree of failure in nurseries. Few of the seed-borne fungi are also capable of causing seedling diseases in nurseries. Taking into account the unusual flowering and seeding cycle of bamboos, the seeds are so valuable that any loss from microbial deterioration is of great concern. Although data on losses owing to biodeterioration of seeds has not been available from any of the bamboo-growing countries in the region, it is evident from some of the disease records that large quantities of bamboo seeds are being deteriorated by microorganisms in different process stages.

Non-infectious diseases, which are caused by abiotic factors, also play a major role in limiting culm production. Severe damage caused by glazed frost has been reported from China and Taiwan. About 3 800 ha of *Phyllostachys pubescens* stands in Qianshan county of Anhui province were severely affected by glazed frost during 1988, causing a loss of 9 million Chinese Yuan from culm damage. Culm rust caused by *Stereostratum corticioides* poses a major threat to *Phyllostachys glauca* and *P. meyeri* stands in Jiangeu, Hunan, Zhejiang and Anhui provinces in China. The incidence of rust infection that affects the production of edible shoots ranged from 30% to 90% in different provinces. Economic losses from the disease are reported to be very high and a serious outbreak of rust infection in Yangzhong county destroying about 200 ha of *P. glauca* stands has been reported.

CONCLUSION

Among the various diseases that affect bamboos in the Asian region, only a few are categorized as potentially serious. Culm blight, culm rust, top blight and culm mosaic are spreading at a very fast pace in bamboo stands in the respective countries and occasional new outbreaks are common, and these affect the bamboo industry as well as rural economy as a whole. Disease management measures practiced rely mainly on silvicultural measures and, to a lesser extent, on prophylactic treatments. In general, etiology and epidemiology of most bamboo diseases are little investigated, and available control measures are often inefficient. Hence, a concerted effort is needed to develop appropriate measures to check further spread of the disease as well as to safeguard against the inadvertent introduction of diseases to new areas. Disease management strategies comprising both short-term and long-term measures – including the broadening of the genetic base of bamboos and introduction of disease resistant species/provenances – have to be drawn up implemented.

ACKNOWLEDGEMENTS

The author is grateful to the International Network for Bamboo and Rattan (INBAR) and International Development Centre (IDRC), Canada, for providing an Internship to carry out this work.

REFERENCES

Anonymous. 1982. China forest diseases. China Forest Publishing House, Beijing, China. 24 pp.

Boa, E.R.; Rahman, M.A. 1987. Bamboo blight and bamboos of Bangladesh. Bulletin I, Forest Pathology Series, Forest Research Institute, Bangladesh. 43 pp.

Gnanaharan, R.; Mohanan, C.; Chand Basha, S. 1993. Post-harvest technology of reed bamboos. BIC-India Bulletin, 3(1), 1–6.

Kumar, S.; Shukla, K.S.; Indira Dev; Dobriyal, P.B. 1994. Bamboo preservation techniques - a review. INBAR Technical Report No. 3. International Network for Bamboo and Rattan, New Delhi, India; Indian Council of Forestry Research and Education, Dehra Dun, India. 59 pp.

Liese, W. 1993. Methods of bamboo treatment and preservation. *In* International Bamboo Festival 1990. Hamburg Forest Research Institute, Hamburg, Germany. pp. 25–28.

Lin, X.; Wu, G. 1987. Study on witches'-broom of fish-scale bamboo - I. Symptoms and pathogens. Journal of Central-South Forestry College, 7, 132–136.

Mohanan, C. 1990. Diseases of bamboos in Kerala. *In* Ramanuja Rao, I.V.; Gnanaharan, R; Sastry, C.B., ed., Bamboos: current research. Proceedings of the International Bamboo Workshop, Cochin, India, 14–18 November 1988. Kerala Forest Research Institute, Kerala, India; International Development Research Centre, Ottawa, Canada. pp. 173–183.

Mohanan, C. 1992. Diseases - potential threat to productivity of bamboos in India. *In* Proceedings of III International Bamboo Congress, Manamata, Kumomoto Perfecture, Japan. pp. 25–26.

Mohanan C. 1993. Epidemiology and control of *Rhizoctonia* web blight of bamboos. *In* K.S.S. Nair et al., ed., Impact of diseases and insect pests in tropical forests. Proceedings of the IUFRO Symposium, Kerala Forest Research Institute, Peechi, Kerala, India.

Mohanan C. 1994a. Diseases of bamboos and rattans in Kerala, India. KFRI Research Report 98. Kerala Forest Research Institute, Peechi, Kerala, India. 120 pp.

Mohanan C. 1994b. Studies on diseases of bamboos and nursery management of *Rhizoctonia* web blight in Kerala. Ph.D. Thesis, Cochin University of Science & Technology, Kerala, India, 316 pp.

Mohanan C. 1995. Methods to control important diseases of bamboos. *In* Silviculture and management of bamboos in India (Phase II). Final Technical Report to IDRC, Ottawa, Canada. (unpublished).

Mohanan, C. 1997. Diseases of bamboos in Asia – an illustrated manual. INBAR Technical Report No. 10. International Network for bamboo and Rattan, Beijing, China. 228 pp.

Mohanan, C.; Liese, W. 1990. Diseases of bamboos. International Journal of Tropical Plant Diseases, 8, 1–20.

Ou, Z.S.; Zhang, W.Q. 1993. A study on the prediction method for the top blight of *Phyllostachys heterocycla.* Journal of Fujian Forest College, 13(2), 141–146.

Rahaman, M.A.; Zethner, O. 1971. An interim report on results obtained in Forest Pathology Section from September 1969 to August 1971. Forest Dale News, 4(1), 45–48.

Varma, J.C.; Bahadur, K.N. 1980. Country report and status of research on bamboos in India. Indian Forest Records (Botany), 6, 28 pp.

Zhu, X.Q.; Zhang, J.N. 1988. A study of the extended experiment for control of culm rust of bamboos. Journal of Nanjing Forestry University, 1, 35–39.

FURTHER READING

Mohanan, C. 1995. Little leaf disease of bamboos. BIC-India Bulletin 4 (1/2) 30–31.

Purushotham, A.; Sudan, S.K.; Vidyasagar, 1954. Preservative treatments of green bamboos under low pneumatic pressures. Indian Forestry Bulletin No. 178.

Wang, S. 1985. Mould of bamboo culm and control methods. Bamboo Research, 1, 1985, 47–51.

Bamboo for Sustainable Development, pp. 323–331
Eds. A. Kumar, I.V. Ramanuja Rao and Ch. Sastry

Technology for large-scale propagation of bamboo through aerial seeding

M. JAGANNADHA RAO
M.S. Swaminathan Research Foundation, Chennai, India

Abstract—The depletion of forests in tropics and sub-tropics is a cause for alarm not only for conservationists but also for all those who live in and work for the development of the countries of the Third World. Already there is a critical shortfall in the local supply of wood in many developing countries.

The obvious answer is to plant more trees, both to reduce the pressure on forests and to meet the increasing demand for food, fuel and fodder. However, large-scale afforestation programs are very expensive and generally beyond the financial means of the Third World Governments. Even when money is available, limited technical knowledge often results in the failure of such schemes. Furthermore, afforestation of extensive tracts of degraded forests and hill slopes by conventional means of raising nurseries and planting out is a time-consuming as well as a highly expensive process. In these conditions, aerial seeding becomes relevant as a handy and effective tool to cover large areas within a short period and taking full advantage of short monsoon periods.

INTRODUCTION

Broadcasting of seed from aircraft, in particular for forest crops, is a recognized technique in Australia, New Zealand, Canada, the United States and China (Abott 1965; McCracken 1969; Cayford 1974; Smith 1984) and pastures (Huang and Li 1985). In China alone, 15 million hectares were aerial-seeded from 1958 to 1985 at an average annual rate of 536 000 ha. In India, aerial seeding was first tried in 1950 in Gujarat State to stabilize sand dunes. Aerial sowing of seeds was also experimented in Rajasthan along the international border in 1953-1954. However, aerial seeding in large scale and in a systematic way was introduced in Chambal ravines of Madhya Pradesh in 1980 and by 1986, more than 50 000 ha of land was reported to have been covered (Ramprasad 1988). Other states – such as Uttara Pradesh, Rajasthan (Joshi 1986) Maharashtra and Karnataka – also adopted this technology with varying degrees of success to afforest degraded areas.

The Palani Reserve Forest in Dindigul Forest Division, Tamilnadu, is classified as a degraded, dry deciduous and thorny forest. The terrain is undulating with small hillocks of 500–750 m elevation. Rainfall is low, with just 700 mm received mostly during 4-month monsoon period from July to October. Apart from the obvious ecological fragility, this area assumes social and economic significance since its 14 villages with a population of 42 000 along with their cattle (numbering 12 000) depend heavily on this resource, leading to massive degradation and denudation of the area. The short monsoon season, the vastness and ruggedness of the land, and scarce resources have made the regeneration of the area through the conventional means an uphill task for the foresters. In view of the urgency to restore the forest cover to meet the ever-increasing needs of the local population, aerial seeding was tried as a restoration method and 2500 ha area was aerial-seeded in July 1989. The results of this aerial seeding operation are discussed in this paper.

AERIAL SEEDING OPERATION

Ground preparation

For the success of aerial seeding, activities such as soil raking, construction of check dams, and digging of percolation ponds and contour trenches were carried out wherever feasible, to prepare the ground for positive reception of the seeds and provide optimum conditions for the seeds to germinate and grow.

Choice of species

Ten species that can meet the day-to-day demands of the local people and are suitable to the eco-climatic conditions of the area were selected. The suitability to aerial seeding – such as prolific seed availability, easy collection and handling, and ability to withstand mechanical pelletization – was also a criterion in the selection of the species.

Pre-treatment of the seeds

Before pelletization, the seeds were cleaned and graded to achieve a minimum 95% purity. Hot water treatment was given to Acacia seeds, which have a mucilage pulp coating over them. Manual depulping was done for *Zizyphus jujuba* and *Tamarindus indica*. These pre-treated seeds were dried under shade for 24 hours, taking care to keep the moisture around 6%. Seeds of *Tamarindus indica, Azadiracta indica, Anona squamosa* and *Dendrocalamus strictus* were kept under regulated temperature and humidity to retain their viability.

Pelletization

After repeated laboratory trials, the research wing of the Forest Department developed a successful pelletization technique. The Genetics Division fabricated five power-operated pelletizers, each with a seed pelletizing capacity of 150 kg per day.

First, the cleaned seeds are fed into the rotary mixer drum of the pelletizer in small quantities (about 10 kg, depending on the seed size). While the mixer drum is rotated, an adhesive solution (Titebond and Fevicol mixed at 1:1 ratio) is sprayed with a knapsack sprayer for a minute. Simultaneously, a fungicide spray (Thiram at 2 g per kg seed) is added to the seeds, followed by a thin coat of filler material (Talc and Dolomite 1:1 ratio). The seeds are then dried for five minutes by blowing dry air over them.

The rotary drum is then allowed free rotation and a micro nutrient mixture (Tracel at 2 g per kg seed) is dusted over the rotating seed. This is quickly followed by an application of filler and adhesive material and drying for few seconds. Finally, a thin coating of pesticide (Aldrin at 2 ml per kg seed) is given. If any residue is found, a fine spray of adhesive and minute quantities of a coloring agent is given. During the entire pelletization process, the mixer drum is rotated at a uniform speed so that coating is uniform over the seed. The treated seeds are then dried under shade.

Although the available literature indicated pelletization on 1:2 to 1:3 ratio (seed:coating) on weight basis, taking into consideration the rainfall pattern and the need for quicker breaking up and faster germination a ratio of 1:1.05 to 1:1.54 was maintained, depending on the species.

Beaver aircraft (fixed wing), with a load carrying capacity of 600 kg per sortie, of the agricultural aviation wing of Vayudoot Air Services was engaged for aerial seeding. To achieve seed broadcasting within the targeted area uniformly, boundaries were demarcated on the ground at frequent intervals with indicative flags.

Data on germination, survival and other parameters were obtained from 320 quadrats of 20×20 m (one square chain) laid out by stratified random sampling.

RESULTS

Particulars of the species used and aerial seeding are given in Table 1.

A thorough examination of the data indicates that the survival performance of the seed sown ranged from 0.11% in the case of *Acacia auriculaeformis* to 19.46% in the case *A. ferruginea,* closely followed by *Tamarindus indica* (17.75%), *Azadirachta indica* (7.95%) and *Anona squamosa* (2.31%). Besides *A. auriculaeformis,* two other species – *Ziziphus jujuba* and *A. leucophloea* – gave poor performance; seeds germinated were less than 0.5% of those sown. The data collected after 8 months of sowing were analysed mainly for species selection for a large-scale aerial seeding in the following year. [1]

The information analysed (Table 2) also revealed a positive correlation between the size/weight of seed and germination percentage, indicating that larger and heavier seeds give better germination.

---

[1] Results of this subsequent operation, which indicated a high performance for *Bambusa arundinacea,* are being published separately.

Table 1.
Aerial seeding at Dindigul Division Palani Hills on 14.7.1989 and 23.7.1989 (Total area covered 2 500 ha)

Name of species	Cleaned seeds (kg)	Treated & pelletized (kg)	Treated alone (kg)	No. of seeds per kg	No. of seeds per ha	No. of seedlings per ha*	Survival percent-age
Acacia auriculaeformis	1000	1000	–	45000	18000	20	0.11
Acacia ferruginea	1000	1000	–	7500	3000	584	19.46
Acacia leucophloea	3039	539	2500	45000	54700	266	0.48
Acacia planifrons	2993	2993	–	30000	36000	298	0.82
Anona squamosa	4375	3175	1200	4100	7175	166	2.31
Azadirachta indica	3000	–	3000	4400	528	42	7.95
Dendrocalamus strictus	1505	805	700	30000	18060	272	1.50
Leucaena leucocephala	2000	2000	–	17000	13600	300	2.20
Tamarindus indica	4200	500	3700	1350	2270	403	17.75
Zizyphus jujuba	3000	3000	–	2800	3360	15	0.44
Total	26112	15012+	11100	–	–	2366	–

Notes: * On 12.03.1990; + 34 348 kg after pelletization

Though the *Zizyphus jujuba* seed is comparatively large, the germination percent was low even after 8 months of sowing. This is attributed to the hard seed coat, which takes considerable time to break. This is evident from Table 2, wherein an increasing trend in germination with advance in time can be observed for some species.

The data on the species composition in the sample plots three years after the aerial seeding (Table 3) indicated that *Acacia ferruginea, A. leucophloea, A. planifrons* and *Dendrocalamus strictus* emerged as the successful species capable of withstanding the limiting factors such as moisture, high temperatures during summer and damage by wild animals.

The high success of *Dendrocalamus strictus* is attributed to its ability to better withstand adverse conditions and to the presence of rhizome that facilitates recovery and reestablishment.

Table 2.
Analysis of data collected after 8 months from sowing

Name of species	Weight of 100 seeds (g)	Average diameter (mm)	Germination percentage *	Number. of seedlings per hectare **		
				I count	II count	III count
Acacia auriculaeformis	1.98	4.60	0.11	–	–	20
Acacia ferruginea	13.78	9.80	19.46	502	518	584
Acacia leucophloea	2.94	4.50	0.48	260	284	266
Acacia planifrons	4.18	3.80	0.82	320	330	298
Anona squamosa	21.50	8.80	2.31	67	85	166
Azadirachta indica	19.42	10.60	7.95	30	30	42
Dendrocalamus strictus	3.22	4.60	1.50	441	326	272
Leucaena leucocephala	6.14	5.70	2.20	454	459	300
Tamarindus indica	85.98	13.10	17.75	253	253	403
Zizyphus jujuba	39.42	10.10	0.44	6	3	15
Total	–	–	–	2333	2288	2366

Notes: * As on 12.03.1990; ** I count on 04.01.1990, II count on 04.02.1990, III count on 12.03.1990

The analysis of the results of the four species, for which both pelletized and non-pelletized seeds were used, indicated significant improved performance of pelletized seeds over the other. This was very evident in case of *Dendrocalamus strictus* (32.23%) *Anona squamosa* (18.61%) and *Tamarindus indica* (12.10%), but not significant in the case of *Acacia leucophloea* (4.28%) indicating that the survival rate in the area would have been better if more proportion of pelletized seeds of these species were used.

There was also significant difference in the seedling establishment in various quadrats. While 18% of them situated on the hilltops did not have any plant in them, those in the valleys showed maximum establishment (2 661 plants). The areas near streams and other waterways, and where soil raking and other ground preparation was carried out properly showed better results.

Table 3.
Establishment of aerial-seeded plants

Species	Seeds sown/ha	No. of plants/ha on 12.03.90	Species composition A	No. of plats/ha after 3 years	Species composition B	Change in species (B–A)
Acacia auriculaeformis	18000	20	00.85	12	5.31	4.46
Acacia ferruginea	3000	584	24.68	68	30.09	5.41
Acacia leucophloea	54700	266	11.24	26	11.51	0.26
Acacia planifrons	36000	298	12.59	46	20.35	7.76
Anona squamosa	7175	166	07.02	8	3.54	-3.48
Azadirachta indica	528	42	01.78	9	3.98	2.21
Dendrocalmus strictus	18060	272	11.50	33	14.60	3.11
Leucaena leucocephala	13600	300	12.68	12	5.31	-7.37
Tamarindus indica	2270	403	17.03	4	1.77	-15.26
Zizyphus jujuba	3360	15	00.63	8	3.54	2.91
Total	–	2366	100.00	226	100.0	–

DISCUSSION

Arial seeding is being practised the world over as the most practical and logical means for ecologically compatible regeneration of areas where natural agencies of seed dispersal have been largely lost or no longer perform the functions they once did because of human interference or other natural factors. Arial seeding was also reported as an effective method of restoration/reclamation of problematic areas such as sand dunes (Suresh Kumar and Sankarnarayan 1998, Gupta 1992) deserts and mountain regions (Li et al. 1990) mangroves (Lahiri 1991), areas exposed to wild fires (Brown 1984), degraded and denuded lands (Vedant 1984) and ravines (Sharma 1985, Ramprasad 1988).

The underlying concern in employing aerial seeding is to move from the relatively safe cocoon of established silvicultural practices (with all their limitations) to innovative (though perhaps risk-prone) and potentially invaluable measures and techniques that could help achieve the desired results far more quickly. However, certain preconditions – such as the selection of suitable species, preparation of the ground, selection of proper sowing time, preparation of the seed (treatment and pelletization) and effective post-seeding care – are essential for the success of the program.

Selection of species

Despite the initial germination being good for most species aerially seeded, only the drought-hardy *Dendrocalamus strictus*, *Acacia ferruginea*, *A. leucophloea* and *A. planifrons* have shown better establishment under the present study. This performance is in conformity with the results obtained in Madhya Pradesh. Out of several species tried in the degraded ravines in this state, *Prosopis juliflora* and *Acacia nilotica,* both of which can withstand adverse weather conditions, were reported to have performed well (Ramprasad 1988). Species such as *Albizzia lebbeck* and *Holoptelia integrifolia*, and grasses were found to be completely absent in this area. Various quantities of seeds, ranging from 20 to 40 kg/ha, were used in seeding and re-seeding operations (0.8–1 million seeds/ha). It was reported that the survival count survey showed a plant density of 1 148 plants/ha on gully beds, 833 plants/ha on gully slopes and 413 plants/ha on gully tops.

Results of aerial seeding in Chambal ravines of Uttar Pradesh during 1982 reveal *Prosopis juliflora* and *Acacia nilotica* as successful species (Joshi 1986). Other species like *Dalbergia sissoo, Holoptelia integrifolia* and *Ailanthus excelsa* sown here were reported to have been complete failures. Vedant (1984) reported a mean number of 1 038 seedlings/ha with 25% of the sown area blank and 16.7% adequately germinated (> 2 000 seedlings/ha). It was also reported that ravine tops were usually without seedlings, while ravine slopes were highly populated.

More seedlings were reported to have been observed on mulched crests, leeward slopes and ploughed inter-dunes in an aerial seeding operation carried out on sand dune areas of Rajasthan (Sankaranarayan and Kumar 1986). A disc chain system for seedbed preparation and aerial seeding were described as a promising combination for regeneration of arid rangelands in Texas, the United States (Wiedemann et al. 1988).

Pelletization

The reasons for the use of pelletized seeds and their advantages and benefits were described by Murthy (1988), citing the use of one method in Andhra Pradesh and other parts of India. While the filler coat improves the ballistic property required for aerial seeding, the pesticide coat acts as a rodent/bird repellent. Application of dolomite adds strength to the pelletized seeds, besides improving porosity. This also helps in dissolving the artificial coat immediately on receipt of moisture. The process of alternate wetting and drying with adhesive, filler, pesticide, fungicide and micronutrients results in the formation of successive thin layers of coating over the seed.

Increased levels of success with improvised pelletizing and seed encapsulation was also reported by Ballie and Elward (1980). Jha et al. (1994) recommended encapsulation of seed with alginate entrapped rhizobia beads as an effective and cheap inoculation technique for possible use in aerial sowing programs.

Increased performance with the pelletized seeds was also observed in the present study especially in the case of bamboo offering great scope for adopting this technique for large-scale use. It was also felt that pelletization, besides providing the benefits mentioned above, increased the size and weight of the seed thereby allowing better anchorage and establishment of seeds, particularly the lightweight seeds like that of bamboo.

Cost

The cost of this aerial seeding, excluding machinery and permanent establishment costs, came to about Rs. 254 per ha, compared with Rs. 8 000 to 20 000 per ha under conventional method of planting. Vedant (1984) reported a cost of Rs. 122.30/ha for an aerial seeding operation in Uttar Pradesh, where the mean number of seedlings/ha was 1 038 two years after sowing and concluded this technique as an appropriate method for afforesting extensive tracts of denuded land.

CONCLUSION

Aerial seeding is best suited to inaccessible areas, the remoteness and ruggedness of which make physical movement difficult and thus render large-scale seeding/planting by conventional methods strenuous. Its success is found to be more dependent on certain prerequisites like ground preparation, choice species, pre-treatment and time of broadcasting. With advancements made in the disciplines of species biology, weather forecasting, seed encapsulation and pelletization, aerial seeding can play a significant role in large-scale plantation programs.

Aerial seeding also holds immense potential, especially for the propagation of species like bamboo confronted with the problems of gregarious seeding and short seed viability. By adapting this technique, surplus seeds in one region can be moved to other similar bio-geographic regions and aerial-sown to act as an early colonizer leading to the process of vegetative succession. Aerial seeding, however, is yet to assume its full potential as an invaluable means for the propagation of bamboo species. Intensive research, testing and development are required before this can be achieved.

REFERENCES

Abbott, H.G., ed. 1965. Direct seeding in the Northwest. Symposium proceedings. Agricultural Experiment Station, University of Massachusetts, Massachusetts, USA. 127 pp.

Ballie, T.S.; Elward, M. 1980. Coated seed – its potential use in land reclamation areas. *In* Proceedings of the Canadian Land Reclamation Association Conference. pp. 189–199.

Brown, G. 1984. Aerial seeding of wildfire areas in the Northwestern Region. *In* Proceedings of Jack Pine Symposium in Timmins, Ontario, Great Lakes Forest Research Centre, Canada. pp. 93–101.

Cayford, J.H., ed. 1974. Direct Seeding Symposium proccedings. Publication No. 1339, Canadian Forestry Service, Department of Environment, Ottawa, Canada. 178 pp.

Gupta, J.P. 1992. Changing the face of the desert land. Indian Farming, 42(3), 27–31.

Huang, W.H.; Li, H.L. 1985. Pasture improvement by aerial seeding in China. *In* Proceedings of the XV International Grassland Congress, 24–31 August 1985, Kyoto, Japan. pp. 610–611

Jha P.K.; Nair, S.; Babu, C.R. 1994. Development of an inexpensive legume Rhizobium inoculation technology which may be used in aerial seeding. Journal of Basic Microbiology, 34(4), 231–243.

Joshi S.C. 1986. Aerial seeding for environmental conservation. Indian Forester, 112(1).

Lahiri, A.K. 1991. Aerial seeding in mangrove swamps. Indian Forester, 117(3), 159–161.

Li, R.A.; Dixon, J.A.; James, D.E.; Sherman, P.B. 1990. Dry land management: economic case studies. National Environment Protection Bureau, Beijing, China. pp. 34–40.

McCracken, I. J. 1969. Direct seeding for watershed revegetation. Protection Forestry Report No. 62, New Zealand Forest Service, Forest Research Institute, Christchurch, New Zealand.

Murthy, A.V.R.G.K. 1988. Pelletization and aerial seeding in forestry. Myforest, 24(4), 279–287.

Ramprasad, 1988. Effectiveness of aerial seeding in reclamation of Chambal ravines in Madhya Pradesh. Indian Forester, 114(1), 1–18.

Shankarnarayan, K.A.; Kumar, S. 1986. Aerial seeding of sand dunes I. Trends in seed germination and seedling distribution. Journal of Tropical Forestry, 2(1), 4–20.

Sharma, D.G. 1985. Ravine reclamation through aerial seeding: an appraisal. Journal of Tropical Forestry, 1(1), 18–28.

Smith, B.W. 1984. Aerial seeding requirements for Jack Pine regeneration. Proceedings of Jack Pine Symposium in Timmins, Ontario, Canada Great Lakes Forest Research Centre, Canada. pp. 78–86.

Suresh Kumar; Shankarnarayan, K.A. 1988. Aerial seeding on sand dunes: seedling survival and growth. Journal of Tropical Forestry, 4(2), 124–134.

Vedant, C.S. 1984. Aerial seeding technique for afforestation of degraded lands an assessment. Myforest, 20(2).

Wiedemann, H.T.; Whitehead, E.E.; Hutchinson, C.F.; Timmermann, B.N.; Varady, R.G. 1988. Engineering systems for vegetative rehabilitation of arid lands. *In* Whitehead, E.E.; Hutchinson, C.F.; Timmermann, B.N., ed., Arid lands: today and tomorrow. pp. 871–881.

Bamboo for Sustainable Development, pp. 333–344
Eds. A. Kumar, I.V. Ramanuja Rao and Ch. Sastry

Micropropagation of woody bamboos: from biotechnological dream to commercial reality

JOHAN GIELIS and JAN OPRINS
Oprins Plant N.V., Rijkevorsel, Belgium

Abstract—Worldwide interest in bamboo as a biomass source in sustainable agriculture and agro-forestry systems has increased rapidly in recent years. However, while several classical propagation techniques are available, shortages of planting material and depletion of natural resources are both increasing rapidly. The potential of micropropagation for mass-scale propagation has raised high hopes and a lot of research has been focused on the development of protocols for large-scale and rapid propagation. Despite numerous publications, however, few published protocols could be transformed into commercially viable propagation systems.

In Oprins Plant company, research focus has been directed towards micropropagation of elite selected genotypes, using mature or juvenile genotypes of temperate and tropical bamboos. It currently produces over 60 different temperate and tropical bamboos commercially, all from selected genotypes. Axillary branching is the preferred *in vitro* method for mass-scale production of bamboos. It is highly efficient for large-scale, even mass-scale, propagation ensuring clonal fidelity and true-to-type propagation.

In the plant production system of the company, special emphasis is placed on the optimization of the process and of the added value of micropropagation. Added values are especially found in the selection procedures for elite plants and the use of micropropagated bamboos either as planting or propagation material in optimized quality bamboo production systems. Forward integration and product mix are also very important. The influence of micropropagation on bamboo availability and prices is discussed, both for ornamental bamboos and for tropical bamboos used in agro-forestry. Other tissue culture techniques – such as *in vitro* flowering of mature bamboos, cryo-preservation and somatic embryogenesis – are also under investigation in the company's laboratory.

INTRODUCTION

Both woody and herbaceous bamboos are found in the subfamily Bambusoideae (Poaceae), but only woody bamboos have economic potential. Historically, woody bamboos have been used for different applications and interest from paper and wood industries has increased recently, both for tropical and temperate woody bamboos. Temperate bamboos of the northern hemisphere are used for agro-forestry, mainly in China, and as ornamental plants in Europe and the United States. Some preliminary trials to use bamboo as a source of biomass are ongoing in the latter regions (it was Thomas Jefferson who said that the best contribution one could give to a culture was to add a valuable crop to its agriculture).

Bamboo is often advocated as an ideal renewable resource for biomass, useful for wood and paper industry. Positive arguments also include ecological arguments; indeed in the future forests and agriculture, water conservation and carbon recycling will become very important criteria. However, the classical economic criteria (profit and added value) will remain very important. Moreover, if bamboo is to be used as a source of biomass, it will have to compete with other plants, not to speak of competitions with industrial materials. This will certainly impose much pressure on bamboo, particularly regarding selection of elite genotypes, silvicultural methods, new approaches for harvesting and the production of quality biomass. Therefore, the time horizon of mass-scale bamboo utilization may be quite beyond what the advocates of bamboo foresee (or hope for) at present.

One of the main problems with bamboo is that it has been regarded as a resource that is simply there to take, as has been done for thousands of years by people in rural economies. However, in industrial economies such practice leads to considerable overexploitation and rapid depletion of bamboo resources nearer to paper mills and factories. This has already happened in many places, to the point that transportation costs have become too high for bamboo to be economical because of the increasing distance between the raw material source and the point of utilization. Estimates regarding future use of bamboo indicate that there will be a huge shortage for bamboo planting material in the medium and long terms (Subramaniam 1994; Nadgauda et al. 1997). To cope with this forecasted shortage, some large or mass-scale bamboo planting has been planned/carried out, such as in Green Sarn in Thailand and the proposed reforestation in India. Several pilot plantations or larger scale plantings are planned or have been carried out throughout in many parts of the world. It is clear that bamboo propagation and silviculture are the pivotal upstream technologies in the whole process. If something fails there, it will have consequences down the line (causing, in the worst case, for the interest in bamboo to fade instead of increase).

The downstream process of bamboo production and transformation consists of many different steps. Each of these steps is important and, like in any successful industrial enterprise or process, optimization in each step, integration of all steps and feedback are important. In general, the consumer has to be satisfied and the whole production scheme should be aimed at the market.

On the production side, propagation is crucial for mass-scale utilization of bamboo. Different propagation techniques are available for bamboo, such as seed propagation, clump division, and rhizome and culm cuttings (Banik 1994). These methods, however, suffer from serious drawbacks when one considers large or mass-scale propagation. Most of the classical techniques for clonal propagation are useful for production on smaller scales (up to 10 000 plants per year), although some techniques such as macroproliferation (Adarsh Kumar 1991) can go higher. Much will depend on the availability of mother plants. Large-scale propagation (between 10 000 and 500 000 plants per year) requires very efficient techniques. For mass-scale propagation (> 500 000 plants per year), classical techniques are mostly insufficient and inefficient, and tissue culture is the only viable method.

Indeed, the volume of the demand for bamboo planting materials indicates that micropropagation will inevitably be necessary for mass-scale propagation (Subramanyam 1994; Gielis 1995). Classical techniques alone can never solve this problem. It is also important to point out that the total world production of all tissue cultured plants in 1995, was estimated at 600 million (Debergh 1981, personal communication), or less than the total need projected for bamboo planting material.

MICROPROPAGATION OF BAMBOO

A historical note

To date, a large number of papers on micropropagation (the use of tissue culture for propagation) of bamboos have been published – original papers, reviews as well as some in which tissue culture was described in a more general way. The main center of research has been Asia. The first paper on successful tissue culture was by Alexander and Rao (1968) who described embryo culture. The 1980s saw a considerable increase in micropropagation: of seedlings in tissue culture (Nadgir et al. 1984), the induction of somatic embryogenesis in bamboo seeds of tropical species (Rao et al. 1985), clonal propagation of *Guadua angustifolia* (Manzur 1988) and other species (Prutpongse and Gavinlertvatana 1992), and the induction of organogenesis (caulogenesis) in mature bamboos (Huang et al. 1989).

Many researchers have focused on somatic embryogenesis of seedlings of tropical bamboos, eventually culminating in the INBAR publication "Propagation of Bamboo and Rattan through Tissue Culture" (Ramanuja Rao et al. 1991). Major research focused also on the clonal propagation of elite genotypes, both juvenile and adult. The literature on this subject, however, is much less, solely because of the lack of success. Technically, the propagation of adult plants via axillary branching is much more difficult than with seedlings. An inventory (BIC 1994) shows that, at one time, there were at least 21 labs in the South and Southeast Asia involved in bamboo tissue culture, mainly in India (reviews on tissue culture of bamboo can be found in Ramanuja Rao, et al. 1991; Saxena and Dhawan 1994; Zamora 1994; Nadgauda et al. 1997).

A critical note

Micropropagation via tissue culture attracted a lot of attention since it was believed that this method could solve most, or at least many, problems that hinder the propagation of bamboo. This biotechnological dream was fuelled by a large number of publications. However, the translation and transformation of these expectations into commercially viable propagation systems has been beset with a number of problems (which is normal to any new technological process). Main problems were either technological or marketing-related. The published research all lacked some crucial elements. They were either not very efficient, not applicable to all bamboos, or both.

The use of starting material (seeds or adult plants) and the choice of the propagation method are crucial. The two major advantages of using seedlings are that it is a new generation, and that the technology is easier. The disadvantages too are considerable:

1. Insufficient or no knowledge of genetic background;
2. Restricted availability of seeds for most species, and rapid loss of germination capacity; and
3. Comparison of *in vitro* to *in vivo* performance has not been thoroughly evaluated.

There are problems when using adult bamboos too, the main being:

1. Endogenous contamination;
2. Hyper hydricity and instability of multiplication rates; and
3. Many problems with rooting, even in bamboos that root readily in nature.

Rooting percentages for adult bamboos ranged from very low to 73% for *Dendrocalamus longispathus* (Saxena and Bhojwani 1993). Rooting percentages of 77% were obtained for adult *D. giganteus* in 3-4 weeks (Ramanayake and Yakandawala 1997). While 77% of success is good on 500 plants in a laboratory trial, it means a loss of 33 000 plants when 100 000 plants are transplanted.

Initial trials for commercialization were carried out in some companies in different parts of the world. An unknown number of commercial laboratories succeeded in propagating one or more bamboo genotypes. However, some of the bamboos produced were of inferior quality, owing either to insufficient identification of the mother plants or to the chase for short-term profit. While it is very important to show positive results, especially for market penetration, the time to market is still far away and feedback times are considerable (for instance, 30 years or longer for flowering observation). It is not easy to convince foresters to invest, even with data that seem very convincing. Tissue culture has to adapt to the market and not vice versa (or at least not in the first instance). It must be remembered that micropropagation is a tool, not an aim.

A personal note

Oprins Plant's response to these problems was that it had to develop a universal system for all bamboos – a system that was very efficient for the propagation of selected genotypes, while ensuring genetic stability. In short, develop tissue culture as the "best available technique". This technology had also to be integrated into the plant production scheme in the company. With regard to commercialization, the strategy was to combine the production of high-quality plants with product mix and market diversification (for both temperate and tropical species), and effect a forward integration in the downstream process of bamboo propagation and transformation.

SELECTION OF ELITE GENOTYPES

Because of the peculiar flowering habits of bamboo, it has been almost impossible to breed for superior traits in woody bamboos. Bamboos are natural polyploids and a lot of variability occurs within seedlings or natural populations (at an infraspecific level, in general) in tropical as well as temperate bamboos (Banik 1995). In seedling populations, variability between genes occurs at the genetic and phenotypic levels. Characteristics and properties are quite variable and, in general, overall characteristics are unfavorable for biomass production. Selection in existing populations to improve overall quality of yields is possible. Clonal forestry based on one or more elite selected genotypes, propagated via clonal propagation, allows a considerable genetic improvement (in this sense, 'genetic' refers to better yields, quality, etc.).

Population enhancement of bamboo is relatively easy in the first approach: selecting the plus clones (e.g. the top 20%) and using mass propagation for upscaling these plus clones will result in more homogenous populations with higher yields. The selection and subsequent vegetative propagation of superior clones allow the improvement of bamboo germplasm considerably, compared to natural populations. Simple selection procedures such as this can also be done for seedling populations. It is important to select clones in such a way that a broad genetic basis is maintained, since many tropical bamboo are natural monocarps (Gielis and Oprins 1998).

Among the large number of species of woody bamboos, the economically most promising ones can be used for a variety of purposes. Selection criteria depend on the biological characters of the bamboo – macromorphological, micromorphological and anatomical – directly related to ultimate transformation or to biomass production. Selection procedures should be refined for two main reasons. Firstly, the quality requirements imposed by industry (or some international standards) will get more severe, especially when bamboo is used as a substitute for (and in direct competition with) existing wood-based biomass transformation. Secondly, refining the selection procedures, incorporating morphological, physico-chemical and anatomical characters, as well as molecular markers, will also allow to considerably speed up selection through early assessment of seedlings.

Molecular markers have been proposed as tools for identification (fingerprinting) of bamboos, assessment of variations and identification of characters or traits. Oprins Plant applies AFLP™, which is considered as one of the most sensitive and most reproducible methods among different laboratories. The advantage of AFLP patterns is that the same fingerprint can be used for identification, taxonomy, the study of natural variability and early assessment. It was shown that differences in DNA isolation procedures could seriously affect AFLP patterns

(Gielis et al. 1997). DNA-based markers are still in their infancy and, in general, they represent data not necessarily connected to phenotypic character or loci in the strict sense, covering only a limited part of the genome. Therefore, incongruities between morphological and molecular data can be expected and any result should be interpreted only with maximum precaution (Gielis et al. 1996; Gielis et al. 1997). Despite these difficulties, however, molecular markers can be very helpful in the detection of useful characters and traits, or in the detection of elite genotypes using multivariate analysis, in careful combination with phenotypic markers.

Selection of elite genotypes is at the basis of clonal forestry and, at the same time, the propagation of selected genotypes is one of the major advantages of micropropagation. In a very short time (about 1 year), 100 000 to 1 000 000 copies can be made from one single elite clone.

GENETIC STABILITY IN MICROPROPAGATION

In tissue culture, three basic methods exist for the propagation of plants:

1. *Axillary branching,* which miniaturizes the natural process of branching;
2. *Organogenesis,* the neo-formation of organs or shoots; and
3. *Somatic embryogenesis,* the regeneration of embryos from somatic cells.

In bamboo, all these methods have been investigated, but both organogenesis and somatic embryogenesis involve an intervening callus phase.

Tissue culture has been associated with molecular aberrations in general (including chromosomal defects), and phenotypic aberrations may show up only some years later. To avoid such problems one should be very cautious in using such techniques, especially when employing methodologies that are prone to higher chances of such genetic aberrations. Even when using axillary branching, the hi-fi technique in tissue culture, caution must be taken. Huang and Huang (1995) reported that the bulbous internodes of *Bambusa ventricosa* were lost after tissue culture. Variegated leaves appear very often during tissue culture in many species.

For large or mass-scale production, efficiency of the propagation methods is important; but perhaps even more important is the genetic stability. The evaluation of clonal stability is very difficult and protracted in bamboos, and since feedback times are very long in bamboos to monitor the genetic stability, ensuring genetic stability is very important. The molecular tools to test genetic stability are, in general, rough approximations. Use of molecular markers may reveal some differences, but detection of a single gene mutation is very difficult and slow. Use of other tools, such as flow cytometry to check ploidy level, does not provide any information at the molecular level. Molecular tools are not a watertight system to detect variations. The best strategy, therefore, seems to be the most prudent also: use axillary branching for clonal multiplication.

From this perspective, it is surprising that most of the efforts in bamboo have concentrated on the use of seeds and somatic embryogenesis for mass-scale propagation (Ramanuja Rao et al. 1991). For tropical bamboos like *Dendrocalamus giganteus, D. strictus* and *Bambusa bambos,* seeds are readily available in India. The ease of somatic embryogenesis in these species has attracted most attention, but somatic embryogenesis is a process with higher chance for aberration at the molecular level. Very few (if any) plants are micropropagated commercially via somatic embryogenesis, especially because of clonal fidelity problem (Cervelli and Senaratna 1995). Surprisingly, few people seem to care about this in bamboo.

In most plants produced on large scale, axillary branching is used since this methodology has a lesser risk of aberration. All shoots originate from preformed buds, and no neo-formation of organs is involved even in nodule cultures of pseudospikelets (Gielis 1998). Oprins Plant, therefore, has selected axillary branching as the method of choice for bamboo propagation.

AXILLARY BRANCHING: A UNIVERSAL TECHNIQUE FOR BAMBOO

Lack of good starting material has been a problem from the beginning. While seeds of some tropical bamboos were available for tissue culture studies, only adult material could be used for temperate bamboos. The development of a universally applicable protocol was thus technically difficult. At present, all remaining bottlenecks for axillary branching have been solved. Basic research has allowed to identify the endogenous factors involved in senescence and browning of explants, in hyper hydricity and variability of multiplication, and in inhibition of rooting of *in vitro* bamboos (Gielis, unpublished results).

In Oprins Plant laboratory, commercially feasible micropropagation for large-scale propagation is practiced for species and varieties of *Arundinaria, Chimonobambusa, Fargesia, Phyllostachys, Pleioblastus, Sasa, Sasaella, Semiarundinaria, Shibataea* and *Yushania* (temperate bamboos), and *Bambusa, Dendrocalamus, Dinochloa, Oxytenanthera* and *Thyrsostachys* (tropical bamboos). With the new technology developed, it will probably be possible to propagate any bamboo (Gielis and Oprins 1998). Over 60 different species and cultivars of bamboo have already been successfully propagated, with emphasis on certain ornamental bamboos (*Phyllostachys, Fargesia, Semiarundinaria,* etc.) and on some tropical bamboos (*Bambusa* and *Dendrocalamus*) for reforestation and plantation.

In the company's greenhouse, *ex situ* conservation of elite plants is combined with stage 0 conditions (Debergh and Maene 1981), which facilitate the initiation in tissue culture. In micropropagation practice, Stage I, the initiation stage, involves all sub-cultures until a constant and stable multiplication rate is obtained. If prepared under proper Stage 0 conditions for about one month, the time for Stage I is reduced to less than 4 months for most tropical and temperate species, though for some it can take up to 8 months. After this period, more than 1 000 plants per species enter the production line (Stage II). Multiplication rates vary according to species from 3 to 10 every 3–4 weeks (in practice, multiplication is kept below 6). Within 6–7 months from initiation, 100 000 plants can be produced, which corresponds to theoretical calculations (or rather earlier expectations) of tissue culture technology. One skilled operator can easily transfer 7 000–10 000 plants daily. By handling clumps of 3–5 plantlets each, the transfer of clumps can be more efficient. New cultures are initiated at least every two years and stay in production for a maximum of 20 cycles.

Rooting is achieved either by a separate rooting medium or by administering liquid medium (Stage IIIb, Debergh and Maene 1981). Thereafter, the plants are transplanted with 98–100% success under conditions of 100% relative humidity. Under such optimal weaning conditions, rooted plantlets of most bamboos, for example *Phyllostachys* and *Bambusa* species, can be potted 10–14 days after transplanting in the greenhouse. Rhizome development is equally fast: *Pleioblastus auricomus* rhizomes are visible in about 8 weeks after transplanting.

In the classical scheme of tissue culture laboratories, the tissue culture phase terminates at Stage III. With regard to marketing of plants, however, one can distinguish subsequent stages:

1. Stage IV, the transplantation stage with the end-product being a rooted plantlet in a tray;
2. Stage V, the production of liners, either for production of saleable plants or for use as mother plant in micropropagation; and
3. Stage VI, the production of saleable plants.

This distinction is important if the complete chain of production is to be integrated in a single company, since this determines the added values.

COMMERCIAL MICROPROPAGATION OF WOODY BAMBOOS

A selected number of private laboratories have succeeded in mass propagating one or more bamboos from either seedlings or mature plants. Micropropagation is much more efficient when developed in private companies, where cost-benefit analyses have to be made and optimal strategies for efficient implementation and integration of micropropagation into a complete plant production scheme have to be developed and adapted as necessary.

While the tissue culture lab is a separate entity in most commercial plant production, forward integration is almost a necessity if a company wants to mass-produce bamboos. Forward integration involves the integration of several steps of plant production in one company, sometimes even going beyond production. This can have different goals:

1. To be able to monitor how the micropropagated plants perform downstream;
2. To get sufficient feedback to improve either selection or technology;

3. To increase the added value of the plant; and/or
4. To improve market penetration by increased know-how of several steps in the downstream process.

Forward integration of bamboo tissue culture can indeed be rewarding; a straightforward comparison can illustrate this. In Europe, plants of *Phalaenopsis* selections and clones are sold at USD 0.25 per piece by a tissue culture lab. After hardening and growing, and through passing through distributors, the ultimate retail price is USD 10–20. Thus, the price of the tissue culture plant is between 1.25% and 2.5% of the retail price. In the case of *Ficus benjamina* the disparity is even more glaring. The tissue culture laboratory sells clumps for about USD 0.25. After hardening, 15–20 cuttings are taken from this plant, rooted and sold as liners to growers at the retail price of around USD 10. The price of the tissue-cultured plant, therefore, makes up only 0.125% to 0.25% of the retail price. To generate sufficient turnover for the laboratory, a large number of plants have to be produced.

In contrast, if the laboratory is integrated in the company, the situation will be very different, especially when one operates in small niches in the market. For a *Hydrangea* plant from tissue culture, wholesale prices after 6 months of growth range from USD 2 to USD 4, depending on the species. For the company to have the same turnover as the large laboratories producing *Ficus* or orchids, only one-tenth of the number of plants have to be produced; indeed the added value is gained on the saleable plant. At the same time, this strategy can create and protect (stabilize) market niches. Oprins Plant tries to apply the same strategy for bamboo. Besides increasing the added value for the plant producer/seller, it also avoids a downward spiralling of bamboo prices in general, and tissue culture bamboos in particular.

The company aims at a product mix: tissue culture bamboos used for (1) production of saleable plants (end of Stage VI) sold only at this stage, (2) production of liners (end of Stage V), and (3) hardened plantlets in trays (end of Stage IV). It targets producing tissue-cultured plants that give a distinct added value to the company (new genotypes, high quantity and quality) and use those in the most profitable way.

With respect to return on investment, the higher added value will of course be for saleable plants, but faster return on investment is obtained for direct sale of tissue-cultured plants. In terms of time to market, earlier sales can indeed be more profitable and also practical: logistically, it is quite difficult to annually grow about two million tissue-cultured bamboos further to saleable plants. Most of the production of tropical bamboos is directly exported to several tropical countries for planting.

THE FUTURE IS HERE AND NOW

The development of micropropagation systems for bamboos initially caused great excitement. New technologies or developments, however, often evolve in a typical way referred to as "double vision forecasting". A new technology can generate mass excitement and its short-term impact is generally overestimated. Because of the slow diffusion of the technology, time lags must be taken into account and short-term expectations are not met. However interesting and promising the new technology may be, the initial excitement will fade and, at the same time, create an under-expectation of medium and long term results when the technology ripens. In bamboo tissue culture, problems of technological and logistic nature have slowed down the development. Logistic factors involve the collection of elite germplasm, the incorporation of new technology in classical plant production and all aspects involved with market penetration.

This paper has examined the evolution of micropropagation of bamboos – starting from a biotechnological dream of tissue culture being the solution, to the development of a universally applicable technology as a practical and commercial reality. The problems involved in transforming such research results into commercial technologies are numerous and include: endogenous contamination; browning; lack of efficiency in rooting and transplanting; the choice of a specific propagation method; and the choice of appropriate starting material. This paper has outlined the approach Oprins Plant adopted to develop a commercially feasible process. It involves three major components:

1. Fundamental research on bamboo physiology and genetics;
2. The development of axillary branching into a universal technique with a high clonal fidelity and very high efficiency; and
3. Forward integration to optimize the added value of micropropagation.

In conclusion, it is now possible to micropropagate almost any selected bamboo on a mass scale in a short time frame. The hypothetical numbers of propagation and short time frames proposed in earlier research papers have been realized. The efficiency and time frames given in this paper are real figures and real productions, not on laboratory scale. Huge numbers of bamboo can be transported from the lab to the site of planting within 48 hours or less anywhere in the world. The quality of young plants derived from tissue culture is generally excellent. The plants are vigorous growers and in many ornamental bamboos, the quality of 1-year-old plants derived from tissue culture is considerably better than of those propagated via conventional methods of propagation.

Micropropagated plants are also price-competitive with other plants (if one includes quality and more long term factors). Moreover, micropropagation is the only technique for the mass propagation of many bamboos. As predicted some years ago (Gielis 1995), micropropagation via axillary branching is a universal technique and, above all, it is the "best available technique" and would become the standard for mass-scale propagation of bamboos. A comparison with *Ficus benjamina* seems appropriate: nobody in the tropics would even consider propagating a weedy fig tree via tissue culture. It is of little value and can be propagated easily via cuttings. Yet, all *F. benjamina* plants produced in Europe today (about 50 million) are obtained from tissue culture; micropropagation has replaced propagation via cuttings completely.

Micropropagation of bamboo has also transformed into a medium technology at the same time the technology matured. Hence, it can be applied for large and mass scales of production. However, diffusion of technology will have to be based on normal market rules, such as maximizing profits (profits are higher on ornamentals than for reforestation), protection of know-how (via patents or black-box), and technology transfer on a contract basis. Private companies are generally much more efficient, especially for production. Bamboo and its future are too important to leave it only to scientists.

REFERENCES

Adarsh Kumar. 1991. Mass production of field planting stock of *Dendrocalamus strictus* through macroproliferation: a technology. Indian Forester, 117, 1046–4052.

Alexander, M.P.; Rao, T.C. 1968. In vitro culture of bamboo embryo. Current Science, 37, 415.

Banik, R.L. 1994. Review of conventional propagation research in bamboos and future strategy. *In* Constraints to the production of bamboo and rattan. Report of an Experts Consultation, Bangalore, India, 9–13 May 1994. INBAR Technical Report No. 5. International Network for Bamboo and Rattan, New Delhi, India. pp. 115–142.

Banik, R.L. 1995. Selection criteria and population enhancement of priority bamboos. *In* Williams, J.T.; Ramanuja Rao, I.V.; Rao A.N. ed., Genetic enhancement of bamboo and rattan. INBAR Technical Report No. 7. International Network for Bamboo and Rattan, New Delhi, India. pp. 99–110.

BIC (Bamboo Information Centre). 1994. Bamboo researchers and projects of South and Southeast Asia: a directory. Bamboo Information Centre India, Kerala Forest Research Institute, Trissur, Kerala, India.

Cervelli, R.; Senaratna, T. 1995. Economic aspects of somatic embryogenesis. *In* Aitken-Christie, J; Kozai, T.; Lila Smith, M., ed., Automation and Environmental Control in Plant Tissue Culture. Kluwer Academic Publishers, The Netherlands. pp. 29–64.

Debergh, P.C.; Maene, L. 1981. A scheme for commercial propagation of ornamental plants by tissue culture. Scientia Horticulturae, 14, 335–345.

Gielis, J. 1995. Bamboo and biotechnology. European Bamboo Society Journal, 6 May 1995, 27–39.

Gielis, J.; Everaert, I.; Goetgheberu, P.; De Loose, M. 1996. Bamboo and molecular markers. *In* Ramanatha Rao, V.; Ramanuja Rao, I.V. ed., Bamboo, people and the environment, Vol. 2, Biodiversity and Genetic Conservation. Proceedings of the Vth International Bamboo Workshop, Ubud, Bali, Indonesia, 19–22 June 1995. International Network for Bamboo and Rattan, New Delhi, India. pp. 45–67.

Gielis, J.; Ngoc, L.H.T.; Angiolillo, A.; Gerats, T.; Van Gysel, A.L.; Breyne, P. 1997. AFLP™ analyses in bamboos: influence of DNA extraction procedures. Mededelingen Faculteit Landbour Universiteit Gent 64/2a, 1471–1475.

Gielis, J.; Oprins, J. 1998. Biotechnological approaches to germplasm improvement in woody bamboos. *In* El Bassam, N. et al., ed., Sustainable agriculture for food, industry and energy. (In press).

Gielis, J. 1998. Micropropagation and in vitro flowering of temperate and tropical woody bamboos. (In press).

Huang, L.C.; Huang, B.L.; Chen, W.L. 1989. Tissue culture investigations of bamboo IV - Organogenesis leading to adventitious shoots and plants excited from shoot apices. Environmental and Experimental Botany, 29(3), 307–315.

Huang, L.C.; Huang, B.L. 1995. Loss of the species distinguishing trait among regenerated *Bambusa ventricosa* McClure plants. Plant, Cell, Tissue and Organ Culture, 42, 109–111.

Manzur, D. 1988. Propagacion vegetative de *Guadua angustifolia* Kunth. Agronomia, 2(3), 14–19.

Nadgauda, R.S.; John, C.K.; Joshi, M.S.; Parasharami, V.A.; Mascarenhas, A.F. 1997. Application of in vitro techniques for bamboo improvement. *In* Chapman, G., ed., The bamboos. Linnean Society, London, UK. pp. 163–177.

Nadgir, A.L., Phadke, C.H.; Gupta, P.K.; Parsharam, V.A.; Nair, S.; Mascarenhas, A. 1984. Rapid multiplication of bamboo by tissue culture. Silvae Genet, 33, 219–223.

Prutpongse, P.; Gavinlertvatana, P. 1992. In vitro micropropagation of 54 species from 15 genera of bamboos. HortScience, 27, 453–454.

Ramanayake, S.M.S.D.; Yakandawala, K. 1997. Micropropagation of the giant bamboo (*Dendrocalamus giganteus* Munro) from nodal explants of field-grown culms. Plant Science, 129, 213–223.

Ramanuja Rao, I.V.; Rao. I.U.; Narang, V. 1985. Somatic embryogenesis and regeneration of complete plantlets in the bamboo *Dendrocalamus strictus*. Plant Cell Reports, 4, 191–194.

Ramanuja Rao, I.V.; Yusoff, A.M.; Rao, A.N.; Sastry, C.B. 1991. Propagation of bamboo and rattan through tissue culture. IDRC Bamboo and Rattan Research Network, International Development Research Centre, Ottawa, Canada. 60 pp.

Saxena, S.; Bhojwani, S.S. 1993. In vitro clonal multiplication of 4 year old plants of the bamboo *Dendrocalamus longispathus* Kurz. In Vitro Cellular and Developmental Biology, 29, 135–142.

Saxena S.; Dhawan, V. 1994. Micropropagation research in South Asia. *In* Constraints to the production of bamboo and rattan. Report of an Experts Consultation, Bangalore, India, 9–13 May 1994. INBAR Technical Report No. 5. International Network for Bamboo and Rattan, New Delhi, India. pp. 101–113.

Subramaniam, K.N. 1994. Bamboos: demand and supply of planting stock. INBAR Newsletter No. 5, December 1994, 24–25.

Zamora, A.B. 1994. Review of micropropagation research on bamboos. *In* Constraints to the production of bamboo and rattan. Report of an Experts Consultation, Bangalore, India, 9–13 May 1994. INBAR Technical Report No. 5. International Network for Bamboo and Rattan, New Delhi, India. pp. 45–100.

Bamboo for Sustainable Development, pp. 345–365
Eds. A. Kumar, I.V. Ramanuja Rao and Ch. Sastry

Flowering of *Phyllostachys pubescens* and germination of caryopses

SHOZO SHIBATA

Kamigamo Experimental Forest Station, Kyoto University Forests, Graduate School of Agriculture, Kyoto University, Japan

Abstract—One strain of *Phyllostachys pubescens* flowered in 1997 at eight places in Japan. It was on the 67th year after seeding, and the second record in the world. The changes in the bamboo grove after flowering were investigated, and germinability of the caryopses was examined. At Kamigamo, Kyoto, none of the flowering culms died and small new leaves grew twice after flowering. Furthermore, small new racemes with a few flowers appeared during the same season again. Therefore, flowering of this strain was not considered as controlled solely by the flowering gene with short-term flowering periodicity. The grain setting rate was low, as generally reported. However, the grain quantity was considered sufficient for the next generation. The results of germination tests suggested that the husk plays an important role in germination.

INTRODUCTION

The species *Phyllostachys pubescens* Mazel ex Houzeau de Lehale in Japan is said to have been introduced from China to Kagoshima in 1746 (Suzuki 1978), although the year of introduction is not confirmed. In Japan, there are few records of gregarious flowering of *P. pubescens,* and only some episodes of partial flowering have been recorded. The flowering record of the strain grown from seedlings has been recorded only once, in 1979, the 67th year after seeding, in Kyoto and Yokohama (Watanabe et al. 1982). Most of the *P. pubescens* groves in Japan have not flowered since they were introduced from China. Therefore, they are considered to have the gene with long-term flowering periodicity. The partial flowering, sometimes observed, is considered to be caused by the mutation on the growth point of rhizomes. Caryopsis harvested from partially flowering groves may have some genes causing short-term flowering periodicity. However, the periodicity cannot be verified until they flower again and their seedlings grow. In Japan, experimental cultivation for this purpose was started about 80 years ago (Konno 1937).

In the summer of 1997, a strain different from the one that flowered in 1979 flowered. This flowering periodicity was also 67 years. This paper reports on that rare occurrence, only the second flowering of *P. pubescens* in the world, and discusses the changes in the bamboo grove after flowering and the results of the germination tests.

DISTRIBUTION AND HISTORY OF *P. pubescens* THAT FLOWERED IN 1997

The strain of *P. pubescens* that flowered in 1997 was raised from the seedlings collected in 1931 from the bamboo grove that flowered in 1930 at Miyata, Yokohama, Japan (Table 1). This grove originated from six clones transplanted in 1924, and all culms derived from one particular clone flowered. This flowering is an example of partial flowering. The seedlings collected from the flowered bamboos were planted at two stations of Tokyo University Forests (Tanashi and Chiba) in 1934 and two stations of the National Forestry Institute (Meguro in 1934 and Akanuma in 1939). The seedlings at Meguro were later lost. The rhizomes were transplanted (number unknown) from Akanuma to Kamigamo Experimental Forest in February 1955. The rhizomes of the clone at Akanuma was transplanted to five locations in Japan, and kept there (Table 2, items D–H). In addition, the rhizomes from Wakayama farm were transplanted to Tochigi Forestry Center. Thus, in the flowering year, 1997, the strain of this clone was growing at 8 locations, three from seedlings and five from transplanted rhizomes.

Table 1.
History of the strain *P. pubescens* Mazel ex Houzeau de Lehale that flowered in 1997 at Kamigamo Experimental Forest, Kyoto, Japan

1924	May	Six rhizomes with culms were transplanted to Miyata, Yokohama, in Kanagawa Prefecture.
1930		One rhizome in six rhizomes showed gregarious flowering (partial flowering as a whole).
1931		Seedlings were collected from flowering bamboo grove floor.
1939	April	Seedlings were transplanted to Akanuma Experimental Station, Forestry and Forest Products Research Institute.
1955	February	Rhizomes with culms were transplanted to Kamigamo Experimental Forest, Kyoto University Forests, Kyoto.
1957	July	Gregarious flowering, but no culms died.
1997	December	Some culms flowered again; flush of new small leaves on all culms.
1998	June	Flush of new small leaves on some culms.
1998	July	Some other culms flowered; one of the 42 remaining flowering culms died.

Source: Ide and Suzuki 1998.

FLOWERING OF *P. pubescens* RAISED FROM SEEDLINGS IN 1930

The *P. pubescens* grown at eight locations in Japan simultaneously flowered in 1997. However, at Godai (A. in Table 2), Akanuma (C), Fudago (D) and Utsunomiya (F), partial flowering on small scale was already recorded in 1996. The flowering in 1979 was gregarious at all flowering sites, and most of the culms died out after flowering. Nevertheless, the flowering pattern at eight planting sites of 1997 was not identical. Flowering was gregarious at four planting sites – Godai (A), Tanashi (B), Fudago (D) and Kamigamo (E) – and partial at other four planting sites. At three of the four planting sites that had gregarious flowering, most of the flowering culms died out, but those at Kamigamo survived even after flowering.

Table 2.
Location of the strains of *Phyllostachys pubescens* that flowered in 1997 in Japan

	Site	Planted year	Flowering pattern	Remarks
A	Godai, Chiba Tokyo University Forests, Chiba	1934	Gregarious	Seedling
B	Tanashi Experimental Forest, Tokyo University Forest, Tokyo	1934?	Gregarious	Seedling
C	Akanuma Experimental Station Forestry & Forest Products Research Institute, Saitama	1939	Partial	Seedling
D	Fudago, Chiba Tokyo University Forests, Chiba	1955	Gregarious	Transplanted from C
E	Kamigamo Experimental Forest, Kyoto University, Forests, Kyoto	1955	Gregarious	Transplanted from C
F	Wakayama Farm, Utsunomiya, Tochigi	c. 1967	Partial	Transplanted from C
G	Headquarters, Forestry & Forest Products Research Institute, Ibaragi	1977	Partial	Transplanted from C
H	Tochigi Prefecture Forestry Center, Tochigi	c. 1975	Partial	Transplanted from C

Source: Ide and Suzuki 1998.

FLOWERING AT KAMIGAMO EXPERIMENTAL FOREST, KYOTO

Kamigamo Experimental Forest, Kyoto University Forests, is located north of Kyoto City at 35°04' N, 135°45' E. The average temperature is 14.5°C, and annual precipitation is 1 600 mm with a little snow in winter. In the 1950s, the late Dr. Koichiro Ueda started to collect several strains of bamboo from seedlings. He also collected the strain that flowered in 1997. The first flowering of *P. pubescens* grown from caryopsis also occurred there.

The strain introduced from Akanuma was planted in the planting bed (5×5 m) set at the most interior part of the bamboo garden, which is in the northern part of the University Forest and located in a small valley opening eastwards and surrounded by pine plantations. When *P. pubescens* flowered in the summer of 1997, 18 culms were growing in this bed. In addition, another 36 flowering culms were observed in neighboring plantation area of *Pinus taeda* and *P. elliottii.* They are considered as the culms from the rhizomes that had spread out from planting bed in the past. The distribution of the flowering culms outside the planting bed is shown in Fig. 1. This strain produced 54 flowering culms in all. The youngest age of flowering culm was two years but only one culm was of this age (No. 6 in Fig. 1). The other culms were between three and eight years of age. In the flowering year, no new culms were produced. Most flowering culms were 8–10 cm in diameter, with an average of 7.7 cm for all (3.1–11.3 cm) and 8.9 cm for those inside of the planting bed (Fig. 2). Their growth condition was poor because of the lack of adequate nutrients and the shade caused by surrounding pine trees.

Although most of the normal leaves were shed just before and after the flowering, none of the culms died. Only a few normal leaves remained on the flowering culms, with a flush of small new leaves before December. In December, small racemes were formed and some spikelets flowered again on some culms. These spikelets did not set any grain. Small new leaves were shed during April–May in 1998, but there was a flush of small new leaves in June in several culms. In July 1998, just one year after gregarious flowering, some small flowering spikelets were observed again on a few culms that flowered in 1997, and one culm died.

PRODUCTION OF RACEMES AND CARYOPSES AT KAMIGAMO

Appearance of flowering culms

Twelve flowering culms, selected from the 36 culms outside the planting bed (Fig. 1), were felled. Their age, height, diameter at breast height (dbh), number of nodes, and height and number of nodes under the lowest branch were measured. Branches (one or two) were collected from each node. The number of racemes and grain-setting caryopses were counted, and the branches and the racemes and caryopses on each branch were weighed. Ten racemes from every 5–6 branches at three positions (lower, middle and upper parts, Table 3) were sampled from each culm, and the number of spikelets and position of grain-setting caryopses were examined.

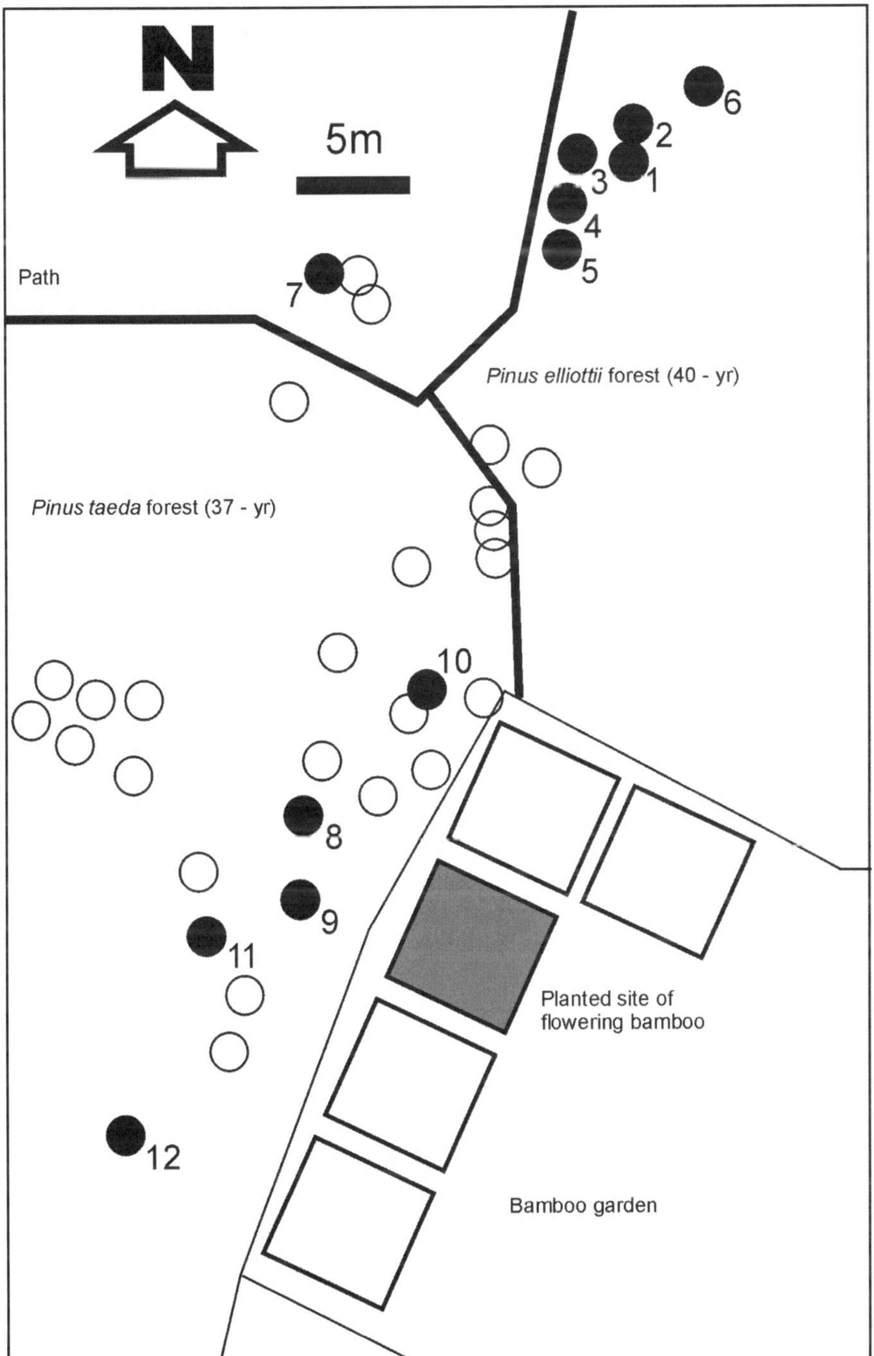

Figure 1. Sites of flowering culms of *P. pubescens* (18 culms) at Kamigamo Experimental Forest, Kyoto (white circles indicate 24 flowering culms outside planted site; black circles indicate 12 sampled culms)

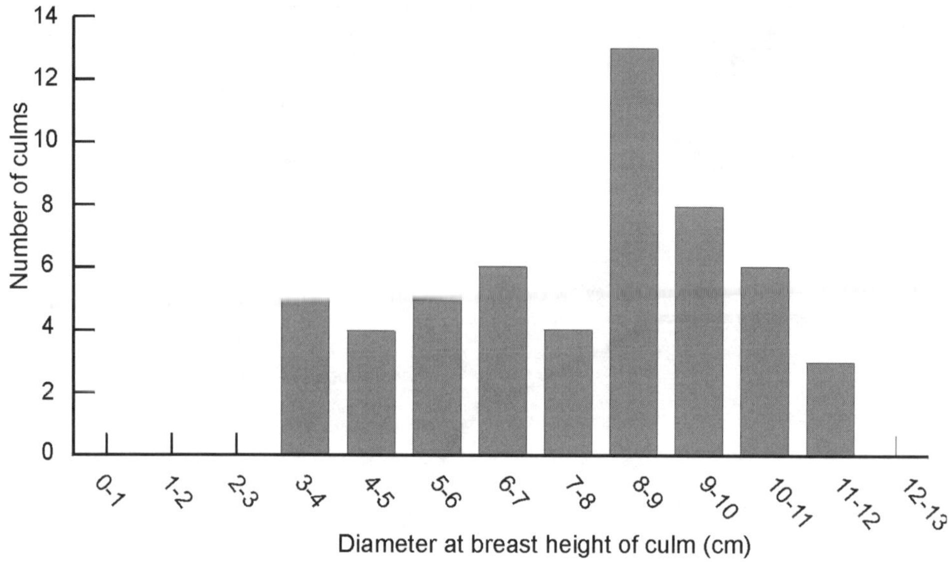

Figure 2. Distribution of diameter at breast height (dbh) of whole flowering culms at Kamigamo Experimental Forest, Kyoto

Table 3 shows the data on sampled culms. The height and dbh were greater in young culms and 2-year-old culms were the tallest. As the size of culm decreases with age, the rhizomes were assumed to have run out from the planting bed within the last ten years. The culm height and number of nodes under the lowest branch varied in proportion to culm height. The number of nodes was 41 to 62. The thicker the culm, the more nodes they had, and the younger the culm, the higher the number of racemes. The total weight of branches and racemes, and the number of grain-setting caryopses per node are shown in Fig. 3. Generally, *P. pubescens* produces two branches from each node on every alternative side. Each value in Fig. 3 on successive node positions tended to increase and decrease alternately. This suggests that some factors determining the number and weight of production of branches, racemes and caryopses on one side of the culm is different from those on the opposite side.

The number of grain-setting caryopses per culm markedly varied with culm age (0–1278, Table 3). Three-year-old culms produced the most caryopses, especially on Culms No. 4 (1 012) and No. 8 (1 278). On the contrary, the 6- and 7-year-old culms had no caryopses. The number of grain-setting caryopses per raceme was also high on 3-year-old culms. The number on Culm Nos. 3, 4 and 8, was 0.124, 0.120 and 0.114, respectively. On these culms, at least one grain-setting caryopsis was harvested per 10 racemes. On the other hand, a 2-year-ol culm (No. 6), which produced the largest number of racemes, produced one caryopsis per about 27 ra-

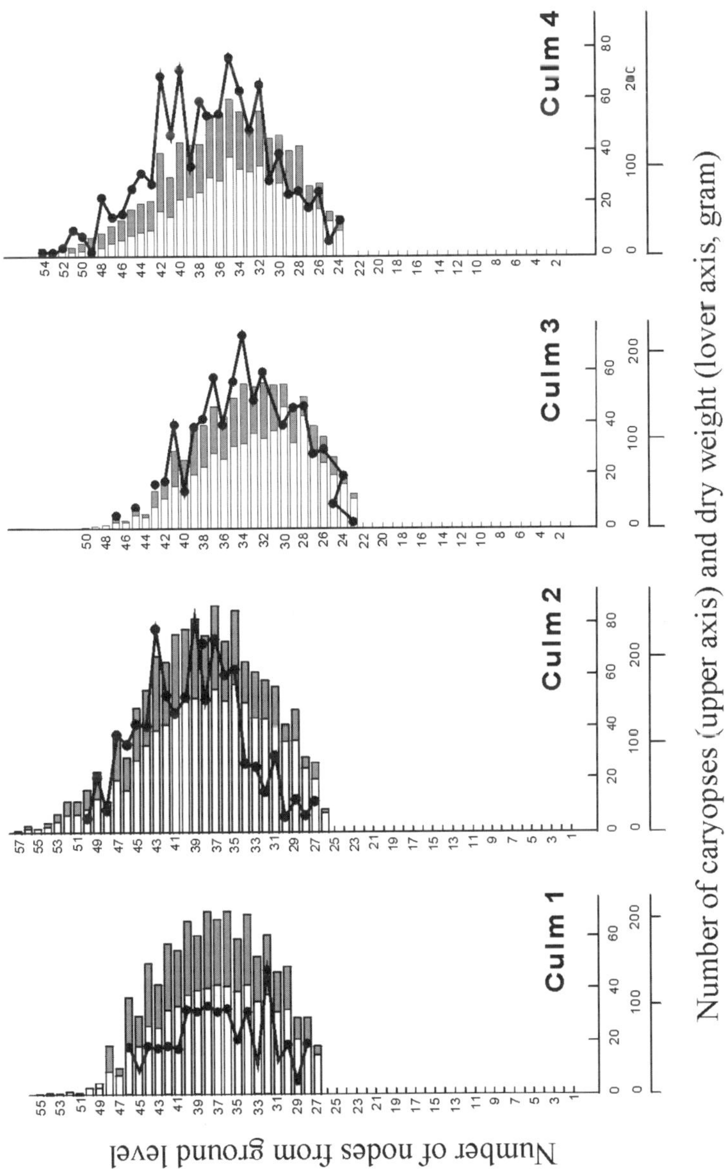
Culm 1
Culm 2
Culm 3
Culm 4
Number of nodes from ground level
Number of caryopses (upper axis) and dry weight (lover axis, gram)

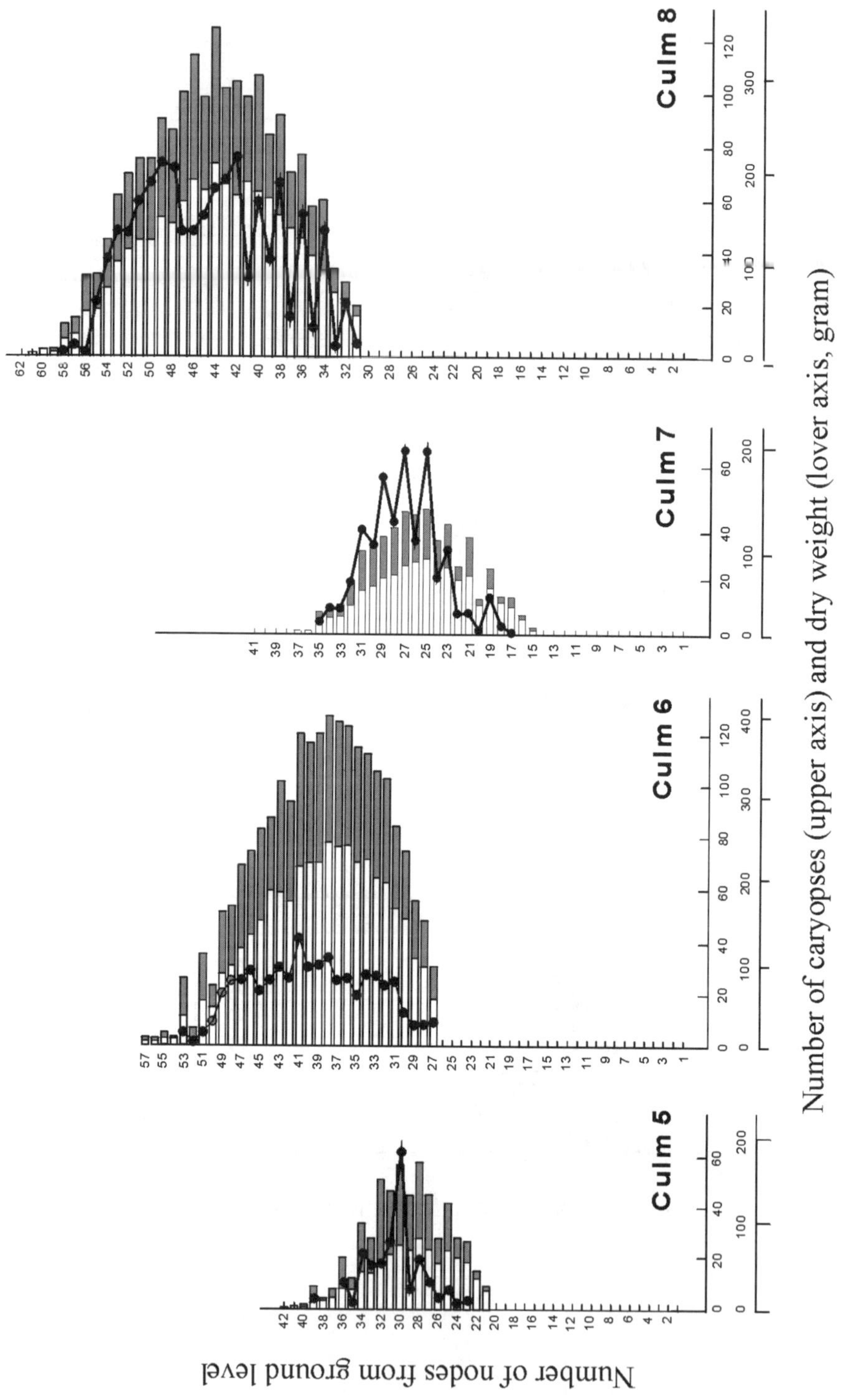
Culm 5
Culm 6
Culm 7
Culm 8
Number of nodes from ground level
Number of caryopses (upper axis) and dry weight (lover axis, gram)

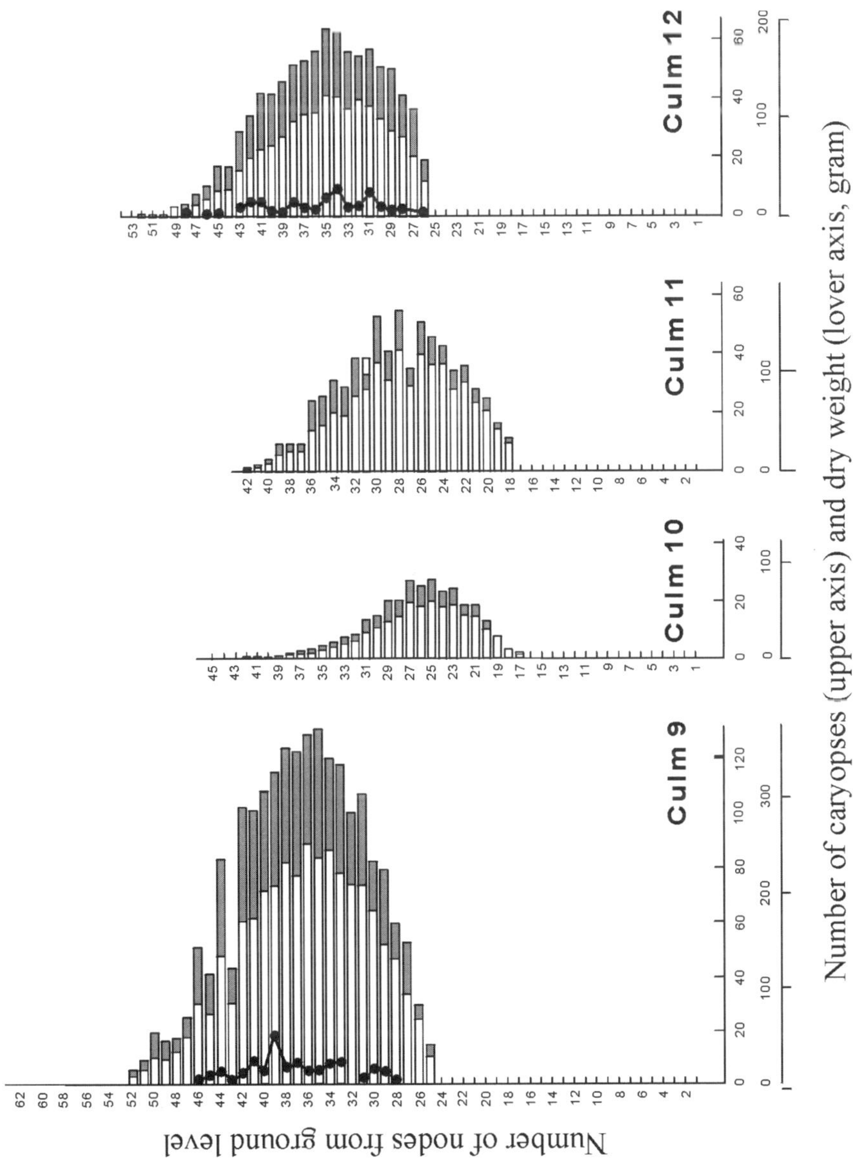

Figure 3. Distribution of dry weights of branches (white bar) and racemes (gray bar), and number of caryopses (black line) on each node of the sampled culms Production of grain-setting caryopses on culms

Table 3.
Morphology of sampled culms of *P. pubescens* that flowered in July 1997 at Kamigamo Experimental Forest, Kyoto

Culm no.	Age (y)	Dbh (cm)	Height (cm)	Height under lowest branch (m)	No. of nodes	Nodes under lowest branch	Harvested		Caryopsis per raceme	Dry weight of			Germination rate		No. of nodes for each raceme sampled		
							Raceme (A)	Caryopsis (B)	(B/A)	Branches (g)	Racemes (g)	Caryopses (g)	Soil (%)	Incubator (20 c.d.%)	Upper	Middle	Lower
1	4	8.3	12.76	5.02	55	26	8467	452	0.053	1992	1217.8	8.44	20.3	–	46	38	31
2	3	9.4	13.00	5.30	57	25	9674	943	0.097	2685	1374.7	16.64	7.9	90.9	52	41	31
3	3	6.9	10.62	4.14	50	22	6535	812	0.124	1618	907.8	15.12	26.7	–	43	35	28
4	3	6.4	10.55	3.55	54	23	8420	1012	0.120	1499	1260.8	18.90	29.6	81.8	49	39	29
5	4	4.5	8.10	2.84	42	20	7975	222	0.028	915	864.4	4.40	17.1	–	35	30	25
6	2	10.6	15.65	6.65	57	26	16030	590	0.037	4070	2789.6	11.18	20.1	–	53	42	31
7	4	4.1	7.30	1.68	41	14	4995	480	0.096	1007	591.2	9.64	15.9	–	32	26	19
8	3	11.1	14.30	6.36	62	30	11181	1278	0.114	3374	2352.9	23.24	13.3	72.7	56	45	34
9	3	11.3	14.25	5.98	52	24	11382	129	0.011	3860	1923.8	2.50	23.3	–	47	38	29
10	7	3.5	7.12	2.10	45	16	2290	0	–	656	232.2	–	–	–	34	27	21
11	6	5.8	9.37	2.88	42	16	4374	0	–	1562	526.2	–	–	–	39	30	21
12	5	7.1	11.78	5.85	55	25	5036	73	0.014	1663	1029.2	1.36	11.9	–	46	37	29
Average	3.9	7.4	11.23	4.36	51	22.3	8029.9	190.8	0.069	2075.1	995.7	11.14	19.5	81.8	–	–	–

cemes. This means that a young culm produces a large number of racemes, but this grain-setting rate is low.

Fig. 4 shows the number of grain-setting caryopses per raceme on each culm node. There were no clear differences among the culms at different ages in caryopsis number on each node. In some culms, the number of caryopses on each node was low, although the total number on the culm was high. Even the old culms produced many caryopses on the nodes at the middle part of the culms.

Production of spikelets and caryopses on racemes

Using ten racemes sampled at random from three nodes of the sampled culms, the number of rachis nodes, rachillae from each rachis node, and caryopses including grain- and empty grain-setting caryopses were counted. The number of caryopses was nearly the same as that of flowering spikelets. The dry weights of caryopses, rachises, rachillae and all the other appendages (glume, palea, etc.) were also measured. In addition, 100 racemes harvested at Wakayama Farm, at Utsunomiya were also investigated.

The number of caryopses including grain- and empty grain-setting caryopses ('a' in Table 4) was low on No. 5 and No. 10 culms. There was no variation with the culm age. On Culm No. 8, which produced the largest number of racemes, the number of caryopses per raceme was also high (about 16). The 5-year-old culm (No. 12) also had a large number of caryopses, especially at upper parts of culm (more than 17). The average number of caryopses in each raceme harvested at Kamigamo was 9.93. On the other hand, racemes from Utsunomiya, where the bamboo grove was maintained in fertilized conditions, had 13.8 caryopses on the average. Generally, no variation in the number of caryopses with the position on culm was observed, except at the upper part of culm No. 12 and the lower part of culm No. 3, where the number was high. The number of rachis nodes on each raceme ('b' in Table 4) was 6.85–9.55, and there was no conspicuous variation with the culm or position on each culm. The average number of rachis nodes on each raceme in the samples from Kamigamo was 8.12 and similar to that from Utsunomiya, 8.14.

The total weight of racemes ('c' Table 4) was 1.00–2.51g and showed a similar tendency to the number of rachis nodes. However, the racemes of Culm Nos. 6, 8 and 9, which produced a large number of racemes, and Culm No. 12 with many caryopses on each raceme tended to have heavy racemes. Table 5 shows the dry weights of each part of raceme, and their percentage. On each culm, appendages such as glume, palea accounted for 45–52%, and caryopses not including grains and empty grains about 40%. Each caryopsis with grains or empty grains accounted for 0.7–1.5%, rachis 7–9% and rahillae 4–5%.

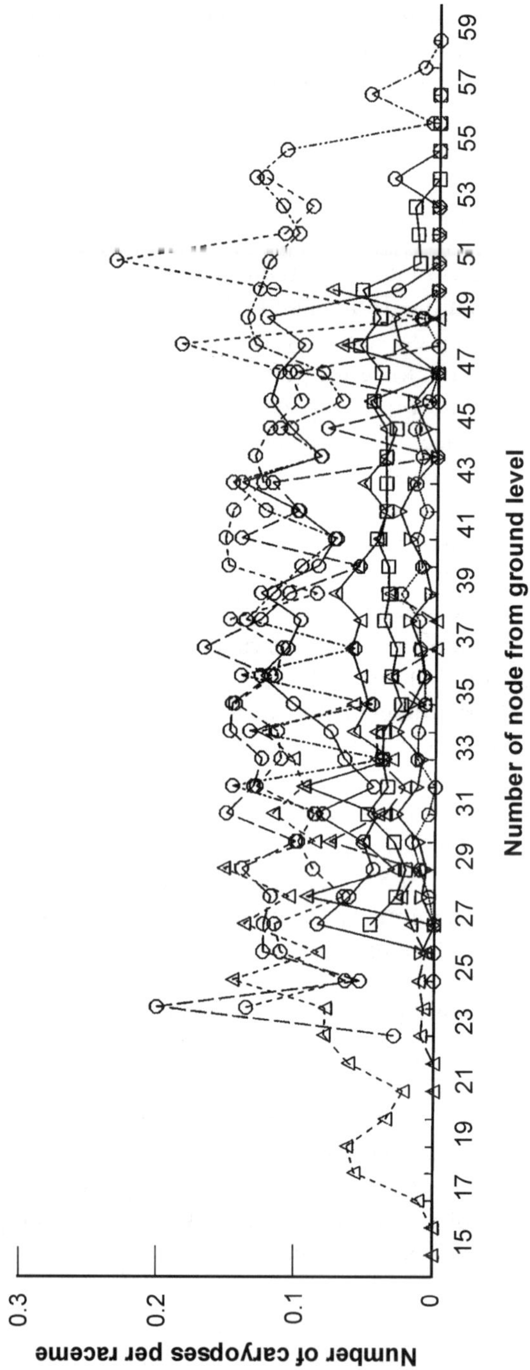

Figure 4. Number of caryopses per raceme on the nodes of each culm sampled (Square = 2-year-old culm; circle = 3-year-old; triangle = 4-year-old; inv. triangle = 5-year-old)

The number of caryopses on each rachis node of raceme is shown in Table 6. The main node producing plural spikelets is generally the second to fourth node from the rachis base. On the raceme having many caryopses, three or five spikelets were produced at each of these nodes. This tendency was common to the caryopses from both Kamigamo and Utsunomiya. The number of grain-setting caryopses on the racemes was not so high. The average position of rachis node bearing grain-setting caryopses is shown in Table 7 by the node position counted from the base of the rachis. Most of the grain-setting caryopses were produced on the third to fifth node on 2- to 4-year-old culms except for the 5-year-old Culm No. 12. These nodes are also the position that produced many spikelets, as shown in Table 6. This shows that the grain-setting rate is also high at these positions.

GERMINABILITY OF CARYOPSES

Generally, the germination rate of *P. pubescens* harvested from partially flowering groves is low. This is commonly explained as follows. Since the partial flowering is caused by mutation at the growing point, whole cells in one clone or culm do not have the same flowering genes. However, in the clone that showed gregarious flowering in 1979, many caryopses were produced – 4 840 grain-setting caryopses from nine culms (Watanabe et al. 1982).

Germination tests for caryopses collected from flowering culms

From 12 culms flowering in Kamigamo in 1997, 5 991 grain-setting caryopses were harvested and dried until February. Then the racemes were detached, and the caryopses collected from March to the beginning of April. The harvested caryopses, except for those used for the germination test in April and those used for seeding, were stored at 5°C. The caryopses harvested from branches from which racemes were sampled were kept at room temperature. Caryopses for germination tests were selected from 3-year-old Culm Nos. 2, 4 and 8, as these culms produced adequate caryopses. In some experiments, the husks were removed just before germination (naked grain). Caryopses harvested at Utsunomiya were also tested.

Table 4.
Characteristics of racemes – (a) number of caryopses; (b) number of nodes on rachis of each raceme; (c) total dry weight of each raceme – average of every 10 racemes from one or two branches at three parts of each sampled culm

Characteristics	Sample culm number												Kamig.*	Utsun.**
	1	2	3	4	5	6	7	8	9	10	11	12		
(a) Number of caryopses														
– lower part	7.35	7.40	10.45	8.75	6.70	10.45	7.05	16.45	12.20	6.75	9.75	11.65		
– middle part	9.20	8.50	9.85	9.90	7.00	9.90	7.45	15.90	12.10	6.80	9.00	13.90		
– upper part	8.45	8.65	8.05	9.20	7.10	9.95	8.70	15.90	11.70	7.10	9.00	17.35		
– average	8.33	8.18	9.45	9.30	6.93	10.10	7.54	15.95	12.00	6.84	9.25	14.30	9.93	13.76
(b) Number of nodes on rachis of each raceme														
– lower part	7.60	7.65	8.30	8.10	6.85	9.00	7.30	9.50	8.65	7.15	7.65	8.70		
– middle part	8.00	7.95	8.20	8.45	7.25	8.60	7.50	9.50	8.80	7.25	8.00	9.15		
– upper part	7.90	7.55	7.45	8.50	7.25	8.45	7.60	9.45	8.25	7.40	7.30	9.55		
– average	7.83	7.72	7.98	8.32	7.12	8.68	7.44	9.48	8.57	7.24	7.65	9.13	8.12	8.14
(c) Dry weight of each raceme (g)														
– lower part	1.279	1.310	1.654	1.494	1.012	2.040	1.024	2.641	1.905	1.003	1.482	2.175		
– middle part	1.532	1.415	1.648	1.623	1.185	1.869	1.265	2.405	1.898	1.163	1.594	2.275		
– upper part	1.384	1.258	1.294	1.525	1.161	1.609	1.260	2.275	1.669	1.150	1.337	2.505		
– average	1.411	1.337	1.530	1.561	1.122	1.845	1.176	2.449	1.824	1.096	1.471	2.329	1.599	–

Notes: * Data from Kamigamo are average of all sampled racemes; ** Data from Utsunomiya are average of 100 racemes

Table 5.
Dry weight of 10 racemes from each sampled culm in Kamigamo Experimental Forest (average of 50 or 60 racemes from 5 or 6 branches at three parts of each sampled culm)

Culm no.	Total	Glumes, etc.*	Caryopses with no seed	Rachises	Rachillae	Empty caryopses			Caryopses		
						Empty grains	Husks	Subtotal	Grains	Husks	Subtotal
1	1.411	0.702(49.7)	0.519(36.8)	0.103(7.3)	0.062(4.3)	0.0037	0.0062	0.0099(0.7)	0.0101	0.0059	0.0160(1.1)
2	1.337	0.655(49.0)	0.488(36.5)	0.103(7.7)	0.062(4.6)	0.0026	0.0062	0.0088(0.7)	0.0123	0.0037	0.0209(1.6)
3	1.530	0.722(47.2)	0.571(37.3)	0.109(7.1)	0.068(4.5)	0.0030	0.0072	0.0102(0.7)	0.0316	0.0136	0.0502(3.3)
4	1.561	0.757(48.5)	0.586(37.5)	0.107(6.9)	0.072(4.6)	0.0050	0.0079	0.0129(0.8)	0.0154	0.0100	0.0253(1.6)
5	1.122	0.528(47.1)	0.440(39.2)	0.085(7.6)	0.046(4.1)	–	–	–	0.0151	0.0032	0.0232(2.1)
6	1.845	0.970(52.5)	0.636(34.4)	0.140(7.6)	0.092(5.0)	0.0033	0.0060	0.0093(0.5)	–	–	–
7	1.176	0.543(46.2)	0.464(39.4)	0.089(7.5)	0.053(4.5)	–	–	–	0.0182	0.0098	0.0279(2.4)
8	2.449	1.141(46.6)	0.966(39.4)	0.167(6.8)	0.131(5.3)	0.0035	0.0056	0.0091(0.4)	0.0289	0.0053	0.0353(1.4)
9	1.824	0.873(47.9)	0.735(40.3)	0.125(6.9)	0.091(5.0)	–	–	–	–	–	–
10	1.096	0.496(45.2)	0.460(41.9)	0.103(9.4)	0.038(3.5)	–	–	–	–	–	–
11	1.471	0.699(47.5)	0.582(39.6)	0.119(8.1)	0.070(4.8)	–	–	–	–	–	–
12	2.329	1.141(49.0)	0.892(38.3)	0.156(6.7)	0.124(5.3)	–	–	–	0.0110	0.0057	0.0167(0.7)
Average	1.599	0.776(48.5)	0.616(38.5)	0.124(5.3)	0.077(4.8)	0.0032	0.0067	0.0100(0.6)	0.0197	0.0095	0.0293(1.8)

Note: * Including all glumes, lemmas and paleae

Table 6.
Number of caryopses on each rachis node of raceme (average of every 10 racemes from 5 or 6 branches at three parts of each sampled culm)

Culm no.	Node numbers*											
	1	2	3	4	5	6	7	8	9	10	11	12
1	0.83	1.35	1.10	1.15	1.08	1.00	0.98	1.00	1.00	1.00	–	–
2	0.80	1.25	1.22	1.22	1.02	1.00	0.98	0.97	1.00	1.00	–	–
3	0.82	1.62	1.53	1.27	1.18	1.10	1.00	0.95	0.94	1.00	–	–
4	0.78	1.42	1.50	1.18	1.08	1.06	1.00	0.98	0.94	1.00	1.00	–
5	0.75	1.03	1.05	1.00	1.00	0.98	1.00	1.00	1.00	–	–	–
6	0.82	1.32	1.50	1.36	1.15	1.18	1.08	1.04	0.97	1.00	1.00	–
7	0.72	1.14	1.08	1.14	1.02	1.00	1.00	1.00	1.00	–	–	–
8	1.17	2.43	2.97	2.57	2.00	1.28	1.03	1.02	1.00	1.00	1.00	1.00
9	1.20	2.08	1.95	1.70	1.38	1.08	1.03	1.00	1.00	1.00	–	–
10	0.56	1.02	1.02	1.02	1.00	0.98	1.00	1.00	1.00	–	–	–
11	1.20	1.65	1.38	1.22	1.10	1.03	1.00	1.00	1.00	1.00	–	–
12	1.18	2.38	2.37	2.12	1.67	1.37	1.02	1.02	1.00	1.00	1.00	1.00
Kamigamo **	0.91	1.57	1.57	1.42	1.23	1.09	1.00	1.00	0.99	1.00	1.00	1.00
Utsunomiya ***	1.81	2.44	2.23	2.01	1.67	1.30	1.04	1.04	1.00	1.00	1.00	–

Notes: * Number from the base of raceme rachis; ** Average of all sampled racemes; *** Average of 100 racemes.

Table 7.
Average position of rachis node bearing grain-setting caryopses on each raceme (average 50 or 60 racemes from three parts of each sampled culm)

Characteristics	Sample culm number												Kamig.*	Utsun.**
	1	2	3	4	5	6	7	8	9	10	11	12		
Average position	6.00	3.00	4.33	6.00	3.40	–	2.40	4.00	–	–	–	1.00	4.13	4.16

Note: * Kamigame – average of att sampled racemes; ** Utsunomiya – average of 100 racemes.

Germination tests in spring

Germination tests were conducted under 12 conditions at three temperatures (25°, 20° and 15°C), and under two light conditions (light and dark), using caryopsis containing grains and naked grains. Three petri dishes, each with two wet filter papers and 33 seeds, were used for each treatment. Tests at 25°C were started on the 27th and those at 20°C and 15°C on the 28th of April 1998. The number of germinated seeds was counted daily. Each test was stopped when new germinated seeds were not observed for three consecutive days.

Germination tests in summer

Germination tests were conducted in light and in darkness at 20°C. The caryopses stored at room temperature and 5°C were used with or without husks (caryopses and naked grains). In the experiments using the caryopses stored at 5°C, three petri dishes, each with 33 seeds, were used. In those using the caryopses stored at room temperature, three petri dishes, each with 22 seeds for the Kamigamo samples and two petri dishes each with 22 seeds for the Utsunomiya samples. The tests were started on 2 July 1998.

Germination in seeding beds

Caryopses from 12 sampled culms in Kamigamo, 2 150 in total, were sown in seeding beds on 24–30 April 1998. The number of germinated caryopses was counted daily. Nine hundred caryopses from Utsunomiya were also investigated. Soil used in the seeding bed consisted of mixed soil of weathered granite soil, volcanic ash and bark compost.

Results of germination tests

The average germination rate in spring (Table 8) was generally high, and the rate was higher for the Kamigamo seeds (67.5–86.4%) than for the Utsunomiya seeds (32.3–85.9%). The duration of the germination tests was 10–23 days for the Kamigamo seeds and 10–28 days for the Utsunomiya seeds. Naked grains germinated earlier than caryopses, and the test period for grain was shorter. In the seeds from Utsunomiya, the germination rate of naked grains was also higher than that of caryopses. The seeds from Kamigamo showed no clear difference in germination rate with the treatment, and the variation was the smallest at 20°C. The period required for germination was the shortest at 20°C, followed by 25°C and then 15°C. Nevertheless, the final germination rates were similar at all temperature. The difference in the germination rates in light and darkness was slight in the seeds from Utsunomiya. In the seeds from Kamigamo, the rate was higher in darkness at 25°C and in light at 15°C.

In summer, the germination rates of the seeds stored at room temperature was extremely low (Table 8). Only one caryopsis out of 132 seeds from Kamigamo and 8 naked grains out of 88 seeds from Utsunomiya germinated. The germination rate of the seeds stored at 5°C was high: 72.7–78.1% for Kamigamo seeds, showing no significant difference with the condition. The Utsunomiya seeds showed a clear difference between caryopses and naked grains. The germination rate of the caryopsis was clearly lower than that in spring. The period required for germination in darkness was shorter than that in light.

In the seeding beds, the germination rate was very low as compared with that in petri dishes. They were 19.5% for the caryopses from Kamigamo, and 7.3% for those from Utsunomiya. Such low germination rates are considered as caused by the covering of caryopses with soil.

DISCUSSION

Characteristics of the flowering of P. pubescens in 1997

Although the strain flowered simultaneously at eight locations in Japan in 1997, it did not show the same flowering pattern at each place. Regardless of whether the plants were grown from the seedlings or after rhizome transplantation, both gregarious flowering and partial flowering were observed. In addition, the culm showing gregarious flowering died at three locations but not at Kamigamo. This phenomenon is quite different from the flowering in 1979. At Kamigamo, not only did the culms after gregarious flowering survive, but also small new leaves and small racemes flushed on some culms in the following winter and early summer.

Mode of flowering and fruiting

Although many racemes were produced on young and thick culms, the number of caryopses was not in proportion to that of racemes. The number of caryopses produced on each culm ranged from 0 to 1 278, although the number of spikelets on each culm was considered to be about 100 000. Thus, grain-setting rate was very low (about 1%). Caryopses were produced the most abundantly on large 3-year-old culms; 4-year-old and 5-year-old culms also produced many caryopses, but their grain-setting rates were high only at the middle of the culm. On the other hand, the caryopses from 2-year-old culms showed grain-setting rates, which were, as a whole, relatively low.

The numbers of racemes and caryopses on each culm node were not uniform. They were in proportion to the branch size, and higher at the middle part of culm than the other parts. In addition, they varied with the branching side on the culm. It is uncertain whether this was a species-specific characteristic or caused by imperfect genetic control by the short-term flowering periodicity gene.

About a half the weight of each raceme was occupied by appendages such as the glume, and about 40% was related to caryopses. The percentage of grain-setting caryopses was only 1%. This means that a large amount of energy is required for the production of caryopses. In many culms, caryopses were produced mainly on the lower part of each raceme, and grain-setting caryopses were also produced on the same part.

Table 8.
Results of germination tests of *Phyllostachys pubescens* seeds (caryopsis grain) harvested at Kamigamo Experimental Forest and Wakayama Farm

Season of germination tests	Place of seed harvest	Temperature (°C)	Light condition	Seed condition	Storage condition	Average germination percentage *3	Period of test (days) *4
Spring *1	Kamigamo *2	25	Light	Caryopsis	Room temp.	73.7	17
				Grain		68.2	13
			Dark	Caryopsis		81.8	15
				Grain		81.1	10
		20	Light	Caryopsis		81.8	15
				Grain		77.3	12
			Dark	Caryopsis		72.7	15
				Grain		86.4	10
		15	Light	Caryopsis		70.7	23
				Grain		81.5	14
			Dark	Caryopsis		74.7	16
				Grain		67.5	14
	Utsunomiya	25	Light	Caryopsis		58.6	16
				Grain		77.8	14
			Dark	Caryopsis		64.6	16
				Grain		80.8	14
		20	Light	Caryopsis		32.3	15
				Grain		85.9	11
			Dark	Caryopsis		47.5	16
				Grain		76.8	10
		15	Light	Caryopsis		39.4	24
				Grain		84.8	19
			Dark	Caryopsis		52.5	28
				Grain		74.1	16
Summer *5	Kamigamo *2	20	Light	Caryopsis	Cold *6	72.7	19
				Grain		78.1	12
			Dark	Caryopsis		76.5	14
				Grain		74.1	09
			Light	Caryopsis	Room temp.	01.5	15
				Grain		00.0	15
			Dark	Caryopsis		00.0	15
				Grain		00.0	15
	Utsunomiya		Light	Caryopsis	Cold *6	30.3	16
				Grain		80.8	13
			Dark	Caryopsis		18.2	16
				Grain		81.8	10
			Light	Caryopsis	Room temp.	00.0	15
				Grain		02.3	10
			Dark	Caryopsis		00.0	15
				Grain		15.9	12

Notes: *1 – Tests in spring were started on 27 (25°C) and 28 (20°C & 15°C) April 1998.
*2 – Seeds were selected from sample culms No. 2, 4 & 8.
*3 – Twenty-two seeds were used for each petri dish, with the seeds stored at room temperature in the summer tests and 33 seeds for the others. Each figure is an average of two petri dishes for Utsunomiya seeds stored at room temperature in the summer tests, and of three treatments for the others.
*4 – Each test stopped when new germinated seeds were not observed for three consecutive days.
*5 – Tests in summer were started on 2 July 1998.
*6 – Seeds under cold storage were kept at 5°C before tests.

Germination of caryopses

In the range between 15°C and 25°C, the germination rate was high, but a long period was required for germination at a low temperature (15°C). The germination rate in petri dishes was not influenced by light, but the low germination rate in the seeding bed suggests that soil cover on caryopses lowered the rate.

Cold storage is important to keep caryopses alive. Most caryopses kept at room temperature lost their germinability within three months after harvest. Only a few naked grains, obtained from Utsunomiya and stored at room temperature until the summer of 1998, germinated. This suggests that husks control the germination of caryopses to some extent. The caryopses from Utsunomiya generally showed low germination rates in the form of caryopsis (with husks). This is probably because they were harvested in November 1997, which was four months earlier than the harvest of caryopses at Kamigamo. Usually, the caryopses of *P. pubescens* ripen in November to December. However, the caryopses kept on the culm or on the detached branches (after-ripening) are suggested to have a high germinability. To grow *P. pubescens* seedlings efficiently on the nursery, caryopses should be harvested after being kept on the flowering culm as long as possible and to be sown in the form of naked grain within a short period after harvest.

Existence of husks decreased the germination rate and extended the germination period. Especially, in the caryopses from Utsunomiya, the existence of husks reduced the germination rate. In nature, *P. pubescens* seeds are dispersed in the form of caryopses (grains with husks) in winter, and germinate in the next spring. At Kamigamo, the first germination of caryopsis at flowering site was observed on 2 April 1998. The bamboo grove floor after the dispersion of caryopses has generally a low litter supply. Therefore, dispersed caryopses are preserved in an exposed condition on the ground. The present results of the germination tests suggest the following strategy of *P. pubescens* caryopses for regeneration.

The husk of the caryopsis exists to protect the grain from physical damage and feeding by animals. This is because *P. pubescens,* with low germinability, has to keep alive as many grains as possible. Furthermore, the husk may also play an important role to cope with the change in weather conditions. In the present experiments, the naked grain germinated earlier than the grain with husk (caryopsis). Since the East-Asian temperate zone where *P. pubescens* is indigenous has severe changes in weather conditions, especially in early spring, too early germination in response to warmth during this season may cause the death of seedlings. The husk is considered to have a role to protect the grain from a rapid response to warm temperatures. The husk may also function to select the caryopsis with strong vitality. The grain without sufficient power to break through the husk will die even in the case of rooting. In nature, weak seedlings without sufficient vitality will not survive. Thus, the husk is considered to have some important function to guard the caryopsis and to select seedlings.

CONCLUSION

The strain that flowered in the summer of 1997 is not considered to have the complete set of genes necessary for flowering in the 67th year after fruiting. Some recombination of genes might have occurred after transplant to each of the eight sites of flowering. Therefore, it is uncertain whether or not the mutant of the long-term flowering gene of *P. pubescens* caused the short-term flowering periodicity to be 67 years. However, the strain that flowered in 1997 produced many grain-setting caryopses, which showed high germination rates. This flowering is considered to have served its function of preserving the species to the next generation. Whether all the caryopses produced in 1997, or at least those produced at the same location, have the same flowering genes should be examined by DNA analyses. We must wait until the next flowering, perhaps in the year 2064, of this strain to examine how the gene that caused flowering in 1997 is succeeded to the next generation.

ACKNOWLEDGEMENTS

My special thanks are due to Dr. M. Suzuki (Tokyo University), Dr. K. Kasahara (formerly from Hosei University), Dr. M. Watanabe (Japan Bamboo Society), Mr. Y. Wakayama (Wakayama Farm), Ms. J. Morimoto (Kyoto University) and the staff of Kamigamo Experimental Forest, Kyoto University, for their help.

REFERENCES

Ide, Y.; Suzuki, M. 1998. Flowering of *Phyllostachys pubescens* in plant site for the recognition of flowering periodicity of bamboo species in Chiba, Tokyo University Forests. Ringyo-Gijutsu, 672, 11–14. (In Japanese).

Konno, E. 1937. Start of experiment for periodicity of bamboo species for three hundred years. Sanrin, 656, 20–24. (In Japanese).

Suzuki, S. 1978. Index to Japanese Bambusaceae. Gakken, Tokyo, Japan.

Watanabe, M.; Ueda, K.; Manabe, I.; Akai, T. 1982. Flowering, seeding, germination and flowering periodicity of *Phyllostachys pubescens*. Journal of Japanese Forestry Society, 64(3), 107–111.

Bamboo for Sustainable Development, pp. 367–372
Eds. A. Kumar, I.V. Ramanuja Rao and Ch. Sastry

Control of *in vitro* flowering in *Bambusa bambos* from multiple shoots and somatic embryos

I. USHA RAO and RAMA RAGHAV
Department of Botany, University of Delhi, Delhi, India

Abstract—*In vitro* flowering has been successfully observed from multiple shoots as well as somatic embryos of *Bambusa bambos*. Somatic embryos and multiple shoots were derived from seeds cultured on MS basal medium supplemented with benzylaminopurine (BAP) at different concentrations. Eight to ten weeks old cultures at 8×10^{-6} M to 10^{-5} M of BAP showed formation of multiple shoots. Subsequent sub-culture of these shoots in the same medium resulted in 10-12 fold increase in the number of multiple shoots. After 12 weeks, more than 40% of the cultures exhibited the formation of flowering shoots in lower BAP concentrations (10^{-7} M and 5×10^{-7} M). Multiple shoots were also sub-cultured on MS medium supplemented with 2,4-dichlorophenoxyacetic acid (2-4,D). Within 3-4 weeks of culture, the multiple shoots showed callusing and formation of somatic embryos. These then formed flowering plantlets. Hence, a procedure has been developed for precocious induction of flowering of multiple shoots and somatic embryoids of *Bambusa bambos*. The potential of this research has great significance in understanding bamboo flowering.

INTRODUCTION

Bamboo is a versatile group of ligneous plants closely linked with the culture of tropical Asia. Given its unrivalled position in terms of diversity, distribution and uses, coupled with vital role it plays in rural economics of several countries around the world, bamboo has emerged in recent years as the most important non-wood forest resource that has the potential to replace wood in construction and other uses. However, increasing demand for bamboo raw material, especially for the pulp and paper industry, is resulting in their overexploitation and a steady decline of their natural stands, making it critical to urgently increase the production of bamboo by all available means.

The conventional methods of propagation, both sexual and vegetative, have numerous problems. Production of seed is irregular and seed viability is poor. Further, large-scale consumption of seeds by wild animals and seed sterility restrict sexual propagation. Poor seed setting during off-season and flowering just once before culm death compound the problem. With respect to vegetative propagation, cuttings, offsets and rhizomes are bulky and non-available in required number. They are also difficult to handle and transport, and plantlet survival is usually low (Ramanuja Rao and Usha Rao 1990). In comparison, *in vitro* propagation of bamboos, if adequately developed, can prove to be a reliable method for establishment of new bamboo plantations. Several successful attempts have been made to micropropagate bamboos. Bamboos have a peculiar flowering and seeding behavior. They grow vegetatively for a long species-specific period, before flowering, seeding and dying (Janzen 1976). This has made propagation by seeds and improvement by hybridization difficult. *In vitro* flowering has been attempted in a number of species of bamboos (Ramanuja Rao and Usha Rao 1990; Chambers et al. 1991; Rout and Das 1994; Lin and Chang 1998). This paper reports a repeatable procedure for large-scale *in vitro* flowering of multiple shoots and embryos of *Bambusa bambos.*

MATERIAL AND METHODS

Seeds (caryopses) of *Bambusa bambos* were obtained from Bannerghatta, Forest Research Institute, Bangalore, India. After dehusking the seeds, these were washed in 2% Teepol (liquid detergent, Shell, India) solution on a magnetic stirrer for 5 minutes. The Teepol solution was removed by washing in running tap water for 15–20 minutes. Surface sterilization of the material was effected by immersing the seeds in 0.01% silver chloride ($HgCl_2$) solution for 5–10 seconds, followed by thorough washing in sterile distilled water in laminar air flow chamber. The seeds were implanted under aseptic conditions in culture test tubes containing nutrient medium.

The nutrient medium consisted of salts and vitamins of MS basal medium (Murashige and Skoog 1962), supplemented with various concentrations of BAP (5×10^{-6}M, 2×10^{-5} M). The pH of the medium was adjusted to 5.8 prior to autoclaving at 15 psi for 15 minutes. Cultures were maintained at 25±2°C under continuous illumination provided by cool white fluorescent lamps. Twenty-four test tubes were inoculated for each treatment and experiments were repeated thrice.

RESULTS AND DISCUSSION

Seed germination and the ability of seeds to form multiple shoots were dependent on the concentration of BAP incorporated in the medium. Seeds inoculated in medium containing 10^{-5}M and 2×10^{-5}M BAP germinated within one week of culture and produced up to 30 shoots within 8–10 weeks of culture (Fig. 1A). Multiple shoots in clusters of 4–5 shoots each were multiplied (sub-culture period of 3 weeks) on MS basal medium supplemented with 5×10^{-6}M BAP. Flowering was observed in nearly 40% of the cultures raised on BAP (5×10^{-7}M) medium (Fig. 1B). The inflorescence was a compound one and consisted of many florets. The basic structural unit of inflorescence – the spikelet – was a spike having an indeterminate axis bearing many florets (Fig. 1C). Each floret was observed to have a single lemma and palea, a pair of lodicules, six stamens and a gynoecium having

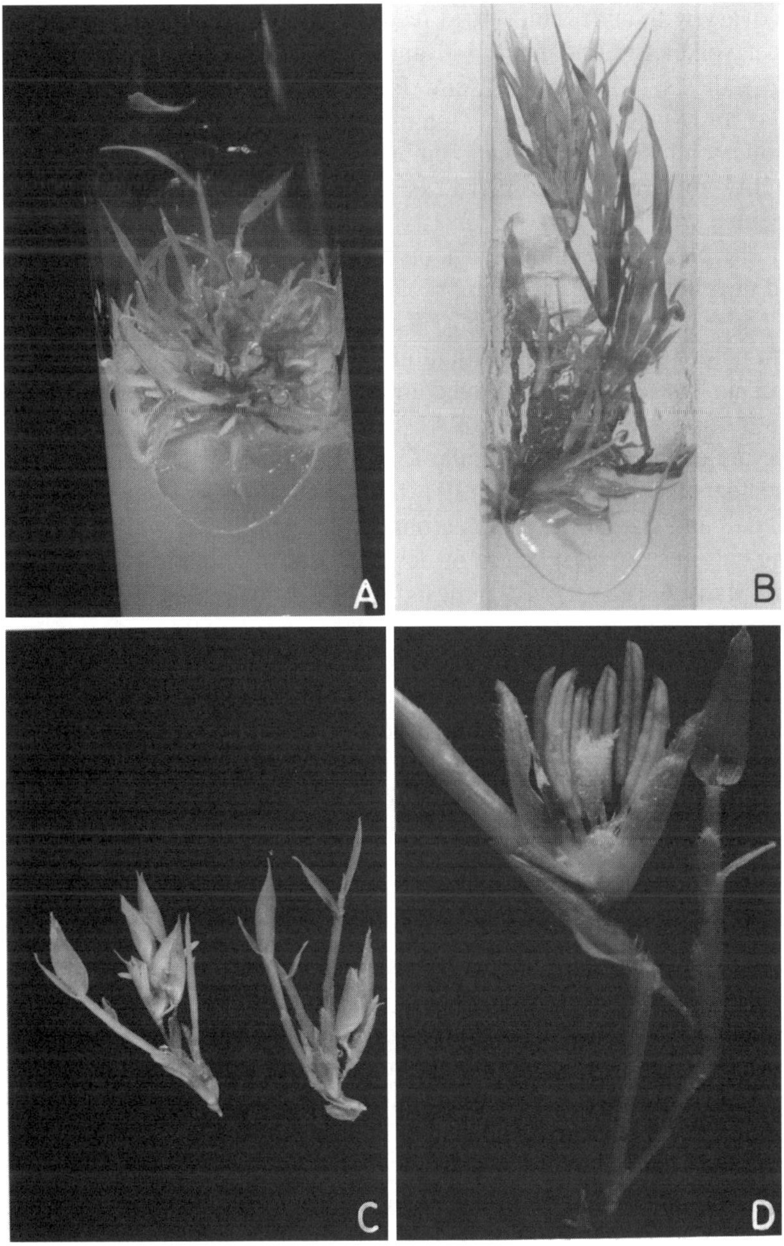

Figure 1. *Bambusa bambos*

A. Induction of multiple shoots on MS basal medium containing 10^{-5}M BAP. Photograph taken after 14 days (×1.8)

B. Flowering of multiple shoots on 5×10^{-7} M BAP after 12 weeks of culture (×1.4)

C. Isolated spikelets (×1.3)

D. Magnified few of a single floret showing lemma, Palea, six stamens and a single ovary. Inflorescence axis is indeterminate (×4)

a single ovary with style and a trifid feathery stigma (Fig. 1D). Mean number of florets per spikelet varied from 4–6 and the mean size of each floret (mature, showing anthesis) was up to 1 cm. Florets had well developed anthers and changed color from green to yellow on maturity.

To induce callusing and somatic embryos, multiple shoots, which were raised on MS+BAP medium for 10–12 weeks, were sub-cultured on MS basal medium supplemented with 2,4-D (10^{-5}M, 2×10^{-5}M). Formation of compact callus was observed at the base of the multiple shoots within 2 weeks of culture. Subsequently localized development of green embryoids was observed in the compact callus in another 2 weeks of culture. About 70% of these embryoids germinated and flowered within four weeks when sub-cultured in a fresh medium having the same composition. The embryoids formed inflorescences without establishing themselves into normal plantlets (Fig. 2A). Florets were stout and more in number. The structure of the floret was similar to the structure of the floret observed from multiple shoots (see details in Fig. 2B). The mean number of florets per spike varied from 6–8 and size of the florets around 1–1.2 cm. Roots were also seen in the flowering cultures. Several stages of development of somatic embryos were also observed. Many florets showed anthesis. Although the pollen was found to be viable, no seed set was observed.

Chang and Hsing (1980) showed that embryoids have the potential to develop into flowers directly in defined conditions. They also reported the production of embryos from root callus of mature *Panax ginseng,* and these embryos formed flowers with fertile pollen within one month of subculture. In a recent report, Jumin and Ahmad (1999) have reported the induction of flowering *in vitro* at high frequency from seeds, stem internodes and segments derived from seedlings of *Murraya paniculata* on half strength Murashige-Tucker basal medium containing 5% sucrose supplemented with 0.1 mgl^{-1} benzyladenine. The present study confirms these reports.

In bamboos, flowering of the juvenile shoots and embryoids provide a valuable system for studying the physiological and molecular events underlying the shift from the stable vegetative phase to the monocarpic floral phase. The results on *in vitro* flowering have far-reaching implications for bamboo improvement programs and micropropagation systems (Ramanuja Rao and Zamora 1995). The research on *in vitro* flowering could lead to a greater understanding of the flowering process in different bamboos, enable *in vitro* hybridization and haploid culture, and double haploid generation. In bamboos, gregarious flowering leads to death of the whole clump and a sharp increase in rodent population, which threatens food stocks. A control on the flowering process would allow the lengthening or shortening of the intermast period according to requirements.

Figure 2. *Bambusa bambos*

A. Culture showing callusing, embryos and inflorescences raised on MS basal medium containing 2,4-D (10^{-5}M) (×3.2)

B. Enlarged view of an isolated floret showing lemma, Palea, six stamens and a single ovary with trifid stigma (×5)

REFERENCES

Chambers, S.M.; Heuch, J.H.R.; Pirie, A. 1991. Micropropagation and *in vitro* flowering of the bamboo *Dendrocalamus hamiltonii* Munro. Plant Cell Tissue Organ Culture, 27, 45–48.

Chang, W.C.; Hsing, Y.I. 1980. *In vitro* flowering of embryoids derived from mature root callus of ginseng (*Panax ginseng*). Nature, 284, 341–342.

Janzen, D.H. 1976. Why bamboos wait so long to flower? Ann. Rev. Ecol. Syst., 7, 347–391.

Jumin, H.B.; Ahmad, M. 1999. High frequency *in vitro* flowering of *Murraya paniculata* (L.) Jack. Plant Cell Reports, 18, 764–768.

Lin, C.S.; Chang, W.C. 1998. Micropropagation of *Bambusa edulis* through nodal explants of field-grown culms and flowering of regenerated plants. Plant Cell Reports, 17, 617–620.

Murashige, T.; Skoog, F. 1962. A revised medium for rapid growth and bioassays with tobacco tissue culture. Physiologia Plantarum, 15, 473–497.

Nadgauda, R.S.; Parasharami, V.A.; Mascarenhas, A.F. 1990. Precocious flowering and seeding in tissue cultured bamboos. Nature, 344, 335–336.

Ramanuja Rao, I.V.; Usha Rao, I. 1990. Tissue culture approaches to the mass propagation and genetic improvement of bamboos. *In* Ramanuja Rao, I.V.; Gnanaharan, R.; Sastry, C.B. ed., Bamboo: current research. Proceedings of the International Bamboo Workshop, Cochin, India, 14–18 November 1988. Kerala Forest Research Institute, Kerala, India; International Development Research Centre, Ottawa, Canada. pp. 151–158.

Ramanuja Rao, I.V.; Zamora, A.B. 1995. In vitro flowering and its implications for bamboo development. *In* Williams, J.T.; Ramanuja Rao, I.V.; Rao A.N. ed., Genetic enhancement of bamboo and rattan. INBAR Technical Report No. 7. International Network for Bamboo and Rattan, New Delhi, India. pp. 61–67.

Rout, G.R.; Das, P. 1994. Somatic embryogenesis and *in vitro* culture of 3 species of bamboo. Plant Cell Reports, 13, 683–686.

Bamboo for Sustainable Development, pp. 373–384
Eds. A. Kumar, I.V. Ramanuja Rao and Ch. Sastry

Plant regeneration via indirect organogenesis in callus derived from young leaves of *Dendrocalamus giganteus*

M. MUÑOZ-FONSECA
Graduate School of Agriculture, University of Costa Rica

E. GUEVARA-BERGER
Center for Grain and Seed Research, University of Costa Rica

M. MONTIEL-LONGHI
Animal Sciences School, Faculty of Agronomy, University of Costa Rica

Abstract—*In vitro* plant regeneration was achieved in *Dendrocalamus giganteus* (giant bamboo) through *de novo* formation of meristems in calli initiated from apex and young leaves from buds, excised from lateral branches, and cultured on MS medium with 3 mg/l 2,4-D and 1 mg/l zeatin, or with 3 mg/l 2,4-D alone. Friable, white, soft and non-organogenic calli were commonly found when calli were initiated in culture medium with auxin alone. They undergo necrosis after 28 days of culture. A second type of calli – white to pale yellow, compact and smooth, and composed of both meristematic and non-meristematic cells – was developed in culture mediums with both 2,4-D and zeatin. Those calli developed roots and shoots upon transfer to an MS medium with 1 mg/l zeatin in the light. Whole plants were developed when transferred to an MS medium with 1.5 % sucrose for four weeks. They were hardened successfully and then transferred to a greenhouse.

INTRODUCTION

Bamboo has been exploited in Asia for hundreds of years: as food, raw material for building, paper making, handicraft, fiber production and even as a species for eroded soil management (Soderstrom and Calderón 1979). In Latin America, it has not been used extensively despite some species, such as *Dendrocalamus giganteus* and *Bambusa Guadua,* adapting well in the tropical environs of the region. Propagules supply represents the main constraint for its cultivation (Montiel and García 1985). Propagation by seeds is also impractical given that giant bamboo flowers only once every 80–120 years, and that most of its seeds are non-viable, with low germination rate and recalcitrance to storage (Banik 1987; Ramanuja Rao et al. 1985). Mass propagation using lateral shoots, culm-nodes or rhizome sectioning is cumbersome because of the bulkiness of the propagules, low availability and difficulty in transportation (Banik 1987).

Propagation by *in vitro* culture has been applied successfully in some species of bamboo. Plants developed from somatic embryos have been obtained in *Bambusa bambos* (syn. *B. arundinacea*) (Mehta et al. 1982) and *Dendrocalamus strictus* (Ramanuja Rao et al. 1985), *B. oldhamii* and *B. beecheyana* (Yeh and Chang 1986a; 1986b) from young inflorescences or seed-derived calli. Most bamboo *in vitro* propagation systems rely on the availability of seeds and inflorescences in order to develop somatic embryos from embryogenic calli. Calli derived from vegetative tissues, such as nodes and leaves, are not embryogenic and they lead to the formation of non-organogenic callus, as observed in *Bambusa*, *Phyllostachys* and *Sasa* (Huang and Murashige 1983), *Schizostachyum brachicladum* and *Thyrsostachys siamensis* (Dekkers et al. 1987).

Some species of the genus *Dendrocalamus* have been already propagated *In vitro:* Ramanuja Rao et al. (1985) developed plantlets by somatic embryogenesis from *D. latiflorus* and *D. strictus*; Pruntpongse and Gavilertvatana (1992) reported the propagation of six species of this genus by culture of shoot segments.

Given the difficulty to obtain seeds or inflorescences in order to produce somatic embryos, this paper presents an alternative method for the micropropagation of *Dendrocalamus giganteus* based on shoot formation from calli developed from lateral-branch node apexes.

MATERIALS AND METHODS

Plant material

Nodes from lateral branches and aerial roots from basal nodes were collected from culms of two-year-old bamboo clumps. Branches were cut into internodes, each section being 2 cm in length and having a node with dormant buds (~2 cm long). They were washed for 30 minutes in water and a detergent (Alconox), dipped in 70% ethanol for 1 minute, immersed for 5 minutes in 0.3% silver chloride ($HgCl_2$) (w/v) plus Tween-20, and rinsed thoroughly with sterile water. Adventitious aerial roots were surface-sterilized likewise, but 1.5% sodium hypochlorite (NaOCl) solution was used instead of $HgCl_2$.

Leaf sheaths covering axillary buds were removed, and the apex (1–5 mm long) was removed and transferred to calli induction mediums. Unfurled leaves were dissected from the apex and cultured separately. Adventitious roots were cut into 2 mm-wide segments.

Culture conditions

The basal medium contained macro and micro elements of Murashige and Skoog (1962), supplemented with 3% (w/v) sucrose, 100 mg/l myo-inositol, 1 mg/l thiamine-HCl, 0.5 mg/l nicotinic acid, 0.5 mg/l pyridoxine-HCL, 2 mg/l glycine. The pH of the medium was adjusted to 5.5 before solidifying with 2 g/l Gelrite. Cultures were maintained at 25°C in the darkness for callus induction, or with a photoperiod regime of 16/8 hours light/dark cycle (4 W/m^2) for plant regeneration.

The first calli induction medium (CIM1) was MS basal medium supplemented with 2,4-D in a range of 2 to 20 mg/l; a second calli induction medium (CIM2) was supplemented with 3 mg/l of 2,4-D and different concentrations of three cytokinins: BAP (0.1, 0.5 and 1.0 mg/l), kinetin (0.1, 1.0 and 2.0 mg/l) and zeatin (0.5, 1.0 and 2.0 mg/l).

The effect of exogenous application of 10 000 ppm CPPU (forclorfenuron) was tested by mixing with lanolin and pasting on dormant buds for one and two weeks. The apexes were then dissected and incubated in the two calli induction media (CIM1 and CIM2).

Three different culture mediums suitable for rice culture were tested with bamboo apexes: N6 (Chu medium) supplemented with 1 g/l casein hydrolizate and 2 mg/l 2,4 D (N6 D), N6 plus 1 g/l caseín hydrolizate, 3 mg/l 2,4 D and 1 mg/l de zeatin (N6-31), c- MS plus 2 mg/l de 2,4-D and 50 mg/l L-tryptophan (MSTf).

For plant regeneration MS basal medium was supplemented with 1 mg/l zeatin, shoots were rooted in basal medium devoid of hormones and supplemented with 1.5 % sucrose (w/v).

Each experiment included six replicates with 10 explants per replicate, and two incubation cycles of 4 weeks each.

Histological studies

Axillary buds, 2, 4 and 10 weeks old calli were fixed in Navashin solution (0.75% CrO_3, 5% glacial acetic acid and 20% formalin) in a vacuum chamber for 30 minutes, followed by 24-hour incubation in the same solution. The tissues were dehydrated by serial incubation in increasing concentrations of terbutilic alcohol (TA). They were incubated for one hour in a mixture of paraffin and TA (1:1 ratio), and then in paraffin-TA (3:1 ratio) for another hour at 65°C. The samples were cooled down to 4°C and cut into 0.5 cm^2 pieces. Serial 12 μm cuts were made with a Reichter-Jung 820a michrotome, and staining was performed following Sharman's method.

RESULTS

Efect of auxin 2,4-D on calli induction

Calli were initiated from axillary buds and unfurled leaves on MS medium supplemented with different concentrations of 2,4-D. Calli induction increases according to 2,4-D concentration in the range of 2 to 20 mg/l when axillary bud apex were used as explant (Table 1). The highest number of explants-forming calli was obtained at concentrations 3 and 4 mg/l of 2,4-D, which also corresponded to the largest size (5–7 mm) and a low number of calli that underwent necrosis. When concentrations above 4 mg/l were used, only small calli were recovered (2–3 mm diameter). For the rest of the experiments, a level of 3 mg/l 2,4-D was used in the calli induction medium.

Table 1.
Effect of 2,4-D concentration on calli induction from axillary buds of giant bamboo after four weeks of incubation in calli induction media

2,4-D (mg/l)	No. of explants	Percentage of explants producing calli	Percentage of calli and explants showing necrosis
0	50	0	54
2	50	54	30
3	50	76	12
4	50	86	8
8	50	50	0
20	50	46	10

Callus proliferation started 10–12 days after culture in calli induction medium. Cell division was first observed in sub-epidermal cell layers (parenchyma) in the foliar primordia and the inner unfurled leaves; the meristem and the rest of the apex underwent necrosis within the first two weeks of culture. Calli initiation took place in the medial part of the foliar primordia and young leaves, but eventually the whole foliar tissue gave rise to different types of calli.

Several types of calli were observed, most of which were white to pale yellow in color, soft and friable. A second type was of mucilaginous appearance, usually interspersed with the first type. A third type of calli was smooth and compact, pale yellow in color.

Friable and mucilaginous calli were composed of large, isodiametric and vacuolated cells. They are usually large and elongated, with thin walls and few or no intercellular connections, which gives rise to a spongy-like appearance of the calli. Compact calli were composed of small cells with conspicuous nucleus and thin cell walls (some were found to accumulate starch grains), organized as meristems surrounded by larger cells and epidermis. Those calli developed into shoots when cultured in a culture medium with 1 mg/l zeatin under light conditions, whereas roots were obtained when compact calli were transferred to 1 mg/l zeatin and 3 mg/l 2,4-D under dark conditions.

Calli growth was maximum during the first 2–3 weeks, until they reach 3–6 mm in diameter. After transferring to fresh medium, the calli tend to grow slower and necrosis was present in most of the calli by the end of the sixth week, especially in friable calli.

Morphogenic response of different explants

Root apexes and root segments

Development of root apexes into roots was low (5%) and no calli were recovered from root segments, even after five rounds of culture in calli induction medium. While most of the root apexes got brown during the second week, few of them started to develop into friable calli by the fourth week of culture. All the calli recovered got necrosed upon transfer to fresh media.

Shoot apexes and young leaves

The outer young leaves were removed from the apex (approximately 0.5 cm in length) and cultured separately on calli induction medium. Calli proliferation took place in 24% of the apexes and 31% of the young leaves. In both cases, cell division started in foliar mesophyll tissues (Figs. 1A & 1B).

Effect of shoot apex size

Shoot apexes were incubated in calli induction media with 3 mg/l 2,4-D and 1 mg/l zeatin. Leaf sheaths covering the axillary buds were removed to get 1 mm apex without leaf sheaths. Also, apexes of 2 mm and 3 mm were prepared with 3 and 5 leaf sheaths, respectively. No significant difference was observed in calli induction among the different explants: 77% of 1-mm apex, 60% of 2-mm apex and 70% of 3-mm apex produced calli. Only 12–20% of the explants necrosed.

Effect of cytokinins on calli induction and plant regeneration

The addition of one of the three cytokinins in different concentrations to the calli induction medium 1 (CM1) increased the number of explants forming calli up to 70% as compared with 46% of the control treatment (3 mg/l 2,4-D alone) (Table 2). However, calli induction was lower than the control when 0.5 mg/l zeatin and 1 mg/l BAP was used: 33% and 37% of calli were recovered in those treatments. The number of explants and calli that necrosed was higher than in the control when 1 and 2 mg/l kinetin and 1 mg/l BAP was used, reaching up to 65% of explants necrosed.

All the calli induced with BAP were of the friable type, and they got necrosed within the third and fourth week of culture. Only 5–10% of the calli recovered in 0.5 mg/l of kinetin and 1 mg/l of BAP were transferred to fresh medium. In the case of calli initiated with BAP, the number of necrosed calli increased with the concentration of BAP and those calli tended to be smaller than the control.

Around 60–70% of the calli induced with zeatin were transferred to fresh media. Twenty percent of those obtained in 1 mg/l zeatin were compact and nodular calli (Figs. 1A & 1B). Roots were developed in those calli when maintained for eight weeks in calli induction medium supplemented with 3 mg/l 2,4-D and 1 mg/l zeatin. When those roots were excised and maintained for over 10 months in MS supplemented with 3 mg/l 2,4-D, proliferation and development of secondary roots as well as the formation of large masses of mucilaginous calli were observed. Reversion of roots to large, white and friable calli was also noticed after 10 months in continuous culture.

Table 2.
Effect of three cytokinins on calli induction using axillary buds apexes

Concentration of cytokinins (mg/l)	Percentage of explants forming calli	Percentage of explants and calli necrosed	Percentage of calli transferred to fresh media
Control (3 mg/l 2,4-D)	46	52	89
Zeatin			
0.5	37	40	77
1.0	70	38	62*
2.0	61	51	59
Kinetin			
0.1	58	48	4
1.0	63	63	40
2.0	63	65	34
BAP			
0.1	56	44	46
0.5	58	55	21
1.0	33	63	10

Note: * 20% (12 out of 60) of the calli were organogenic

Plant regeneration from compact organogenic calli

Compact calli obtained in MS supplemented with 3 mg/l 2,4-D and 1 mg/l zeatin for 8 weeks were transferred to MS plus 1 mg/l zeatin for another 8 weeks under dark (Fig. 1B). Calli that are 10–12 weeks old showed greenish areas where shoots started to emerged upon transfer to fresh medium and under light (Fig. 1C). Each callus gave rise to 2–6 shoot primordia, forming a cluster with a dominant shoot and axillary inhibited smaller shoots (Figs. 1D-F). Shoot clusters developed roots in the same medium and they were transferred to MS medium without hormones and supplemented with 1.5% sucrose (Fig. 1G). They were hardened and transferred to a greenhouse, where they started to develop new leaves after two weeks and three new shoots per plant after 4 weeks in the greenhouse (Fig. 1H).

Evaluation of different calli induction mediums

Two calli induction mediums used above (CM1 and CM2) were compared with the following calli induction mediums commonly used for the induction of embryogenic calli from rice zygotic embryos:

1. N6 supplemented with 1 g/l casein hydrolizate and 2 mg/l 2,4-D (N6);
2. N6 plus 1 g/l casein hydrolizate, 3 mg/l 2,4-D and 1 mg/l de zeatin (N6-31);
3. MS plus 2 mg/l de 2,4-D and 50 mg/l L-tryptophan (MSTf).

Figure 1. Plant regeneration from organogenic calli induced from the culture of *Dendrocalamus giganteus* axillary buds. (A) organogenic calli cultivated on a MS with 3 mg/l of 2,4-D and 1 mg/l of zeatin in dark conditions. (B and C) Organogenic calli transferred to light and showing meristematic zones. (D, E and F) successive growth of axillary buds cultivated on a MS medium with 1 mg/l of zeatin. (G) single shoot with roots formed. (H) Plant transferred successfully to greenhouse conditions.

Almost 100% of the explants incubated in MSTf formed calli with no necrosis after 4 weeks of culture as compared with 26% of calli induced in MS supplemented with 2,4-D and zeatin. In other experiments, the maximum calli induced with 2,4-D and zeatin was 80% under optimum conditions, which indicated that further combinations of 2,4-D, tryptophan and zeatin may lead to even higher calli induction rates. On the other hand, only 1% of those incubated in N6 plus 2,4-D and casein developed calli. Dormancy of axillary buds was ten times lower when both 2,4-D and zeatin were added to calli induction medium compared to the use of 2,4-D alone (data not shown). All the calli induced in MSTf were friable, non-organogenic type.

To induce shoot formation, these calli were transferred to a similar culture medium used for calli induction with and without zeatin (Table 4). Calli initiated in N6 medium plus casein did not survive for more than two weeks; the same was observed with those initiated in MS plus zeatin and 2,4-D. Of the calli formed in MSTf, 92% necrosed during the first week of culture in culture medium supplemented with zeatin. However, two out of three calli that survived did develop into roots, indicating that meristem was formed *de novo* in small organogenic sectors of the calli. On the other hand, 20% of those calli transferred to the medium without zeatin survived during the second subculture, but no roots or shoots develop-

Table 3.
Effect of de different culture media on calli induction after four weeks of culture*

Calli induction mediums	Percentage of explants forming calli	Percentage of explants and calli necrosed
MS Tf	99	9
N6 Cs	1	13
N6 Cs 31	25	60
MS 31	26	19

Note. * Each treatment included seven replicates with 10 explants each

Table 4.
Effect of the addition of zeatin to calli initiation media for induction of organogenic calli

Calli induction mediums	Recovered calli	Necrosed calli	Calli transferred to fresh media	Calli surviving after four weeks
N6 Cs –> N6 Cs + Zt *	17	17	0	0
N6 Cs –> N6 Cs	15	15	0	0
MSTf –> MSTf + Zt	36	33	8	3 (2)
MSTf –> MSTf	35	27	23	8 (1)
MS31 –> MS31	10	10	0	0
MS31 –> MS1 **	10	10	0	0

Notes: * Zt indicates 1 mg/l zeatin; ** 31 indicates 3 mg/l 2,4-D and 1 mg/l zeatin, 1 indicates 1 mg/l zeatin; Numbers in parentheses (last column) indicate the number of roots per calli.

ment was observed. Therefore, it could be inferred that most of the calli were non-organogenic, but those able to form organs did so because of the presence of zeatin.

Effect of exogenous application of cytokinins to axillary buds in lateral branches

Since the incubation of axillary apexes in culture media supplemented with cytokinins leads to the formation of organogenic calli, tests were conducted on the effect of exogenous application of CPPU mixed with lanolin on dormant buds for 1–2 weeks. The apexes were then dissected and incubated in calli induction medium supplemented with 3 mg/l 2,4-D and 1 mg/l zeatin or 3 mg/l 2,4-D alone.

As shown in Table 5, buds treated with CPPU for 1–2 weeks showed 100% increase in calli induction, as compared with the control when cultured in MS medium supplemented with 3 mg/l 2,4-D.

Table 5.
Effect of exogenous application of forchlorfenuron (CPPU, 10 000 ppm in lanolin) on the formation of calli from dormant axillary buds [a]

Incubation period (weeks)	Percentage of explants forming calli	Percentage of necrosed calli and explants
MS 2,4-D [b]		
0	42	48
1	86	6
2	94	0
MS 2,4-D + ZT [c]		
0	64	19
1	44	20
2	68	12

Notes: a – each treatment included 50 explants; b – MS medium + 3 mg/l 2,4-D; c – MS medium + 3 mg/l 2,4-D and 1 mg/l zeatin

DISCUSSION AND CONCLUSION

Calli induction and plant regeneration in bamboo largely depends on the type of explant used, its physiological state and its morphogenic competence. Different bamboo species have been propagated *in vitro* via somatic embryogenesis and indirect organogenesis from calli derived from zygotic embryos, immature inflorescences, meristems, leaf sheaths, apex and root (Ramanuja Rao et al. 1985; Yeh and Chang 1986; El-Hassan and Debergh 1987). Somatic embryogenesis seems to be the easiest way for micropropagation; however, it relies on the availability of embryogenic tissues, which are not easily available in species like giant bamboo. Since few reports have demonstrated the feasibility of inducing somatic embryos from mature tissues, any method that can regenerate plants from axillary buds will be of much benefit. In this case, calli from mature tissues (axillary buds, young leaves and foliar primordia) were obtained which represents an advantage since axillary buds from lateral branches are the most abundant source of vegetative material. Moreover, even the smallest apex used in this experiment bore at least three pairs of foliar primordia, from which calli can be easily initiated. A system for micropropagation of giant bamboo was reported by Ramanayake and Yakandawala (1997), based on the incubation of nodal segments from lateral branches and the formation of shoots out of the dormant bud in media supplemented with BAP, kinetin and adenine sulphate. In previous unreported experiments, the authors tried to incubate nodal cultures but establishing axenic cultures was hampered by contamination with vascular bacteria.

Three different type of calli noticed, as previously described by other authors in different Gramineae and Poaceae species, including bamboo (Ramanuja Rao et al. 1985; El-Hassan and Debergh 1987). Most of the calli were friable and translucent. A second kind was mucilaginous and a third type was compact and nodular calli, which were recovered at low frequency. Friable and mucilaginous calli did not show any morphogenic competence, although it has been reported that friable calli are able to develop roots (Dekker et al. 1989). However, it is possible that small sectors in the friable calli were compact and thus able to develop meristems *de novo* under a suitable culture medium. It was also observed that the formation of roots from friable calli initiated in MS medium supplemented with 2,4-D and trytophan, but histological analysis revealed that friable calli are unable to develop any organ. It is interesting to note that even compact organogenic calli can also have interspersed small friable sectors.

Compact and nodular calli showed a much higher morphogenic competence, however they were recovered at low frequency from the explants evaluated in this experiment. Induction of compact calli seems to be determined by the combination of auxins (2,4-D) and cytokinins (zeatin). In other species of bamboo cytokinins such as BAP, 2ip, kinetin or zeatin have been added to the culture media in order to induce calli formation, differentiation and regeneration of both shoots and roots (Huang and Murashige 1983; Yeh and Chang 1986).

Calli induced and maintained in 3 mg/l 2,4-D and 1 mg/l de zeatin were able to develop roots after 8 weeks of culture, whereas those transferred to 1 mg/l zeatin were able to develop shoots 12 weeks after calli initiation. It is remarkable that organogenic calli were recovered from media supplemented with zeatin as the only cytokinin, while BAP and kinetin were ineffective in inducing the formation of compact calli, at least at the concentrations evaluated in this experiment. Since the activity of different cytokinins even at the same concentration is not comparable, it is possible that zeatin is more active than the other cytokinins. Moreover, exogenously application of cytokinins (CPPU) to dormant axillary buds increased the formation of calli when their apexes were dissected and incubated in a medium without cytokinins, indicating the need of disrupting the dormant state before calli induction.

However, this is not the only factor, given the complex interactions between different growth regulators that are required during the dedifferentiation and formation of meristems *de novo* as well as somatic embryos. The same results can be obtained by different combinations of auxins, cytokinins and even abscisic acid, as demonstrated by Jullien and Tran Thanh (1994). In their case, friable calli were recovered from semi-thin sections of *Bambusa glaucescens* leaves incubated in calli induction medium supplemented with 2,4-D only, whereas the combination of 2,4-d, IAA, BAP and ABA gave rise to organogenic compact calli and to embryo-like structures. Further research in calli induction mediums should be done in order to determine which combinations could increase both calli induction and morphogenic response. In the present experiments, the combination of 2,4-D, zeatin and tryptophan proved to lead to a substantial increased in calli formation as compared with previously used culture mediums.

This experiment succeeded in developing whole plants via indirect organogenesis using calli derived from vegetative tissues, bypassing the need to rely on seed or inflorescence for calli and plant formation. However, there is always the risk of somaclonal variation when plants are derived from calli.

Recently, it was reported that the production of giant bamboo can be achieved using seedlings from *in vitro* germinated seeds, from which axillary shoots can be multiplied under the presence of BAP or kinetin (Rout and Das 1994; Maity and Ghosh 1997; Arya and Arya 1997). Somatic embryogenesis has also been achieved using calli derived from zygotic embryos incubated in MS plus 2,4-D, kinetin and adenine sulphate (Rout and Das 1994).

REFERENCES

Arya, I.; Arya, S. 1997. *In vitro* culture and establishment of exotic bamboo *Dendrocalamus asper*. Indian Journal of Experimental Biology, 35 (11), 1252–1255.

Banik, R.L. 1987. Techniques of bamboo propagation with special reference to pre-rooted and pre-rhizomed branch cuttings and tissue culture. *In* Rao, A.N.; Dhanarajan, G.; Sastry, C.B. ed., Recent Research on Bamboo. Proceedings of the International Bamboo Workshop, Hangzhou, China, 6–14 October 1985. Chinese Academy of Forestry, Beijing, China; International Development Research Centre, Ottawa, Canada. pp. 160–169.

Dekkers, A.J.; Rao, A.N.; Loh, C.S. 1987. *In vitro* callus in bamboos *Schizostachyum* and *Thyrsostachys* species. *In* Rao, A.N.; Dhanarajan, G.; Sastry, C.B. ed., Recent Research on Bamboo. Proceedings of the International Bamboo Workshop, Hangzhou, China, 6–14 October 1985. Chinese Academy of Forestry, Beijing, China; International Development Research Centre, Ottawa, Canada. pp. 170–174.

El-Hassan, A.A; Debergh, P. 1987. Embryogenesis and plantlet development in the bamboo *Phyllostachys viridis* (Young). Plant Cell, Tissue and Organ Culture, 10, 73–77.

Huang, L.C.; Murashige, T. 1983. Tissue culture investigations of bamboo. I. Callus cultures of *Bambusa*, *Phyllostachys* and *Sasa*. Botanical Bulletin of Academia Sinica, 24, 31–52.

Huang, L.C.; Chen, W.L.; Huang, B.L. 1989. Tissue culture investigations of bamboo. IV. Organogenesis leading to adventitious shoots and plants in excised shoot apexes. Environmental and Experimental Botany, 29 (9), 307–315.

Jullien, F.; Tran Thanh, V.K. 1994. Micropropagation and embryoid formation from young leaves of *Bambusa glaucescens* Golden goddess. Plant Science, 98, 199–207.

Maity, S.; Ghosh, A. 1997. Efficient plant regeneration from seeds and nodal segments of *Dendrocalamus strictus* using *in-vitro* technique. Indian Forester, 123 (4), 313–318.

Mehta, U.; Ramanuja Rao, I.V.; Mohan Ram, H.Y. 1982. Somatic embryogenesis in bamboo. *In* Fujiwara, A., ed., Proceedings of the Vth International Congress of Plant Tissue & Cell Culture, Tokyo, Japan. pp. 109–110.

Montiel L.M.; Garcia, P.R. 1985. Programa Bambú: Análisis de una alternativa de desarrollo. Vicerrectoría de Investigación, Universidad de Costa Rica, Costa Rica. 16 pp.

Murashige, T.; Skoog, F. 1962. A revised medium for rapid growth and bioassays with tobacco tissue culture. Physiologia Plantarum, 15, 473–497.

Prutpongse, P.; Gavinlertvatana, P. 1992. *In vitro* micropropagation of 54 species from 15 genera of bamboos. HortScience, 27, 453–454.

Ramanayake, S.M.S.D.; Yakandawala, K. 1997. Micropropagation of the giant bamboo (Dendrocalamus giganteus Munro) from nodal explants of field-grown culms. Plant Science, 129, 213–223.

Ramanuja Rao, I.V.; Rao. I.U.; Narang, V. 1985. Somatic embryogenesis and regeneration of complete plantlets in the bamboo *Dendrocalamus strictus*. Plant Cell Reports, 4, 191–194.

Rout, G.; Das, P. 1994. Somatic embryogenesis and *in vitro* flowering of three species of bamboo. Plant Cell Reports, 13, 683–686.

Soderstrom, T.; Calderon, C. 1979. A commentary on the bamboo (Poaceae: Bambusoideae). Biotropica 11(3), 161–172.
Yeh, M.L.; Chang, W.C. 1986a. Plant regeneration through somatic embryogenesis in callus culture of green bamboo (*Bambusa oldhamii* Munro). Theoretical and Applied Genetics, 7, 161–163.
Yeh, M.L.; Chang, W.C. 1986b. Somatic embryogenesis and subsequent plant regeneration from inflorescence callus of *Bambusa beecheyana* Munro var. beecheyana. Plant Cell Reports, 5, 409–411.

FURTHER READING

Chu, C.; Wang, C.; Sun, C.; Chan, H.; Yin, K.; Chu, C.; Bi, F. 1975. Establishment of an efficient medium for anther culture of rice through comparative experiments on the nitrogen sources. Scientia Sinica, 18, 659–668.
Manzur, D. 1988. Propagación vegetativa de *Guadua angustifolia* Kunth. III Simposio Nacional Bambú Guadua. Colombia, Octubre 1988, pp. 14–19.
Saxena, S. 1990. *In vitro* propagation of the bamboo (*Bambusa tulda*) through shoot proliferation. Plant Cell Reports, 9, 431–434.

Bamboo for Sustainable Development, pp. 385–395
Eds. A. Kumar, I.V. Ramanuja Rao and Ch. Sastry

Origin and development of embryogenic callus, somatic embryos and secondary somatic embryogenesis in *Dendrocalamus strictus*

I.V. RAMANUJA RAO
International Network for Bamboo and Rattan, Beijing, China

VIBHA NARANG and I. USHA RAO
Department of Botany, University of Delhi, Delhi, India

Abstract—Cultures of caryopsis of *Dendrocalamus strictus* raised in B_5+2,4-dichlorophenoxyacetic acid (2,4-D, 3×10^{-5}M) were fixed and sectioned to determine the origin of the embryogenic callus and embryoid development. Callusing is restricted to the embryonal axis without the involvement of the scutellum. Examination of the callusing embryo revealed that the callus masses arose from vascular bundles. Before callusing, accumulation of starch is noted in the tissues adjoining the bundle. Callusing initiates from the xylem parenchyma cells towards the base of the bundle. This subsequently spreads all around the bundle, except where the fibrous cap is located. A cambium-like layer then forms which cuts off cells both towards the inside as well as outside. The origin of somatic embryos appears to take place from the cambium-like layer. Once there is externalization of the cambium like layer, embryoid initiation takes place. Once embryoid development is initiated at one point in the cambium-like layer, it rapidly spreads to the rest of the callus, and results in the formation of a large number of somatic embryoids, which are often complete with a shoot axis, a coleoptilar region and a scutellum. All embryoids are superficial in origin.

Secondary somatic embryogenesis has been commonly observed in B_5+2,4-D (3×10^{-5}M)+ BAP (10^{-6}m). The epidermal cells of the developing embryoids divide to give rise to globular masses that later develop into embryoids. If continually maintained in this medium, these embryonal masses proliferate and give rise to a culture that consists of secondary embryoids.

INTRODUCTION

Bamboos are one of the world's most renewable resources. These giant arborescent grasses have woody culms arising from rhizomes. They are extraordinarily useful plants, for which myriad applications have been devised. The woody material has bundles that remain separate in the long internodes, giving bamboo an easy splitting characteristic that makes it so amenable to handicrafts. Bamboo propagation is by and large through offsets and seeds.

Plant tissue culture not only offers the unique opportunity for realizing the totipotency of cells into whole plants, but also provides conditions under which physiological manipulations can be carried out to overcome endogenous controls inherent in the intact plant. *In vitro* methods present an attractive alternative to the traditional planting materials – such as offsets, cuttings and seeds – for the propagation of bamboos. One method is the formation of plantlets from somatic embryos, while the other involves micropropagation. Somatic embryogenesis is the development from somatic cells of structures that follow a histo-differentiation pattern, which leads to a body pattern resembling that of zygotic embryos (Emons 1994). *In vitro* somatic embryogenesis can occur either directly from cells of an organized tissue or indirectly from callus or suspension culture (Williams and Maheswaran 1986). Somatic embryos are organized bipolar structures arising from a single cell and having no vascular connection with maternal tissue (Haccius 1978). According to Williams and Maheswaran (1986), "there appears to be universal formation of compact clump of embryogenic cells, the proembryonal complex from which one to many embryoids develop". This paper describes the origin and development of embryogenic callus, somatic embryos and secondary somatic embryogenesis in the bamboo, *Dendrocalamus strictus.* These studies have also been performed on somatic embryos of different species and two pathways of origin – unicellular (Vasil and Vasil 1982; McCain et al. 1988; Jones and Rost 1989) and multicellular (Wernicke et al. 1982; McCain and Hodges 1986; Taylor and Vasil 1996) have been given.

MATERIAL AND METHODS

Caryopses (seeds with husk) of *D. strictus* were used as experimental material for raising embryogenic cultures. Seeds were washed in 2% Teepol (liquid detergent, Shell, India) solution on a magnetic stirrer for 5 minutes. The Teepol solution was removed by washing in running tap water for 15–20 minutes followed by a rinse in distilled water. Sterilization of the material was effected by a 5-minute immersion in chlorine water (saturated chlorine water diluted 5 times with distilled water) followed by thorough washing in sterile distilled water. The procedure used was essentially similar to that described earlier by Rao et al. (1985). The seeds were implanted aseptically in test tubes containing the nutrient medium.

The medium consisted of B_5 salts and vitamins (Gamborg et al. 1968), sucrose (2%) and 0.8% agar agar (w/v) supplemented with 2,4-D (3×10^{-5}M) for embryogenesis. For achieving secondary embryoids, the cultures already growing on B_5+2, 4-D (3×10^{-5}M) were sub-cultured on $B_5+3\times10^{-5}$ 2, 4-D+10^{-6}M BAP and incubated for 3–4 weeks. The pH of the medium was adjusted to 5.8 prior to autoclaving at 1.05 kg/cm^2 for 15 minutes.

Dissections of the cultures at different developmental stages were carried out under the stereomicroscope and photographs taken. In addition the origin and development of embryoids was studied by cutting wax sections at various stages of development. Photographs were taken in a Zeiss ICM 405 inverted microscope.

Histological Techniques

Embryogenic calli (raised from caryopsis) on B_5+2, 4-D (3×10^{-5}M)+BAP (10^{-6}M) were fixed in 10% neutral formalin (4% aqueous solution of formaldehyde at pH 7.0). The pH was adjusted using pyridine at 4°C for 48 hours and transferred to 70% ethyl alcohol. The material was dehydrated through TBA-Xylene series and embedded in paraffin wax. Serial sections were cut at a thickness of 8–10 μm. The sections were stained with 0.05% aqueous solution of Toluidine Blue.

OBSERVATION AND DISCUSSION

In *D. strictus* (present work), it was observed that somatic embryos originated from compact calli (embryogenic) and not loose, friable ones. When seeds are inoculated in 2,4-D (3×10^{-5}M), the embryo germinates after 3–4 days and the radicle first emerges out. Callusing starts soon after germination. Most of the callusing is restricted to the embryonal axis while the scutellum largely remains as such. Some studies have found scutellum of immature embryos as being the tissue from which embryogenic calli originate (McCain et al. 1988; Ryschka et al. 1991; Oka et al. 1995).

The initial callus is largely restricted to the coleoptile and its removal reveals the embryonal axis as a white, solid, central core. Tubular callus masses were seen arising from the vascular bundles. Callusing was observed to be taking place from only one side of the bundle and not on the other side (Fig. 1 A-E).

When a section of 14-day-old callusing embryo was cut, several deeply staining vascular bundles were seen scattered in the ground parenchyma of the embryonal axis (Fig. 2 A-B). Prior to callusing there is accumulation of a large amount of starch around the bundle. Deeply staining cells (xylem parenchyma) can be seen located towards the base of the bundles (Fig. 2 C-E): callusing initiates from these cells. Cell division activity spreads to the other cells on the periphery of the bundle resulting in a circular mass of divisions radiating away from the bundle. This zone of cell division activity that originates from the vascular bundle spreads radically outwards and assumes the nature of a cambium-like layer, a few cells thick. This meristematic layer consists of stacked flattened and rectangular cells with deeply staining nucleus (Fig. 3 A-D).

The origin of the somatic embryos appears to take place from the cambium-like layer that is derived from the bundles. The initiation of embryo formation in the growing compact callus takes place when the meristematic activity of the cambium-like layer, which in the initial stages is deep-seated owing to the production of cells towards the outside (in addition to those towards the inside), spreads to the surface of the callus. Once embryoid development is initiated from the cambium-like layer, this activity rapidly spreads to the rest of the callus resulting in the formation of several embryoids. Thus, in a section of the embryogenic callus various stages of embryoid development can be seen, with the youngest at the

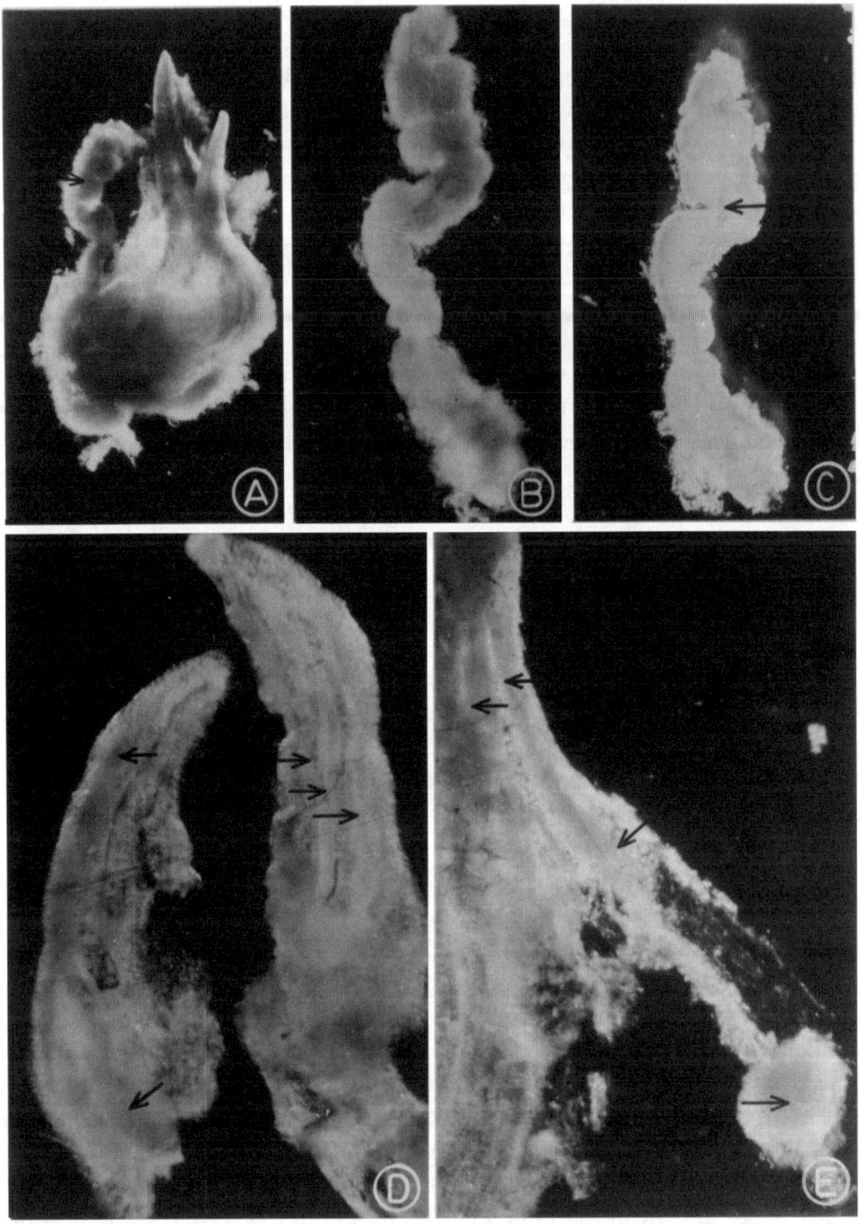

Figure 1. Dissections of *D. strictus* embryos isolated from seed cultured in B_5+2,4-D (3×10^{-5}M)
A. Embryo showing callusing of axis. The coleoptile has been removed to show the tubular callus mass arising from the core of compact callus (arrow). This callus has formed along one of the vascular bundles of the coleoptile.
B-C. Isolated tubular compact callus mass. Note remnant of the fibrous cap of the bundle as a line traversing the length of the callus (arrow).
D-E. Coleoptile isolated from (A) showing callusing vascular bundles (arrows). Note accumulation of starch around vascular bundles in E (dark field photograph), starch seen as shining masses.

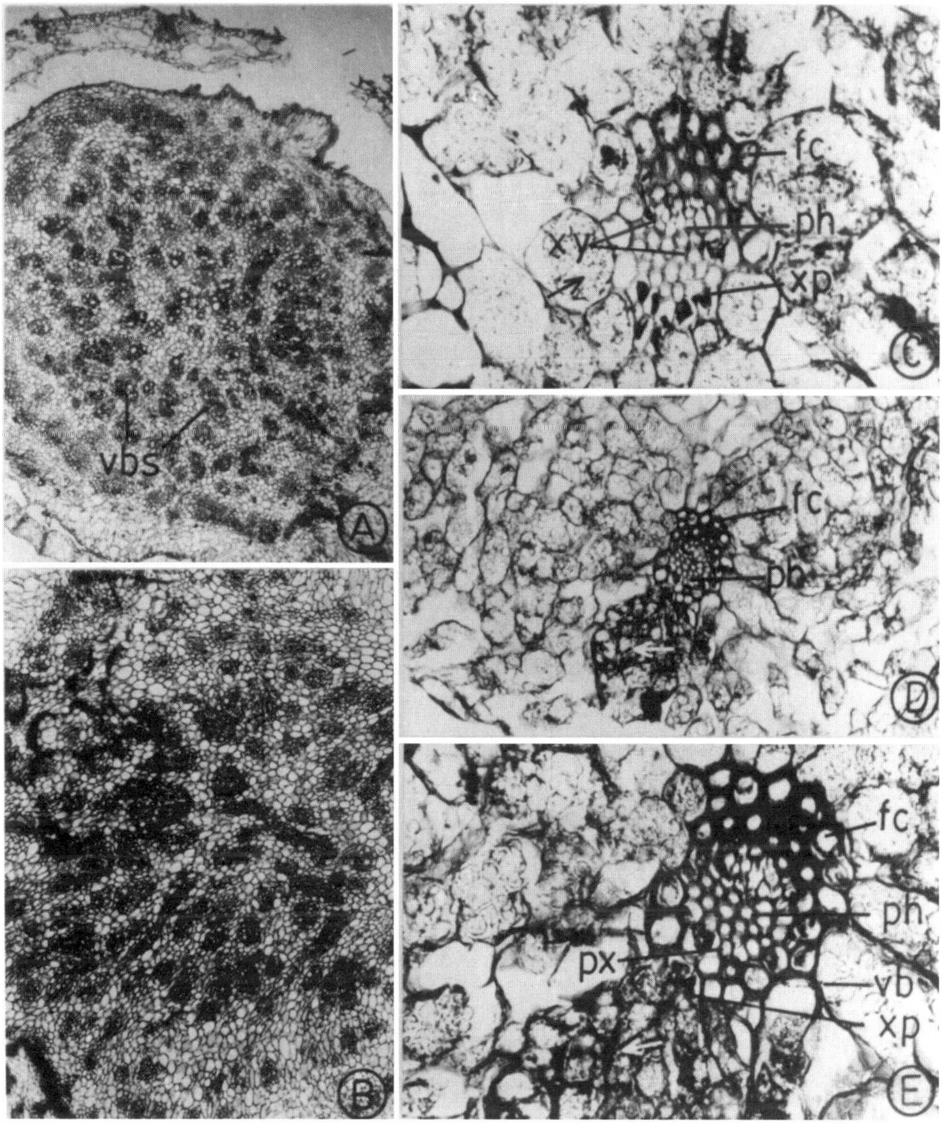

Figure 2. *D. strictus:* transverse sections of 14-day-old mature embryos in B_5+2,4-D (3×10^{-5}M)
A-B. Transverse section of embryonal axis showing the presence of many scattered vascular bundles (vbs). The vascular bundles are deeply stained indicating meristematic activity.
C. Magnified view of a single vascular bundle showing the fibrous cap (fc), xylem (xy), deeply stained cytoplasmic xylem parenchyma (xp) and phloem (ph). The adjacent cells have starch grains (arrow).
D. Transverse section of embryo showing callus (arrow) originating from the xylem parenchyma (xp) of the vascular bundle (vb).
E. Enlarged view of D.

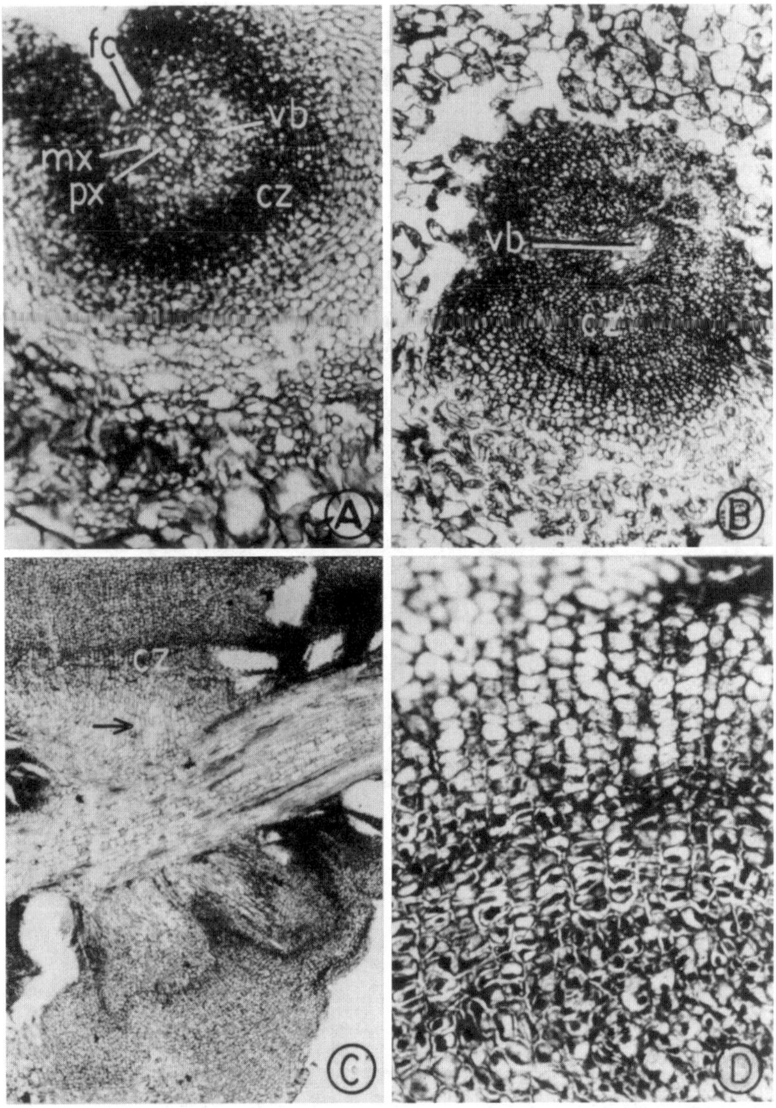

Figure 3. *D. strictus:* transverse and longitudinal sections of callusing embryo obtained from seeds cultured in B_5+2,4-D (3×10^{-5}M) for three weeks.

A. A portion of the transverse section of callus showing meristematic activity in the vascular bundle (vb). No divisions are seen in the region of the fibrous cap (fc). The division initiate from the cells of the bundle and this results in the formation of a meristematic cambium-like zone (cz). Cells of this zone are small, ajtoplasmic and deeply stained. Protoxylem (px) and metaxylem (mx) are seen.

B. A portion of the transverse section of the callus showing meristematic activity around the vascular bundle. The cells of the meristematic cambium-like zone (cz) are deeply stained.

C. L.S. of the callusing embryonal axis. Note the formation of cambium-like layer (cz), and the elongated cells formed towards the inside and the smaller cytoplasmic cells on the outside of the layer.

D. Magnified view of C showing the cambium-like layer (cz) having rectangular cells with deeply stained nuclei.

edge of the cambium-like layer where it meets the surface of the callus and the oldest away from it. During the later stages of embryoid development, the apical part expands out such that it comes to have a broader upper region attached to the underlying tissue by a narrower part (Fig. 4 A-D).

The somatic embryoids that are formed from the embryogenic callus are often complete with a shoot root axis and coleoptilar and scutellar regions (Fig. 5 A-C).

Secondary somatic embryogenesis

This has been commonly observed in B_5+2,4-D (3×10^{-5}M) by the sixth week and also in B_5+2,4-D (3×10^{-5}M)+BAP (10^{-6}M). The epidermal cells of the developing embryoids initially divide radially, and later tangentially, to give rise to globular masses that later develop into embryoids. If continually maintained in B_5+2,4-D (3×10^{5}M)+BAP (10^{-6}M), these embryonal masses proliferate giving rise to a culture that entirely consists of secondary embryoids (Fig. 6 A-F). Broadening of the apical regions of the embryonal masses takes place and several daughter embryoids form from these broader apical regions, giving the entire structure a branched appearance.

In *D. strictus,* callusing initiates from the xylem parenchyma cells, which stain deep and are located towards the base of the bundles. However, oil palm histological studies showed that calli grow from perivascular cells (Schwendiman et al. 1990). Proliferation of calli from cells adjacent to the vascular bundles has been recorded particularly in Agavacea (Hunault 1978), Coconut (Pannetier and Buffard-Morel 1986) and *Hevea* (Michaux-Ferriere et al. 1992). The embryoids arise as globular masses from the outer layer.

Research work on immature embryos of wheat (Magnusson and Bornman 1985) and barley (Ryschka et al. 1991) gave evidence of different origins of somatic embryos. The periphery of the scutellum becomes slightly enlarged owing to elongation of cells from the abaxial scutellar epidermis. Single cells of the scutellar epidermis start dividing periclinally and give rise to pseudothalus-like structures. The parenchymatous cells of the scutellum and cells of the coleorhiza enlarge and vacuoles coalesce, leading to separation of cell groups. Divisions take place in sub-epidermal cell layer, resulting in the formation of distinct meristematic areas without polarity described as meristematic units (Fransz and Schel 1991). The proembryogenic or meristematic areas may be considered a result of coalition of densely crowded proembryoids loosing their bipolarity, resulting in the formation of compact embryogenic calli (Ryschka et al. 1991). Nonohay et al. (1999) describe that in *Hordeum* somatic embryo development could also occur through an intermediary stage, in which the meristematic region grows without embryo differentiation and unknown number of sites of somatic embryo origin (SSEO) fuse and then differentiate to form an aggregate of embryos with independent vascular systems. According to Ryschka et al. (1991), the embryoids could originate from each of the three basic tissues of scutellum – the epidermis,

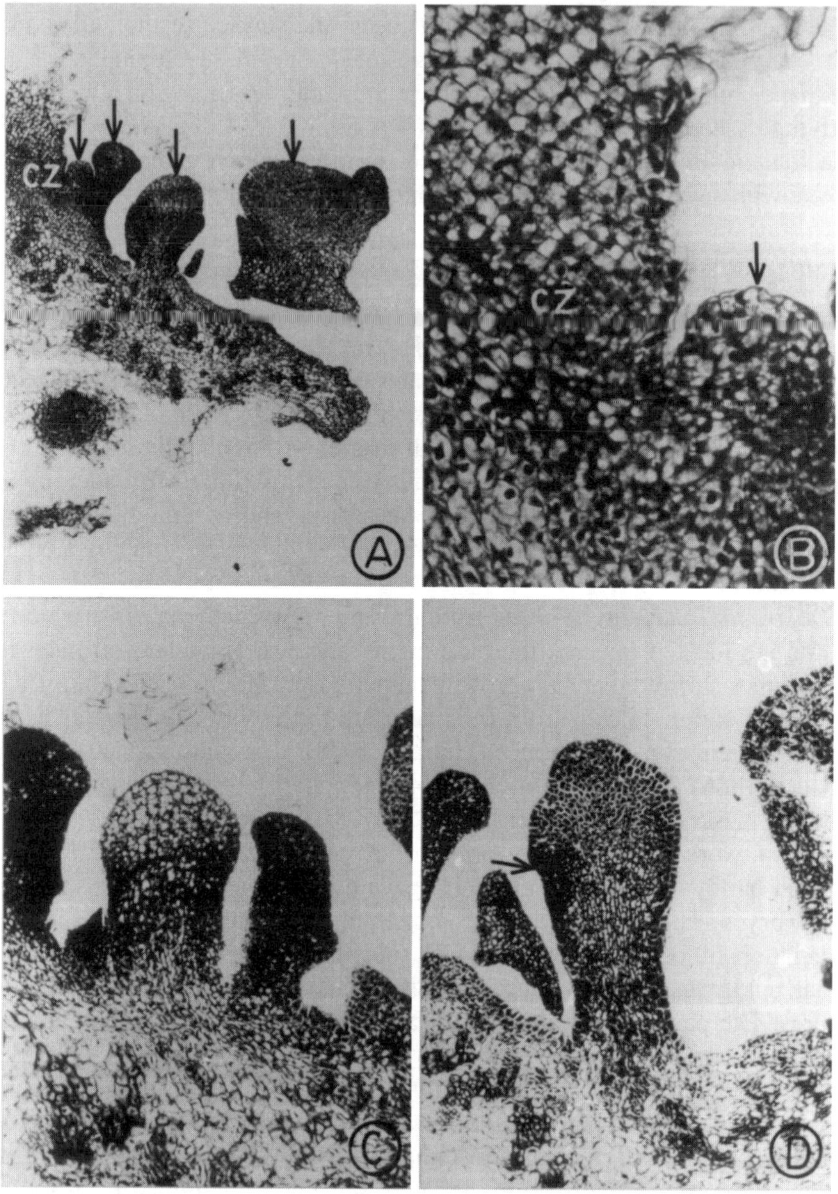

Figure 4. *D. strictus:* Section of embryogenic compact callus raised from seed cultured on B_5+2,4-D (3×10^{-5}M) for three weeks

A. Section of the embryogenic callus showing the origin of embryoids. The cells of the young embryoids take up dense stain and are highly meristematic. Note the progressive conversion of the callus into embryoids (arrow), with the youngest adjacent to the callus and older ones away from it.

B. A magnified portion of (A) showing embryoids (arrow) developing adjacent to the cambium-like zone (cz).

C. Embryoid at the globular stage of development. The cells are highly meristematic.

D. Magnified view of a developing embryoid. Note expanded upper part and narrower stalk. The arrow indicates the notch at which the shoot apex will form.

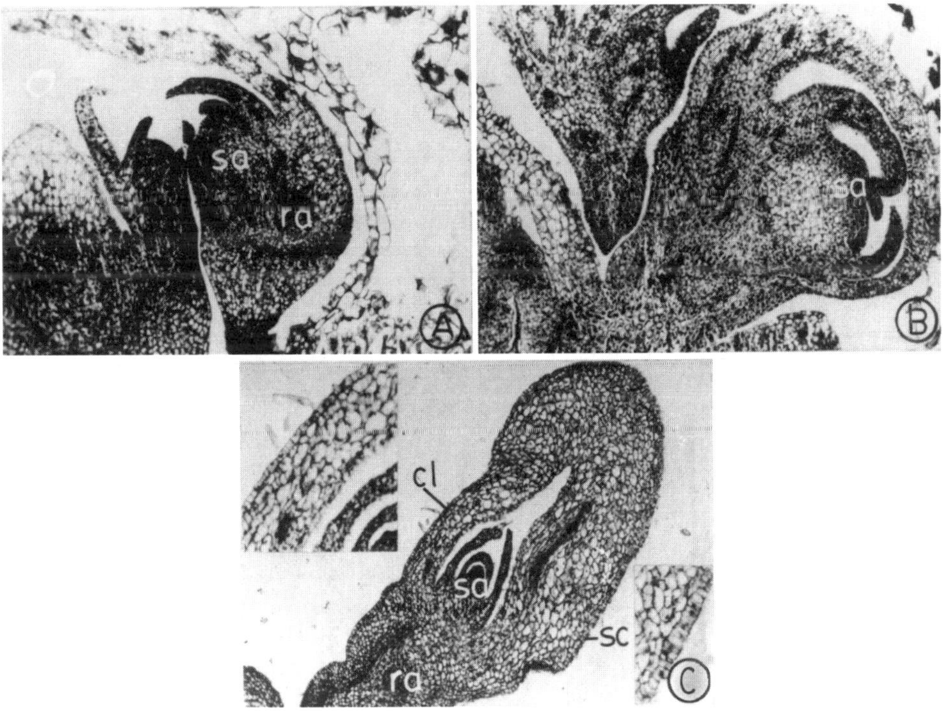

Figure 5. *D. strictus:* Sections of embryogenic callus with embryoids obtained from seeds cultured B_5+2,4-D (3×10^{-5}M) for five weeks.
A-C. L.S. of somatic embryos. Note the presence of the shoot apex (sa) and root apex (ra). The coleoptilar (cl) and scutellar (sc) regions can be seen clearly in C. These are enlarged in the insets.

sub-epidermal layer or procambium. In our work, the origin of somatic embryos has been observed from cambium-like layer derived from vascular bundles related to procambium.

REFERENCES

Emons, A.M.C. 1994. Somatic embryogenesis: cell biological aspects. Acta Botanica Neerl, 43, 1–14.

Fransz, P.F.; Schel, J.H.N. 1991. An ultrastructural study on the early development of *Zea mays* somatic embryos. Canadian Journal of Botany, 69, 858–865.

Gamborg, O.L.; Miller, R.A.; Ojima, K. 1968. Nutrient requirements of suspension cultures of soybean root cells. Experimental Cell Research, 50, 151–158.

Haccius, B. 1978. Question of unicellular origin of non–zygotic embryos in callus cultures. Phytomorphology, 28, 74–81.

Hunault, G. 1978. Recherches sur le comportement des fragments d'organes et des tissue de monocotyledones cultives in vitro I. Considerations historiques. Revel Generale de Botanique, 85, 305–351.

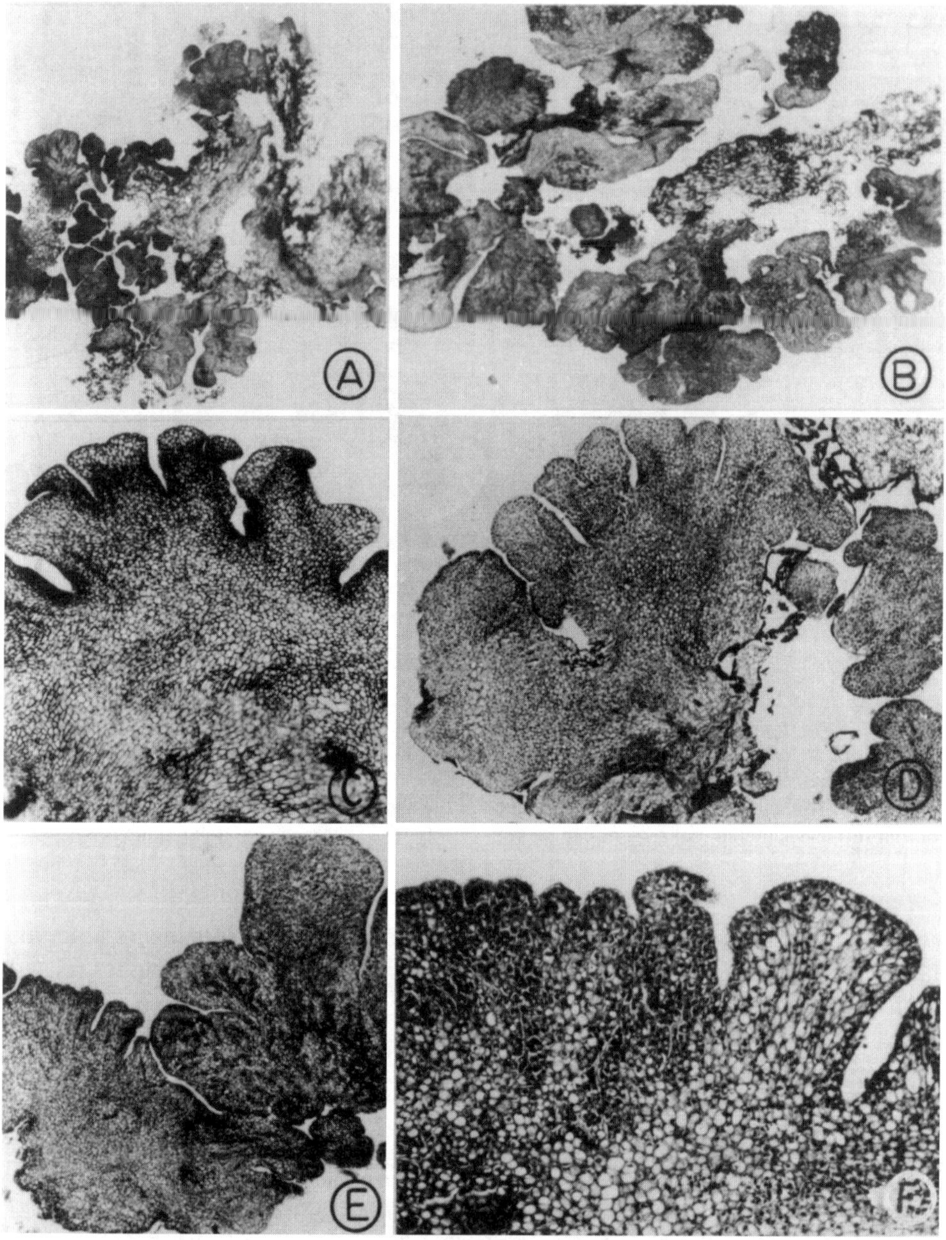

Figure 6. *D. strictus:* Sections of secondary embryoids obtained on B_5+2,4-D (3×10^{-5}M)+BAP (10^{-6}M) after 3 weeks of culture (Sub-cultured from same medium)
A-B. Low magnification photographs of embryogenic callus showing numerous secondary embryonal masses.
C-F. Sections of the embryogenic callus showing secondary embryonal masses.

Jones, T.J.; Rost, T.L. 1989. The developmental anatomy and ultrastructure of somatic embryos from rice (*Oryza sativa* L.) scutellum epithelial cells. Bot. Gazette, 150, 41–49.

Magnusson, I.; Bornman, C.H. 1985 Anatomical observation on somatic embryogenesis from scutellar tissues of immature zygotic embryos of *Triticum aestivum*. Physiologia Pl., 63, 137–145.

McCain, J.W.; Hodges, T.K 1986. Anatomy of somatic embryos from maize embryo cultures. Bot. Gazette, 147, 453–460

McCain, J.W.; Kamo, K.K.; Hodges, T.K. 1988. Characterization of somatic embryo development and plant regeneration from friable maize callus cultures. Bot. Gazette, 149, 16–20.

Michaux–Ferriere, N.; Grout, H.; Carion, M.P. 1992. Origin and ontogenesis of somatic embryos in *Hevea brasiliensis* (Euphorbiaceae). American Journal of Botany, 79, 174–180.

Nonohay, J.S.; Mariath, J.E.A.; Winge, H. 1999. Histological analysis of somatic embryogenesis in Brazilian cultivars of barley, *Hordeum vulgare* vulgare, Poaceae. Plant Cell Reports, 18, 929–934.

Oka, S.; Saito, N.; Kamaguchi, H. 1995. Histological observation on initiation and morphogenesis in immature and mature embryo derived callus of barley (*Hordeum vulgare* L.). Ann. Bot., 76, 487–492.

Pannetier, C.; Buffard-Morel, J. 1986. Coconut plant (*Cocos nucifera* L.). *In* Bajaj, Y.P.S., ed., Biotechnology in agriculture and forestry I. Trees. Springer-Verlag, Berlin, Germany. pp 430–450.

Rao, I.U.; Ramanuja Rao, I.V.; Narang, V. 1985. Somatic embryogenesis and regeneration of complete plantlets in the bamboo *Dendrocalamus strictus*. Plant Cell Reports, 4, 191–194.

Ryschka, S.; Ryschka, U.; Schulze, J. 1991. Anatomical studies on the development of somatic embryoids in wheat and barley explants. Biochem. Physiol. Pflanz., 187, 31–41.

Schwendiman, J.; Pannetier, C.; Michaux-Ferriere, N. 1990. Histology of embryogenesis from leaf explants of the oil palm *Elaeis guineensis* Jacq. Ann. Bot., 62, 43–52.

Taylor, M.G.; Vasil, I.K. 1996. The Ultrastructure of somatic embryo development in pearl millet (*Pennisetum glaucum,* Poaceae) American Journal of Botany, 83, 28–44.

Vasil, V.; Vasil, I.K. 1982. The ontogeny of somatic embryo of *Pennisetum americanum* (L.) K. Schum. in cultured immature embryos. Bot. Gazette, 143, 454–465.

Wernicke, W.; Portykus, I.; Jhamas, E. 1982. Morphogenesis from cultured leaf tissue of *Sorghum bicolor* – the morphogenetic pathways. Protoplasma, 111, 53–62.

Williams, E.G.; Maheswaran, G. 1986. Somatic embryogenesis: factors influencing coordinated behavior of cells as an embryogenic group. Ann. Bot., 57, 443–462.

Bamboo for Sustainable Development, pp. 397–405
Eds. A. Kumar, I.V. Ramanuja Rao and Ch. Sastry

Development of sustainable supply and use of tropical bamboos

R.J. MURPHY
Department of Biology, Imperial College of Science, Technology and Medicine, London, The United Kingdom

Abstract—A project, supported by the European Commission under its INCO-DC program, is under way to assist in increasing the sustainable supply, use and quality improvement of selected bamboo species in Southeast Asia. It is being undertaken by the following institutes: Imperial College, London; Science University, Malaysia; Forest Research Institute Malaysia; Forest Research and Development Center, Indonesia; University Los Baños, the Philippines; BFH, Hamburg, Germany; and Oprins Plant NV, Belgium. Emphasis is placed on the use of bamboo in higher technology applications, as these are likely to benefit the most from improvements in raw material quality through silviculture and management. The project has the following, general objectives:

1. To determine the influence of management practices on basic culm properties in managed, natural stands and plantations;
2. To optimize the processing and performance of higher technology bamboo products based on 'improved' raw material; and
3. To disseminate the information to user groups.

The approach of this project is to conduct integrated research on raw material production, culm quality and product technology for selected bamboo species. Several different experimental methodologies will be used, ranging from fundamental studies of culm anatomy (including electron microscopy) through molecular biology methods of PCR for genetic variability assessment, to preservative treatment methods and mechanical property testing. Preliminary results are reported here.

INTRODUCTION

In 1996, a consortium of institutes, universities and a commercial company developed a joint research project proposal on sustainable management and quality improvement of tropical bamboos. The proposal was submitted to the European Community (EC) Research and Technical Development (RTD) Program 'Cooperation with Third Countries and International Organizations' (INCO-DC) via DGXII of the Commission of the European Communities. The project was reviewed favorably, awarded, and commenced in November 1997. The purpose of this paper is to outline the aims and objectives of the project and to summarize the progress to date. Preliminary results are also presented.

The INCO-DC Program is intended to add value to EC-RTD via targeted RTD cooperation and synergy with other EC activities, to strengthen the EC's science/technology base and to support implementation of other EC policies. Three sectors of general importance were identified in the INCO-DC program within the Fourth Framework program: Sector I - Sustainable Management of Renewable Natural Resources; Sector II - Sustainable Improvement of Agricultural and Agro-industrial Production; and Sector III - Health. The project addressed issues relevant to Sector II in respect of cooperation with partners in Southeast Asia.

Bamboo culms normally attain maximum height within the first year of growth. The culm then undergoes a further maturation period of several years, followed by senescence (Parameswaran and Liese 1976; Murphy and Alvin 1992; Liese and Weiner 1996; Murphy and Alvin 1997). During this time, several important material changes occur, such as increases in density and cell wall thickness, lignification and fluctuations in starch content. These changes are related to location in the culm (Alvin and Murphy 1988) and probably contribute significantly to changes in properties of the material – for example, mechanical strength and durability with maturation (Sulaiman 1993; Sulaiman and Murphy 1995). A more detailed scientific knowledge of the maturation process and its relationship with material properties of the culm and its products is urgently needed.

Traditional and modern developments in bamboo utilization have been reviewed by Liese (1985) and, more recently, with regard to Malaysia by Abd. Razak et al. (1995). However, such publications and the activities of the International Network for Bamboo and Rattan (INBAR) indicate that many aspects of bamboo silviculture, genetics, structure and processing warrant continued investigation (for instance, see INBAR 1996). The great potential for cultivation of bamboos as plantation species is recognized but appropriate silviculture regimes for the region need further development. Bamboos have also been typically considered as a low-quality material owing to variability and problems with biodeterioration. Genetic variability in the resource is also probably great and the degree variability and heterozygosity warrant further investigation (Gielis et al. 1996).

As a fiber material, bamboo has clear potential for widespread use and products of high quality, produced sustainably, are considered to have good potential for export success in markets such as the EC. Improved, sustainable utilization of bamboo for such products will allow future diversification of the uses of this renewable resource.

The partner organizations in the project are listed in Table 1, and the scientific teams are indicated in Table 2. Professor Waiter Liese, University of Hamburg and BFH, has an active role in the project as Senior Adviser.

It can be seen from Table 1 that there are equal numbers partners from EC countries and from Southeast Asia. This appeared most appropriate for the present project, as it ensured a reasonable geographical spread and R&D support in the growing regions for tropical pachymorph bamboos, as well as a diversity of input from EU-based organizations. The project commenced on 1 November 1997 and is scheduled for a four-year period in the first stage. Completion will be in October 2001 with final reporting by December 2001.

Table 1.
Partner Organizations in INCO-DC Project on tropical bamboos

No.	Organization	Type	Country	Role in the project
1.	Imperial College (IMPCOL)	University	UK	Coordinator, liaison with EC, research
2.	Universiti Salns Malaysia (USM)	University	Malaysia	Research
3.	Forest Research Institute Malaysia (FRIM)	Research institute	Malaysia	Research, dissemination
4.	Forest and Nature Conservation R&D Center (FNCRDC)	Research institute	Indonesia	Research, dissemination
5.	University of the Philippines at Los Baños (UPLB)	University	Philippines	Research, dissemination
6.	Bundesforschungsanstalt für Forst und Holzwirtschaft (BFH)	Research	Germany	Research
7.	Oprins Plant N.V. (OPRINS)	R&D, commercial	Belgium	Research

Table 2.
Scientific staff on INCO-DC Project on tropical bamboos

Organization	Scientist/official responsible	Team members
IMPCOL	Dr Richard Murphy r.murphy@ic.ac.uk	Mrs. Cristina Grirsch
USM	Dr Othman Sulainman othman@usm.my	Dr Rokiah Hashim
FRIM	Dr Azmy Mohamed azmy@frim.gov.my	Dr Rashid Malek, Dr Abd. Latif, Dr Razak Wahab
FNCRDC	Ir Sutiyono slitbang@indo.net.id	Ir Dwiatmo Siswomartono, Dr A. Ng Gintings, Mr. Mohammad Rosid
UPLB	Dr Elvira Fernandez elcf@mudspring.uplb.edu.ph	Dr Armando Pajilon
BFH	Dr Rolf-Dieter Peek rpeek@aixh0401.holz.unihamburg.de	Dr Hans Leithoff
OPRINS	Dr Johan Gielis joprins@oprins.be	Mr. Jan Oprins

AIMS AND OBJECTIVES

The overall objective of the project is to assist in increasing knowledge on the sustainable supply, use and quality improvement of selected bamboo species in Southeast Asia. The project will address this through the following specific actions:

1. Assessment of the effect of fertilizer on the productivity and quality of natural and planted clumps of *Gigantochloa* and *Dendrocalamus* species in comparison with controls;
2. The influence of the above on culm properties such as density, maturation (cell wall thickening, lignification), fiber/ground tissue ratios, and internode length. Molecular markers will be used to study the degree of genetic variability in natural stands;
3. Improve the processing and performance of technical bamboo products based on selected raw material;
4. Assessment of the natural durability and effectiveness of preservative treatments (including novel approaches) for the resistance of culm raw material and laminates/composites against biodeterioration. Field trials will be established in Southeast Asia and EU and durability will also be assessed in accelerated laboratory tests; and
5. Disseminate the information to potential users (such as growers, manufacturers and R&D organizations).

In order to achieve the aims and objectives, the activities in the project have been organized into five major work packages, each broken down further into subtasks. These are summarized in Table 3.

INITIAL RESULTS

During the first year of the project, most of the research effort has gone into organizing the natural stands of *G. scortechinii* and *D. pendulus* in Malaysia for subsequent fertilizer application and harvesting, establishment of plantation material of *G. levis* and *D. asper* in the Philippines, and management of *D. asper* and *G. apus* in Indonesia.

The experimental design for the field sites is: for each species, the block is divided into 4 replicates, each subdivided into 3 subplots (fertilizer type) in which are 4 sub-subplots (application rate) in a split plot design. The fertilizer types and rates are slightly different between sites but are designed to be comparable. For example:

Malaysia (natural stand)	3 fertilizers (palm oil mill effluent, 'Avanti' (chicken dung base) and 'PeatGro' – applied at 0, 4, 8 and 12 kg per annum per clump
Philippines (planted)	3 fertilizers (dried cow dung, chicken manure and rice hull ash), applied at 0, 3, 6 and 12 kg per annum per clump

Characterization and soil analysis have been undertaken as shown in Table 4.

Results from the plantation establishment of *G. levis* and *D. asper* in the Philippines are given in Table 5.

Table 3.
Organization of work packages in INCO-DC Project on tropical bamboos

Work package	Lead partner	Others	Sub-task	Status (as on Oct. 1998)
1. Silviculture	FRIM	FNCRDC	1.1 Set up plots	Commenced
		UPLB	1.2 Characterize sites	Data available
		OPRINS	1.3 Culm production	Commenced
			1.4 Natural variability	Commenced
			1.5 Management guides	–
2. Raw material properties	USM	IMPCOL	2.1 Anatomy investigation	Commenced
		FRIM	2.2 Culm physical properties	Commenced
		FNCRDC	2.3 Culm chemical properties	Commenced
		UPLB	2.4 Culm mechanical properties	–
		OPRINS	2.5 Molecular markers for superior traits	–
3. Processing	FRIM	USM	3.1 Laminate manufacture	–
		UPLB	3.2 Composites	Commenced
			3.3 Pulping	Under revision
4. Durability & preservation	FRIM	IMPCOL	4.1 Field trials on durability	–
		USM	4.2 Lab. trials on durability	Commenced
		BFH	4.3 Preservative treatments	–
			4.4 Novel heat treatments	Commenced
5. Dissemination	FRIM	All	5.1 Publications, etc.	–
			5.2 Dissemination seminar	–

Several of the other tasks have been initiated (see Table 3) and preliminary information is available. However, much of this has been concerned with establishing the research techniques and material is not yet available from fertilized clumps. For example, the anatomy studies have been commenced as part of Work package 2, sub-tasks 2.1 & 2.3. Culms of *G. scortechinii* aged 0.5, 1.5, 3.5, 5.5 and 6.5 years have been collected from Nami. The initial basic anatomical properties – such as cell wall thickness and fiber length – and chemical properties of unfertilized culms are being evaluated. Within this, a transmission electron microscopy (TEM) study of internode 7 has commenced (see Fig. 1).

The light and electron microscopy studies will be used, in conjunction with other data on culm properties, to assess the influence of the differing fertilizer treatments on culm development and maturation. Such data are also necessary to begin developing the definition of 'superior' culms and associating these characteristics with genetic variability within the natural populations. Ultimately, it is intended that characterization of the extent of genotypic variation in natural populations of bamboo species, coupled with data on the 'quality' of culms for end uses and molecular markers for 'superior' traits, will permit rapid selection of material

Table 4.
Site characteristics Malaysia and Philippines

Characteristics	Malaysia	Philippines
Location	Nami, Chebar Besar Forest Reserve, Kedah	Laguna-Quezon Land Grant
Area	2 ha	2.2 ha
Stand type	Natural *G. scortechinii/D. pendulus*	Plantation of *G. levis* and *D. asper*
Vegetation type	Previous dipterocarp forest	Coconut, *Imperata cylindrical,* bush vegetation, fern (*Dicranopteris* sp.), trees (mainly *Syzygium* sp.)
Elevation	200 m above sea level	500 m above sea level
Annual rainfall	2 500 mm	4 000 mm
Soil type	Acrisols consolidated clastic rocks (red and yellow latosols of variable fertility)	Antipolo, Luisiana, macolod clay series (very acidic with low nutrient content)
Soil analysis		
- pH	5.4	4.8
- organic matter	1.1% (organic carbon)	6%
- nitrogen	0.1%	0.19%
- phosphorus	8.2 ppm	1.6 ppm
- potassium	–	0.17 me/100 g soil
- CEC	–	27.78 me/100 g soil
- sand	–	18%
- silt	–	24%
- clay	–	58%
- textural grade	–	Clay
- bulk density	–	0.77
- WHC	–	99%

Table 5.
Survival of propagules in plantation establishment of *Gigantochloa levis* and *Dendrocalamus asper*

Species	Collection date	Total no. of propagules	Number potted	Rooting	Surviving number	Survival rate
G. levis	Feb-May	1 288	827	64%	611	74%
D. asper	Feb-May	723	543	75%	389	72%

for propagation and bamboo plantation management, or to inform management of natural bamboo stands. Clearly, for such a long-range goal to be achieved, it is necessary to have an integrated knowledge base on the species concerned, its propagation and management and its potential applications. This project is an attempt to begin developing such knowledge.

There are also a number of fundamental questions that are raised by these objectives such as, how to define culm 'quality' and when are culms and clumps 'mature'? The scientists working in this project will have to address these and similar questions throughout the project and are keen to discuss such issues with colleagues in the research and bamboo community through meetings like this International Bamboo Congress and Workshop.

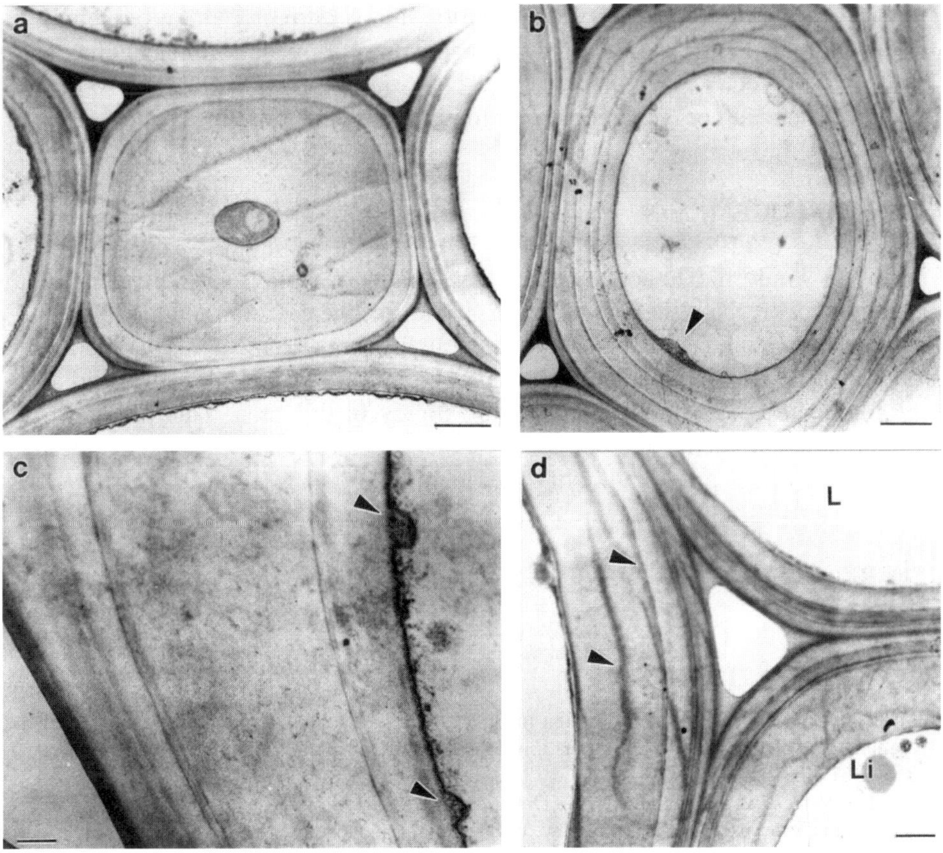

Figure 1. TEM micrographs of fibers from free fiber strand, 0.5 year old culm
a) Tip end of a fiber from a free fiber strand from a vascular bundle, showing a very thick wall with few lamellae and a very small lumen. Surrounding fibers appear to be larger and have much thinner walls. Scale bar: 2 μm
b) Small fiber cell exhibiting polylamellated cell. Cytoplasmic remnants can be seen in the periphery of the cell (arrowheads). Scale bar: 2 μm.
c) Detail of the cell wall of a fiber from a free fiber strand. Arrowheads show vesiculating plasma membrane. Scale bar: 0.25 μm.
d) Detail of the cell walls of three large fiber cells from the same specimen shown in (a) and (b). A lipid droplet is present (Li). Arrows indicate folds in the resin. Scale bar: 1.0 μm.

CONCULSION

The Commission of the European Communities, through its INCO-DC RTD program, supports the project. The project aims to begin developing an integrated body of information for selected species of tropical pachymorph bamboos in Southeast Asia. The information being generated concerns:

- production of culm raw material;
- influence of growing practice and genetic variability on culm properties; and
- relationships between culm properties and performance of potential products.

The project is initially of four years duration and is currently at the beginning of its second year. Both natural stand management and plantation management are being investigated. Dissemination of results and communication with other researchers, users and organizations is an important element of the project.

ACKNOWLEDGEMENTS

The author would like to acknowledge that this research has been carried out with financial support from the Commission of the European Communities INCO-DC Program, 'Sustainable management and quality improvement of bamboos and products'. The content does not necessarily reflect the Program's views and in no way anticipates the Commission's future policy in this area. The author would also like to acknowledge the help of all project partners and Prof. Walter Liese in the preparation and presentation of this paper.

REFERENCES

Abd. Razak, O.; Abd. Latif, M.; Liese, W.; Haron, N. 1995. Planting and utilization of bamboo in Peninsular Malaysia. FRIM Res. Pam. No. 118, Forest Research Institute Malaysia, Kuala Lumpur, Malaysia. 118 pp.

Alvin, K.L.; Murphy, R.J. 1988. Variation in fiber and parenchyma wall thickness in culms of the bamboo *Sinobambusa tootsik.* International Association of Wood Anatomists Bulletin, n.s., 9, 353–361.

Gielis, J.; Everaert, L.; Goetghebeur, P.; De Loose, M. 1996. Bamboo and molecular markers. *In* Ramanatha Rao, V.; Ramanuja Rao, I.V. ed., Bamboo, people and the environment, Vol. 2, Biodiversity and Genetic Conservation. Proceedings of the Vth International Bamboo Workshop, Ubud, Bali, Indonesia, 19–22 June 1995. International Network for Bamboo and Rattan, New Delhi, India. pp. 45–67.

INBAR (International Network for Bamboo and Rattan). 1996. INBAR Newsletter (several items), Volume 3, No 4. International Network for Bamboo and Rattan, New Delhi, India.

Liese, W. 1985. Bamboos - biology, silvics, properties, utilization. Deutsche Gelleschaft für Technische Zusammenarbeit (GTZ), Nr. 180, TZ Verlagsges., Roßdorf 1, 132 pp.

Liese, W.; Weiner, G. 1996. Ageing of bamboo culms. Wood Science and Technology, 30, 77–89.

Murphy, R.J.; Alvin, K.L. 1992. Variation in fiber wall structure of bamboo. IAWA Bulletin, n.s., 13, 403–410.

Murphy, R.J; Alvin, K.L. 1997. Fiber maturation in bamboos. In G. Chapman, ed., The bamboos. Linnean Society, London, UK. pp. 293–303.

Parameswaran, N.; Liese, W. 1976. On the fine structure of bamboo fibers. Wood Science and Technology, 10, 231–246.

Sulaiman, O. 1993. The development of soft rot decay in bamboo cell wall. Ph.D. thesis, University of London, London, UK.

Sulaiman, O.; Murphy, R. 1995. Ultrastructure of soft rot decay in bamboo cell walls. Material u. Organismen, 29(4), 241–253.

Bamboo for Sustainable Development, pp. 407–419
Eds. A. Kumar, I.V. Ramanuja Rao and Ch. Sastry

Criteria for the selection and management of bamboo shoot species for Australia

VICTOR CUSACK

Bamboo World, Wadeville, New South Wales, Australia

Abstract—Established viable edible shoot plantations use species chosen for their high productivity in the chosen environment. The species also produce timber culms as a second crop, because of the management system chosen. Different species are traditionally used in different countries. The reasons for these choices must be considered when selecting species and management systems for new growing areas. Environmental aspects, production capacity and quality will be the deciding factors in these choices.

INTRODUCTION

Australia is establishing bamboo plantations for high-quality edible shoots and timber culms Cusack 1997). To ensure that the most suitable and productive species for the diverse climatic conditions in Australia are planted, the author studied different species grown in large plantations in different countries. These include areas in China, Thailand and, to a lesser degree, Taiwan. This paper discusses the conclusions reached after visiting research centers and plantations in those countries a number of times. It covers some of the significant factors to be considered when choosing species and management techniques for large plantations.

SPECIES SELECTION AND CRITERIA

Three species stood out in the final analysis, mostly because they exist in larger plantings than any other species – *Dendrocalamus asper*, *Dendrocalamus latiflorus* and *Phyllostachys heterocycla* var. *pubescens* (moso bamboo). They are also by far the most researched edible shoot-producing bamboos and the most important to the economies of the regions or countries where they are grown (Cusack 1997). All three are capable of both high productivity and yield superior quality crop.

Two other species will be discussed because they have advantages applicable to specific environments despite being less relevant to Australian conditions – *Dendrocalamopsis oldhamii* (commonly referred to as *Bambusa oldhamii)*, which is a less productive sympodial species with a higher cold tolerance, and *Phyllostachys praecox,* which has gained recent economic importance to Anji province in China.

While there are many other species cultivated in places where shoots are highly prized (such as *D. brandisii, D. hamiltonii, D. beecheyana, Phyllostachys prominens*, etc.), the three species mentioned initially represent, in the author's view, the most important and relevant to Australia. It must also be mentioned that there appears to be very little relevant research on alternative species.

The most important factor in choosing an appropriate species is the plantation environment; more specifically, the lowest annual temperature expected, and the extent and time of rainfall. This, to a great degree, dictates the selection of a sympodial or monopodial species, or both, because of the following reasons.

1. Sympodial bamboos suitable for high productivity of superior quality edible shoots have low minimum temperature limits (-4° to -5°C), and sustain damage if the temperature drops even once below those limits. The qualified exception to this is *Bambusa oldhamii* (see later text), which has a lower limit of -9°C, but is only half as productive as some larger sympodial species.
2. Monopodial species are far more cold tolerant and can be grown productively even where temperatures drop to -15°C.
3. Because of their natural yearly dormant-to-active cycle, all bamboo species require a major rainfall or water intake before and during their natural shoot season. Their productivity will fall unless this happens. Sympodial species suit monsoon rainfall patterns, late summer through early winter. Monopodial species require their major rainfall, in contrast to sympodial species, in late winter and spring, with some rain in early summer for rhizome development.
4. Subject to low temperature limitations, it is possible to grow sympodial and monopodial bamboos in rainfall areas not suited to their natural cycle, but this requires massive irrigation (except in pockets not typical of the general environment of the area). To maximize the crop, both sympodial and monopodial bamboos require preferably a minimum of 1 500 mm rainfall per year with more than 60% falling during the active shoot period specified earlier. A large plantation would need a substantial river source and quite expensive equipment to constantly supply 150 mm/ha/month if it does not receive its natural rainfall.

Areas fortunate in having high rainfall spread over all seasons (or with access to abundant irrigation) may benefit from planting both sympodial and monopodial species because they will have two separate harvest periods. Taiwan grows both sympodial (*D. latiflorus*, *B. oldhamii* and *B. edulis*) and monopodial (moso) species, the latter mostly grown in cold areas, but does not grow *D. asper.* The choice of a priority species to plant should take into account the following factors:

1. Monopodial species have a very short shoot season – about five weeks peak period and eight weeks total – whereas sympodial species have a shoot period ex-

tended by the management system to about 25 weeks, and a less longer peak period.

2. Large monopodial species like moso have shoots that require harvesting by digging down and cutting off fairly close to the rhizomes, which are located on average about 250–700 mm below ground level (Oshima 1931)[1]. Each shoot averages 1.5 kg (Oshima 1931). Digging is not a problem for smaller monopodial shoots of species such as *Phyllostachys praecox*, but they require more labor to harvest the large number of small shoots needed to make up the tonnage. Each 10 tons of *P. praecox* shoots harvested, for example, will have about 50–60 thousand shoots compared with 6–8 thousand moso shoots. In countries with high labor costs, this needs to be considered.
3. Sympodial bamboo shoots are harvested much closer to ground level. The species mentioned earlier as most productive have much larger shoots on average than the monopodial species (averaging 1–5 kg or more depending on cultivar). The combination of more weight per shoot and easier harvesting (no digging) results in significantly less labor needed to harvest the crop (an average of 3 kg per sympodial shoot easily cut off versus twice as many moso shoots averaging 1.5 kg to dig up).
4. *P. pubescens* (moso) takes twice as long to produce a commercial crop of equal volume as the sympodial *D. asper* (approximately 5–6 years for sympodial and 8–10 years for the large monopodial moso). *P. praecox* reaches a high tonnage faster than moso, but is disadvantaged by the number of shoots to harvest and the short shoot season.
5. The three selected species (both monopodial and sympodial) appear to be capable of producing much the same high shoot capacity/hectare/year (and timber crop when mature). Conservative annual shoot crop estimates would be 10 tons/ha (Thammincha et al. 1997)[2]; some published figures put this at 15 to 20 tons/ha/y with intensive cultivation and lots of water (Dransfield and Widjaja 1995). The management system developed for sympodial species, in particular the Thai system used for *D. asper*, does not require the clump or plantation to be renewed by replacement. Moso is normally progressively deep-ripped and re-established in 6-m wide bands in 20 years after planting (Lou Yiping and Xiao Jianghua 1997). Considered together with its longer establishment period and its tendency towards alternate crop bearing (Wu Liangru and Xiao Jianghua 1997), moso does have its apparent annual productivity reduced if those factors are taken into account. It appears that this need to renew an aging stand would apply to all monopodial species, as the long monopodial rhizomes must constantly renew themselves because of their limited (2–3 years) shoot producing life (Ding Guohua and Ding Yulong 1997). The ground eventually becomes clogged with non-shoot producing rhizomes, useful only as energy storage and collection systems, leaving little room or energy for new rhizomes.

---

[1] Also, Fu Maoyi in personal communication 1997.
[2] Also, Fu Maoyi in personal communication 1997.

Sympodial rhizomes, being very short, develop a slow progressive growth that remains within a tight circle. The Thais have developed management and feed systems that renew and sustain each clump without having to destroy the rhizome system (see later text). This enables the life of a plantation to be almost infinite (the author saw one *D. asper* cv. "Dam" plantation in Thailand claimed to be more than 100 years old). In China, management systems applied to *D. latiflorus* are different and more intensive. They may not be more productive, although *D. latiflorus* capacity claims are a little higher when planted closer than *D. asper.*

6. Another factor that may influence species choice is plantation access and maintenance access, both during the establishment and maturing period and after maturity. Using vehicles or tractors in monopodial plantations is likely to damage rhizomes and developing shoots. Hence, weed clearing, cultivation, harvesting of shoots and culms, mulching and fertilizing must be by manual labor, which is expensive in developed countries. This is particularly time-consuming during the longer establishment time of moso because there is no canopy to retard weed growth and it has to be cleared by hand or hand-driven light machinery. With sympodial species, subject to planting distances being sufficient (which is recommended for the selected species) a tractor can be used throughout the plantation area for tasks such as cultivation and harvesting.

When establishing any bamboo plantation, it is critical to study the flowering cycle of the bamboo chosen for planting, to protect (by planting either the right cultivar or species) the plantation from the massive destruction that most bamboos suffer when these rare flowerings occur.

PLANT CHARACTERISTICS

Dendrocalamus asper (Schultes f.) Backer ex Heyne

D. asper is a large-leafed and fairly vertical sympodial species. The largest cultivars are capable of growing 30 m in high and 30 cm in diameter, though not normally as large in plantations. It is one of the strongest and most researched bamboos in the world, and of major importance as a large structural bamboo in Southeast Asia. The current recommended planting density for *D. asper* is about 200 plants/ha, planted on a 7×7 m or 8×6 m grid.

There were 54 000 ha planted in Thailand in 1994 (Thammincha 1996), consisting of (more than) five different Thai and Indonesian cultivars that have distinctly different characteristics, shoot sizes, flavors, culm diameters and productivity, as Table 1 indicates.

It is vital that the correct cultivar of this bamboo is planted, or the plantation might be destroyed by a gregarious flowering cycle before its economic potential could be realized (see notes later this chapter). The only cultivar that has gregari-

Table 1.
Characteristics of Thai and Indonesian *D. asper* cultivars

Characteristic	Pai tong mo	Pai tong dam	Pai tong keo	Pai tong nu	Indonesian
Shoot weight	5–10 kg	3–6 kg	1–4 kg	1–3 kg	3–7 kg
Shoot color	Dark brown to purple	Dark brown and shiny	Black	Dark brown and shiny	Dark brown
Shoot flesh	white, sweet minor fiber	white, sweet and coarse	creamy with fine and crisp	white, sweet, minor fiber	creamy with fine and crisp
Culm diameter	12–16 cm	9–12 cm	5–12 cm	3–6 cm	9–16 cm

ously flowered is *pai tong keo* (see above), which is unlikely to flower for the next 100 years or so. It has some advantages over the other cultivars:

1. It produces shoots about one month earlier than the others;
2. It is considered the most productive for shoots;
3. It also produces a range of small-sized shoots (1–4 kg), which may help marketing (large shoots to restaurants and smaller, less expensive shoots for households); and
4. The culm has a comparatively smaller diameter, and provides a higher quantity and tonnage than the other cultivars (the very large diameter timber produced by some cultivars may not be as useful as the 50–120 mm diameter culm range produced by *pai tong keo*).

Although *pai tong dam*, with its larger shoot and culms, provides shoots of a slightly better quality, the difference is small enough to ignore. All bag-grown *D. asper* shoots are of superior quality and can be eaten raw in a salad or introduced into a recipe without pre-cooking. *Pai tong keo* was the most popular shoot species established in Thailand before its gregarious flowering and death in 1994 (40 000 out of 54 000 ha), and is certainly the cultivar being currently planted there, with 35 000 ha being planted this year alone![3] The maximum production in one large area of Thailand was achieved from *pai tong keo* was 18.7 tons/ha/y of edible shoots, and 23 tons of 12% dried timber (Thammincha et al. 1997)! A 10 ton/ha/y shoot production and a little more in timber each year should certainly be achievable with rational management.

As with edible shoots of all species, only half or even less of a shoot can be claimed for edible purpose; sometimes, this is down to 38%, depending on the growing technique.

[3] This particular cultivar was successfully tissue-cultured by Bamboo World and was produced in thousands for plantation establishment.

Dendrocalamus latiflorus (Munro) syn. Bambusa latiflorus (Munro) Kurz & Sinocalamus latiflorus (Munro) McClure

D. latiflorus is a large, very vertical bamboo, capable of growing 25 m high with culms of 20 cm diameter. There is more than 90,000 ha in Taiwan and an even larger amount planted in mainland China (statistics not available).

D. latiflorus culms are marginally straighter than those of *D. asper,* but culm wall is a little thinner and not quite as strong as a structural bamboo. *D. latiflorus* produces a slightly bitter shoot, but still of high quality. Some people, such as the Vietnamese, prefer bitter shoots. Perhaps *D. latiflorus* statistics claim a slightly larger yield than *D. asper* in Taiwan, but that may be because of different management techniques. Planting costs are somewhat higher than for *D. asper* because the recommended 6×6 m *D. latiflorus* planting requires 277 plants/ha as against about 200 plants for *D. asper.*

D. latiflorus shoots average 4 kg each and have basically the same diameter as the culms they produce (about 120–170 mm), unlike *D. asper* that produces shoots larger in diameter than the culms that would grow from them.

Available records going back more than 100 years indicate that *D. latiflorus* is a sporadically flowering bamboo. It is not expected to create a disaster by flowering gregariously and dying in a new plantation; however, dealing with its flowering habit does require a management system. In Chinese plantations, about 6–8% of *D. latiflorus* clumps produce flowers in a 4–6 years cycle on one or more culms dedicated to flowering. Those clumps will not produce marketable shoots that year, causing a small reduction in the yield. Usually, those flowering culms are cut off to help the bamboo produce a normal shoot crop in the following year. Being a sporadically flowering bamboo, almost all the seeds produced are non-viable.

Phyllostachys heterocycla var. pubescens (Mazel) syn. Phyllostachys edulis (Carr.) Riviere

Commonly known as moso in Japan and the western world, *Phyllostachys heterocycla* var. *pubescens* is the bamboo shoot most consumed in Japan, Korea and the mid to northern half of China. It is the largest of the monopodial species. Approximately 2.6 million hectares of mostly natural stands exist in China (Ding 1996)[4] and significant plantings in Japan, Korea and Taiwan, where it grows mostly above 900 m elevation (Oshima 1931). This essentially cold-climate bamboo is slow and difficult to establish by comparison with the sympodial species, taking 8–10 years to produce a mature crop, and about 5–6 years for a first small crop. Its somewhat aggressive, invasive habit also requires a management control system. Its culms, measuring 160–180 mm in diameter, can reach a height of 26 m but are often cropped to prevent damage from heavy snow.

---

[4] Also, Fu Maoyi in personal communication 1997.

Moso is a gregariously flowering bamboo with a 70–75 year cycle,[5] but with many mutant variations or cultivars. It is often flowering in different parts of China, with consequential damage to plantations or forests. The moso originally planted in Australia came from seed introduced in the 1980s, and therefore is unlikely to gregariously flower before 2050 or so. Planting stock should preferably come from a cultivar that has flowered as recently as possible to ensure a maximum plantation life before another gregarious flowering takes place.

There is substantial evidence that a well-run moso plantation can produce more than 10 tons of shoots/ha/y (an extrapolation of most common overseas yield figures of 8–15 tons/ha, with some claims that 20 t/ha/y is possible). The shoots produced are generally 75–125 mm in diameter and weigh about 1.5 kg each (6 600 shoots per 10 tons). Both the Chinese and the Japanese prefer smaller early harvested *moso* shoots that weigh 0.5–1 kg each. Early dug winter shoots or shoots harvested from a young plantation (leaving the larger ones to develop the plot) will be considerably smaller.

Winter shoots are highly valued but expensive to harvest. They have to be dug out carefully from well below ground level. These very early shoots, which have just begun to swell, are covered in a hard casing of yellow culm leaf, but have very tender meat and a special flavor (Fu Maoyi 1996). Care must be taken in harvesting winter shoots because if the rhizome is damaged, it causes the shoot crop from that rhizome system to be reduced for up to three years. About 70% of moso rhizomes are 30–70 cm beneath the surface (Fu Maoyi 1996) [6]. One cannot see winter shoots harvesting being viable in Australia because of the cost of harvesting them, but normal moso shoots are certainly very viable. The Chinese pay about USD 2.00/kg for winter shoots' or smaller underground shoots, but only USD 0.70/kg for shoots harvested after they emerge from the ground (1997 prices).

The following figures apply to an intensively cultivated, efficient moso plantation run for edible shoots. Culms that are 5–7 years old (less vigorous) are felled at the rate of 250–350 culms/year to leave a fairly constant number of mature, healthy culms (1 600 to 2 500 culms/ha), giving a space of 2–2.5 m between each culm. The spacing depends on soil quality and growing conditions, and may be decreased to 1.5×1.5 m (4 440 culms/ha) in poor soil to produce more small-diameter shoots.

In a very intensively cultivated and efficient moso plantation run for timber in Anji by a research instititute, the culms are spaced at about 1.3×1.3 m (5 917 culms/ha). Mature culms harvested (aged 5 years for moso timber harvesting) yield 900 to 1 275 culms/year (about 37 tons of dried wood/year), but the edible shoot yield is reduced to only 1.5 tons/year instead of 10 tons or more/year. This plantation is apparently yielding about five times as much as the average yield from natural forests. A more realistic annual yield for a properly managed, inten-

---

[5] Fu Maoyi in personal communication 1997.

[6] Also, Fu Maoyi in personal communication 1997.

sively cultivated plantation will be about 20–30 tons/ha of timber (Shi Quantai et al. 1987), or about 75% of the figure achieved in the research institutes well managed plantation.

Bambusa oldhamii (Munro) syn. Dendrocalamopsis oldhamii (Keng f.)

There are an estimated 5 million *B. oldhamii* plants in Fujian Province, but because it is a relatively small bamboo (100 mm in culm diameter) planted on a 3×4 m grid, this equates to only about 6 000 ha. It is unlikely to produce more than 6 000 kg/ha/year of shoots. That would mean harvesting about 15 000 to 20 000 shoots/ha because average shoot weight is less than 400 g. The timber is straight and attractive, but too thin walled for structural work. The bamboo has only one advantage when compared with *D. asper* or *D. latiflorus* – its apparent cold tolerance up to -9°C. If the site is too cold for the larger and more productive sympodial bamboos, and the grower does not want to plant the monopodial moso, *B. oldhamii* is an option; otherwise, it is not a practical choice.

Phyllostachys praecox (C.D. Chu et C.S. Chao)

This monopodial species is mentioned because 1 000 ha of land in Anji Province, China, was recently planted with *P. praecox* and the planted area is being rapidly expanded. It is a high-quality, cold-climate bamboo that realizes its full shoot crop potential faster than moso (though not as quickly as the sympodial bamboos). It is reported to be yielding shoots at the annual rate of 20 tons/ha. Most of the shoots are being exported fresh to Japan. Its disadvantages are the intensive labor required to cultivate and harvest more than 100 000 shoots per hectare within the short monopodial harvest period. *P. praecox* is unlikely to be a practical commercial alternative for a labor-expensive country.

MANAGEMENT TECHNIQUES

The management techniques vary in each country according to custom and species planted, with colder climates being forced to grow cold-tolerant monopodial species. In some cases, management choices may be dictated by established practice or custom, rather than by an analysis of techniques for efficiency or productivity for that growing environment. One example is the reluctance of some Thai growers to adopt the bag-growing technique, but going to the length of bleaching shoots after harvest to disguise their lesser quality.

In Thai plantations, growers of high-quality shoots cover them as they emerge from the ground with a black plastic planter bag of a specific size stuffed with straw or carbonized rice husk. The shoot, isolated from light, grows longer, fatter and fuller, with a higher percentage of edible meat, a whiter color and a less-fibrous texture. When the shoot is tall enough to reach the top of the bag, it begins to lift the bag off the ground, indicating that the shoot is ready to harvest! The bag is removed and placed over another emerging shoot, which causes the plant to both transfer its energy towards fattening and growing the next emerging shoot, and towards producing a replacement shoot. Normal annual mulching is applied only as deep as the tops of the rhizomes, which means the shoot is cut off, using a light flat-ended spear, virtually at ground level at an angle sloping downwards. Shoots grown in a bag can be harvested in a pristine, clean condition (a little like the Tasmanian apples grown in paper bags for the Japanese market). The only post-harvest requirement is for the slightly soiled end to be cut off cleanly before packing for market presentation.

Instead of the bag, Thai farmers sometimes use a tin of the correct height, with the top cut out. After placing the tin over the emerging shoot, it is filled up with carbonized rice husk. When the shoot begins to project from the filled tin, it is ready to harvest. The bag system is preferable because of its simplicity.

Chinese farmers cover the rhizome area with mulch about 50 cm deep to exclude the light so that the shoots have to grow through the mulch before they are ready to harvest. This system, still being used by some Thai farmers for *D. asper,* apparently has some drawbacks according to Thammincha.[7] It causes the clump to climb out of the ground more quickly because all the rhizome necks reach upward, and it is hard to see the shoot size and position properly when choosing healthy appropriately located shoots to grow into culms. Thai experts recommend that the deep mulching system should be used sparingly, certainly not more than each second year. To harvest such shoots one has to remove the mulch cover down to the rhizome to expose the shoot. In Taiwan, the deep mulch is also covered with black plastic to increase heat and humidity. Further work is needed to establish whether the bag system can be applied to *D. latiflorus* to advantage.

Sympodial shoots are harvested from each clump in a 3–4 day cycle during warm weather, and a 4–6 day cycle during cooler weather during a harvest period of about 25 weeks. More shoots are harvested during peak harvest periods, which can vary with rainfall irregularities.

The Chinese have developed more elaborate systems than the Thais for controlling the tendency of bamboos to climb out of the ground as they grow older. This phenomenon occurs because the shoot buds are all located in the side of each rhizome, causing each new rhizome and shoot to be higher than its mother. One researcher explained that they dig out about 25% of the rhizomes in a clump when it is 10–12 years old, and re-plant an offset deeper into the ground. This has the effect of re-establishing a portion of the clump deeper, with its new shoots starting deeper in the ground than the older plant section, which by that age has its rhizomes close to or projecting from the ground. Is it possible that the Thai system could replace that practice to advantage, or is it necessary because *D. latiflorus* rhizome forms are different to *D. asper*?

---

[7] Songkram Thammincha in personal communication 1998.

Fertilizers and application techniques

According to Chinese and Thai experts, fertilizer application during the shoot season causes the shoots to be soft-textured and darker-colored, lacking the crisp texture so valued in the market and given to easy discoloration and bruising. It also causes large inferior shoots to develop with excessive culm leaf content and reduced edible flesh.

In terms of total fertilizer requirements, the different species grown in different countries all appear to have the same amount applied. Total nitrogen applied in Thailand to *D. asper* is about 150 500 kg/ha/year (Thammincha et al. 1997, Cusack 1998). In intensively cultivated moso plantations in China, total nitrogen can be as high as 600 kg/ha/year.[8]

The Thais apply total nitrogen in the following manner:[9]

A. In October (2–3 months before the shoot season), 6–9 tons of manure or compost (3% nitrogen) plus 420–840 kg of 15:15:15 NPK fertilizer would be distributed per hectare. A portion of it is mixed with mulch and deposited into a 20 cm deep trench dug in a square about 1–1.5 m outside the rhizome circle. The remainder is scattered over the entire ground area surrounding the clumps (outside the square trench) and the ground disc-ploughed to a depth of about 15 cm, thereby ploughing in the fertilizer and the heavy leaf litter dropped during the dormant period.

B. About 4 weeks later, in November (one month before shoots are about to appear), 1–2 kg per clump (200 to 400kg/ha) of urea (40:0:0) or 16:20:0 NPK fertilizer is distributed inside the trenched-in area. Part of the fertilizer is thrown onto the old cut-off rhizome area at the center of the clump. This has the effect of causing shoots to appear up to four weeks earlier than otherwise expected. In the case of *pai tong keo,* shoots appear one month earlier than other cultivars, and can be induced to shoot just as far into the winter as other cultivars, thus giving that cultivar a slightly longer shoot season.

Clump management and culm harvest

Most culms harvested in the *pai tong keo* plantations are between 50–130 mm in diameter, with some a little larger thus giving an unusually large range. It is necessary to harvest culms each year, partly as good plantation management and partly to sell as a second crop. To train a *D. asper* clump, the Thais recommend the following procedure:

---

[8] Fu Maoyi in personal communication 1997.

[9] Information sourced from two booklets compiled by the Thai government as recommendations to growers and translated into English for the author by Dr Songkram Thammincha of Kasetsart University. One booklet also advocated the additional, progressive application of 1 kg per clump of both 13:13:21 NPK and 46:0:0 urea divided into small increments over the whole shoot season.

- 1 year old – Leave all culms.
- 2 years old – Cut off very small culms.
- 3 years old – The plant should have about 10–12 culms of various sizes by now. Remove inner shoots that appear. Leave large shoots at even intervals spaced around the outside of the clump, and harvest all other shoots, all small ones, and any that dry up. Keep about 5–10 culms, depending on the clump. Late winter to early spring that year, cut off all the 3-year-old culms at a node above ground level.
- 4 years old and each year thereafter – Repeat the Year-3 procedure, but building up to having 8–12 culms in each clump depending on its vigor. There should be about the same number of one-year-old, two-year-old and three-year-old culms in each clump.

Using this system, all culms are harvested in the dormant (mostly dry) season, and the harvested culms would have been growing for at least 3 years. Such wood is less prone to powder beetle attack because of reduced starch and moisture contents (Liese 1985), and would have developed acceptably high fiber strength by that age. When the clump is 7 years old, it should be possible to reliably harvest 2–4 culms per year per clump (400–800 culms/ha/year at a density of 200 clumps/ha).

CONCLUSION

To establish high-quality edible shoot plantations, the two sympodial bamboos *Dendrocalamus asper* and *Dendrocalamus latiflorus* seem to be the most suitable choices for planting in the warmer, wet, monsoon rainfall areas of Australia. The monopodial bamboo *Phyllostachys heterocycla pubescens* (moso) seems the most suitable choice for the southern winter-rainfall cooler areas of Australia. While it is possible to grow both monopodial and sympodial in some areas, the temperature and rainfall patterns plus the availability of irrigation will be the deciding factors.

Already established young plantations of *D. asper* and *D. latiflorus* in north-eastern Australia are showing spectacular growth in the warm monsoon rain areas, and now represent at least 90% of the species being planted. The monopodial moso will become popular and established in the southern part of Australia, though at present there is little awareness of its potential in those areas and perhaps some hesitation because of its longer establishment time.

Australia is now well on the way to establishing a viable edible shoot industry, with more than 150 hectares planted or currently being planted this season. An incorporated marketing organization was recently voted into existence (Australian Commercial Bamboo Corporation) by a dedicated group of existing growers.

Note: Some data have come from yet to be published papers presented at the workshop "Bamboo towards the 21st Century" at Anji, China, in 1997.

ACKNOWLEDGEMENTS

Thanks are due to: Dr Songkram Thammincha of Kasetsart University, Bangkok; Prof. Fu Maoyi and Assoc. Prof. Lou Yiping of the Research Institute of Subtropical Forestry, Chinese Academy of Forestry; Prof. Zhang Guangchu of the Forest Research Institute of Guangdong Province; and Liu Hui Tang of Forestry Department of Guangzhou.

REFERENCES

Cusack, V. 1997. Bamboo rediscovered. Earth Garden Books, Australia.

Cusack, V. 1998. Observations on Thai *Dendrocalamus asper* plantations. Paper presented at the ACBC - CQU - Rural Industries Research & Development Corporation Workshop, Brisbane, Australia, August 1998.

Ding Guohua; Ding Yulong. 1997. Preliminary study on the dynamics of auxin in below-ground organs of moso bamboo. Paper presented at the Bamboo towards the 21st Century International Bamboo Workshop, Anji, China, 7–11 September 1997.

Ding Yulong. 1996. *Phyllostachys* in China and its utilization. Belgian Bamboo Society Newsletter 12, 1996, 7–15.

Dransfield, S.; Widjaja, E.A. ed. 1995. Bamboos. Plant Resources of South-East Asia No. 7. Backhuys Publishers, Leiden, the Netherlands. 189 pp.

Fu Maoyi, 1996. Cultivation and utilization of bamboos. Subtropical Forestry Research Institute, Fuyang, Zhejiang Province, China.

Liese, W. 1985. Bamboos - biology, silvics, properties, utilization. Deutsche Gelleschaft für Technische Zusammenarbeit (GTZ), Nr. 180, TZ Verlagsges., Roßdorf 1, 132 pp.

Lou Yiping; Xiao Jianghua. 1997. Research needs and strategies on long-term site productivity of moso bamboo forests. Paper presented at the Bamboo towards the 21st Century International Bamboo Workshop, Anji, China, 7–11 September 1997.

Oshima, J. 1931. The culture of moso bamboo in Japan - Part 1 & 2. (Reproduced in the American Bamboo Society Newsletter, Vol. 3, No. 1, 1982.

Shi Quantai; Bian Yaorong; Wang Yongxi. 1987. Study on the application of chemical fertilizer to the timber and paper-pulp stands of *Phyllostachys pubescens. In* Rao, A.N.; Dhanarajan, G.; Sastry, C.B. ed., Recent Research on Bamboo. Proceedings of the International Bamboo Workshop, Hangzhou, China, 6–14 October 1985. Chinese Academy of Forestry, Beijing, China; International Development Research Centre, Ottawa, Canada. pp. 87–90.

Thammincha, S. 1996. Bamboo shoot industry and development. *In* Belcher, B.; Karki, M.; Williams, J.T., ed., Bamboo, people and the environment, Vol. 4, Socio-economics and Culture. Proceedings of the Vth International Bamboo Workshop, Ubud, Bali, Indonesia, 19–22 June 1995. International Network for Bamboo and Rattan, New Delhi, India. pp. 33–39.

Thammincha, et al. 1997. The final report to INBAR on the INBAR-Kasetsart University project on bamboo (Thailand). (Unpublished).

Wu Liangru; Xiao Jianghua. 1997. Dynamics of the endogenous phytohormones in on-off year moso bamboo stands. Paper presented at the Bamboo towards the 21st Century International Bamboo Workshop, Anji, China, 7–11 September 1997.

FURTHER READING

Cusack, V. 1997. The establishment of commercial bamboo plantations in Australia. Paper presented at the Bamboo towards the 21st Century International Bamboo Workshop, Anji, China, 7–11 September 1997.

Cusack, V. 1998. Bamboo World. Simon & Schuster, Kangaroo Press, Australia.

Cusack, V. 1998. Bamboo shoot and timber plantation: comparative data across species. *In* Proceedings of the Rural Industries Research & Development Corporation Workshop, Brisbane, Australia, October 1997.

Cusack, V. 1998. *Dendrocalamus asper:* establishment costs and cash flow. *In* Proceedings of the Rural Industries Research & Development Corporation Workshop, Brisbane, Australia, October 1997.

Bamboo for Sustainable Development, pp. 421–429
Eds. A. Kumar, I.V. Ramanuja Rao and Ch. Sastry

Analysis of growing parameters in the vegetative propagation of *Bambusa vulgaris, Gigantochloa apus* and *Guadua chacoensis* in the neotropics

MAYRA MONTIEL LONGHI, ANA ISABEL RUIZ C. and LUIS MURILLO R.
Universidad de Costa Rica, San José, Costa Rica

Abstract—Several studies were conducted to evaluate the effect of age and culm bud position on the propagation of *Bambusa vulgaris* and *Gigantochloa apus,* and the effect of age, culm sectioning and grass cover (*Pennisetum purpureum* var. King Grass) on the propagation of *Guadua chacoensis*. Besides, the effect of Naphtalene-acetic acid (NAA) on the rooting of culm sections of *B. vulgaris* was also tested. It was statistically determined that for *B. vulgaris* the culm age does not affect the growing variables, but for *G. apus* three years proved to be the best. Basal and middle sections of both species produced more shooting and rooting than the distal ones. Three-year basal culms of *G. chacoensis* produced less shoots growing higher and larger with lower death rates. Middle shoots reached intermediate values for every variable. The use of grass cover produced higher shoots and benefited the survival rate during the dry season. Culm sections of *B. vulgaris* treated with different concentrations of NAA showed no significant effect. The control group produced the greatest number of shoots and roots, followed by the distal and middle sections using 1 000 ppm NAA, and the basal sections using 2 000 ppm NAA.

INTRODUCTION

The fact that Costa Rica (as well as other countries of Central America and northern South America) belongs to the tropical ecological belt benefits the country, because it allows a great variety of soils and convenient relative humidity and precipitation. Costa Rica can be termed as a fertile field, capable of assuring a promising future to bamboo as a naturally renewable and highly profitable crop (Montiel and García 1985).

In Asian countries and in the Neotropics, bamboos are one of the most visible and important components of the flora. According to studies on bamboo ecology, the plant's rapid growth helps it to perform a cicatrizant role for damaged ecosystems (Montiel 1998).

Guadua chacoensis, a native of northwestern South America, has a wide distribution, including in Central America (McClure 1966). The culms reach 18–30 m height and 10–18 cm diameter, with hollow internodes. The species has been recognized as a tropical plant of great adaptability, capable of growing from 40 to 2 400 meters above sea level, and in temperatures ranging from 20°C to 26°C.

The minimum acceptable growing temperature is 13°C (Faivre 1981; Soto and Valencia 1981).

For this study on *G. chacoensis*, the objectives were: to evaluate the effect of two ages, three culm sections and a cover system (*Pennisetum purpureum* var. King Grass). The experimental design used was of divided parcels in time, in random complete blocks with four repetitions in a factorial arrangement 2×2×3 (2 ages, the use or not of grass cover and 3 culm sections). Each parcel was 2.5 m in length, and five rows wide with a 0.5 m gap. The ages of the culm were one year and three years, and the sections used were basal, middle and distal for each one.

With *Gigantochloa apus* and *Bambusa vulgaris*, both Asian bamboos but well adapted to the tropical conditions in Costa Rica, the objectives were to evaluate the effect of age and bud position in the culm for the vegetative propagation using the following variables:

1. Size of the higher shoot produced;
2. Number of nodes with buds per parcel;
3. Number of buds per parcel capable of surviving a transplant; and
4. Number of nodes with root.

It was used a random blocks experimental design with eight treatments in a factorial arrangement and eight repetitions.

The methods of asexual reproduction are commonly used for the propagation of bamboo and it can be affected by climate factors for each species. However, the dormancy of the buds can be activated to produce roots and shoots by the manipulation of the internal physiological processes. Because there are scarce studies about this topic, another objective of this work was to evaluate the effect of different concentrations of NAA on the vegetative propagation of *B. vulgaris.*

RESULTS AND ANALYSIS

In *G. chacoensis* (Fig. 1), the interaction between the grass cover and age has an effect on the height of the shoots. The age does not influence the height when no cover was used: the shoots of one-year-old mother culm (planting material) had an average height of 54 cm, while the shoots of three-year-old culm (planting material) were 53 cm high. However, when the cover was used, the difference between the shoots that were born from culms of one and three years old was 11 cm. The use of grass cover favored shoot height for both ages, but the larger values were for those from the three-year-old culm.

Cell division and enlargement in periods of growth depends on an efficient transportation system. It is probable that the three-year-old culms presented this condition and also took advantage of a more humid environment, by making better use of the available water, which is essential for different physiological processes involved in growth. Thus, an accelerated longitudinal growth resulted.

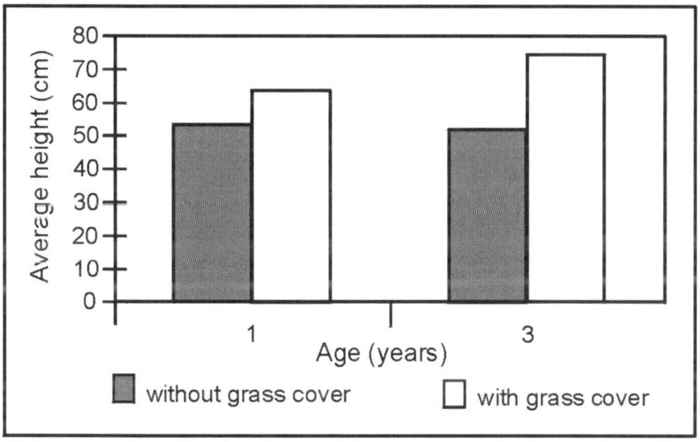

Figure 1. Effect of age-cover interaction on the height of shoots

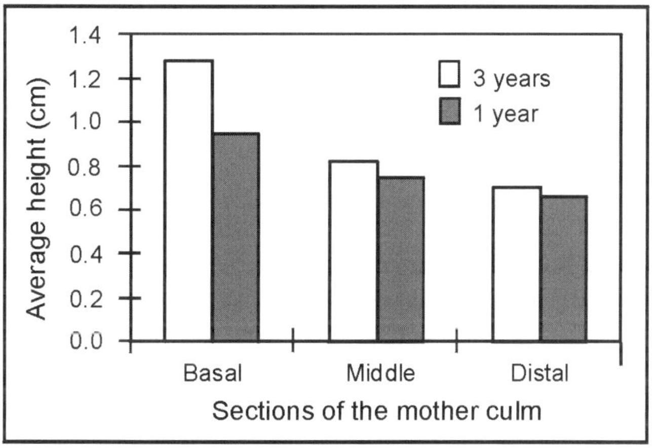

Figure 2. Effect of age-section interaction on the diameter of shoots

Fig. 2 shows the results of age-section interaction. The shoots of the basal section planted reached large diameters for both ages: 0.95 cm for shoots from the one-year-old and 1.27 cm for those from the three-year-old. They were followed by shoots from the middle section, with 0.74 and 0.80 cm for shots from the one and three-year-old culms, respectively. The difference between shoot diameters from the basal and middle sections was larger for shoots from three-year-old mother culms than for the ones from one-year-old mother culms. Shoots of smallest diameters were from the distal section.

The three-year-old culms were thicker than the one-year-olds. The effect of mother culm's age on shoot mortality is showed on Fig. 3. It can be seen that the shoots of one-year-old mother culm had a higher mortality rate than the shoots from the three-year-old culm. An explanation can be found in the fact that the

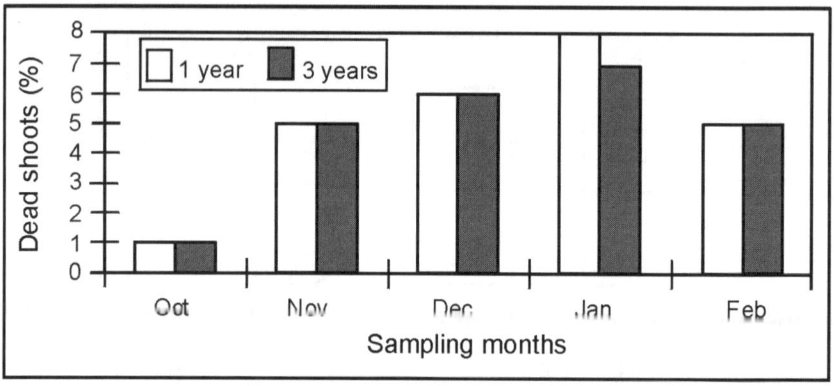

Figure 3. Effect of the mother culm's age on shoot mortality

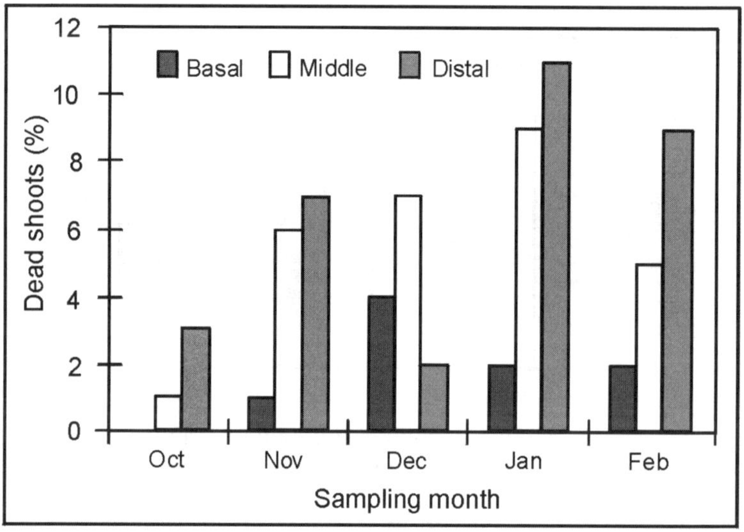

Figure 4. Effect of the mother culm section on shoot mortality

one-year-old culms produced more shoots per bud, while the three-year-old culms produced only one shoot per bud. Also, the plants were damaged by *Fusarium* (Figs. 4 & 5) during the rainy season. The distal section presented the highest mortality rate.

One requirement for the success of vegetative propagation is that each axis should produce more roots (Vela 1982; Castaño et al. 1982). Fifty percent rooting was obtained in the rainy season, and 100% in the beginning of the dry season; the rooting was successful in the middle and basal sections, while the distal section did not react in the expected way. This is because the middle and basal sec-

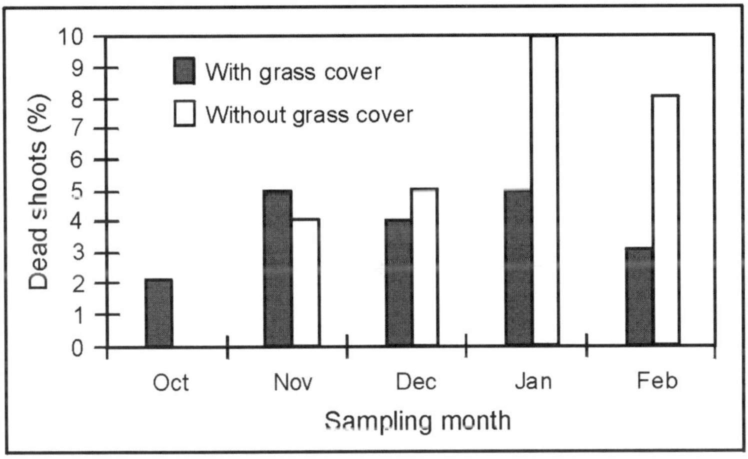

Figure 5. Effect of using and not using grass cover on shoot mortality

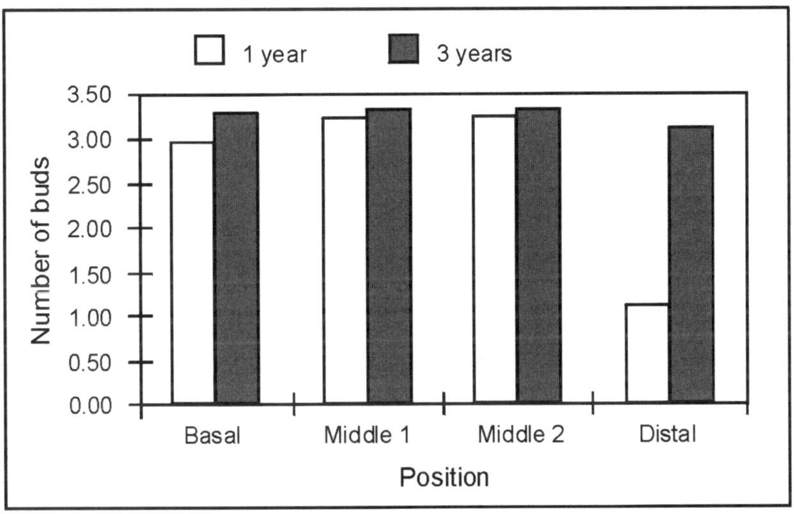

Figure 6. Effect of the interaction between the culm position from where *Bambusa vulgaris* section was taken and the age of the culm, on the number of buds per parcel

tions have a larger reservoir of water than the distal section. The grass cover promoted fungal growth during the rainy months. The cover modifies the environmental conditions around the plants, especially humidity and temperature, by decreasing the evaporation rate from the soil, protects the surface against water and wind erosion, and reduces weed growth during the rainy season.

The emerging of the shoots from the culms of both species began two weeks after the planting was done. At the fourth week (1 month), the tests using *G. apus*

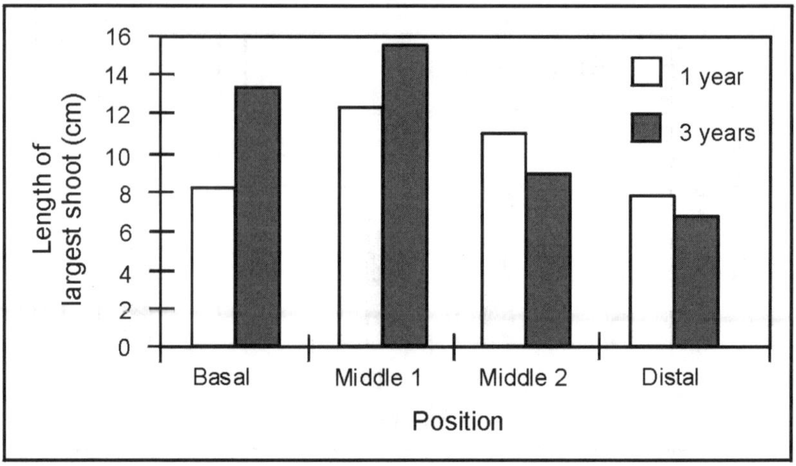

Figure 7. Effect of the interaction between the position from where the *Gigantochloa apus* section was taken and the age of the culm, on the length of the largest shoot

and *B. vulgaris* presented an average of 70% and 71% of culms with shoots, respectively. Some treatments obtained more than 80% of shooting.

It is presumable that the bud latency in *G. apus* (Fig. 6) broke more easily at the base than at the apex, since the one-year-old distal section presented a lower number of buds compared with the others and three-year-old sections, though they recovered at the age of three. As a result, it was assumed that at the moment of the evaluation, the young distal section was just starting to break the bud latency. It has been described in literature that the tendency of the buds to lose their latency basipetaly or acropetaly along the culm, depending on the species (Banik 1980).

In *B. vulgaris* (Fig. 7), the quantities of buds produced by one-year-old culms were similar from the basal, middle 1 and middle 2 sections. The interaction of culm age and position affected the length of the shoots of *B. vulgaris*, but not the ones of *G. apus*. For both ages, the distal section was the most inadequate. The basal section showed the best response when age changed from one to three years.

The effect of the position over the length of the larger shoot increased in the basal and middle 1 sections when the age of the mother culm changed from one to three years. The parameters used to evaluate the treatments constitute, at the same time, different methods of transplantation, depending on the decision to plant: only the larger shoots; all the shoots irrespective of the size; or only the nodes with roots, with or without shoots.

In the case of *B. vulgaris*, the number of shoots and roots produced was quantified to evaluate the effectiveness of NAA (in 20% alcohol) in the rooting of buds in different culm sections, since NAA has been known as a shooting inhibitor and a rooting stimulator (Weaver 1976). As the hormone concentration or the time of

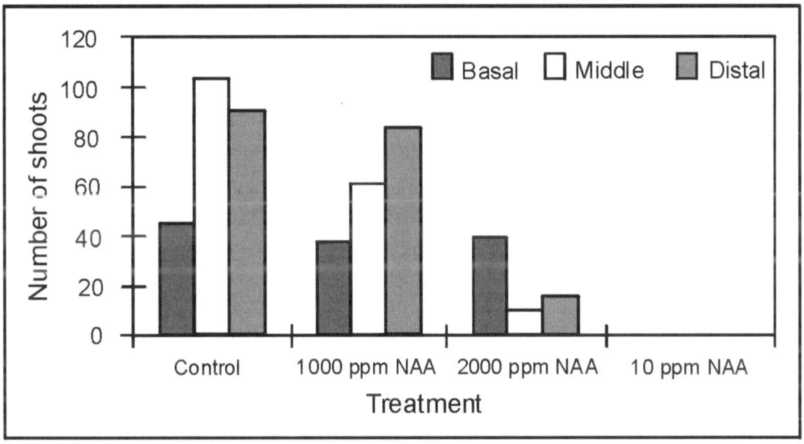

Figure 8. Number of shoots produced after treatments with NAA

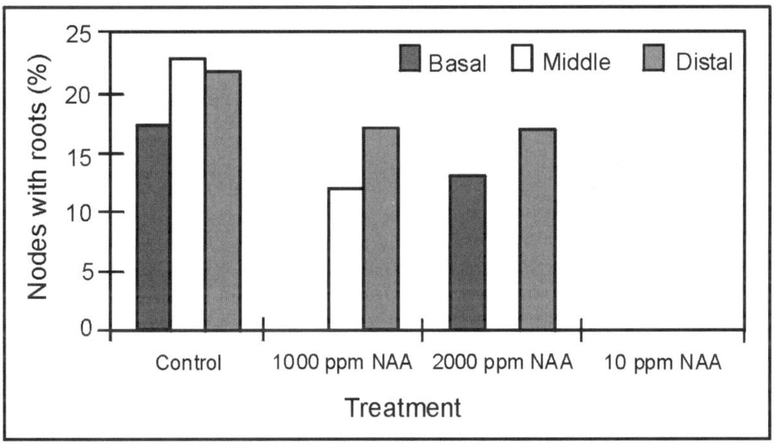

Figure 9. Percentage of rooting produced after treatments with NAA

exposure increased, the number of shoots decreased. However, the decrease in the number of shoots was also affected by the bud position along the culm, whether it was on the basal, medial or distal section (Fig. 8).

The rooting percentage decreased in all the treatments, compared with the control group. The use of NAA also affected each culm section in a different way, depending on the concentration used or the immersion time (Fig. 9).

CONCLUSION

By the analysis of all the considered variables that give information about the effect of age and position from where the vegetative material was taken for the propagation of *B. vulgaris* and *G. apus*, the following was determined:

1. For *G. apus*, three-year-old culms were the best propagules.
2. In the case of *B. vulgaris,* the age of the culm did not affect any variables.
3. The basal and middle 1 and 2 positions of *G. apus* were the ones that give the best results for planting
4. In the case of *B. vulgaris,* the bud position affected only the number of nodes with shooted buds and the number of nodes with roots; the basal and middle 1 sections give the best results.
5. Considering the effect of culm age and position, it is advisable to use the middle 1 and 2 sections of one-year-old culms to obtain the largest number of budding nodes.

In *G. chacoensis*, to determine the age and section of the culm that will give the most suited vegetative propagule, it is necessary to first identify the need. If quantity is required, one-year-old culms and the middle and basal sections provide good shoot production, in terms of quantity, average shoot height and diameter, and low mortality rate. However, if the need is to produce shoots with large diameter and height, the use of three-year-old culms of the basal section is recommended. Grass cover favors the height but have almost no effect on the emerging shoots.

The effect of hormones neither increased the number of shoots in any of the treatments nor stimulated the rooting. This agrees with the results of White (McClure 1966) and Delgado (Hasan 1980). The control group (without NAA) presented a greater number of nodes with shoots and roots. The treatment at 1 000 ppm of NAA improved substantially the number of shoots of the distal section. Without NAA, the most adequate section to use as a propagule is the middle, followed by the distal.

ACKNOWLEDGEMENTS

The authors are grateful for was support provided for this investigation by the Vice-Rectoría de Investigación of the University of Costa Rica. Thanks are also due to the International Journal of Tropical Biology and Conservation (University of Costa Rica).

REFERENCES

Banik, R. L. 1980. Propagation of bamboos by clonal methods and by seed. *In* Lessard, G.; Chouinard, A. ed., Bamboo research in Asia. Proceedings of a workshop held in Singapore, 28–30 May 1980. International Development Research Centre, Ottawa, Canada. pp. 135–150.

Castaño, N.; Londoño, X.; Prieto, L. 1982. Introducción al estudio fitoecológico de los guaduales del valle geográfico del Río Cauca en Colombia. 2° Simposio Latinoamericano, Ecuador.

Faivre, Y. 1981. Cultivo y explotación del bambú malayo para la producción de pérticos de cosecha. Palmas, Colombia, pp. 40–42.

Hasan, S. M. 1980. Lessons from partial studies on the propagation of bamboos. *In* Lessard, G.; Chouinard, A. ed., Bamboo research in Asia. Proceedings of a workshop held in Singapore, 28–30 May 1980. International Development Research Centre, Ottawa, Canada.

McClure, F.A. 1966. The bamboos: a fresh perspective. Harvard University Press, Cambridge, Mass., USA. 347 pp.

Montiel, M. 1998. Planting and using of bamboo in the neotropics. International Journal of Tropical Biology and Conservation, Vol. 46, Suppl. 3, June 1998.

Montiel, M.; García, R.. 1985. Bambú análisis de una alternativa de desarrollo. Editorial de la Universidad de Costa Rica. San José, Costa Rica.

Soto, H.; Valencia, H. 1981. La guadua. Corporación regional Autónoma. Boletín de Colombia. División Técnica. Sección Conservación de Suelos y Agua.

Vela, G. 1982. Los bambues forestales SFF y SHH. Boletín Técnico N° 50, México.

Weaver, R. 1976. Reguladores de crecimiento de las plantas en la agricultura. Trilla Ed., México.

Bamboo for Sustainable Development, pp. 431–442
Eds. A. Kumar, I.V. Ramanuja Rao and Ch. Sastry

Prospects of research on sustainable management techniques of bamboo forests in China

FU MAOYI and LOU YIPING
Subtropical Forestry Research Institute, Fuyang, Zhejiang Province, China

Abstract—Since the 1950s, research and extension of high-yielding bamboo management techniques have contributed greatly to an increase of bamboo productivity in China. The emphasis of these activities was on stand composition and structure control techniques, such as: regulating species composition, standing culm density, diameter-class and age-class composition; raising harvesting intensity of shoots and culms; shortening harvesting cycle for culms; and soil management techniques on tending, loosening soil, fertilizer application, and target-specific chemical herbicide treatment. Following these management techniques, there has been marked raise in the productivity of low-yielding bamboo forests in a relatively short term.

However, some Chinese researchers have realized that some of the existing management techniques have caused negative influences on sustainable productivity by degrading the function of bamboo ecosystem. While high output was achieved in a stand for short-term economic benefits, maintenance of long-term productivity of the managed stands was neglected without any foresight. Some of the management practices – such as shifting mixed stands into pure bamboo stands, intensive harvesting in a short rotation, whole-bamboo harvesting, and frequent loosening of soil, chemical herbicide treatment and commercial fertilizer application – probably caused major negative impacts on the function of bamboo ecosystems. Loss of soil nutrition and reduction of undergrowth vegetation resulted, thereby undermining the biodiversity and deteriorating the physical, chemical and biological properties of soil, and eventually causing the decline of land and stand productivity.

It is proposed that the future research goal should be to rectify the negative consequences improper actions pertaining to maintenance of ecological systems and sustainable productivity, and to ensure that management practices favor the stability, resistance and resilience of bamboo forest ecosystem.

INTRODUCTION

Bamboo forest covers over 4 million ha, accounting for about 4% of the total area of the forests in China and ranking third in area in terms of forest species of southern China. Bamboo forest not only plays an important role in rural economy as a traditional multiple-use tree species but also has, as an important part of the forest ecosystem, a great impact on natural environment, especially in the regions dominated by bamboo forests. In the past, bamboos were considered as non-wood forest product, mainly because there was adequate supply of wood from natural forests. At present, forests are shrinking in area and timber tree species of good quality or value are diminishing in numbers. Hence, interest is increasing in fast-growing species to enhance the biomass and produce wood for general purposes.

Bamboos grow faster than most hard wood species, and their harvest cycle begins after 3–4 years of planting. Bamboo clumps can continue to grow for many years. For this reason, bamboo forests have been managed widely and intensively in the past few decades, using high-yielding cultivation techniques, in some countries such as China. In most cases, the productivity of such managed bamboo stands were highly enhanced by these techniques. These high-yielding cultivation techniques, however, were aimed mainly at achieving economic benefits from bamboo forest; the negative impacts on ecological function and sustainable productivity of these forests were not considered. Some evidences of declining ecological function and hampered long-term productivity have been found in bamboo forests in the past two decades; consequences of massive negative disturbances caused by improper management practices (Fu 1997; Lou and Xiao 1998).

The impact of undesirable management practices on the composition and function of forest ecosystems have evoked worldwide concern in recent times. Many research findings have proved that current management practices may greatly disturb forest ecosystems; for instance, it happens when mass nutrients are removed and soil is disturbed during heavy harvesting (Dyck et al. 1994). Even though at times some of these practices have brought about high productivity and resultant economic benefit in a comparatively short term, some of them probably have a strong negative effect on sustainable productivity resulting from undermining the stability and resilience of forest ecosystems, as the natural ecosystem function was changed (Chen and Zhang 1998).

Research studies on the effects of management practices on sustainable site productivity have paid much attention to the "second rotation effect", harvesting effects and subsequent site preparations. Researchers, however, found that the impact of management is a very complex issue and that in most cases it is difficult to differentiate the effects of various practices on site productivity. Some practices often produced different effects on short-term and long-term site productivity. For this reason, many researchers are now focusing on the effects of multi-practices on sustainable site productivity.

This paper discusses the impacts of certain current management practices on ecological function and long-term site productivity of bamboo forest ecosystems, and suggests some technical strategies for future research on ecological maintenance of sustainable productivity, and restoration and reconstruction of degraded bamboo ecosystems. The goal of future research is to rectify the negative consequences of past and current management actions on the maintenance of ecological function and sustainable site productivity and to ensure that the management practices favor the stability, resistance and resilience of bamboo forest ecosystem. It is expected to improve upon the current management practices so that the long-term consequences of bamboo forest management can be controlled. This paper intends to not only outline the information needs but also list some specific studies.

IMPACTS OF CURRENT MANAGEMENT PRACTICES ON SUSTAINABLE PRODUCTIVITY OF BAMBOO STANDS

Most of the current silvicultural practices related to the management of bamboo forests probably have strong impacts on the function of bamboo ecosystem function and sustainable site productivity. These practices – such as the maintenance of pure stands, nutrient removal (especially removal of mass organic matter) in intensive shoot and culm management as well as during whole-bamboo harvesting, deep tillage, and chemical herbicide treatment – have significant effect on soil erosion and soil nutrition loss, elimination of undergrowth vegetation, and undermining of biodiversity. The effects of chemical fertilizer application on the fertility, compactness and biodiversity of soil are well documented. Research studies have proved that the productivity decline of bamboo forest occurs in cases such as when a mixed bamboo forest is transformed into a pure stand for long-term management (Lou et al. 1997a). Lou et al. (1997b) also reported site degradation in such cases, or the productivity can be maintained only when frequent and intensive tillage and fertilizing are adopted. Hence, the effects of existing management practices on the maintenance of bamboo forest ecosystem function and on the sustainable land productivity of managed stands should be the focus of future research and management programs.

The negative effects of current management practices have caused concern among researchers and these require a discussion. This does not imply that these are the only factors, which have impact on the maintenance of ecological function and sustainable productivity of managed bamboo forests, worthy of consideration. Indeed, any management practice would have some impact, positive or negative, on the maintenance of ecological function and sustainable productivity of bamboo forests. However, assuming some of these practices to continue, and to even intensify, in many commercial bamboo forests in the world, the study of this topic is timely.

Impact of pure bamboo stands management on sustainable productivity

The effect of species change on forest productivity, concerning the effects of different stand compositions on soil degradation, go back at least several decades in many tree crops. The significant declines of site and stands in pure coniferous forests in second rotation were observed in Norway spruce (Prichett and Fisher 1987) in Germany, radiata pine (Evans 1990) in Switzerland and fir plantations (Sheng 1992) in China. Nutritional losses were suspected to result in site declines, but could not be conclusively proven (Prichett and Fisher 1987). Otherwise, several studies have documented the significant effects of species composition on the rate of nutrient uptake (Johnson et al. 1988).

In the early of 1950s, most of the Moso bamboo forests in China were unmanaged forests, where bamboos grew mixed with trees and abundant undergrowth vegetation. Till the middle of 1960s, all mixed bamboo forests were encouraged, as per the high-yielding cultivation techniques prevalent at that time, to be changed into pure stands by removing trees and undergrowth vegetation to raise the productivity. Since then, most of these bamboo forests were managed as pure stands. Meanwhile, the area of bamboo forests was increased by rehabilitating the soil. Until the early 1990s, for instance, Anji County alone had about 4 700 ha of high-yielding bamboo forests raised through fertilizer application and soil rehabilitation. This is about one-tenth of the total bamboo forest area in the County (Xu 1998). Until now, most of the managed bamboo forests in China are being maintained as pure stands.

High productivity and economic benefits have been achieved from most of the managed pure stands over a relatively short period. However, the ecological environment of these bamboo forests has also altered greatly because of the vast area of pure stands. Some of the evidences of ecological imbalance observed are discussed in the following paragraphs.

Frequent and serious pests and diseases attack

As reported by Xu (1998), pre-1970 historical records have shown very few pests and diseases breakout in bamboo forests in Anji County. Consequently, bamboo stands have been considered as a kind of forest not commonly affected by pests and diseases. The only available record was about a small area of bamboo damaged by bamboo shoot weevils (*Otidognathus davidis*) in the former Xiaofeng County. However, pests and diseases infestations occurred very frequently and caused serious damage to bamboo forests from the 1970s. In 1971, for instance, there was a serious breakout of *Ceratosphaeria phyllostachydis*, resulting in the death of over 300 000 culms died in 50 000 Mu (1 Mu = 1/15 ha) of affected bamboo forests. Most recent case occurred in 1987, when *Hippotiscus dorsalis* broke out in 81 000 Mu of bamboo forest, causing the death of 70 000 culms. A recent survey has found that there are at least 5 pest species that pose serious threat to the bamboo forests, with the possibility of frequent breakouts in Anji County. Some unidentified pests and diseases found in recent times have also caused extensive bamboo production losses. For instance, an unidentified disease affected 12,000 Mu of bamboo forests in 1993 and destroyed over 100 000 culms.

As mentioned above, harmful pests and diseases are occurring more frequently and seriously than ever before, causing a great deal of loss to bamboo production. Most probably, the reason for this is the ecological imbalance in the bamboo forest ecosystem, caused by mono-population of pure bamboo stands, less undergrowth vegetation and destabilized biodiversity.

Site degradation

In pure bamboo stands, mixed trees were removed and some undergrowth vegetation cover was reduced as part of the management practices. Some reports showed that the simple canopy composition and low undergrowth coverage led to serious loss of water and soil during heavy rains. As the soil nutrients were washed away, the physical and chemical properties of the soil in the forests steadily worsened. Consequently, the productivity of bamboo forests was greatly affected.

The forest coverage in Anji County increased from 54.1% in 1989 to 60% in 1994 because of afforestation. However, the capacity of forests for water retention and soil conservation was reduced, which probably resulted from the simplified forest composition (over 60% pure bamboo stands in Anji County, for example) (Xu 1998) and less undergrowth vegetation in managed pure bamboo stands. Large areas of managed pure stands will have negative impacts on sustainable productivity of bamboo forests.

Impact of intensive harvesting in a short rotation and whole-bamboo harvest

Nutrient removal from biomass re-emerged as a major concern in harvesting, especially whole-tree harvesting (WTH), in the 1970s (Boyle et al. 1973; Leaf 1979) and remains a significant concern to date. The problem with WTH is that, in many cases, the nutrient export per unit of biomass increases dramatically with little gain in useable product.

In China, to meet the strong demand for bamboo products, intensive harvesting of shoots and culms in a short harvest cycle and whole-bamboo harvest have been practiced in the past few decades. This increased harvesting intensity, wherein the age of harvested culms fell from 7 years to 5 years, resulted in a decrease in the culm density of bamboo stands.

A lot of organic matter was removed from the bamboo ecosystem by the intensive harvesting of culms and shoots in a short rotation and the clearance of vegetation coverage. Not only were culms, shoots and branches harvested to remove mass nutrients off site, but bamboo leaves were also removed, amounting to 15 000 kg per ha in some cases (Xu 1998). Intensive harvesting of shoots and culms once every year or two not only removed mass organic matter and nutrients from the bamboo ecosystem through energy and substance flow, but also altered the biological composition of the bamboo ecosystem and negatively affected the soil ecosystem.

Since adding organic matter to the soil under bamboo stands is difficult, nutrient removal during intensive harvesting and whole-bamboo harvesting leads to site degradation. This loss of organic matter from the soil alters the habitats of plants, animals and microbes, leading to a reduction in the soil biodiversity and the resultant fall in productivity.

Impact of frequent soil loosening

Soil rehabilitation by loosening is an efficient practice to achieve high productivity of bamboo forests. This method improved the chemical and physical properties of soil in a short time and reduced the competition of other vegetation. Since most bamboo forests are located on mountainous and hillside fields, however, this practice caused loss of soil and water from the land, resulting in mass soil and nutrient removal.

Moreover, following the loosening of the soil, a significant decrease in the green cover of the earth has been observed. The direct result of this is that the soil becomes exposed and porous, producing more surface soil runoffs than before, especially in bamboo stands on steep slopes. As soil is thinned, its nutrients are gradually washed away and its ability to conserve water decreases. The degradation of environmental components results in a negative impact on the bamboo stands: their productivity falls. All these lead to poor shooting, ability to retain water and capacity to adjust to the temperature and humidity in a canopy (Chen and Zhang 1998).

Impact of chemical herbicide treatment

There is a lot concern about the pollution of water bodies, such as lakes and streams, caused by chemical herbicide treatments. Studies by Likens et al. (1969) showed major increases in stream-water nitrate concentrations following herbicide treatment at Hubbard Brook, New Hampshire. Recent research studies show that the harmful impact of chemical herbicides on the function and productivity of forests is much greater. No doubt, herbicides are an economical way to clear shrub and weeds; but its negative impacts on the ecosystem of bamboo forests cannot be ignored.

It was found that there was hardly any vegetation growing in bamboo forests where chemical herbicides were frequently applied; bamboos were nearly the only plant population present. The mono-ecosystem in pure bamboo forests is very simple in species and community components. Bamboos growing in such conditions will have poor capacity to resist pests and diseases, and the function of soil and water conservation, water retention and nutrient storage worsen significantly.

Chemical herbicide treatments also cause sharp decrease in biodiversity, and the consequent deterioration in the environment conditions reduces the productivity of bamboo stands. For this reason, though chemical weeding is a new commercial method to tend bamboo forests, it is necessary to control the frequency of application, and to rehabilitate undergrowth vegetation and maintain of biodiversity in bamboo forests.

Impact of tending

Tending (removing undergrowth vegetation), usually used in multi-story bamboo forests that are rich in undergrowth vegetation, is a traditional silvicultural practice. Tending alters the biological components of bamboo forest, as the vegetation is cleared away temporarily. Improved land fertility resulting from the decomposition of the felled brushes branches and leaves promotes bamboo growth.

On the other hand, if tending reduces the undergrowth vegetation – in terms of species and quantity – beyond the desirable level and if the vegetation does not regenerate, the decline of the green cover and the loss of biodiversity will eventually cause the deterioration of ecological function and sustainable productivity.

Impact of fertilizer application

As mass nutrients are lost through management practices – such as harvest of shoots and culms and removal of branches and leaves – and the capacity of the land to regenerate nutrients on-site is weakened by activities – such as chemical herbicide application and soil loosening – it is necessary to input some nutrients to maintain the productivity of bamboo forests. Commercial fertilizer application is a traditional way to add nutrients, but it has the disadvantages of short-term efficiency and heavy pollution. Organic fertilizer is a good solution to this problem, but its sources are limited and costs are high. For this reason, chemical and organic fertilizer application is considered only as a supplementary process to maintain the balance of nutrients in a bamboo ecosystem.

According to recent research, frequent commercial fertilizer applications damage the physical and chemical properties of soil, and thereby the habitat of living things in the soil. The resultant low biodiversity in the soil ecosystem causes deficiency of soil nutrition.

POSSIBLE TECHNICAL STRATEGIES AND RESEARCH PRIORITIES FOR SUSTAINABLE MANAGEMENT OF BAMBOO FORESTS

As mentioned earlier, the current management practices have achieved significant gains in bamboo forest productivity and economic returns. In the context of ecologically sustainable management of forests, the purpose of intensive bamboo management must not only be to enhance economic benefits but also to maintain the eco-system and sustainable productivity so as that high outputs can b maintained in the long term rather than the short term. Unfortunately, most management practices in bamboo forests have, to date, targeted only economic benefits.

Every management practice has some positive impacts, especially on short-term productivity, and some negative effects, mostly on long-term productivity. What is desirable for future research, in terms of sustainable management, is to retain the positive impacts while eliminating the negative ones. Although it is almost impossible to overcome all the disadvantages of any management system, it is expected that some technical strategies and research directions be adopted to address these concerns.

Complete recovery of a natural bamboo forest ecosystem, which has been disturbed by management practices, is impossible in the short term. This is particularly so, when the strong demand for bamboo products does not let any reprieve to intensive cultivation. However, it would be possible to recover and restore the bamboo ecosystem to some extent by regulating the stand composition and protecting the existing undergrowth vegetation. The possible technical and research strategies directed towards achieving sustainability of bamboo forests are discussed in the following paragraphs.

Management of bamboo stands by cultivation category

Considering the current status in bamboo management, it is proposed that bamboo forests be managed by category – such as high-yielding culm-producing stands, shoot-producing stands, both culm and shoot-producing stands, or biodiversity conservation stands. The principle of target management in bamboo forest is to cultivate stands in terms of site, techniques and specific products, and to view both economic benefits and eco-system maintenance as a single target.

In high-yielding bamboo plantations, the high output is achieved by high input, and the capacity to maintain the ecological function and to rehabilitate land damaged by management disturbances is weak. All ecological management practices should be conducive to maintaining sustainable productivity. Future research should focus on ways and means to achieve both economic and ecological benefits through keeping stable and diverse biological and environmental components under the various management practices. Such practices could include fertilizing bamboo clumps with organic or biological fertilizers, keeping coniferous and deciduous trees at a rational proportion to optimize the composition of a stand, etc.

Techniques on regulating culm density and canopy composition

With regard to future research on regulating culm density and canopy composition of pure bamboo stands, the following techniques should be emphasized, so that a sustainable management strategy could be evolved.

1. Culm density control
Culm density should be such that it will not only produce high stand productivity but also favor recovering biodiversity by promoting undergrowth, increasing organic matter in soil, and produce a favorable environment for animals, insects and microbes in bamboo forest ecosystem. The density should be studied and determined in terms of specific site and stand conditions.

2. Canopy component regulation
The purpose of regulating canopy component is to establish a favorable environment for recovering depleted biodiversity and site productivity by pruning.

3. Planting of selected trees
In some pure stands, where land degradation is serious and long-term, it is possible to recover the ecological resilience by planting some selected trees. The tree species selection, planting density and spacing should be determined in accordance to the stand and site in future studies.

For managed bamboo forests with degraded ecosystem, the practices to maintain sustainable productivity should be focused on increasing biodiversity and sustainable productivity, so that stability and resilience of bamboo ecosystems can eventually be recovered.

New techniques of rehabilitating low-yielding mixed bamboo forests

Bamboo productivity is low in most of the mixed bamboo forests, even when the soil fertility is good and abundant undergrowth exists. The management target in this kind of bamboo forest is to raise output by clearing some trees to create room for bamboo growth. A new ecosystem may thus be established, losing the original composition that perpetuated low productivity. However, this change in the ecosystem may be dramatic enough to affect the biodiversity, which may negatively influence the overall ecological function and sustainable production of the bamboo stand. Hence, the ecological function and spacing of trees should be closely studied to find out new techniques for managing mixed stands.

Techniques for managing undergrowth vegetation and maintaining biodiversity

There is no doubt that undergrowth vegetation is an important part in bamboo ecosystem; they have some specific functions in maintaining biodiversity, ecological functions and site productivity. Non-crop vegetation can, as reported by Beets (1994), play a key role in maintaining long-term productivity by: 1) Stabilizing soil to minimize erosion, mass movement, and rainfall impacts; 2) Acting as a nutrient sink to reduce leaching; 3) Cycling nutrients to crop species; 4) Fixing nitrogen; and 5) Contributing to a diverse and beneficial soil biota. These essentially beneficial processes must be appropriately balanced against the negative impacts on the allocation of site resources – light, nutrients and water – to crop trees.

Some studies have demonstrated, for a range of sites and species, that herbaceous and woody vegetation can have a significant negative impact on early plantation growth (Carrier and Dobkowski 1991). The situation in bamboo plantations, however, is very different from that in young tree plantations because the negative competition usually does not occur in young plantations with big transplanted mother-bamboos or in matured bamboo forests with tall culms. For this reason, in bamboo forests, the benefits from vegetation growth will overcome the negative impacts of competition. On the contrary, the most direct cause of site degradation in a bamboo plantation is low biodiversity resulting from loss of undergrowth vegetation and bamboo monoculture. For this reason, it is crucial to protect existing undergrowth vegetation. Hence, it is important undertake studies to establish techniques to manage undergrowth vegetation in managed bamboo stands.

Techniques to control the balance of nutrients and organic matter

For sustainable productivity of bamboo stands, it is important to balance the input and output of nutrients, especially in view of nutrient removal owing to heavy harvesting and soil disturbance. It is equally, if not more, important to guard against reduction of organic matter in the soil ecosystem in managed bamboo stands. Abundant organic matter will not only improve the chemical and physical properties of soil, but also help improve biological properties in soil to maintain the biodiversity of soil ecosystem.

The function and importance of organic matter in soil ecosystem of bamboo forests are topics that are worthy of study, as they probably relate closely to the population of microbes and consequently to the biodiversity of soil ecosystem and soil fertility.

New fertilizers and fertilizing techniques for sustainable productivity

It is known that frequent application of chemical fertilizers has a negative impact on the environment and soil properties. Organic fertilizer, especially microbe fertilizer, for bamboo forests should be produced to replace chemical fertilizers for maintaining sustainable productivity. Besides, the fertilizing techniques should be further studied to increase the fertilizing efficiency and reduce waste and pollution.

As mentioned earlier, current management practices – soil loosening, chemical weeding, tending, shoot and culm harvest in short cycles, whole-bamboo harvest, chemical fertilizing, etc. – regulate the components of bamboo ecosystem so as to improve its short-term productivity. Unfortunately, the long-term sustenance of ecological function and site productivity is often neglected as emphasis is on short-term economic benefits. For achieving sustainable productivity and ecological function of bamboo forests in the 21st century, however, emphasis must be on management practices that help maintain ecological function and site productivity on the long term, while providing attractive economic returns.

Given this, the goal for future research is to establish a comprehensive sustainable management regime that recovers and reconstructs degraded bamboo ecosystems and maintains ecological function and long-term site productivity. This regime may include:

1. Management of bamboo forests by category in terms of target output;
2. New techniques on regulating culm density and canopy composition of bamboo;
3. New techniques on rehabilitating low-yielding bamboo forests with abundant undergrowth;
4. Techniques on managing undergrowth vegetation and maintaining biodiversity;
5. Techniques to control balance of nutrient and organic matter; and
6. New fertilizers and fertilizing techniques for sustainable productivity.

Avoiding or correcting the negative consequences of today's plantation management practices is the general goal for future research. Degradation of ecological function and site quality can be avoided by accepting sustainable management techniques that do not adversely affect site quality. If a certain negative effect is an unavoidable consequence of the management practice, then future ecological and sustainable productivity problems need to be averted by incorporating the appropriate ameliorative techniques into the silvicultural regime prescribed for the site. This can be a very logical and cost-effective way. To do so, however, requires a full understanding as to: 1) what has changed owing to the current management practices; 2) what are the effect of these changes on ecological function and long-term productivity; and 3) what corrective practice is needed to minimize the long-term negative consequences to the ecological function and site productivity of managed bamboo forests.

From the research perspective, the current knowledge is a long way from such a comprehensive understanding of the bamboo forest ecosystems and the management techniques necessary to accurately control the consequences of alternative corrective measures that might be studied. This is the challenge before the scientific community in the 21st century.

ACKNOWLEDGEMENTS

The authors would like to thank the Chinese government and the International Network of Bamboo and Rattan. Without their support, the presentation of this paper would not have been possible.

REFERENCES

Beets, P.N.; et al. 1994. Management system for sustainable productivity. *In* Dyck, W.G.; Cole, D.W.; Comerford, N.B. ed., Impacts of forest harvesting on long-term site productivity. Kluwer Academic Publishers, Dordrecht, The Netherlands.

Boyle, J. R.; Phillips, J.J.; Ek, A.R. 1973. "Whole-tree" harvesting: nutrient budget evaluation. Journal of Forestry, 71, 760–762.

Carrier, B.D.; Dobkowski, A. 1991. The effects of three different vegetation management regimes on sixth-year Douglas fir growth in Oregon and Washington. Technical Report 050–3910/31, Weyerhaeuser Company, Centralia, WA, USA. 32 pp.

Chen, L.C.; Zhang, P.X. 1998. Bamboo Ecology and Management. *In* Fu Maoyi; Lou Yiping; ed., Proceedings of International Workshop on Bamboo towards 21st Century, Anji, China, 7–11 September 1997. China Forestry Publishing Press, Beijing, China.

Dyck, W.G.; Cole, D.W.; Comerford, N.B. ed. 1994. Impacts of forest harvesting on long-term site productivity. Kluwer Academic Publishers, Dordrecht, The Netherlands. 384 pp.

Evans, J. 1990. Long-term productivity of forest plantation: status in 1990. *In* Corcoran, T.J.; Linehan, P.E.; Liu, S. ed., Proceedings of the XIX IUFRO World Congress, Montreal, Canada, 5–11 August 1990. Maine Agricultural Experiment Station, Maine, USA. 296 pp.

Fu, M.Y. 1997. Current bamboo research dynamics and strategies in the world. Forest Science and Technology, 322(3), 34–35.

Johnson, D.W. 1994. Reasons for concern over impacts of harvesting. *In* Dyck, W.G.; Cole, D.W.; Comerford, N.B. ed., Impacts of forest harvesting on long-term site productivity. Kluwer Academic Publishers, Dordrecht, The Netherlands.

Johnson, D.W.; Henderson, G.S.; Todd, D.E. 1988. Changes in nutrient distribution in forests and soils of Walker Branch Watershed, Tennessee, over an eleven-year period. Biochemistry, 5, 275–293.

Leaf, A.L. ed. 1979. Impact of intensive harvesting on forest nutrient cycling. State University of New York, Syracuse, USA.

Likens, G.E.; Bormann, F.H.; Johnson, N.M. 1969. Nitrification: importance to nutrient losses from a cut-over forest ecosystem. Science, 163, 1205–1206.

Lou, Y.P. et al. 1997a. Stand dynamics of pure *Phyllostachys heterocycla* var. *pubescens* stands transformed from mixed stands, Forest Research, 10(1), 35–41.

Lou, Y.P. et al. Effect of long-term management of pure *Phyllostachys heterocycla* var. *pubescens* stands on soil fertility, Forest Research, 10(1), 125–129.

Lou, Y. & Xiao, J. Decline Evidence of Long-Term Site Productivity of Moso Bamboo Forests and Its Research Needs. *In* Fu, M.Y.; Lou, Y.P.; ed., Proceedings of International Workshop on Bamboo towards 21st Century, Anji, China, 7–11 September 1997. China Forestry Publishing Press, Beijing, China.

Pritchett, W. L.; Fisher, R.F. 1987. Properties and management of forest soils. John Wiley and Sons, New York, USA.

Sheng, W.T. ed. 1992. Researches on declines of tree plantations in China. China Science and Technology Publishing House, Beijing, China.

Xu, R.S. et al. On the protection of undergrowth vegetation in managed moso bamboo forests in Anji of China. *In* Fu, M.Y.; Lou, Y.P.; ed., Proceedings of International Workshop on Bamboo towards 21st Century, Anji, China, 7–11 September 1997. China Forestry Publishing Press, Beijing, China.

Bamboo for Sustainable Development, pp. 443–448
Eds. A. Kumar, I.V. Ramanuja Rao and Ch. Sastry

Starting a commercial bamboo plantation in the Yucatan jungle

ROBERT H. GOW

President/Owner, Hacienda Xixim, Houston, the United States

Abstract—The author, an entrepreneur, has planted 22 000 *Dendrocalamus strictus* clumps at a 16th century hacienda in the jungles of Yucatan. This paper explores his reasons for doing this, what he is learning about environmental factors that affect growth, major cost elements, and bamboo product development.

INTRODUCTION

The author is not a bamboo researcher; rather, as the magazine *Mexico Today* put it, "an entrepreneur with a conscience", living in a 16th century hacienda in the jungles of Yucatan. In 1995, he embarked on a project of studying the possibility of growing *Dendrocalamus strictus*. The decision to grow bamboo was prompted three reasons:

1. The growing of bamboo has a desirable effect on the environment. At the very least, if the bamboo is used in place of wood, trees are saved.
2. Near the hacienda (named Hacienda Xixim), there is a small village of Mayan Indians. The University of Texas believes that this village was in existence before Columbus. Today there is no employment in the village. The young people drift into city slums, where they are ill equipped to cope. It would be desirable if bamboo could find a way to let these full-blooded, indigenous, Mayan-speaking young people remain gainfully employed in their ancestral village.
3. Use bamboo to make lots of money.

The author has seen many people come to the jungles of Guatemala and Yucatan from the United States and other countries. Some of these men and women have spent years patiently telling the local people to use knowledge to change their economic lives for the better. Often these intelligent, dedicated advisors feel ignored and frustrated when their advice is only partially followed.

When anyone raises a new crop – for example, one that clearly makes a lot of money – everyone rushes to copy him or her. The author feels that there is a chance to change the economy of Yucatan and provide employment to Mayan Indians in many small villages. However, there is only one way that the author knows to do this: raise lots of bamboo plants and establish that cultivating bamboo is extremely profitable. Hundreds of thousands of bamboo plants can be made to grow in the Yucatan one day.

It is very important, not only to the author but also to the future of bamboo growing, to prove that growing bamboo is a very worthwhile endeavor from an economic point of view.

TWO INITIAL QUESTIONS

In late 1994 and early 1995, the author was in possession of a batch of *Dendrocalamus strictus* seed and was faced with two questions. First, could *Dendrocalamus strictus* or possibly another useful species of bamboo be grown in Yucatan, given Yucatan's relatively dry agricultural conditions? Second, could a profitable market be found for the bamboo that would grow in Yucatan?

Obviously, a negative answer to either question would doom the project. However, the search for a market by far would be the most important and most doubtful. Who would want bamboo enough to pay a good price for a large quantity of it? It became clear very early that the greatest obstacle to growing bamboo was that any existing, large, profitable market for it could not be found. A market would have to be created – always a much more difficult endeavor.

If bamboo were to be profitably cultivated in the Yucatan, it would have to be sold to a high-end market. Serviceable, nesting plastic chairs can be bought in the local market for a few dollars. There is no problem with rain or with powder-post beetles. Bamboo is superior aesthetically and is very strong, but for a modern profitable operation, bamboo culms cannot be sold against timber, which easily takes nails, much less be sold against plastic. The author could not find a market for bamboo as a "poor man's timber", but bamboo could be a wonderful "rich man's timber".

Others have shown the feasibility of making bamboo into laminated timber to use as a substitute for expensive hardwood products, such as flooring. Once rectangular cross-sectioned bamboo strips are created, they can be made into a large variety of different products. A beam, for example, could be made to be very strong if it is made from rectangular cross-sectioned bamboo strips. It could be made to any dimension of thickness, width and length. Clearly, the larger the cross-sectioned area of each strip, the fewer strips required for making a large beam.

OBJECTIVES OF THE BAMBOO PROJECT

The author next worked with a consultant specialized in the development of machinery for processing small-diameter logs into rectangular cross-sectioned timber. As described later in the text, it had to be assumed that bamboo of different sizes and characteristics would be processed. A variety of books and experts indicated that *Dendrocalamus strictus* would produce culms with a range of dimensions. It was also clear that the bamboo has to be straight, thick-walled and large in diameter, and *D. strictus* had all these characteristics. All these confirmed the good fortune in having thousands of tiny *D. strictus* seedlings growing in small plastic bags filled with water-retaining bagasse.

During the International Bamboo Congress and Workshop in Bali, Indonesia, the author received varying answers to a host of questions on key economic aspects related to *D. strictus:* the diameter ranged from 3¼ to 5 inches; the wall thickness ranged from solid to ¾ inch; 8 to 35 culms (3-year-old) could be expected per plant per year; the height or length of the culm ranged from 30 to 70 feet. In terms of making rectangular cross-sectioned laminates, 35 long straight culms per year of 5 inches diameter solid bamboo would obviously present an entirely different economic picture than 8 shorter, curved culms per year of 3¼ inches diameter with a 1¾ inches cavity!

GROWING BAMBOO

Costs advantage

In terms of growing costs, there were two advantages. First, the total cost of growing *D. strictus* seedlings was only about eight cents apiece. As there were more seedlings than one could afford to put in the ground and raise to maturity, they could be considered "free". Second, as there already was land to grow bamboo, that too could be considered as free. Thus, land and plant costs were out of the cost equation.

Planting and spacing

It was already late June 1995, when the author returned from Bali, and decided to plant 22 000 plants before the end of the rains in September in 11 groups of approximately 2 000 plants each. Each group would be treated to a different combination of environmental conditions. The principal conditions that would vary in a variety of combinations were:

1. Amount of water
2. Shade/sunlight
3. Quantity of high-nitrogen fertilizer
4. Changes in the trace elements the plants were receiving
5. Degree of weed control.

The importance of genetics could be seen by observing variations within each of the 11 groups. Since almost all the 22 000 plants were grown sexually (that is from seeds) rather than by vegetative propagation, there would be plenty of genetic combinations to observe. If one or more of these plants produce the desirable characteristics, one could consider tissue-culturing those plants.

From late June to September 1995, the land was worked with the help of 54 Mayan Indians. Old stone roadbeds were rebuilt, and the bamboo plants were planted in rows, perpendicular to these stone roads. The rows were put 4.5 m apart, with 4 m between each plant within the rows, even though most people the author consulted in Bali had suggested 6×6 m spacing. There were three reasons for this:

1. It was clear that until the bamboo could keep itself free of weeds, weeding would be a major expense. Bamboo with 4.5×4 m spacing was expected to become weed-free sooner than bamboo spaced 6×6 m.
2. It was hoped that having the bamboo closer together would yield straighter culms. The thinking was that the bamboo would not have as great a tendency to curve out towards the light. With closer spacing, this might occur when the bamboo was shorter.
3. A very expensive irrigation system was planned to water approximately 12 000 of the plants. This would cost less, if fewer pipes and less ditches were required.

In one part of the plantation, *D. strictus* was planted on 6×6 m spacing for use as "control" for comparison. The story of planting 22 000 bamboo plants in about three months is filled with drama, frustration, loneliness, heat and exhaustion. Poisonous snakes, burrowing rats and constant breakdown of the 30-ton trench-digging machine had to be overcome. By the end of the rains of 1995, however, all the bamboo seedlings were in the ground; some were even beginning to shoot. About 95% had survived transplanting from their bags, and the 5% mortality was easily replaced.

Growing program

The rainy seasons of the summers of 1996 and 1997 produced progressively superior (in terms of the objectives – large diameter, thick-walled, straight and tall) culms. It was soon clear that the best fertilizer was the "footsteps of the farmer". Since the planting, the author has walked the bamboo rows thousands of times, trying to look at all bamboo rows (if not all plants) at least three times a week. It also helped that this was not an experiment. As soon as it was absolutely clear that something was beneficial, all the plants in all the groups got it. For example, supplementing zinc as a trace element was clearly an improvement. Since the objective was to produce a profit before too much money is invested, and not determining the differences in the output of plants that received zinc and those that did not, all plants received zinc as soon as it was shown to be beneficial for them.

THINGS LEARNT ABOUT GROWING BAMBOO

One important thing that the author learnt in Bali is that *how* bamboo should be grown is very dependent upon *where* it is being grown. Some other things learnt, in the Yucatan, about growing bamboo are listed below with a cautionary note that these too are subject to change.

1. Environmental considerations appear to dwarf genetic considerations. For example, at the end of the first rainy season, bamboo that had received the most favorable environmental conditions had probably 20 times the plant biomass of those plants receiving the least favorable environmental conditions (water, fertilizer, etc.). In contrast, genetically the best plant within a group under similar environmental conditions had less than twice the mass of genetically the worst plant.
2. There was no indication that *D. strictus* loves water and fertilizer. Anyone interested enough in bamboo to be reading this paper would know that. It was found, however, that with lots of water it is possible to gain incremental benefit from the use of larger amounts of ammonium nitrate. Without knowing the value of a culm, it has not been possible to get an economic optimum on fertilizer application. Without sufficient water, too much ammonium nitrate just burns up the plants.
3. Approximately, the principal costs associated with growing bamboo in Yucatan are (in approximate order of their magnitude):
   a. Weeding
   b. Watering
   c. Clump thinning
   d. Initial clearing
   e. Fertilizing
   f. Control of tusas (large burrowing rats that eat rhizomes).
4. In the future, it might be possible to grow *D. strictus* clumps to maturity in Yucatan for about USD 20 a clump (this figure does not include cost for land, plants, the time value of money or the years of effort the author is putting into it). It will not be possible to achieve as low a cost as USD 20 a clump with the present group of plants.
5. Constant concern with trace elements is very desirable. Chemical analysis of soil often varies from one part of the land to the other. In Yucatan, it was not sufficient to deal only with NPK needs.
6. There are no plants in the plantation that have grown all their life under optimum conditions. After two rainy seasons in the ground, the better *D. strictus* plants had solid or almost solid culms 3–4 cm in diameter and about 10 m tall. Early in the third (1998) rainy season shooting, most culms appear to be coming up about 6 cm in diameter within a range of 5–7 cm.
7. Propagation by digging up a ball of rhizome, keeping this moist and cutting back leaves, appears to be a very desirable propagation method. The mother

plants, from which the balls are taken, appear to suffer little or no damage. The new plants grow much more rapidly than tissue-cultured plants or seedlings. It appears that it would be possible to take at least two new plants per year from each plant. Over 80% of these plants survive if planted within hours of taking the rhizome bundle.

In all likelihood, *D. strictus* is not the most rapidly growing bamboo species for this region. Recently, a number of *D. giganteus* plants were planted on 8×8 m spacing. Some of the same rules were found to apply, as with *D. strictus*, though *D. giganteus* seems to require more water and grow faster.

MARKETING CHALLENGES

On the marketing side, the right product for the bamboo is still to be developed. The author has been working with David Flanagan of "The Bamboo Fencer" and other contacts to sell fencing. There are several beautiful fence designs, made from the small-diameter culms that first grew when the plants were very young and which are being cleared from the clumps to thin them. These bamboo culms are compressed into panels enclosed in frames made from a wild bamboo locally called "jimba" (*Guadua amplexifolia*). These fence panels have been selling well, and work is going on to make them more beautiful, easier to ship and less expensive. *G. amplexifolia* will soon be replaced by home-grown *D. strictus*.

In the meanwhile, the author has talked to several companies and entrepreneurs, and many people have come up with some really good ideas. However, markets have not yet been clearly identified, and the question remains: "Who will buy our bamboo?"

Bamboo for Sustainable Development, pp. 449–457
Eds. A. Kumar, I.V. Ramanuja Rao and Ch. Sastry

Silviculture of guadua bamboo in Costa Rica

WILLIAM GARCÍA
Engineer, Forest Department, FUNBAMBU, San José, Costa Rica

Abstract—This paper discusses the response of Guadua bamboo (*Guadua* spp.) to diverse silvicultural works carried out in bamboo plantations in the Atlantic zone of Costa Rica.

Asexual propagation method was found to be the most economically beneficial because of its operational simplicity, ease of transport of planting materials and facility to ensure quality of the material. Equally important is the silvicultural management of groves that permits the penetration of air and light into the midst of the clumps, stimulating the growth of new buds with a larger diameter, as observed in bamboo plantations in Guápiles, Limon province. These plantations of FUNBAMBU are 8 years old, and into the 2nd year of profitable commercial use. About 45% of the Guadua culms are mature enough for harvest on a sustained basis annually, maintaining the harvest cycle and silvicultural management of the plantation. The results are encouraging, although not definitive yet as the stabilization of the production starts only after the tenth or twelfth year of seeding.

INTRODUCTION

The term "Guadua" applies to several woody bamboo species that belong to the largest, most economically important and autochthonous genus of America: *Guadua* spp. The uncontrolled felling of plants belonging to this genus has put some of the species of socio-economic value in serious danger of extinction.

In view of the lack of appropriate research on bamboo propagation techniques and management in South and Central Americas, the National Bamboo Project of Costa Rica undertook in 1988 a massive *Guadua* spp. afforestation program, and a simultaneous silvicultural research project that comprised:

1. Identification of suitable species for planting;
2. Methods for adequate propagation of each species;
3. Determining the density of seeding in different site conditions;
4. Fertilizer application;
5. Establishment of permanent plots in different sites; and
6. Thinning and other silviculture techniques.

There are no known natural stands of giant Guadua bamboos in Costa Rica; there are, however, medium-sized Guaduas – such as *Guadua paniculata* – in the South of the country. In the North of the country, the species *Guadua amplexifolia* is found.

The two giant species of Guadua were brought to Costa Rica from Brazil and Colombia. *Guadua atlantico* from Brazil was planted in the Atlantic zone of the country. The other, *Guadua angustifolia,* was introduced from Colombia by native merchants and planted parallel to the Camino de Mulas in the South zone of the country.

This paper aims to share with people interested in Guadua the experiences of FUNBAMBU regarding the propagation, seeding, management and profitable use of this plant, in order to contribute to the development of cultivation of this valuable species.

PROPAGATION OF GUADUA SPECIES

In general, there are two methods of propagation for bamboo plant: sexual and asexual. In the first method, the seeds are the means of propagation. In the second method, the vegetative parts – such as rhizomes, stems, branches – are employed as planting material. Meristems and some other plant parts are used for *in vitro* propagation in laboratories.

Sexual (seed) propagation

Bamboo is reproduced naturally through seeds. The benefit of this method of reproduction is the genetic variations obtained within a grove. In nature, the variability is important since it offers the genetic material that ensures continued adaptation of a specific species to the habitat.

The sexual reproduction of is seldom used or practiced for Guadua species, as it is difficult to collect seeds because of the lengthy interval between two flowerings. Although sporadic flowering occurs from time to time among individual plants in a plantation, it is not a dependable system as majority of the spikelets turn out to be empty and a very large percentage of the seeds prove unviable.

Asexual (vegetative) propagation

Asexual propagation consists of obtaining individuals from the vegetative sections of the plant. It is the most common propagation system used in the reproduction of bamboo, especially Guadua, wing to the options it presents. Propagation could be done through the rhizome, a section of the culm or branch, or meristem cultivated in laboratories.

Vegetative propagation permits the reproduction of desirable clones (genetically uniform material arising from the same individual). Millions of cellular divisions happen in the vegetative reproduction of a clone and if the clone perpetuates for a long of time, the probability of a spontaneous change or mutation is significant.

Likewise, mutations or chromosomal changes could be induced artificially or accidentally. In general, for the asexual propagation of Guadua, culms from 2 to 3 years old plants have to be selected. This is an intermediate age when the culms can develop roots easily.

Propagation by means of rhizomes
Rhizomes are part of the culms, and they grow underground constituting the mechanical support of the bamboo. The plant uses them to store nutrients, and they also function as the organ in charge of multiplication of stems through their buds.

Guadua has a rhizome of the pachymorph (sympodial) type, which induces the plant to grow as clumps that radiate outwards from one point, with older culms in the middle and newer ones in the outer periphery.

Culms from 2 to 3 years old plants are selected, cutting them 1 meter above the ground. The rhizomes are dug out of the ground using a shovel, taking care not to damage the buds. This method is very onerous and is recommended only for ornamental transplantations or small groves; but never for large plantations as an established grove could be destroyed in trying to start a new one. It is recommended that the transplantation be started early in the morning and at the beginning of the rainy season.

Propagation by culm sections
Culms from 2 to 3 years old plants with vigorous and healthy buds are selected. They are cut in 3-meter lengths to facilitate transportation and avoid dehydration. The soil is then prepared, with furrows dug 2 meters apart, and the culms are placed into them taking care not to damage the buds.

Budding occurs between the second and ninth month after planting. It is recommended that the plants be left for a period of more than 1 year at the site, so that the buds grow and develop into thick plantlets; thin and small plantlets are not suitable for the plantation as they might be smothered by the undergrowth. In about 1 year or so, the rhizomes will be stronger, and the culms taller and their diameter larger. Cutting for transplantation has to be done within 2 years, as the rhizomes will be large and heavy after that, making the job difficult.

In very dry places, it is necessary to make a cut on each internode, pour in some water through the cut, cover the cut with the same piece of bamboo or with a plastic sheet, and place the culm over a 5–7 cm layer of soil. The base portion is more suitable for planting, as it has shorter internodes and larger and vigorous buds. It is recommended that the culms be taken from the interior of the clump since they would have higher carbohydrate content, as less light falls on them, which promotes better rooting.

Propagation by branches
The branches at the basal section of Guadua may be used as reproductive material, using pre-rooting or cutting method. The main or primary branch is selected, and a bud available at a height of about 5 cm is covered with moss or soil rich in humus. This is then covered with aluminum or black plastic sheet, leaving a hole to add water. In a few months, rooting takes place. Once rooted, the branches are separated from the culm and transplanted in the field, nursery or greenhouse.

Reproduction by means of cuttings consists uses lower branches, separated from the stem and buried in prepared propagation beds at the greenhouse or in plastic bags containing soil and nutrients. This operation should not be carried out under sun to prevent dehydration of the branches; it is advisable to carry out this procedure in shade. Chemical growth stimulators may be used for rooting.

Propagation using shoots
The planting material here is the shoots or young culms, of about 1–3 cm in diameter and a height up to 1 meter, which appear at the center of the groves after the large culms have been harvested. These are removed with their roots and rhizomes intact. The leaves are cut off to avoid dehydration. These shoots could be kept in plastic bags, where they can remain for several months until they develop to a stage adequate for transplantation.

In cases where there is excess development of roots, it would be convenient to trim them. When planting Guadua, it is necessary to press down the soil well so that the roots remain in optimum contact with the soil, preventing the formation of air-pockets.

ESTABLISHMENT OF PLANTATIONS

In establishing a plantation, ecological factors – such as precipitation, soil and land type, altitude and wind speed – as well as the purpose of the plantation have to be given due consideration before the planting time and density and the silvicultural work required can be determined.

Land conditions, rainfall and altitude are of great importance in the development of the species. Although bamboo has a good adaptability and can grow in the majority of lands, it flourishes when the soil is medium deep and well drained. The best time to start planting is at the beginning of the rainy season, as the plantlets require a lot of moisture. It is estimated that 5 months of rain would be enough to guarantee the success of a plantation. Heavy precipitation, soils of heavy textures, high altitude and thick undergrowth are factors that contribute to the failure of a plantation. Guadua is very susceptible to undergrowth, mainly grass and climbing plants, which may smother young plants.

Under adverse conditions – such as eroded, hard or shallow soils, extremely dry conditions, irregular felling, and altitudes of over 1 200 meters above sea-level – the plants would show many undesirable characteristics such as high mortality, slow growth and low reproduction of buds. In windy regions, deterioration of buds could occur owing to the constant wind-induced rubbing of culms and branches against each other.

Preparation of land

Land is prepared – loosening the soil, removing weeds, digging planting holes, etc. – to provide the soil with the best conditions for the development of bamboo plants. Careful land preparation will help increase plant survival rate and promote rapid growth of young plants.

Planting density

Because of its tendency for rapid multiplication, it is crucial to control the density of planting during the cultivation of Guadua species. A rope grid may be used to make optimum use of the land in terms of orderly rows and correct density of the plants.

The density of the plantation will depend directly on the site. The initial density, defined by the distance between two plants, will be a function of the objectives of production and of environmental factors such as altitude, precipitation, soil type, etc. Densities from 5×5 m to 5×10 m would be adequate to manage the competition for space between the plants and to ensure optimum production per hectare.

Planting

As said earlier, the land has to be prepared before planting can start. A weedicide may be applied on the soil to prevent undergrowth appearing before the plants establish themselves. The land can be cleared totally or partially in 2-m wide rows. Once the land is prepared, the next step is to dig holes for planting the bamboo. The size of the holes will depend on the size and age of the planting material and the soil. Although holes of 30×30 cm size will suffice in most cases, in compact soils it is better to make larger holes and fill them later with loose soil to permit better root penetration.

Maintenance

The maintenance of a bamboo plantation involves a series of activities or cultural work that has the following objectives:

1. To guarantee the survival of the plants.
2. To help obtain adequate phytosanitary conditions.
3. Speed up the growth and production of the plantation.

A newly planted grove should be protected against weeds, cattle, fire, pests and diseases. It is also important to ensure timely supply nutrients to ensure rapid establishment and best development of the plants.

Control of undergrowth

One of the first and important tasks of maintenance of a plantation is the elimination of plants of other species that compete with bamboo plants for space, water, nutrients and light. In the first 12 months after their planting, the bamboo plants will not in a position to compete with these weeds. Hence, it is necessary to remove the weeds that sprout without damaging the bamboo plants. This weeding activity needs considerable investment and therefore, a judicious decision has to be taken by balancing the most economical method with the most effective.

Weeding can be carried out manually or mechanically using small tools and equipment, or using a chemical weedicide.

Clearing of the site and weeding may be undertaken fully, or by patches, rows or plots, depending on the kind of undergrowth present and the environmental conditions. However, it is not advisable to weed several consecutive plots and then leave some areas untended, because of the likelihood of the undergrowth strengthening in those untended areas making it necessary to carry out weeding over the entire area.

In very humid climates, climbing plants, such as those belonging to the *Convolvulaceae* family, pose a serious problem. Tall grasses can also block the light reaching the bamboo plants to the point of choking them. These weeds should be removed immediately on their detection.

Control of pests and diseases

Pests and diseases of economic importance have not yet been detected in the Costa Rican bamboo plantations, possibly because of the small size of the plantations. A strict surveillance, however, is in order to detect them at the first instance. Establishment of live barriers of tall plant species is also recommended in order to prevent infection spreading from or to neighboring groves.

Fertilizer application

Bamboo, particularly a species like Guadua with its rapid growth characteristics, requires adequate amount of nutrients and certain elements for uptake from the soil. If there is deficiency of any of these in the soil, it should be rectified by adding the lacking component in the right amount and at the right time. Thus, the type and quantity of fertilizer to be applied depends on the results of a chemical analysis of the soil. However, this could be estimated to some extent based on observations of the behavior of the plantation.

In general, after 3 months of planting the bamboo, 50–75 g of 10:30:10 NPK should be applied at the base of the plant. When the rainy season is over, 75 g of nitrate or urea needs to be applied to each plant. At the beginning of the rainy season, 60–75 g of the complete formula must be applied.

Association of Guadua with other cultivations

Normally, bamboo is planted in corn, beans or pumpkin fields. This is extremely advantageous because the other plants keep the land free of underbrush and the bamboo plants benefit from the care given to the other crops. However, owing to its fast growth, its capacity for rapid multiplication and its long thorny branches at the bottom, Guadua cannot share its site with other cultivations after its 2nd year of growth.

Silvicultural management of plantations

The management of Guadua plantations is next only to propagation in importance, as the quality of bamboo is almost entirely dependent on it.

Establishment of a plantation is an economic investment, and it should be seen as any other business that requires an adequate return on investment. Silvicultural treatments form an essential part of the management regime that rewards investments with culms and shoots of high quality and quantity. The success of the management regime relies on its planning and coordination. The development of a plan has to consider a detailed examination of the resource in terms of:

1. Composition per species (if there are more than one species, variety);
2. Quantity of culms according to diameter;
3. Quantity of culms in different age categories;
4. Vegetative cycle of the species; and
5. Duration of each of the different stages (juvenile, seasoned, mature) in the life cycle.

This information would help determine the percentage of thinning and/or extraction required, as well as the cutting cycle or interval of time between two harvests.

Branch cutting

This involves the removal of superfluous branches on the culms. The main objective is to permit circulation of air and penetration of light to the center of the clump. This will stimulate the birth of buds, avoid very humid atmosphere that induces rotting of buds, improve the quality of the final product and improve accessibility within the grove.

A curved handsaw or a motorized one may be used for branch cutting. The saw must be very sharp so that it makes clean cuts and does not damage the outer skin in the remaining portion. The idea is to avoid giving an entry point for any pathogen that might deteriorate the quality of the culms.

If a machete is used, the first cut should be made by an upward stroke and then another one downwards. The branches should be pruned to a height of at least 2 m from the culm base.

Thinning

The objective of thinning – selective removal of culms – is to concentrate the potential for growth on the better culms and thus favor the development of plantation. The remaining culms will get the opportunity to grow in good shape and size.

There are two types of thinning:

a. Sanitary thinning. It consists mainly of the elimination of weak, damaged, bent or broken culms and branches.
b. Commercial thinning. It consists of cutting of culms, with diameters below 8 cm, for use in the manufacture of furniture and other products.

Harvesting

It is necessary to plan the harvesting of culms by first taking stock of the plantation to know the total number of culms, number of mature culms and their diameter. It is advisable to harvest only 33–45% of the mature Guadua in order to maintain a continuous cutting cycle. The idea is to maintain a maximum sustainable yield.

When the felling is carried out, no large spaces or clearings must be left within the plantation, as this could induce bending in developing culms. The cutting should be done close to the first node, in order to avoid water collecting in the stump's internode and promoting rotting. The culms should be felled in such a way that they do not damage the remaining culms.

After each cutting, Guadua plantation should present such a density that the apical branches of the remaining culms serve as support to the neighboring ones. Besides being a guide for harvesting, this will help avoid lodging of the culms in strong winds.

Commercial plantations

A model of a profitable *G. atlantico* plantation is presented below. It is assumed that the plantation is 6 years old and has a planting density of 5×5 m. The data are per hectare.

Number of buds	: 450
Juvenile culms	: 850
Seasoned culms	: 900
Mature culms	: 976
Total	:3,176

Harvesting of 45% of the mature Guadua (876 culms) will yield 439 culms.

Tables 1 and 2 provide estimates of the annual extraction of culms from six-year plantations in the Atlantic region of the country (265 m above sea level, 4 500 mm annual precipitation and sandy soil).

Table 1.
Estimate of production per hectare of *Guadua atlantico* culms with three planting densities and harvest rates of 70%, 45% and 33% of mature culms

Planting density (m)	Harvest		Culm yield (number/ha)				
	rate (%)	Year 6	Year 7	Year 8	Year 9	Year 10	Year 11
5 × 5	70	515	1030	2060	3000	3090	3090
	45	331	662	1324	1986	1986	1986
	33	243	486	972	1458	1458	1458
6 × 6	70	358	1074	1790	2506	3090	3090
	45	230	690	1150	1610	1986	1986
	33	169	507	845	1183	1458	1458
7 × 7	70	263	789	1578	2104	2630	3090
	45	169	507	1014	1352	1690	1986
	33	124	372	744	992	1240	1458

Table 2.
Estimate of production per hectare of *Guadua sur* culms with three planting densities and harvest rates of 70%, 45% and 33% of mature culms

Planting density (m)	Harvest	Culm yield (number/ha)					
	rate (%)	Year 6	Year 7	Year 8	Year 9	Year 10	Year 11
5 × 5	70	1080	2160	3240	5400	5400	5400
	45	792	1584	2376	3960	3960	3960
	33	578	1156	1734	2890	2890	2890
6 × 6	70	750	1150	2250	3750	5400	5400
	45	550	1100	1650	2750	3960	3960
	33	401	802	1203	2406	2890	2890
7 × 7	70	551	1102	1653	3306	4408	5400
	45	404	802	1212	2424	3232	3960
	33	294	588	882	2058	2352	2890

CONCLUSION

Guadua is a promising species in America, with several potential uses. With widespread cultivation and rational utilization, it will make a very valuable contribution in relieving the pressure of the Latin American forests.

For plantations to be an economic success, their objectives, silvicultural aspects and environmental conditions of the site have to be well understood. These data would decide the techniques to be followed in the selection of the propagation material, planting density and the management regime for the plantation.

A Guadua plantation forms, by itself, a whole ecosystem and it is necessary to manage it in a way that satisfies the socio-economic objectives without breaking the equilibrium established by nature. Undoubtedly, Guadua is a species with a great afforestation potential. It is, therefore, urgent to continue research on the following silvicultural aspects of this underutilized genus:

- Planting densities in sites of different qualities;
- Fertilizer application;
- Morphological differences among species;
- Optimum harvest percentages and cycles for different species and sites;
- Population dynamics of different species; and
- Vegetative cycle of different species.

It is also recommended that research be also directed towards the profitable use of the resource and industrialization of production.

Bamboo for Sustainable Development, pp. 459–462
Eds. A. Kumar, I.V. Ramanuja Rao and Ch. Sastry

Bamboo plantation on a land after shifting cultivation in Laos

ETSUZO UCHIMURA
Application Forum for Bamboo Resources (AFOBAR), Laos

Abstract—Natural distribution of bamboo in Lao P.D.R. (Laos) is at an altitude from 200 to 1 000 m above sea level. There are, however, quite a few natural bamboo forests in the northern part of Nam Ngum Reservoir, at Vangvieng district, Vientiane province, where the area of this study is located. There are also many small bamboo clumps grown on land in fallow, as soon as the shifting cultivation is completed. Clumps of second growth start on the bare land in fallow, which has good conditions – for example, no shade or weed – to grow bamboo. During the season, the rains make it very easy for bamboo seeds to germinate or culm cuttings to sprout.

This type of bamboo cultivation is increasing in the tropical Asian countries, and needs to be encouraged in Laos also. If such kind of bamboo cultivation is not practiced when the land is in fallow, the productivity of land is bound to fall. Further, the dwindling natural bamboo forests calls for urgent remedial measures. Therefore, for the conservation of natural bamboo resources and to keep the land productive, a model plan for bamboo plantation is proposed for slash-and-burn agriculture areas.

INTRODUCTION

Any land use must be rational and optimized in terms of sustainability, taking into consideration aspects such as land productivity, gradient and workability, in addition to environmental conservation needs. In Lao P.D.R. (Laos), agriculture and forestry are identified as the two key land use categories. Area under bamboo spreads from small hillocks to mountains. Much of the productive land is reserved for agriculture, the predominant economic activity in the country. Much of the land on gentle slopes and at altitudes lower than 500 m above sea level is used for cultivating rice (upland rice fields) by slash-and-burn method. These areas are called symbiosis areas, as people must live there in accord with the prevailing natural environment.

Farmers in much of the study area at Vangvieng district, Vientiane province, still practice slash-and-burn agriculture. Annual rainfall in this area is about 3 000 mm and the average annual temperature is 26°C. Land productivity has reduced because of short-rotation land use in recent years. Several years without proper tending deteriorate most of the land, and small bamboo clumps begin to grow there within one year after shifting cultivation. Farmers, however, dislike these bamboo clumps since their culms are short and small in diameter, which reduces the scope of their use.

BAMBOOS IN THE STUDY AREA

In general, distribution of bamboo in Laos is at altitudes ranging from 200 to 1 000 m above sea level. In the study area, a few large-sized bamboos can be seen near farmers' houses or riverbanks. Some of these are growing naturally, while others are those planted by farmers for their own consumption. According to a survey conducted, secondary growth of bamboos – bamboos that have grown after slash-and-burn cultivation – is occupying a very large area (Table 1): about 85% of the total area of bamboo in the study area. However, the culms of bamboos in this secondary growth – such as mai sout (*Oxytenanthera parviforia*) and mai hea (*Cephalostachyum virgatum*) – are very small in the first year and have no economic value.

Table 1.
Distribution area of bamboo in the study area (%)

		Village	Watershed area	Others	Total area
Natural bamboo forest	(a)	5.89	602.54	1,243.24	1,851.67
Secondary growth	(b)	3,505.15	2,942.05	3,579.50	10,026.60
(a) + (b)	(c)	3,511.04	3,544.59	4,822.64	11,878.27
(a) ÷ (c) × 100	(%)	0.17	17.00	25.78	15.59

It can be understood from this that small bamboos and fast-growing coppice start to establish on the fallow land soon after slash-and-burn cultivation. Although mai hea (*C. virgatum*) can be used for wall, roof and other construction application, large-sized bamboos – such as mai hok (*Dendrocalamus brandisii*), maiphai ban (*Bambusa blumeana*) – are the more useful bamboo species. The area can be improved by making it more high productive and planting more land with large-sized bamboo species.

METHODOLOGY OF BAMBOO PROPAGATION

Asexual propagation

There are two methods used for this kind of propagation.

(1) Offset: Commonly used with non-clump forming type of bamboo species that grow in the temperate zone because these tend to have long rhizomes. Unfortunately, clump-forming types of bamboo that grow in the tropical region have short rhizome, and culms grow directly from buds attached to the bottom of

the mother culms. In such bamboos, cutting of parts for offset method might damage the clump. Also, transportation of large-sized offsets is difficult and their handling is laborious. Offset planting, therefore, is not recommended in the case of clump-forming type of bamboo species.

(2) Culm cuttings. Typical propagation method adopted for tropical bamboo species. Young culms sections, cut horizontally to have two nodes per section, are planted horizontally in pits dug in the ground to about 30 cm depth. Watering is necessarily one or twice a day if no rainfall. New sprout and roots come out from the node within one month. Out-planting can be done in about 3–4 months after nursery practice.

Sexual propagation

Typical sexual propagation is using seedlings. Bamboo flowers in long intervals and it often occurs without any warning. In general, clump-forming types of bamboo come into flower more often than non-clumping bamboo and also yield more seeds.

NURSERY PRACTICE

Seeds are sown in a pot or seedbed as soon as possible after seeding, because germination rate of decreases rapidly on storage, even in cold storage. Normally, seeds germinate within a week or so. The soil of nursery bed or pot is the most important factor in nursery practice. Seedlings and cuttings use their own nutrient at the first stage; thereafter, they need more nutrient supply from the soil, sunlight, fertilizer, etc. Growing seedlings need shade in the beginning but this is not necessary with cuttings. Regular watering, and easy drainage, is needed for the growth of bamboo shoots. Seedlings can be out-planted when they reach a height of 40 cm or more or 1 cm in diameter. Usually this takes about three to four months after sowing.

PLANTING

Planting primarily involves land preparation. Large tree stumps and bushes are dug up from the planting site (small stumps and bushes may be just trimmed to save labor). The size of planting hole is about 50×50×50 cm and planting distance is kept at 5×5 m or 5×6 m. The planting density (number of seedlings or cuttings planted per hectare) is maintained between 300 and 400.

Plantation

Plantations or artificial bamboo forests are raised from seedlings or cuttings. For sustainable and profitable plantations, it is necessary to have a rational management system that includes all necessary cultural measures such as density control, clearing land of weeds, stand improvement, trimming of branches, etc. Bamboo stands need such tending every year. The age of standing culms in the stand should be one to four years; culms over five years of age, as well as the damaged ones, should be felled. The culms should ideally be three to four years or above when harvested; below that age, they would have high moisture content that makes them soft and unsuitable for any worthwhile use.

Natural regeneration

Clumps that grow in the first year consist of four to six culms. As the study area has much rainfall and a long shooting season, bamboo shoots will be able to sprout three times a year. Seedlings that grow in natural bamboo forests after flowering and seeding must be used to the extent possible for regeneration of natural forests. In warranted, the seedlings may be brought to a nursery and grown there till they are old enough for out-planting.

It has been observed that the number of clumps that grow naturally after shifting cultivation is so high that small-sized or poor clumps could be removed without causing any harm to regeneration. It would be ideal to select 400–500 clumps per hectare and then whittle down the finally density of clumps to 350–400 clumps.

CONCLUSION

People in Laos do not use many bamboo products. However, the experiences of many Asian countries that plant and use bamboo point to a direction that Laos must take for managing bamboo as a valuable and sustainable natural resource.

Large-sized bamboo species need to be raised in plantations, both to generate income from the products that can be made from the culms and to ensure sustainable and productive land use after shifting cultivation.

Section 3:

Bamboo Construction and Engineering

Bamboo for Sustainable Development, pp. 465–480
Eds. A. Kumar, I.V. Ramanuja Rao and Ch. Sastry

Effects of smoke drying as a pretreatment for bamboo

TAKAYA NOMURA
Wood Research Institute, Kyoto University, Kyoto, Japan

Abstract—Smoke drying is an effective technique for processing bamboo and bamboo materials, and for bamboo charcoal making. It prevents bamboo culm from splitting through stress and moisture conditioning. The soot and other pyrolysis chemical components that adhere to the surface of bamboo culm protect it from mold and other biological agents of deterioration. Furthermore, smoke drying of bamboo used in charcoal making helps shorten the moisture conditioning time of the carbonizing materials, protect the material from biological deterioration, retain good form and increase the rate of charcoal production.

INTRODUCTION

Smoke drying has been known in Japan since early times as one of the wood drying methods. The same method is used for preserving and flavoring food. However, smoke drying of wood has been almost neglected up to the present because the drying efficiency and precision of the method were claimed to be low. The drawbacks reported for the method are that: (1) it is difficult to control temperature and humidity; (2) it generates uneven drying owing to circulation of smoke; (3) the surface color of the material is turned black by adhesion of soot; and (4) it is a fire hazard since temperature control is difficult. These demerits, however, were based on observations of the existing smoke drying kilns, considered as non-mechanical, and the pre-modern process of charcoal making; they were not based on proper scientific evidences. The non-existence of literature on smoke drying method makes this point amply clear.

In most existing wood drying techniques, air and steam have been used as the heat transfer mediums, and conduction and convection have been used as heat transfer methods. As a heat transfer medium, air has very low efficiency, while both conduction and convection are time-consuming and difficult to control as methods of heat transfer.

In addition to heat conduction and convection, heat radiation can be used in smoke drying as the method of heat transfer. As a medium of heat transfer, mixed gas – unburned gases – can be used in place of air. Unburned gases are the product of the incomplete combustion of a heat source, such as waste wood or bamboo.

Smoke drying will be an effective technique to process all kinds of bamboo. Smoke-dried bamboo materials can be used in construction, agriculture, fishery, articles of daily use, interior and exterior decoration, furniture and other applications. The process is also suitable for pretreatment in the production of bamboo charcoal. Smoke drying is an effective process to ensure a certain level of quality.

This paper describes how heat is transferred in smoke drying, the mechanism of smoke drying and the effective results of this process in bamboo drying and bamboo charcoal making.

HEAT TRANSFER DURING SMOKE DRYING

Smoke drying has some special features. It uses heat radiation in addition to heat conduction and convection. Another feature is that in place of air, an inefficient heat transfer medium, it uses unburned gas, a product of the incomplete combustion of a heat source, such as waste wood or bamboo. The main components of this gas are air, steam, carbon particulates, and a pyrolysis product called wood or bamboo vinegar.

Let us check the heat value and thermal conductivity of each component. Specific heat is the quantity of heat (expressed in calories) required to raise 1 gram of a material by 1°C in temperature. Specific heat of air is 0.2366 (20–440°C); specific heat of water is 1 (at 15°C); and that of carbon is 0.204 (20°C). The mean specific heat of wood is 0.324 in the range of 0°C to 100°C.

Green wood log or green bamboo culm has high moisture content (100–200%). Because water has a high specific heat, when the kiln temperature rises in the smoke drying process, the moisture in the log or culm serves to prevent rapid increase in log or culm temperature.

As the specific heat of carbon particulate floating in the kiln is lower than that of air, the mixed gases of carbon and air have a larger heat accumulation effect than air alone.

Thermal conductivity of air is 5.33×10^{-6} (Cal* $cm^{-1}s^{-1}deg^{-1}$) at 0°C and 6.81×10^{-5} at 100°C. Thermal conductivity of carbon is 8.49×10^{-3}, and thermal conductivity of water is 1.64×10^{-3}. It can be seen that the thermal conductivity of carbon is about 125 times as large as that of air at 100°C, while thermal conductivity of water is about 24 times as large as that of air. This means that the thermal conductivity of the mixed gases is greater than that of air alone, and greater the quantity of carbon or steam in the kiln, the greater the thermal conductivity.

Heat conduction is a phenomenon in which heat is transferred, through a material, from a hotter part to a cooler. Convection is the process by which heat is transferred by the flow of a fluid, such as gas and liquid. Radiation is a phenomenon in which energy is directly transferred to a material in the form of electromagnetic waves, and the electromagnetic waves are absorbed by the material and changed into thermal energy. This radiation does not require heating of mediums such as air and water. One merit of smoke drying process is that it can effectively utilize heat of blackbody radiation of carbon contained in smokes. In the temperature range of smoke drying process (20–150°C), electromagnetic waves radiated are of wavelengths 7–10 microns, in the range of far-infrared rays. Thermal radiation of these electromagnetic waves is absorbed by the log or culm to raise the inner temperature.

MECHANISM OF SMOKE DRYING

The next aspect to examine is how the dimensions of wood and bamboo are stabilized and their warping, cracking and distortion reduced by smoke drying. An understanding of the field of forces acting on a tree or bamboo growing on the ground will be helpful in this: (1) the tree or bamboo must stand vertically against the gravity; (2) it must support its own weight; and (3) it must resist against external forces, such as wind. To meet these requirements, trees and bamboos have various contrivances, and growth stress is one such.

In the case of timber, stresses change at one-third of the radius from tensile stresses to compressive ones in the direction perpendicular to the cross section (direction of fibers) of the trunk. The largest compressive stress is at the pith. In addition to this, there are forces that act in the circumferential direction of the trunk to clamp the entire trunk, just as the forces exerted by the hoops of a barrel. When these forces are more complex, tissue structures are formed as if long pieces of cloth are bound over the surface of the trunk spirally, and in S-shape and Z-shape alternately. These growth stresses cause warping, distortion, twisting and cracking of timber. Hence, removal of these growth stresses is the key to successful utilization of timbers.

It is said that for any logs of the same kind of tree, the magnitudes of growth stresses in the outer circumferences of the trunks are virtually identical. This means that the shorter the diameter of the log, the greater the distortion of the timber.

A large growth stress would amount to several hundred kilograms per square centimeter. In the case of tropical woods, when a board that shows a bent owing to growth stresses is to be made straight, a large load of ten tons may be needed. If such large growth stresses are released when a tree 1–2 m in diameter is cut down, the log may split from one cut-end to over 10 m longitudinally.

Such large growth stresses are generated by groups of cells that are several millimeters in length, and from ten to several tens of microns in width. The major chemical components of these cells are cellulose, lignin and hemicellulose. Cellulose forms the framework. Lignin has a shapeless matrix and serves as cement. Hemicellulose binds cellulose and lignin together. The respective cells are formed by these components, and distortions generated by the individual cells are integrated to appear as growth stress of the entire trees.

The Wood Research Institute of the Kyoto University has been working hard to find an effective way of controlling the growth stresses of trees and bamboos. Smoke drying of woods and bamboos attempts to control this growth stress by controlling the very root of the problem – the distortions that are generated in the individual cells. Cell walls of plant cells of woods and bamboos are made of cellulose, lignin and hemicellulose. Cellulose gathers in the form of fibers to form a framework of cell walls. This framework has the same role as the reinforcing bars in concrete. Lignin, which surrounds the framework of cellulose, plays the role of concrete. Hemicellulose works as a binder that makes cellulose and lignin compatible with each other. At the formation of these cell walls, forms that originate various distortions are also made by cellulose fibers. Lignin deposits around the fibers to fix these distortions. As a result, the individual cells have a variety of internal stresses. The magnitude of stress is proportional to the magnitude of distortion. When this fixed distortion is eliminated or moderated, the stress will be reduced.

Lignin, which fixes the distortion, has an interesting property. Components of woods and bamboo have their thermal softening points. When wood and bamboo are in a dry state, the thermal softening point of lignin is in a range of from 134–235°C. However, when wood and bamboo have much moisture (like in green wood), the thermal softening point of lignin drops to 77–108°C. Smoke drying of woods and bamboos uses this effectively to regulate distortions at low temperatures.

SMOKE DRYING OF BAMBOO

While using bamboo materials, two issues assume importance: how to prevent bamboo culm from splitting and how to protect it from biological deterioration. The most important problem is splitting. Growth stresses, and hygro stresses owing to different moisture contents in the circumferential direction of each internode, are the main causes of splitting. Hence, it is important to know the moisture content and the internal stresses in bamboo culm.

Moisture content in bamboo

Fig. 1 shows the distribution of moisture content in each internode of the culm of Moso bamboos (*Phyllostachys heterocycla*). The difference in the moisture content between basal part and upper part in the culm is two or three times. It is known that moisture content in the basal internodes reduces with age. Fig. 2 shows the distribution of moisture contents in 100 equal parts of the same internodes. That is, one internode is divided into 10 parts in the circumferential direction (radial direction), each part 36°. The internode is also divided into 10 parts in the longitudinal direction. As shown in Fig. 2, moisture content in the culm is distributed periodically in the circumferential direction, and the periodical changes in moisture content in each longitudinal portion are nearly the same. The average difference in moisture content between maximum and minimum is 16 points.

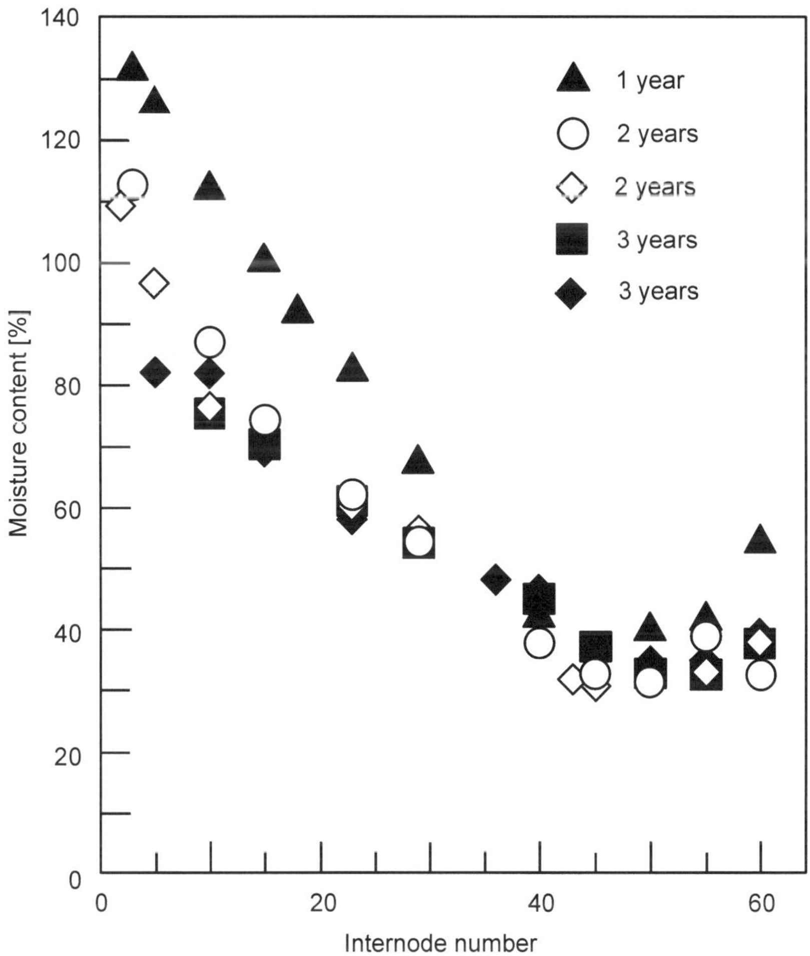

Figure 1. Internodal distribution of moisture content (*Phyllostachys heterocycla*)

The variation in moisture content in the same internode may be related to the cell turgor pressure and may be affected by the formation of internal stress, which is formed against external force.

Growth stress of bamboo

As mentioned earlier, bamboo must stand vertically against the gravity, support its own weight and resist against external forces. Growth stress is one of the devices that help bamboo to meet these requirements. It is well known that the structure of the bamboo culm helps it to resist external forces; however, the formation of prestress in bamboo culm and the mechanical formation of prestress as growth stress have not been clearly understood. For the utilization of bamboo materials, knowledge of this stress conditioning is as important as moisture conditioning.

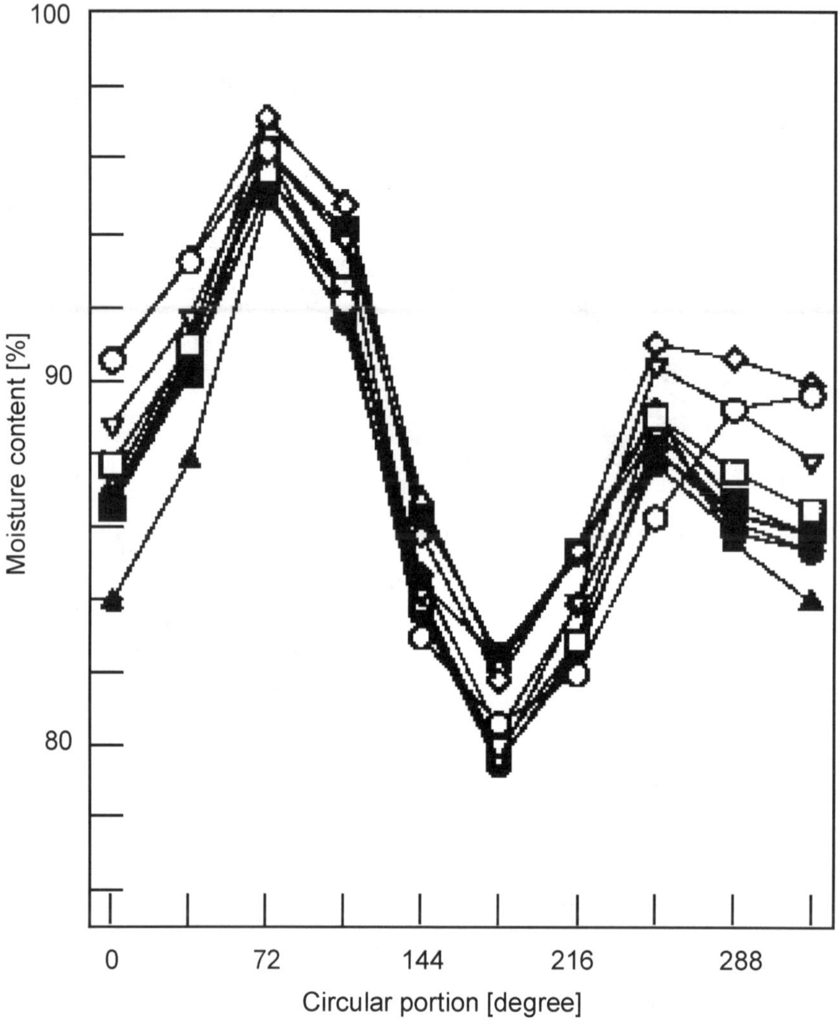

Figure 2. Moisture content, circular portion's distribution in 20th node

Notes: Each mark in the figure shows the longitudinal portion divided into 10 parts. The order from top to bottom is: ○, ◇, ▽, □, ▼, ●, ◆, △, ■, ▲

The changes of growth stresses were estimated as a residual stress by released strains. Distributions of these released strains on the peripheral surface of the bamboo culm were measured by strain gauge method. The peripheral surface of each sample was divided into 10 equal parts and the released strains in each part were measured. Strain gauges were stuck on the surface of each internode in directions parallel and perpendicular to the longitudinal direction. Figs. 3 and 4 show the results.

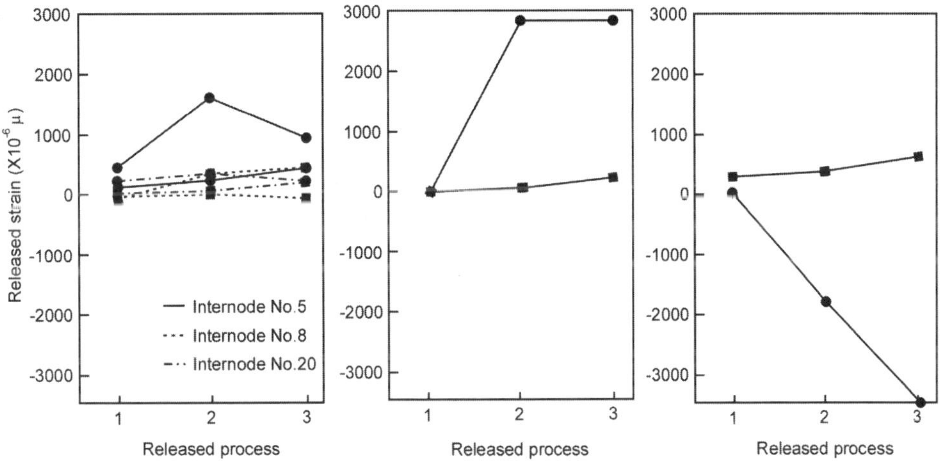

Figure 3. Comparison of growth stresses as released strains in *Phyllostachys heterocycla* (left), *Phyllostachys bambusoides* (center) and *Dendrocalamus giganteus* (right)

Notes: 1 = Released longitudinal stresses; 2 = Released radial stress; 3 = Released strain stress when separated 10 parts in radial direction. ■: Longitudinal direction ●: Radial direction

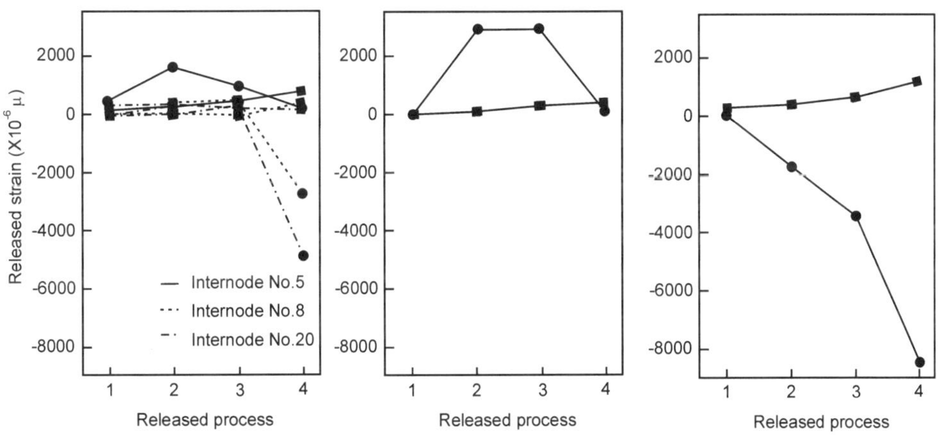

Figure 4. Comparison of changes of released strains when growth stresses released in *Phyllostachys heterocycla* (left), *Phyllostachys bambusoides* (center) and *Dendrocalamus giganteus* (right)

Notes: 1 = Released longitudinal stresses; 2 = Released radial stress; 3 = Released strain stress when separated 10 parts in radial direction; 4 = Released strain after growth stress is freed. ■: Longitudinal direction ●: Radial direction

Figure 3 shows the comparison of growth stresses as released strains among three different bamboo species – *Phyllostachys heterocycla* (Moso), *Phyllostachys bambusoides* (Madake) and *Dendrocalamus giganteus* (DG). Fig. 4 shows the changes of released strains when the growth stress released. As shown in Fig. 3, growth stress in the basal internode differs qualitatively among species. As *Phyllostachys heterocycla* and *P. bambusoides* are flora of the Temperate Zone, there is apparent compressive stress on the longitudinal and the circumferential directions of the peripheral surface. The released strain of *P. bambusoides* is about twice that of *P. heterocycla* in the circumferential direction and about one-third in the longitudinal direction. *D. giganteus* being a tropical plant, there is apparent tensile stress in the circumferential direction and apparent compressive strain in the longitudinal direction. The released strain of it is about 3.8 times that of *P. heterocycla* and 1.8 times of *P. bambusoides* in the circumferential direction, and 1.4 times of *P. heterocycla* and 4 times of *P. bambusoides* in the longitudinal direction in comparison with absolute values. As stress is proportional to the modules of elasticity and strain, the growth stress in the basal parts of *D. giganteus* is higher than those of *P. heterocycla* and *P. bambusoides*. This means that the use of *D. giganteus* as a material is more problematic. After releasing of growth stress, the change of the released strain after 24 hours is clearly different between species and portions of the culm.

This change of the released strain is considered a combination of stress relaxation change and drying stress. As the moisture content reduces, shrinkage in the circumferential direction increases and the released strain goes into negative values. When the structure of bamboo is considered, shrinkage in the longitudinal directions is not very large and shrinkage with drying mainly appears in the circumferential direction. In this process, compression stress following shrinkage appears on the outer surface and tensile stress on the inner surface. As moisture content decreases, these stresses increase and splitting occurs on the surface of the culm, which cannot bear against these stresses. How to prevent this phenomenon is the most important issue in using bamboo materials. These growth stresses have a particularly adverse effect on charcoal making: when bamboo is exposed to high temperature, splitting and crushing of bamboo charcoal occur.

Smoke drying of bamboo

It is important to control moisture content and internal stress (as growth stress) in order to protect bamboo materials against splitting and warp. Smoke drying is the most excellent method for the conditioning of moisture and stress in bamboo and wood, not only as the method of stress relaxation but also to bring out larger prestresses without splitting or warping the bamboo culm.

The construction of the smoke drying furnace is shown in Fig. 5. The green bamboos are piled up in a smoke drying chamber (R1), which can be sealed, located adjacent to the combustion-gas generator (R2). The upper section of R2 is connected to the upper portion of R1 by means of a combustion-gas passage (A). The lower section of R2 forms a combustion chamber (B) in which suitable fuels, such as wood wastes, are burned to produce hot combustion gas and unburned gas. R2 is connected to the air-inlet conduit (C) for introducing fresh air into B.

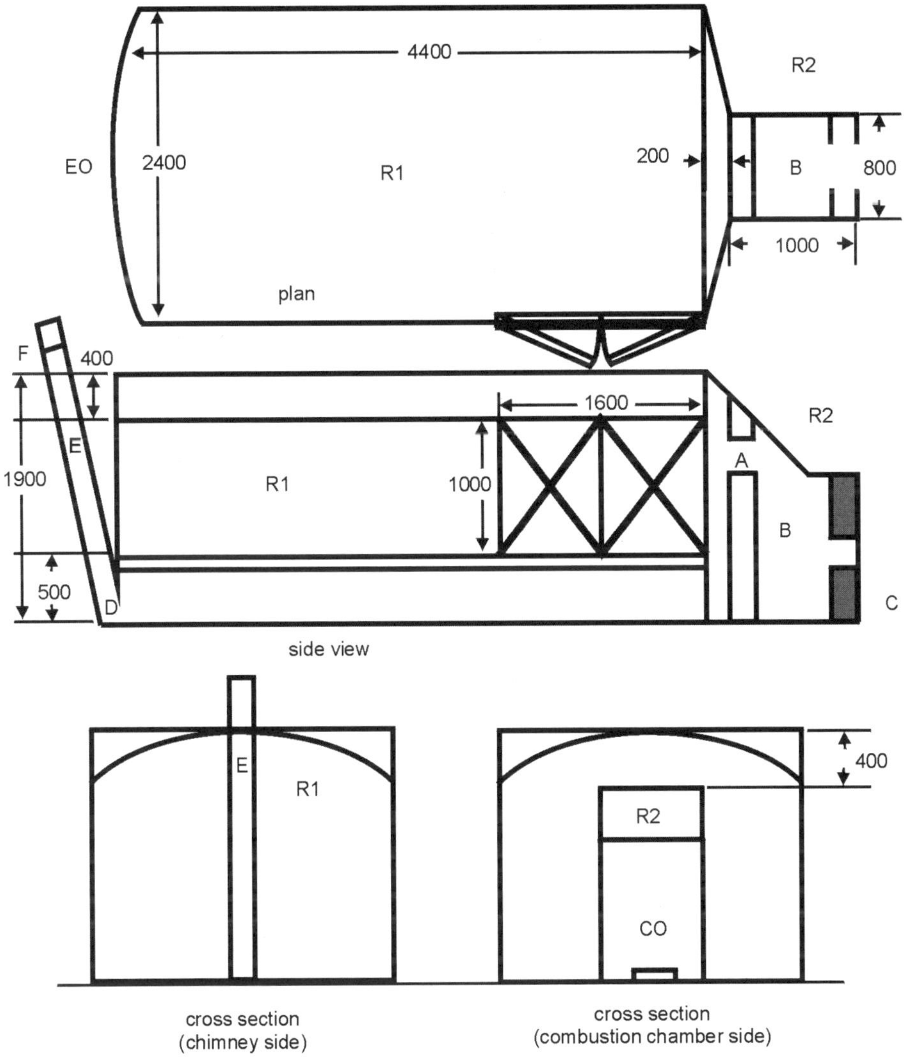

Figure 5. Schematic drawing of smoke drying furnace

Notes: R1 = Smoke drying chamber; R2 = Combustion gas generator; A = Combustion gas passage; B = Combustion chamber; C = Air inlet conduit; D = Exhaust gas discharge conduit; E = Chimney; F = Damper

An exhaust-gas discharge conduit (D) is connected to a corner of R1 to discharge gas from A in the diagonal direction of R1. The end of D which is connected B is positioned at the remotest point of R1, while the other end extends to a chimney (E) through which the exhaust gas passing through D is discharged into the atmosphere. A circulator-damper (F) is provided at the chimney end of D for controlling the flow rate of the exhaust gas to be discharged.

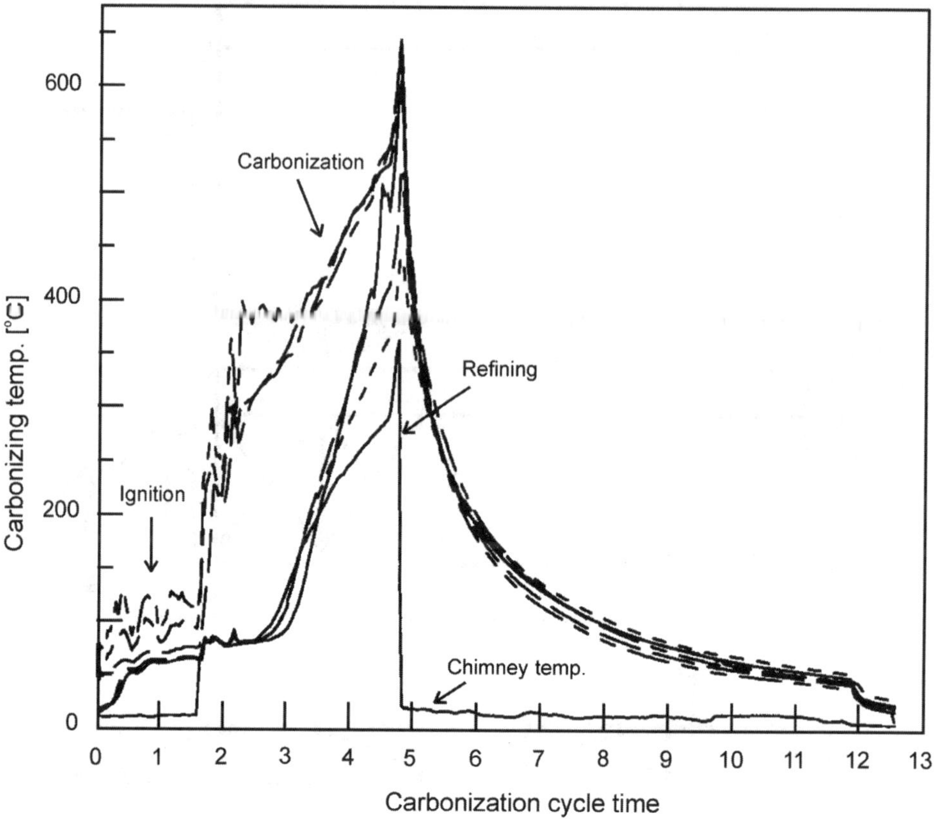

Figure 6. Carbonizing temperature curves

The maximum temperature in R1 is in the range 150–200°C. Fig. 6 shows an example of temperature change in a heat treatment process. The temperature distribution in R1 is stratified by natural convection and the difference of temperature between the upper space and the lower space in R1 is in the range of 50–60°C. A question to examine is whether this temperature differentiation brings the variation in the finishing moisture content.

Fig. 7 shows the state of the treated bamboo samples after smoke drying and Fig. 8 shows the differentiation of moisture content (in terms of weight loss) of bamboo samples. The samples were placed at different positions inside R1. After smoke drying, very little splits and cracks are seen on the surface of the treated culm.

Fig. 9 shows the distribution of the average moisture content in an internode treated by smoke drying. The average moisture content is reduced to 40%. Fig. 10 shows the released strain of the smoke dried sample in comparison with the control.

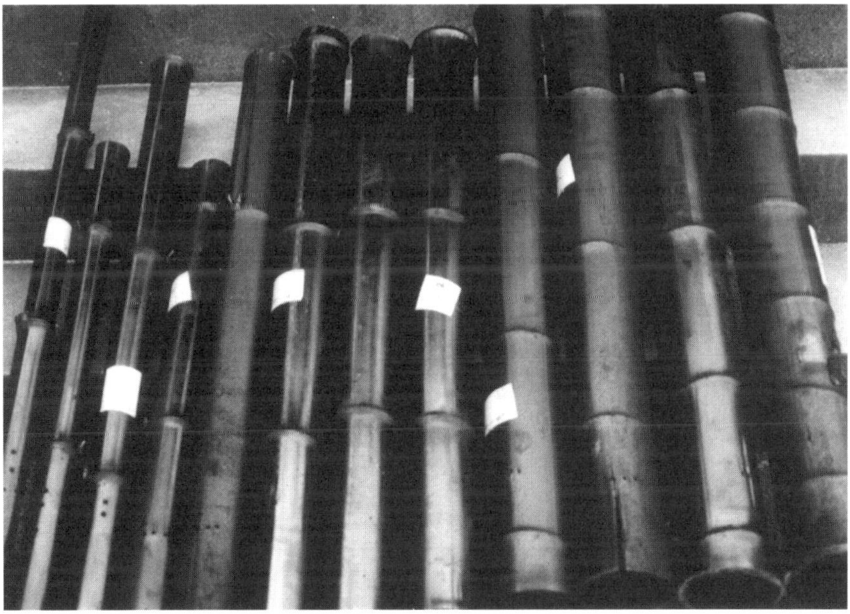

Figure 7. Bamboo samples after smoke drying

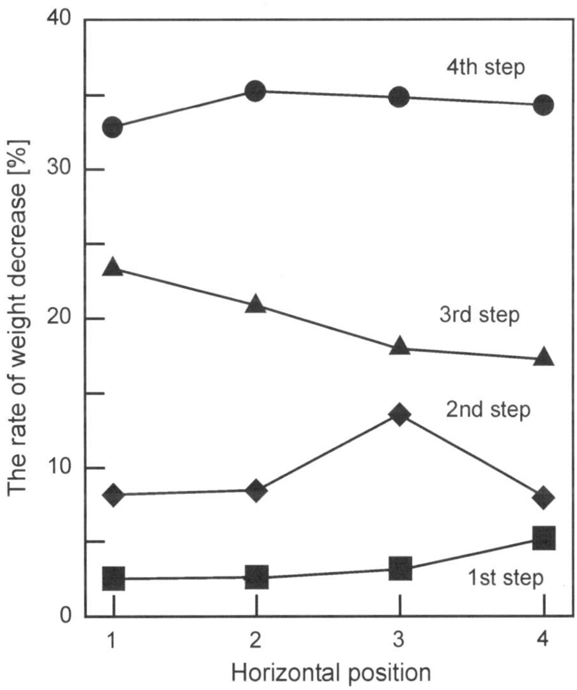

Figure 8. Weight loss of bamboo samples placed at different positions in the smoke-drying chamber
Note: The positions are numbered from the innermost part outwards

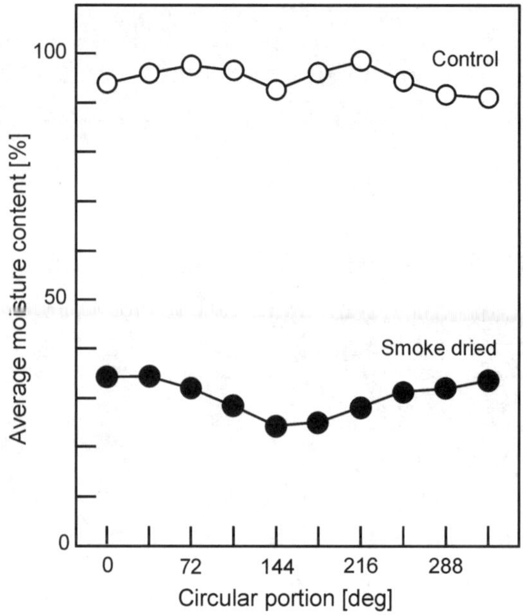

Figure 9. Circular distribution of average moisture content in an internode of a smoke-dried sample compared with the control

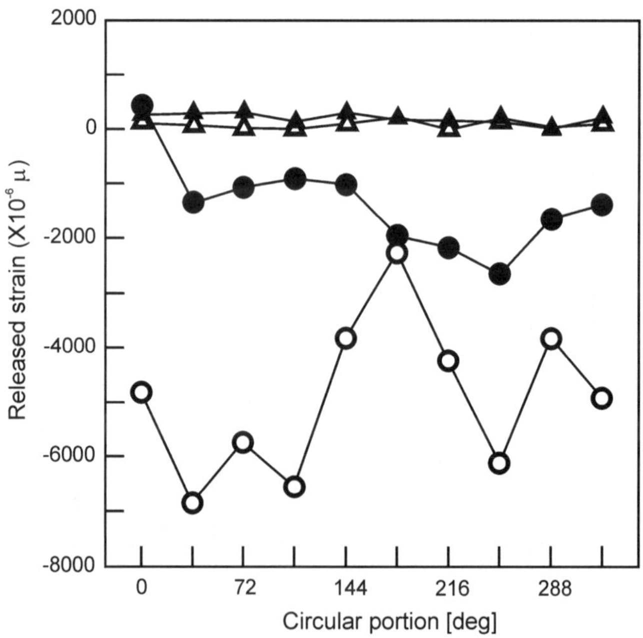

Figure 10. Circular distribution of released strain

Notes: △ = Control (longitudinal direction); ▲ = Smoke dried sample (longitudinal direction); ○ = Control (radial direction); ● = Smoke dried sample (radial direction)

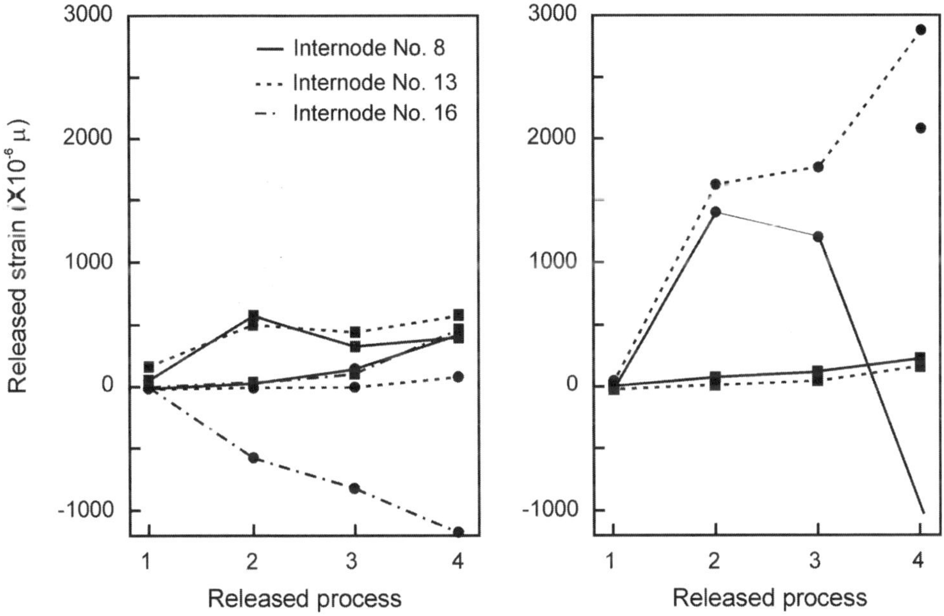

Figure 11. Change of released strain in smoke dried samples of *P. heterocycla* (left) and *P. bambusoides* (right)

Notes: 1 = Released longitudinal stresses; 2 = Released radial stress; 3 = Released strain stress when separated 10 parts in radial direction; 4 = Released strain after growth stress is freed. ■: Longitudinal direction ●: Radial direction

The difference between released strains in the longitudinal direction before and after the smoke drying is very small. In the radial direction, however, the released strain is clearly diminished by smoke drying. This result shows the effectiveness of smoke drying in stress conditioning of bamboo.

Fig. 11 shows the change of released strain after 24 hours from growth stress release, showing a comparison between the control and the smoke-dried sample. As the moisture content in bamboo decrease in air-drying process, the shrinkage in the radial (circumferential) direction increases in the control bamboo. The shrinkage in the smoke dried sample is not so large because the culm has been shrunk already by about 10% against diameter.

In the case of *P. heterocycla*, it has already been shown that the compression stress of the control in the radial direction is larger than that of *P. bambusoides*. The radial shrinkage of *P. bambusoides,* which was treated by smoke drying, increased after 24 hours. This result proves that the prestress in the radial direction in the culm of *P. bambusoides* increased by smoke drying.

The above-mentioned facts show that the effects of smoke drying on bamboo materials are different not only among bamboo species but also among the different internodes of a bamboo culm. In general, however, smoke drying brings about desirable effects to eliminate the defects of bamboo. When smoke-dried, bamboo culm is prevented from splitting and the prestresses are reinforced in the culm, making it possible to use bamboo culm as a whole.

Effect of smoke drying as a pretreatment to bamboo charcoal making

As mentioned above, smoke drying reduces the moisture content in green bamboo and ensures that the moisture content in the internodes is uniform. It also relaxes the internal stresses in bamboo culm and makes them uniform. Fig. 12 shows a comparison between charcoal made from bamboo culms that have been treated and the untreated by smoke drying.

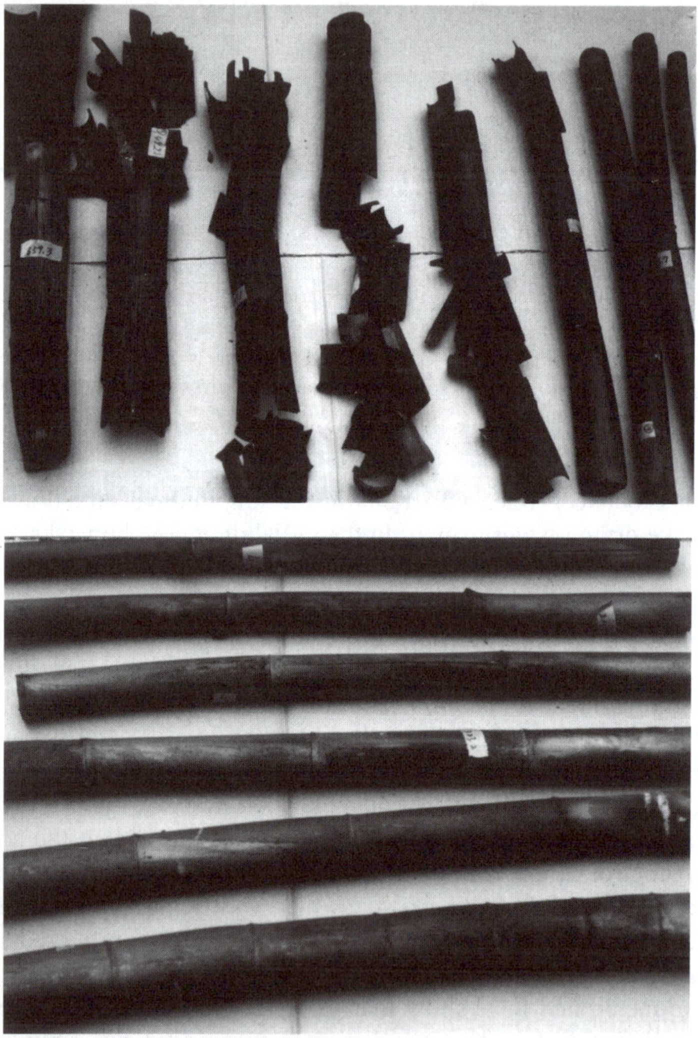

Figure 12. Bamboo charcoal from culms that have been untreated (top) and treated (bottom) by smoke drying

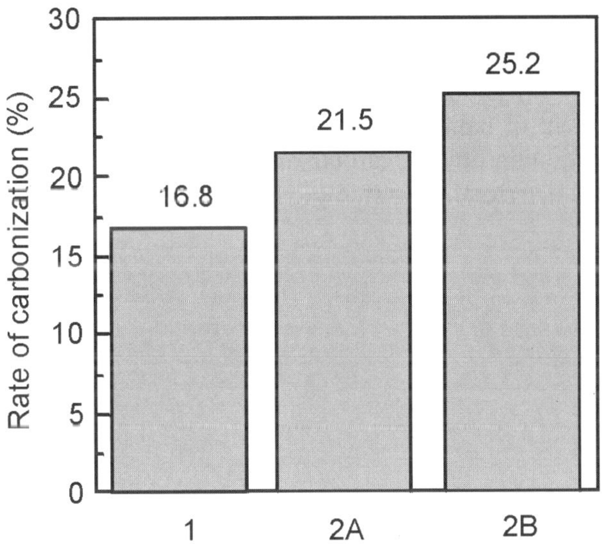

Figure 13. Comparison of the rate of carbonization between pretreated bamboo and untreated bamboo

Notes: 1 = Untreated by smoke drying; 2A = Treated by smoke drying, the rate based on green weight; 2B = Treated by smoke drying, the rate based on smoke drying.

Samples 1 m in length were cut from the basal part to upper part of a culm, and then packed into a bundle after splitting it into four or six pieces. One set was pretreated using smoke drying, while another set was kept green. Both samples were then carbonized to make charcoal. Bamboo charcoal, which was not made from smoke-dried samples, cracked up in the middle parts of the culm, while the charcoal from smoke-dried bamboo kept the original form with limited splitting and cracking. It is clear that the pretreatment by smoke drying is important to make good quality charcoal. Fig. 13 shows a comparison of the rate of carbonization of bamboo culms that have been smoke-dried and not smoke-dried. The rate of carbonization of smoke-dried bamboo is 25.2% based on the weight of smoke drying and 21.5% based on the green weight. In the case of untreated bamboo, the rate of carbonization is only 16.8%. Thus, pretreatment by smoke drying not only improves the quality of bamboo charcoal but also increases its quantity.

CONCLUSION

Smoke drying is an effective technique in the processing of different kinds of bamboo and bamboo materials, and in bamboo charcoal making. The effects of smoke drying are:

1. It prevents bamboo culm from splitting through stress and moisture conditioning;
2. The adhesion of soot and other pyrolysis chemical components on the surface of culm protects it from biological deterioration; and
3. As a pretreatment of bamboo in charcoal making, it not only of shortens the moisture conditioning time of carbonizing materials but also protects charcoal from biological deterioration and improves the quality and the rate of charcoal production.

Bamboo for Sustainable Development, pp. 481–494
Eds. A. Kumar, I.V. Ramanuja Rao and Ch. Sastry

Preservation of bamboo for the construction of houses for low-income people

WALTER LIESE
Chair for Wood Biology, University Hamburg, Germany

JORGE GUTIÉRREZ
National Laboratory for Materials and Structural Models, University of Costa Rica

GUILLERMO GONZÁLEZ
Bamboo Foundation (FUNBAMBU), San José, Costa Rica

Abstract—Bamboo can be destroyed by fungi and insects in a relatively short period. Therefore, it is necessary to protect and preserve bamboo culms when good quality and durability are required. This paper presents the scientific and technical experience with preservation methods developed in Costa Rica by the National Bamboo Project (PNB) and the Bamboo Foundation (FUNBAMBU). Anatomical considerations as well as techniques for drying, storage and quality control are discussed. Performance of treated bamboo structures used for the building of houses for low-income people, furniture and other building components is examined.

INTRODUCTION

The Costa Rican National Bamboo Project (PNB) was created to offer low-cost houses to help reduce the housing deficit. Initially, in 1988, PNB worked under the administration of the United Nations Center for Human Settlements (Habitat) and the United Nations Development Programme (UNDP) – (Projects COS 87/001 and COS 90/005) – with financing from the Royal Government of the Netherlands (Chaves and Gutiérrez 1990). As of January 1996, all professional, technical and material resources of these projects were transferred to the Bamboo Foundation (FUNBAMBU) – a non-profit, non-government organization presided over by the Minister of Housing and Human Settlements (MIVAH) and dedicated fundamentally to the construction of housing for low-income people. This Foundation has the function of perpetuating in a self-sustaining manner the gains and benefits of the National Bamboo Project (Gutiérrez, this volume).

Since then, nearly 3 000 houses have been built and more than 300 ha of bamboo planted. The government plans to extend the plantations up to 2 000 ha in the near future.

The need for bamboo preservation

When the project started, insect attack of the bamboo in use was a very serious problem. Today, the durability of bamboo is no longer in question, since it was possible to find effective, low-cost preservatives and to develop simple and efficacious systems for its treatment. In recent years, the demand for bamboo culms for the construction of houses has considerably increased. Besides, it has become necessary to treat several thousand of culms and laths for furniture making, handicrafts and construction components, for the local market as well as for exports.

Bamboo culms provide an excellent structural material for many purposes. Its use, however, is restricted because of the low resistance against degradation by fungi and insects. Whereas fungal decay can be prevented by architectural and construction details that keep the bamboo dry, insect attack can be prevented only by a suitable chemical treatment. The possible treatment techniques for bamboo have recently been summarized in numerous papers such as, for example, Liese (1980, 1997), Sulthoni (1987) and Kumar et al. (1994).

A wider application of bamboo preservation is, however, rare. This is mainly because bamboo is a low-cost material, which would become more expensive and less accessible to the poor after proper preservation, although it may be more economical in the long term. In addition, bamboo culms, when compared with timber, show a refractory behavior towards penetration, which makes simpler preservation methods like immersion difficult. Furthermore, a constant market demand for durable material is required to merit the necessary treatment installations and to develop the technical skills.

BAMBOO PRESERVATION RESEARCH IN COSTA RICA

The methods and preservatives available in the country were analysed, based on the results of consultations by the first author during 1988–1992 (Liese 1988, 1989, 1990a, 1990b, 1992) and the research done at the University of Costa Rica and the Technological Institute of Costa Rica. The technical personnel of PNB, with the help of researchers and laboratory facilities at the above-mentioned universities, conducted tests for adapting to the treatment of bamboo in the rural areas of the country. It was necessary to treat bamboo culms as well as *esterilla* (culms spliced open with an axe and flattened into a board) for the construction of houses as well as the production of furniture, handicrafts and decorative elements for buildings. Later on, it became necessary to treat, besides *esterilla*, also *regilla* (small boards or laths obtained by longitudinally sawing bamboo culms). First trials for the application of boron preservatives were done by either soaking or sap displacement for round bamboo pieces, and dip diffusion was tested for the treatment of *esterilla* and *regilla*.

Selection of the treatment methods

After preliminary trials, it was decided to concentrate research on the treatment of culms to be used as structural members by the sap-displacement process. This method appeared to be effective, environment-friendly and particularly suitable for the treatment of bamboo for houses in the rural areas of Costa Rica. For the treatment of *Guadua* culms, the sap-displacement method, on comparison with other methods such as long-period immersion, is less time consuming and gives the best results in terms of penetration, distribution and retention of the preservative.

For preservative treatment of *esterilla* and *regilla*, a dip-immersion/diffusion process was tested both in the laboratory and in the field. Considering the great potential of contamination and misuse of the preservatives obtained from the first trials, it was recommended that the dip-diffusion method be applied only at a few centralized places as close as possible to the plantations, and to have the work done only by skilled labor under proper supervision. A preservative treatment for laths was developed later as an improved, effective and safe way to treat the material.

The sap-replacement treatment, also called Boucherie process after its French inventor A. Boucherie in 1839 (Willeitner and Liese 1992), has been applied since long for the treatment of wooden poles. Its use for bamboo was first tried in India by Narayanamurti et al. (1947) and was further developed into the "modified Boucherie treatment" by Purushotham et al. (1954) and Liese (1959).

Although encouraging results were obtained from these experiments, the practical use of the method remained rather limited, because further clarification on several parameters was needed before it could be used on a semi-commercial scale. Based on the literature about bamboo preservation and practical experiences in several Asian countries, the research at PNB was directed to the testing of several parameters, which had to be defined previously to a commercial operation at a larger scale. It was especially necessary to answer the following questions:

- What kind of preservative is to be used?
- What will be the response of *Guadua* bamboo to sap-displacement treatment?
- What are the parameters necessary to obtain a good penetration?
- What kind of equipment will be necessary to treat *Guadua* culms?
- What kind of protection will be obtained for the treated material?
- What will be the cost and time for the preservation of culms and laths by this process?

Selecting the preservative

As the final price of the treated bamboo culms for the construction of low-cost houses has to be maintained as low as possible, the cost of bamboo preservation must also be low. At the same time, the treatment should be reliable to ensure the expected duration and performance of the material. From the first tests, this seemed to be very difficult to achieve.

About a dozen products nationally available in the market that the manufacturers claimed to be "good" for the preservation of timber were tested. However, most of these products had to be rejected when the following criteria of selection were adopted:

- A competent and responsible manufacturer or distributor of the product.
- A reasonably priced product.
- Good preservation properties, especially against *Dinoderus* spp. and other insects.
- Good distribution of the preservative within the culm by diffusion.
- Low toxicity to human beings and low environmental risks.

After many facts were considered, a preservative with boron as an active component was selected. This type of preservative is quite economical and has, in general, good preservation properties, good diffusion rates and low toxicity for humans and animals. The main disadvantage is that it does not possess chemical fixing properties and, in consequence, the treated material can be leached out if it is not well protected against humidity.

The preservative TIMBOR showed most of the requirements needed. However, since this product was not imported into the country, preliminary trials were done using the product "Menorel-8" (SOLUBOR). This product is manufactured as a foliage fertilizer and is widely distributed in Costa Rica. It consists mainly of disodium octoborate tetrahydrate ($Na_2B_8O_{13}\cdot 4H_2O$), with 66.1% active boron content expressed as B_2O_3.

ANATOMICAL CONSIDERATIONS FOR SAP DISPLACEMENT TREATMENT

Impregnation stage

As mentioned, since the hard skin of bamboo culm is resistant to lateral penetration, the idea of the Boucherie treatment for an axial liquid exchange was adopted. Along the culm length, both at the inner and outer parts, vessels run straight without any branching or obstructing connections of single vessel members (Fig. 1). Nodes, however, interrupt the easy fluid transport. Observations of the nodal structure have revealed that many vessels turn lateral, changing into shorter cells with many perforations for an intensive interconnection with other vessel members. Thus, the diaphragm, as the inner part of the nodal area, provides a distribution of liquid and sap in the vessels and sieve tubes of the phloem to all sides of the culm (Ding and Liese 1995). Obstacles between these cells reduce an easy flow. Apart from the twisting and realignment in the nodal part, some vessels do not turn in but run straight through into the following vessel, as shown in Fig. 2.

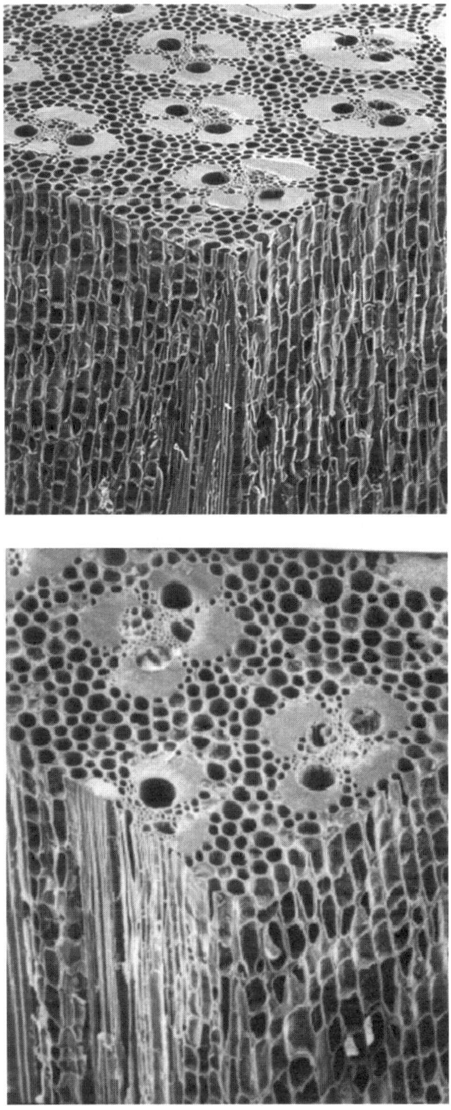

Figure. 1. Cell tissue of a bamboo culm with vascular bundles surrounded by sclerenchyma sheaths and parenchyma as ground tissue (100x)

The long-stretching vessels are thus perfect pathways for the exchange of sap by the preservation solution. However, a major obstacle has to be addressed for a successful treatment: the vessels occupy only a rather small area of the whole cross section to be protected. Whereas for an axial penetration of softwoods, the lumina of tracheids provide 60–70% cross-sectional area, and the vessels in diffuse porous hardwoods amount to 20–30%, bamboo has only about 5–8% vessel area (Liese 1994).

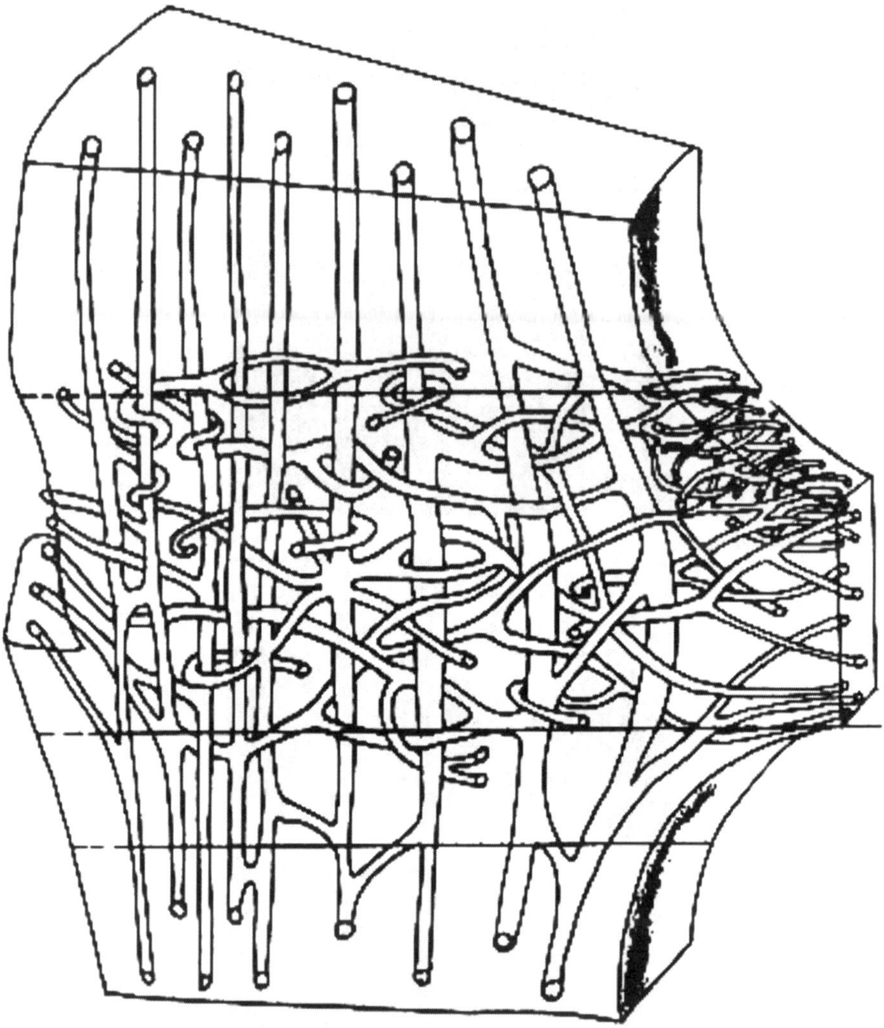

Figure 2. Twisting and realignment in the nodal part (Drawing after Ding and Liese 1995)

Total vessel area available for liquid flow

To get more information about the vessel area available, some detailed investigations were made for *Guadua* "Sur" and *Guadua* "Atlantico", the two main *Guadua* bamboos that have been planted in Costa Rica. Internodal parts were cut from the bottom, middle and top portions of the culm. Samples were taken from each disk from two opposite sides, divided into their outer and inner half. The measurements were made with an Image Analyzer (CUE-3, Olympus) at 40 times magnification. After a delicate preparation, the metaxylem vessels could be clearly differentiated from the fibers and parenchyma. The results are summarized in Table 1.

Table 1.
Vessel area of *Guadua* Sur and *Guadua* Atlantico

Portion of bamboo	*Guadua* Atlantico Vessel area (%)			*Guadua* Sur Vessel area (%)		
	Outer	Inner	Average	Outer	Inner	Average
Top	5.2	13.7	9.4	4.4	9.8	7.1
Middle	5.3	10.7	8.0	3.8	7.8	5.8
Bottom	3.9	6.4	5.2	3.1	5.0	4.0

The data on the vessel area demonstrate considerable differences, since the vessel area at the inner half is about double than that of the outer half. Moreover, the vessel size increases considerably from the bottom to the top, both at the inner and outer halves.

Table 2.
Average vessel size (μm^2)

Portion of bamboo	*Guadua* Atlantico Cross-sectional area		*Guadua* Sur Cross-sectional area	
	Outer	Inner	Outer	Inner
Top	12.400	38.700	10.800	29.100
Middle	11.300	34.200	9.400	25.700
Bottom	12.200	35.600	12.200	29.000
Average	12.000	36.200	10.800	27.900

The average vessel area for the two *Guadua* bamboos reveals different mean values, since *Guadua* Sur showed distinctly lower values. As both bamboos differed considerably in their flow behavior in the preliminary testing, their average vessel size was further analysed as shown in Table 2. The results of such measurements demonstrate that *Guadua* Sur has considerably smaller vessels, which could account for the lower vessel area and the longer treatment time experienced.

Independent of the respective vessel area, the bamboo culms must be treated as fresh as possible, because soon after their felling the vessels get blocked by cellular outgrowths (tyloses) and slime (Fig. 3). Therefore, both ends have to be cut off immediately before the treatment.

Diffusion phase

For an effective treatment of the bamboo culm by the sap-replacement process, it is important to realize that even a complete filling of all the vessels covers less then 10% of the whole tissue to be treated and that the remaining large portion stays liable for degradation. Consequently, the sap replacement phase has to be followed by a second treatment phase: the diffusion of the preservative from the

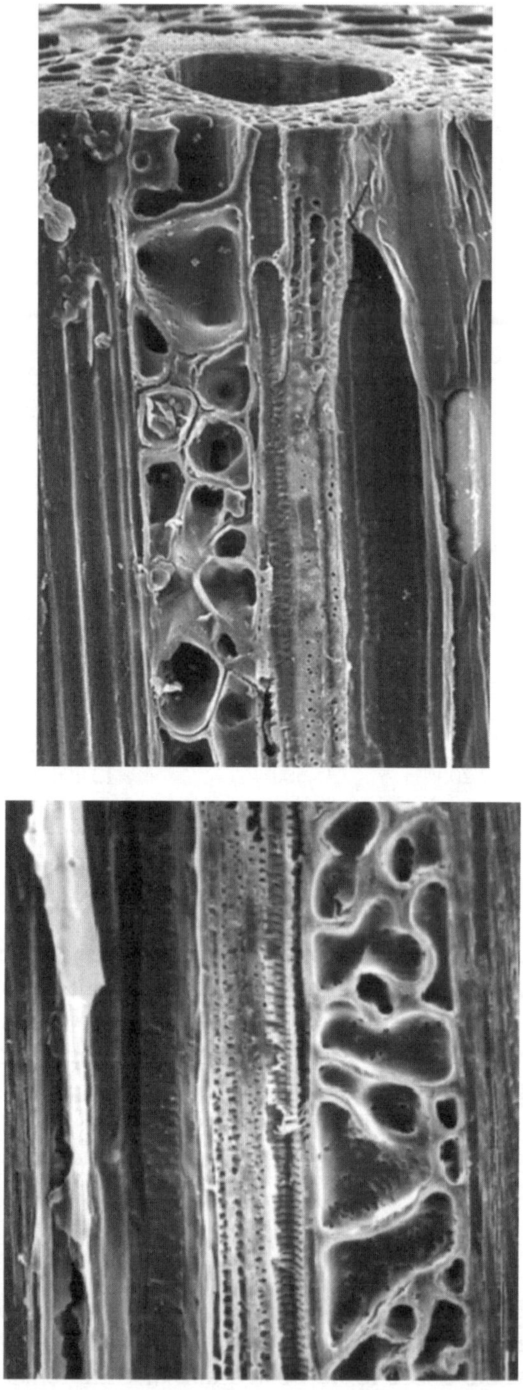

Figure 3. Tyloses blocking a vessel of a vascular bundle (400x)

vessels into the surrounding fiber sheaths and bundles as well as into the parenchyma. To ensure a maximum-size reservoir for such diffusion, nearly all the vessels have to be completely filled to a sufficiently high concentration. Filling will happen first in the larger vessels at the inner part of the culm and later in the smaller vessels at the outer zone. The time necessary for the treatment will hence be governed by the time necessary for a complete exchange of the sap by the preservative solution.

BAMBOO PRESERVATION AT FUNBAMBU

For the commercial production of treated bamboos using the sap-displacement method, several technical matters have to be resolved or improved, which require a very tedious trial and error process.

- Some variables of the process are sensitive to the bamboo species being treated and therefore have to be adjusted accordingly.
- Harvesting and transport need to be well coordinated in order to keep the culms as fresh as possible (otherwise water storage facilities are necessary).
- Design, construction and operation of the treatment caps at the butt end are critical. They must be flexible and durable to accommodate repetitive deformations and displacements without brittle fracture of parts and spilling of liquid. Good connecting devices between culm and cap are essential.
- The caps must include a special device to allow for the complete air removal from the system.
- An adequate pressure should be maintained at all times of the treatment process.
- Clean water must be used for the preservative solution and it must be kept clean in order to avoid suspended particles that will produce obstruction of the flow through the vessels of the bamboo culm.
- Permanent and strict quality control of the process as well as of the treated bamboo products is mandatory.

The sap-displacement preservation plant has been described before (González and Gutiérrez 1996). Since then, both sap-displacement and the dip-diffusion methods have been improved as described below.

Commercial sap-displacement pressure treatment plants

In the treatment plants being commercially used at FUNBAMBU, the rubber caps embracing the culm ends were improved to simplify and speed up the process. They allow rapid fixing of the cap to the culm with a tight joint that prevents leakage. It is possible to have 12 bamboo culms ready for treatment within 8 minutes. The problem of the air remaining in the upper part of the cap during the preservation process and thus preventing the flow of the liquid in this part of the culm was eliminated by adding a bleed valve (González and Gutiérrez 1996).

Even though clean water was used in the preservative solution, the liquid flow was found decreasing with time. Particles of soil and dust, which got mixed in the preservative solution during the process, were plugging the end of the culm where the liquid was entering. To overcome this problem, two kinds of filters were installed. One was a coarse one, made out of cheese cloth, to filter the mixture of solution and sap coming from the top end of the culm as well as from the reservoir, collecting the liquid from the bleed valves. The other was a fine in-line filter with a paper filter element, which was installed just before the entrance of the preservative solution to the manifold where the culms are connected.

The pressure system was improved for a larger capacity with an 80–liter pneumatic tank. With this larger capacity tank, more culms could be treated simultaneously and the pressure kept constant for longer periods. The pressure plant was further improved by placing corrugated galvanized iron sheets underneath the culms being treated, in order to collect the liquid spilling from the nodes. In this way, almost 100% of the preservative could be collected for recycling (Fig. 4).

The dip-diffusion treatment plant

The dip-diffusion treatment plant was built to treat *esterilla* and *regilla*. It consists of a tank made of four used steel oil barrels, which have been cut into halves and welded to produce a tank of approximately 300 liters capacity (Fig. 5). A steel frame, which allows a proper height to facilitate the immersion of bamboo laths, supports this tank. After a brief period of immersion, the laths are placed over corrugated iron sheets and left for a brief time to collect the excess liquid and return it into the tank. The laths are removed, wrapped with a plastic sheet and stacked for the preservative to diffuse into the material. Finally, the plastic sheet is removed and the material placed in a well-ventilated covered place for drying. Since plastic sheets are expensive and do not last long, a permanent covered place has been proposed that can be closed for the diffusion stage and opened later for drying.

Figure 4. Improved sap-displacement pressure treatment plant

Figure 5. Bamboo laths dip-diffusion treatment plant

QUALITY CONTROL

To ensure that the material is properly treated, it becomes essential to strictly follow the different steps of the treatment process. First, the preservative solution has to be prepared. For this purpose, the required amount of preservative is weighed and the water is measured by volume and mixed in the working tank, according to the required concentration. When the liquid is clear, a sample is taken into a cylinder where, with the help of a hydrometer and a thermometer, the specific gravity and the temperature of the solution are noted.

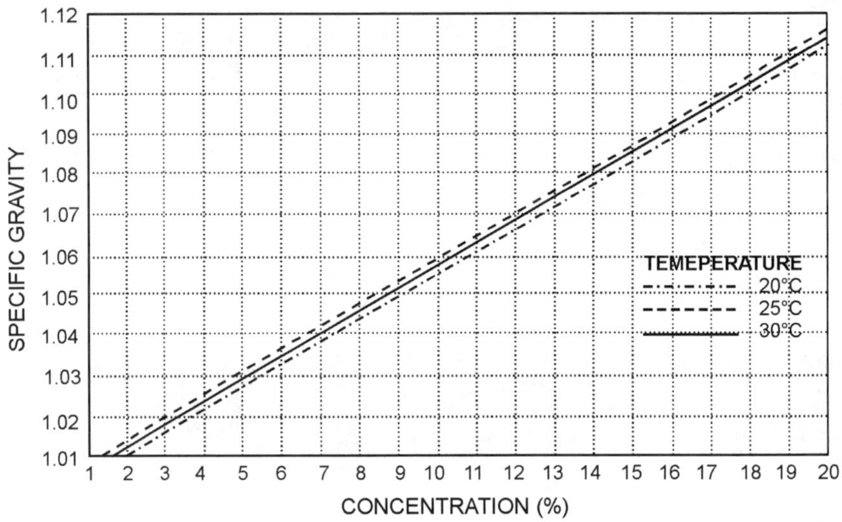

Figure 6a. Variation of specific gravity with concentration for disodium octoborate tetrahydrate solution at 20°, 25° and 30°C temperatures

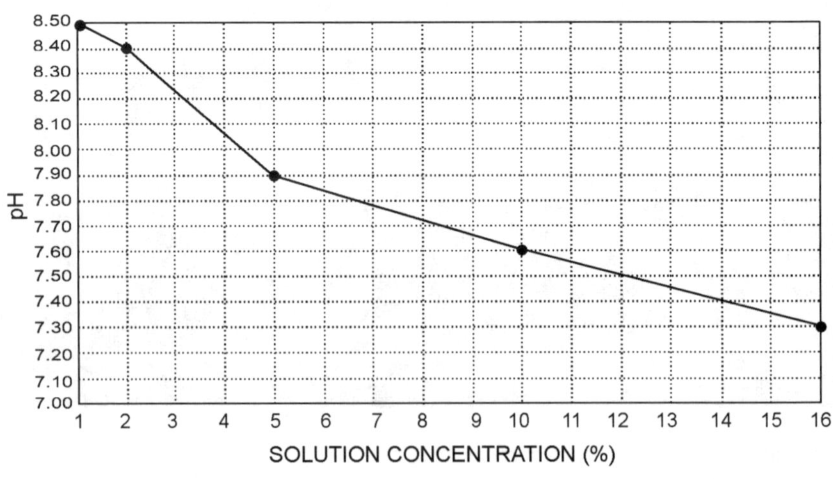

Figure 6b. Variation of pH of disodium octoborate tetrahydrate solution at 23°C temperature

With the help of chart in Fig. 6, the concentration of the solution can be read directly. For example, 1.068 specific gravity at 25°C gives 12% Boron-salt concentration, which is the concentration usually needed to treat *Guadua* bamboo culms. The concentration is checked regularly and adjusted as needed by adding more preservative or more water until the required concentration is reached again. Afterwards, the culms are connected to the rubber caps and the system is filled with the liquid preservative, bleeding each cap to ensure that all air is expelled out of the system. Next, each one of the inlet pressure liquid valves is opened and the treatment process started. The sap will start dripping at the other end of the culms almost immediately. The complete replacement of the sap by the preservative may take from a few minutes to a few hours, depending on the length and wall thickness of the culms and the amount of time elapsed since it was cut. For pieces 3–4 meters long, the average time is below 20 minutes. To check that the culms have reached the final point of treatment, a simple test with a pH sensitive paper is carried out. A change of color is observed because there is a relation between boron concentration and the pH of the liquid coming out of the culm.

After treatment, the culms are stored, and allowed to dry in a covered place to avoid direct exposure to moisture from rain and heat (and UV rays) from the sun. Moisture may produce leaching of the preservative and subsequently fungi infestation and insect attack may take place, while heat will cause splitting and structural collapse of the cross section.

Drying in a very humid tropical climate takes a lot of time. Therefore, it is recommended that the bamboo pieces be stored at an adequate distance from each other and from the wet soil – not less than 40 cm – for good ventilation.

When the preservative has completely diffused in and the material is dry, thin cross-sectional samples from several representative culms are cut near the end where the liquid emerged. These samples are further dried for 3–5 hours in a kiln, heated to a temperature of no more than 60°C. The sample is covered throughout with fine drops of a curcumin solution (Solution 1)[1] and, 15–30 minutes later, with an acid solution (Solution 2)[2]. Within 10 minutes time a bright red color should develop, indicating that the culm has being adequately treated and contains a sufficient amount of boron. A yellow color indicates insufficient amount of boron.

CONCLUSION

The sap-displacement treatment of bamboo culms, as developed by the Costa Rican National Bamboo Project, is a well-proven technology that has already been transferred to, among others, the Environmental Bamboo Foundation, Bali, Indonesia, since 1994 (EBF 1994) and to the Bamboo World of Victor Cusack, Australia (Cusack 1997).

---

[1] Solution 1 = 1% alcoholic extract of curcumin.

[2] Solution 2 = 6.0 g salicylic acid in 20 ml of concentrate hydrochloric acid diluted to 100 ml with ethanol.

ACKNOWLEDGEMENTS

The authors wish to thank the FUNBAMBU staff for their continuous support that made this work possible.

REFERENCES

Chaves, A.C.; Gutiérrez, J.A. 1990. The Costa Rican Bamboo National Project. *In* Ramanuja Rao, I.V.; Gnanaharan, R.; Sastry, C.B., ed., Bamboos: current research. Proceedings of the International Bamboo Workshop, Cochin, India, 14–18 November 1988. Kerala Forest Research Institute, Kerala, India; International Development Research Centre, Ottawa, Canada. pp. 344–349.

Cusack, V. 1997. Bamboo rediscovered. Earth Garden Books, Trentham, Victoria, Australia, 95 pp.

Ding, Y.L.; Liese, W. 1995. On the nodal structure of bamboo. Journal of Bamboo Research, 14(1), 21–27.

EBF (Environmental Bamboo Foundation). 1994. Training manual for bamboo preservation using the Boucherie system. Trainer's Guide. Environmental Bamboo Foundation, Bali, Indonesia. 24 pp.

Gonzales, G.; Gutierrez, J. 1996. Bamboo preservation at the Costa Rican National Bamboo Project. *In* Ganapathy, P.M.; Janssen, J.A.; Sastry, C.B. ed., Bamboo, people and the environment, Vol. 3, Engineering and utilization. Proceedings of the Vth International Bamboo Workshop, Ubud, Bali, Indonesia, 19–22 June 1995. International Network for Bamboo and Rattan, New Delhi, India. pp. 121–129.

Kumar, S.; Shukla, K.S.; Dev, I.; Dobriyal, P.B. 1994. Bamboo preservation techniques - a review. INBAR Technical Report No. 3. International Network for Bamboo and Rattan, New Delhi, India; Indian Council of Forestry Research and Education, Dehra Dun, India. 59 pp.

Liese, W. 1959. Bamboo preservation and soft-rot. FAO Report to Government of India, No. 1106, 37 pp.

Liese, W. 1980. Preservation of bamboos. In Lessard, G.; Chouinard, A. ed., Bamboo research in Asia. Proceedings of a workshop held in Singapore, 28–30 May 1980. International Development Research Centre, Ottawa, Canada. pp. 165–172.

Liese, W. 1988, 1989, 1990a, 1990b, 1992. Bamboo preservation in Costa Rica. Progress Report Nos. 1, 2, 3, 4 and Final Report for the United Nations Centre for Human Settlements (Habitat). Project COS-87–001, Hamburg, Germany.

Liese, W. 1994. Biological aspects of bamboo and rattan for quality improvement by polymer impregnation. Folia Forestalia Polonica, Series B-Drzewnictwo, 23, 43–56.

Liese, W. 1997. The protection of bamboo against deterioration. In Jena, P.K. ed., Proceedings of the 2nd International Conference on Non-Conventional Building Materials (NOCMAT-97), Bhubaneswar, India, 17–19 June 1997. Natural Resources Development Foundation, Bhubaneswar, India. pp. 111–117.

Narayanamurti, D.; Purushotham, A.; Pande, J.N. 1947. Preservative treatment of bamboo. Part I. Treatment of green bamboos with inorganic preservatives. Indian Forest Bulletin, No. 137, 14 pp.

Purushotham, A.; Sudan, S.K.; Vidya, S. 1954. Preservative treatment of green bamboos under low pneumatic pressures. Indian Forest Bulletin, No. 178, 21 pp.

Sulthoni, A. 1987. Traditional preservation of bamboo in Java, Indonesia. *In* Rao, A.N.; Dhanarajan, G.; Sastry, C.B. ed., Recent Research on Bamboo. Proceedings of the International Bamboo Workshop, Hangzhou, China, 6–14 October 1985. Chinese Academy of Forestry, Beijing, China; International Development Research Centre, Ottawa, Canada. pp. 349–357.

Willeitner, H.; Liese, W. 1992. Wood protection in tropical countries: a manual on the know-how. GTZ, Eschborn, Germany, 228 pp.

Bamboo for Sustainable Development, pp. 495–510
Eds. A. Kumar, I.V. Ramanuja Rao and Ch. Sastry

Preliminary study on the drying of bamboo (*Bambusa blumeana*) in a wood waste-fired kiln

YOSIAS GANDHI

Graduate Student, Department of Forest Product and Paper Science, University of the Philippines at Los Baños, Laguna, the Philippines

Abstract—Drying bamboo in a kiln is rarely done by a small and medium-scale bamboo industry because of the limited knowledge of the physical and mechanical properties of bamboo and its drying behavior in relation to its end-use.

The study of bamboo (*Bambusa blumeana* J.A. & J.H. Schultes) drying behavior in a wood waste-fired kiln was conducted with emphasis on the rate of drying and shrinkage behavior.

There were three levels of drying time: 24 hours, at 30% relative humidity and temperature of 100°F; 24 hours, at the same temperature; and equalization period of 8 hours at a temperature of 120°F. There is a sharply reducing of moisture content at the first 24-hour drying time, followed by gradually reducing during the next 24 hours, and then slowly reducing during the last 8-hour drying time. Tangential, thickness and volumetric shrinkage of bamboo at nodes and internodes tend to increase from the initial to final drying periods. The base portion shows the greatest shrinkage, followed by the middle and then the top portion. The results gathered from this experiment are suitable only for split bamboo. It is suggested that similar experiment be carried out for whole bamboo.

INTRODUCTION

Among the three bamboo-growing regions of Asia-Pacific, Americas and Africa, the former has the largest areas (Fangchun 1992). Much research has been conducted with the aim of improving the use of bamboo in view of the possibility of bamboo replacing wood as a raw material. Current bamboo products include bamboo furniture, bamboo mat plywood, bamboo plywood, bamboo particleboard, bamboo fiber-cement board, bamboo parquet, peeled bamboo veneer and bamboo plastic.

Drying is a general production process in the manufacture of bamboo products. Most furniture industries in the Philippines, especially the small, medium and cottage-scale ones, lack facility to dry their raw materials, and this reduces their competitiveness in the export market (Palisoc and Bello 1995). A study conducted by Roxas (1990) in the Philippines identified two methods of drying employed by bamboo furniture makers: sun and air-drying.

Drying bamboo in a kiln is rarely done because of the limited knowledge of the physical and mechanical properties of bamboo in relation to the end-uses. Therefore, it is important to conduct a preliminary study on the drying behavior of bamboo.

A wood waste-fired kiln is appropriate for small-scale industry since it is cheap and easy to operate. This study on the drying behavior of bamboo in a wood waste-fired kiln emphasized the rate of drying and shrinkage behavior.

WOOD WASTE-FIRED KILN

General description

Semi-commercial size timber drier in the Wood Machining Laboratory, Department of Forest Product and Paper Science, University of the Philippines at Los Baños, is a wood waste-fired kiln (Fig. 1). It is compartment-type, cross-shaft, with internal fan-and-furnace or electrical heating. It has a dry-and-wet bulb mechanism to maintain humidity at the desired level, and operates at the temperature range of 40°-100°C.

The kiln is designed for seasoning timber and timber products, veneer, rattan, bamboo, seeds and other forest products in carefully controlled conditions of temperature and humidity above ambient conditions. It may also be used in accelerated air-drying of timber. The kiln has a capacity of 700 board feet of timber 1 inch thick, 6 inches wide and 7 feet long, or a load size 4 feet wide, 4–5 feet high and 7 feet long (1.2 m wide, 1.2 m high and 2.3 m long).

Figure 1. Wood waste-fired kiln drier

Parts of the kiln

Structure: The kiln housing measures 2.57 m (width) × 1.30 m (height) × 1.42 m (length). Both outer and inner walls are made of steel. The loading doors are of wood, swing type, 99 cm (W) × 135 cm (H), fastened to the door by means of heavy steel strap and hinges. The internal dimensions are 2.49 m (W) × 1.22 m (H) × 1.30 m (L). The floor is made of 12-mm plywood. Two steel rails, 76 cm center to center, are embedded on the floor extending outside of the kiln to facilitate loading and unloading. One vent, 14 cm square is located on top of the kiln.

Heating system: The furnace heating system, capable of producing temperatures up to 40°C, uses 1 m^3 of sawdust for 8–9 hours. It is located at left side of plenum chamber. The electrical heating coils are located vertically at the right plenum chamber and consist of two plain copper heaters, each 8 mm in diameter. Electrical heating is usually used for achieving the equalization temperature (above 60°C).

Air circulation system: There is one fan, 34 inches in diameter, flat blade, propeller-type and reversible. It can produce air speeds up to 4.5 m/s. The motor is 4 kWh, 220 VAC.

METHODOLOGY

Material and specimen preparation

Two bamboo culms collected from the Makiling Forest, College, Laguna, were used for this study. The length, diameter and number of internodes of each culm were recorded.

The culms were cut into 2.2-m lengths as representative sections of the base, middle and top portions. Each portion was divided into six parts (Fig. 2). The segments were number-coded, to indicate their original position in the culm. Each segment was divided into six strips of 2×170 cm for shrinkage specimens and two oven-dry weight specimens of 2×2 cm taken from the end of the shrinkage sample.

Surface of the specimens were cleaned and ensured that they are free from decay and other defects that affect the result of treatments. Lines were drawn across the specimen to indicate position in measuring shrinkage at node and internode. Each specimen was numbered according to its position in the culm and coated at the ends with paint to prevent evaporation from longitudinal directions.

Seasoning and monitoring the moisture content

Shrinkage

Shrinkage values from green to oven-dry condition in width (tangential direction) and thickness (radial direction) of the specimen at node and internode were de-

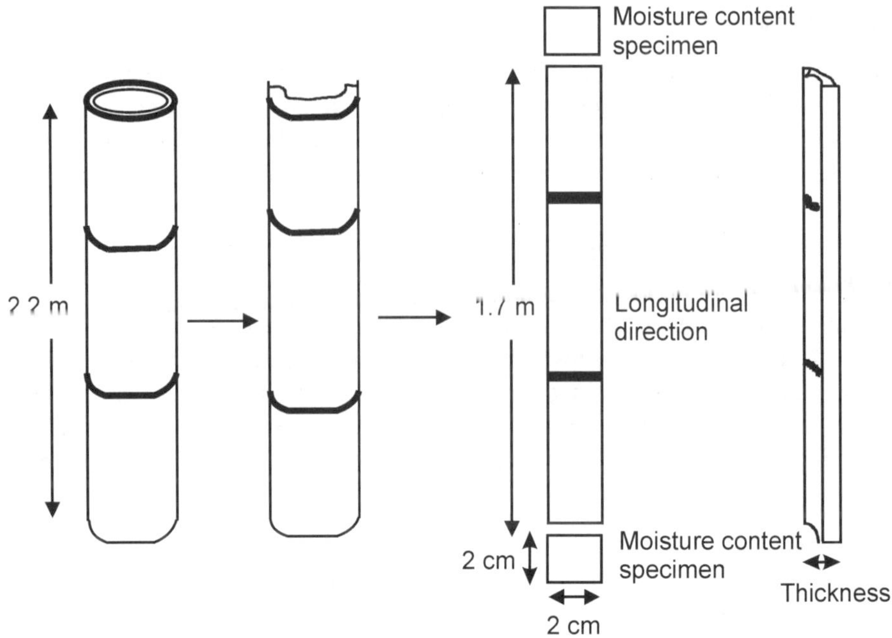

Figure 2. How the culms are split, and (right) one half designated for kiln drying

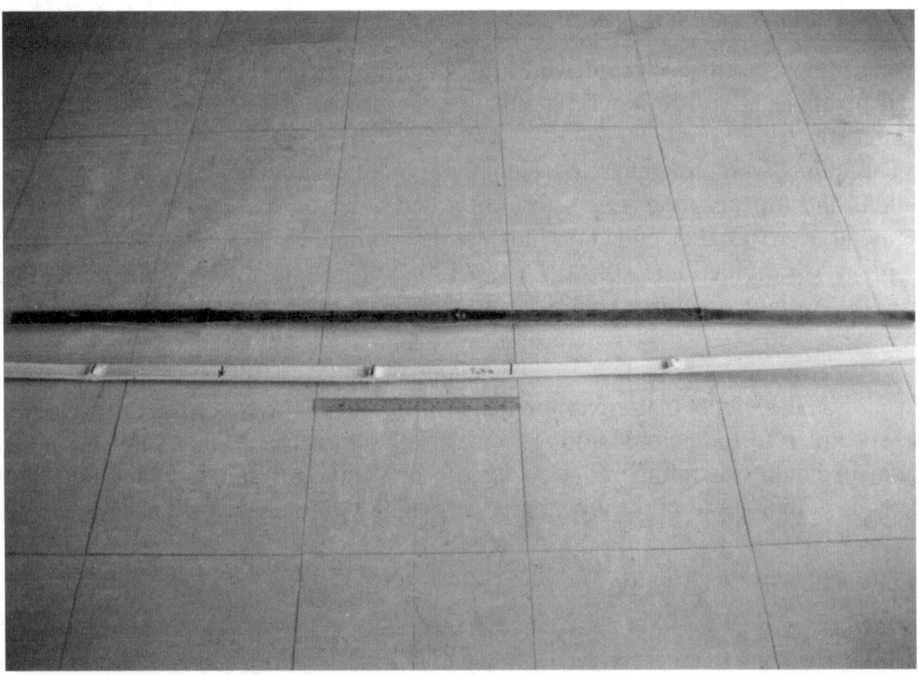

Figure 3. Specimen for shrinkage variables

termined every day. As dried bamboo usually cup, it was necessary to measure the shrinkage value on the inner and outer sides of the specimen. The longitudinal length of the specimen was assumed as constant. The percent volumetric shrinkage was determined using the following formula:

$$Sh_r = 100\ (V_g - V_d) \div V_g$$

Where, Shr = volumetric shrinkage in percent

V_g = green volume

V_d = dry volume.

Initial moisture content

The specimens were weighed from green to dry condition every day in order to obtain the current weight. Initial moisture content was determined as the difference of current weight and oven dry weight of the specimens. The initial moisture content of the specimen was then presented in percent of oven dry weight.

Procedure of the experiment

Preparation of the drier

The rack was arranged in three levels so that the air can freely circulate over all the specimens at the constant speed of 4.5 m/s. The specimens were then put on the rack; top portion on the upper level, middle portion on the middle level and base portion on the lower level as can be seen in the Fig. 4.

Figure 4. The arrangement of specimens inside the kiln

Sawdust was filled in a container, pressed to make it compact with a hole along the longitudinal axis. The filled container was placed in the furnace to run the drier at a low temperature of 40°C (100°F). The burning time of the sawdust in the container is 8–9 hours, after which time the container needs to be refilled. A motorized fan is used to circulate the air within the kiln. The temperature is controlled by reading the dry-and-wet bulb thermometer.

Drying schedule

Drying began with dry bulb temperature of 40°C and moisture content was recorded every 24 hours. Once the moisture content reached 20–30%, the temperature was increased to the equalization temperature of about 60°C until the moisture content of the specimen reached about 10–15%.

RESULTS AND DISCUSSION

Data of moisture content and volumetric shrinkage were analysed using variance analysis, so that the effect of the treatments could be determined.

Drying schedule

The 36 specimens of split bamboo – each piece measuring 2 cm wide, 1 cm thick and 170 cm long, or having a volume of 340 cm^3 – were first dried in the wood waste-fired kiln at low temperature of 40°C (100°F) to attain the moisture content of 20%. They were dried then at the equalization temperature of 60°C (120°F) to reach the moisture content of 10–15%. The drying schedule, taking totally 56 hours, of the experiment is given in Table 1.

There were three levels of drying time recorded in this experiment. In the first 24 hours, moisture content reduced from the average of 61.21% to 29.27% at the relative humidity of 30%; in the second 24 hours, there was only about 10% moisture content extracted from the specimens to reach around 20% moisture content. The final level, equalization period, was only for 8 hours, to gain 13.62% moisture content on the average. Because temperature was increased to 120°F, the relative humidity decreased to 25%.

Table 1.
Drying schedule of bamboo (*Bambusa blumeana*) in the wood waste-fired kiln

Moisture content (%)	Temperature (°F)		Relative humidity (%)	Drying time (hours)
	dry bulb	wet bulb		
61.21 - 29.27	100	75	30	24
29.27 - 19.18	100	75	30	48
19.18 - 13.62	120	85	25	56

Moisture content

The average moisture content of six specimens measured at base, middle and top portions from culm I and II at green condition and three drying levels are presented in Table 2.

The base portion of the two culms has the highest moisture content at every level of drying, followed by the middle portion and then the top portion. Table 3 shows summary of analysis of variance on moisture content between the two culms and among the three part of the bamboo. The result indicated that there is a significant difference among the three portions of the bamboo at all levels of drying time. There is no significant difference, however, between the two culms at all levels of drying time.

The reducing moisture content in base, middle and top portions of bamboo as the drying time increases is graphically demonstrated in Fig. 5. The graph shows that all portions of bamboo have the same trend of the moisture content reducing with the increasing drying time. There is a sharp reduction of moisture content – from 61.21% to 29.27% on the average – in the first 24 hours of drying time, followed by a gradual reduction at 48 hours (19.18%) and at 56 hours (13.62%).

Starting at the initial moisture content of 77.64%, the base portion contained the highest moisture content at all levels of drying time (43.71% at 24 hours, 32.77% at 48 hours and 25.51% at 56 hours). The middle part had initial moisture content of 54.94%, which dropped to 25.11% at 24 hours, 14.77% at 48 hours and

Table 2.
Moisture content (%) of the specimen during the drying schedule at base, middle and top portions of bamboo

Drying time	Base		Middle		Top		Average
	I	II	I	II	I	II	
Green	71.64	83.64	54.41	59.54	46.35	51.66	61.21
24 hours	42.65	44.80	26.71	23.50	19.84	18.12	29.27
48 hours	33.16	32.37	16.85	12.68	10.38	9.66	19.18
56 hours	26.77	24.25	11.80	6.16	6.50	6.24	13.62

Table 3.
Summary of analysis of variance in moisture content

Drying time	Culm (1 & 2)	Part (Base, Mid, Top)	Mean	Standard error
	Mean square and statistical significance			
Green	502.432^{ns}	1333.45*	61.21361	3.73583
24 Hours	7.645^{ns}	1008.31*	29.26972	0.46083
48 Hours	32.281^{ns}	869.87*	19.18694	0.94694
56 Hours	70.840^{ns}	657.28*	13.62111	1.40278

Notes: * – Significant at 99% level of probability; ns – Not significant.

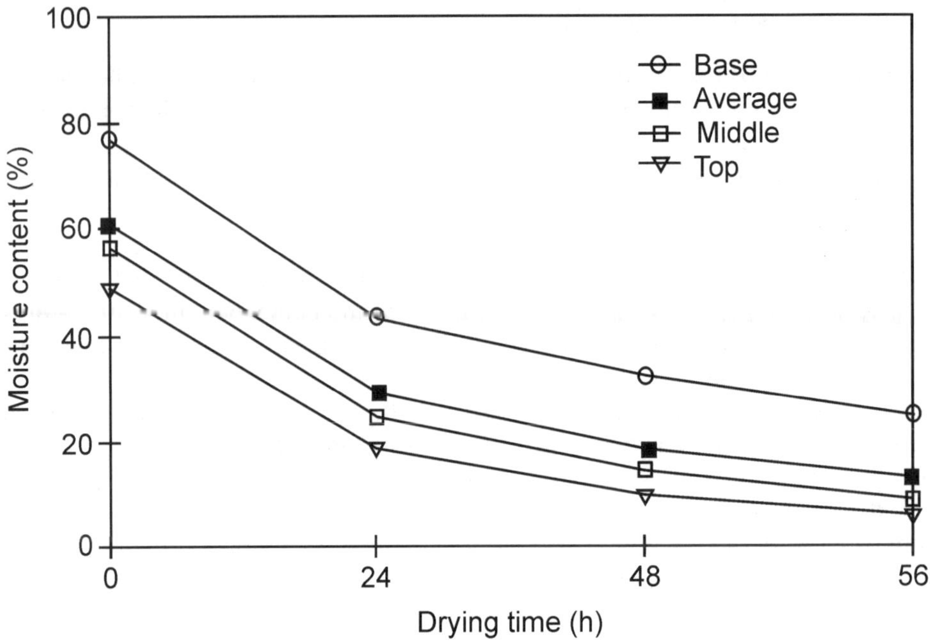

Figure 5. Relationship of moisture content (%) and drying time (hours)

8.98% at 56 hours. The lowest moisture content was in the top portion, 49.01% at the start, 18.98% at 24 hours, 10.02 at 48 hours and 6.37% at 56 hours.

Shrinkage

Tangential shrinkage

The average tangential shrinkage of 12 specimens measured at the internode (outer and inner part) and node (outer and inner part) of base, middle and top portions are presented in Tables 4 and 5.

As can be seen in Table 4, tangential shrinkage at internode tends to increase from the initial to final drying time, both at the outer and inner parts. The base portion showed the greatest shrinkage at every stage of drying, followed by the middle and the top portion, respectively.

Table 4.
Tangential shrinkage (%) of bamboo at internode (inner and outer)

Drying time	Moisture content (%)	Internode (outer part)				Internode (inner part)			
		Base	Middle	Top	Average	Base	Middle	Top	Average
24 hours	61.21-29.27	4.82	4.71	3.30	3.93	7.69	5.00	3.17	5.29
48 hours	29.27-19.18	7.36	6.09	4.43	5.66	11.90	7.11	3.78	7.59
56 hours	19.18-13.62	8.16	7.67	5.20	6.04	13.57	8.27	4.91	8.92

Table 5.
Tangential shrinkage of bamboo (%) at node (inner and outer)

Drying time	Moisture content (%)	Node (outer part)				Node (inner part)			
		Base	Middle	Top	Average	Base	Middle	Top	Average
24 hours	61.21 - 29.27	2.20	2.25	2.35	2.27	4.25	3.90	2.62	3.59
48 hours	29.27 - 19.18	3.25	3.77	3.88	3.63	7.52	4.90	3.61	5.34
56 hours	19.18 - 13.62	4.09	4.79	4.42	4.43	9.27	7.67	4.60	7.18

The tangential shrinkage of bamboo at the node tends to increase from the initial to final drying levels, both at the outer and inner parts (Table 5). The base portion displayed the greatest shrinkage at every level of drying, followed respectively by middle and top portions.

It was determined statistically that there is significant difference in tangential shrinkage between the inner and outer parts at both internode and node, and also between internode and node. The results of the t-test on tangential shrinkage are presented in Table 6.

According to the mean of each variable, tangential shrinkage of the inner part is greater than that of the outer part at both internode and node. The results also showed that tangential shrinkage at node is greater than that at internode.

In general, the difference in tangential shrinkage between each variable tends to increase with the drying time, as demonstrated in Fig. 6.

Thickness shrinkage

The average thickness shrinkage of 12 specimens measured at node of the base, middle and top portion are presented in Table 7. It was found that thickness shrinkage at node tends to increase from the initial to final drying level, and that the basal portion shows the greatest and the top portion the least shrinkage throughout the drying stages. As to be expected, thickness shrinkage increased with drying time (Fig. 7).

Table 6.
Summary of t-test for paired sample of tangential shrinkage

Variable	Degree of freedom	Mean	Standard error of mean	T-value
Inner and outer at internode	107	7.27 & 5.31	0.383	5.12*
Inner and outer at node	107	5.68 & 3.48	0.304	7.26*
Internode and node	215	6.29 & 4.58	0.214	8.01*

Note: * – Significant at 95% level of probability

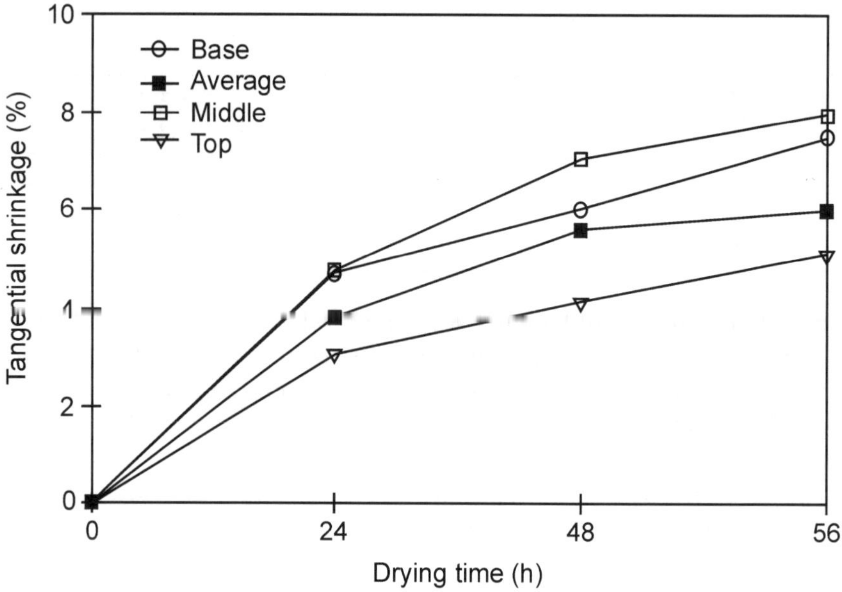

Figure 6a. Relationship of tangential shrinkage to drying time at outer internode

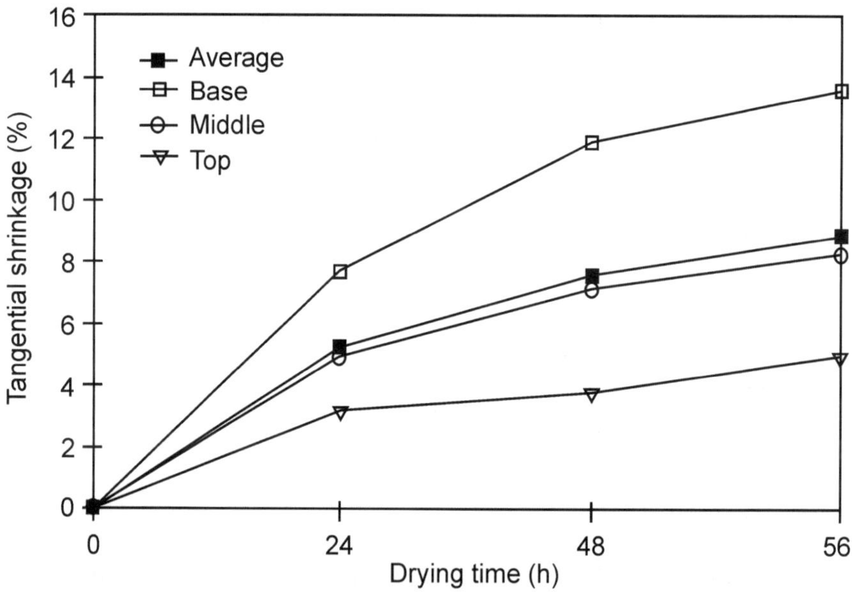

Figure 6b. Relationship of tangential shrinkage to drying time at inner internode

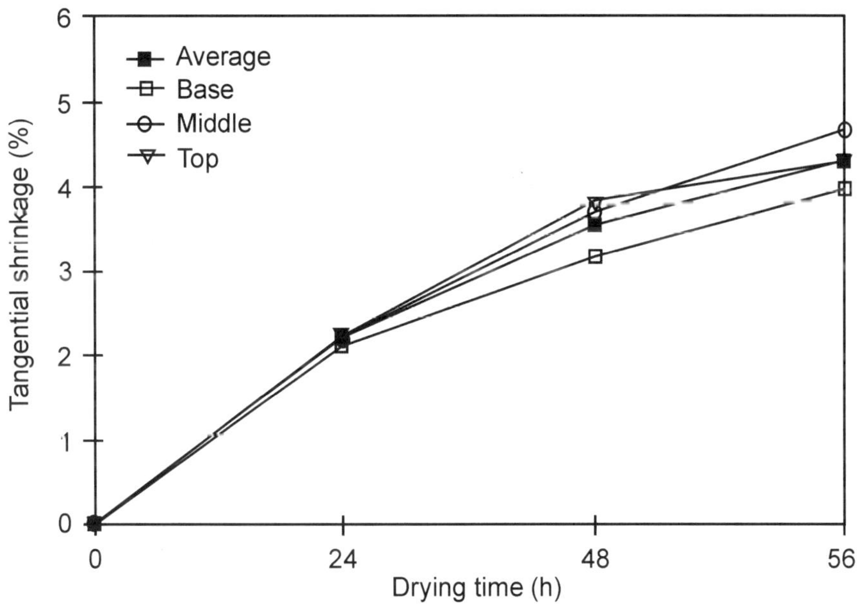

Figure 6c. Relationship of tangential shrinkage to drying time at outer node

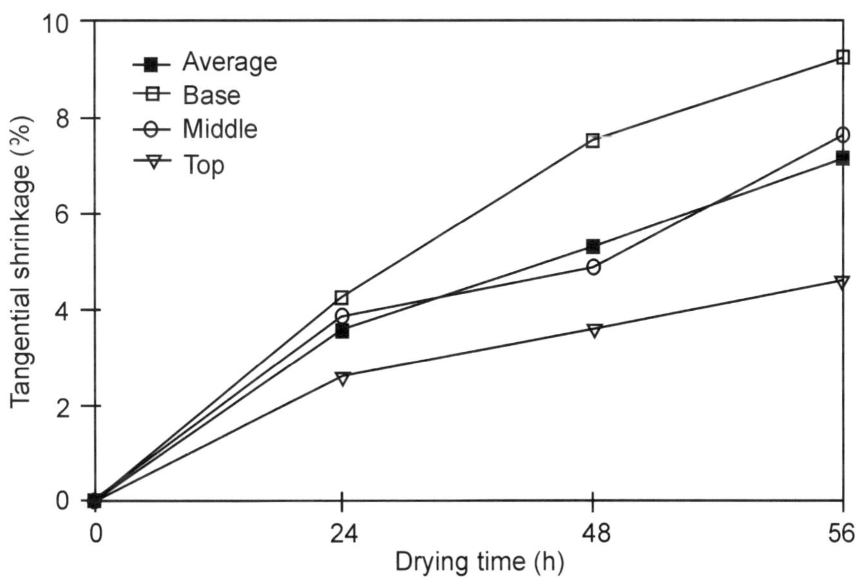

Figure 6d. Relationship of tangential shrinkage to drying time at inner node

Table 7.
Thickness shrinkage (%) of the specimens during drying

Drying time	Moisture content (%)	Thickness shrinkage (%)			
		Base	Middle	Top	Average
24 hours	61.21-29.27	6.698	5.716	3.150	5.188
48 hours	29.27-19.18	8.706	7.664	6.097	7.489
56 hours	19.18-13.62	12.351	8.215	8.072	9.546

Volumetric shrinkage

The average volumetric shrinkage of six specimens measured in base, middle and top portions from Culm I and Culm II at three levels of drying time are presented in Table 8. The basal portion of both culms showed the greatest volumetric shrinkage, followed respectively by middle and top portions at every stage of drying.

Table 8.
Volumetric shrinkage of bamboo (%) during drying

Drying time	Moisture content (%)	Base		Middle		Top		Average
		I	II	I	II	I	II	
24 hours	61.21-29.27	10.93	2.31	9.74	0.90	7.01	7.62	9.75
48 hours	29.27-19.18	16.28	16.57	13.31	14.11	9.31	11.08	13.44
56 hours	19.18-13.62	19.25	2.16	4.43	6.38	2.54	2.85	16.27

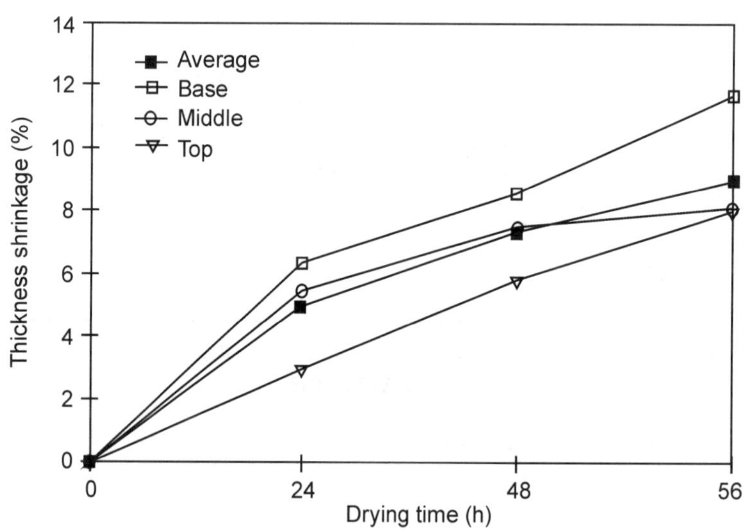

Figure 7. Relationship of thickness shrinkage in bamboo to drying time

It was determined statistically that there is a difference in volumetric shrinkage among the three portions of the bamboo. Table 9 shows the summary of analysis of variance in volumetric shrinkage between the two culms and among the three portions.

The result indicated that there are significant differences in volumetric shrinkage among the three portions of the bamboo at all levels of drying time. However, there was no significant difference between the two culms at any level of drying and hence, the data from Culm I and Culm II can be used together in determining the trend of volumetric shrinkage as the drying time increases.

The trend of increasing volumetric shrinkage in the base, middle and top portions with drying time is demonstrated in Fig. 8. The graph shows that all parts of bamboo have the same linear increase in volumetric shrinkage as the drying time increases. At the first 24 hours, the average of volumetric shrinkage was 9.75%, which then went up to 13.44% at 48 hours and reached 16.27% at 56 hours drying time.

Table 9.
Summary of analysis of variance in volumetric shrinkage

Drying time	Culm (1 & 2)	Part (Base, Mid, Top)	Mean	Standard error
	Mean square & statistical significance			
24 Hours	21.778^{ns}	41.456*	9.53444	0.77778
48 Hours	8.160^{ns}	59.34**	13.44278	0.47611
56 Hours	26.694^{ns}	102.02**	16.2717	0.86111

Notes: * – Significant at 95% level of probability; ** – Significant at 99% level of probability; ns – Not significant.

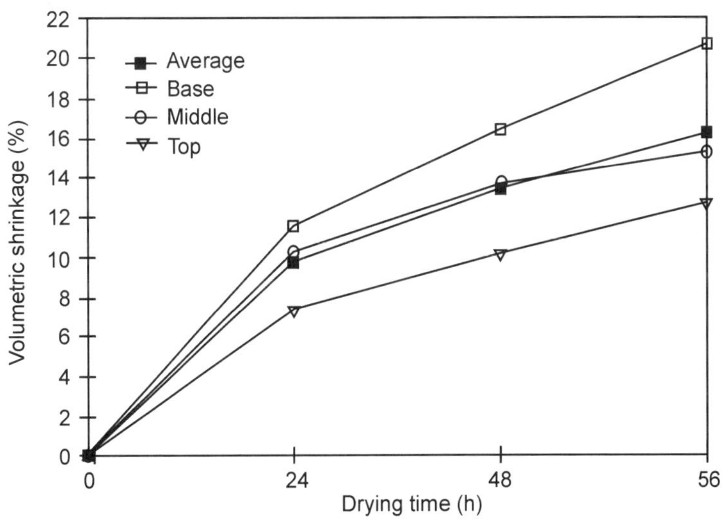

Figure 8. Relationship of volumetric shrinkage (%) and drying time (hours)

The base portion of bamboo displayed the largest volumetric shrinkage at all levels of drying time (11.62% at 24 hours, 16.43% at 48 hours and 20.71% at 56 hours). The middle portion showed more volumetric shrinkage (10.32% at 24 hours, 13.71% at 48 hours and 15.41% at 56 hours) than the top portion (7.32% at 24 hours, 10.20% at 48 hours and 12.70% at 56 hours).

Relationship of shrinkage and moisture content

Moisture content reduction and volumetric shrinkage in split bamboo specimens are related, as indicated by the data plotted in Fig. 9.

As can be seen from the scatter plot of the data, when the moisture content reduces, the volumetric shrinkage increases. To determine their relationship, the data was analysed using linear regression as presented by linear line in Fig. 9. The formula for the linear line is:

$Y = 15.6198 - 0.2016X$, where X is the moisture content (%) and Y the shrinkage (%).

CONCLUSIONS

Three levels of drying time were employed in drying the bamboo (*Bambusa blumeana* schultes *f.*) specimens in the wood waste-fired kiln. There was a sharp reduction in moisture content in the first 24 hours, from 61.21% to 29.27% on the average, followed by a gradual reduction in the next 24 hours (19.18%) and the last 8 hours (13.62%). The basal portion was proved to have the highest moisture

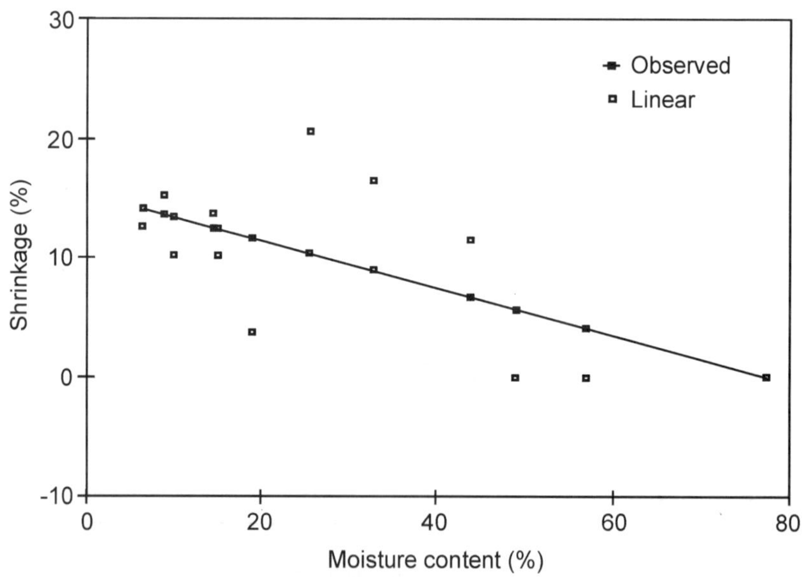

Figure 9. Relationship of shrinkage (%) and moisture content (%)

content at all drying stages (43.71% at 24 hours, 32.77% at 48 hours and 25.51% at 56 hours). The lowest moisture content was recorded in the top portion (18.98% at 24 hours, 10.02 at 48 hours and 6.37% at 56 hours).

Shrinkage (tangential, thickness and volumetric) of bamboo at internode and node tend to increase from the initial to final drying stage, both at the outer and inner parts. The base portion contains the greatest shrinkage at every level of drying time, followed by the middle and the top portions. Shrinkage of the inner part is greater than that of the outer part at both the internode and the node, and the shrinkage at the internode is greater than that at the node.

There is a linear relationship (Y = 15.6198 – 0.2016X) between the moisture content (X) and the volumetric shrinkage (Y), indicating that loss of moisture from bamboo increases volumetric shrinkage: the volume of water lost equals the volumetric shrinkage of cell wall and the bamboo culm.

The results gathered from this experiment are for split bamboo. It is suggested that similar experiments be conducted for round bamboo.

The wood waste-fired kiln is more suited for small-scale operations with split bamboo because it can dry faster without any damage. A similar experiment should be carried out with air-drying for data comparison. Different drying schedules – such as higher temperatures with shorter drying times – may also be tried out, taking into account important physical and mechanical properties of bamboo.

ACKNOWLEDGEMENTS

The author expresses sincere gratitude and appreciation to Dr. D.M. Lantican for his advice, encouragement and guidance in conducting the research, and to SEARCA for financing the research.

REFERENCES

Fangchun, Z. 1992. The state of utilization, trend and strategies for the world bamboo industry. World Forestry Research, Vol. 5, No. 1, 56.

Palisoc, J.G.; Bello, E.D. 1995. Bamboo furniture industry in the Philippines. *In* Ganapathy, P.M.; Janssen, J.A.; Sastry, C.B. ed., Bamboo, people and the environment, Vol. 3, Engineering and utilization. Proceedings of the Vth International Bamboo Workshop, Ubud, Bali, Indonesia, 19–22 June 1995. International Network for Bamboo and Rattan, New Delhi, India. pp. 263–265.

Roxas, J.O. 1990. Manpower training needs analysis of the bamboo furniture industry in Luzon Island, Philppines. M.S. Thesis, University Pertanian Malaysia, 67 pp.

FURTHER READING

Espiloy, Z.B.; Ella, A.B.; Floresca, A.R. 1986. Physico-mechanical properties and anatomical structure relationships of two erect bamboo species. Philippine Timberman, Vol. 32, No. 4, 1986.

Isidro, G. 1988. Fiber-length variability in relation to the anatomical structure of bamboo. Technical Note, Forest Product Research and Industries Development Commission (FORPRIDECOM). National Science Development Board College, Laguna, The Philippines.

Tamolang, F.N.; Lopez, F.R.; Semana, J.A.; Casin, R.F.; Espiloy, Z.B. 1980. Properties and utilization of Philippine erect bamboo. FORPRIDE Digest, Vol. IX, No. 3 & 4.

Ueda, K. 1987. Looking at bamboo again and its practical use. Bamboo Journal, No. 4, 1987.

Usui, H. 1987. Phylogeny of Japanese bamboo based on nodal anatomy. Bamboo Journal, No. 4, 1987.

Bamboo for Sustainable Development, pp. 511–522
Eds. A. Kumar, I.V. Ramanuja Rao and Ch. Sastry

A structural jointing system for bamboo

L.E. MOREIRA
Department of Structural Engineering, UFMG School of Engineering, Belo Horizonte, Brazil

K. GHAVAMI
Department of Civil Engineering, Pontifícia Universidade Católica, Rio de Janeiro, Brazil

Abstract—This paper discusses part of an extensive research program on non-conventional materials, technology and structural forms developed at the School of Engineering, Federal University of Minas Gerais, initiated in 1994. The thinness of bamboo wall considerably limits the structural use of bamboo. For example, space truss structures are the most optimum form to cover a large space without using many columns, and the strength and form of bamboo are well suited and economical for the construction of such structures. However, joining different elements is the most difficult part in the assembly and dimensioning of space structures, and bamboo's thin wall poses a problem in this. To address this issue, different jointing systems have been studied under the research program. This paper presents a joint using the "Basic System of Bamboo Reinforcement" (BRBS) developed for *Dendrocalamus giganteus*, as well as its dimensioning and axial tests.

INTRODUCTION

As we are approaching the 21st century, the building science and technology lag behind other major disciplines, although civil engineering has been developed thousands of years ago. In this century, significant developments are taking place in many technological areas such as transportation, electronics and computers. These developments started principally in industrialized countries with a cold to moderate climate and limited raw materials. The main feature of this fast process of change is characterized at present time by one major objective: the improvement of performance.

Surprisingly, all new technologies have developed in different societies without following any particular tradition or culture. However, when the usefulness of a new technology is realized in a "developed" country, usually accompanied by vast publicity, it readily gets adopted by others without taking into consideration the local conditions, which might lead wastes generation in more than one way. Because of this, import of whole structures – factories, bridges, nuclear reactors, etc. – has become highly questionable. Is it important to participate in this race for modernity when local conditions offer other, and often better-suited, possibilities?

Any new approach in civil construction must incorporate a high-performance design. It should provide new levels of physical performance and environmental attributes, using locally available materials efficiently. It should consider aspects such as natural light ventilation, so that optimum use of environmental forces – such as wind and solar energy – is effected to minimize the need for non-renewable energy. Material and structural design aspects must integrate to create a pleasing solution to complex architectural design problems, taking into account the need for buildings with a long service life and low maintenance and capable of change over time. Responsible engineering involves a permanent search for better use of structural forms and materials.

NATURAL AND STRUCTURAL PERFORMANCE OF BAMBOO

The optimized natural mechanical performance of bamboo needs an equivalent engineering solution. Bamboo culm is conical, with a mean diameter decreasing from bottom to top, and hollow in majority of the species. The ratio of the ring area to total area of the transversal section has a mean value of 30%.

Bamboo is also an optimized tube in terms of mechanical strength to mass ratio, and it has no volume to contain natural defects. Hence, in contrast to wood, bamboo has no knots or deviation of the fibers in the internode regions. However, fiber deviation occurs at nodal regions in radial direction, which is responsible for the enlargement of these regions. This increase in wall thickness t, as well as the presence of a diaphragm, gives bamboo a natural reinforcement that permits the development of branches.

Tension and compression tests of the nodes reveal that bamboo material is less rigid in these regions than in internodal regions. Axial and bending rigidities of the nodes are not well defined for many species of bamboos. Therefore, it is necessary to conduct more investigations to eliminate contradictory results.

The continuity of the fibers in the nodal region is the guarantee of resistance to tension in bamboo strips used in reinforced concrete. The resistant sclerenchyma tissue is immersed in the soft parenchyma tissue. This composite configuration is characterized by high values of Young's modulus E and low values of transversal modulus G.

The low values of G allow for the large twists that bamboo is subjected to under the action of wind. In the same way, the low values of the moment of inertia I of bamboo's sections allow the large displacements under the strongest winds. So, the low G and I are the natural resources that help bamboo withstand load, support its foliage on adjacent foliage or reduce lever arm through displacements.

This natural characteristic of bamboo, however, has been of little value in structures, owing to the lack of efficient joints between structural elements. This drawback has been reported in most literature on the subject (Janssen 1981, 1991; Moreira 1991, 1998).

Thus, bamboo can be safely used structurally only in compression as columns, or in bending as beams. The majority of traditional joints are based on penetration through slots or holes in bamboo wall, and the low support surfaces of these systems induce high stress concentration on joints. The consequences of these solutions are relatively low efficiency and impossibility of application in modern structures, such as space truss.

The "Basic System of Bamboo Reinforcement" (BRBS), shown in Fig. 1, solves this problem, as it basically works through shear stress in the adhesive line. A large glued surface allows asymmetric distribution of shear stress caused by axial forces.

THE CONCEPT OF BRBS

The concept of BRBS is based on the following premises:

- The analogy between chemical composition of bamboo and wood. The lignin-cellulose tissues allow efficient gluing between the two materials. The adhesives suitable for wood-wood gluing are suitable also for bamboo-bamboo gluing (Carrasco 1984; Carrasco et al. 1995) and consequently, for bamboo-wood gluing.
- The possibility of adjustment between bamboo and a cylindrical wood that is introduced at one end of bamboo.
- The presence of strong fibers at the inner part of bamboo wall.

THE EXECUTION OF BRBS

Wooden cylinders, with a length calculated according to the axial force *N*, are shaped in a turning lathe. The axis of the cylinder has to be parallel to the wood fibers.

The fitting between bamboo and wood is made possible through equally spaced cuts on the bamboo wall (Fig. 1). The length l_c of the cuts is greater than the glued length l_g. This procedure allows the bamboo to be fitted over the entire surface of the cylinder.

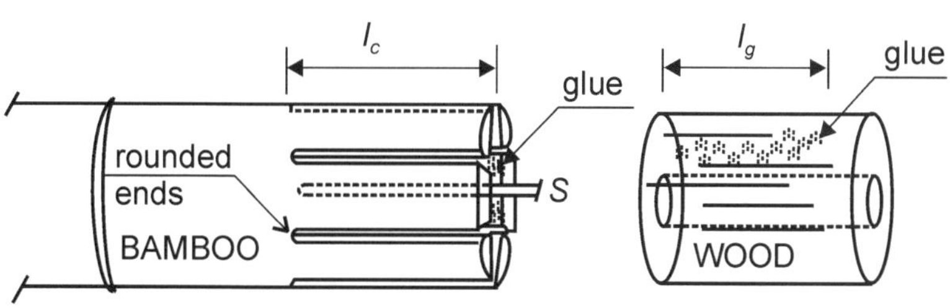

Figure 1. BRBS system

The bamboo wall is then deformed over the wood cylinder using leather straps. A thin layer of approximately 0.4 mm thickness has to be removed from the inner part of bamboo before gluing. This removal makes inner surface rough and exposes bamboo fibers, increasing friction and bonding between bamboo and wood. The adhesive has to be spread over the cylinder surface as well as the inner wall of bamboo for better adhesion.

THE GLUING OF CELLULOSE TISSUES

Carrasco (1984) carried out a large amount of work related to the selection of an efficient adhesive, which is strong and compatible with woods components. The chemical composition of bamboo and wood is similar, but bamboo tissue is more uniform than wood tissue. Bamboo has no coupling of modes of strains, which is common in wood and cause warping and other drying defects. Among the many important conclusions of this work, two are very important for the safety of the bamboo-wood joint:

- When resorcinol glue is used, shear stress, strain of gluing, relative displacement between glued elements and glued area in service are not important functions of the temperature and humidity of air.
- The natural variation of temperature and moisture of the air are not relevant to the strength of wood in service.

STRENGTH OF GLUED JOINTS

When two materials are glued together, the adhesive line has at least 5 components: adherent 1; interface 1; adhesive; interface 2; and adherent 2 (Fig. 2). The lower resistance among these 5 components will define the resistance of the joint. The compatibility of adhesive and adherent is responsible for a strong joint. One of the parameters of compatibility is the solubility of the adhesive; a similar solu-

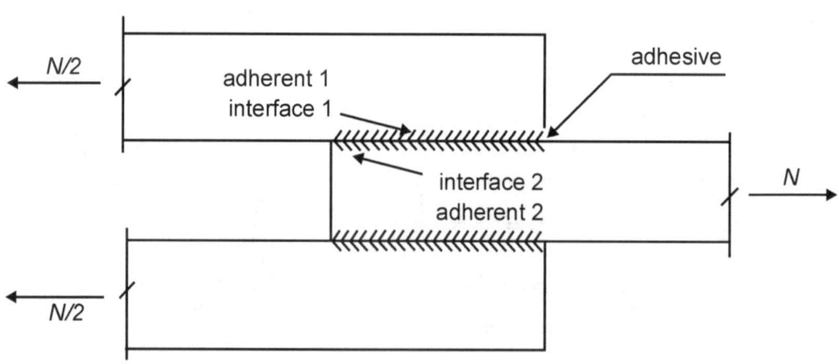

Figure 2. Components of the adhesive line

bility parameter among the adhesive and the adherent will contribute to the efficiency of the joint.

Fig. 3 presents a complete BRBS for absorption of axial forces *N*, which may be used as space or plane truss elements. Radial stress f_c and f_t arise in adhesive line, owing to eccentricity between the pressure *p* and the reaction τ. The force *N* will be resisted by shear stress *t* acting on interfaces. The equilibrium in axial direction produces:

$$N = \int_0^{l_c} \int_0^{\pi} \tau \, R_i d\theta \, dl \therefore \tau = \frac{N}{2\pi \times R_i \times l_g} \tag{1}$$

Literature available has little information about the normal stress acting over adhesive line at the instant of the collapse of joint. The stress distributions, according to Leicester's experimental results (Carrasco 1984), have triangular form in compression zone f_c, and parabolic form in tension zone f_t (Fig. 3 & 4). The application of Leicester models to asymmetric distribution of stress produces the differentials shown in Fig. 4.

The eccentricity of the application of the axial force *N* in relation to adhesive line (Fig. 5) produces moment *M* (Equation 2), which produces f_c and f_t (Fig. 4).

In Fig. 5, the figure on the right shows the adhesive line resistant to half of the force *N*. This force is transmitted over the wood cylinder by a washer with radius r_c. This force is transmitted to bamboo by adhesive line and goes through bamboo, theoretically, at the center of the thickness wall *t*. The resultant force is 0.5 *N* over the washer, whose resultant actuate in G_2 is $\frac{4r_c}{3\pi}$ of the diametrical line affecting half the circumference of bamboo wall, whose resultant actuate in G_1 is $\frac{2(R_i + 0.5t)}{\pi}$ of the same line. So, we have the moment *M* given by Equation (2):

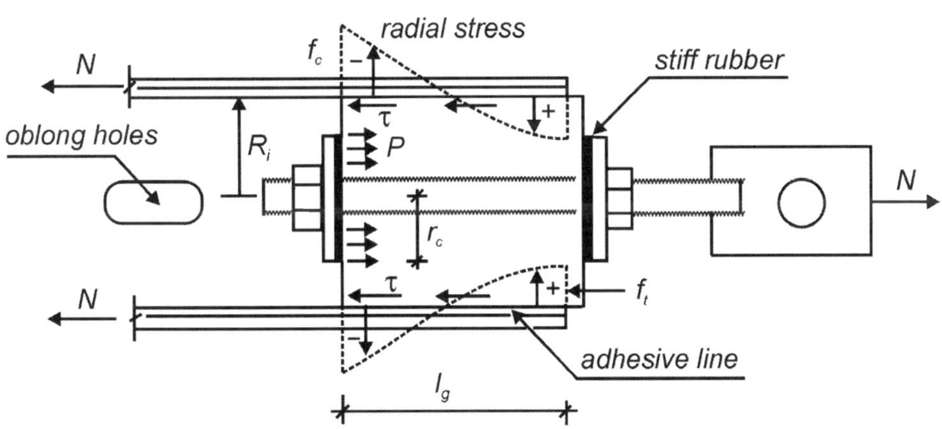

Figure 3. Longitudinal section of BRBS and accessories for space truss

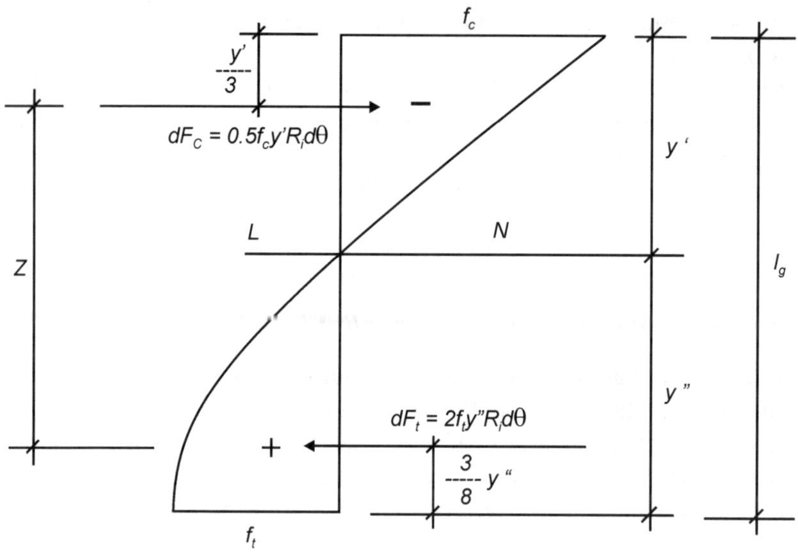

Figure 4. Differentials of the radial forces normal to adhesive line

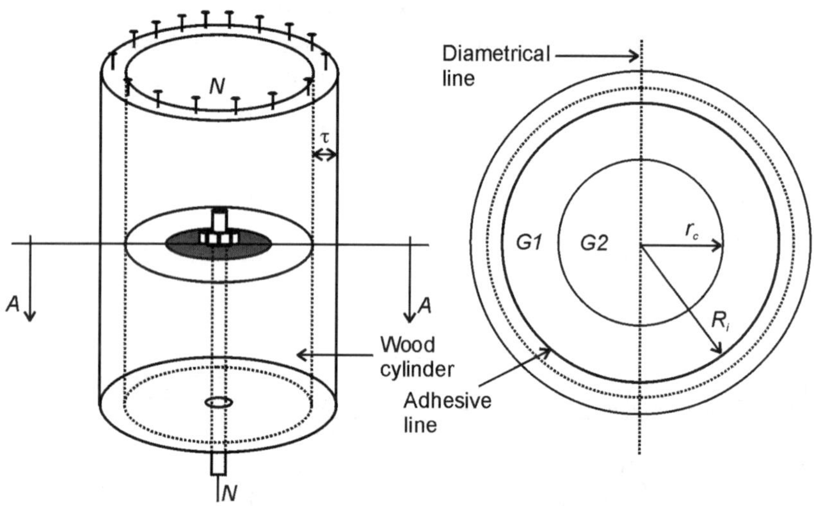

Figure 5. Geometrical characteristics

$$M = \frac{N}{2} \times \left[\frac{2(R_i + 0.5t)}{\pi} - \frac{4r_c}{3\pi} \right] \tag{2}$$

This moment has to be resisted by the binary

$$M = F_t \times z = F_c \times z \tag{3}$$

Taking the normal component of the radial force over the diametrical line of the wood cylinder (Fig. 4) and making the integration:

$$F_c = \int_0^{\pi} dF_{cr} \sin\theta = \int_0^{\pi} 0.5 \times f_c y' R_i \sin\theta \times d\theta \therefore F_c = f_c y' R_i \tag{4}$$

$$F_t = \int_0^{\pi} dF_t \sin\theta = \int_0^{\pi} \frac{2}{3} \times f_t y'' R_i \sin\theta \times d\theta \therefore F_t = f_t y'' R_i \tag{5}$$

The equilibrium over diametrical line produces:

$$F_t = F_c \therefore \frac{4}{3} f_t y'' = f_c y' \tag{6}$$

According to Fig. 4,

$$l_g = y' + y'' \tag{7}$$

$$z = \frac{5}{8} y'' + \frac{2}{3} y' \tag{8}$$

According to Leicester $f_c > f_t$, but in this paper it is considered $f_c = f_t$ in favor of safety. The combination of the given equations results in the gluing length l_g, given by Equation (9), which assures that the failure will occur by shear of the adhesive line before the failure of the wood subjected to tension normal to the fibers f_t.

$$l_g \geq \sqrt{2N_d \left[\frac{R_i + 0.5_t}{\pi} - \frac{2r_c}{3\pi} \right] \times \frac{1029}{763 R_i f_{t90,d}}} \tag{9}$$

Where,

R_i = radio of the wood cylinder
N_d = axial force of calculus
r_c = radio of the washer
$f_{t90,d}$ = resistance of calculus of the wood normal to the fibers
t = mean thickness of bamboo wall
l_g = gluing length.

MATERIALS AND METHODS

Tension test was carried out on 7 specimens soaked in water for 48 hours and on 14 specimens in dry condition (mean moist content of 12%). Compression test was performed on 8 specimens in dry condition.

A turned wood cylinder of *Angelim araroba*, with a length (l) of 8 cm was used. It was adapted to a length (l_g) of 6 cm for the different specimens. Because of this, some bamboo specimens had spliced the wood cylinder by action of tension stress f_t, before the shear neared the adhesive line.

For tension tests of the specimens, two oblong holes of 5/8 inch diameter were drilled over the longitudinal axis of the wood cylinder (Fig. 3 & 5). A washer r_c (radius 30 mm) and a nut were introduced through the holes for fixing the threaded rod (Fig. 3). The mean tension force was 16.5 kN/min.

The failure modes are listed in the last column of Table 1: shear (S), tension (T), rotation (R) or explosion (E). In some cases, failure was due to a combination of two or more modes, and it was difficult to identify the mode really responsible for the loss of resistance of the system.

Table 1.
Tension and compression tests results of BRBS

Spec. no.	*Ri* (mm)	At (cm^2)	Ac (%)	*N* (kN)	f_t (MPa)	τ (MPa)	Test	Failure
1	94.0	177.2	49.0	34.0	3.40	3.96	Comp.	S
2	93.5	176.2	84.0	44.3	4.40	3.00	Comp.	S
3	95.0	179.0	100.0	72.5	7.20	4.05	Comp.	S
4	95.0	179.0	96.0	88.6	8.80	5.13	Comp.	S
5	91.5	172.5	100.0	48.3	4.85	2.80	Comp.	S
6	96.0	181.0	100.0	56.1	5.57	3.10	Comp.	S / R
7	98.0	184.7	82.5	44.3	4.40	2.90	Comp.	S
8	102.9	194.0	100.0	51.7	5.07	2.66	Comp.	E / S
1	97.6	184.0	82.0	46.8	7.50	3.10	Tension	S / T
2	85.0	160.0	87.5	39.0	4.00	2.79	Tension	S / T
3	90.0	170.0	100.0	39.0	6.00	–	Tension	S / T
4	85.0	160.2	93.0	43.0	6.40	2.89	Tension	S / T
5	94.0	177.0	100.0	43.0	6.76	2.43	Tension	S / T
6	102.0	192.3	95.0	38.0	6.20	2.07	Tension	S / T
7	93.0	175.3	100.0	42.0	6.60	2.40	Tension	T
8	82.0	155.0	84.5	50.0	7.30	3.82	Tension	S
9	78.5	148.0	100.0	52.5	7.42	3.50	Tension	S
10	85.0	160.2	100.0	52.1	7.75	3.25	Tension	E / S / T
11	67.0	126.0	100.0	32.5	4.50	2.58	Tension	S
12	83.8	158.0	98.0	47.5	7.00	3.06	Tension	S / T
13	100.0	188.5	100.0	38.0	6.20	–	Tension	S / R
14	110.5	208.3	100.0	32.5	5.50	–	Tension	S / R
1	70.0	132.0	100.0	36.0	4.70	2.73	Tension	S / T
2	86.3	162.6	100.0	44.0	6.60	2.71	Tension	S
3	99.0	186.6	88.0	29.2	4.71	1.79	Tension	S
4	100.0	188.5	96.0	48.2	7.82	2.66	Tension	S
5	90.4	170.4	89.0	48.2	7.43	3.03	Tension	S / T
6	105.0	198.0	76.0	35.8	5.84	2.38	Tension	S
7	95.0	179.0	81.0	33.0	5.22	2.28	Tension	S

Shear (τ) and tension stress (f_t) in Table 1 were calculated using Equations (10) and (11), respectively:

$$\tau = \frac{N}{\pi R_i l_g} \tag{10}$$

$$f_t = \frac{2.697 N_d}{l_g^2 R_i}\left[\frac{R_i + 0.5t}{\pi} - \frac{2r_c}{3\pi}\right] \tag{11}$$

The pressure p in compression test is distributed over the entire section of the wood cylinder, or $r_c = R_t$. In compression test, the speed of load application was not constant owing to manual control of the equipment.

ANALYSIS OF RESULTS

As shown in Table 1, the 8 specimens in dry conditions that underwent compression test were subjected to a mean shear stress (τ) of 3.45 (±0.85) MPa. The mean value for tension stress (f_t) in the wood cylinder, calculated through Equation (11), was equal to 5.46 (±1.74) MPa at failure.

For tension tests of dry specimens, the mean shear stress (τ) was 2.9 (±0.52) MPa and tension stress (f_t) 6.33 (±1.07) MPa.

For compression test specimens saturated with water, τ was 2.51 (±0.4) MPa and f_t was 6.06 (±1.27) MPa. The shear always has occurred in bamboo, for both tension and compression specimens.

It may be observed from Fig. 5 that when the axial force N is of tension, the radial forces of tension f_t will be greater than when the force N is of compression, because r_c in the latter case is equal to the radius of the cylinder R_i. In the former case, the value is greater for M, given by Equation (2), which explains the different performance of tension and compression. Thus, owing to the differences in r_c, the tension stress over the adhesive line, combined with shear stress, causes the collapse of specimens under tension more easily than in specimens under compression. This explains the ambiguity about the failure mode of some tension test specimens.

The comparison between dry and wet specimens shows for the latter a loss of resistance of 13% in relation to shear stress. This loss of resistance is probably caused by stress concentration in the adhesive line, owing to hygroscopic strain between bamboo and wood.

The asymmetric stress distributions over the glued surface of this system are responsible for the large efficiency of this joint, as proved by the tests.

Although it is recommended that the cut ends of specimens be rounded to alleviate stress concentration at these points, this procedure was not adopted during the tests, so as to test the specimens in their worst conditions.

During the tests, there was no shear at the cuts made on the bamboo before joint collapse occurred in the adhesive line. Some specimens showed shear rupture, but not over the glued surface. It is, hence, assumed that a twist of the wood cylinder, making it press laterally against the bamboo wall, must have caused cracks in the direction of the cuts.

SOME DERIVED SYSTEMS OF BRBS

Some accessories may be added to the basic BRBS (Fig. 1) for different applications, according to the requirements of the structural elements. Fig. 3, for example, shows a complete system for use in tension and compression as an element for space or plane truss, in which rubber pieces and metallic washers have been added at both ends of the wood cylinder for proper distribution of the compression pressure *p*. Fig. 6 shows the jointing of two bamboo culms, using a wood cylinder.

In Fig. 7, the top of a bamboo column terminates in a wood reinforcement. Ease of work, proper stress distribution and larger support surface are the advantages of this arrangement. Besides this, it also eliminates the need of nodes near the joining point of elements, as recommended by traditional methods of bamboo jointing (Hidalgo 1981).

In Fig. 8, the wood reinforcement system is used for slanting joints, as used in plane trusses. In all these different types of joints, the wood cylinder must be sized according the approved standards.

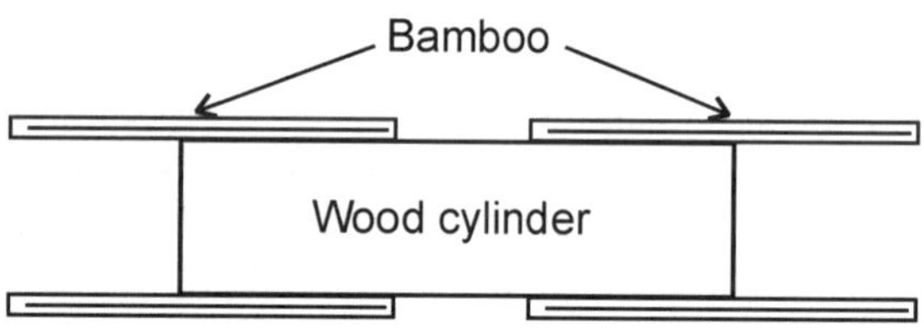

Figure 6. Jointing of round bamboo pieces

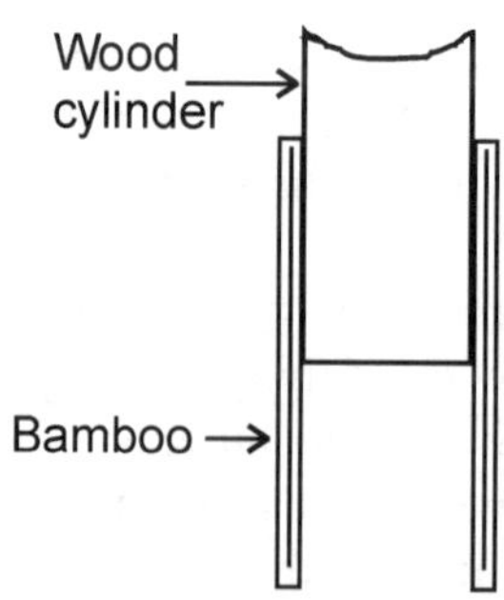

Figure 7. Wooden top for bamboo columns

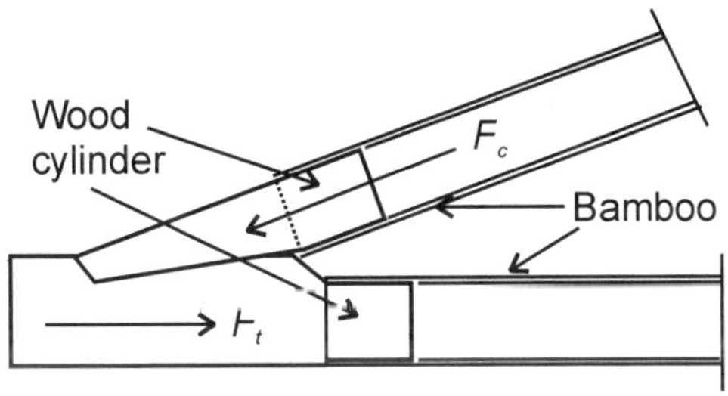

Figure 8. Wood-reinforced joint in a plane truss

CONCLUSION

The science and technology of civil engineering is strongly ruled by a complex, traditional, social hierarchy. Engineers are trained to calculate the structures chosen by architect, who in turn considers only conventional forms and materials. The contractors build according to the plan given to them. It is not surprising, then, that each part of this hierarchy plays only within its limited scope, and there is very little mutual understanding between them. Therefore, it is hard to optimize building design and construction under the present system.

This paper is an attempt to show that modern and traditional bamboo structures may be successfully used in building construction. This new system for joining elements has proved very efficient, and it also has the advantage of being low cost.

The efficiency of other accessories as appendages at bamboo's ends and around some points of the elements has been investigated with the intention of preventing the collapse of the elements from exposure to sun (Moreira et al. 1995).

A comparison between BRBS and another system based on the transmission of the axial force through holes in the bamboo wall, showed a 100% gain in resistance in favor of BRBS (Moreira 1995; Moreira 1998).

These significant results, together with the ease of fabrication and low cost of BRBS, make this system a major promise in the field of modern bamboo structures.

REFERENCES

Carrasco, E.W.M. 1984. Ligações estruturais de madeira. Dissertação de Mestrado, São Carlos, USP, 436 pp.

Carrasco, E.W.M.; Moreira, L.E.; Xavier, P.V. 1995. Bambu laminado colado. *In* Anais do V EBRAMEM, E.E.U.F.M.G., Belo Horizonte, Minas Gerais, Brazil. pp. 246–259.

Hidalgo, O. 1981. Manual de construccion con bambú - construcción rural. Estudios Técnicos Colombianos Ltda, Bogotá, Colombia. 72 pp.

Janssen, J.J.A. 1981. Bamboo in building structures. Ph.D. Thesis, Eindhoven University of Technology, Eindhoven, the Netherlands. 235 pp.

Janssen, J.J.A. 1991. Mechanical properties of bamboo. Kluwer Academic Publishers, The Netherlands. 134 pp.

Moreira, L.E. 1991. Desedvolvimento de treliças espaciais de bambu. Dissertação de Mestrado, PUC/Rio de Janeiro, Rio de Janeiro, Brazil. 172 pp.

Moreira, L.E. 1995. Análise pelo M.E.F. de ligações com pinos circulares em tubos de bambu solicitados axialmente. *In* Anais do Encontro Brasileiro em Madeiras e Estruturas de Madeira V, Belo Horizonte, Minas Gerais, Brazil. pp. 260–275.

Moreira, L.E. 1998. Aspectos singulares das treliças de bambu - flambagem e conexões. Tese de Doutorado, PUC/Rio de Janeiro, Rio de Janeiro, Brazil. 300 pp.

Moreira, L.E.; Ripper, J.L. Ghavami, K 1995. Estudo analitico de uma nova ligação para tubos de bambu solicitados axialmente. *In* Anais do Encontro Brasileiro em Madeiras e Estruturas de Madeira V, Belo Horizonte, Minas Gerais, Brazil. pp. 276–289.

Bamboo for Sustainable Development, pp. 523–533
Eds. A. Kumar, I.V. Ramanuja Rao and Ch. Sastry

Adaptation of wood machining processes for glued bamboo board production

M.T.T. GONÇALVES, M.A. DOS R. PEREIRA, C. AIRES and G. MARIANO
Departamento de Eng. Mecanica, Universidade Estadual Paulista (UNESP), São Paulo, Brazil

Abstract—Bamboo, although widely known, is little used in Brazil owing to lack of knowledge about its useful characteristics and possible applications. Bamboo's rapid growth and great utilization potential, together with its notable strength properties, led us to carry out the present study. The study involved the following aspects: analysis of the giant bamboo species *Dendrocalamus giganteus* for the manufacture of glued boards; proposal of manufacturing processes, equipment and tools; production of pieces of glued boards and analysis of their performance when subjected to conventional wood machining processes.

INTRODUCTION

Some large bamboo species have been employed with great success in the industrial-scale production of durable goods, such as: ply-boards, curtains, various types of containers, furniture, etc.

The specific objective of the present work was to analyse the industrial processes and equipment employed for the production of "Plybamboo" (glued bamboo board) developed in the eastern countries and already implanted in Europe, and based on these analyses, to consider the use of conventional wood machining processes so to adapt them for the preparation of composite products of bamboo.

The giant bamboo species *Dendrocalamus giganteus* was selected for the study because of its widespread availability in the region, its fast growth, and the good mechanical strength properties of its large-sized culms (Pereira 1992).

ANALYSIS OF ALTERNATIVES

D. giganteus species produces large culms, measuring 15–25 m in length, with an average external diameter of 14 cm, 10–30 mm wall thickness and 20–45 cm long internodes. These dimensional features provide good possibility for working the bamboo with conventional wood machining processes.

Bamboo culms are normally processed in the "green" state, i.e. before drying is carried out. In the manufacture of products such as boards, drying is carried out after cutting the cylindrical culm sections longitudinally to form laths or slivers and after removing the external and internal (diaphragm) nodes. Bamboo pieces, thus, are submitted to machining processes in an air-dried state in which they contain as much moisture as the environmental humidity. This is a major difference with the conventional woodworking process in which only well-dried timber is used in the machining process.

Schematically, Fig. 1 represents the conventional method of production of glued, laminated bamboo boards for application as flooring elements (Chin 1994). At the end of the alternative analysis phase, it was concluded that the optimum forms for glued bamboo board in industrialized production are laths and plywood-type composite boards. Products of these forms allow the optimum use of the culms, with the least waste generation during machining operations.

The subsequent proposal developed for the large-scale production of bamboo boards specify the operations, equipment and tools, and process parameters for the different production stages such as harvest, treatment, cross-cutting of culms, removal of nodes, drying, longitudinal cutting, planing, cutting into strips, planing of the four sides, glue application, pressing, sanding, finishing and handling.

The pieces produced in the form of laths and composite boards were subjected to conventional machining processes for evaluating machining performance and the surface finish obtained. Based on this evaluation, a comparative analysis could be made with commercial wood processing using identical machining conditions (Gonçalves 1990).

PROPOSAL FOR THE MANUFACTURING PROCESSES

As can be observed from the method for the production of flooring elements (Fig. 1), the bamboo culms are cut into laths when still in the green state.

A great part of the wood processing industries does not make use of forced drying installations. Hence, bamboo culms in the naturally air-dried condition were used in the study aimed at adapting wood machining processes (Gonçalves 1993) for the production of bamboo composites.

According to the objectives of the study, the sequence shown was considered as the best method for plybamboo production:

1. Harvest of bamboo culms with diameters of 10 cm and lengths of 2.20 m.
2. Cutting off the branches with manual tool.
3. Packing up in culms to protect it from direct sunrays.
4. Preservation by immersion in a tank filled with preservative (or using Boucherie sap displacement method).
5. Stocking in covered and ventilated place for drying.
6. Transversal cutting the bamboo culms to uniform size.
7. Piling up for natural drying in a covered and ventilated place.
8. Longitudinal cutting for the strips production.
9. Planing off the strips to remove the internal soft layer and the internal nodes.

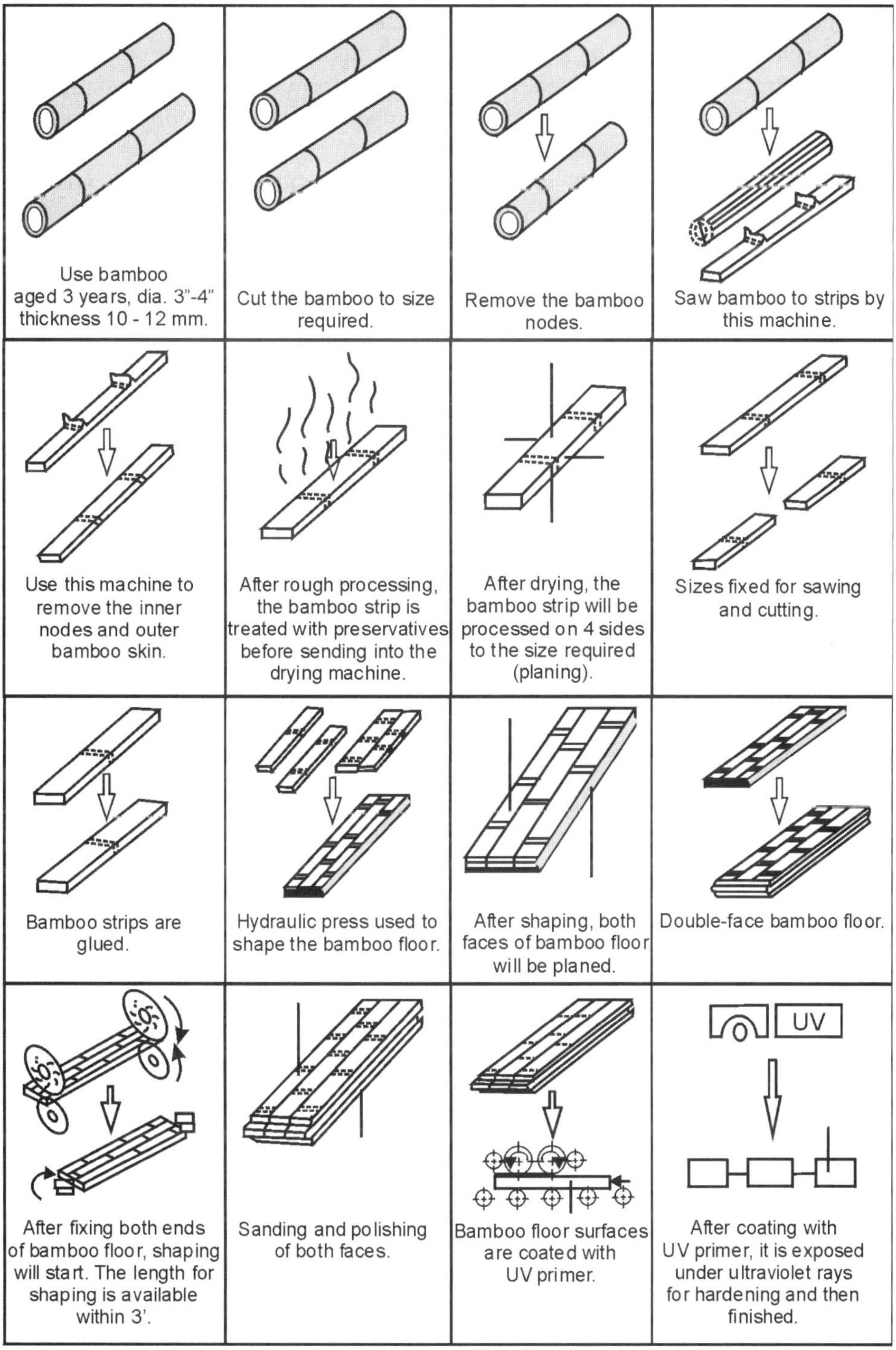

Figure 1. Conventional production flow for laminated bamboo board

10. Lateral planing of the strips for lateral face alignment.
11. Transversal cutting of the strips to uniform length.
12. Spreading adhesive on the sides of the strips to form one layer of plybamboo.
13. Planing off the outer surface of the plybamboo layer (which contains bamboo skin), to obtain layers of uniform thickness.
14. Longitudinal and transversal cutting of the plybamboo layers for standardizing width and length to the final dimensions.
15. Spreading adhesive on the layers and pressing them together to compose bamboo board of a specified thickness.
16. Milling the four sides and the top, and cutting the required rabbets and grooves for fitting.
17. Sanding and finishing.

EQUIPMENT AND TOOLS

Harvesting of bamboo culms

Sawing with a chain saw, with a 4.4-hp motor, 15-inch blade, 3/8-inch chain.

Removal of branches

Cutting manually with a large knife.

Transversal cutting of bamboo culms

Sawing with a crosscutting circular saw (radial, articulate and pendulum or a crosscutting circular saw with sliding table) with a minimum height of cut of 17 cm, hard metal blade of 500 mm diameter and 60 teeth, and a 5 hp motor.

Preservation

Either by immersion in a preservation tank or by sap displacement using the modified Boucherie method, using CCB preservative.

Removal of internal nodes (optional)

The diaphragm may be removed at this stage using a manual tool or a mechanized equipment for internal machining.

Drying

The preserved culms are piled up for natural drying in a covered and ventilated place. Optionally, force-air circulation through fans or exhaust fans may be used for speeding up the drying process.

Longitudinal and transversal cutting

A fixed circular saw, equipped with special devices such as bracket, feed adjustment and safety guards is used. The saw has a cutting height of 0–80 mm, and a 3-hp motor that gives a spindle rotation of 3 800 rpm. For longitudinal cutting, a 300-mm circular blade (36 alternating teeth and 20° exit angle) is used. For crosscutting, a 300-mm, anti-backlash rip saw blade (48 alternating teeth and 10° exit angle) is employed.

Surface planing and side cutting

Surface planing is carried out using an automatic-feed thicknessing planer, while a vertical milling machine is employed for cutting male-female joints on the sides. The 5-hp, 3 400-rpm thicknessing planer has a table with 600 mm width and use knives with steel-alloy insert (500 × 80 × 8 mm). The 5-hp vertical milling machine has a spindle rotation speed range of 3 500 to 6 000 rpm. The milling table measures 900 × 1 000 mm, and the tools used are standard 150 mm and 120 mm (8 mm thickness) milling cutters.

Sanding and finishing

A belt sander machine with continuous feed is used for fine sanding and a coat of varnish is applied to the product as finishing using a pistol-type sprayer with compressed air system.

The belt sander can work with sizes up to 230 mm width and 100 mm height. The feed motor has a rating of 1.5 hp, while the sanding operation is powered by a 7.5 hp, 1 300 rpm motor. The 1-hp varnishing system can accommodate a width of up to 1 000 mm.

EXPERIMENTAL PRODUCTION OF BAMBOO BOARD

The experimental production of bamboo board undertaken involved the following steps and equipment.

1. Harvesting of bamboo culms – of 10 cm diameter and 5 m length – and cutting of branches were carried out using a motorized chainsaw, handsaw and a large knife.
2. A 3-hp crosscutting circular saw (Fig. 2) and a band saw (Fig. 3) were used to cut the culms into uniform length. The circular saw has a fixed table and a guard for the saw blades. The three blades used had the following specifications:

Diameter (mm)	Thickness of tooth (mm)	Number of teeth
300	4.4	48
350	4.4	54
400	4.4	60

Figure 2. Crosscutting circular saw (articulated model)

The band saw (Fig. 3) was powered by a 5 hp, 1 720 rpm motor, and had a wheel diameter of 70 cm. The fixed table was provided with a feed device and a brace for fixing the culm. The saw blade had a width of 32 mm, a thickness of 0.7 mm and 10 mm teeth.

Figure 3. Band saw machine

3. The modified Boucherie process (sap displacement by capillary action), using a CCB preservative, was employed for the preservation of the culms.
4. Treated bamboo sections were covered and stocked in a ventilated place for natural drying.
5. A rotary cutting machine (Pereira 1997) was employed to remove the inner nodes (diaphragm).

6. The sections were then cut longitudinally into strips of uniform size using a circular saw (Fig. 4), powered by a 3 hp, 3 500 rpm motor. Four blades with straight teeth (FZ) and alternating inclined teeth (WZ) used had the following specifications:

Diameter (mm)	Thickness of tooth (mm)	Number of teeth
300 (FZ)	3.1	20
300 (WZ)	3.2	28
300 (WZ)	3.2	36
300 (WZ)	3.2	48

7. The strips were then planed by removing the soft inner layer using a thicknessing planer (Fig. 5) driven by a 5hp motor, running at 2 800 – 3 400 rpm. The planer had a table with 60 cm width and used four knives with steel alloy inserts measuring 600 × 80 × 8 mm.

Figure 4. Circular saw with fixed table

Figure 5. Thicknessing planer

8. Lateral cutting and crosscutting of the strips were then performed, using the circular saw mentioned under Item 6, for obtaining pieces with closely aligning sides and uniform length.
9. A resin-based adhesive (vinyl acetate: resorcinol resin) was then spread on the sides of the strips, and the strips joined to form a layer of plybamboo. A workbench with assembling and pressing devices was employed for the purpose.

10. The top surface of the plybamboo layer (the face that contains skin) was planed off using the thickness planer mentioned under Item 7.
11. The circular saw mentioned under Item 6 was then employed for cutting the layer to size longitudinally and transversally.
12. The adhesive mentioned under Item 9 was applied on the surfaces of the layers and the layers pressed together, using a workbench with assembling and pressing devices, to form plybamboo board of specified dimensions.
13. Milling of the four sides and the top was done for cutting jointing grooves. The machine used was a vertical milling machine powered by a 5 hp, 3 500 rpm motor, and having a table measuring 900 × 1000 mm. Standard milling cutters were used for milling.
14. A belt sander performed the final sanding. The sander, powered by a 3 hp machine, had a working speed of 1 100 mm/min and a working dimension of 7 100 × 130 mm. The final finishing of the product involved varnish application using a compressed air, pistol-type sprayer.

DISCUSSION AND CONCLUSION

The giant bamboo species *Dendrocalamus giganteus* proved appropriate for the purpose of manufacturing glued bamboo board, because of its good physical, mechanical and dimensional characteristics.

The bamboo culms were processed in the green state, owing to technical difficulties and the high cost of artificial drying. In certain applications, this may make the use of bamboo as a wood substitute unfeasible.

The sequence of operations proposed as the better method for the production of plybamboo showed the technical viability in terms of using conventional wood processing equipment. However, the work with green bamboo culms implies a series of safeguards to be adopted in terms of protection against fungal and insect attacks, as well as in terms of drying and storage procedures.

Preserving the culms with CCB preservative implies altering their appearance, as the chemical tend to impart a green coloration to the bamboo.

The best process for drying culms should be determined for each particular situation, in function of the volume of plybamboo production, owing to the long drying period required by the natural method.

The milling of grooves on the top surface of the bamboo board/bamboo floor pieces can be eliminated if the crosscutting that precedes this operation can be accomplished with tools that give fine finishing conditions within appropriate machining speeds.

The process of gluing bamboo strips can be optimized using a machine that would perform automatic jointing (gluing and pressing) of the pieces. The manual assembly and pressing operations reduce the production volume, thereby resulting in lesser revenue.

The height of cut selected for crosscutting bamboo culm sections must be more than the largest diameter of the selected culm sections, to obtain better finish and ensure safety to the operator.

While performing longitudinal cutting, a device for supporting the culms should be used on the machine table. A V-profile, with angle of approximately 120°, proved adequate during the experimental production. In this process, height of cut should be more than the culm thickness.

The planing of the sides of strips for alignment of the faces, together with the planing of the upper and lower surfaces, can be made in a single operation if a planing machine is used.

In the sanding operation, the grade of the sandpaper employed would depend on the intended usage of the product.

The technical data, equipment and tools specified for the operations matched the conditions set for wood machining operations (Koch 1964; Kolmann and Cote 1984).

The product made in this experiment showed good characteristics in terms of finish, strength and appearance, demonstrating a good scope for acceptance by specialists who work with wood.

REFERENCES

Chin, Y. 1994. Machine works. Catalogue of complete set making machinery for bamboo floor/bamboo board. Taiwan.

Goncalves, M.T.T. 1990. Contribuicao para o estudo da usinagem de madeiras. Dissertacao (Mestre-Engenharia Mecanica), EESC - USP, Sao Carlos, São Paulo, Brazil. 241 pp.

Goncalves M.T.T. 1993. Proposta de metodologia para medicao dos esforcos de corte da Madeira. Tese (Doutor - Engenharia Mecanica), EESC- USP, Sao Carlos, São Paulo, Brazil. 148 pp.

Koch P. 1964. Wood machining process. Ronal Press Company, New York, USA. 530 pp.

Kolmann, F.F.P.; Cote, W.A.J. 1984. Principles of wood science and technology, Vol. I. Springer-Verlag, Berlin, Germany. 592 pp.

Pereira, M.A. dos R. 1992. Viabilidade da utilizacao do bambu para fins de irrigacao-aspectos tecniocos. Dissertacao de mestrado (master degree in agronomy), FCA/UNESP, Botucatu, São Paulo, Brazil. 103 pp.

Pereira, M.A. dos R. 1997. Caracteristicas hidraulicas de tubos de bambu gigante *Dendrocalamus giganteus*. Tese (doctor degree in agronomy), FCA/UNESP, Botucatu, São Paulo, Brazil. 161 pp.

Bamboo for Sustainable Development, pp. 535–539
Eds. A. Kumar, I.V. Ramanuja Rao and Ch. Sastry

Plybamboo sheets as housing construction material

JULES J.A. JANSSEN
Faculty of Building and Architecture, Eindhoven University of Technology, The Netherlands

GUILLERMO GONZÁLEZ JR.
Postgraduate student, Eindhoven University of Technology, The Netherlands

Abstract—The aim of this research proposal is to study the use of plybamboo sheets as structural elements (walls) in prefabricated housing in developing countries. One problem major with plybamboo sheets is that each batch could be different. The proposed project, therefore, starts with a research on the properties of sample plybamboo sheets. In the next step, joints between walls will be designed and tested, mainly to resist earthquakes and hurricanes. The result of the study will be a theory on plybamboo sheets as structural element in housing, validated for one sample batch of plybamboo sheets.

INTRODUCTION

Until now, bamboo is largely being used as it is: full culms. Many new plantations are maturing within a few years, and it has been calculated that in Costa Rica alone an annual crop of about 40 000 tons dry material can be expected. If this is "translated" into sheets like plywood, it works out to about a million square meters of 12-mm thick sheets each year. This means that it is time to think about an industrial process, which needs applied scientific research, the subject of this proposal. The most promising item is a sheet like plywood – or plybamboo in the case of bamboo.

Many researchers, like Tandon (1988) and Damodaran and Jagadish (1994), have already research on this product, but all the studies refer to fruit boxes, furniture and the like. These items are good, but not appropriate for a large-scale industrialized production. A more appropriate line of though would be the use of bamboo in housing construction. Developing countries need to develop their economic self-reliance by exploiting their own resources to the full.

The thesis by Arce "Fundamentals of the design of bamboo structures" (1993) gives an appropriate philosophy on the structural design of joints. The International Network for Bamboo and Rattan (INBAR), the international coordinating agency in the field of bamboo and rattan, has recognized the importance of this subject and made it a part of their research program for the immediate future.

Plybamboo sheets do exist as a product, and is being used for products like boxes and doors. The aim of this research proposal is to study the use of these sheets as structural elements (walls) in prefabricated housing in developing countries. One major problem with plybamboo sheets, however, is that each batch could be different. Therefore this project starts with a research on the properties of sample plybamboo sheets. Next, joints between walls will be designed and tested, mainly to resist earthquakes and hurricanes. The result of the study will be a theory about plybamboo sheets as structural element in housing, validated for one sample batch of plybamboo sheets.

AREA OF RESEARCH

Production

This research will work with plybamboo panels 12 mm thick, 1.20 m wide and 2.10 m high, to be used as walls in housing and able to withstand vertical loads (the roof) and horizontal loads (earthquakes, hurricanes). Positive aspects of the project include the possibility of prefabrication and industrialization of plybamboo panels with little capital investment, and the number of jobs created by such a venture. According to Nijhof (1992) the lamination process of bamboo can be as follows:

1. Cutting culms into boards	Bamboo culm is crosscut at the desired length and the inner and outer surface layers are scraped with special equipment. The external nodes are removed. The culms are then split open into two or three pieces. The internal nodes are removed.
2. Preservation	Preservation treatment by soaking the pieces in a cooking vat for several hours.
3. Pretreatment	The key to increase plasticity of bamboo is softening. By means of steam treatment in an electrical heating box at the temperature 130–150°C the spreading behavior of bamboo materials is improved and their splitting property decreased.
4. Flattening into boards	The treated pieces are then spread out, flattened and stabilized in a hot-press and breathing drier, specially used for processing bamboo sheets. The pieces are planed smooth and edged straight on both sides.
5. Bonding	Applying the glue (250 g/m^2), assembling, preparatory pressing (2 N/mm^2), hot-pressing (1.2 min/mm thickness, 130°C).

6. Cooling — Hot-stacking; same as for plywood.
7. Finishing — Edging and shaving.

The research will use commercial plybamboo already stocked in the storeroom of Eindhoven University of Technology (EUT).

Structural design approach

In practice, while working with well known materials, the steps involved in the design process can be as given below:

1. Make a model of the structure, that is, idealize the real structure by a theoretical model that could be analysed with the available computation procedures.
2. Determine the design actions such as earthquakes and hurricanes, which are generally defined by codes; but the designer has to study the local environmental data in order to define the design action.
3. Determine the effects caused by the design actions on the model.
4. Define the strength and stiffness of construction and joints (dimensioning).
5. Determine the quality of the structure, to verify that the structure fulfills the security requirements.

This research, working with plybamboo sheets, will follow the process the other way round:

1. Determine the physical and mechanical properties sample panels.
2. Design joints with these panels, and test them.
3. Determine the strength and stiffness of the construction.
4. Define the design actions (earthquakes and hurricanes) that the structure can resist.

The method must be generally valid, though the validation is with one sample of sheets only. Consequently, the research area can be divided into two parts: material properties and design of joints. Research on both parts will be carried out simultaneously.

The first part deals with material properties of the panels in stock, to be tested according to, for example, CEN standards for plywood. These properties have to be tested to the extent needed for the joints, which are also under study. For instance, if the design of the joint uses nails, then the nailing properties of the plybamboo sheets have to be tested. Similarly, other properties like bending, shear, nailing properties, screwing and nail-holding capacity can become relevant.

With these tests it will be known how to make joints between panels, like T or L types for meetings of internal walls and external walls. The second part, which is scientific research and the most important, is structural design of joints. It will study:

1. The corner of a house – strong and stiff against wind and earthquake, and safe against breaking in.

2. Wind load perpendicular to a wall (bending of the panel).
3. The stability of walls, complete with windows and doors, corners, and connections with roof and foundation.
4. Other loads perpendicular to a wall (such as a break-in).
5. Shear in a joint between two panels.
6. Buckling by vertical loads (roof).
7. Adequate tests to verify the hypothesis and to formulate design rules.

Timetable

Months of the first 2 years	Activities
1–3	Study of literature; design of experiments related to the sheets; development of ideas about joints between walls.
4–11	Prepare specimens, observe them in moist environment during the remaining part of the year; tests on mechanical properties of sheets and T and L type joints.
12	Evaluation of behavior in moist environment; developing new and better ideas about joints, and a series of tests.
13–21	Same as in months 4–11.
22–24	Evaluation and interim report on properties and joints.
Months of the next 2 years	
1–2	Literature study; design of test program.
3–7	Build wall panels to be tested perpendicular to the plane and test them (wind load, break-in), and same procedure for vertical loads (from roof) and buckling.
8–14	Shear in the plane of a panel, by earthquake, with connections to foundation; stability; shear in a joint between panels.
15–19	Corners between two panels; tests on wind, earthquake and break-in.
20–24	Thesis writing, defense.

CONCLUSION

Results of these studies will be published as a thesis, and as a handbook on technical and physical properties of plybamboo. The results will be distributed to groups in developing countries by means of the Information System of INBAR (INBARIS).

The application of bamboo suffers from a tremendous drawback owing to insufficient applied scientific research. Bamboo and bamboo-based products should be brought to the level of familiar wood-based products. This would enable the bamboo-sector to benefit from the enormous design expertise and product range available in the wood-based industries. A western university could play a leading role in this research on bamboo sheets as a housing construction material.

This high-profile use of bamboo would gain the material, as an environmental-friendly material, an entry into the market of building materials and products. The initial processing of the material would be done by the poorer sections of the society and rural households, thus creating jobs in the villages without high capital investment. The final processing would be at the industry-level. The product might contribute to prefabricated housing in developing countries.

ACKNOWLEDGEMENTS

The authors are grateful to the International Network on bamboo and Rattan (INBAR), Eindhoven University of Technology (EUT) and Bamboo Foundation (FUNBAMBU) for financial support and scientific guidance.

REFERENCES

Arce, O. 1993. Fundamentals of the design of bamboo structures. Ph.D. thesis Technical University of Eindhoven, The Netherlands. 260 pp.

Damodaran, K; Jagadish, H.N. 1994. Application of bamboo mat composites in construction and packaging. *In* Bamboo in Asia and the Pacific. Proceedings of the 4th International Bamboo Workshop, Chiangmai, Thailand, 27–30 November 1991. International Development Research Centre, Ottawa, Canada; Forestry Research Support Programme for Asia and the Pacific, Bangkok, Thailand. pp. 299–304.

Nijhof. 1992. Feasibility study regarding the production of plybamboo in Costa Rica. M.Sc. thesis, (Tema) Technical University of Eindhoven, The Netherlands.

Tandon. 1988. Wirebound boxes from bamboo. Indian Forester, October 1988, 693–700.

Bamboo for Sustainable Development, pp. 541–548
Eds. A. Kumar, I.V. Ramanuja Rao and Ch. Sastry

Bamboo mat composite panel – A potential construction material

T.S. RANGARAJU, S.S. ZOOLAGUD and H.N. JAGADEESH
Indian Plywood Industries Research & Training Institute, Bangalore, India

Abstract—In tropical countries like India, wood and wood-based panel products constitute important building materials for structural and non-structural applications. Of late, wood is becoming scarce and costly and is being gradually replaced by high-energy materials like steel, RCC and plastics. There is no hope of increasing supplies of hard wood, which has been a major raw material for panel materials owing to conservation-oriented forest management practices. The building industry, therefore, has to find alternative sources of raw materials, mainly through raising plantations of fast-growing woody species.

Bamboo is a very fast-growing lignocellulosic material possessing excellent mechanical properties. It has been traditionally exploited for it being a strong building material, mainly for rural low-rise house construction. However, in its traditional form, bamboo has low durability. This paper deals with research experience gained in the development of highly durable panel materials from bamboo under a project carried out by the Indian Plywood Industries Research & Training Institute (IPIRTI) and sponsored by International Development Research Centre (IDRC) of Canada. IPIRTI is in a position to offer technologies for manufacturing bamboo mat composite panels – bamboo mat boards, bamboo mat veneer composite panels and corrugated roofing sheets. Technology for corrugated roofing sheet from bamboo has been developed at IPIRTI under another project funded by the Building Material & Technology Promotion Council, (BMTPC), Ministry of Urban Development, Government of India. Development of these materials has led to value addition and enlargement of utilization scope of bamboo for specific applications in fields such as housing construction and furniture manufacture. These materials have clearly demonstrated the potential of bamboo as an excellent alternative to wood and plywood in specific applications. This paper describes the production technology, properties and application of bamboo mat composite panels, with emphasis on their use in construction.

INTRODUCTION

The wood-based panel industries in India, which traditionally depend on natural forests for their raw material requirements, are in a state of crisis now. This is mainly because of the dwindling supplies of prime quality hardwood that was favored for making panel products. The situation has compelled the wood-based panel industries to orient itself towards a larger raw material base of renewable resources, such as plantation timber and other lignocellulosic materials like bamboo and agricultural residues. Various strategies have been evolved in different countries for the conservation of prime quality wood and its judicious use for the production of wood-based panels, in combination with lesser-known secondary wood from plantation or fast-growing species and agricultural residues to comply with functional requirements of the panels while keeping in view the techno-economical viability and user acceptance. In this context, bamboo, with its excellent mechanical and strength characteristics, holds a major promise.

Bamboo is a very important construction/building material in most bamboo-growing countries. For low-rise and low-cost houses, the use of bamboo for walling, roofing and even flooring is very well known and does not need elaboration. Engineers have been attracted towards the excellent mechanical properties and easy workability of bamboo. In particular, the high tensile strength and modulus of elasticity of bamboo have motivated engineers to experiment with bamboo as a reinforcing material in cement concrete. There has been mixed success in the utilization of bamboo as reinforcement, as the material has the tendency to shrink and swell, and perishes easily if not treated. Round bamboo, although characterized by high tensile strength, is prone to splitting and insect attack. In order to extend the structural possibilities of bamboo, work is being carried out in countries such as China and India to develop composite panel materials to overcome limitations of bamboo such as swelling, shrinkage and splitting.

The limitations of bamboo have been overcome in the bamboo mat board (BMB) developed by the Indian Plywood Industries Research & Training Institute (IPIRTI). It has excellent mechanical properties, plane rigidity, stiffness, weatherproofness, and resistance to decay and insects. Experiments have shown that bamboo mat composite panels – BMB and bamboo mat veneer composite (BMVC) – are suitable for roof and wall cladding, non-load bearing partitions/ceiling, etc. Besides, because of their high modulus of rigidity, the material has potential application as web material in Box and I-beams, arch panels, shear walls, storage bins, etc., and is equally suitable as infill material in door shutters.

Corrugated bamboo mat board is another promising material recently developed at the Institute. It is light, strong, stiff and weather-resistant, and appears to be a good alternative to Asbestos-cement sheets. These derivative materials from bamboo can revolutionize house construction, particularly in disaster prone areas and prefabricated houses.

DEVELOPMENT OF BAMBOO MAT COMPOSITES

Bamboo mat board (BMB)

Process technology has been developed for the manufacture of BMB at IPIRTI, and several entrepreneurs in India have started using the technology developed at IPIRTI to manufacture exterior grade BMB. Studies conducted at the Institute have clearly established that the manufacture of BMBs up to 6 mm thick is economically viable and commercially feasible (Zoolagud and Rangaraju 1994).

BMB and BMVC are examples of high-quality materials with high internal bond strength, surface strength, modulus of rigidity and durability (preservative chemicals are mixed into the resin system itself). Experiments have shown that PF-bonded BMB, BMVC and corrugated bamboo roofing sheets can be used for low-cost construction, especially in rural housing.

Test methods were developed at IPIRTI to evaluate bond quality of bamboo mat board (Naidu and Aswathanarayana 1991). Subsequently, the Bureau of Indian Standards has brought out an Indian Standard Specification on BMB for general purposes (IS: 13958 – 1994). Physical and mechanical properties of BMB were evaluated by adopting standard test procedures for plywood (IS: 1734 – 1983).

Extensive studies, conducted at the Institute to identify applications for BMB, have established its suitability for non load bearing applications similar to plywood in areas like housing, packaging, storage and transportation (Damodaran and Jagadeesh 1994). Physical and mechanical strength properties of BMB are given in Table 1. Utilization of BMB for load-bearing structural applications similar to structural plywood and concrete shuttering plywood were found unfeasible because of its lower strength and stiffness ratio to density (Jagadeesh and Damodaran 1993). Strength properties of BMB are comparable with plywood made with lesser-known timber species as well as plantation-grown species. Table 2 gives physical and mechanical properties of plywood using lesser-known species such as *Duabanga grandiflora* (Lampati), *Tetrameles nudiflora* (Maina), *Artocarpus chaplasha* (Chaplash), *Parishia insignis* (Red dhup), *Alnus nepalensis* (Indian alder), *Paraserianthes falcataria* (Albizzia) and *Hevea braziliensis* (Rubber wood).

Table 1.
Physical-mechanical properties of BMB bonded with PF resin (Species used: *Ochlandra travancorica*)

Property	No. of mats used for making board			
	2-layer	3-layer	5-layer	7-layer
Density (kg/m^3)	751	766	771	790
Internal bond strength (N/mm^2)				
– Dry state	2.23	2.18	2.42	1.97
– Wet state	1.79	1.98	2.14	1.73
Surface strength by torque wrench (N/mm^2)				
– Dry state	12.14	11.42	11.23	9.47
– Wet state	11.01	11.42	10.47	9.10
Tensile strength (N/mm^2)	–	22.69	26.59	29.54
Compressive strength (N/mm^2)	–	16.77	30.35	35.30
Modulus of Rupture (N/mm^2)	–	50.74	56.31	59.35
Modulus of Elasticity (N/mm^2)	–	3678	3220	3114
Modulus of Rigidity (N/mm^2)	5401	5881	6050	6066

Table 2.
Physical and mechanical properties of plywood using lesser-known as well as plantation-grown species

Property	Wood species							
	Lampati	Maina	Chaplash (Assam)	Red dhup	Indian alder	Chaplash (Andamans)	Albizzia	Rubber wood
Density (kg/m^3)	559	407	502	552	450	458	416	520
Tensile strength (N/mm^2)								
–Along	43.2	19.8	25.3	51.6	35.3	49.5	21.2	48.3
–Across	30.4	19.3	29.6	32.7	28.4	31.9	20.2	37.3
Compressive strength (N/mm^2)								
–Along	28.1	18.6	27.5	30.0	27.9	30.9	19.1	30.3
–Across	29.7	17.3	27.0	28.9	26.0	30.0	16.2	22.4
Modulus of rupture (N/mm^2)								
–Along	51.8	30.1	46.0	50.6	45.1	49.9	27.5	62.5
–Across	44.2	22.0	32.0	29.9	30.9	30.4	21.5	37.9
Modulus of elasticity (N/mm^2)								
–Along	6418	4564	6195	5415	5697	5143	4056	7713
–Across	3552	2734	2729	2592	2576	2485	2498	3572
Modulus of rigidity (N/mm^2)	536	445	459	504	556	514	424	504

Bamboo mat veneer composite (BMVC)

In view of the lower physical and mechanical strength properties of BMB in comparison with the requirements of structural grade plywood and non-availability of BMB beyond 6 mm thickness, attempts were made to overcome these deficiencies by developing a BMVC under the IDRC-sponsored project. Process technology developed for the manufacture of BMVC makes use of bamboo mats in combination with veneers from plantation species (Rajeev et al. 1996). The physical mechanical strength properties of BMVC are given in Table 3. Strength data of BMB, plywood from lesser-known and plantation-grown wood species, and BMVC reveal the following aspects.

Strength properties of BMB are comparable to plywood of lesser-known and plantation-grown wood species. However, BMB has a significantly high modulus of rigidity in comparison with plywood. Low strength properties are expected, as BMB is made of layers of bamboo mat with herring-bone type weave arranged in the same direction with respect to adjacent mat, unlike plywood which has cross-grained layered construction.

Table 3.
Physical-mechanical properties of bamboo mat veneer composites

Thickness of panel (mm)	Density (kg/m^3)	Modulus of rigidity (N/mm^2)	IB dry (N/mm^2)	IB wet (N/mm^2)	SS dry (N/mm^2)	SS wet (N/mm^2)	Tensile strength (N/mm^2)		Comp. Strength (N/mm^2)		MOR (N/mm^2)		MOE (N/mm^2)		Wood panel replacement (%)
							Along	Across	Along	Across	Along	Across	Along	Across	
6	771	6050	2.42	2.14	11.23	10.47	26.59	26.59	30.35	30.35	56.31	56.31	3220	3220	100
8	750	6066	1.97	1.73	9.47	9.10	29.54	29.54	35.30	35.30	59.35	59.35	3114	3114	100
12	708	3135	2.14	1.55	12.65	9.05	50.20	29.30	43.40	34.80	75.65	44.10	5725	3015	40
16	783	2282	2.28	1.84	13.80	11.0	41.60	22.80	49.20	38.30	72.40	35.1	7750	2240	72
19	827	3240	2.36	1.70	14.20	10.0	38.40	26.10	47.10	42.60	59.90	55.60	6740	3730	75
16	760	2324	1.56	1.54	11.80	8.80	34.80	32.50	43.20	34.70	72.20	39.30	8030	2410	57
19	752	2490	2.2	1.60	13.50	10.00	41.90	33.40	45.70	36.30	72.70	42.40	7400	2620	63
21	802	3316	2.30	1.65	13.00	9.60	36.40	35.80	43.90	40.20	68.50	55.40	7820	3210	67

Introduction of wood veneer in the panel has increased most properties considerably, except the modulus of rigidity, depending on the number of veneers and their position and pattern in the assembly with respect to grain direction of adjacent veneers. Though modulus of rigidity of composite panels has reduced when compared with the values of BMB, the composite panels still possess much higher modulus of rigidity (4–9 times) than the values prescribed for structural plywood. Introduction of bamboo mat in the composite panel has substantially increased the strength properties of plywood made entirely from plantation-gown species.

Density of BMVC panels is comparable to that of BMB; however, it is higher when compared with the values of plywood made entirely with veneers of plantation-grown wood species.

Considering the strength values of composite panels, manufacturing parameters and applications, it is desirable to manufacture panels using veneer as face plies and mid-core interspersed with bamboo mats equally on either side.

There is scope to carry out further work in the manufacture of composite panels by changing the assembly pattern and thickness of veneers, and substituting wood veneers with bamboo mats to the extent possible. This would help meet the required strength properties for conforming to Indian Standard Specification for structural plywood.

Corrugated bamboo mat roofing sheets

Technology for corrugated roofing sheets from Bamboo mat has been developed at IPIRTI under a project funded by the Building Material & Technology Promotion Council (BMPTC), Ministry of Urban Development, Government of India. Development of corrugated roofing sheet has led to value addition and enlargement of scope of bamboo utilization for specific applications, especially in the field of housing. A modified process has been developed for the manufacture of corrugated roofing sheets, which are highly resistant to weather, borer and decay. Studies are in progress to upscale and standardize the technology and evaluate the roofing sheets. Preliminary tests have shown that corrugated sheets exhibit good impact strength coupled with excellent resilience.

Application of bamboo mat composite panels in construction

Studies at IPIRTI have clearly demonstrated that BMB can be used as infill panels in furniture and panel doors and as skins in flush doors. Designers can take advantage of BMB's high modulus of rigidity in designing furniture, partition cubicles, etc. BMB of 4–6 mm thickness is sufficient for such applications. It is also possible to use BMB as a structural component – web beam, trussed shed walls, stressed skin roofing, arch and vault roofing, etc. BMVC may also be suitable for such applications. By combining veneer and bamboo mat, it is possible to produce thicker panels suitable for panel furniture without using any wood.

The technology for manufacture of BMB has now been extended to the manufacture of corrugated roofing sheets. A roofing sheet of only 3 mm thick with suitably designed corrugations possesses high stiffness strength and resilience. The technology is in the development stage.

In general, BMB and BMVC panels are suitable for making prefabricated shelters because of their high modulus of rigidity and weather resistance. The sheets, however, have to be reinforced with wooden or bamboo studs. It is expected that with proper design, such shelters would be suitable for withstanding high winds and earthquakes. BMB and BMVC are also suitable to be used as wall and roofing panels in wood-frame housing, though more research and developmental work are needed in the design of these shelters. They can also be used in low-cost housing in combination with other materials like stone, bricks, and split and round bamboos.

CONCLUSION

Based on the technologies developed at IPIRTI, it is possible to manufacture high-strength bamboo mat boards and bamboo mat veneer composite panels, which have immense potential for different end-uses including structural/constructional applications. There is ample scope to substitute wood veneers to the extent of 50–60% and to improve strength properties by changing the veneer-mat assembly pattern and veneer thickness in the manufacture of composite panels, so that they meet the standards of Indian Standard Specification for Structural Plywood.

ACKNOWLEDGEMENTS

The authors are grateful to the International Development Research Centre (IDRC) of Canada for funding the project on wood substitutes, and to BMTPC, New Delhi, for sponsoring work on corrugated roofing sheet. The authors are also indebted to Shri Arun K. Bansal, Director, IPIRTI, for his advice and encouragement. Authors wish to express their thanks to several others who rendered assistance in the investigation.

REFERENCES

Damodaran. K; Jagadeesh, H.N. 1994. Application of bamboo mat composites in construction and packaging. *In* Bamboo in Asia and the Pacific. Proceedings of the 4th International Bamboo Workshop, Chiangmai, Thailand, 27–30 November 1991. International Development Research Centre, Ottawa, Canada; Forestry Research Support Programme for Asia and the Pacific, Bangkok, Thailand. pp. 299–304.

Jagadeesh H.N.; Damodaran, K. 1993. Applications of bamboo mat board. Paper presented at the National Workshop on Bamboo Mat Board, 12 February 1993. IPIRTI, Bangalore, India.

Naidu, V.M.; Aswathanarayana B.S. 1991. A study on test methods for evaluating bond strength of bamboo mat board. *In* Proceedings of the National Seminar on Bamboo, 19–20 December 1990.

Rajeev, B.N.; Naidu, V.M.; Rangaraju, T.S.; Zoolagud, S.S.; Aswathanarayana, B.S. 1996. Development of BMB wood veneer composite as wood substitute. Paper presented at the National Seminar on Bamboo, Bangalore, India, 28–29 November 1996.

Zoolagud, S.S.; Rangaraju, T.S. 1994. An improved economical process for the manufacture of bamboo mat board. *In* Bamboo in Asia and the Pacific. Proceedings of the 4th International Bamboo Workshop, Chiangmai, Thailand, 27–30 November 1991. International Development Research Centre, Ottawa, Canada; Forestry Research Support Programme for Asia and the Pacific, Bangkok, Thailand. pp. 292–298.

FURTHER READING

Bureau of Indian Standards. 1983. Indian Standard: 1773–1983, specification for methods of tests for plywood. Bureau of Indian Standards, New Delhi, India.

Bureau of Indian Standards. 1983. Indian Standard: 10701–1983, specification for structural plywood. Bureau of Indian Standards, New Delhi, India.

Bureau of Indian Standards. 1994. Indian Standard: 13958–1994, specification for bamboo mat board. Bureau of Indian Standards, New Delhi, India.

Naidu, V.M.; Aswathanarayana, B.S; Raghunath Rao, D.M. 1987. Determination of physical-mechanical properties of wood. IPIRI Research Report No. 27, IPIRTI, Bangalore, India.

Zoolagud, S.S.; Rangaraju, T.S. 1990. Phenol-formaldehyde adhesive for bamboo mat board. *In* Proceedings of the National Seminar on bamboo, 19– 20 December 1990. pp.106–126.

Bamboo for Sustainable Development, pp. 549–558
Eds. A. Kumar, I.V. Ramanuja Rao and Ch. Sastry

Industrialized use of bamboo in structures

L.E. MOREIRA
Department of Structural Engineering, School of Engineering of UFMG, Belo Horizonte, Brazil

K. GHAVAMI
Department of Civil Engineering, Pontifícia Universidade Católica, Rio de Janeiro, Brazil

J.L.M. RIPPER
Department of Arts, Pontifícia Universidade Católica, Rio de Janeiro, Brazil

Abstract—According to Maturana (1997), a prominent researcher in neurobiology, works in knowledge can be defined as an operational system of distinction. This operation system, which is expressed in different languages, permits access to living beings – human beings, in particular – in a consensual way inside an existing space and transforms itself during existence. Following these views on cognitive mechanisms, this paper presents certain ways that have been considered adequate for transforming bamboo into structural elements with applications compatible to industrially produced elements. The attempt sums up a 10-year study on the transformation of bamboo into a structural element, with the possibility to manipulate using mechanical engineering concepts. Tests carried out in different laboratories of UFMG and PUC-Rio later confirmed the results of the study. The paper also describes the basic characteristics of a pre-fabricated model of a bamboo structure.

INTRODUCTION

Intensive exploitation of the Brazilian forests had existed since colonial times, with the main objectives of supplying firewood and enlarging agricultural area. As these activities resulted in destruction of the forest areas, extraction of timber was occasional. In the 1920s and 1930s, however, the efforts at industrialization made the first timber industries appear. By the 1950s, the timber industry was already representing 7.4% of the total industrial production in the country, providing 5.2% of the national income. Prodigious industrial development in all areas in following two decades also caused indiscriminate exploitation of timber and the exhaustion of traditional forest areas in the South of the country. Since the 1970s, the greater part of the national timber industry has moved to the Amazon region. Currently, the timber industry in Brazil can be divided into the following segments: plywood industry; chipboard industry; furniture, joinery and carpentry industry; and civil construction industry.

Possibilities of large-scale use of bamboo

The population of Brazil will have reached 179.5 million in the year 2000, according to the Brazilian Institute of Geography and Statistics, causing an increased demand for housing. Already, there are hundreds of thousands of people living in cardboard boxes in the streets, pavements, street squares and beaches. As the population in cities is increasing rapidly owing to migration from the rural areas, it is becoming impossible to provide basic services such as housing, education and public health. Further more, the rate unemployment is growing at the same rate at which the cities are growing. The complexity of the problem, which is more of historical and socio-political origins than just economical, is evident.

On the other hand, deforestation has devastated the country in the last few decades, leading to desertification across the whole national territory, principally in the southern region of the country where a large percentage of fertile areas has been laid waste (Moreira 1991).

These facts necessitate an emphasis on searching for socio-economically appropriate and environment-friendly solutions to overcome the grave crisis that faces the country. This is where bamboo comes in.

Bamboo is, of course, not a panacea; however, its large-scale utilization appears in this context as an option parallel to the use of timber in a different situation. Bamboo presents a very valuable resource with a wide range of uses in countries like China, India and Indonesia. It is, however, practically unknown in the Occident, with the exception of Colombia and Costa Rica where bamboo has recently found application in housing construction (Hidalgo 1985; Chaves and Gutierrez 1990).

Bamboo has several advantages such as:

- A resource with high rates of production (2.5 to 3 years after appearance of shoot) and renewal.
- High mechanical resistance, superior to most timbers (Janssen 1991; Moreira 1991; Moreira and Ghavami, this volume).
- Tubular form, structural stability and ready-to-use finish, with rigidity to deflection superior to metallic tubes and composites, concrete structural elements and timber (Wegst et al. 1993).
- Low specific mass (6–9 kg/m^3) and low structural weight owing to its hollow form, which promotes workability.

In the past 8 years, the authors have been studying two types of bamboo structures for civil construction: a double-layer space structure and a single-layer geodesic structure. For the first structure, a system of metallic pin connectors was earlier used. This was later replaced with the modified connector BRBS. This system (see Moreira and Ghavami, this volume) presented a much higher mechanical efficiency than the former system, at lower cost and easier fabrication.

The double-layer grid could be employed for grain silos, stables, chicken pens, community centers etc. in the rural areas, and supermarkets, restaurants, workshops, churches, sport halls, etc. in the urban areas. The geodesic structure could serve in rural areas as coverings of green houses, dwellings, among others and in urban areas as coverings of dwellings, meeting places, shelters in general, stalls etc.

Bamboo can be used well in simple structures such as stiffeners, scaffolding, trusses, beams and furniture as long as derivations over BRBS are used. All these applications, which are frequently required, are usually resolved using other materials such as steel, aluminium and timber.

APPLICATIONS OF BAMBOO STRUCTURES

During the past 8 years, the authors have been studying two types of bamboo structures for civil construction: a double-layer space structure and a single-layer geodesic structure. Fig. 1 shows the prototype of the first double-layer space structure, which was tested until rupture (Moreira 1991). A system of metallic pin connectors was used in this structure, which was later modified.

Fig. 2 presents a joint of the same structure, where the modified BRBS connector (Fig. 7) is used. This system, described in this article, presents a much higher mechanical efficiency, at a lower cost and easier fabrication, than the earlier system. Fig. 3 shows a single-layer geodesic structure with a diameter of 7 m. Other researchers have also reported on the use of bamboo as a structural element and on various tests carried out with the prototypes of the structures (Janssen 1981; Villalobos 1993).

Figure 1. Prototype test of the double-layer bamboo space truss

Figure 2. BRBS node of a double-layer bamboo space structure

Figure 3. A single-layer geodesic bamboo structure

The double-layer space structure could be used in the rural areas to cover grain silos, stables, chicken coops, community centers, etc. and in the urban areas for supermarkets, restaurants, workshops, churches, recreation halls, etc. The geodesic structure could serve as coverings for green houses, dwellings, meeting places, vendor stalls and other types of shelters.

Bamboo can also be used in simple structures such as stiffeners, scaffolding, trusses, beams and furniture as long as derivations over BRBS are used. All these applications may require the use of other materials – such as steel, aluminium and timber – also to optimize the structures.

PREPARATION OF BAMBOO FOR STRUCTURAL ELEMENTS

Basic preparations

Machinery, tools, gadgets, accessories and basic materials employed in the fabrication of bamboo structural elements include the following:

- Electric machines – vertical saw, vertical circular saw, plane, electric horizontal and vertical drill, mechanical lathe, welding machine, slicing and cutting machines, thinning machine.
- Manual tools – hacksaw, chisels, screwdrivers, hammers, scraper, steel ruler, knives, a set of angles, drill for the internodes.
- Special gadgets – jigs and fixtures.
- Basic accessories – nails, nuts and bolts, cloth, nylon and sisal ropes, galvanized wire, brushes and sandpaper, metallic braces, steel or aluminium plates, drill bits, vises.
- Basic materials – bamboo, timber, varnish, glue, preservatives, electrodes, paper.

The factory needs to have the following areas, defined in such a way as to make the fabrication process most efficient, for: (1) preservation treatment of bamboo; (2) drying of treated bamboo; (3) cutting and transforming bamboo into structural elements; (4) finishing the structural elements; and (5) storage of structural elements.

When harvesting the culms, care must be taken to follow suitable operational procedures and good management practices, as mentioned by Hidalgo (1981). The length of the culms should be determined in terms of ease of transportation, treatment and the end-use requirements. Ends of the culms should not be nodes, as this usually crack during the drying process. Open ends (internodes) can be strengthened after drying using BRBS. Care should be taken while handling, as any minor crack will get amplified during drying.

The first step in the process is perforation of the nodes (diaphragm), using a tool such as the one mentioned by Pereira (1997), a twin-blade tool that works in both longitudinal directions (Fig. 4). Preservation treatment is required for bamboo to increase its durability, and Boucherie or modified Boucherie method could be used for this (Hidalgo 1981). After treatment the bamboo is dried, ensuring that drying is as uniform as possible to avoid strong internal tensions caused by different moisture levels. A hothouse, with a controlled humidity and temperature, may be used for the purpose. Although relatively costlier, reduced losses from cracking and accelerated drying (natural drying process takes about 3 months, depending on the physical characteristics of the species) compensates for this. Dried bamboo is ready to be transformed into structural elements.

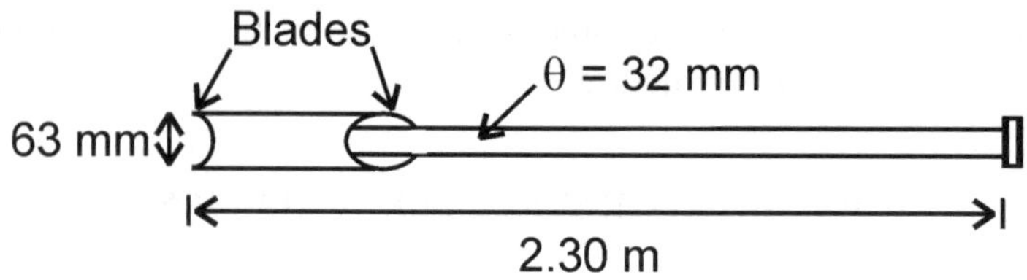

Figure 4. The manual impact tool for perforating the diaphragm

Strengthening of the ends of structural elements

BRBS is a finish given to the extreme ends of bamboo to make it more resistant and more workable (see Moreira and Ghavami, this volume, for details). It consists of fitting a wood cylinder into the opening at the end of a bamboo culm by means of an adhesive and a set of braces (Fig. 5). Six equally spaced slits cut parallel to the bamboo axis facilitate the introduction and fitting of the wood cylinder. The length of the slits depends on the length of the joint, which is determined by structural calculation in function of required load (Moreira and Ghavami, this volume).

The wood cylinder must not contain any knots or cracks. Before applying the adhesive, a thin layer is removed from the inner bamboo wall to expose the fibers and provide a rough surface for better adhesion. A good fitting between bamboo and wooden cylinder is essential for good adhesion and consequently good strength of the system (Fig. 6). After glue is spread on the inner wall of bamboo and the outer surface of the wood cylinder, the latter is inserted into the former. Clamps are used to maintain pressure over the glued area. Once the glue sets, the clamps can be removed.

After the ends of the bamboo are strengthened by BRBS as described above, various applications can be considered in connecting the bamboo poles to other elements, using various accessories. The accessories could be threaded rod, bolt, discs and washers, etc. BRBS has been subjected to several tension and compression tests (Fig. 7) and its efficiency and versatility confirmed (Moreira and Ghavami 1997; this volume).

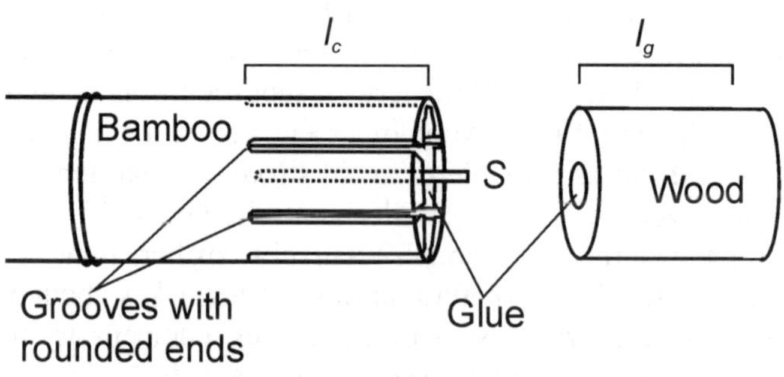

Figure 5. Basic components of the BRBS system

Figure 6. Bored wooden cylinders fitted into bamboo sections

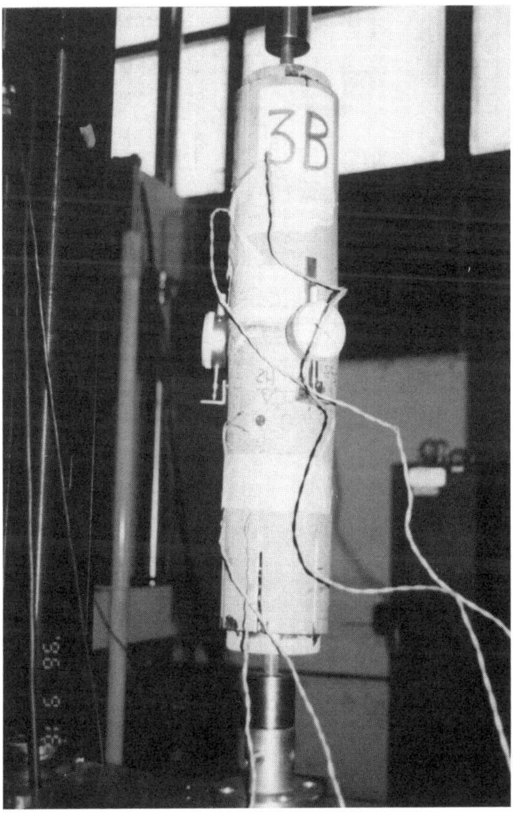

Figure 7. Bamboo element with BRBS under tension test

PRODUCTION OF BAMBOO STRUCTURAL ELEMENTS

Layout of factory

Fig. 8 provides a layout for a factory for fabricating bamboo structural elements. Optimum number of workers, time of execution for each step, energy consumption, etc. will depend on factors such as the demand, the type of structure, the funds available and local conditions. In the figure, arrows indicate the flow of the main operational movements.

Cost estimation of elements strengthened with BRBS

For bamboo element with an internal diameter of 10 cm and a length of gluing (joint) of 10 cm, the total area of gluing – considering the two ends of the bamboo

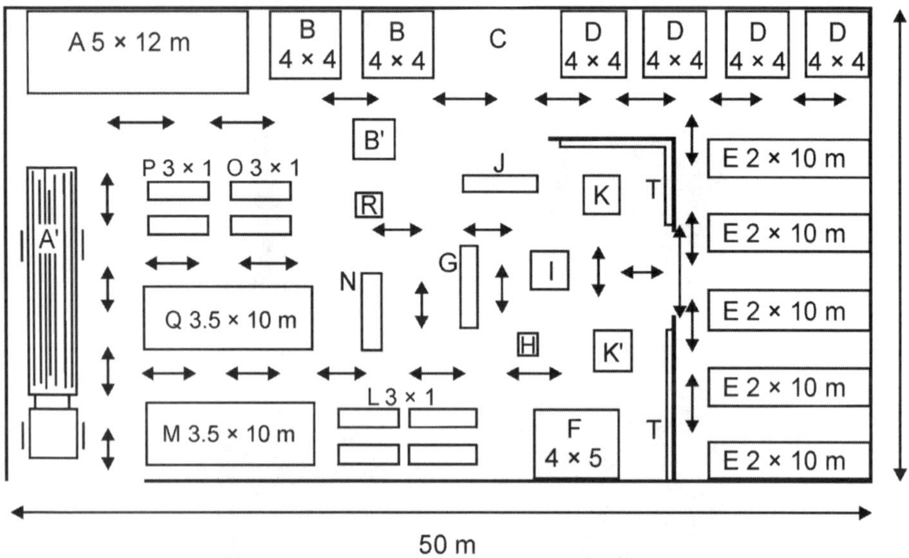

A' : Loading/unloading area
A : Storage for fresh bamboo
B' : Vertical saw (vertical blade)
B : Storage for bamboo culms
C : Area for washing bamboo
D : Tanks for immersion treatment
E : Drying shelves
F : Storage for dry bamboo
G : Plane
H : Thinning machine
I : Vertical saw (circular blade)
J : Mechanical lathe
K : Vertical saw
K' : Horizontal saw (circular blade)
L : Benches for glue application
M : Shelves for reinforcing elements
N : Benches for boring wooden cylinders
O : Benches for fitting accessories
P : Benches for fitting accessories
Q : Shelves for storage of completed elements
R : Multiple cutter
S : Sliding door
T : Storage for tools and accessories

Figure 8. A proposed factory layout for bamboo structural element production

culm as well as the wood cylinder – is 0.126 m^2, which accounts for 25–44 g of the adhesive per element. At industrial level, this would reduce considerably because of the reduction in assembly time (from glue application to curing).

The costs of inputs for the fabrication of a structural bamboo element with the above specified dimensions, strengthened at the ends with a wooden cylinder, may be worked out as follows:

$$\text{Wooden Cylinders} - 2\,\frac{cylinders}{element} \times 0.5\,\frac{U\$}{cylinder} = 1.00\,\frac{U\$}{element}$$

$$\text{Cascophen RS 216 M} - 30\,\frac{g}{elem} \times 17.5\,\frac{U\$}{1000g} = 0.53\,\frac{U\$}{element}$$

$$\text{Labor to manufacture the wooden cylinders and gluing} - 0.50\,\frac{U\$}{element}$$

$$\text{Total} - 2.03\,\frac{U\$}{element}$$

The cost of bamboo may be estimated in relation to the annual productivity. A hectare of *Dendrocalamus giganteus* can produce annually about 4 200 new culms. Assuming that each culm would be sold for US$ 2 in the plantation, without any additional cost for the up-keep, there could be a turnover of US$ 8 400 annually.

It can be assumed that the cost of maintaining of bamboo plantation and transportation of harvested bamboo to the factory, which is charged to the customer, would add US$ 0.50 per bamboo culm. As the length for *Dendrocalamus giganteus* is about 15 m, several sections of different lengths could be cut from it, depending the utility, and the cost per element will vary accordingly.

In the case of spatial grid elements of 2 m length, for instance, up to 6 elements could be obtained from one culm (leaving out the narrower top end). The cost of bamboo per element would thus be only US$ 0.41. Adding the production cost of US$ 2.03, the cost of an element with strengthened ends (with BRBS) would be approximately US$ 2.45, not taking in account the indirect costs such as the costs of equipment and electric energy. If the factory has its own bamboo plantation located nearby, the transportation and material costs would reduce to the tune of about 17% of the final cost of the element.

CONCLUSION

It might be said that bamboo is yet to get a defined space in Brazil. The way to achieve this, in the opinion of the authors, is through constructing structural models wherein not only the technological aspects but also the aesthetic and health aspects are given equal importance. Once introduced into the market, bamboo structures will be in a position to diversify without harming the status of bamboo.

The structure proposed here permits the fabricator to make, within the general body of construction, any alterations required. Construction details of the super structure have been studied and lateral covering will be studied in the near future.

The factory model proposed in this paper may be taken as the basis for laying out facilities smaller or larger in size, in accordance with the type of structure to be manufactured and the characteristics of the industrial process adopted.

REFERENCES

Chaves, A.C.; Gutiérrez, J.A. 1990. The Costa Rican Bamboo National Project. *In* Ramanuja Rao, I.V.; Gnanaharan, R; Sastry, C.B., ed., Bamboos: current research. Proceedings of the International Bamboo Workshop, Cochin, India, 14–18 November 1988. Kerala Forest Research Institute, Kerala, India; International Development Research Centre, Ottawa, Canada. pp. 344–349.

Janssen, J.J.A. 1981. Bamboo in building structures. Ph.D. Thesis, Eindhoven University of Technology, Eindhoven, the Netherlands. 235 pp.

Janssen, J.J.A. 1991. Mechanical properties of bamboo. Kluwer Academic Publishers, The Netherlands. 134 pp.

Hidalgo, O. 1981. Manual de construccion com bambu. Estudios Tecnicos Colombianos Ltda., Universidade Nacional de Colombia, Bogotá, Colombia.

Hidalgo, O. 1985. Constructing with bamboo in South America. *In* Gass, S.; Drusedau, H.; Hennike, J., ed., IL31 Bamboo. Institute für Leichte Fachentragwerke, Stuttgart, Germany. pp. 288–293.

Maturana, H. 1997. Ontologia da realidade. UFMG, Belo Horizonte, Brazil.

Moreira, L.E. 1991. Desenvolvimento de treliças espaciais de bambu. Dissertaçao de Mestrado, PUC/Rio de Janeiro, Brazil. 172 pp.

Moreira, L.E.; Ghavami, K. 1997. Wood reinforced joint for bamboo space structures and for general use in bamboo structures. *In* Jena, P.K. ed., Proceedings of the 2nd International Conference on Non-Conventional Building Materials (NOCMAT-97), Bhubaneswar, India, 17–19 June 1997. Natural Resources Development Foundation, Bhubaneswar, India.

Pereira, M.A. dos R 1997. O uso do bambu na irrigação: montagem de um sistema de irrigação por aspersão de pequeno porte, utilizando tubulação de bambu. *In* Materiais não convencionais para construções rurais, UFPB, Brazil. pp. 293–323.

Villalobos, O. 1993. Fundamentals of the design of bamboo structures. Ph.D. Thesis, Technical University of Eindhoven, Eindhover, The Netherlands. 261 pp.

Wegst, U.G.K.; Shercliff, H.R.; Ashby, M.F. 1993. The structure and properties of bamboo as an engineering material. Department of Engineering, University of Cambridge, United Kingdom.

FURTHER READING

Baudrillard, J. 1973. O sistemas dos objetos. Perspectiva S.A., São Paulo, Brazil.

Kollmann, F.F.P.; Côte Jr., W.A. 1968. Principles of wood sciences and technology – solid wood. Springer Verlag, Berlin, Germany.

Moreira, L.E.; Ghavami, K. 1997. Dimensionamento de estruturas de bambu. *In* Materiais não convencionais para construções rurais. UFPB, Brazil. pp. 237–291.

Moreira, L.E.; Ghavami, K.; Costa, P.C.S. 1995. Analise pelo MEF de ligações com pinaos circulares em tubos de bambu solicitados axialmente. *In* Proceedings of the V EBRABEM, EE-UFMG, Belo Horizonte, Brazil, Vol. II. pp. 160–175.

Bamboo for Sustainable Development, pp. 559–573
Eds. A. Kumar, I.V. Ramanuja Rao and Ch. Sastry

Development of bamboo tension structures

JOSÉ LUIZ MENDES RIPPER and MARCELO DA FONSECA E SILVA
Pontifícia Universidade Católica do Rio de Janeiro, Rio de Janeiro, Brazil

LUIS EUSTÁQUIO MOREIRA
Universidade Federal de Minas Gerais, Belo Horizonte, Brazil

Abstract—During the last two decades, a great deal of research has been carried out to study the use of locally available materials, such as bamboo or natural fibers, in producing different engineering elements or products. The development of these products is an interdisciplinary study and requires knowledge and know-how from different areas of science and technology. The main products developed by the group of industrial designers are for physically disabled people. The studies about physically disabled people are carried out at the Center for Independent Living (CVI), which is directly connected to the Laboratory for Training and Development of Prototypes (LOTDP). Several objects recently developed at the LOTDP, are described in this paper.

INTRODUCTION

One of the main problems in developing countries is the lack of systematic research into locally available low-cost materials; bamboo is one such material. It is an ideal material that can be used together with a variety of other materials such as plastics, rubber, nylon, metals, steel and wood to manufacture an object economically. Every new object produced allows a choice of materials, which is usually made by the stock of materials available in the laboratory where its prototype is created. Despite the use of a wide variety of materials, current research has managed to have a focus on the study and application of bamboo.

Bamboo, besides being a low-cost material, offers other advantages such as its resistance to impact load, aesthetic appeal, lightness and fineness. When held, it reminds one of toys. Thus, a walker made of bamboo is not just a walking device for the physically disabled but also a toy that a child would want use to stand up and walk, perhaps for the first time in its life. Bamboo domes, besides offering an economical advantage, have a psychological effect on the users of the space covered by the structure.

Bamboo stands for nature, flexibility, lightness and beauty. These values, ignored by the modern humans, together form an extremely important way to a good and healthy life.

LOTDP

The Laboratory for Training and Development of Prototypes (LOTDP) occupies an area of 400 m^2 and was designed for the tropical climate in Rio de Janeiro. Its inside is totally open to the outside, in that there are no dividing walls, doors or enclosed rooms that would separate the students from the teachers (Fig. 1).

This design of LOTDP made it easy to have contacts with the neighboring institution, the Center for Independent Living (CVI), a not-for-profit organization that offers social services to physically disabled people. Today, LOTDP actively cooperates with CVI to find solutions that meet the very special needs of physically disabled people. Designers, engineers and architects collaborate in the development of bamboo structures, while students, teachers and prospective clients work together on solutions that make living with a certain physical disability a lot easier. Various orthoses – devices that help correct an orthopedic problem – have been developed in recent years such as: devices for people with paralyzed hands (Figs. 2 and 3); and sliders (Fig. 4) and walkers (which are still in research stage) (Figs. 5, 6 and 7) for people with difficulty in walking.

Besides these objects a light bamboo structure for civil construction, such as the geodesic dome (Ripper et al. 1995) shown in Fig. 8, and a spatial structure made by lashing joints (Fig. 9) have been developed.

Figure 1. Inside LOTDP

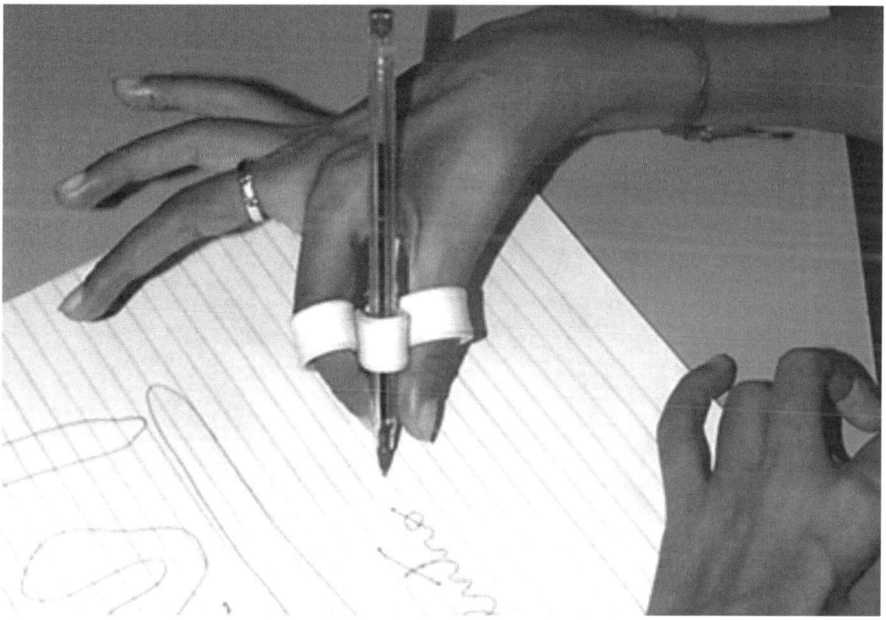

Figure 2. An orthosis for writing

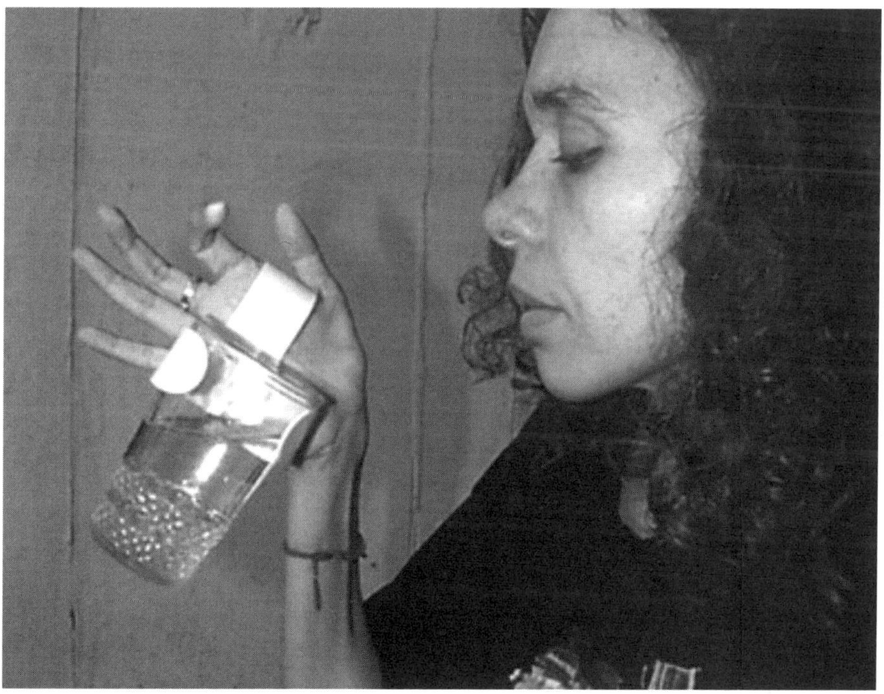

Figure 3. An orthosis for drinking

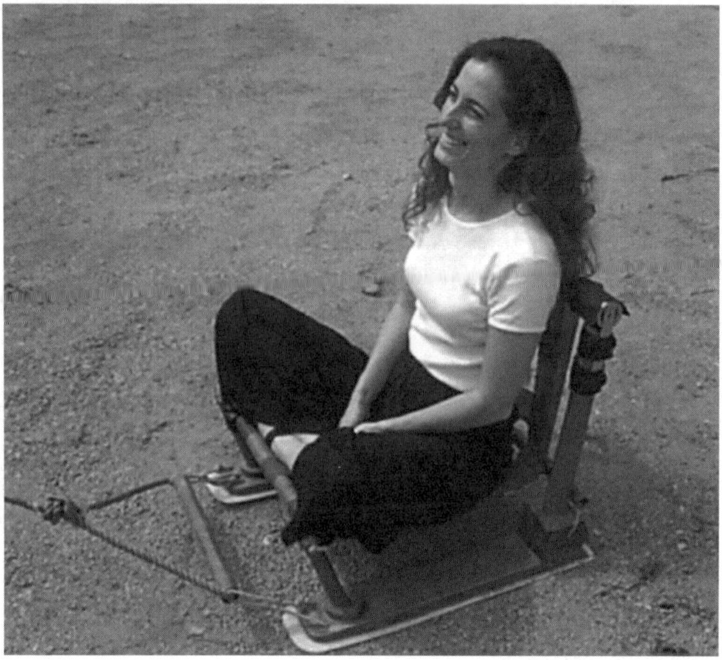

Figure 4. A slider

Figure 5. A walker

Figure 6. Another type of walker

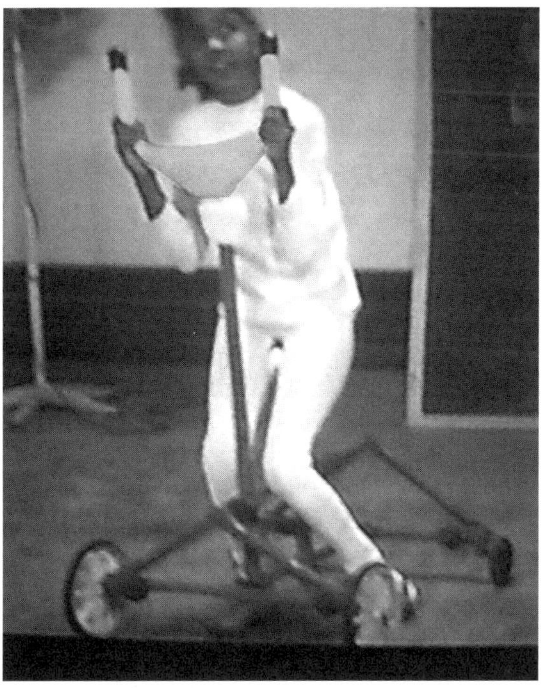

Figure 7. Another design of walker

Figure 8. A geodesic structure

Figure 9. A spatial structure

Working methodology

Under the methodology followed at LOTDP, the students and their teachers receive a complete feedback on object development, which is a concrete process of optimization of the structural project. Thus, LOTDP has the following basic working scheme:

The problem
↓
Aesthetic and geometric analysis by specialists in terms of biomechanics and engineering
↓
Development of the object (orthosis) as small-scale models
↓
Revision of the adopted solutions, aiming at low cost, rational manufacture, etc.
↓
Development of full-scale model
↓
Direct test of the object (orthosis) by physically disabled people or direct test in the environment for light structures in civil construction → Research result.

TENSION STRUCTURES

During the process of object development, users, recipients, students and teachers, are constantly exchanging information.

Among the different solutions found at LOTDP for the joints between bamboo elements – such as making slits in the ends of bamboos, coning these ends and fixing solid wooden cones (Moreira et al. 1995), etc. – this paper concentrates on a system specially created for tension structures (Borrego 1968). These structures have proved to be an excellent case for the use of bamboo because:

- They are flexible, which is compatible with the mechanical performance of bamboo;
- The ease of assembling the tension structures; and
- Bamboo is always subject to compressive forces, and this has a positive consequence for the joints in terms of efficiency, low cost and ease of manufacture. This means affordability of the structure to potential users, since it can be developed through relatively simple tools.

In Figs. 10–17, different types of tension structures are presented, showing that they have played a historical part in the process of industrial development. Fig. 10 shows solutions by Santos Dumont, one of the aviation pioneers, who created the structures of the first aircrafts from bamboo and cables. Fig. 11 shows a bamboo exposition panel based on meteorological kite made as a tension structure.

Fig. 12 shows a reduced model of a bamboo geodesic dome with tension structures created at LOTDP, and the detail of the joining of the elements can be seen in Fig. 13.

Figs. 14 and 15 show a bamboo bicycle frame developed at LOTDP and the detail of the joint. Fig. 16 shows a support for a platform developed at LOTPD.

A bathing chair for physically disabled people developed at LOTDP is shown in Fig. 17. The novelty of this work in relation to tension structures is in the types of objects created and the system of tension developed.

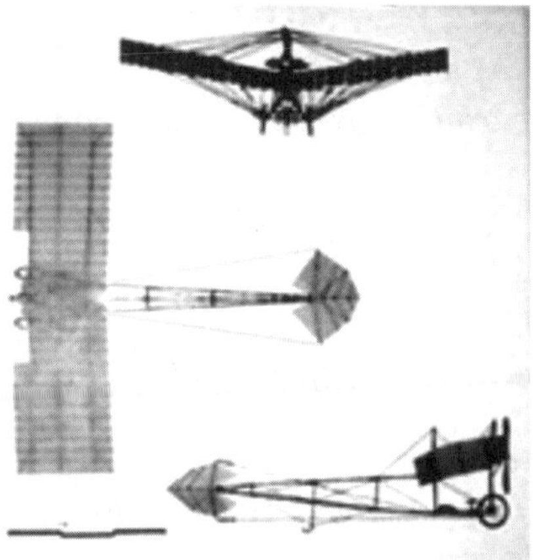

Figure 10. Santos Dumont's demoiselle

Figure 11. Exposition panel

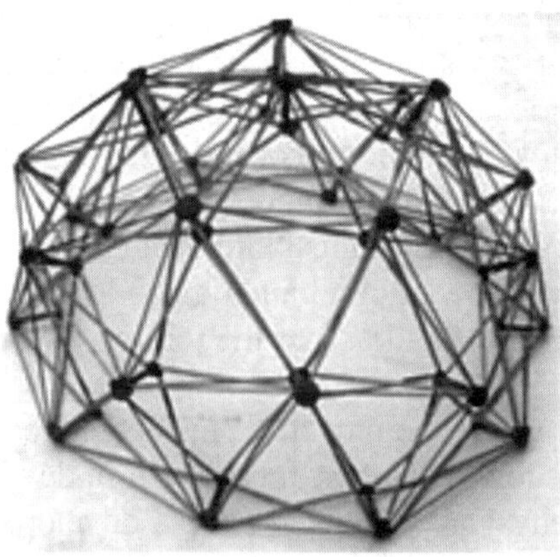

Figure 12. Model of a bamboo geodesic dome

Figure 13. Joining of the elements

Figure 14. Bamboo frame for a bicycle

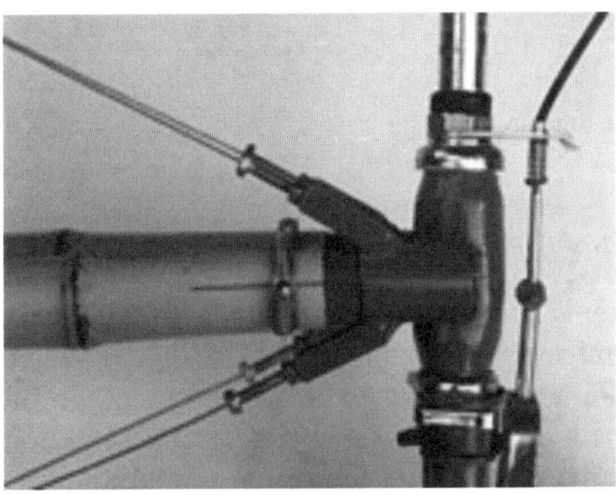

Figure 15. Detail of the joint

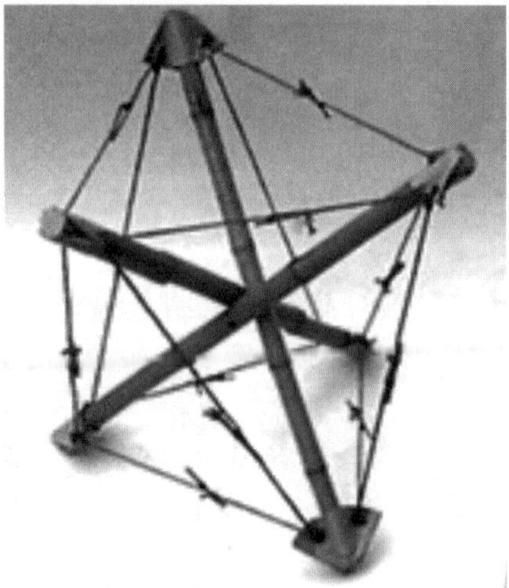

Figure 16. Support for a platform

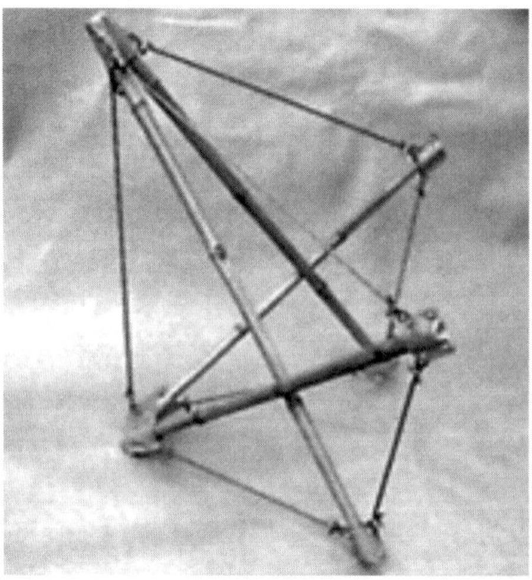

Figure 17. Bathing chair structure

Adaptors responsible for the working of the system

Of the different systems studied, the one finally approved was an adaptor that can be made of rigid, semi-rigid or flexible materials (Fig. 18 & 19).

The concept of the adaptor consists of encasing the end of the bamboo culm, blocking its movement and, at the same time, conveying compressive forces to the bamboo. This is the great advantage of tension structures. In these structures, the rigid element works in compression on a continuous basis. In the case of bam-

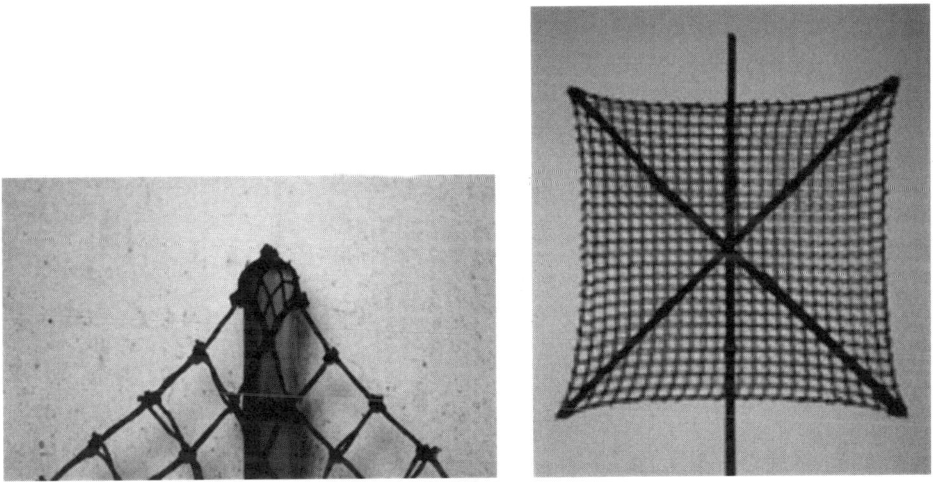

Figure 18. Adaptor made of a rope web, and the structure used in it

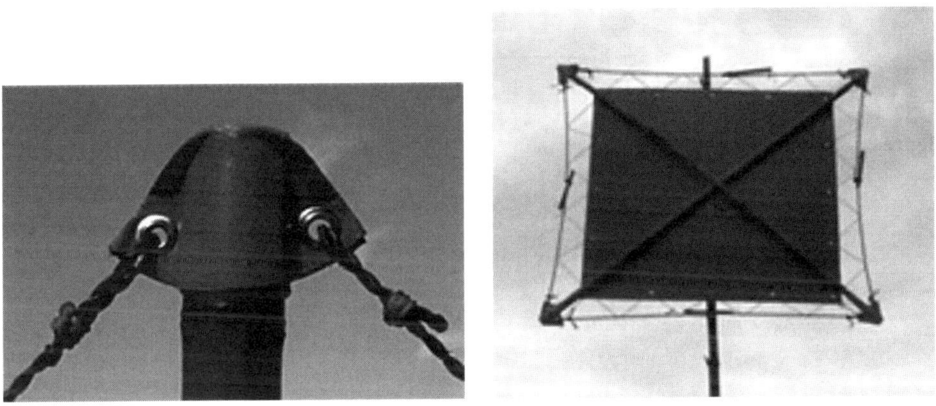

Figure 19. Adaptor made of semi-rigid materials and the structure used in it

boo, this is a definite advantage because, as it is known, the major difficulty in keeping bamboo in tension lies in the development of joints, which do not concentrate tension in their walls (Moreira and Ghavami 1997). Fig. 20 illustrates the distribution of forces in the system shown in Fig. 19.

Force τ, exerted on the cable, compresses the end of the bamboo with a force f_c in such a way that

$$f_c = \tau \cos\theta 1 + \tau \cos\theta 2$$

$$f_c = \tau (\cos\theta 1 + \cos\theta 2)$$

The problem of stress concentration, which is transferred to the holes made in the adaptor, needs to be studied in more detail. However, in this case, more flexibility is gained because the adaptor can be made of a material that has a suitable resistance to the stress applied.

Fig. 21 shows the stages of semi-rigid adaptor development, while Fig. 22 shows the conjugation of adaptors to be used in another point of the bamboo culm.

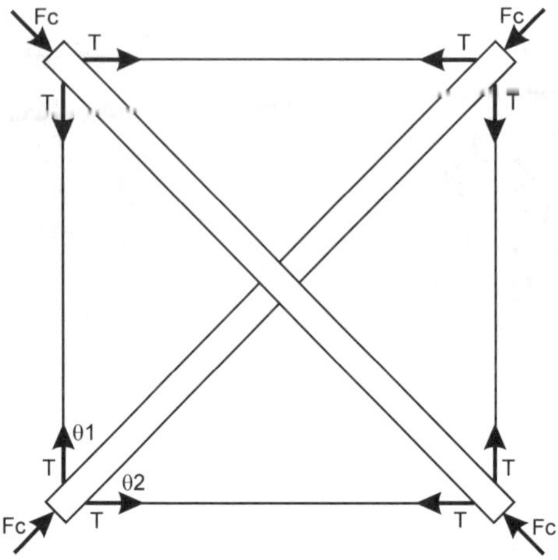

Figure 20. Distribution of forces in the system

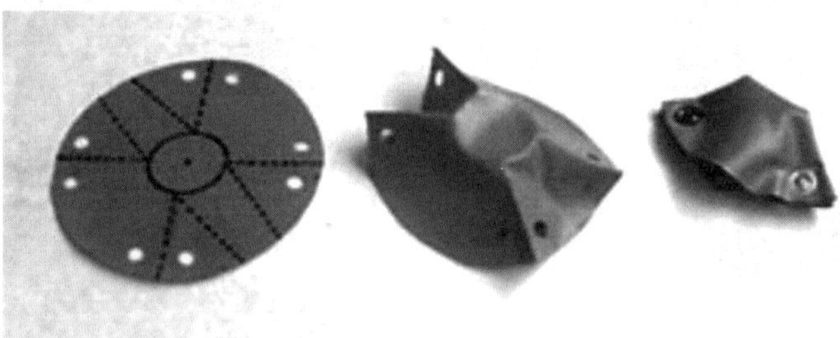

Figure 21. Stages of semi-rigid adaptor development

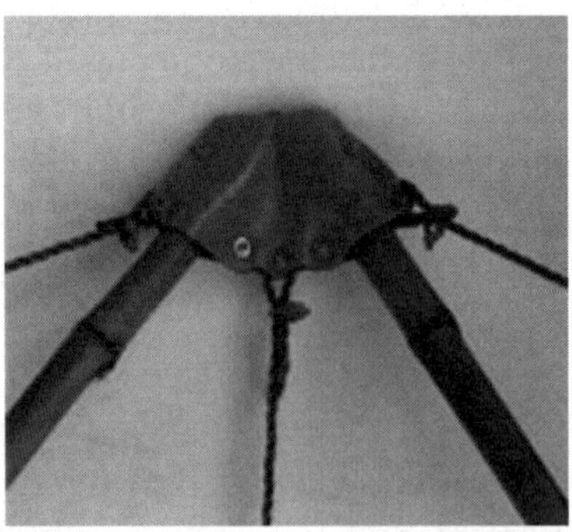

Figure 22. Conjugation of adaptors

The adaptor presented is a system for the transmission of compressive forces, which requires another complementary element – the “blocking end” shown in Fig. 23. This blocking structure will transmit the compressive force to the bamboo, compressing its ends with a tension τ; at the same time, the blocking structure is also subject to critical shearing stress. Fig. 24 illustrates the forces acting on the adaptor.

Another system of transmission of compressive forces is shown in Fig. 25, this time a rigid one, also using the basic idea of encasing the end of the bamboo culm. Again, the ends of the bamboo are subject to the compressive stress τ. To strengthen this end, depending on the acting force, a cylindrical wooden reinforcement may be used.

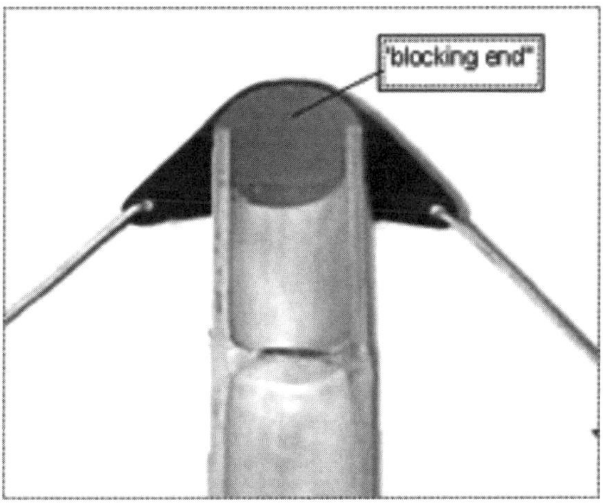

Figure 23. Semi-rigid adaptor (longitudinal section)

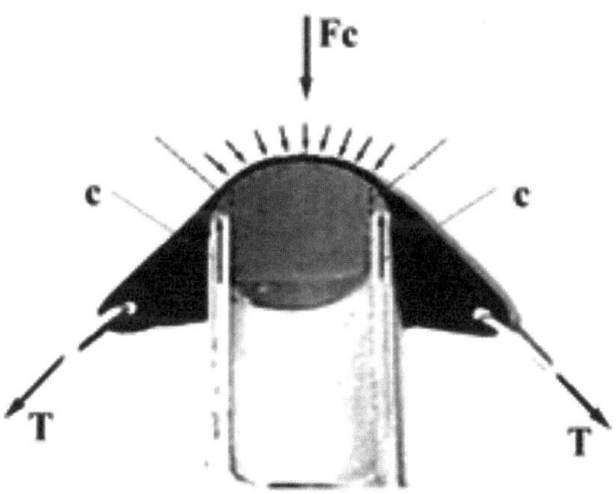

Figure 24. Action of forces on the adaptor

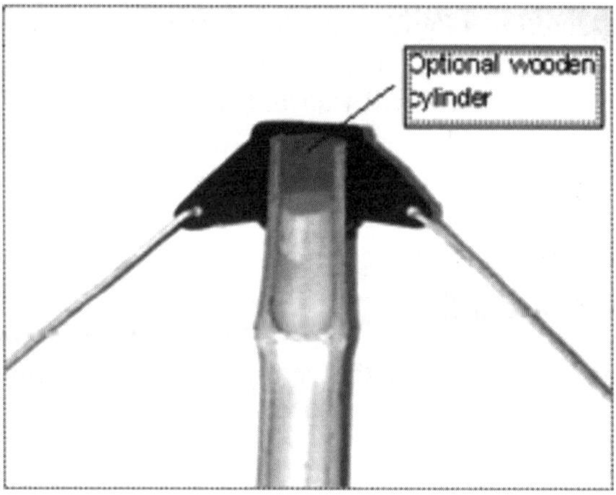

Figure 25. Rigid adaptor (longitudinal section)

CONCLUSION

LOTDP's objective is to conceive structural systems, such as joints and their fundamental principles, based on structural mechanics and biomotor aesthetics. At this level of research and in the area of industrial design, no attempt has been made to experimentally define mechanical strengths. This has been a specific stage of the work conducted at the laboratories for mechanical tests. Therefore, there was no intention to show the mathematical evaluation of the structures nor to build systems of force transmission between the elements.

LOTDP is mostly concerned with the conception of construction systems and their physical creation. Simple tests are done by the users themselves in the case of vehicles, and by the action of wind and rain in the case of structures.

In doing this, LOTDP hopes to contribute towards the development of bamboo tension structures, using different solutions. The novelty offered by the system of tension structures is the idea of encasing the end of the bamboo.

The simplicity of the solutions described above allow for the accessibility of the general population to engineering solutions, making it possible for the needy to afford modern and useful structures, both from economical and health standpoints.

ACKNOWLEDGEMENTS

The authors are indebted to Prof. Khosrow Ghavami, Civil Engineering Department PUC-Rio, and Prof. Shigeyasu Amada, Department of Mechanical Engineering, Gunma University, Japan, for their competent collaboration. They also thank Flavio Leão Deslandes, student, Industrial Design Program PUC-Rio, for his dedication and assistance.

REFERENCES

Borrego, J. 1968. Space grid structures. The MIT Press, Massachusetts Institute of Tecnology, Cambridge, USA.

Moreira, L.E.; Ghavami, K. 1997. Wood reinforced joint for bamboo space structures and for general use in bamboo structures. *In* Jena, P.K. ed., Proceedings of the 2nd International Conference on Non-Conventional Building Materials (NOCMAT-97), Bhubaneswar, India, 17–19 June 1997. Natural Resources Development Foundation, Bhubaneswar, India. pp. 1–17.

Moreira, L.E.; Ripper, J.L.; Ghavami, K 1995. Estudo analítico de uma nova ligação para tubos de bambu solicitados axialmente. *In* Anais do Encontro Brasileiro em Madeiras e Estruturas de Madeira V, Belo Horizonte, Minas Gerais, Brazil. pp. 276–289.

Ripper, J.L.; Ubésio, A.; Moreira, L.E. 1995. Cúpula geodésica de bambú (CGB). *In* Anais do Encontro Brasileiro em Madeiras e Estruturas de Madeira V, Belo Horizonte, Minas Gerais, Brazil.

Bamboo for Sustainable Development, pp. 575–584
Eds. A. Kumar, I.V. Ramanuja Rao and Ch. Sastry

Use of bamboo as structural material in flooring slabs

LUCY AMPARO BASTIDAS PASSOS and EDGARD FLORES BORJA
Independent Architects, Popayan, Cauca, Colombia

Abstract—Seeking alternatives to lower the construction cost, using materials associated with the culture of place and, at the same time, making good use of contemporary materials and advanced technologies, has been a preoccupation in the construction sector over the last two decades. For this combination can produce good results in construction, with excellent structural, seismic and aesthetic characteristics. So has it been with bamboo, its application in the construction of houses being based on historical and technical knowledge of the material. This investigation emphasizes the use of the *Guadua angustifolia* as a structural material in flooring slabs.

INTRODUCTION

The authors' explorations and applications with bamboo began in Colombia and Ecuador for two reasons. First, these two countries have a millennial tradition in the use of bamboo, which forms part of not only the architectonic heritage but also the cultural heritage. Second, the Colombian Guadua (*Guadua angustifolia* Kunth) is a bamboo that has one of the most desirable physical and mechanical properties, which make it an excellent material for construction. It is a woody bamboo that possesses some extraordinary properties in terms of tensile, bending and compression strengths, and they are comparable to the strength properties of steel. Apart from these, it is also a lightweight and flexible material. All these properties provide Guadua bamboo with a high potential for a number of uses in the construction sector and elsewhere.

The main objective of this paper is to give an account of the history of use of bamboo in Latin America, and to discuss the use of Guadua bamboo in the construction of housing, particularly in flooring slabs, for the middle strata of the society. It also reports on the results of laboratory tests on Guadua bamboo, during various stages of its application in the construction of flooring slabs.

GUADUA THROUGH HISTORY

In the history of architecture in Latin America, there are some threads of information, though broken at places, on use of bamboo and other indigenous materials and techniques. A major emphasis is given in the universities on the study of primitive constructions that employed “durable materials”, such as the use of stone constructions by the pre-Inca Machupicchu or the Inca buildings of Sacsaguaman, Pisag and Ollaitaitambo, among others.

In Cuzco, besides constructions in stone, some earthen, wooden, bamboo, straw or mixed construction elements were also found, although they fast deteriorated and disappeared owing to the action of weather or other natural phenomena (Cartagena and Cartagena 1977). It is believed that these vestiges from the past are probably ruins of temporary houses that the Incas built for workers, soldiers and artisans.

The constructions for the upper sections of the society were made with stone and many of them still exist. For the lower classes, perishable materials were used in constructions and nothing much is left of them today. This is not unique of the Americas; the same trend can be observed in other places such as in India, where a great number of houses for the poorer classes are still being made with bamboo. In such houses, there is hardly any protection for the bamboo material and hence, these construction have a life of 12–15 years only.

In Flores Island, Indonesia, stone tools were found in 1998, proving that the *Homo erectus* crossed vast sèas some 800 000 years ago (Anonymous 1998). The archeologist Mike Morewood opines that these long crossings were probably done from Africa, using rafts of bamboo, much like the ones being used even today in this part of the world.

Several researchers have concluded that the use of bamboo was prevalent in the Americas – principally in Colombia, Ecuador and Panama – since historic times, and had continued through the pre-Inca, Inca and Colonization periods to the contemporary epoch.

Pre-Inca period

In the Americas, Guadua stands are found from Mexico to Argentina, and very abundantly in Colombia, Panama and Ecuador. If we start with the assumption that the natives extended the principle of using “what the earth gives” in their constructions also, it can be safely deduced that they must have used bamboo in their constructions also.

There are, in fact, definitive indications that bamboo was utilized in construction in the remote past. It is reported that traces of bamboo were found in recent archeological excavations in Ecuador, where it was used in constructions about 9 500 years ago (Hidalgo 1978). More recently, remnants of small bamboo logs used in construction about 3 500 years ago were found in Santa Elena Island, near Guayaquil City of Ecuador.

Inca period

According to the botanical history of the Americas, large bamboo forests were prevalent in a good part of the region in ancient times. The Incas were well versed in nautical knowledge, which they put to good use for the development of trade and expansion of territory. It is known that they used bamboo, a light material that floats in water, for these long crossing of the seas for commerce and conquest. They utilized it in the construction of hanging and anchored bridges, demonstrating interesting civil engineering techniques. Similar bridges existed until few years ago in Paez, Cauca, a unique area in Colombia where this bridge-building technique is still alive. The Inca civilization is responsible also for the development of bamboo-based musical instruments such as the flute, quena, zampoña and marimba.

It could be concluded that bamboo was utilized alone or in combination with earth by almost all the tribes of the Americas; however, the utilization was more widespread among the tribes located in the pacific coast – Panama, Colombia and Ecuador – where large Guadua stands existed (Hidalgo 1978).

Colonization period

When Cristobal Colon stated that he found canes "as thick the thigh of a strong man" in Colombia and Ecuador (Cartagena and Cartagena 1977), it is clear that the conqueror was speaking about the Guadua. The huts in the native settlements were made of cane and earth, in the *bahareque* technique. The bridges were of Guadua and they served like strategic points in the defense against the Spaniard conqueror, as the bridges could be erected and dismantled in a few days. It is also known that some Spaniards used to burn Guadua stands to prevent the natives from hiding in the cover provided by these stands.

The use of Guadua in the Antiguo Caldas region (currently the Departments of Caldas, Risaralda and Quindío) has converted Colombia into a unique country; it is Colombia that developed durable techniques for construction with Guadua (Robledo 1992). In the last century, nearly a hundred villages in the coffee-growing areas had only *bahareque* style houses built with Guadua, taking advantage of bamboo's high seismic resistance in a zone under constant threat from earthquakes.

Contemporary period

In Colombian cities, two parallel applications of Guadua can be noticed. One is the use of bamboo by poor people as a material for marginal housing. Many urban centers in Colombia are surrounded by such marginal houses built with Guadua. The other is the use of bamboo in secondary structures, such as formworks and scaffolds, in the construction of large buildings. The image of Guadua as a fragile material is the reason for the provisional and disposable character given to it.

Another aspect in this period is the advancements in construction processes achieved with Guadua in Colombia, where it is fully recognized as an important material and utilized together with new materials and modern techniques. The country is home to a good number of professionals with sound knowledge on the material and its applications. Bamboo is used in the furniture industry and in a large variety of crafts. In rural areas, it finds us in almost every work. Diverse government programs aimed at environmental protection and rehabilitation are utilizing bamboo, for example, in reforestation, for protecting riverbanks and for controlling erosion.

In concluding this brief discussion on Guadua's history in the Latin America, one can clearly see that to help solve housing deficit and contribute to the necessities of the modern life, it is not necessary to replace traditional materials with new ones. Such materials can well serve the purpose if they are properly understood and improved techniques for their processing and use are developed.

USE OF GUADUA IN HOUSING

If the belief that Guadua is a material fit only for provisional housing for the lower-income sections of the society is to be erased, bamboo needs to be used in durable houses for the middle and upper-income classes. One such effort was realized in a ward for middle class in Popayán City, Colombia. The technique used here was an improved form of *bahareque*, with walls of Guadua poles and *esterilla* (plates of bamboo made by longitudinally cutting open the culm and flattening it) and sand-cement mortar. This system is employed not only in Colombia but also in several other countries in the region. Research for further improvements, such as the construction of various types of floors for houses of two and three floors, is continuing at present.

Different types of floors

Floors built traditionally have some variations:

1. Floor of wood is the more traditional and older type employed in the construction of houses in several regions of the Colombia (Fig. 1). The system consists of placing floorboards over joists fixed on a tie-beam. All the elements are made of wood. The floorboards, placed side to side or joined in tongue-and-groove arrangement, are nailed to the joists. Electrical wires run below the floorboard.
2. Floor with mat decking and Guadua joists (Fig. 2) also have a timber tie-beam, over which Guadua poles are placed across 40 cm apart. A mat of bamboo laths is nailed over this (Hidlago 1978). Concrete is then poured over the mat to a thickness of 4 cm. Over this mat and concrete decking, a flooring material of choice (cement or ceramic tile, parquet, etc.) is placed. In this technique, the Guadua poles are left exposed.
3. Floor with covered Guadua decking (horizontal *bahareque*) is another type of flooring technique (Fig. 3). In this, *esterllia* is placed over the tie-beam and then Guadua poles set 40 cm apart. Bamboo mat is nailed to the bamboo joists and concrete is poured over the mat to 4 cm thickness. The floor is finished with paint or a material, such as tiles or parquet.

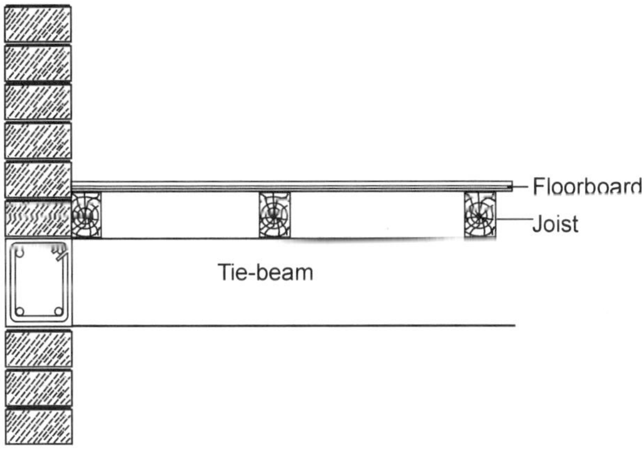

Figure 1. Floor made of wood in a farm house in Chachagüi, Nariño, Colombia

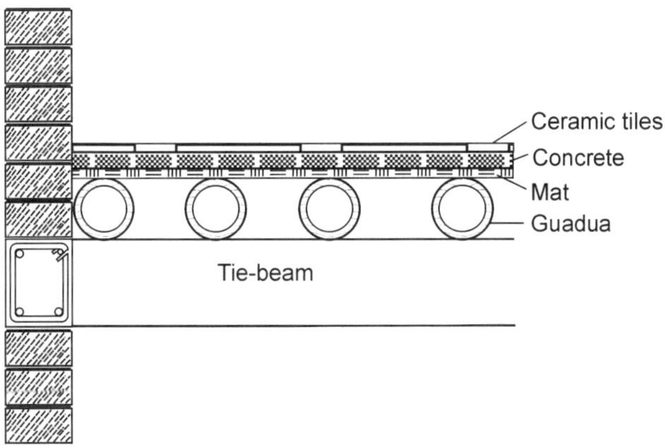

Figure 2. Floor of mat-and-concrete decking used in country houses in Cauca, Colombia

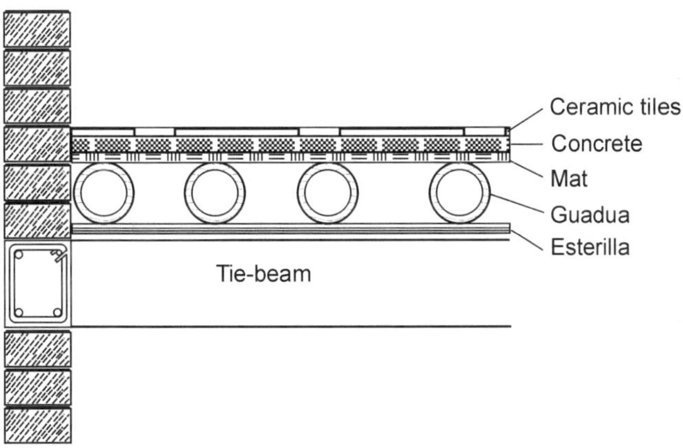

Figure 3. Floor with Guadua joists covered on both sides (horizontal *bahareque*) found in houses in some parts of Popayán City, Colombia

4. Combination of Guadua joists and plank is used for floor and ceiling, though the floor will have concrete between the joists to prevent the bamboo poles getting crushed (Fig. 4). Guadua poles are placed over planks nailed to the tie-beam at 40 cm distance to each other. Concrete is poured among the Guadua joists. Planks are then nailed to the joists to complete the work.

   One major problem in this system is that it is difficult to get Guadua poles with the same diameter. Poles of different diameters will make the planks nailed to them to be on different levels.
5. Reinforced concrete and Guadua are employed in another type of flooring technique. The steps involved in the construction of this type of floor are almost similar to the one mention under item 4, with the difference that some Guadua poles and are replaced with steel sections to reinforce the concrete

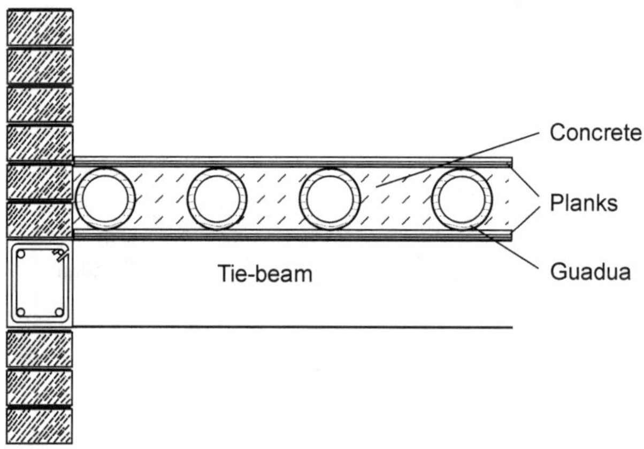

Figure 4. Floor with Guadua and plank decking as used in flooring in houses in some parts of Popayán City, Colombia

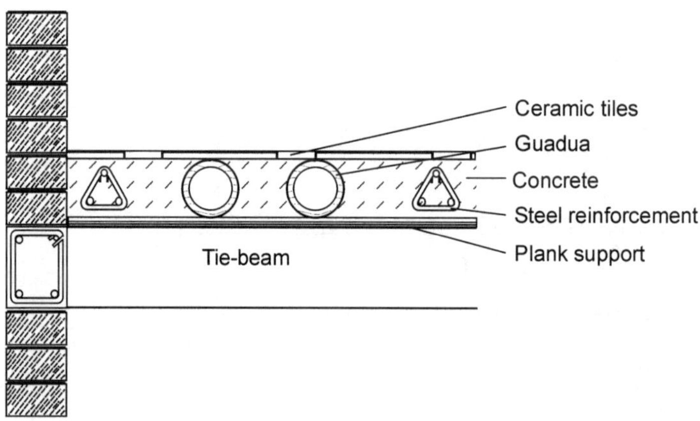

Figure 5. Floor with slabs of reinforced concrete and Guadua, seen in some parts of Popayán City, Colombia

(Fig. 5). As the use of Guadua is more for lightening the floor than for any structural use, poles of smaller diameters and otherwise unusable are employed.

A variation to this is the use of slabs of reinforced concrete. The steps are the same, but concrete slabs are used instead of poured concrete. Here too Guadua plays only the role of lightening element.

Use of Guadua as a structural material in concrete slabs

This part discusses a flooring technique that uses Guadua to reinforce concrete. The concept of this system, however, is different from that of conventional reinforced concrete slab in which the two materials (concrete and steel) work in a mutually dependent way, with both fulfilling structural functions. In this system, the load is exerted on the Guadua, with the role of concrete restricted to making the floor rigid. A wide variety of materials – ceramic tile, cement tile, vinyl tile, parquet, marble, carpet, etc. – could be used for the finish.

It is important to note that this system Guadua bamboo, besides being the material that contributes to the strength, also lightens the slab owing to its hollow form. No other natural or artificial material can simultaneously fulfill these two roles.

The process

The principal steps involved in its construction are the following. A bamboo mat or light planks are fixed over the tie-beam. Short Guadua poles, with diameters smaller than normally employed, are laid out in sections that are horizontal to each other. Electrical and telephone cables, as well as sanitary and water pipes, are installed at this stage.

Concrete is then poured over the Guadua poles to approximately 4 cm thickness, using formwork and levelled. Once it sets, finishing is carried out using materials of choice. If this floor is above another, then the ceiling of the latter is covered with sand-cement plaster and painted.

There are several houses that such use this system at Los Guayacanes and Guayacanes del Rio of Popayán City (Fig. 8). At the latter location, there are also two commercial establishments built using this technique.

Figure 6. Guadua laid out in sections (note that the poles at the upper part of the picture are horizontal to those at the lower part)

Figure 7. Concrete being levelled

Figure 8. Completed houses at Los Guayacanes in Popayán City

Laboratory tests

Tests of the floor were conducted two, three and five years after it was laid to observe changes in behavior, strength, etc.

In the first samples, the pH was high while in the last sample the pH was neutral. The flexibility index (F) values showed that the mechanical strength of the fibers had improved. The cement in the concrete had acted on the lignin in the fibers of Guadua and replaced it, increasing the strength of the fiber. The microbiological resistance of the Guadua was also seen to have improved. No variation was observed in the natural starch content of Guadua.

Advantages of the system

- It replaces the costlier steel for reinforcement (about 45% cost reduction).
- It lightens the concrete slabs.
- The formwork is required only during the pouring of concrete.
- It reduces the volume of concrete required.
- It provides excellent acoustic insulation.
- Its execution is fast.
- In Popayán City, it has found ready acceptance among the users, as well as the Municipal Planning Office of the Mayor.

This type of floor has been utilized for heights of up to 3.5 m, with pillar supports (Fig. 9). For higher constructions, a steel beam is placed at the center of the pillars in such a way that it supports the Guadua poles that run across at points where they could bend.

This floor construction method is ideal for buildings that employ mixed techniques, like the 200 houses in three urban centers in Popayán City. To date, the authors have built over 6 000 m^2 of slab floors using this system for houses of two and three floors.

A variation of this technique, used in a house of games for children, is to leave the Guadua poles exposed in the ceiling (Fig. 10).

CONCLUSION

It is important to know the historical background of traditions and traditional materials to arrive at solutions that are in harmony with culture of place, its environment and its architectonic heritage.

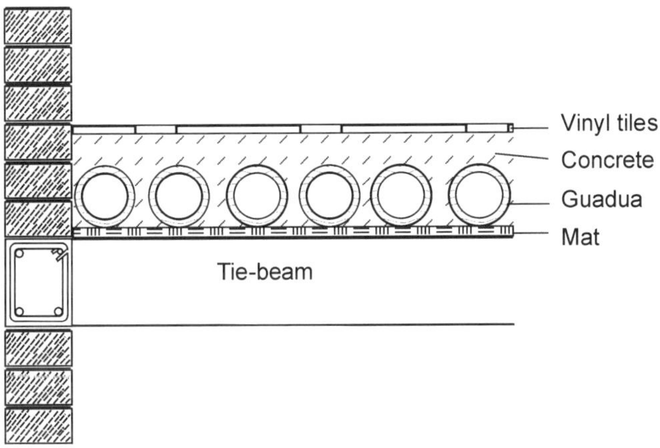

Figure 9. Floor with structural Guadua and concrete

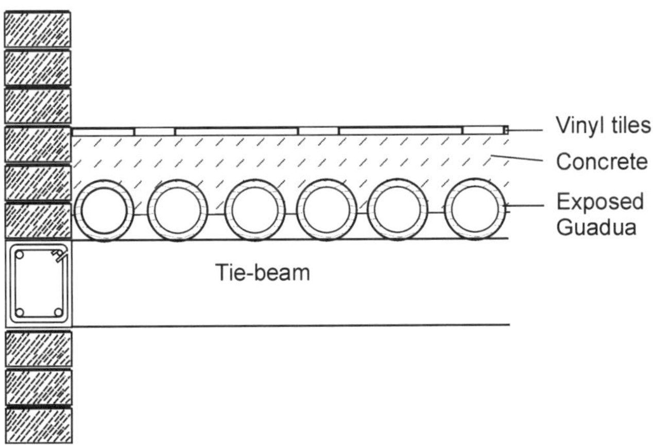

Figure 10. Ceiling of 1st storey (floor of 2nd storey) with exposed Guadua and concrete

The use of bamboo in construction must have adequate technical basis, obtained through the analysis and application of traditional construction systems and verified by laboratory tests. When other materials are used in combination with bamboo, it is critical to examine the short and long-term influence of these materials on bamboo. Bamboo and other materials together should form a construction system that offers security, durability, strength and economy.

ACKNOWLEDGEMENTS

The authors wish to acknowledge the work of Oscar Hidalgo López, the Colombian architect who pioneered contemporary research on the structural use of bamboo in Latin America and whose contributions to this field have converted many Latin Americans into bamboo lovers.

REFERENCES

Anonymous. 1998. Article in El Liberal, May 1998. Popayán City, Colombia.

Cartagena, N.; Cartagena, H. 1977. Sur la piste des Incas (On the trail of the Incas). R. Laffont, Paris, France. 317 pp.

Hidalgo, O. 1978. Nuevas técnicas de construccion con bambu. Estudios Técnicos Colombianos, Colombia. 137 pp.

Robledo, J.E. 1992. Un siglo de bahareque en el Antiguo Caldas. Memorias 1er Congreso Mundial de Bambu/Guadua, Pereira, Colombia, 8–15 August 1992.

Bamboo for Sustainable Development, pp. 585–597
Eds. A. Kumar, I.V. Ramanuja Rao and Ch. Sastry

Mechanical structure of bamboo based on natural, functionally graded materials

SHIGEYASU AMADA
Department of Mechanical Engineering, Gunma University, Japan

Abstract—Bamboo has remarkably flexible and tough characteristics as compared with other woody species, even though it grows very fast. These properties have been obtained to adapt to our natural environment for a long time. Functionally graded materials (FGMs), which have been developed recently, allow their microscopic structure to be varied so that a physical quantity induced into this material is maximized or minimized. Bamboo has two kinds of functionally graded structures, macroscopic and microscopic ones, each of which has different objectives. These structures can lead to very useful mechanical characteristics. This report presents the mechanical structure of bamboo as a functionally graded and composite material.

INTRODUCTION

Functionally graded materials (FGMs) have been developed to maximize (or minimize) a function by tailoring a microscopic material structure and combining different material elements (Niino et al. 1990). The first generation of FGMs has been applied for relaxing thermal stresses (STA 1986). At present, FGMs have become more versatile and their applications are expanding to different fields (Shiota and Miyamoto 1997).

Bamboo is a natural functionally graded and composite material. This structure leads to excellent characteristics such as high flexibility, high strength, lightness and resistance to buckling. Bamboo has reinforced fibers called vascular bundles that, together with the bundle sheaths that surround them, play the same role as fibers in composite materials. Furthermore, the fiber distribution in bamboo constructs a functionally graded structure. The bamboo used in this study is 2-year-old *Phyllostachys pubescens* Mazel ex H. de Leh (Moso bamboo), with a height of 16–20 m and a maximum diameter of 16–13 cm.

This paper presents an overview on bamboo as a natural composite (Oda 1980, 1985; Chuma et al. 1980; Jain et al. 1992) and functionally graded material (Nokata and Seo 1990; Amada et al. 1996, 1997). The functionally graded structures of bamboo display a hierarchical system that can lead to strength and flexibility characteristics that are much better than those of trees.

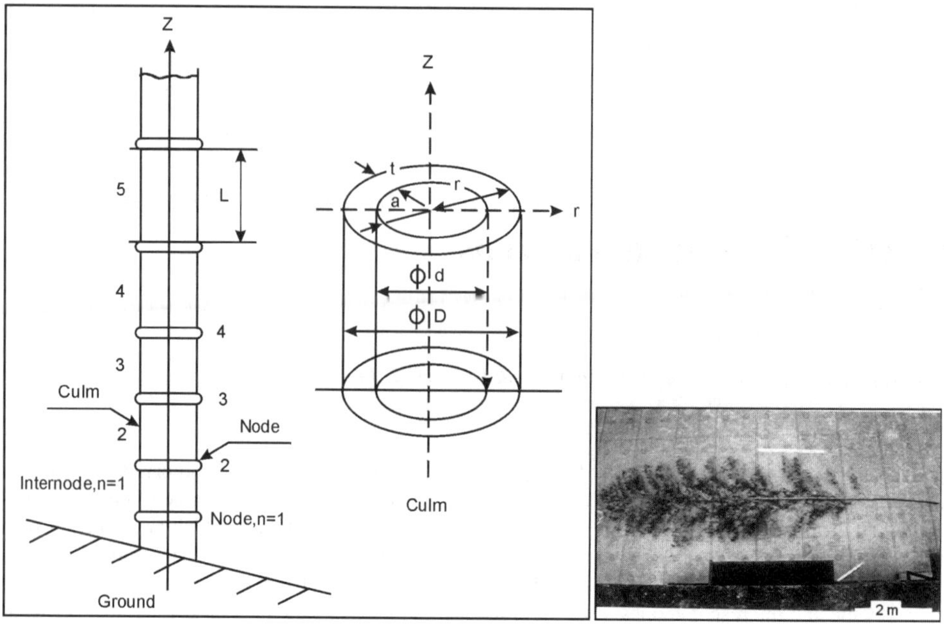

(a) Coordinate system (b) Bamboo

Figure 1. Coordinate system (a) and shape of bamboo (b)

SHAPE: MACROSCOPIC FUNCTIONALLY GRADED STRUCTURE

The coordinate system, as shown in Fig. 1, is introduced to represent the bamboo shape. Fig. 2 shows the changes of diameter D and thickness t with respect to culm internode number n. These quantities correspond to the structure of FGMs called the macroscopic functionally graded structure (Hirai 1996). The question is why bamboo has such a FGM structure.

Bamboo is subjected mainly to wind loads. To predict the load, the projection area of the bamboo was measured using image analysis of the photograph as

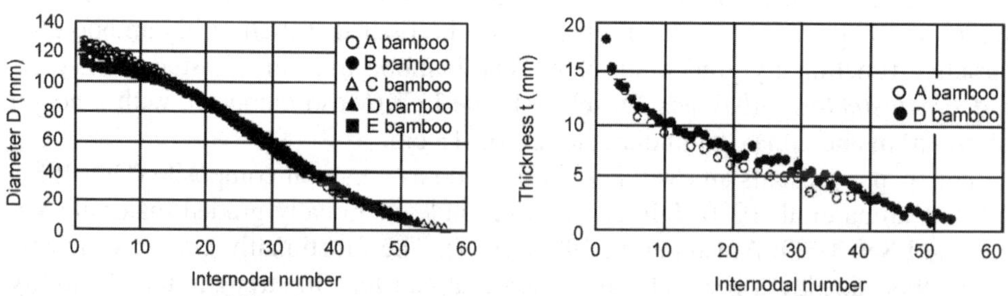

(a) Diameter (b) Thickness

Figure 2. Diameter (a) and thickness (b) of bamboo studied

shown in Fig. 1b. The wind load distribution, shown in Fig. 3, is obtained as the product of wind pressure and the measured projection area. Assuming that bamboo is modelled like a tapered cantilever beam with a variable circular cross-section and is subjected to a horizontal load, the bending stress on the surface can be easily calculated based on the beam theory. The result is shown in Fig. 4, where non-dimensional surfaces stress is normalized by the surface stress at root. The results indicate the surface stress at every height is almost constant, except for the top region. Therefore, the macroscopic functionally graded structure leads to a constant surface stress design; in other words, an optimum design.

FIBER DISTRIBUTION: MICROSCOPIC FUNCTIONALLY GRADED STRUCTURE

The volume fraction of fibers changes with respect to radius of the culm cross-section, which corresponds to a structure called the microscopic functionally graded structure. Using image analysis, the volume fraction *Vf* of fibers has been evaluated and the results plotted are shown in Fig. 5. There is so much difference in the position of fibers; *Vf* is less than 20% in the inner layer and more than 60% in the outer layer.

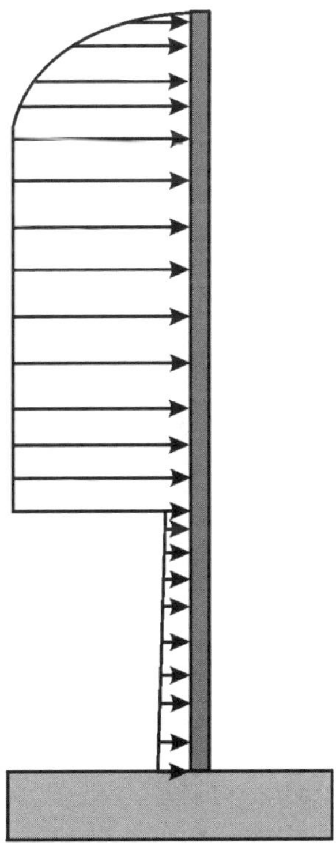

Figure 3. Wind load distribution on bamboo

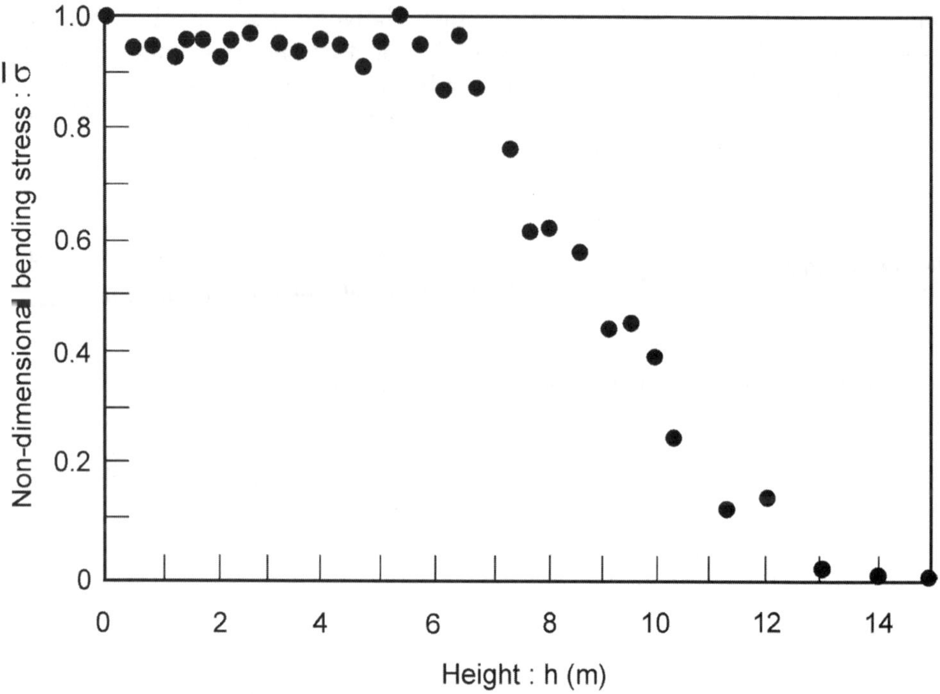

Figure 4. Surface stress distribution on bamboo

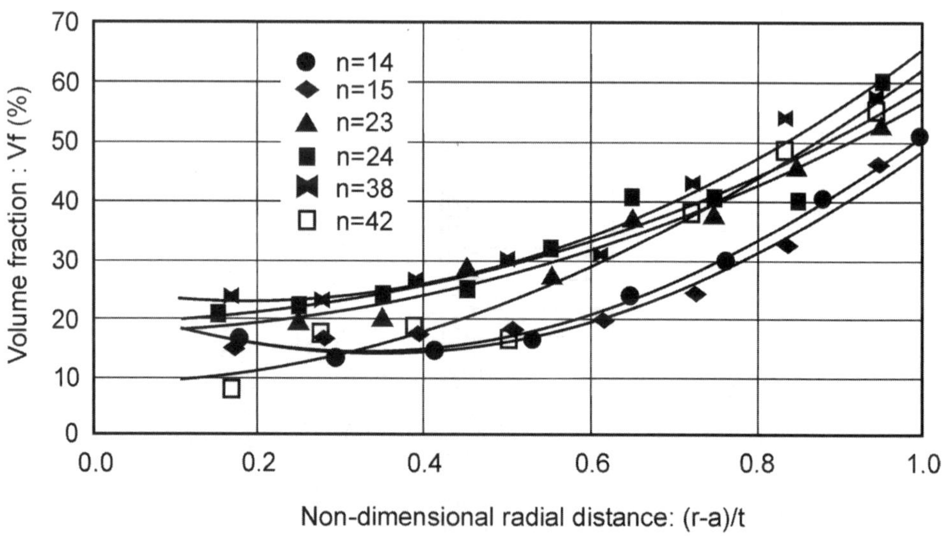

Figure 5. Volume fraction of fiber in a bamboo culm

Tensile tests

To determine the mechanical properties of the bamboo under study, tensile tests were carried out for the specimens sliced out from the different radial positions in the culm cross-section. Fig. 6 shows the tensile strength σ_Z and Young's modulus E with respect to radius. For the culm $n = 14$, σ_Z is about 80 MPa in the inner layer and 450 MPa in the outer layer; that is, the tensile strength of the outer layer is about 5.5 times the strength of the inner layer.

Let the fiber and matrix strengths be denoted by σ_{Zf} and σ_{Zm}, respectively, and their Young's modulus by E_f and E_m, respectively. Assuming that the mixture principle equations

$$\sigma_Z = \sigma_{Zf} \cdot V_f + \sigma_{Zm}\,(1\text{-}V_f) \tag{1}$$

$$E = E_f \cdot V_f + E_m\,(1\text{-}V_f) \tag{2}$$

can be applied to bamboo culm, the tensile strength and Young's modulus were obtained through their plot with V_f as given in Table 1. The fiber strength reaches up to 610 MPa, which is 12.2 times higher than the matrix strength and corresponds to the strength of steel. The value of 46 GPa was obtained for the fiber Young's modulus, which is less than that of steel. Judging from these data, it can be seen that the bamboo fiber has almost the same strength as steel but is more flexible.

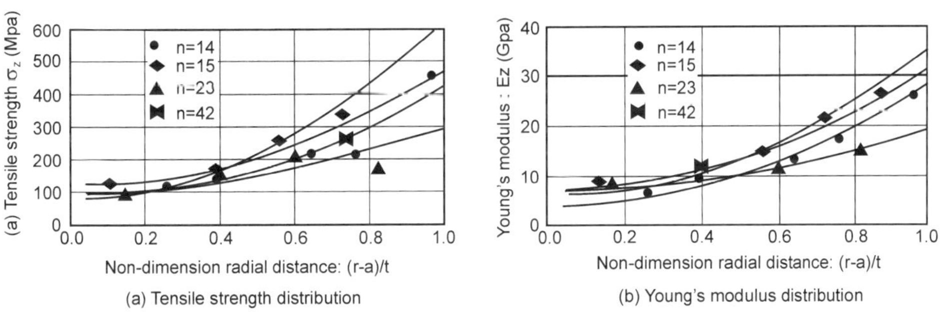

Figure 6. Distribution of tensile strength (a) and Young's modulus (b)

Table 1.
Measured material properties

Method	Material	Tensile strength	Young's modulus
Tensile test + Mixture principle	Fiber	610 (MPa)	46 (GPa)
	Matrix	50	2.0
	Composite	140–230	11.17
Scanning acoustic microscope	Fiber	–	60.2

Measurement by scanning acoustic microscope

Recently, a scanning acoustic microscope (SAM) has been developed to non-destructively measure properties of materials (Ishikawa et al. 1996; Canumalla et al. 1997; Nishida et al. 1997), which has been possible because of the V(z)-effect (Atlar 1978; Welgein and Wilson 1978; Kushibiki and Chubachi 1985). Fig. 7 shows the schematic configuration of a reflection type SAM, which can measure speed of the elastic surface wave. It consists of an acoustic lens with a piezoelectric film, a pulser, receiver, scanning control system and image construction system.

Scanning in the direction of height, an image called the XZ-mode image is obtained. The contrast in the image output by SAM varies in a periodic fashion owing to the interference of two acoustic waves as the distance between the lens and the specimen is decreased. The periodicity in the intensity of the signal received at the transducer is referred to as the V(z) effect (Welgein and Wilson 1978). The interference curve, or the V(z)-curve, generated gives the speed of the elastic surface wave, which is closely related to the elastic properties of the scanned material. A typical XZ-mode image is given in Fig. 8 (for n = 11), where the bright band corresponds to the surface layer of the specimen. Next dark islands and bands stand for the interference pattern.

The interference condition claims the following relation (Welgein 1985):

$$V_R \cong V_{SAW} = \frac{V_W}{\sqrt{1-(1-V_W/2f\Delta z)^2}} \tag{4}$$

where V_R is the Rayleigh wave speed, V_{SAW} is the speed of the surface acoustic wave, V_W is the speed in the coupling fluid (water in this experiment), f is the frequency and Δz is the period of the oscillations in the V(z)-curve.

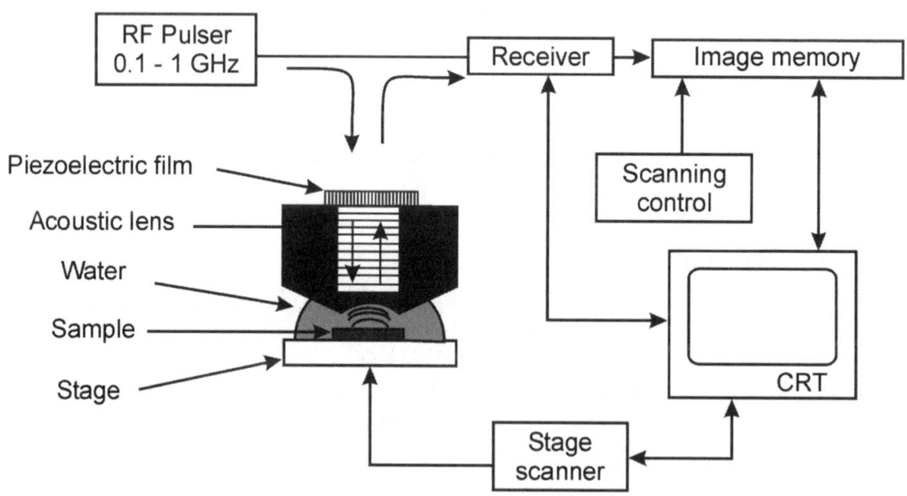

Figure 7. Scanning acoustic microscope

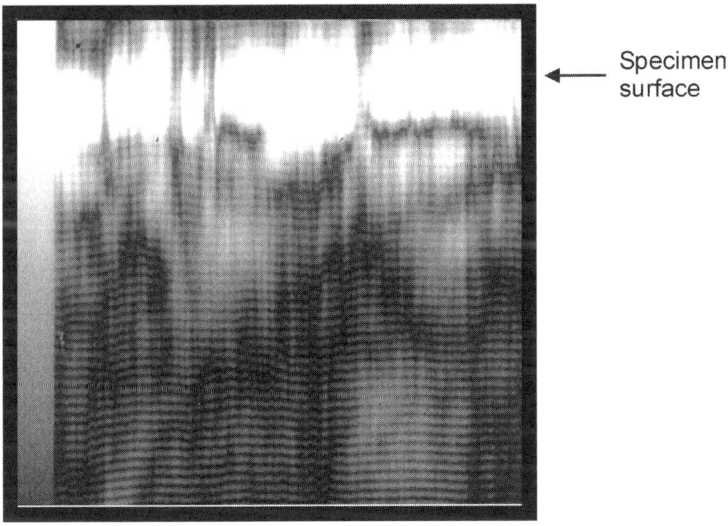

Figure 8. X-Z mode image for n = 11

V_R is related to the speed of the transverse elastic wave V_T (Fung 1965)

$$V_R = 0.9194 \cdot V_T$$

Assuming that the fiber is isotropic, the following holds:

$$V_T = \sqrt{\frac{1}{2(1+V)} \cdot \frac{E}{\rho}} \tag{5}$$

where V is the Poisson's ratio and ρ is density of the fiber. The adopted value (Amada et al. 1996) $\rho = 1.16 \times 10^3$ kg/m^3 and the assumed value $V = 0.3$ are a little controversial. The Poisson's ratio of bamboo has not been obtained yet and it may considerably change with the relative orientation to the fiber direction. Taking the Poisson's ratio $V = 0.25 \sim 0.35$, the Rayleigh speed is influenced only by 3.8%. Therefore, the Poisson's ratio makes little influence on V_R.

Solving equation (5) on E and using the relation equation (4), one gets the Young's modulus E:

$$E = \frac{2(1+V)}{0.8453} \cdot \rho \cdot V_R^{\ 2} \tag{6}$$

If V_R is measured, then the Young's modulus can be evaluated from equation (6).

Using SAM with a frequency 200 MHz, measurements were taken on the exposed tube wall at several different radii and different internodes of the bamboo culm. Scanning the XZ-mode image as shown in Fig. 8 on a vertical line, a V(z)-curve is obtained as shown in Fig. 9, where the upper curve is as measured and the lower curve is obtained by smoothening it. The curve periodicity gives the Rayleigh wave speed of 3 554 m/s from equation (1).

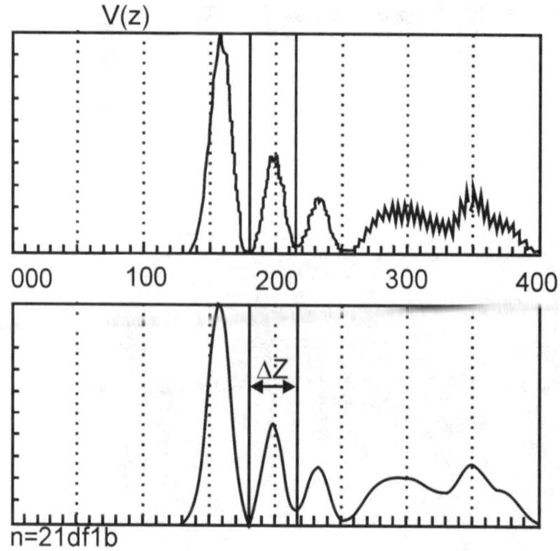

Figure 9. V(z) curve at r = 0.38 for n = 21

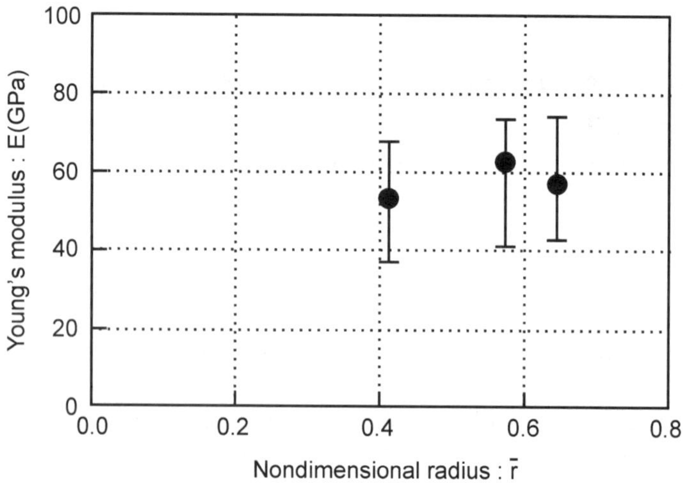

Figure 10. Measured Young's modulus for n = 11

Fig. 10 gives the measured Young's modulus with respect to non-dimensional radius for n = 11. There were only three fibers fully exposed on the specimen surface and the measurements were carried out on these fibers. The average value of the measured E is represented by the symbol and the vertical line segment stands for its scattered range. The average value spans 52.1~61.7 GPa. The measured data for n = 21 and n = 41 are shown in Figs. 11 and 12, respectively. Those figures suggest the average Young's modulus around 60.1 GPa. These data indicate that the fibers have approximately the same Young's modulus for the average value with respect to radius.

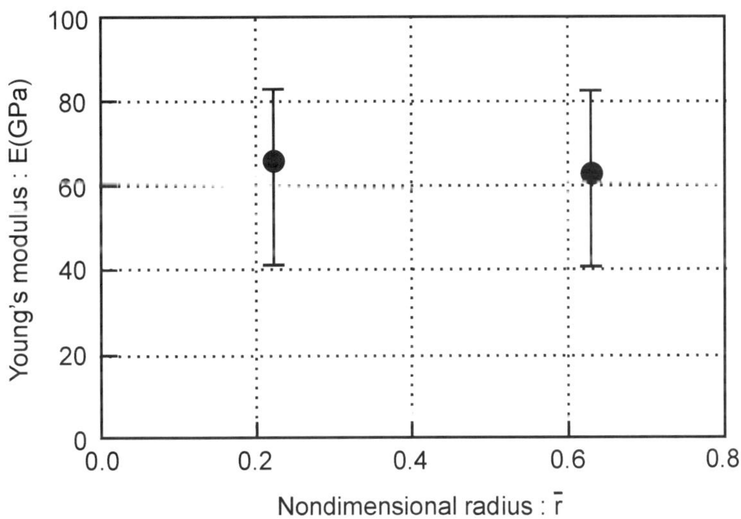

Figure 11. Measured Young's modulus for n = 21

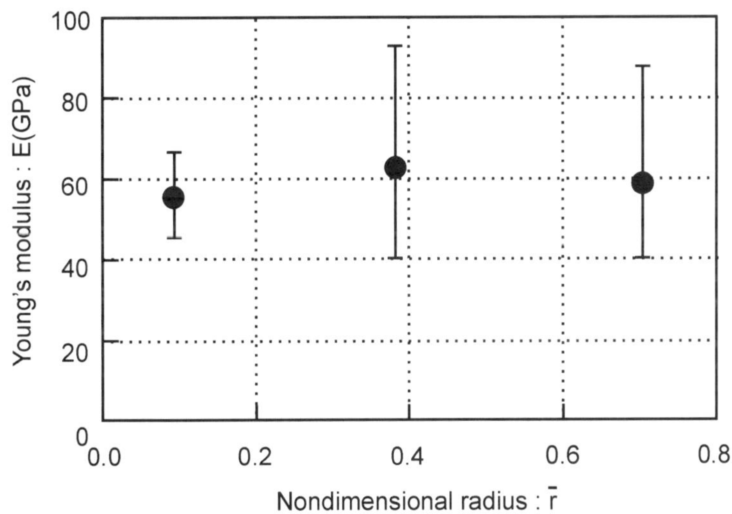

Figure 12. Measured Young's modulus for n = 41

Taking an average value of all measured data, the Young's modulus becomes 60.2 Gpa, as listed in Table 1, which is higher than the previous result 46.0 GPa (Amada et al. 1996, 1997) based on the plot of the mixture principle and image analysis. There were two kinds of errors introduced into the tensile test for the evaluation of *E* from the stress-strain curve and for the evaluation of the volume fraction of fibers through image analysis. Judging from those errors introduced into the latter approach, the result obtained by the SAM technique seems to be the more reliable value.

GRADED STRUCTURE OF FIBERS

Since the Young's modulus distribution varies with radius, the stress distribution in bamboo with respect to radius is different from that of a homogeneous hollow cylinder under bending moment M. Assuming that the bamboo culm can be modelled as a multi-layered cylinder (Fig. 13) and the theory of composite cylindrical beams applies to the bamboo culm, the stress in the 'j' layer is expressed by:

$$\sigma_j = \frac{E_j \cdot R_j \cdot M}{\sum_{n=1}^{N} E_n \cdot I_n} \tag{7}$$

where E_j and R_j are the Young's modulus and radius, respectively, of the 'j' layer and its real moment of inertia is I_j. Introducing the non-dimensional stress $\bar{\sigma}_j$, which is normalized by the stress at the outer surface:

$$\bar{\sigma}_j = \frac{E_j \cdot R_j}{E_n \cdot I_n} \tag{8}$$

where E_n and R_n are the Young's modulus and radius, respectively, at the outer surface. Using the data given by Fig. 6(b), the non-dimensional bending stress was calculated from equation (8) and the results for four different culm numbers are plotted in Fig. 14.

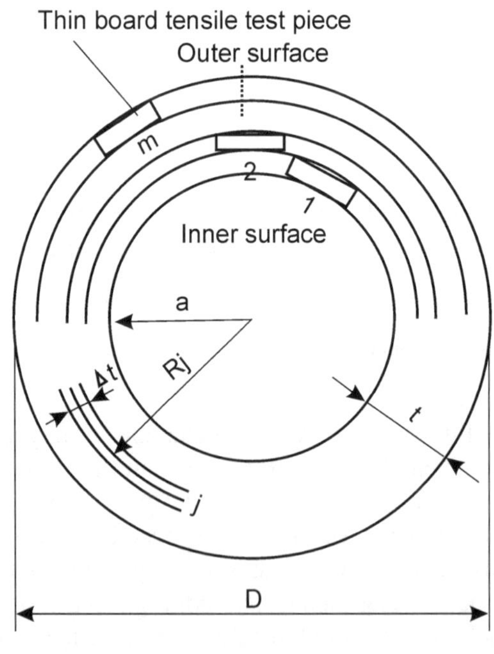

Figure 13. Multi-layered cylindrical model

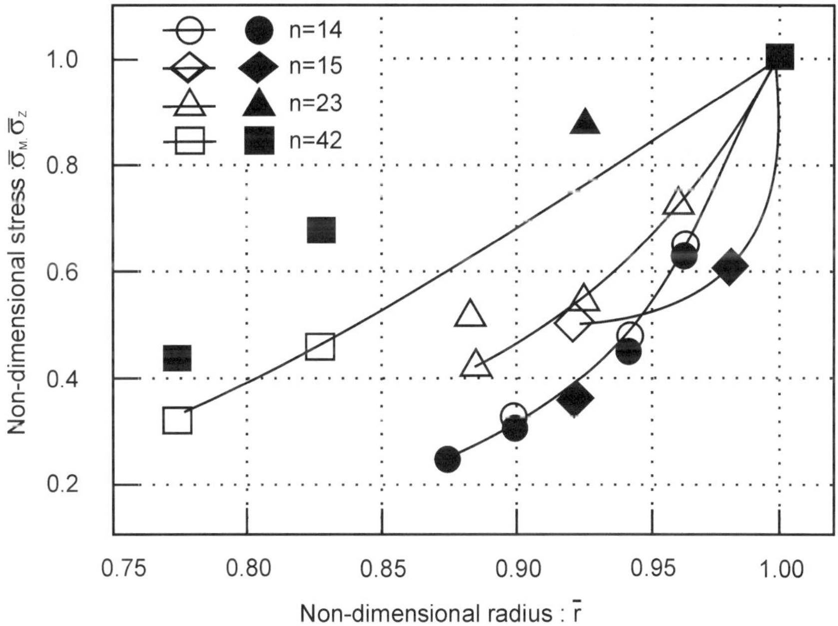

Figure 14. Comparison of bending stress and strength

The strength distributions with respect of radius were shown in Fig. 6(a). Normalizing them by the strength σ_N at the outer surface, they are defined by:

$$\bar{\sigma}_f = \sigma_f / \sigma_N \tag{9}$$

Taking into consideration the individual differences in the bamboo culms, the two kinds of values calculated from equations (8) and (9) are approximately close. In other words, the distribution of the fiber adapts to the bending stress under environmental loads. This may be an objective of the graded structure of bamboo fiber.

CONCLUSION

Bamboos are a natural, functionally graded and composite material, reinforced axially by fibers. Focusing on their mechanical characteristics, the following conclusions can be made.

1. Bamboos have a hierarchical graded structure; that is, a macroscopic graded structure in culm diameter, thickness and internodal length, and a microscopic one in the fiber distribution.
2. The tensile tests and image analysis derived the strength and the Young's modulus of the fiber and matrix, respectively. The fiber strength is about 12 times higher than that of the matrix.

3. The Young's modulus of the bamboo fiber measured non-destructively by a scanning acoustic microscope (SAM) was 60.2 GPa, which was higher than the 46.0 GPa obtained from tensile tests and image analysis.
4. Bamboos have a structure that is well adapted to environmental loads, the characteristics of which are as follows:
   a. The maximum surface stress at different heights under wind loads is almost a constant along the entire length.
   b. The distribution of fibers with respect to radius approximately fits the bending stress generated by the wind loads.

ACKNOWLEDGEMENTS

This project is supported by the Grant-in-Aid for Scientific Research on Priority Area, No. 274, the Ministry of Education, Science, Sports and Culture of Japan.

REFERENCES

Amada, S.; Ichikawa, Y.; Munekata, T.; Nagase, Y.; Shimizu, H. 1997. Fiber texture and mechanical graded structure of bamboo. J. Composite Eng. Pt. B. 28B, 13–20.

Amada, S.; Munekata, T.; Nagase, Y.; Ichikawa, Y.; Kirigai, A.; Zhifei, Y. 1996. The mechanical structures of bamboos in view point of functionally graded and composite materials. J. Composite Materials, 30, 800–819.

Atlar, A. 1978. An angular spectrum approach to contrast in reflection acoustic microscopy. Journal of Applied Physics, 49, 5130–5139.

Canumalla, S.; Gordon, G.A.; Pangborn, R.N. 1997. In situ measurement of the Young's modulus of an embedded inclusion by acoustic microscope. J. Eng. Materials and Technology, 119, 143–147.

Chuma, S. et al. 1990. Composite structure and tensile properties of mousou bamboo. J. Soc. Materials Sci. Japan, 39, 847–851.

Fung, Y.C. 1965. Foundation of solid mechanics. Prentice-Hall.

Hirai, T. 1996. Functionally gradient materials. Materials Science and Technology, Processing of Ceramics, Pt.2, VCH VerlagsgesellschaftmbH, 293–341.

Ishikawa, K.; Ogura, Y.; Ikeda, S.; Suganuma, M.; Narita. 1996. Measurement of sound velocity near a crack by an acoustic microscope. J. Non-destructive Testing, Japan, 45, 749–755.

Jain, S.; Kumar, R.; Jinda, U.C. 1992. Mechanical behavior of bamboo and bamboo composites. J. Mat. Sci. 27, 4598–4604.

Kushibiki, K; Chubachi, N. 1985. Material characterization by line-focus beam acoustic microscope. IEEE Trans. on Sonics and Ultrasonics, SU-32, 189–211.

Niino, M.; Hirai, T.; Watanabe, R. 1990. Functionally graded materials. Composite Materials, 13, 257–264.

Nishida, M.; Endo, T.; Adachi, T.; Matsumoto, H. 1997. Measurement of anisotropic elastic moduli in local area by acoustic microscope (the case of cubic single crystal). Trans. Japan Soc. Mech. Eng. Ser. A 63, 1467–1474.

Nokata, F.; Seo, K. 1990. Functionally graded structure in nature. FGM News, 8, 1–5.

Oda, J. 1980. Feature and mechanical characteristics of bamboo. Trans. Japan Soc. Mech. Engr. Ser. A 409, 997–1002.

Oda, J. 1985. Minimum weight design problems of fiber-reinforced beam subjected to uniform bending. Journal of Mechanisms, Transmission and Automation in Designing, Trans. ASME, Ser. R. 107, 88–93.

Shiota, I.; Miyamoto, K. ed. 1997. Functionally graded materials: 1996. Elsevier, 1997.
STA (Science and Technology Agency). 1986. Review of a new functional material for thermal stress relaxation. Survey Report of the Science and Technology Agency, Japan.
Welgein, R.D.; Wilson, R.G. 1978. Characteristic material signatures by acoustic microscopy. Electron Lett. 14, 352–354.
Welgien, R.D. 1985. Acoustic micro-metrology. IEEE Trans. on Sonics and Ultrasonics, SU-32, 225–234.

Bamboo for Sustainable Development, pp. 599–606
Eds. A. Kumar, I.V. Ramanuja Rao and Ch. Sastry

Testing of a bamboo school building using tilting table

PURWITO
Research Institute for Human Development, Cileunyi Wetan, Bandung, Indonesia

Abstract—Bamboo has several characteristics that support its use in the construction of seismic-resistant building in seismic zones. After the 1992 earthquake in Flores Island, Indonesia, a research was conducted to examine the feasibility of bamboo constructions for school buildings. A scale model was constructed using bamboo, and the specimen was tested on a tilting table with 0–42° tilting capacity. The tests revealed that bamboo could be safely used in anti-seismic constructions of similar design parameters.

INTRODUCTION

Bamboo is one of the oldest materials used for the construction of houses and other structures. As an excellent building material, it is relatively cheap, easy to work with and readily available in most villages in bamboo-growing countries.

Laboratory tests and filed experiments have proved the technical characteristics of bamboo as a construction material to be good. In arcas where earthquakes are frequent, the lightweight and resilient bamboo frame houses will provide a longer service life than any other type. Bamboo holds a high potential as the construction material for building temporary structures and reconstructing damaged buildings in seismic zones after an earthquake.

In 1995, three years after the Flores Island earthquake, the Indonesian Government developed a plan to utilize bamboo to support the reconstruction of school buildings in the island. Some Institutions in Indonesia – such as the Department of Public Works and some universities and consultation companies – jointly carried out this project with financial assistance from the World Bank.

The school building was designed by Bina Karya, a consultant, the Directorate General of Human Settlements and the Research Institute for Human Settlements (RIHS).

The testing of the bamboo structure designed was conducted at RIHS by using local bamboo. The following sections describe the testing method employed.

FIELD RESEARCH DATA FROM SOME INSTITUTES

Some research data obtained from field observations in Flores Island are given below. These played an important role in the building design.

1. Flores Island lies within active seismic zone, classified in very high-risk category.
2. Between 1981 and 1990, there were 14 earthquakes of 4–6.5 magnitude.
3. Of the 1.4 million people inhabiting Flores Island, over 80% are engaged in agriculture and fishery.
4. The 1992 earthquakes occurred on a Saturday, 12 December, with epicenter in the Flores Sea and had a magnitude of 3.0 to 6.0 in the Richter scale. The effects of earthquakes were, tidal wave (tsunami) along the northern coast, ground shaking, landsides, liquefaction and sand boils. Seriously damaged area was around 6 000 km^2 (40% of Flores Island) and at least 31 villages were completely destroyed. More than 25 000 houses, 500 schools, 135 health facilities and 6 000 ha of irrigation areas were either destroyed or severely damaged.
5. Analyses of the impact of earthquake revealed that damage to buildings varied according to soil conditions, nature of construction and the quality of construction.

USE OF BAMBOO FOR SCHOOL BUILDING

The type of construction most common in Flores is brick masonry with heavy roof structure. Typical damages in such structures result from:

- Lack of reinforcement in masonry walls;
- Failure or lack of collar beams;
- Large openings (for doors and windows); and
- In wooden structures, the failure of the foundation and lack of adequate anchorage of the posts.

Apart from its excellent mechanical strength characteristics, there are several other reasons for using bamboo for construction in Flores Island. The major ones are:

- Bamboo is available in about 9 000 ha of area in Ngada District.
- The elasticity of bamboo is better than wood for seismic-resistant housing, as has been proved in the case of several small houses;
- With proper treatment, bamboo provides a service life of up to 30 years; and
- Bamboo has acceptability because it is part of traditional architecture;
- Bamboo is affordable by most people;
- Constructing with bamboo is easy;
- Post-construction maintenance is easy; and
- Bamboo is highly adaptable.

The mechanical properties of bamboo make it an ideal material for building construction. These include:

- Tensile strength varying from 1 000 to 4 000 kg/cm^2
- Compression strength varying from 250 to 1 000 kg/cm^2
- Bending strength varying from 700 to 3 000 kg/cm^2
- Modulus of elasticity varying from 100 000 to 300 000 kg/cm^2
- Pressure test results – at 4 Atm perspiring at node, at 5 Atm serious leaking and at 10 Atm bursting.

MODEL OF SCHOOL BUILDING

The design of the school building (Fig. 1) was made conforming to the Standard set by the Education and Cultures Department for school buildings, with some modifications in the building materials (bamboo structures) used in the site.

Because the size of the testing device (tilting table) is standardized, a scale model (specimen) of the building conforming to the actual condition needed to be used.

The specimen was erected on the tilting table using a Distance Variable Scale or Length Dimension, employing the formula:

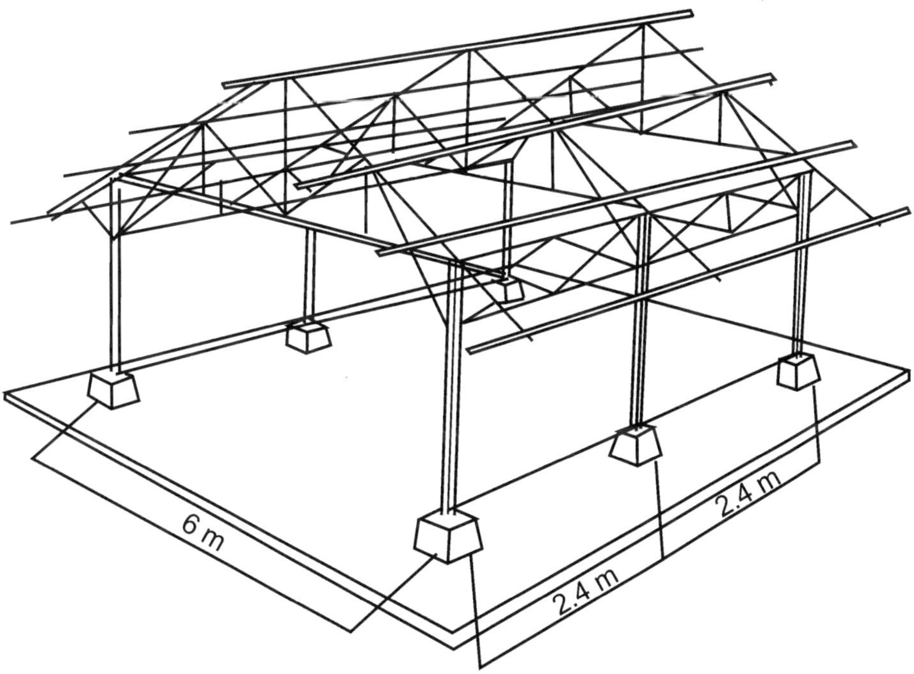

Figure 1. Basic design of the building

$$fm = \frac{rm}{rp}$$

Where: r = length of the table
m = model
p = prototype
fm = length of model plan.

By substituting values in this formula, the result obtained was:

$$fm = \frac{6.00}{7.35} = 0.816$$

This means that all dimensions of the specimen must be multiplied by 0.816 for them to conform to design specifications. Accordingly, the dimensions of the specimen building become:

Length = 6.00 m
Width = 6.00 m
Height = 2.20 m.

The materials used in this specimen building were as follows:

1. Column – bamboo pole of 12 cm diameter
2. Truss – bamboo pole of 12 cm diameter
3. Purlin – bamboo pole of 12 cm diameter
4. Bracing beam – bamboo pole of 10 cm diameter
5. Doors and widow frames – bamboo pole of 8 cm diameter
6. Foundation (to connect column and table a steel plate of 8 mm thickness was used)
7. Jointing – *injuk* rope of 10 mm diameter (Fig. 2)
8. Roof cover – corrugated zinc sheet of 0.29 cm thickness (Fig. 2).

The six columns (each a set of two or three poles fastened together with bolts, Fig. 3) were set in concrete blocks (60 cm high) set at 2.4 m apart (center to center) on the side with three columns and 6 m apart on the side with two columns (Fig. 1). The building specimen was completed, from the preparation of all components to erection on the tilting table, in 26 days by four workers.

METHODOLOGY OF TESTING

The testing started 30 days after the erection of the model was completed. Loading of the specimen was carried out by 3° increments to the right side and the left side, starting from 0° up to 42°, the maximum capacity of the equipment (Fig. 4).

The different loading steps involved are shown below:

1. Step 1 = 0– 6° (2 cycles per each direction)
2. Step 2 = 0–12° (2 cycles per each direction)

Figure 2. Jointing with *injuk* rope, and the galvanized zinc roof

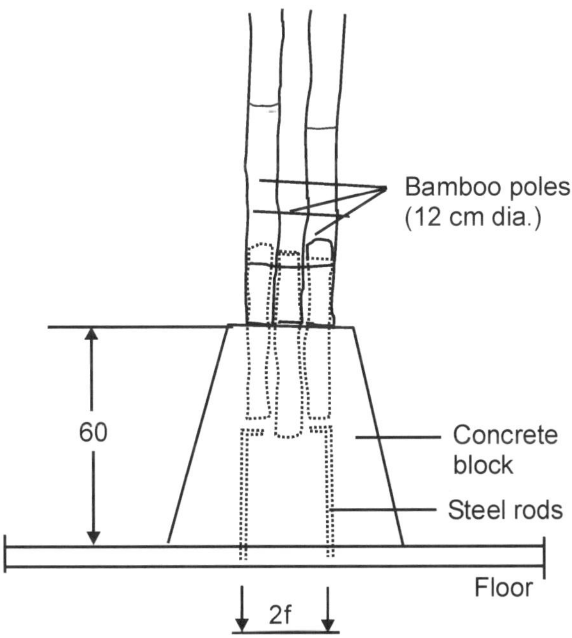

Figure 3. Setting of columns

Figure 4. The specimen under test on tilting table – 30° right

3. Step 3 = 0–18° (2 cycles per each direction)
4. Step 4 = 0–24° (2 cycles per each direction)
5. Step 5 = 0–30° (2 cycles per each direction)
6. Step 6 = 0–42° (1 cycles to one direction).

The scheme of loading is as shown in Fig. 5.

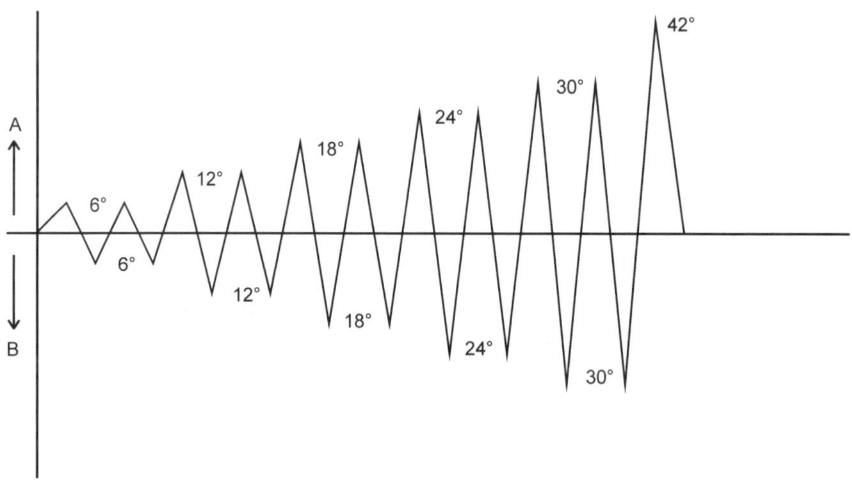

Figure 5. Scheme of loading of specimen

Transducers were installed at 12 locations in the frame to monitor deformations of the specimen, and the data collected by these were transmitted to a computer for analysis.

The safety factor used in building specimen is 3, in view of the building characteristics that will be used and the erection of the construction.

TEST RESULTS

Monitoring of the specimen before tests showed no cracks, although the tests started 30 days after erection of the specimen construction. This meant that the bamboo used in the specimen was of good quality. The test results are summarized below.

- No large deformations were noticed even when the building was tilted 42° on the tilting table.
- The response capacity of the specimen building was found to be 0.67 G.
- Jointing between the components (mainly in trusses and column parts), using *injuk* rope and bolts, was still in good condition and strong at the conclusion of the tests (Fig. 6). No damage was noticed either in the joints or in the jointing material.
- Height and width of span was in normal condition.

Figure 6. Joints and jointing materials in good condition after tests

Tests were carried out in cross direction, same as employed in structural analysis calculation. In this direction, the shear strength of the specimen building was reinforced by 4 portal trusses.

The quality of bamboo used in this building (8–12 cm diameter and 3 years old) was good enough (no crack or deformation), except that some bamboo poles aged less than 3 years showed some shrinkage.

CONCLUSION

Based on the test results in cross direction, some conclusions could be drawn:

1. The selection of materials is important, especially regarding the age of the bamboo used. It is recommended that only culms aged three years or above be used.
2. The result of structural calculations was not found different on comparing with the results of tests with scale model.
3. The design and structure of the building was suitable for seismic areas.

Bamboo for Sustainable Development, pp. 607–613
Eds. A. Kumar, I.V. Ramanuja Rao and Ch. Sastry

Physico-mechanical evaluation of mineralized particles in bamboo-Portland cement composite

A.L. BERALDO and J.C.T. LEITE [1]

Agricultural Engineering, Universidade Estadual de Campinas, São Paulo, Brazil

Abstract—This paper presents the physical and mechanical evaluation of composites made from natural or washed mineralized bamboo particles and Brazilian Portland cement. *Bambusa tuldoides* and two Brazilian Portland cements – Type II (CP II E 32) and Type V (CP V ARI, a high initial strength cement) – were utilized for the study. Mineralization involved two salt solutions: sodium silicate and aluminium sulfate. Changes in the weight particles after mineralization and the moisture content of the samples were determined. It was followed by sample tests for structural changes by means of non-destructive evaluation (NDE). While natural bamboo particles inhibit the settling and hardening of Type II Portland cement, mineralization enhanced their performance with cement paste. Young's dynamic modulus of the composites ranged from 1 800 MPa to 3 700 MPa.

INTRODUCTION

Bamboo is a very widely used raw material in many developing countries. It has many important characteristics, such as renewability, easy availability, good workability, etc. Bamboo can be utilized directly or combined with other materials (clay, concrete, plastics or metals) for construction purposes. Particles from bamboo culms can be used for the fabrication of resin-bonded boards. However, technical information on bamboo-inorganic bonded materials is very scarce in literature.

Wood-cement bonded boards were known even in the early 1930s. Several researches showed the importance of this composite in construction applications. At the same time, the researchers also detected serious problems with Portland cement and some wood species. Complex chemical reactions develop when wood particles are added to the Portland cement paste. Sometimes cement setting is delayed; in other cases, set cement failed in use.

Accelerators (chlorides, sulfates and aluminates) are generally used to overcome the wood-cement chemical incompatibility. For each wood species used in this process, there is a specific efficient accelerator identified. Calcium chloride, however, is the most used accelerator in the fabrication of wood particle-cement boards.

---

[1] Undergraduate student

Theoretical wood-cement chemical compatibility can be evaluated by analysing the composite hydration curve (time against temperature), according to Moslemi and Lim (1984). Other possibilities are the compression strength test of the samples (Lee and Hong 1986) and by non-destructive evaluation method (Beraldo 1994).

Setting and hardening of a lignocellulosic-cement composite can be improved by mineralization technique. Impermeable layers envelop the lignocellulosic particles and prevent the displacement of undesirable substances on to the particle surface. According to Furuno et al. (1991), mineralization can also reduce the wood hygroscopicity and enhance its fire performance.

Non-destructive evaluation (NDE) was initially proposed to detect defects in metals and alloys. Results obtained were normally reproducible and the modifications found were due to the internal cracks. For concrete, it is very interesting to find a relationship between ultrasonic wave speed and compression strength of the samples. For the same aggregate and the same ratio of the constituents, the changes in ultrasonic wave speeds can indicate changes in the composite structure.

MATERIAL AND METHODS

Bamboo particles

Bamboo particles smaller than 2 mm were washed in water (for 2 hours at 80ºC). For washing 0.1 kg of bamboo particles, 1 liter of water was used. After washing, bamboo particles were dried under natural conditions for one week.

Mineralization

Natural or washed and dried particles were treated with two chemical solutions: sodium silicate at 2% and 4% concentrations and aluminium sulfate at 10%, 20% and 30% concentrations (hereafter, the percentage concentrations of sodium silicate and aluminium sulfate will be denoted, in that order, by two values separated by '/'). This mineralization procedure was adapted from Furuno et al. (1991).

Samples preparation

Samples (three replication by treatment) were prepared using natural and washed bamboo particles (both mineralized). Two Types of Brazilian Portland cements were employed: Type II (CPII E 32 - NBR 11578) and Type V (CP V ARI - NBR 5733, a high initial strength cement). For three cylindrical samples (Ø 50 mm and 100 mm height), 300 g of cement and 150 g (before washing) of the bamboo particles were used. Water used ranged from 250 to 400 ml.

A mix of cement-bamboo particles was prepared. Water was added and the mix was placed into cylindrical molds. After 24 hours, the samples were removed from the molds. Mass of the samples was measured over a period of 44 days. Compression test was conducted in a Versa Tester apparatus for about three minutes.

Non-destructive evaluation

Ultrasonic waves were utilized to survey changes in the structure of the composites. A surgical gel was utilized to enhance the surface contact of the samples with the probes of a Steinkamp BP-5 apparatus (sensibility 0.1 μs). Measurements were conducted over a period of 14 days.

RESULTS

Natural bamboo particles were found to inhibit cement setting (Beraldo 1994). Therefore, it was used only as reference, and the sample was prepared with washed bamboo particles (hereafter, 'N' represents natural bamboo particles and 'W' denotes washed bamboo particles).

At the same moisture content, mineralization treatment increased the weight of bamboo particles. Values ranged from 14% (treatment N2/10 – natural bamboo particles mineralized with 2% sodium silicate and 10% aluminium sulfate) to 58% (treatment N4/30).

Drying curves of composites show different behavior for mineralized and non-mineralized particles (Figs. 1, 2). It can be observed that the moisture content of a reference composite (W0/0 – washed bamboo particles without mineralization) decreased faster than other composites made from mineralized particles. This proved that mineralization modifies the structure of the composites.

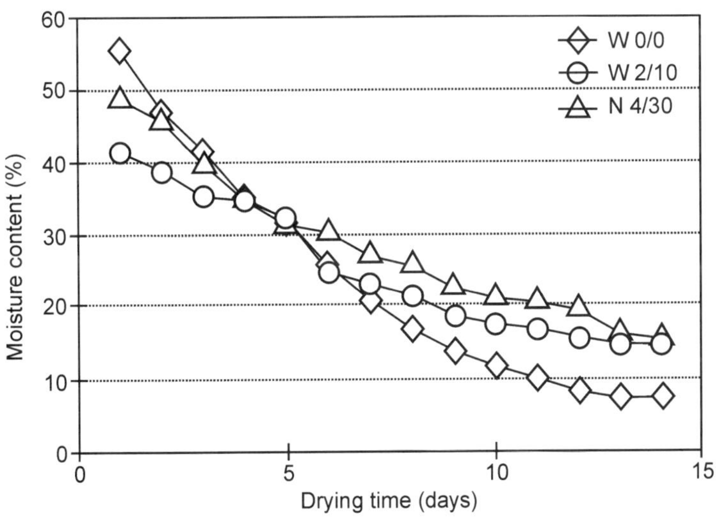

Figure 1. Drying curves for CP VARI

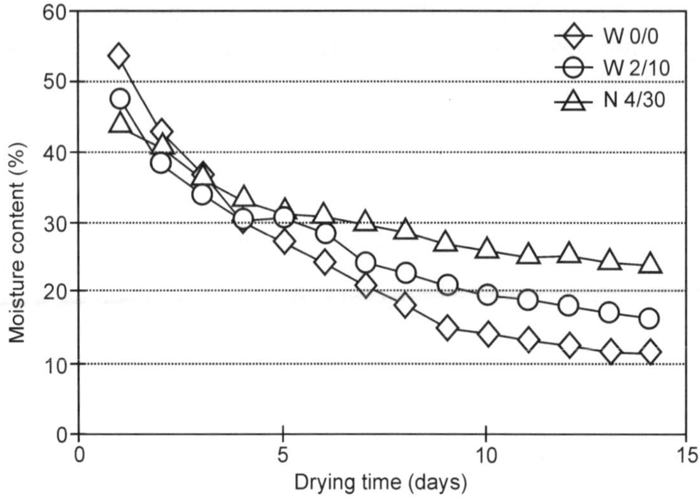

Figure 2. Drying curves for CP II E 32

Mineralization overcame the inhibitory effect of the substances in natural bamboo particle, enhancing the setting and hardening of Portland cement. However, this procedure was very sensitive to the mix constituents, mainly the water quantity. Table 1 shows the results of the compression strength of the samples.

For Type II Portland cement, the composite compression strength of washed and mineralized particles (W) increased, when compared with a composite from natural mineralized particles (N). For Type V Portland cement, an opposite effect was observed. Cement Type II is therefore more sensitive to the harmful substances in natural bamboo particles.

Higher salt concentrations produced lower compression strengths for the composite. This, though unexpected, can be explained by the rapid exothermic reaction that occurs between cement and other chemicals. When water is added to the dry mix, temperature rises suddenly (15°C in 5 minutes for N4/30). Increase in the mix temperature lowers composite workability, and this can damage the structure of the material.

Table 1.
Compression strength (MPa) for bamboo mineralized-Portland cement composites

Treatment AS/SS (%)	Cement CP II				Cement CP V			
	SS 2		SS 4		SS 2		SS 4	
	N	W	N	W	N	W	N	W
AS 10	3.7	6.6	3.6	4.9	10.4	6.4	9.9	5.3
AS 20	3.8	4.4	3.7	4.7	6.7	5.0	8.0	4.1
AS 30	2.2	2.6	2.2	3.1	2.8	2.6	4.6	2.9

Note: AS – aluminium sulfate; SS – sodium silicate; N – natural particles; W – washed particles.

It was not necessary to utilize a high salt concentration for composite fabrication; good results were obtained for 2/10 treatment. However, caution needs to be exercised, as changes in water quantity drastically reduce the workability of the mix. Table 2 shows the compression strength for two kinds of treatments (N/2/30 and N/4/30).

For 2% sodium silicate concentration, compression strength of the samples increased with the amount of water; for 4% concentration, however, the effect was reverse. For both mixtures employed, compression strength of the samples was less than 5.0 MPa.

Non-destructive evaluation

Composite surface was very rough and a good contact between the samples and the probes was difficult. Hence, a surgical gel was used to ensure contact. Ultrasonic wave speed across the composites increased during the first days, mainly during the initial hours. After a variable time (5–8 days), wave speeds were observed to stabilize.

An exponential relationship between ultrasonic wave speed and the time can be proposed as per the following equation:

$$V = Vmax\ [1\text{-}exp\ (\text{-}kt)]$$

Where Vmax = maximum ultrasonic wave speed (supposed constant); and k = rate of speed increase by time.

Figs. 3, 4 show the ultrasonic wave speed for the samples.

For Type II cement, the maximum value ranged from 0.84 km/s (W0/0) to 1.68 km/s (W2/20). For Type V cement, the maximum value ranged from 1.01 km/s (W0/0) to 1.80 km/s (N4/30). The coefficient k ranged from 0.31 to 0.44.

For many materials, compression strength and ultrasonic wave speed are dependent on density. Mineralization, however, affects the density of the composite as it deposits substances into the bamboo cells. Probably, mineralization creates heterogeneous regions in the composite structure without invoking favorable modifications in terms of mechanical strength (Table 3).

Table 2.
Compression strength (MPa) of natural mineralized bamboo-cement composite.

Treatment	Water quantity (ml)	MOR (MPa)	
		CP II	CP V
Natural	250	2.2	2.8
SS2	300	3.0	4.4
AS30	400	–	4.2
Natural	250	2.2	4.6
SS4	300	2.1	4.2
AS30	350	1.6	2.2

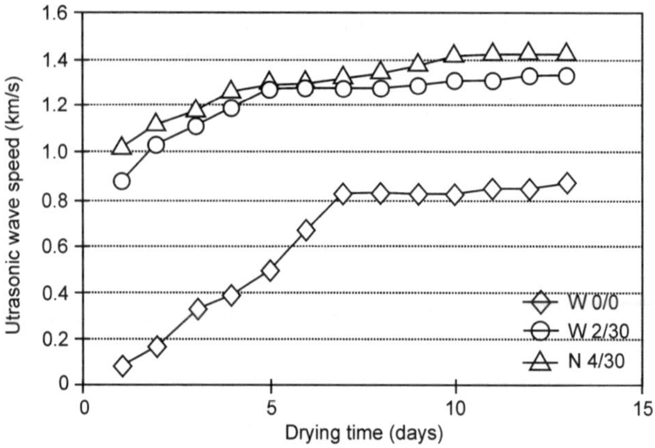

Figure 3. Ultrasonic wave speed for Type II cement

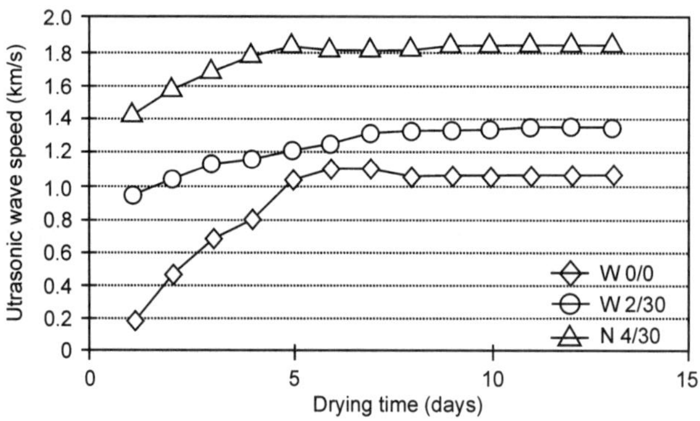

Figure 4. Ultrasonic wave speed for Type V cement

Table 3.
Physico-mechanical characteristics of mineralized particles in bamboo-cement composite

Treatment SS/AS (%)	CP II			CP V		
	Speed (km/s)	Density (g/cm^3)	MOR (MPa)	Speed (km/s)	Density (g/cm^3)	MOR (MPa)
W 0/0	0.84	0.97	1.0	1.01	0.99	3.1
N 2/30	1.40	1.07	2.2	1.57	1.06	2.8
N 4/30	1.34	1.03	2.2	1.80	1.15	4.6
W 2/10	1.60	1.14	6.6	1.65	1.12	6.4
W 2/20	1.68	1.15	4.4	1.63	1.06	5.0
W 2/30	1.33	1.03	2.6	1.37	0.96	2.6
W 4/10	1.41	1.08	4.9	1.46	1.06	5.3
W 4/20	1.50	1.07	4.7	1.41	1.04	4.1
W 4/30	1.36	1.06	3.1	1.39	1.00	2.9

Note: AS – aluminium sulfate; SS – sodium silicate; N – natural particles; W – washed particles

Statistical analysis indicates that ultrasonic wave speed depends strongly on the mix employed: average speed values were 1.48 km/s for Type V and 1.38 km/s for Type II. Average densities were 1.05 g/cm^3 for Type V and 1.07 g/cm^3 for Type II, while compression strength averaged 4.10 MPa for Type V and 3.52 MPa for Type II.

Young's dynamic modulus of the composites could be evaluated based on the density and ultrasonic wave speed measurements. For the composites, reference (W0/0) values were 700 MPa (Type II cement) and 1 000 MPa (Type V cement). For mineralized bamboo particles, values ranged from 3 700 MPa (N4/30, Type V cement) to 1 800 MPa (W2/30, Type II cement).

CONCLUSION

Natural bamboo particles greatly inhibit the setting and hardening of Type II Brazilian Portland cement. Mineralization of bamboo particles enhances the performance with cement paste. This procedure, however, is heavily dependent on the water quantity in the mix. For each mineralization treatment, it was necessary to optimize the water amount. Salts in small concentrations were the most effective for composites. A non-destructive evaluation test (NDE), using ultrasonic wave speed, can survey the structural modifications of the mixes in the early stages. However, ultrasonic wave speed cannot be correlated with compression strength, owing mainly to the different salt concentrations during mineralization procedure. Young's dynamic modulus of bamboo particles-Portland cement composites ranged from 1 800 MPa to 3 700 MPa.

REFERENCES

Moslemi, A.A.; Lim, Y.T. 1984. Compatibility of southern hardwoods with Portland-cement. Forest Products Journal, 34 (7/8), 22–26.

Lee, A.W.C.; Hong, Z. 1986. Compressive strength of cylindrical samples as an indicator of cement-wood compatibility. Forest Products Journal, 36 (11/12) 87–90.

Beraldo, A.L. 1994. Generalisation et optimisation de la fabrication d'un composite biomasse vegetale-ciment a variations dimentionsnelles limitees vis-a-vis des variations de l'humidite. Ph.D. Thesis, Universite Henri Poincare, Nancy, France. 222 pp.

Furuno, T.; Uehara, T.; Jodai, S. 1991. Combination of wood and silicate I - Impregnation by water glass and application of aluminium sulfate and calcium chloride as reactants. Mokuzai Gakkaishi, 37 (5), 462–472.

Bamboo for Sustainable Development, pp. 615–632
Eds. A. Kumar, I.V. Ramanuja Rao and Ch. Sastry

The development of non-traditional bamboo technologies in Costa Rica*

JORGE A. GUTIÉRREZ
National Laboratory for Materials and Structural Models, University of Costa Rica

Abstract—Bamboo, the companion of humanity even before the dawn of civilization, has a very important role to play as a key material of the 21st century if the world is going to move into a Sustainable Development Scheme. For that reason, it has to change its status from a local, low-value material used for traditional to a global, high-value industrial material employed for modern products. In that context, the experience of the Bamboo National Project and the Bamboo Foundation of Costa Rica are described and some promising possibilities for East-West cooperation and North-South commercial exchange are presented in this paper.

INTRODUCTION

Bamboo has been serving humanity from cradle to grave in many countries since ages in many different and ingenious ways. China, India, Japan, Indonesia and Colombia, to name a few, are countries where bamboo is firmly ingrained in the popular culture. In Vietnam, bamboo is considered a "brother". Few other living beings can even aspire to such closeness.

In most developing countries where bamboo plays such an important role, it is considered a local-level community resource. The traditional techniques are passed from generation to generation with hardly any change, and the final products are usually sold locally and have very little market value. Under this scheme, each community or region develops its own unique traditions and techniques, and hardly any technology transfer occurs between nearby regions, not to mention different countries or far-away continents.

For this reason, there is an urgent need to build bamboo bridges and paths between the East and the West. These links should interconnect all the developing countries where bamboo could and should be growing, in order to fully develop the diverse existing techniques and be able to innovate in an environment of cross-fertilization of ideas and products. Research and development on bamboo and bamboo products should be a priority of developing countries in order to de-

* A version of this paper has been earlier published "Traditional Techniques of Bamboo in Modern Life", Proceedings of the International Seminar for the Safeguarding and Promotion of Traditional Techniques of Bamboo in Modern Life, Ho Chi Minh City, Vietnam, 17–20 December 1997.

velop new products, with better technologies and improved quality control, to increase its added value. Scientific and technological exchanges should be intense in both ways in this much needed bamboo paths between the East and the West.

Another equally important bamboo link is needed to interconnect the North and the South. The developed countries of North America, Europe and Asia, where bamboo is scarce or the land for cultivation is very expensive, could be natural consumers of industrialized bamboo products, as well as potential providers of high technology that could be transferred to the bamboo producers of the South. It may sound paradoxical but the best way for safeguarding the traditional bamboo techniques and their promotion into modern life is through an intense and open exchange of traditional and technological know-how and high value-added industrial and crafted products between the East and the West as well as the North and the South.

Within this framework of open exchange of ideas, the author would like to present the experience of Costa Rica. Unlike other Latin American countries, like Colombia or Ecuador, where bamboo has been a traditional material for centuries, Costa Rica did not have a bamboo culture. Rich in tropical forests with plenty of extraordinary timber species, the country exploited them with such intensity and recklessness that, twenty years ago, one per cent of the total territory was being deforested every year. Today, almost all the remaining rain forests, about 25% of the country, are under some type of protection. Timber, which was extensively used as a construction material, has become expensive, scarce and of low quality, giving way to new construction materials, such as masonry or precast concrete, and their related technologies. The total percentage of timber houses in the country dropped dramatically from 86% in 1963 to 60% in 1984, and the plunge continues. Together with the use of the material, timber building techniques and labor skills also diminished from the excellent levels reached half a century ago.

The introduction of more complex housing construction technologies produced three main negative effects:

1. It increased the cost of the houses, because self-building was no longer possible and the materials were more expensive;
2. It stimulated the migration from rural to urban areas, where most of the construction projects were being developed; and
3. It increased the foreign component on construction materials, affecting the trade balance.

To overcome these negative effects, it became necessary to stimulate adequate construction programs in the rural areas, using low-cost materials and appropriate technologies for self-help construction projects that could be afforded by the people of the rural areas, where quality of life is so easily attainable when basic needs are met. Whenever one visits a rural community with good standards of living, one cannot but think about Mexico City, Sao Paulo, Calcutta, Cairo and all the other megalopolises of the developing world, saturated with people that gladly and willingly would have stayed in the countryside, had their conditions been different.

In the case of Costa Rica, since timber was scarce and expensive, there was a need for a substitute material that has similar strength characteristics, and is economical, resistant and locally produced. Bamboo satisfied all these requirements and an ambitious project was conceived to introduce bamboo as a construction material and to develop a complete bamboo culture in the country, practically from scratch. This paper describes the development of a process that was initiated ten years ago, and is about to be reproduced in several countries in Central America and the Caribbean; a process that is ready to start travelling along the East-West path.

Some of the most notable achievements of the project are:

- An effective, non-toxic and economic preservation technique;
- A prefabricated low-income housing system that uses some timber as well as bamboo laths and cement mortar;
- More than 3 000 prefab houses built (30 of them near the epicenter of a 7.5 Magnitude earthquake survived without any damage);
- A recently developed prefabricated 100% bamboo house, that may reduce the construction costs by as much as 20%; and
- Bamboo furniture with original design and relatively low price.

Furthermore, a scheme was developed for introducing bamboo as part of the global Joint Implementation Programs. The scheme could easily provide the needed incentive for extensive cultivation areas, as well as produce the additional raw material for high added-value industrialized bamboo products – such as laminated bamboo, plybamboo, parquet and fiber board – with excellent domestic and international market opportunities.

Ten years of bamboo in Costa Rica

The National Bamboo Project (PNB) was officially endorsed by the Costa Rican Government through the Ministry of Housing and Human Settlements, the Ministry of Planning and the Ministry of Foreign Affairs. The original funding came from a grant of US$ 5.7 million from the Government of the Netherlands and the United Nations Development Program (UNDP). The administration was carried through UNCHS-Habitat and UNDP. The funds lasted for eight years and at that time a private non-profit foundation, called FUNBAMBU, was created to inherit all the activities and the know-how developed by PNB. It has been operating as a self-sustaining business enterprise since January 1996.

PNB had divided its activities into four main areas: Cultivation, Construction, Training, and Research and Development. The main accomplishments in these eight years were as follows.

Cultivation
Guadua bamboo, native of Colombia and Ecuador, was strategically planted in 200 ha (Fig. 1) in different sites in the country to supply the required material without a significant increment in the transportation costs. The total cultivated area was later increased to 350 ha. This amount of planted bamboo, when mature, will suffice for the local needs on housing, furniture and a limited amount of industrialized products (such as laminated bamboo, plybamboo, bamboo parquet, fiberboard, etc.) produced on an experimental scale. There are government plans already for a significant increase of the plantation areas.

Construction
More than 3 000 dwellings were built in the rural communities and Native American reservations throughout the country (Fig. 2). A technology of prefabrication was introduced to significantly reduce the construction time and cost. Since bamboo was not readily available, their walls use a combination of a timber frame, bamboo laths or small diameter canes, and cement mortar. In 1997 alone, FUNBAMBU built about 1 500 units, competing on a very competitive commercial market. This figure represents a notable 6% of the country's total output, providing employment to more than 500 technicians all year round.

Training
Being a new technology, it became necessary to initiate an ambitious training program in cultivation, processing and utilization of bamboo in house construction and furniture making. More than 1 000 persons were trained in some of these techniques.

Research and development
Being a new material and a new technology to Costa Rica, it was obvious that research and development would be necessary. Whenever a traditional construction material is used and something goes wrong, hardly anybody will blame the material. On the contrary, when a new material is introduced, the smallest problem can endanger the complete process.

An ambitious and comprehensive program was conceived covering almost all activities but concentrating on cultivation, preservation, material properties and construction techniques.

The author was the Director of the R&D department since its beginning. The percentage of its budget was very small when compared with that of the cultivation and construction programs. Hence, he perceived the whole National Bamboo Project as an R&D project and, as a consequence, it became possible to collect very valuable information from activities and processes that otherwise would have been quite sterile from the research point of view.

That the Project started from scratch in a country without any bamboo tradition, in a way, should be considered a strength rather than a weakness because it is easier to introduce new techniques and technologies than it is to change ancient traditions. However, a very well conceived R&D program is required to provide all the technical answers necessary to create a sense of confidence among users and program personnel.

Figure 1. A *Guadua* bamboo plantation in Costa Rica

Figure 2. A bamboo house built by FUNBAMBU

CONSOLIDATED ACHIEVEMENTS

This section describes some of the most relevant consolidated achievements of the National Bamboo Project of Costa Rica. Most processes related to these are mature enough to start travelling along the East-West path between developing countries.

Bamboo preservation

According to Liese (1980): "Bamboo's low natural durability means that the service life of structures made from it is often short. One way to extend its life is through processing and construction methods that minimize attack by fungi and insects. Conventional methods, like soaking in water, deplete the starch content of the bamboo and thus may reduce insect infestation. However, better preservatives and treatment techniques, if applied, will extend the service life of bamboo structures considerably, even in hazardous situations, and their increasing application throughout the world proves their economic value." With this philosophy in mind and the key assistance of Prof. Liese, an R&D Program on Preservation was initiated in 1988 to develop the technical capabilities required to commercially preserve bamboo with an effective, non-toxic, low-cost and easy-to-apply procedure. The method based on a sap-displacement technique has been a complete success and can be easily installed in rural areas close to the plantation sites (Fig. 3). Details of this process have been reported elsewhere (González and Gutiérrez 1996).

Figure 3. Bamboo preservation (sap displacement) plant

Physical and mechanical properties of material and structural components

As with any structural material, the determination of the physical and mechanical properties of bamboo and its structural components is an essential requisite for its acceptance as an engineering material, that is, a material that allows for the calculation of the strength, stiffness and stability of the structures built of them, using established design procedures.

All structural materials do have strengths and weaknesses. For the structural designer, the challenge with the material is to fully exploit its strengths while avoiding, at the same time, all negative effects derived from its weaknesses. For instance, unreinforced masonry is a construction material that is strong in compression but weak in tension. With the development of the geometric forms of the arch, the barrel vault and the dome – which basically produce compressive stresses under gravity loads – the material was fully exploited since the Romans some 2000 years ago. It was the predominant construction material of the western world until the advent of iron and steel in the 21st century.

One of the main difficulties of living structural materials like timber or bamboo is that their physical and mechanical characteristics are species-dependent and, even within a species, dependent on site and climate as well. These properties must be determined for each species and, whenever possible, for each particular region. For this, it is essential to adopt standard tests that are unambiguous and universally accepted. The next step should be the preparation and adoption of a set of International Building Codes. Significant steps in this direction are already

Figure 4. A mechanical test under progress

under way (Janssen 1996). A word of caution on this type of international documents is that they must be flexible enough to allow for particularities and traditions as long as they are technically safe; neither should they limit the innovative search for new structural forms and solutions.

A comprehensive test program was defined by PNB for the determination of the physical and mechanical properties of the *Guadua* bamboo, and the Forest Products Laboratory of the University of Costa Rica (Sotela 1990) carried it out. The test procedures (Fig. 4) were based on the pioneer work of Janssen (1981), who acted as a consultant and technical supervisor. A publication describing each of the test procedures was also issued (Sotela 1991).

Costa Rican bamboo houses have walls made of bamboo laths in combination with a timber frame, and cement mortar that completely covers the bamboo. This combination creates a very interesting composite material with unique mechanical characteristics. Being a country in a very active seismic region, there was quite justified a concern regarding the strength, stiffness and modes of failure of this type of walls, especially under the effect of lateral forces. A complete set of tests was carried out by the author for this purpose (Gutiérrez 1993). The results were more than satisfactory because they clearly proved the excellent composite behavior of the three components, which provided strength much higher than the demand induced by earthquake forces. These results were later confirmed for real houses (Fig. 5) built in the very epicenter of a Magnitude 7.5 earthquake in the

Figure 5. A bamboo house that survived intact the Limón earthquake

Limón Province. A group of 30 houses survived intact, without even a single crack in the cement mortar, whereas in the surrounding areas there was extensive damage and collapse of traditional timber and masonry houses.

Prefabricated houses

The need to reduce construction time and cost leads almost invariably to prefabrication, where lines of production can be established for standardized components such as walls and roof trusses.

For the walls, panels of variable horizontal sizes, usually equal to the size of a particular room in the house, are prefabricated and stored in a panel factory. The procedure is very flexible and allows for a great deal of architectural distributions, based on the needs of the client. The total number of each specific panel is then determined for each house. These are then transported in groups to the construction site, where they are easily erected and fixed on top of concrete foundation.

Even if the panels are quite large (typical size can be 2.4 m × 2.7 m), the timber and bamboo lath construction makes them very light. Cement mortar is applied on site to the assembled panels once the roof is in place, allowing for electrical and mechanical work at the same time. The total construction time for a group of low-income houses is less than three months.

As mentioned, the system is price-competitive with traditional construction materials, as the costs are 10–20% lower. This factor, together with the proven structural safety and good construction quality, has given a clear competitive edge to this prefabricated housing. Whole villages are being built in this fashion in many rural areas in the country (Fig. 6).

In addition to prefabricated houses, there has been an emphasis on the architectural development and construction of bamboo structures that can be used as provisional shelter, communal facilities at the village level or as part of large facilities like ecological hotels or resorts. This type of projects greatly contribute to the social acceptance of bamboo as a material that goes beyond the "poor man's timber" image. Some of these structures are inspired by the local traditions (Figs. 7, 8), while others respond to innovative concepts (Figs. 9, 10). In both cases, the knowledge of the material properties allows for reliable engineering analyses and designs of very diverse architectural possibilities. It can be rightly said that the limits are set by imagination.

Furniture

This has been a very active and successful independent line of production and commercialization (Erickson 1992). Original techniques have been developed using bamboo laths and small-diameter culms, resulting from the thinning in the plantations and are useless for construction purposes. The quality and affordable price of these products have led to an increase in their acceptance and demand. Their beautiful designs have allowed them to be exhibited in museums and galleries as pieces of art (Fig. 11).

Figure 6. A village with prefabricated bamboo houses

Figure 7. A conical roof structure: following local tradition

Figure 8. Interior detail of the conical roof

Figure 9. Bamboo icosahedron (during construction)

Figure 10. Bamboo icosahedron (after construction)

Figure 11. Bamboo furniture in a museum

FUNBAMBOO, a non-profit foundation

If it was difficult to successfully achieve the goals of the National Bamboo Project, it was much more difficult to achieve permanence and economic sustainability as a non-profit commercial enterprise. As mentioned, the transition from a project to a business enterprise occurred in January 1996, when FUNBAMBU, a foundation created for this specific purpose, started its activities as heir of the know-how, assets and personnel of the Project. After almost two years of commercial activities, FUNBAMBU has been able to consolidate itself as an important option in the housing market, employing more than 500 workers all year round. To guarantee the required coordination with the government, the five-member Board is chaired by the Minister of Housing or a personal delegate. The Ministry of Agriculture also has a seat, as does a representative from the local government. There was some hesitation at the beginning about such a strong presence of government, especially in a country that has presidential elections every four years, mostly with a change of the ruling political party. However, in the author's opinion, this has been beneficial because low-income housing is highly subsidized (up to a US$ 5 000 subsidy, enough to purchase a 40 m^2 bamboo house) and in consequence, a great deal of coordination with the political sector is required. Besides, the Foundation continued the policies set up in the previous phase, providing the necessary stability in the technical aspects and administration.

NEW DEVELOPMENTS

There are two recent developments still at an R&D stage but promising enough to be shared and subjected to discussion.

PREFABAM: a prefabricated 100% bamboo house

As it was mentioned in the previous section, prefabricated panels with a timber frame and bamboo laths have been successfully used in the construction of thousands of houses in the last ten years. It was also mentioned that timber was used because there was no *Guadua* bamboo available until the plantations started producing the large-diameter (8 cm or more) culms necessary for construction.

One of the main advantages of sawn timber is its dimensional regularity, which allows for prefabrication and modular construction. However, it was always clear to everybody involved in the project that timber frames were just a temporary solution and would be substituted by bamboo, once the material became available.

At first, the idea of prefabricated panels was maintained, with a simple substitution of the timber by bamboo culms. This substitution may seem to be obvious and trouble-free. However, the variable diameters of the bamboo preclude any effective standardization or modular construction. Additionally, the bamboo culms are larger and heavier than the 5-cm square timber sections, making transportation less efficient and handling more difficult. Because of these reasons, the bamboo-frame alternative for prefabricated panels was deferred until such a time when laminated bamboo pieces, with standard and regular sections, can be developed as an economic alternative to sawn timber. The goal was set in the development of a prefabricated 100% bamboo house – that is, no timber in the panels or the roof structure – that allowed for standardization and modular construction using bamboo culms with its inherent variations in diameter.

The solution, in the prototype stage, has been named PREFABAM (PREFAbricated BAMboo). It has a radically different approach that eliminates the panels and substitutes them by individual bamboo poles placed on top of prefabricated concrete cylinders, which serve as a foundation and raise the bamboo from the ground to protect it from humidity. These concrete cylinders are then integrated at the ground level by a continuous 6–mm steel bar embedded in concrete, and the bamboo poles are connected at the top by horizontal bamboo beams, which provide the required structural integrity (Figs. 12, 13). From these beams, prefabricated bamboo curtains, made of laths and flexible wire, are hung all the way down to the concrete at the base (Fig. 14). These curtains are then joined to the poles. The system is ready for application of concrete mortar, once the roof is built and the required spaces for windows and doors are opened with a saw.

This project is now being tested, construction details improved, and special tools to facilitate the construction process developed. A technical manual with precise directions for assembly and construction will be issues afterwards. This technology has a very promising future in Costa Rica and should be easily transferred to others regions along the East-West path.

Figure 12. PREFABAM structure

Figure 13. Detail of PREFABAM structure roof

Figure 14. Bamboo curtains and posts in PREFABAM structure

Bamboo for joint implementation programs

This is an interesting idea along the North-South path that merits exploration. To compensate for their atmospheric carbon emissions, the polluting industries of the developed world are willing to pay for the cleaning effect of the carbon being captured and fixed by the growing trees of cultivated forests. These payments amount to an additional income to the commercial plantations and may represent the difference between profit and non-profit. Obviously, very rigorous auditing processes and some kind of international stock market exchange system are to be established.

Although the idea has been conceived for timber trees, bamboo is an excellent alternative for carbon fixation because of its rapid growth and the fact that it is individually extracted in a process that does not disrupt the plantations. It is, in this sense, the archetypal renewable resource. In Costa Rica, there has been some reluctance to include bamboo in the Joint Implementation Programs. The reason is more bureaucratic than scientific: if a grass is included into the club, it will very difficult to define criteria to accept or reject other non-timber alternatives. FUMBAMBU is convinced that the idea is worth a fight. It is, however, obvious that it should be an international fight because its outcome will have profound effect on the large bamboo-producing countries, which may receive significant capital resources, encouraging plantation expansion and maintenance programs. This surplus may represent one of the most significant ways to promote bamboo in modern life, if industrially processed and transformed into high value-added products like laminated bamboo, plybamboo, parquet and fiberboard.

CONCLUSION

The World Commission on Environment and Development (1987) defined Sustained Development as development that "meets the needs of the present without compromising the ability of future generations to meet their own needs". Sustainable development is a process requiring simultaneous progress in four dimensions: economic, human, environmental and technological. There are close links among these different dimensions, and actions in one area can reinforce goals in another.

The author is positively convinced that bamboo, the archetypal renewable resource, must abandon the seclusion in which it has quietly and graciously lived, in the company of mankind, even before the dawn of civilization, and step forward to play its mission as a key material of the 21st century.

ACKNOWLEDGEMENTS

The author wishes to recognize the spirit of enthusiasm, generosity and open-mindedness that has been the permanent characteristics of all the people in the Costa Rican National Bamboo Project and FUMBAMBU, since their beginning.

REFERENCES

Erickson, B. 1992. Experimental Guadua furniture in the context of Costa Rica. *In* Zhu, S.; Li, W.; Zhang, X.; Wang, Z. ed., Bamboo and its use. Proceedings of the International Symposium on Industrial Use of Bamboo, Beijing, China, 7–11 December 1992. International Tropical Timber Organization; Chinese Academy of Forestry, Beijing, China. pp. 180–191.

Gonzales, G.; Gutierrez, J. 1996. Bamboo preservation at the Costa Rican National Bamboo Project. *In* Ganapathy, P.M.; Janssen, J.A.; Sastry, C.B. ed., Bamboo, people and the environment, Vol. 3, Engineering and utilization. Proceedings of the Vth International Bamboo Workshop, Ubud, Bali, Indonesia, 19–22 June 1995. International Network for Bamboo and Rattan, New Delhi, India. pp. 121–129.

Gutiérrez, J.A. 1993. Comportamiento estructural y resistencia sísmica de las viviendas de bambú. Memorias, 1er Congreso Mundial de Bambú/ Guadua, Pereira, Colombia, 8–15 August 1992. pp. 161–171.

Janssen, J.J.A. 1981. Bamboo in building structures. Ph.D. Thesis, Eindhoven University of Technology, Eindhoven, the Netherlands.

Janssen, J.J.A. 1996. Building codes for bamboo housing. *In* Ganapathy, P.M.; Janssen, J.A.; Sastry, C.B. ed., Bamboo, people and the environment, Vol. 3, Engineering and utilization. Proceedings of the Vth International Bamboo Workshop, Ubud, Bali, Indonesia, 19–22 June 1995. International Network for Bamboo and Rattan, New Delhi, India. pp. 130–139.

Liese, W. 1980. Preservation of bamboos. *In* Lessard, G.; Chouinard, A. ed., Bamboo research in Asia. Proceedings of a workshop held in Singapore, 28–30 May 1980. International Development Research Centre, Ottawa, Canada. pp. 165–172.

Sotela, J. 1990. Determinación de propiedades físicas y mecánicas de bambú del género Guadua de las zonas de Costa Rica". Final Report, Laboratorio de Productos Forestales, University of Costa Rica.

Sotela, J. 1991. Manual: ensayos físico-mecánicos para bambú. Laboratorio de Productos Forestales, University of Costa Rica, informe INII 60–91.

World Commission on Environment and Development. 1987. Our common future. Oxford University Press, New York, USA.

Bamboo for Sustainable Development, pp. 633–638
Eds. A. Kumar, I.V. Ramanuja Rao and Ch. Sastry

Industrial bamboo-based panels in China

JIANG SHENXUE

Bamboo Engineering Research Center, Nanjing Forestry University, Nanjing, China

Abstract—This paper discusses the bamboo resource and the different types of bamboo-based panels manufactured in China. Bamboo-based panel products are classified according to their components. The properties and manufacturing process of each of them and present status of production are described.

INTRODUCTION

China, a country with a long history of bamboo cultivation and utilization, is rich in bamboo resources and stands in the forefront of global development of the modern bamboo industry. China has approximately 4 million hectares of bamboo forest, and the total green culm stocks amount to 80 million tons. *Phyllostachys pubescens* (Moso bamboo), a species with a larger culm diameter and thicker culm wall than most other bamboo species, is the most valuable among all bamboo species in cultivation and utilization. *P. pubescens* spreads to more than 2.6 million hectares and 5 billion green culms, distributed mainly in southern China.

It is well known that bamboo is a regenerative natural material with high strength and hardness, good flexibility, and low degree of moisture absorption when compared with wood. Bamboo, however, has some obvious disadvantages when compared with timber such as hollowness of culm, diameter limitation, thinner wall and relatively larger taper. This is why bamboo has not found wider use as a modern industrial and construction material. A breakthrough came in China when Prof. Zhang Qisheng, a member of Chinese Academy of Engineering, invented the plybamboo and its manufacturing process in the 1980s. A variety of bamboo products, especially bamboo-based panels, and factories sprang up because of the priority and support in terms of policy and capital provided by the Chinese government (in China, there is a special office for bamboo industry management). Bamboo-based panels have played an important role in alleviating poverty in villages in bamboo forest regions and in developing the rural economy in southern China.

CLASSIFICATION OF BAMBOO-BASED PANELS

Bamboo-based panel products can be broadly classified as follows according to their components:

- Plybamboo made from softened and flattened bamboo strips;
- Bamboo curtain and mat board formed from bamboo curtains and woven mats;
- Bamboo mat board made of layers of bamboo mats (slivers of uniform size woven to form a flat surface);
- Bamboo floor made from sawed and planed strips;
- Bamboo particleboard; and
- Bamboo and wood composite products.

MAJOR BAMBOO-BASED PANELS

Plybamboo (bamboo plywood)

Plybamboo is usually made from larger bamboo culms (diameter at the height of eyes is more than 30 cm). The manufacturing process is as follows:

- Whole bamboo culms are cut into required lengths using a rotary saw.
- Outer nodes on the sections are removed with a cutter.
- Each section is cut longitudinally into three strips.
- Inner nodes (diaphragm) are removed.
- The strips are placed in boiling water for 4 hours for increasing the moisture content and temperature, and to eliminate some of the nutrients present.
- The hot strips are placed on the conveying chain of a softening machine and fed into the machine. The air temperature in the machine is maintained at a minimum of 160°C to enhance the inner temperature of the strips and increase their plasticity.
- The softened bamboo strips are fed into a flattening machine in which they are unfolded and flattened.
- The flattened bamboo strips are planed on both surfaces to eliminate both outer and inner layers and to form strips of uniform thickness.
- They are then dried to suitable moisture content.
- Both edges of each strip are straightened and phenol-formaldehyde (PF) adhesive is spread on the strips. Laying-up is completed according to plywood manufacturing method (three layers are the most common).
- The bamboo layers are hot-pressed, sawed and graded.

Most plybamboo pieces are 4–6 m long, 0.9–1.0 m wide and 15 mm thick, and are used as floors of trucks and buses. Some are made 8 ft long, 4 ft wide and 10–16 mm thick for use as the baseboards for a type of high-strength concrete formwork coated with PF or MF film. This type of concrete formwork consists of three materials: the base material is plybamboo or bamboo curtain board, the surface is melamine resin sheets, and there is a layer of wood veneer between the surface and the base material. This is considered as the best concrete formwork in China, as famous as the Finnish 'WISA' formwork. The Xihu Bamboo Concrete Form Company in Hangzhou, Zhejiang Province, has been producing it at the rate of 100 000 m^3 per year.

Plybamboo is a sort of high-strength outdoor engineering material with good hardness and durability. Its modulus of elasticity (MOE) is more than 9 000 MPa and modulus of rupture (MOR) is over 98 MPa (15 mm thick). Currently, there are over 20 plybamboo factories producing more than 20 000 m^3 (apart from the concrete formwork board mentioned above) per year in China. Companies such as Nanjing Automobile Group Company No.1 and No.2 Automobile Group Companies of China extensively use plybamboo for making truck floors.

Bamboo curtain and mat board

This product is made from bamboo slivers (thin strips of 0.8–1.2 mm thickness and 1.0–1.6 cm width) prepared by cutting culms, removing outer and inner nodes, splitting, layering along bamboo wall thickness and drying. The slivers are woven into mats, and bamboo curtains are formed by weaving parallel slivers with thread. Both mat and curtain are dipped in PF adhesive. After drying, layers of curtains are placed in cross-wise orientation, and sandwiched between sheets of bamboo mats. Hot-pressing and trimming follow to complete the process.

Bamboo curtain and mat board is usually used for concrete formwork. To improve the surface and increase the service life, Kraft paper impregnated with thermosetting resins can be overlaid on the surfaces. This not only imparts a longer service life but also gives a better concrete formwork with high strength and low swelling ratio. At present, there are 30 plants in China producing about 40 000 m^3 of bamboo curtain and mat board per year.

Bamboo mat board

Bamboo mat board is made by dipping woven mats in PF or urea formaldehyde (UF) adhesive, drying in belt veneer drier and pressing. In general, the thickness of the mat board is 2–9 mm. It is mostly used as a packing material, especially the thinner boards. Some decorative bamboo mat boards are used as material for furniture items such as dining table, chair, bench, etc. When making furniture, the slivers are bleached or dyed before they are woven. There are many small-scale factories producing bamboo mat boards in Sichuang and Hunan Provinces of China.

Bamboo floor

This is a kind of high-end product that sports a unique aesthetic appeal. The production process has the following main steps:

- After cutting, bamboo culm sections are split using a ripsaw into strips of the same width (about 25 mm).
- The strips are planed (rough planing) and straightened to a uniform size with rectangle cross-section (22 mm wide, 6 mm thick).
- The strips are subjected to preservation treatments (high-temperature steaming, bleaching, anti-fungal treatment and borer-proofing).
- The treated strips are dried to suitable moisture content.
- Dried strips are planed for a second time (fine planing).
- Grading is done according to strip color and quality.
- UF adhesive is spread at the inner layers and at both edges of each strip.
- Hot-pressing is carried out in a multi-direction press.
- Sanding is followed by and making tongue-groove on each floor strip.
- Painting, grading and packing are the final steps.

The dimensions of a bamboo floor strip are 90–120 cm length and 9–12 cm width. Bamboo floor has a high strength and hardness, good abrasion resistance and elegance. The price is, however, higher than most wood floors because of low volume use. There are more than one hundred plants in China at present, and the annual production capability of the largest one is approximately 100 000 m^2 and that of the smallest one is over 20 000 m^2.

Bamboo particleboard

This product is made from small-diameter bamboo culms, irregular culms and residues from plybamboo manufacture. Bamboo resource use can be optimized by manufacturing bamboo particleboard, as the low-quality and waste bamboo materials find use in this product. Except particle preparation and PF adhesive, the process of it is similar to that of wood particleboard. It is usually employed in concrete formwork, especially in civil construction, because of its high strength and low swelling ratio compared with wood-based particleboard. To enhance its strength, bamboo mats can be placed on the surfaces or bamboo curtains incorporated. At present, there are four bamboo particleboard plants in China.

Bamboo-wood composite products

As the development of bamboo-based panels gathers pace, the cost of bamboo is increasing. On the other hand, the prices of timber, especially the fast-growing species, are falling. The properties of fast-growing wood species, however, do not meet most of the requirements. Therefore, the prudent way to economically develop high-strength engineering materials would be to take advantage of the desirable aspects both bamboo and wood to form a scientifically designed composite material. At present, a few such bamboo-wood composite products have made their appearance in China.

Bamboo and wood composite container floor

Earlier, container floors were being made only from tropical hardwood because of the need for high MOR (over 75 MPa) and MOE (over 10 000 MPa). To protect tropical forests and the global environment, the Bamboo Engineering Research Center of Nanjing Forestry University, and Techni-con International Development Ltd. Hong Kong, jointly researched and developed a bamboo-wood composite material for container floors. It is composed of Chinese red pine (*Pinus massoniasa* L.) and bamboo sheets (5 mm thick face and back layers) such as plybamboo, bamboo curtain or bamboo mat. The properties of this product meet the requirement for container floors. It is currently being produced in a small scale. Three pine plywood factories are about to start manufacture of this product in China.

Bamboo-wood laminated board

This is a thick (45 mm) outdoor product. Dried and planed thin boards of Chinese red pine are used as the core material, usually in three layers, and softened and flattened bamboo strips or bamboo curtain form the face and back layers as the surfaces. The properties of the product are excellent: MOR is more than 70 MPa and MOE is over 6 000 MPa. It will be used also as floors of trains in place of Douglas fir timber (70 mm thick). Small-scale production of bamboo-wood laminated boards is scheduled to commence soon.

Bamboo-wood composite floor

Chinese fir (*Cunninghamia lanceolata* Hook) is a fast-growing species that had been used to establish several plantations in the south of China. Up to now, the wood of Chinese fir has not been utilized by the industry. In bamboo-wood composite floors, the thin boards of Chinese fir are placed at the core layer and thin (2.5 mm) bamboo strips are placed on the surfaces. The product has a fine exterior similar to bamboo floor, good strength and low weight. The most significant point is that this composite floor greatly reduces the cost. Therefore, the hardwood resource and natural forests of China could be protected.

THE FUTURE OF BAMBOO-BASED PANEL INDUSTRY IN CHINA

The wood resource in China is scant and this situation will continue for a long time. It was this situation that prompted the development of Chinese bamboo-based panel industry. How will it develop further and what needs to be done to ensure a healthy development of the sector?

First of all, existing factories need to be improved and larger companies need to be formed to strengthen resource management and enhance production, because the actual output of bamboo panels is much less than the production capacity. Second, the industry should aim at overseas markets to increase export of bamboo-based panels, such as bamboo floor, bamboo-wood composite floor, high-strength bamboo concrete formwork and plybamboo. Third, it is important to popularize new research results on products such as bamboo-wood composites. Fourth, it is important to attempt at comprehensive utilization and development of products that have a high utilization percentage. The utilization percentage of bamboo resource is below 40%, which is quite low for the health of the industry. One approach is to locate bamboo particleboard factories at strategic points. Plants or companies that produce bamboo-based panels should enhance their capability to develop new products with higher added value. The bamboo-based panel industry will be entering a new stage of development not only in China but also around the world in the coming millennium, and rational management and innovative handling of the sector are necessary to realize its full potential.

Bamboo for Sustainable Development, pp. 639–644
Eds. A. Kumar, I.V. Ramanuja Rao and Ch. Sastry

Modular system for bamboo house construction

ALEJANDRO UGARTE
Design and Technological Innovations Unit, FUNBAMBU, San José, Costa Rica

Abstract—One must use bamboo wisely to derive its best services. This means that its nature, its ways, its being must be respected. In 1987, a bamboo project was established in Costa Rica that later became the Bamboo Foundation. It began with international transfer of technology to which it later added its own experiences, giving birth to an alternative low-cost housing method consisting of prefabricated components. Various technologies were developed to suit different site conditions. All that knowledge together has now helped place bamboo on central stage. This alternative is a "Universal Platform" because it can be used to solve the most diverse applications in different situations. This paper discusses the steps involved in this modular system for the construction of bamboo houses.

INTRODUCTION

When people talk about bamboo, very often views such as "bamboo is a substitute for steel" or "bamboo can substitute steel in this wall" are heard. They speak so in an attempt to raise the status of the bamboo as a material. However, to be frank, bamboo actually is much more than that. Bamboo does not "substitute" anything. Steel cannot be substituted, the same way that bamboo cannot be substituted.

Each material has its own features and traits, with its own personality, its own technology. It is true that bamboo sometimes behaves like steel; but to insist that it remains so is to minimize the possibilities and characteristics of bamboo.

Louis Kahn's phrase is well known: "Let the form be what it wants to be". This seems paradoxical. If one thinks about a form, by virtue of this action alone one shall have already granted it. How then is one going to let it be what it wants to be? The answer is that form "is" already there before being thought of; one must just have the wisdom to unveil it and let it appear as itself. At the sight of architectural masterpieces, one often thinks: "This is exactly what had to be here".

This is the case with bamboo too, which wants to present itself to us, to be worked on and worked with. We must have the wisdom to unveil it, so that it may be revealed in its true form; we are only a facilitator for it to show itself. We must allow it to be itself. For it is a wise plant.

The experience the Bamboo Project and the Foundation have had in research and development all these years has been very rich and alluring. Costa Rica had no real traditions in using bamboo; in fact, the forests were scarce in bamboo.

There were no known practices in building with bamboo. Far from being a problem, this turned out to be beneficial as the work was not bound by any habit or custom that should be corrected, amended or improved. It allowed an adventure without any burdens, ties or advantages, without preconceived ideas or positions to defend.

The response to this situation was a technology for modular construction based on prefabricated components. Using one of these components, namely panels, a specific type of single-floor house can be built. If certain other pieces, such as bamboo columns and trusses, are added to the system it can be a house type that may even have two floors.

The design for the houses was the outcome of the interrelation and interaction of diverse factors such as:

- Physical and mechanical characteristics of the material;
- Adequate spatial dimensions;
- More common plot dimensions; and
- The workability of materials and components.

THE BAMBOO PROJECT

In Costa Rica, there was no native, autochthonous innovation in the field of construction (this is specially true in the case of roof covers: ribbed galvanized iron sheets are used by the rich and the poor alike). The block and slab and column walls existed, but that was all. About 30–40 years ago, more diversity existed in arts and crafts technologies that used wood very widely and concrete. The construction sector had little to choose from.

This is particularly true in the case of low-cost housing (also called "social interest housing"), since it was not possible to lower costs based on the construction method (colones per square meter) to be applied. The path followed was to lower costs by reducing the constructed area (fewer square meters) and materials, with the end result being few colones for few square meters. Let this be called, based on one's viewpoint, "minimum finish house" or "dignified house".

This was the situation in 1987 when the International Year of Housing for the Homeless was proclaimed. In Costa Rica, this was the beginning of a new era in the field of housing: laws were enacted, projects were drawn up, new institutions were created, and new viewpoints, new functions and even new actors emerged in the field of social interest housing.

It was also when the author began to be part of the group that started to work on the subject of construction technology. This group originated from government negotiations with the Royal Government of the Netherlands and the United Nations with the aim of introducing bamboo into construction. Self-help construction, mutual assistance, native and local materials, cost reductions, adequate housing were the themes of talks.

The primary objective was to build social interest houses and the work began with self-help construction and on-site erection. Parallel to this activity, a research and development process was initiated with the purpose of establishing and furthering an integral or holistic construction language.

The trend at the National Bamboo Project (PNB) was towards using bamboo as timber or in round form and in matting. Following the international transfer of technology, the project started by using bamboo columns of approximately 12 cm diameter every 0.45 m, 3–m prefabricated panels formed on site, bamboo beams, and bamboo nail driver boards. This made for a demand of around ninety culms per house and an intensive use of labor. The use of matting was considered impractical for several reasons:

1. The physical effort needed to build matting was very high (labor-intensive) and the labor in the country was particularly expensive;
2. The overall cost of production was high; and
3. Initially, the availability of bamboo was inadequate.

The use of "round" bamboo was not viable either, at the beginning because:

1. Non-availability of the material;
2. The rather insufficiently evolved preservation and drying techniques available at the time; and
3. The difficulty and complexity of joint technology, which would have made the labor input onerous for massive construction work.

In solving these issues, and to give an economical response to the country's need for housing, PNB evolved through several stages of technology.

First stage

To initiate the processes of mass construction, PNB resorted to the on site use of timber, bamboo matting and the wild reed (*Caña brava*). A "balloon frame" structure was placed on a concrete floor slab. Then the matting or wild reed was nailed manually onto this structure. The truss was made of wood.

Second stage

The research process came up with a hybrid between the prefabricated round bamboo panels and matting, and other materials most readily at hand (wood or wild reed). After the wood panels and wild reed were tested and proven theoretically and practically, the "on site" construction technology was substituted with prefabrication. However, the practice did prevail of placing wild reed manually with hammer and nails. Panels of 0.90 m were built from 5 × 7.5 cm frame pieces and the wild reed nailed in place on site before mounting the panel frame on the concrete slab.

Third stage

The practice of placing the balloon frame on site was abandoned soon and prefabricated panels started to be built in a small rudimentary facility, which gradually was improved. For these panels, 5 × 5 cm pieces were used, as a structural analysis showed that the 5 × 7.5 cm pieces are not necessary. Prefabrication was necessitated as a result of the 1991 earthquake, which called for a much faster assembly to house people rendered homeless by the calamity. Foundations were improved so that panels could be raised farther above the ground to protect them from humidity. The design of the panel, however, was still very rigid.

Fourth stage

After this panel proved to be effective and economical, the lack of housing problem was solved for a large number of families (around 1 000 families). Significant progress – such as integration of crews, including hired labor, and improvement of the characteristics of foundations – was achieved in the construction processes.

Fifth stage

The following step consisted of eliminating the panel bracings, thanks to structural tests that demonstrating that each plaster layer functioned like a rigid diaphragm. Also, the panel measurements were made more flexible, thus achieving a greater variety of housing types.

Sixth stage

When the volume of weekly panel production exceeded the installed capacity of the small panel factory, and it became impossible to expand the facility, it became necessary to improve staff efficiency to increase output. The technology implemented was an adaptation of nails and staples driven in with nail/staple guns (the installed capacity at the time, November 1995, was 80 houses per week). This also caused the unit cost to drop significantly. It became necessary to change the handsaws for machine saws, which originally were being used for something else but were reconstructed for new functions. A modular design method was then prepared and dimensions standardized to 0.15 m increments to obtain panels with a large diversity of measurements ranging from 0.30 m to 3.30 m.

Seventh stage

Later, a panel was developed which does not use wood on the inside of the panel, but only on its periphery or rims to ensure that it stays together, carrying a kind of double lining of bamboo strips and wild reed. This panel allowed: the use of gunite (sand and cement mortar) spraying by mechanical means (also with compressed air), thus achieving significant savings in terms of time and money during plastering; and the placement of a soundproofing barrier in the panel to disallow the transmission of sound from one panel face to the other (transmission depends on the material used as noise barrier). The 0.15 m increment was changed to the 0.1 m, as recommended by the International Organization for Standardization (ISO), starting to use the ISO standardization concepts. The panel was also then designed with a 0.1 m thickness. Duly treated (for preservation) and dried beams were introduced, using screws to fix and anchor them to one another.

Eighth stage

The introduction of round bamboo allowed the integration of all parts. A new technology developed at this time used the best physical and structural characteristics of each one of the building components (elements) created by FUNBAMBU (panels, strips, round bamboo), so as to offer optimum advantage to the building. This system, called a "Universal Platform", is for constructions that may be up to two floors. The holding structure, the shell structure, the second floor structure, and the panels for the walls are all prefabricated. These elements make up a harmonious blend by not only providing a specific housing unit, but also giving scope for solving any other construction-related issue within the ambit of FUNBAMBU's work. It virtually offers a means that work with prefabricated elements to erect different types of constructions.

The system is a combination of elements devoted to receive the gravitational load and lateral stresses. A revolutionary concept is used for the joints between bamboo pieces. The bamboo receives a piece of wood perpendicular to its longitudinal axis, to which any other element of the system – trusses of bamboo laths, other bamboo culm beams or girders, or a prefabricated panel – can be jointed. Pieces that reach this "platform" can be parallel, perpendicular or at an angle to the bamboo beam axis, depending on the designer's objective.

In this system, the structural pieces – such as panels or trusses – achieve minimum dimensions owing to their specialized function; their work in the set or group is limited to doing a single task well. One can find bamboo carrying in a similar role in its natural environment, standing up, alone, erect, and parallel to its fellow bamboos and working well to the welfare of all.

In must be mentioned here that these types of joints have already been tried out in other places. However, they have been isolated approaches without any major possibility of having impact on integrated construction systems, nor have they received due continuity in development.

INTEGRATED DEVELOPMENT IN FUNBAMBU

While the pre-assembled construction elements of FUNBAMBU are very important, these are really one more activity or stage in a chain of processes that have to be carried out to successfully assemble a building.

The development of the Universal Platform technology was guided from the beginning by a strong will and a belief that all prefabrication processes have common elements, without which a failure is certain. Bamboo cannot ignore this knowledge, which can be summarized as follows:

1. Standardization of the parts and processes;
2. Common language for all individuals involved;
3. Finite number of prefabricated elements; and
4. Establishment of and respect for "conventions", which are arbitrary initially but with practice are amended and improved in consultation with all individuals (for example, nomenclature, sequence of the processes, sequence of assembly, laws of construction of the form, etc.).

The prefabrication plants, the storage systems, transportation and handling shall be included in this "coordination". A major conclusion arrived at is that coordination among the designer, parts producer, and those in charge of storage, transport, handling and final assembly is the most important aspect, rather than just the coordination of the modular design from the drawing table of the Design Department.

Modular design coordination is not an activity that is carried out in isolation in an office, but rather an attitude that pervades through the entire process and in which all actors play a role to accept the system as theirs.

The panel production, on-site assembly and work control have also evolved quality control procedures. Manuals have been written and are used widely for quality control of the raw materials and finished products, as well as processes (work with pneumatic tools, stowage and loading, reception of raw material, loading and unloading of trucks, casting of foundation plates, plasters, construction of trusses, monitoring of work progress, control of payments, budget calculations, etc.)

The philosophy of the FUNBAMBU has been, at all times, the ease of use of the systems for all actors, use of practical conventions, adaptation of the systems to reality, and acceptance of new conventions and changes to the systems. In brief, accept and even promote changes to the paradigms established. This has brought about an open-ended technology in continuous evolution, easy to grasp by the users, who become involved and adopt the system as their own, thus making easier their work in building their own houses.

ACKNOWLEDGEMENTS

This is the result of 10 years of teamwork in research and the author wishes to express his gratitude to all his colleagues, who have been in charge of coming to terms with the realities concerning the ideas herein expressed. Without them, it would have been impossible to accomplish what has been described here; without their wisdom there would not have been any progress in the work. The author also expresses his appreciation to his friend John D. Cauty who, in recent months, has contributed in an outstanding way to the development of various ideas.

Bamboo for Sustainable Development, pp. 645–652
Eds. A. Kumar, I.V. Ramanuja Rao and Ch. Sastry

Bamboo: A viable alternative for popular housing

RUBENS CARDOSO JÚNIOR, EDSON DE MELO SARTORI, FÁBIO VERÍSSIMO GONÇALVES and SUZANA MARA FERNANDES
Universidade para o Desenvolvimento do Estado e da Região do Pantanal, Mato Grasso de Sul, Brazil

Abstract—This paper presents a two-year research project, in the process of development, on the use of bamboo in large structures and in the civil construction. The aim of the project is to construct a house of 54 m^2 in which bamboo is used as a structural element and paneling material. It also aims to study the feasibility of pre-fabricated ceiling made of concrete beams, reinforced with bamboo strips made into cables or bamboo lath grid. The finishing of the wall panels will be carried out by the application of mud as plaster over bamboo mats, followed by whitewash.

INTRODUCTION

The search for environment-friendly materials and technologies has become the focus of many researchers. It can be safely assumed that this period of industrial growth, in which the search for new materials of affordable costs has been renewed, would result in technological options in the construction of housing for a growing population that, in certain cases, lack access to resources.

In the areas of engineering, architecture and civil construction, there has been a constant endeavor to study materials and construction techniques that focus on reducing the cost of housing. One can choose bamboo as a potentially material in this category, for possible use on a wide scale, owing to some remarkable characteristics of the material:

- It is available in a wide variety of species, extremely adaptive to several soil types and climates;
- It requires much less energy for production than the several other materials commonly used in civil construction (Ghavami 1992); and
- It is a super-productive material, easy to propagate and manage.

This study presents results of the research developed by the authors, from March 1996 to the present, studying and characterizing some bamboo species found in Mato Grosso do Sul State, as well as some traditional technologies developed by the natives of Pantanal (a swampy region) in the use of this material.

PHYSICAL AND MECHANICAL PROPERTIES OF THE BAMBOOS USED

Tables 1 and 2 present some characteristics of relevance to the use of bamboo as a construction material, resulting from tests carried out in our laboratories. There were four species chosen for the study, based on their availability, size and extent of use. They are:

- *Dendrocalamus giganteus* – giant or pale bamboo;
- *Bambusa vulgaris* Schrad – green or common bamboo;
- *Bambusa vulgaris* var. *vittata* – imperial or yellow bamboo; and
- *Bambusa guadua* – Taquarussu.

Table 1.
Physical properties of the bamboos used in the study

Species	Color	No. of nodes	Diameter (cm)			Thickness (cm)			Weight	
			Base	Middle	Top	Base	Middle	Top	Culm (kg)	Sp. weight (kgf/m^3)
D. giganteus	Dark green	20	19.45	17.20	14.94	3.45	2.46	1.47	82.20	886
B. vulgaris	Green	29	9.40	7.90	6.40	1.90	1.40	0.80	24.75	84
B. vulgaris var. *vittata*	Yellow with green stripes	30	7.70	6.00	4.30	1.60	1.10	0.60	68.30	652
B. guadua	Green with White rings	28	12.20	8.50	4.70	2.40	1.55	0.70	30.60	1000

Table 2.
Main mechanical properties of the bamboos used

Species	Tensile strength (MPa)	Compression strength (MPa)	Bending strength (MPa)
D. giganteus	138.10	64.80	125.10
B. vulgaris	169.00	41.60	136.00
B. vulgaris Var. *vittata*	128.80	38.90	110.50
B. guadua	132.40	49.90	102.00

PRESERVATION OF THE BAMBOO USED IN CONSTRUCTION

To use bamboo as a reliable construction material, care should be taken to maintain its integrity and extend its durability. For this, some simple prescriptions need to be followed. “Bamboo for sale is cut at any time of the year; but for own use, the phase of the moon is observed,” goes a Colombian popular saying (Salgado 1994). Culms cut during the waning of moon have smaller amount of sap and hence, they are less susceptible to attacks by insects. There are also some other traditional practices such as not allowing the cut culm to remain in the stand for some time with the branches and foliage intact, and immersing cut culms in water to leach out the soluble starch. While these practices may have some merit, the use of chemical preservatives is the one adapted mostly for treating bamboo used in construction.

Copper sulphate salt is one chemical that is employed in preservation as its solution has anti-fungal properties. Sodium bicarbonate and boric acid in solution ward off insects, such as the borer (Hidalgo 1981). The treatment can be accomplished by immersing the culm in the preservative solution for a period of time, or using the process called Boucherie method (or its modified version), which consists of displacing the sap with the preservative solution maintained under pressure.

PROJECT DESCRIPTION

The basic objective of this project is to determine, specify, quantify and develop technical conditions for the construction of a popular or rural low-cost house, which can be replicated, with their pattern and finish being similar to existent habitations of conventional construction.

The architectural plan was for a residence with two rooms, a bathroom, a kitchen, a living room and a balcony, with a total built-up area of 54 m^2. The construction process and finish are described below (Fig. 1).

Foundation

Executed using soil-cement blocks measuring 30 × 30 cm to a height of 50 cm, the foundation supports bamboo pillars that rise up to ceiling height. A bamboo lath mat is set at a distance of 5 cm from the ground. The foundation wall is made of bamboo culm sections of 12 cm diameter, placed longitudinally on a bracket of 20 cm width and 15 cm height, with a mat of bamboo laths at the bottom.

Superstructure

The superstructure is made of bamboo pillars with a diameter of about 12 cm. The top portion of the pillars is kept in position using a bamboo belt tied to the pillars using No. 14 galvanized wire.

Wall panelling

It is made of five basic types of modular panels, which can be plain, with a window or with a door (Fig 2). These pre-fabricated panels are 1 m or 1.5 m wide and 2.70 m high. They have an inner core of bamboo culms with 6 cm diameter, with lateral covering of *esterilla* (plates of bamboo made by longitudinally cutting open the culm and flattening it). After mounting, the panels are covered with a 2-cm thick layer of soil-cement plaster on both sides, levelled and smoothened (Fig. 3).

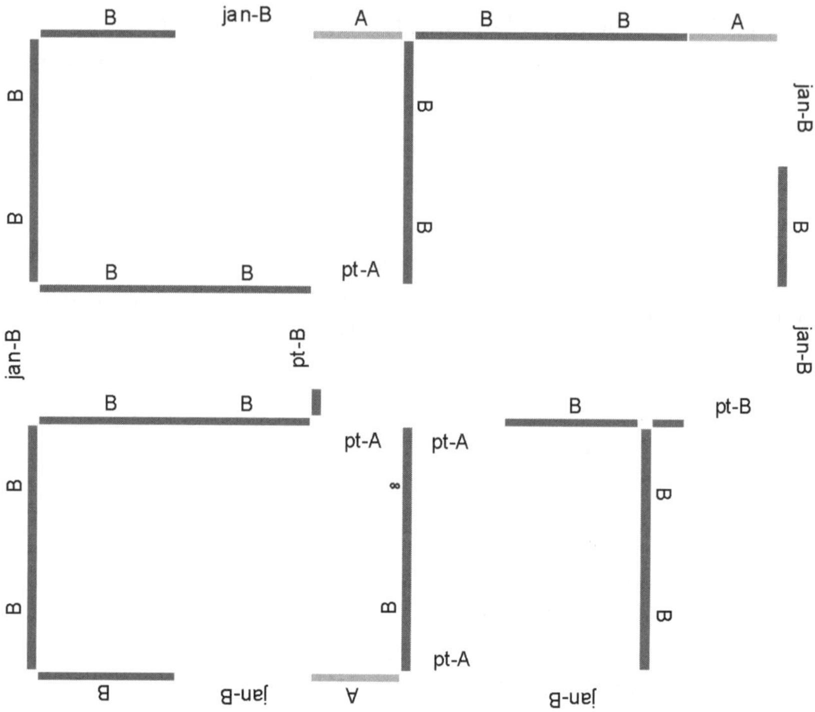

Figure 1. Floor plan of the house

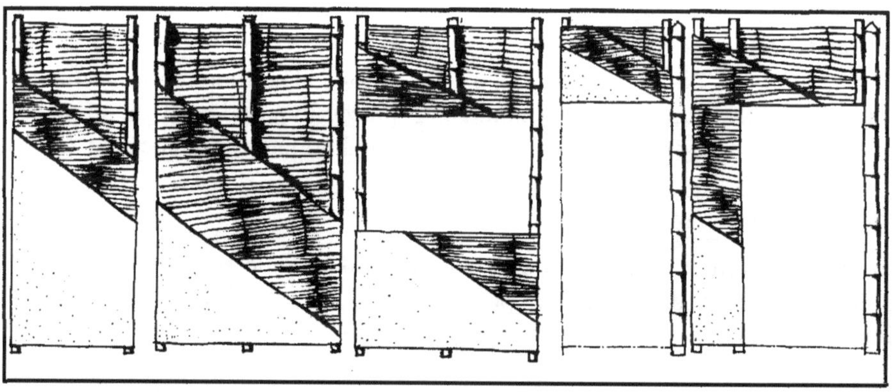

Figure 2. Basic types of modular wall panels

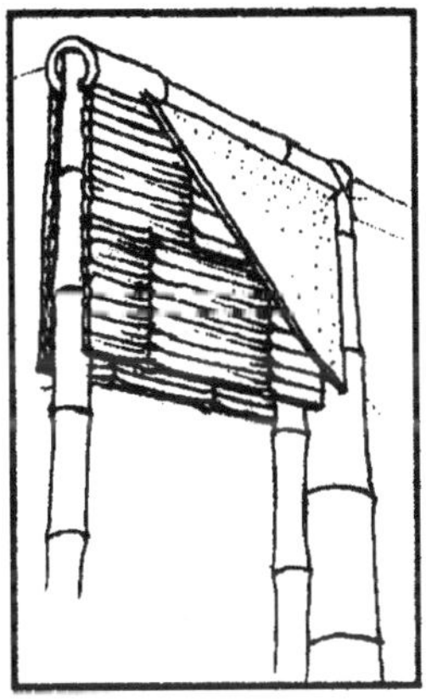

Figure 3. Section of the mounted panel ready for plastering

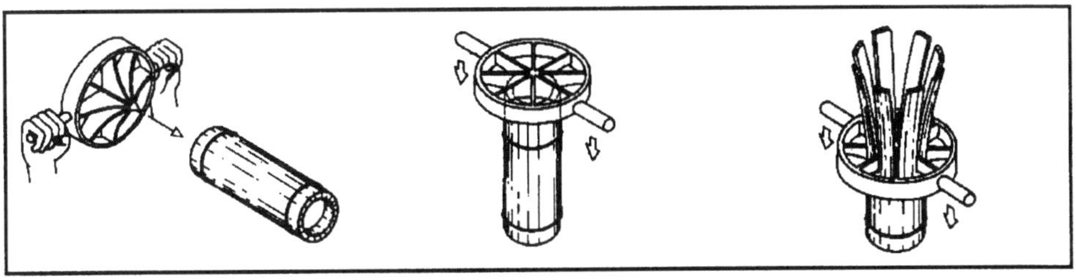

Figure 4. Making laths by cutting a culm with a multiple knife

Roofing

The roof structure is made of bamboo trellis; bamboo culms of 6 cm or 12 cm diameters are used for tiers, rafters and laths also. Iron bolts and nuts with washer are used for structural joints, while galvanized wire is used for secondary joints. Fibrocement tiles of 5 mm thickness will be the material for roofing shingles. The inner ceiling will be made of parallel bamboo laths made by cutting bamboo culms longitudinally into strips (Fig. 4).

Floor

The floor will be made of hand-ground soil-cement mixture, spread, levelled, smoothened, and painted or colored.

Doors and windows

Sliding or awning iron windows will be used in the living room, bedrooms and kitchen, while bathroom will have casement window. The windows will be fixed on to the panels with screws. The doors, as well as headers and studs, will be made of wood and painted over. The hardware (handles, locks, etc.) will be screwed on to the panels. The glass used will have a thickness of 4 mm.

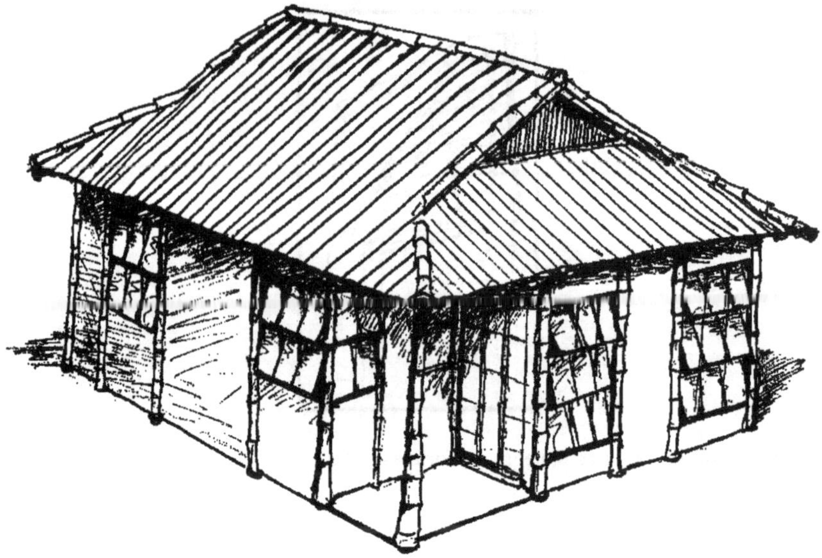

Figure 5. Appearance of a completed house

Electric and plumbing installations

Both plumbing and electrical installations will be on the surface of the wall panels. This will facilitate the work and avoid damage to the panels, in case they need repair.

Painting

Water-based (PVA) paint will be used for the panels, while oil-based paint will be used for doors and windows. A coat of varnish will be applied on bamboo pillars, for protection from sun, and the ceiling.

Fig. 5 gives an idea of how the finished house will look.

BUDGET ESTIMATE

The costs of conventional construction materials were estimated in Campo Grande, the capital of Maro Grosso de Sul. Variations in prices occur owing to differences in transportation costs and other reasons.

Labor costs were calculated taking into account indirect costs, such as taxes and social security expenses. The construction costs will be lower if a larger number of units are built at the same time, using the "mutirão" construction system (a self-help system where the beneficiaries assist in the construction under technical supervision).

The price of bamboo includes the cost of work involved in the selection, cutting and transportation of the material from farms or ranches in or near the city of Campo Grande. There is no regular trade of bamboo in Brazil.

The following budget estimate is not descriptive, and is used only as an approximation, although it takes into account all items involved.

Table 3.
Costs of various components

Component	Cost (US$)	Percentage of total cost
Preliminaries (preparation and demarcation of land)	366.56	6.80
Foundation, structures and panelling	1 023.12	18.90
Covering (structures and roofing)	276.08	11.95
Floor	646.12	5.10
Plastering of walls (render and set)	742.40	13.70
Doors and windows	941.42	17.45
Facilities (plumbing, electrical, etc.)	696.00	12.90
Painting	713.40	13.20
Total	5 405.10	100.00

Table 4.
Comparison of material and labor costs

Item	Cost (US$)	Percentage
Material	3 025.28	56.00
Labor	2 380.32	44.00
Total	5 405.10	100.00

CONCLUSION

The task of building a low-cost housing unit, using alternative materials and taking advantage of available traditional technologies, necessarily involves the search for information. The central concept of this project is to get reliable data on the efficient and durable use of bamboo for building houses for low-income people and the population of Pantanal region, where the material is readily available. The use of bamboo applying the technological process outlined will ensure quality housing at low costs for the people. Furthermore, this might induce professionals in the fields of engineering and architecture to take advantage of the remarkable characteristics of this woody species.

REFERENCES

Ghavami. K. 1992. Revista do Instituto de engenharia. No. 192, Engenho Editora Técnica Ltda., São Paulo, 13–27.

Salgado, A.L.B. 1994. Instruções técnicas sobre o bambu. Boletim Tecnico n. 143, Institute Agronômico, Campinas, Brazil.

Hidalgo, O. 1981. Manual de construccion com bambu. Estudios Tecnicos Colombianos Ltda., Universidade Nacional de Colombia, Bogotá, Colombia.

FURTHER READING

Dedeca, D.M. 1958. A identificação dos bambus. O Agronômico, Vol. 10, No. 9, 8–14.

Dunkelberg, K. 1985. Bambus als bastoff. Karl Kriimer, Stuttgart, Germany.

Engel, H. 1981. Sistemas estrutnrais. Hemus Editora, São Paulo, Brazil.

Farrelly, D. 1984. The book of bamboo. Sierra Club Books, San Francisco, USA. 332 pp.

Graça, V.L. 1988. Bambu: técnicas para o cultivo e suas aplicações. Ícone Editora, São Paulo, Brazil.

Ghavami, K. 1990. Application of bamboo as a low-cost construction material. *In* Ramanuja Rao, I.V.; Gnanaharan, R; Sastry, C.B., ed., Bamboos: current research. Proceedings of the International Bamboo Workshop, Cochin, India, 14–18 November 1988. Kerala Forest Research Institute, Kerala, India; International Development Research Centre, Ottawa, Canada. pp. 270–279.

Ghavami, K. et al. 1995. Propriedades dos bambus e suas aplicações nas obras de engenharia, arquitetura e desenho industrial. CTC/PUC-RIO, Rio de Janeiro, Brazil.

Hidalgo, O. 1974. Bambu, su cultivo y aplicaciones. *In* Fabricraeion de papel, construccion, arquitetura, ingenieria e artesania. Estudios Tecnicos Colombianos Ltda., Cali, Colombia.

Jena, P.K., ed. 1997. Proceedings of the 2nd International Conference on Non-Conventional Building Materials (NOCMAT-97), Bhubaneswar, India, 17–19 June 1997. Natural Resources Development Foundation, Bhubaneswar, India.

Villegas, M. 1989. *Bambusa guadua.* Villegas Editores, Japão. 175 pp. (Published in English as Tropical Bamboo, Rizzoli, New York, USA, 1990).

Bamboo for Sustainable Development, pp. 653–665
Eds. A. Kumar, I.V. Ramanuja Rao and Ch. Sastry

Methodologies and processes for the construction of bamboo houses

FRANCISCO RODRÍGUEZ RODRÍGUEZ
Construction Department, Bamboo Foundation, San José, Costa Rica

Abstract—Based on a decade of experience in building houses combining bamboo, wood and wild reed (*caña brava*), this paper describes the organization of the work to carry out housing projects, utilizing labor of the beneficiary families together with hired labor. It also explains the different stages involved in the construction of one of these houses. Variations that the Bamboo Foundation is introducing in construction technologies to meet the challenge of building 1 000 houses annually, substituting bamboo for wood, are briefly explained.

INTRODUCTION

Traditionally, most houses in Costa Rica were built with wood, adobe and *bajareque* (two types of walls made out of earth, mud, dung and woody matter). Later, baked clay bricks appeared to substitute and supplement these construction systems. Presently, most houses are built with concrete blocks and, of late, prefabricated concrete slabs.

In 1988, the National Bamboo Project (PNB) started to build houses combining wood, *caña brava* (wild reed) and bamboo, as another alternative to help solve the increasing housing deficit in the country. By mid-1995, PNB had built a total of 1 490 houses in several regions of the country. In July 1995, the PNB was transformed into the Bamboo Foundation (FUNBAMBU) and has built 1,500 houses since then.

METHODOLOGIES

Several systems of labor organization have been used for the construction of bamboo houses. The first used, and described below, was self-and-mutual help construction, whereby one individual or family joins other individuals or families to build their houses with their own hands.

Generally, typical housing projects would be 20 houses, located preferably not too dispersed or far apart, within a radius of less than 2 kilometers.

One permanent (full-time) construction technician is assigned for each group of 20 houses. This technician is in charge of training and organizing the families into construction teams for their own houses, and to manage and control the materials storerooms, tools and machinery. A civil engineer or architect, as job director, supplements this fieldwork and a social worker is in charge of motivation and good relations with the beneficiaries.

For any given typical project, beneficiaries are organized into four crews of five persons each, and each crew would build five houses. A coordinator, appointed for each crew, is in charge of transmitting instructions given by the construction technician or the job director.

A housing committee is appointed for each project to serve as liaison between the beneficiaries, the technical staff and local government agencies. It functions are: overview the construction work and, together with the technical advisors, pursue any assistance needed from local governments and associations.

The construction process is gradual and simultaneous for all the five houses being built by each crew. First, the five foundations are laid, the walls of all the five units are then completed, and so on until all the houses are completed.

Training of these amateur builders is carried out following the principle of "learning through doing". These concept and practice are reinforced with simple, easy-to-understand printed materials – such as booklets and pamphlets – that explain in detail the different construction stages.

A typical project involving the aforementioned work organization takes approximately 8.5 months.

With funds from bonds of the National Housing Finance System coming to PNB, and further on to the Bamboo Foundation, completion of these houses was expected to take no more than 4 months. As it was necessary to build more housing units without significantly increasing the field inspection labor and within the 4-month time limit, a reduction in the use of self-help building was effected and hired labor incorporated. Thus, today, the labor composition to build these houses is 20% self-construction labor and 80% hired labor.

Housing bonds are part of the government financing policy to endow a dwelling unit to each poor family. This funding program is aimed at families with incomes that are less than four times the minimum wage of an unskilled construction worker. The financing package is made up of a free bond or non-reimbursable grant (equivalent to 30 minimum wages of an unskilled construction worker) and a loan, the amount of which depends on the family's ability to pay. Both grant and loan amounts vary according to total family incomes: the higher the income, the lower the free bond and the higher the loan base or principal. The loan base is paid off in monthly installments, over a twelve-year period at variable interest.

Inspection and wageworker training plays a vital role in preparing the new workforce composition within the new organization, and this is being applied right through till the present.

Housing projects under this new work organization are for 100 to 150 houses. An engineer or an architect serves as the job director, assisted by two or three construction technicians, and two or three warehouse keepers are employed for two or three work sites.

The work performed by the construction technicians and the engineer is mainly administration and inspection, rather than training.

Families who are the recipients of a housing bond have the duty to perform on a self-help basis the following:

- Clear and level the land of the site where the house will be located;
- Dig ditches and make trusses and frames for foundations and sidewalks;
- Dig holes and ditches for septic tank and drainage system; and
- Painting.

Construction technicians provide the training for these activities. The contractor performs the remaining construction work.

About 100–150 houses are apportioned to a variable number of contractors. According to accumulated house building experience in the rural areas, the number of houses per contractor should not exceed five. In hiring wageworkers, the Foundation prefers local people. The practice of gradual and concurrent (or stepwise) progress is adopted for all houses built at the same time by each contractor. The average completion time for a project of 100–150 houses is 4 months, of which 1 month is taken up by self-building tasks.

THE CONSTRUCTION PROCESS

The construction process reflects all the concepts that need to be followed at the time of engaging in construction of houses of the Bamboo Foundation (Fig. 1). A description of the activities during the construction process of a typical house, in the sequence of execution out in the field, is given below.

Preliminary activities

The activities or sub-activities involved in this stage are closely interrelated and have as their starting point the land on which construction will take place and its specific features.

Land clearing

This consists of removing from the land obstacles – such as stubbles, underbrush, shrubs, roots and rocks – that in one way or the other impair the execution of work. Nests of insects, such as termites, that damage wood and bamboo are also eliminated from the site.

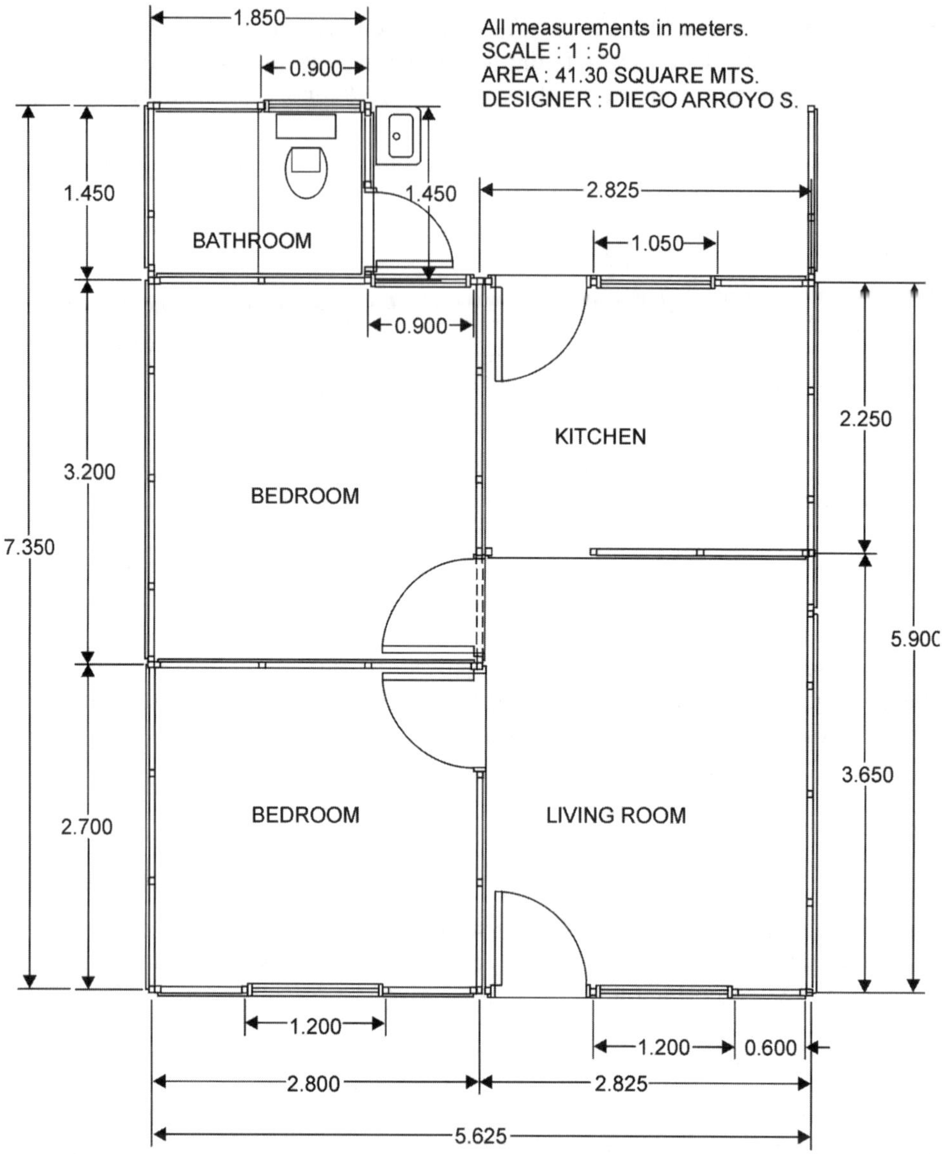

Figure 1. An architectural drawing of a 41 m^2 house designed for bamboo foundation

Location of plot and house

The exact demarcation of the plot will give the area where the house is to be laid out. References on land plan and the existing boundaries are obtained from the concerned offices.

After the plot is defined, the house location is established on the land, as per the wishes and tastes of beneficiaries while respecting technical criteria, such as spacing from neighboring property, streets, waterways, right-of-ways, etc.

Land preparation
After the land is cleared and the housing plot defined, the area where the house is going to be built is levelled. Through shallow cuts and fillings, a flat and even surface is prepared to initiate the layout of the house. To avoid the land settling or caving in, landfills must not be more than 20 cm.

Temporary facilities (storage shack)
This sub-activity consists of building a temporary storeroom in a strategic place of the plot to store materials that are to be used in building the house. The main criteria for the shed are that it should ensure that the materials do not get in the way of construction activities and they are protected from the elements. Ideally it could be placed in such a way that when the beneficiaries start living in the house, the shed could be adapted as a storeroom.

Layout
This consists of transferring the construction blueprint measurements onto the land. With cords on yokes (provisional wooden frames built with wooden bars to make alignments and levels), the plot is laid out in terms of location of rooms, and measurements and levels for foundations and walls.

Foundations

Laying the foundations is one of the important activities in construction process, as the house walls will rest on these foundations. There are five main steps involved.

Ditch digging
The land must be dug deep enough to ensure that, with the designed foundations, base soil will be able to bear the weight of the house without experiencing settling or cave-in that may damage the house. In the case of lightweight houses, such as those made of bamboo, it will be enough to remove the topsoil, provided expansive clays or landfills are not present, to define the depth of the ditches. Ditch width is according to blueprint measurements, offset onto the site during foundation layout.

Making trusses
Design of trusses is based on technical criteria that consider loads to which the soil will be submitted, soil strength and seismic characteristics of the area. The most commonly used truss is the triangular truss or Warren truss. This has three longitudinal No. 3 steel rods, and rings made of No. 2 steel rod placed every 25 cm with hooks made of No. 2 steel rod that later will serve as anchor pins for the lower wall soleplate.

Placing trusses
Placement and fitting of trusses in ditches are done leaving the appropriate spaces between the steel and the soil to avoid steel oxidation from insufficient concrete embedding. Also, when two truss frames are joined, either in the same direction or perpendicular to each other, the overlap of the longitudinal rods must be adequate to eliminate any possible separation of the trusses.

Pouring continuous base plate
This consists of laying a 22-cm thick layer of concrete to cover the foundation truss. The concrete consists of cement and good quality aggregates to a volume ratio of 1:2:3 (cement, sand and rock). During its preparation, sufficient water is added to obtain a pasty and homogenous mixture. In case no machine (hand-driven or motorized mixer) is available to mix, quality control during preparation of the mixture needs to be thorough and meticulous, particularly in the process of folding in the aggregates and the cement to obtain a homogeneous mixture uncontaminated by soil. This procedure help obtain a concrete that has a minimum compression strength of 210 kg/cm^2 after 28 days.

Placement of concrete blocks
This consists of placing two rows (courses) of concrete blocks over the foundation. Block dimensions are: 20 cm (h) × 40 cm (w) × 12 cm (d). The bond between the blocks and the foundation is made with a layer of mortar. The exact and correct prefabricated panel assembly erection will depend upon the vertical and longitudinal levels and alignment of blocks during their placement. All block cells are filled with concrete of the same quality as the one used for the foundation. These two courses of blocks provide a barrier against humidity and termites so that it is not easy for them reach to the lower wall soleplate or the rest of the house. Fig. 2 shows foundation/ground beam construction details.

Wall structure

Before introducing the use of prefabricated panels, the entire construction process used to take place at the house site. First, the wooden skeleton or main frame would be erected, followed by lining with bamboo or wild reed (*caña brava*) matting for plastering thereafter. Today, with prefabricated panels, made in factories specifically set up for this task, execution is greatly eased by just erecting and fastening the factory-made panels in place.

Making and lining wooden panels
The development of a prefabrication system to make wall panels for houses has been one of the most valuable experiences for FUNBAMBU. The advantages this offers, as compared with the on-site construction system, are:

- Different panel types that compose the walls of a house are produced on a production line basis using wood or metal.

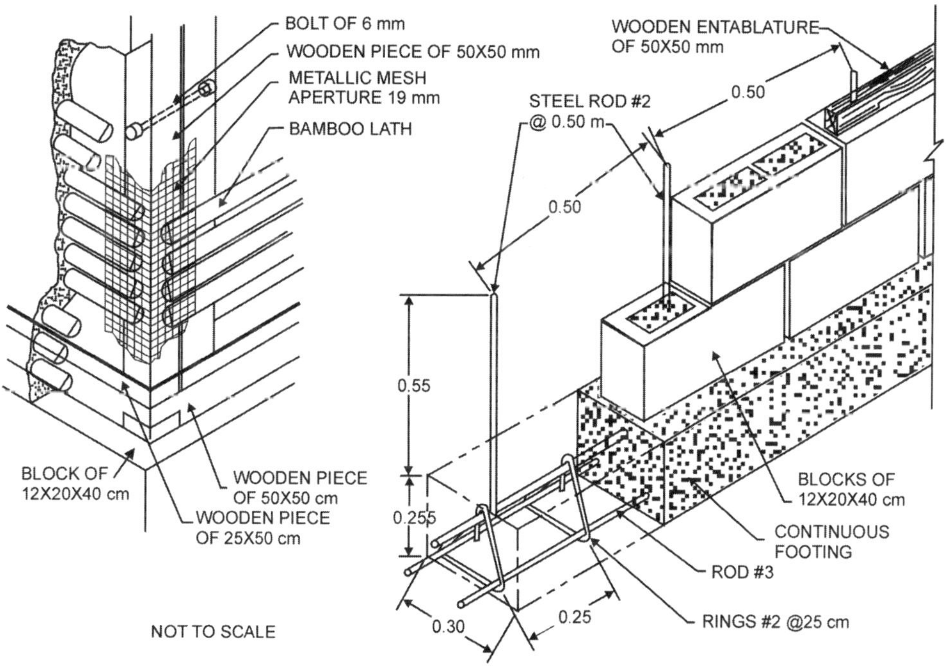

Figure 2. Details of corner panel joint (left) and continuous footing (right)

- Labor output is increased and consequently execution costs and times for this activity are reduced.
- The quality control process is made easy and resource optimization improved, both in materials and labor, as the production process takes place in a factory where tools, machines, materials and labor are exclusively devoted to this activity alone.

The process, in brief, consists of making the wooden structure or frame – composed of a lower wall soleplate, 4–5 verticals and a top wall soleplate – within wooden forms that ensure the dimensions and the squaring (right angles) of the component pieces. The vertical pieces are 5 × 5 cm in their cross section, while the soleplates are 2.5 × 5 cm, joined together with corrugated nails 75 mm long. Once the wooden skeleton is made, one of the faces is lined with wild reed or bamboo laths. The lining is fixed to the structure with nails and No. 20 galvanized wire or clamps. Care is taken to leave openings of at least 1.5 cm between wild canes or bamboo laths to allow plastering mortar penetration.

To build the walls of a 41 m^2 floor area type house (see Fig. 1) 16 panels, each of 2.4 m, height are needed.

Placing lower wall soleplate

This consists of placing wood pieces of 5 × 5 cm cross-section over the top row of blocks. These pieces are fixed or anchored to the blocks with No. 2 steel rod hooks coming from the foundations and placed at a distance of 50 cm from each other. The soleplate receives the panels and for this reason, its placement must conform exactly to the wall axis, so that at the time wall panels are received and set in place there will be no misalignment that may complicate the assembly.

Panel installation

The procedure developed consists of placing and fixing the panels to the lower soleplate and to each other with nails and screws (Fig. 3). After placing the panels in the order specified by the construction blueprints, they are aligned, straightened

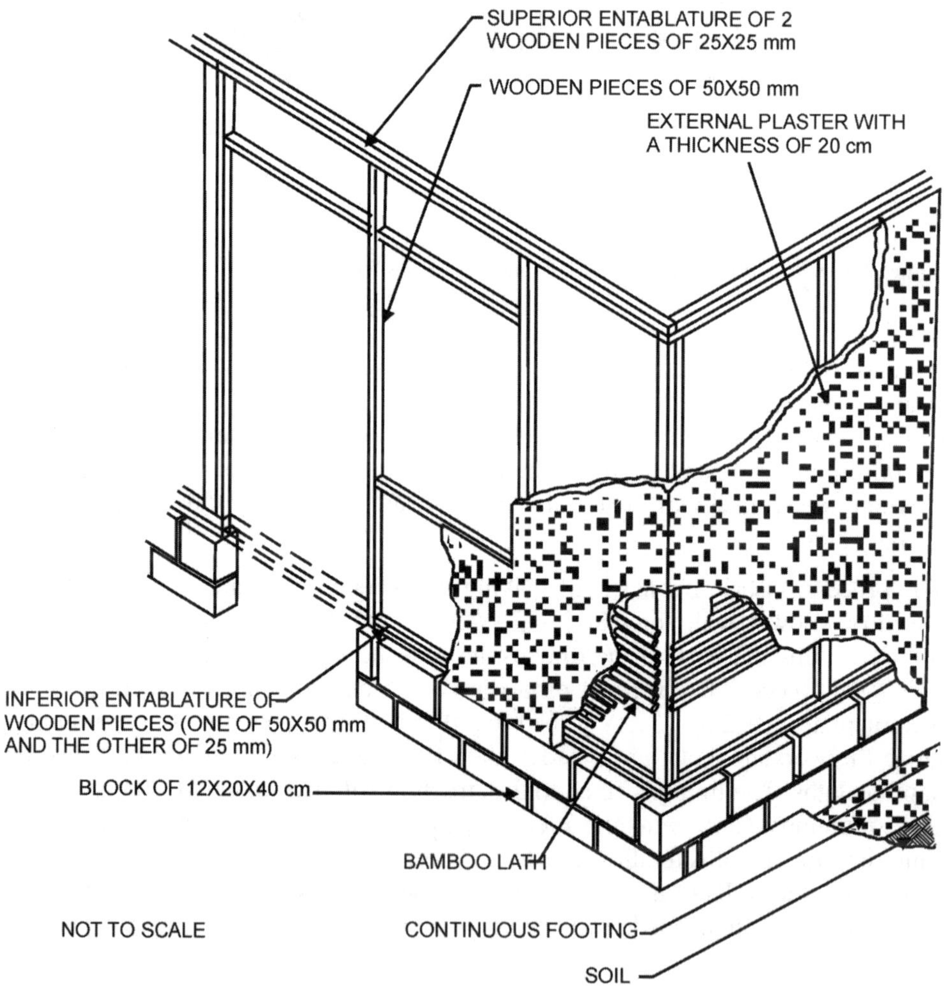

Figure 3. Detail of a typical wall structure

to plumb line and nailed to the lower soleplate, and then fixed onto each other with automobile-body bolts of 6 mm diameter and 10–20 cm long. After the vertical pieces of paneling at the ends are bolted, metal mesh screen strips are attached (see Fig. 2). This screen helps avoid cracking or seaming of the plaster at the panel joints.

Roof structure

The roof structure basically consists of conventional wooden truss beams and stringers (wooden beams with 5.0 × 7.5 cm cross section), placed on the walls at every 90 cm, and nailing rails (purlins with 2.5 × 5.0 cm cross-section) placed on them at 70 cm intervals. These nailing rails receive the nails that fix the galvanized iron sheet roof (Fig. 4).

Truss beams are made out of wood rails with 2.5 × 5.0 cm cross-section. The roofing sheet is No. 28 gauge galvanized iron. Both, truss beams and stringers are fixed to the upper or top soleplate with No. 2 steel rod hooks.

Wall plastering

Wall plastering consists of placing three layers or coatings of mortar – called plaster, splash and finish – on the bamboo lath (or wild reed) lining. Plaster is the first layer applied, by dabbing and spreading the mortar with a trowel or wooden smoothing plane onto the bamboo sticks or wild cane. After hardening, a second thinner and rough layer of mortar is splashed onto the plaster. Two days later, the third and last layer of mixture (finish) is applied, aided by No. 3 steel rod guides, over the splashed layer using a trowel and spreader over until a uniform surface is

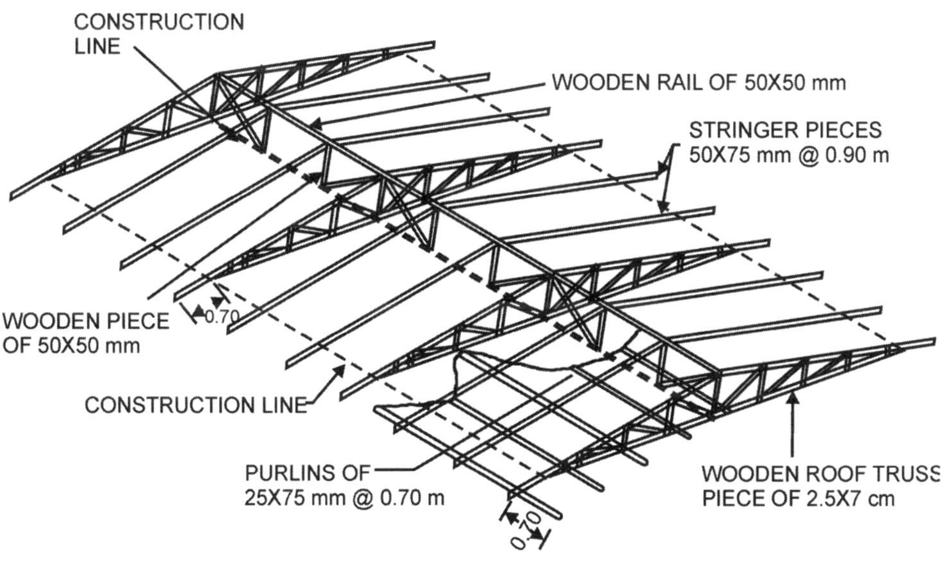

Figure 4. Structural detail of a typical roof

achieved. The surface is further smoothened with a wooden plane. Non-lined wall sides get only two layers of mortar – splash and finish.

The plaster mortar is prepared on a sand-cement ratio of 1:3 by volume for the first two layers and a ratio of 1:4 for the last finishing layer. To obtain good quality, the plastering needs to be wetted with water for setting at least twice every day for the first seven days.

Mechanical installations and fixtures

These installations comprise placement of plastic (PVC) pipes, drinking water fixtures, and wastewater and sewage pipes. Pipe diameters vary from 12 mm (for drinking water) to 50 mm and 100 mm for wastewater and sewage. These installations are carried out before flooring is done. Provisions with fittings are made for later installation of lavatory, toilet, shower and sink.

Floors

Flooring comprises three sub-activity components, carried out after the plastering and pipe installation are finished.

Floor filling and levelling
For filling, a 10-cm thick layer of gravel or pebbles is laid over firmed up soil and compacted. Levelling is then carried out until a smooth, uniform surface is achieved.

Floor pouring
With concrete or wooden guides, an 8-cm thick concrete slab is made over the gravel evenly to one single level. Mixing ratio by volume for this concrete is 1:2:4 of cement, sand and crushed rock.

Floor smoothening
To give the floor the final finish, pure cement is mixed with coloring powder (red ocher) on a ratio of 5:1 by volume. The powdery mixture is sprayed over a 2-cm deep layer of fresh mortar, and then gently rubbed and blended into the mortar with a metal finishing trowel. While using the trowel, care is taken not to leave any streaks or marks.

Sidewalks

The sidewalk around the house insulates against humidity and works as a barrier against termites from entering the house. It consists of delimiting the perimeter of the house to a width of 60 cm with wooden formwork. The area is filled with gravel and topped with an 8-cm poured layer of concrete. The concrete is then smoothened with trowel and plaster. The sidewalk is given a 1% incline perpendicular to the walls for easy water run-off.

Doors and windows

Doors are made with a dressed lumber structure, having a 2.5 × 5.0 cm cross-section, lined with wood slats or plywood. Doors are set on hinges fixed into the wood frame.

Windows – double hung, hinged door-type or latticed – are made of dressed lumber having a 2.5 × 7.5 cm cross-section.

Installation of windowpanes, is a sub-contracted activity carried out by specialized labor. Primary care to be taken is in duly placing the windowpane guards (wooden pieces measuring 12 × 12 mm) that will fasten the windowpanes. After windowpanes have been installed, final inspection is done to determine whether the job was executed according to the quality standards established in the subcontract.

Sanitary and water fixtures

The jobs involved are placing the toilet and lavatory, making check boxes, and installing the sink. Work carried out is monitored closely and in detail to prevent water leaks owing to poor connections between pipes and fixtures, sink and pipes, etc.

Electrical installation

Electrical installation has three circuits that leave the control panel – one for power sockets, another for lighting and a third for the kitchen range (stove). No. 12 electric cable is used for the power socket circuit, No. 14 for the lighting circuit and No. 8 for the kitchen range (stove) circuit. Cables are run through plastic tubes, which are embedded in the walls and covered by plaster mortar. Quality circuit breakers are used in the controls and proper grounding is ensured.

Septic tank and drainage

When a house does not have access to a sewerage system, a septic tank and drainage lines are used to dispose of the sewage and wastewater. The septic tank most commonly used in projects is made of two sets of pipes and a tank (75 cm inner diameter and 150 cm length), both made of concrete. The pipes are placed one after the other and joined to each other by a tube nipple 10 cm long and 10 cm diameter. Both sets of concrete tubes are provided with concrete lids. A 10-cm level differential is maintained between the inlet tube going into the septic tank and the outlet tube going out to the drainage. A 10-cm diameter plastic (PVC) "T" is connected to each one to avoid passage of solids into the drainage.

The drainage consists of a system of perforated plastic piping (10 cm diameter) placed in a rock bed 50 cm wide and 60 cm thick at a depth of 90 cm below the floor or soil surface. The length of the piping depends on the number of family members and the permeability or (seepage capacity) of the earth. The incline (slope) of the piping is kept under 3%.

Painting

Painting involves two coatings of water-based paint (acrylic) to the walls, and reinforcing the outer wall perimeter with one coating of oil-based paint to a height of at least 60 cm to avoid excessive wear of the paint by dirt from splashing rain water. Oil-based paint is used on frames, doors and windows.

VARIATIONS IN THE TYPICAL CONSTRUCTION PROCESS

In view of production from its own bamboo plantations in the near future, the FUNBAMBU has built some model houses with bamboo substituting for most of the wooden pieces in the wall panels and roofing.

In some of the models, prefabricated curtains of bamboo laths, joined to each other with galvanized wire clamped or nailed to bamboo logs as columns, are used instead of prefabricated wall panels. The bamboo logs are placed at 1 m distance from each other. The roofing structure is also basically made of bamboo logs, placed over these columns. The bamboo curtains are plastered with hydraulic mortar. Some inner walls are built with double lining of bamboo laths closely bound together and without any mortar plastering.

In other models, prefabricated panel walls are substituted by a double-layer of bamboo laths, joined together with other bamboo laths placed vertically at distances not exceeding 60 cm. These panels are framed with wooden pieces having a cross-section of 2.5 × 5 cm. Joints between the laths and from these to the wooden frame are made with clamps. One of the layers can be substituted with wild reed. Roofing, composed of trusses made with bamboo laths or bamboo log stringers, is placed over these panels. Plastering of the panels is done with hydraulic mortar.

CONCLUSION AND RECOMMENDATIONS

An inexpensive and easy-to-execute construction system facilitates the assistance by institutions and governments to poorer families with very low incomes. Most of these families are faced with unemployment and under-employment difficulties, and such construction systems may help solve their housing problem through self-construction and mutual help, with or without recourse to hired labor.

If, in addition to solving the housing problem, community organizations could be created or strengthened to take interest in solving other problems that confront these poorer sections of the society. Furthermore, if some of the beneficiaries, after participating in the construction of their homes, learn part of a skill or trade that would give them the opportunity to enter the construction labor market, that would be another good reason to implement such projects in Costa Rica and other countries.

The low weight and ease of building with bamboo, wild reed and wood allow for the participation of women, children and the elderly (senior citizens) in building houses. This provides a reciprocal effect of appreciation of and attachment to their home. Care must be taken that the work they do is a supplementary effort to the work done by the construction crews to avoid the projects taking more time than warranted.

The use of self-construction and mutual help efforts favor housing programs in terms of savings in labor cost. However, the construction time is increased, along with costs of management, operations, technical and social guidance, as well as purchase of tools and machinery, when compared with projects carried out using skilled labor.

The Bamboo Foundation is making good all the experience acquired from the National Bamboo Project, wherein most houses were built in the first stage through self-construction and mutual help. Presently, the Foundation employs much more skilled labor and prefabricated panels with the purpose of improving its construction capacity to 1 500 houses per year.

As more bamboo is used in place of wood in wall panels and roofing, costs of materials diminish; but the need for skilled labor and quality control in manufacturing, placement and fastening of panels and roofing tend to increase. The joining of bamboo log pieces is more complex than joining pieces of sawed wood, because the bamboo pieces have a cylindrical cross section that varies with their length and from culm to culm. Another problem is that bamboo tends to split when nailed.

In addressing the difficulties of working with bamboo, the Foundation is undertaking research in the use of glued bamboo laths, nailed or clamped, to substitute for most wood pieces in the wall panels.

The service life of bamboo constructions depends basically on the protection given against the attack by insects and fungi. The Foundation has solved this problem by treating bamboo with a boron-based chemical preservative to ward off insects, and with an appropriate architectural design that protects the building from being in continuous contact with water and thus prevents fungal growth.

Bamboo for Sustainable Development, pp. 667–671
Eds. A. Kumar, I.V. Ramanuja Rao and Ch. Sastry

Application of bamboo in construction in China

JIANG SHENXUE and ZHANG QISHENG
Bamboo Engineering Center, Nanjing Forestry University, Nanjing, China

Abstract—This paper briefly introduces the bamboo resource and its characteristics in China. The use of bamboo in construction, such as bamboo building, bamboo scaffolding, bamboo floor, bamboo concrete form, bamboo furniture and handicrafts, are described in detail.

INTRODUCTION

The present population of China is over one billion, and its forest resources are scant. Owing to the development of the economy, the Chinese people are increasingly requiring more and more wood products in their daily work and life. Consequently, forestlands in China have depleted sharply, and the quality of forest resources has decreased from overexploitation. In many areas, ecology and environment are rapidly deteriorating. In view of these alarming trends, the Chinese government banned early this year logging in all nature forests and has given directives for their conservation.

On the other hand, the bamboo resource of China is rich in both species and yield. Bamboo not only has many characteristics and advantages – such as fast growth, short rotation period, high yields, high mechanical strength, diverse uses and higher fiber content – but also wide adaptability to varied ecological conditions. Apart from homesteads, they can also be planted on farm boundaries, riverbanks, hill slopes, degraded lands, etc.

In general, intensely managed bamboo plantations need only 4–8 years to be harvested, and will continue to yield benefits for years after planting. The bamboo resource of China is not only extensive – approximate 4 million ha – but also increasing rapidly. In 1950s, there were only 2 million ha of bamboo forests, which increased to 2.34 million ha in 1965, 3.4 million ha in 1980, and 3.55 million ha in 1988. The stocks of *Phyllostachys pubescens* (moso bamboo), a dominant bamboo species in China, have reached 3.4 billion culms. Currently, over 360 million culms (0.3 million tons) of other bamboo species are harvested each year.

Owing to the variety of species of bamboo and their abundance, the Chinese people have extensively used bamboo in their daily life. For example, they wrote on the bamboo strips woven together using thread before they invented paper. Even today, the people in the southern mountain regions, especially some of minority communities, live in bamboo houses, eat bamboo shoots, walk on bamboo bridges, cross rivers in bamboo rafts, and farm with bamboo tools and implements. At present, the Chinese people widely use bamboo in construction applications, including buildings, scaffolding, flooring, formwork, furniture and handicrafts.

BAMBOO BUILDINGS

As said earlier, some minority communities of China still construct bamboo cottages to avoid heat and to enjoy cool air. Notable among these are the Dai people, who live in the south of Yunnan Province, which is a tropical zone with plenty of rain. In general, the cottages consist of two floors. The ground floor is used to store farming tools or raise poultry or livestock, while the people inhabit the first floor. The frames and trusses of the cottages are made of *Dendrocalamus giganteus* or *D. sinicus*, two of the lager bamboo species that can grow over 30 cm in diameter and 30 m in height. The roofing shingles, walls and floors are made of bamboo species with culms of smaller diameter.

Bamboo is also common in landscaped gardens and resorts, and there are a few bamboo towers, pavilions and corridors made in the unique style of the East. In West Lake in Hangzhou, there is a pavilion with its roof, beams, pillars, eaves and rails all made of bamboo. In Xuanwu Lake in Nanjing, a row of bamboo pavilions stand above water surface. These bamboo buildings are carefully structured and painted to last for a long time.

BAMBOO SCAFFOLDING

In China, bamboo scaffolding is widely used in the construction of civil buildings that are usually less than ten floors. Whole round bamboo is fastened to each other to form a scaffold grid around a building being to support wooden work platforms. Bamboo is also used as platforms to support workers during construction. The platform is made of bamboo pieces, following a simple process: the round bamboo culms, usually the top part, are first split into pieces with about 4 cm width; nodes are removed from these pieces; and double-ended steel bolts are used to fasten the pieces to form a scaffold platform. The advantages of bamboo scaffold are its lightness and low cost, when compared with steel.

BAMBOO AND BAMBOO-WOOD COMPOSITE FLOORS

Bamboo floor is a refined product that has a unique aesthetic appeal. The manufacturing process is described elsewhere (Shenxue, this volume). Its price is higher than most wood floor because of the low utilization ratio of bamboo. Currently, there are more than a hundred plants in China that make this product with annual production capacities ranging from approximately 100 000 m^2 to 20 000 m^2. The Bamboo Engineering Research Center, Nanjing Forestry University, has discovered a new bamboo-wood composite to lower the cost of bamboo floors. The process uses boards made of Chinese fir, a fast growing species, as the core layer and a thin layer (2.5 mm) of bamboo strips is used as the facing. The product retains the strength, lightness and fine exterior of bamboo floor, but costs much less than the bamboo floor.

BAMBOO CONCRETE FORMWORK

Bamboo concrete formwork can be divided into three types according to their qualities and value: low grade, middle grade and high grade. Although all the three are made of bamboo, they have different manufacturing processes and different structures.

The low-grade concrete formwork is a bamboo particleboard that has higher strength (1.5 times that of wood particleboard), higher density (0.85–0.95 g/cm^3), and lower moisture swelling ratio compared with wood particleboard. It is less expensive and can last five cycles of use. However, it was found in application this formwork was not tough enough to survive falling down from heights, which negatively influences its service life. There are four plants, with an annual production of over 10 000 m^3, in China that make formwork of this grade. Recently, by incorporating bamboo curtains or mats, an improved bamboo particleboard formwork has been developed.

There are two kinds of middle grade bamboo concrete formworks. One is the bamboo particleboard concrete formwork overlaid with paper impregnated with a PF adhesive. This overlay enhances the surface quality of a building or a bridge. There is a factory in Fuyang, Zhejing Province, which manufactures this product. The other is made from bamboo curtain and mat board. Bamboo curtains, formed by weaving parallel bamboo slivers with thread, are across placed as the core layers and mats impregnated with a PF resin form the facing and backing. When hot-pressing, two paper sheets impregnated with PF adhesive are put on the facing and backing to form films on the surfaces of the board. This formwork has high strength and stiffness, long service life and low water swelling ratio. The only disadvantage is the varying thickness of the product. At present, there are 30 plants that produce more than 40 000 m^3 a year.

The high-grade bamboo concrete formwork is mainly made of bamboo baseboard and several layers of paper impregnated with PF and MF adhesives. Bamboo baseboard is a plybamboo board made of softened and flattened bamboo strips or bamboo curtain board. After the baseboard is sanded to uniform thickness, two sheets of veneer are put at the top and bottom surfaces, and several adhesive-impregnated paper sheets are then placed on the veneer and hot-pressed to form a high-strength concrete formwork. This product is the best, and is comparable to Finish WISA concrete formwork for its excellent qualities. Its MOE parallel to board length is about 10 000 MPa (12 mm thick) and 8 000 MPa (18 mm thick), while the MOE perpendicular to the board length is 7 000 MPa (12 mm thick) and 5 600 MPa (18 mm thick). Its MOR reaches 98 MPa (12 mm thick) and 90 MPa (18 mm thick). The wearing value is less than 0.08 g/100 r. The service life for double-face formwork is over 150 cycles of use. Because of its uniform thickness, it can be inserted into a steel frame, which facilitates assembling of formworks for larger area. The Xihu Bamboo Concrete Form Company in Hangzhou, Zhejiang Province, is manufacturing 100 000 m^3 high-grade formwork per year.

BAMBOO FURNITURE AND HANDICRAFT

Bamboo is widely used in furniture making in China. Bamboo furniture is light, fine, beautiful and less expensive. The process involves splitting the culms, removing the nodes and heating over fire for making them strips pliable enough for bending. The formed pieces are fumigated, drilled, notched or tenoned, and planed before assembly by gluing and splicing. Many furniture items – such as desks, chairs, beds, cupboards and tea tables – are made using split bamboo. A later technology involves boiling the bamboo strips to bend them for molding to required linear or curved shapes. Then, following modern designs and manufacturing processes, they are made into various types of furniture.

Bamboo handicraft can be divided into two types, according to the process employed. One is made by weaving bamboo splits, while the other is made by carving bamboo strips, culm sections and roots. Bamboo weaving has long history in China. Generations of masters of bamboo weaving have created hundreds of patterns, most popular of them being cross weaving, V-weaving and circular weaving. Woven bamboo articles are made both for daily use and decoration. Articles of daily use include baskets, plates, trays, dishes, bottles, jars, boxes, mats, curtain and fans, decorative items include figures of animals, birds, ancient buildings, etc. The color of woven articles varies, but the most preferred are those that maintain the natural finish. Landscapes, pavilions, towers, figures, trees and birds are sometimes carved on bamboo strips. Penholders, fan ribs, walking sticks, and even hairpins are fashioned out of bamboo.

THE STATUS AND THE FUTURE OF BAMBOO APPLICATION IN CONSTRUCTION

Because of the limit of bamboo distribution, bamboo buildings and bamboo furniture appear mainly in southern China. However, bamboo concrete formwork, bamboo floor and bamboo boards are increasingly being accepted and used in construction throughout the nation. In Beijing, Shanghai, Nanjing, etc., many skyscrapers, bridges and highways have been constructed using high-strength bamboo concrete formwork. In Shanghai, the famous Nanpu Bridge and Yangpu Bridge were built using high-strength bamboo concrete formwork. The usage of bamboo formwork has since then spread to other parts of the country. Bamboo floor, owing to its higher cost, is used only by a few. However the development of bamboo-wood composite floor is expected to change this. As it is now becoming fashionable to use decorations made of natural materials, the market for bamboo handicrafts is expanding rapidly. In general, the application of bamboo and bamboo products is becoming increasingly important in the developing economy of China.

Bamboo for Sustainable Development, pp. 673–678
Eds. A. Kumar, I.V. Ramanuja Rao and Ch. Sastry

Use of bamboo in the huts of rural Mexico

GILBERTO R. CORTES and ROSA INES AGUILAR
Instituto Tecnologico de Chetumal, Chetumal, Quintana Roo, Mexico

Abstract—Bamboo finds many uses in Mexico, although not to the extent or as frequent as in Asian or South American countries. One of the prominent uses of bamboo is in the construction of huts, mainly in the rural areas of the country. This paper describes the use of the culms of *Guadua aculeate, Guadua velutina* and *Otatea acuminata* for this purpose.

INTRODUCTION

Nature offers the materials – in variety, abundance and quality – needed for the construction of simple dwellings in rural settings. These materials, in most cases, are easily available locally and only the efforts for their collection and application are required. It can be seen that indigenous people have astutely taken advantage of materials found in the region where they live, choosing with certainty those that offer better results as means for protection against weather and atmospheric changes.

Rural and indigenous communities of Mexico have several species of wood and other plant products suitable for the construction of their huts. Use of non-plant materials is increasing rapidly, threatening the loss of traditions and customs in the use of locally available natural resources.

Among plant products used in the construction of houses, *Guadua aculeate* (tarro), *Guadua velutina* (caña mansa) and *Otatea acuminata* (otate) are rationally used by populations in different regions of Mexico. The predominant users are: the indigenous "Huastecos" in northern Veracruz, southern Tamaulipas and Hidalgo; the "Totonacos" of central Veracruz and Puebla; different rural communities in the States of Jalisco and Puebla (Moya 1988); and some communities of the indigenous "Lacandones" in southeastern Chiapas (Fig. 1).

MATERIALS AND METHODS

This study presents the native species of bamboos of Mexico – *Guadua aculeate, G. velutina* and *Otatea acuminata* – used in construction of traditional huts, describing the form of use and their cultural importance. The method used in this

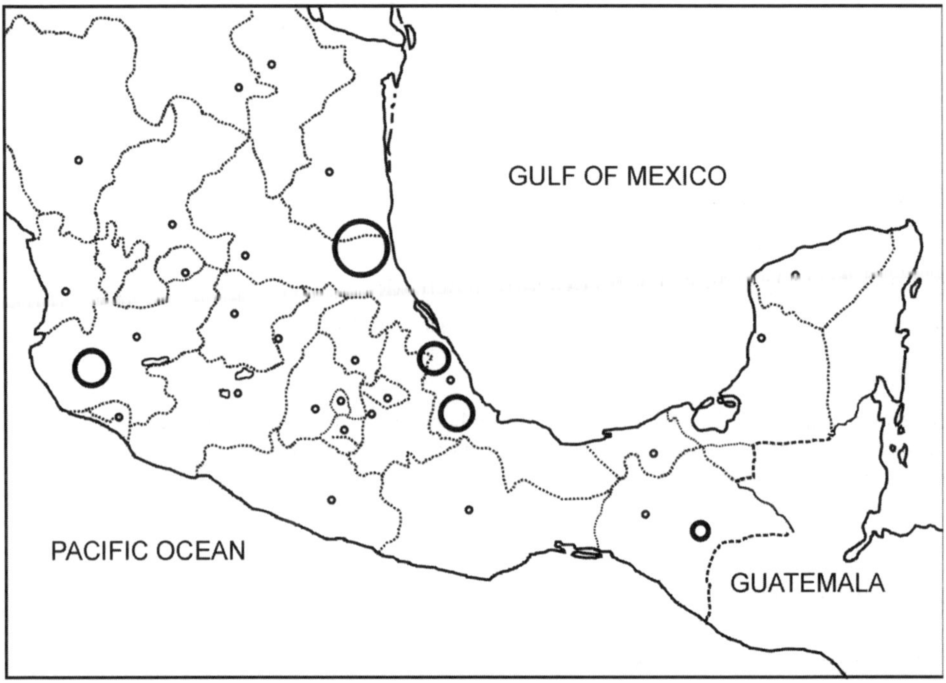

Figure 1. Different areas of Mexico where bamboo is used

brief study basically consisted of carrying out observations in different rural communities, mainly in the state of Veracruz where the three species mentioned above are used for construction of huts. Informal interviews with the people enriched the information, which added to first-hand observations made on site.

RESULTS

Three species of woody bamboos native to Mexico are used by rural communities and indigenous groups for the construction of their dwellings. These are:

1. *Guadua aculeata* Rupr. ex Fournier
2. *Guadua velutina* Londoño & L. Clark
3. *Otatea acuminata* (Munro) C. Calderón & Soderstrom

Guadua aculeata

Characteristics: *Guadua aculeate* is considered as the most robust and strongest of native Mexican bamboos, reaching heights up to 25 m and a diameter of 20 cm. The branches have 3–5 spines grouped at the nodes. The culms can be recognized by their bright red color when mature. The species was described by samples collected from the surroundings of the ethnic community of Colipa and Vega de Alatorre in central Veracruz state. Its normal habitat is disturbed forested areas of medium density (Gomez-Pompa 1980). The species grows better on the banks of water bodies such as rivers. It is known in northern Veracruz by its common name "tarro".

Uses: The culms are cut down with machetes, the branches are removed and transported to the hut construction site. The culms are used in two ways: 1) the entire culm is used without any modification; and 2) the culms are split in half and then hammered into sheet form. The whole culms find use as posts and beams. The sheets are placed in a horizontal or vertical position and used with mud to form the walls of huts. The outer part of the culms forms the exterior side of the wall, while the inner part form the interior side.

Some of the huts observed were notable for the perfect harmony of their construction; halved culms of different diameters were placed in precise spots according to the work it develops.

Ethnology: The Huastecos live to the north of the Maya-Quiché community, and the majority occupies large zones on the west of San Luis Potosi (SLP) and northern Veracruz State (Fig. 1). Typically, their huts are – practically the entire structure of the house (except the roof) – is made with *Guadua aculeate.* This ethnic group also uses *G. velutina*, but less frequently and to a lesser extent. It is interesting to note that traditional hut construction using *G. aculeata* has been maintained for several generations. Several families still prefer to use natural resources in hut construction.

Guadua velutina

Characteristics: *G. velutina* is a recently described species located in southern San Luis Potosi State. This species, which grows in isolated groups, can be found also in Tamaulipas, Puebla and Veracruz. The culms may grow up to 15 m in height and have a diameter of 15 cm at the base. Although nodal spines are considered a characteristic of *G. velutina,* it is common to see large populations without them. The species is known locally as "caña vaquera" and "caña mansa". The grayish-green colored internodes contrast with the white nodes.

Uses: *G. velutina* culms are generally used in their entire lengths in majority of huts, in horizontal position and fastened with bolts, nails, screws, or ropes of "bejuco" vine obtained from the forest. They are frequently used along with clay, carton or zinc sheets. This species has the advantage of being easy to work with because many culms lack spines.

In locations near the City of Papantla, Veracruz, it is possible to see a wall construction technique using this species, in which the hollow internodes are filled with clay. This results in high-quality insulation. The people of this region have developed a construction method wherein culms of different diameters are util-

Figure 2. A house made with the culms of *Guadua velutina* in Veracruz

ized as structural elements to support different loads and forces of different magnitudes. Larger and thicker stems are used as posts and roof beams.

Ethnology: The "Totonacos" of the Totonaca ethnologic community inhabit half of northern Veracruz and its borders with the State of Puebla. It is a well-defined ethnic group in its customs, rites and traditions, but its use of bamboo in hut construction is less extensive than the previous group. Use of "modern" materials is becoming more common in rural construction.

Otatea acuminata

Characteristics: *Otatea acuminata* is a woody bamboo that is most abundant in Mexico and in some regions in southern Jalisco. Often, it is the only species found in that area. This bamboo reaches heights of 2.5 m and its diameter is less than 3 cm. The culms are strong and regular, and the durability is 30–45 years. Botanical samples used in its original description are from the region of Jalcomulco in central Veracruz but revision of samples cites them as from Jalisco, Nayarit, Colima, Puebla, Oaxaca and Guerrero. The species is commonly known as "otate" or "caña de otate". This species usually grows in dry regions with calcareous soil.

Figure 3. A wall made with *Otatea acuminata* in Veracruz

Uses: In central Veracruz, there are two communities that have traditionally used *Otatea acuminata* as supports for clay walls: Jalcomulco and Santa Maria Tatetla, where the majority of huts were built with clay mixed with grass and supported by lengths of otate. In general, it is a widely used material because of its uniform structure and diversity in sizes. The culms are placed at the corners of the house as columns. They are also used to cover walls, make interior partitions and provide support for roofing sheets made of carton or zinc. A common method in some communities consists of making the walls with intertwined otate branches and plastering them with clay, ensuring spaces on the plaster to serve as ventilation.

Ethnology: Communities mentioned above belong to the "Totonacos" group with influences from the "Nahautl" ethnic group of central Mexico. It is important to note that the visit to Jalcomulco Village showed that only the houses on the outskirts of the village still have traditional construction with otate. Some people cite the reason for this as the view of village authorities that these houses don't speak well of the village when centrally located.

CONCLUSION

The use of bamboos, as an element of the ethnic traditions on using natural resources that are part of the local environment, is slowly disappearing in the rural areas of Mexico. Records of the authors show variations in bamboo use among communities and among families: while the Huastecos still consider the tradition as of primary importance, in Jalcomulco village inhabited by the Totonacos, the tradition is being lost rapidly. As with all villages, where traditions and customs are closely related to the region, the groups observed maintain a high respect for local flora – a rational respect to its use and a cultural respect to it as a living factor that forms part of their lives.

ACKNOWLEDGEMENT

The authors are grateful to the people of Veracruz State for the cooperation and assistance extended during this study.

REFERENCES

Gomez-Pompa, A. 1980. Ecologia de la Vegetacion de Veracruz CECSA.

Moya, R.V. 1988. La vivienda indigena en Mexico el Mundo. Universidad Nactional Autonoma de Mexico, Mexcio. 251 pp.

Bamboo for Sustainable Development, pp. 679–688
Eds. A. Kumar, I.V. Ramanuja Rao and Ch. Sastry

Potential of bamboo for building construction in Western India

KANWARJIT NAGI
Independent Architect, Pune, Maharashtra, India

Abstract—This paper makes a case for using bamboo as an alternative to timber in housing construction, especially in roof structures. It goes on to providing technical/practical tips on handling bamboo in construction projects.

Rural housing relies heavily on hardwood timbers that are extracted from already over-stressed forests. There is a rudimentary tradition of building with bamboo in western India. It is used at present for cattle sheds and temporary structures, and for rafters and batons. There is sufficient stock of both cultivated and naturally occurring bamboo in the area. Hence, it seems appropriate to encourage building with bamboo in this region.

The ideas contained in the paper have evolved over the last five years work on different aspects of bamboo in western India. In particular, insights have been sharpened by a recently completed project that the author was involved with: an experimental roof structure in bamboo.

The paper covers the following broad areas:

- Overall design principles
- Structural design principles
- Building components
- Estimation and costing
- Management of the building process
- Preservative treatment

In addition, the paper deals with some critical issues that indirectly affect/facilitate building with bamboo. These are: an enabling legal and policy environment, improvement of genetic stock of bamboo, bamboo plantation management, harvesting and seasoning methods, improved tools and techniques for handling bamboo, and the budding of a new profession – that of "bamboosmiths".

INTRODUCTION: PRESENT STATUS OF BAMBOO IN CONSTRUCTION AND DIFFICULTIES

Western India (mainly western Maharashtra) has three seasons – summer, monsoon and winter. The highest rainfall is 3 500 mm, and the temperature varies between 18°C and 40°C. The light soil ranges from fractured basalt to laterite. The forests in the region are moist deciduous or semi-evergreen.

In urban areas, bamboo is primarily used as a scaffolding material and to build temporary shelters at construction sites. These are put together by the unskilled laborers employed at the sites. Of late, however, bamboo is being increasingly used to decorate the interiors of restaurants, pubs and highway resorts, as well as for garden fences of large plush bungalows. This work is usually executed by a contractor, who employs carpenters and *Buruds* (the traditional bamboo craft community of the region).

In rural areas, bamboo is used mainly for roofing – as batons and rafters. However, the primary beams and columns continue to be made of hardwood timber. In villages, cattle-sheds and extensions to an existing house are often made of bamboo. Bamboo is also a favored material for building temporary pavilions for religious and social functions. Such constructions are done by family members and neighbors. Sometimes, the local *Burud* community may also be involved.

Bamboo available in urban markets usually comes from long distances, but for rural uses it is always obtained locally. The commonly available species are listed in Table 1.

Table 1.
Common bamboo species available in the region

Local name	Botanical name	Characteristics
Manya	*Dendrocalamus strictus*	Dia (2.5), Ht (25), solid or small cavity, hard fiber
Mes/Mesi	*Oxytenanthera stocksii*	Dia (2.5), Ht (22), solid, soft fiber
Kalak	*Bambusa bambos*	Dia (4.0), Ht (35), small to medium cavity, hard fiber, thorny
Udha/Chiwa	*Oxytenanthera monostigma*	Dia (1.0), Ht (12), solid, soft fiber
	Bambusa vulgaris	Dia (3.0), Ht (30), medium to large cavity, hard fiber

Notes: Dia: basal diameter of the culm in inches; Ht: total height of the culm in feet.

Difficulties

To begin with, none of the bamboo available in the market is either seasoned or treated and as a result, the culms do not last beyond 1–2 years. Secondly, as no effort has been made to detail the overall design and joints, the end products are crude and weak. Since the structures are extremely flimsy and temporary, customers are unwilling to pay a good price. Consequently, only unskilled/semi-skilled workers are drawn to this activity. This has become a vicious circle, preventing any innovation or improvement in techniques and design. If adequate attention is paid to overall design, details, the building process and preservative treatment, very elegant and durable bamboo structures are possible.

NOTES ON DESIGN: SITTING LIGHTLY ON THE GROUND

Forms and spaces

A variety of organic forms other than the usual rectangular and square shapes commonly used can be explored for living spaces. Like timber and mud, bamboo allows the freedom to be dismantled and reshaped in contrast to modern cement-based materials.

Traditional roof-forms of this region are pyramidal, two-side sloping and flat (Fig. 1). Even these, when made in bamboo, turn out to be lightweight, open and expressive structures. All natural materials including bamboo tend to enhance the quality of light and acoustics when used for openings and space-dividers.

Assembling the parts

Based on the experience recently gained by the author in building with bamboo, a few do's and don'ts are listed here for the benefit of novices.

The stability of any bamboo frame improves substantially if it is divided into triangles. If these triangles are more or less like equilateral triangles, the performance of the frame will be still better (Fig. 2).

Given the strength of local bamboo species, no member in a truss should be left unconnected or unsupported for more than about 10 feet of its length. When used in structural frames, bamboo poles should have a minimum basal diameter of 2.25 inches.

Bamboo members may be fastened together using nuts and bolts or lashed using wire and/or rope. Rope of natural fibers, like cotton and jute, perform better than others. It is often better to impregnate the rope joint with a resin adhesive for added stiffness to the joint.

It is quite easy to join 2 or 3 bamboo members at one point. If more than 3 members meet at one point, the joint tends to become complicated and cumbersome. This can be avoided by staggering joints or allowing for some eccentricity.

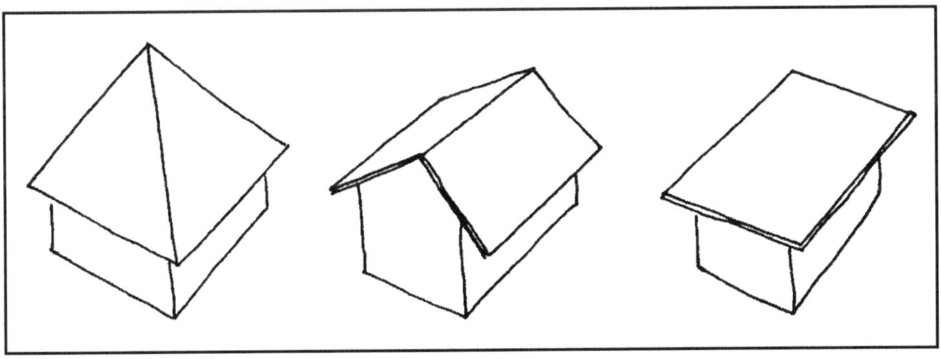

Figure 1. Traditional roof forms

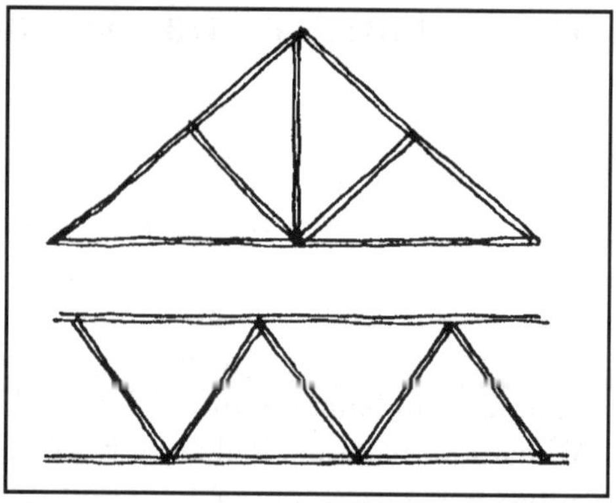

Figure 2. Triangulation in frames

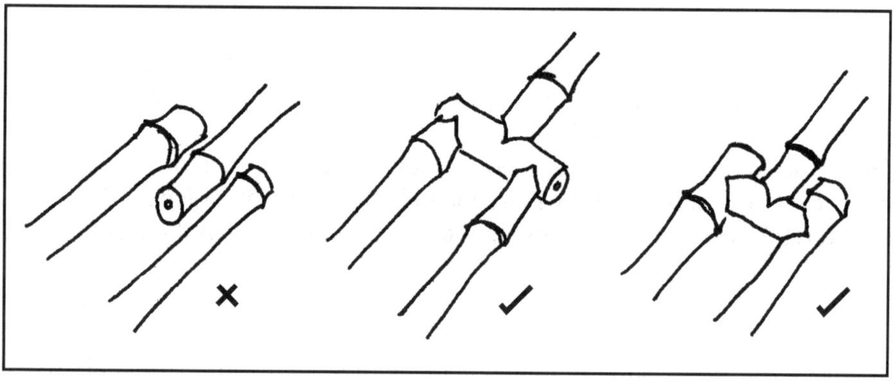

Figure 3. Joints: some do's and don'ts

Depending on the loading conditions, it is sometimes useful to use two bamboo poles instead of one for certain members in a structural frame (Fig. 3). Such double members help resolve stresses and also simplify joints.

If members are carefully cut, shaped and made to sit properly against each other at the joints, the transfer of loads and stresses within the structure improves. If the members sit loosely against one another, there is a danger of the bamboo ends splitting. In this case, the strength of the joint becomes completely dependent on the strength of the binding material.

The size of the cavity in a bamboo pole varies from the base to the tip. Sometimes, a bamboo member at the joint might have a large cavity; this weakens the joint and makes it prone to splitting. Making a plug from a solid piece of bamboo 6–10 inches long, filing it down to fit the cavity and fixing it in place with resin adhesive can prevent this problem (Fig. 4). At joints, the cavity of any bamboo member should not exceed 0.25 of its area of cross-section to avoid the risk of splitting.

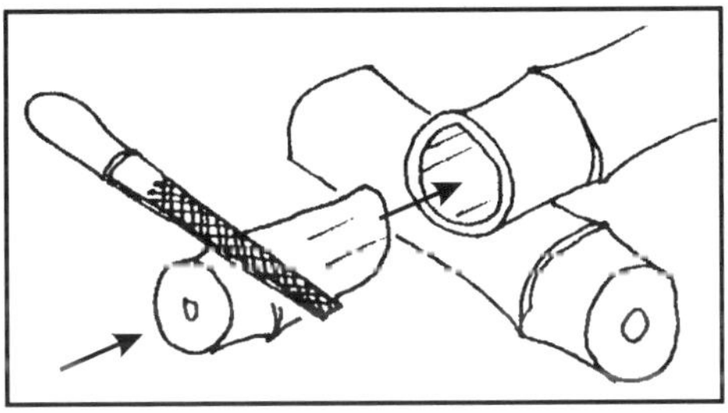

Figure 4. Plug for joints

Building components

A variety of building components – such as doors, windows, partitions and screens, lattice, and shelving – can be made from bamboo. Traditionally, all these components were made from hardwood timber. Today, good-quality timber is very expensive and not easily available; hence, there is a shift towards making these components using steel or aluminium sections.

At this point, bamboo can step in and provide a good natural alternative to steel/aluminium components. The initiative to boost bamboo mat-board production in India seems to be a very timely step in this direction.

Bamboo can be used for these applications in the form of whole culms, split sections, woven mats and/or bamboo ply. Elements can be fabricated either completely from bamboo or in combination with timber.

Proper detailing and the use of high-quality fasteners – such as brass or powder-coated screws, clamps, latches, etc. – will enhance the aesthetic appeal and improve the service life of the components.

Like all natural materials, bamboo also requires protection from sun and rain. So, sufficient roof overhang and/or protective finishes and polishes are necessary.

NOTES ON ESTIMATION: A MEANS TO AN END

Bamboo and labor requirements

Once the design of the building has been finalized, the first step is to estimate the quantity of material required for the execution of the work. One square foot of wall or roof area requires 0.2 bamboos or conversely, one bamboo culm can make 5 sq. ft. of roof or wall.

Local bamboos, for which the above estimates have been made, are usually 18–20 feet long and 2.25 inches in diameter. The multiplication factors will need some correction if the bamboos used are larger or smaller.

Once the material requirement is calculated, the second step is to estimate the amount of labor needed. Experience has shown that a team of six craft workers is an ideal size for construction work. On an average day, one bamboo worker can make about 15 sq. ft. of wall or roof.

It must be noted that there is considerable variation in the skill level of bamboo workers. Some are highly experienced and are able to manage nearly the whole construction process, while others may be more like semi-skilled carpenters.

Estimating costs

The cost estimation method described in Tables 2 and 3 uses a basic rate (B), which is the cost of one bamboo pole at its growing site multiplied by the total number of poles needed for the project. The cost of each step in the construction process can be calculated by multiplying this basic rate (B) with the given factor. This gives the total cost of that particular operation.

Table 2.
Cost of treated bamboo at the building site

Item	Cost
Harvesting and transport to the treatment site	0.5 × B
Cost of treatment (equipment, chemicals, labor and water)	3.0 × B
Taxes, cost of permits, commissions, etc.	0.75 × B
Transport to the building site (within 100 km and in truckloads of about 400 poles)	0.5 × B
Sourcing, handling and supervision costs	0.5 × B
Total Cost	**5.25 × B**

Table 3.
Cost of fabrication and erection of a structure

Item	Cost
Cost of working the bamboo to make walls, roof and components	5.25 × B
Cost of hardware (nut-bolts, wire, rope, adhesive, paints and polishes)	2.25 × B
Rent for hand tools and equipment	2.25 × B
Facilitation, supervision, and management charges	0.25 × B
Total Cost	**9.75 × B**

Therefore, the total cost of a project = 5.25 + 9.75 = 15.00 × B, where (B) is the cost of one bamboo pole multiplied by the total number of poles needed for a project.

The building process and time management

Bamboo construction requires a team of skilled yet versatile craft workers. Traditional bamboo craft consists primarily of woven products. In addition to tradi-

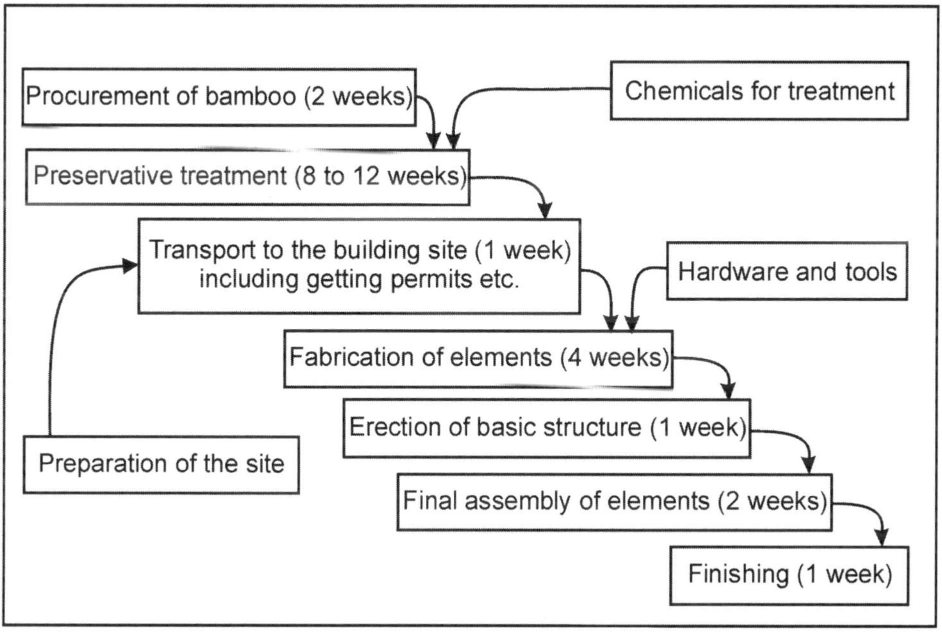

Figure 5. Various tasks in the building process and the time requirement for each

tional skills, construction requires a rudimentary knowledge of carpentry and an understanding of the building process.

Teams of about six craft workers usually work together at a site. During the first phase of fabricating different elements of the structure, they work in groups of two. While erecting the basic structure all need to work together and later, during final assembly and finishing, there are usually four craft workers working on the structure, with two of them passing the material and tools around. In this team, at least two must be highly skilled bamboo craft workers, two or three could be of average experience and one or two could be semi-skilled.

In addition to this team, a supervisor/facilitator is necessary to ensure that the work done is of desired quality and is proceeding at the right pace, and to make available materials, tools, etc. at the right time. A schematic diagram of the tasks involved in the building process is given in Fig. 5.

Other processes – such as acquiring permits, purchase of hardware and collection or purchase of tools – must continue while the preservation treatment is in progress. If these tasks are carried out serially, they may add up to 6 weeks to the total time taken for building.

From January to May craft workers are relatively free from agriculture and other domestic tasks. January–February is also the best time to harvest bamboo. Hence, the process of harvesting must be initiated during January–February to ensure that the construction is completed before the monsoon begins in June.

PRESERVATIVE TREATMENTS: KEEPING THE BUGS AWAY

Once a completely neglected subject, preservative techniques and treatments have now become critical to the survival of the bamboo sector. Many chemical preservative treatments have been prescribed and used with good results. However, their toxic effects, both at the treatment site and at the point of application, are increasingly being questioned.

Preservative treatments are most effective when they are applied on bamboo in the green state. Once the bamboo dries, the treatment acts only like a surface application that does not penetrate into the fibers, which require preservation most. Hence, it is very important that facilities for treatment be available near the bamboo plantation.

During discussions with tribal craft workers, the author came across some traditional non-toxic preservation methods. A few of these techniques, which are worth exploring, are listed below.

- Harvest bamboo in the low-sugar period (December to April).
- Harvest bamboo during the dark phase of the moon.
- Allow culms to dry in the clump after harvest with leaves and branches intact.
- Treat culms or splits with a decoction of tobacco leaves or the fruits of *Zizyphus jujuba* (Ber) or the heartwood of *Acacia catechu* (Khair).
- Apply the oil of *Semecarpus anacardium* (Biba) seeds on bamboo splits and then smear with wood ash.

CRITICAL ISSUES: THE BOTTLENECKS

There are a few more issues that require exploration and further study, especially in the Indian context. These issues affect the bamboo sector directly and can play a crucial role in bringing bamboo use into center stage.

Enabling legal and policy environment

In India, bamboo is classified as a forest product and its harvest and transport needs certification from the forest department.

Most forest laws related to bamboo are over a hundred years old and do not take into account the current social conditions and realities. They are so stringent that local people do not have access to forest bamboo for even non-commercial uses.

It is high time bamboo cultivation was treated as a commercial cash crop. Reduction in taxes, relaxation of rules concerning inspection and certification, and incentives to cultivators will be some steps that would encourage bamboo cultivation and use.

Harvesting and seasoning methods

Proper harvesting and seasoning methods will go a long way in making products more durable and long lasting and thus change the common perception that bamboo products are temporary and therefore cheap.

Today no systematic seasoning of bamboo is practiced. After harvest, poles are often just left out in the sun to dry. The ill effects of drying bamboo poles too fast and their subsequent collapse are well known. The development of low-cost solar kilns and other ways of controlling the loss of moisture in bamboo poles need to be devised.

When bamboo is purchased from the market, nearly 30–40% is deformed, cracked and unfit for use. Seasoning and curing, if done with care, will control deformation and cracking of poles, thus making more material available for construction.

Improvement of genetic stock and plantation management

Traditionally, bamboo has been considered "poor man's timber", an inferior material. No attention has been paid to improving the quality of the harvest.

Research in China and Southeast Asia over the last 15 years has demonstrated the importance of enhancement of genetic stock and proper management to improve the quality and quantity of bamboo produced. However, the gloomy situation still persists in India. Bamboo products continue to be of low quality, fetching low prices. Consequently, the craft workers can pay only a limited amount for raw material. Even if good, well-seasoned and treated bamboo is available, they are unable to pay for the cost increase result from improvement in the quality of bamboo.

This situation can be broken if value-added high-quality products, which fetch a better price in the market, are designed and produced. This will put pressure on the craft workers to look for better raw material.

Improvement in techniques and tools

New ways need to be found for harvesting and cleaning culms at the growing site to prevent damage to the outer skin, as such damage becomes the starting point for pest and fungal attacks.

There is need to develop small, low-power machines that produce even and smooth bamboo slivers of different sizes. This will bring down the cost and improve the quality of matting. Devising jigs and fixtures will speed up repetitive tasks and help standardize products.

Adequate attention is required to improve finishes like lacquer, melamine and polishes, which will enhance the product.

'Bamboosmiths': a new profession

As the use of bamboo increases, it will be critical to find people who are ready to take on the challenge of using bamboo for newer applications. Bamboo construction encompasses a range of skills from building design to structural load calculation, carpentry, fine detailing, supervision, account-keeping and costing, sourcing of material and dealing with clients and responding to the market.

Present day craft workers, carpenters, masons, etc. cannot handle such a variety of tasks, as their own knowledge base is very specific and focused. There is need to diversify their abilities through short-duration practical training courses.

It may be worthwhile to institute a specialized course in bamboo construction, which provides the students with the skills required to handle the entire process, from sourcing of raw materials to completing the building.

CONCLUSION

In this paper, the author has attempted to trace the entire process of transformation bamboo goes through from a culm growing in a field to a usable, comfortable house. Whatever literature he has seen has discussed different aspects of bamboo in fairly great detail but in isolation to each other. For example, there are learned treatises on the botany of bamboo, preservative treatments, manuals on jointing, etc. but hardly any writing that dwell upon the connections between these different aspects and how they interlink in practice.

Decades (and millions of dollars) worth of research has gone into the development of cement-based construction techniques, leading to such widespread application as we see today. A similar quantum of rigorous effort, meticulous research and testing needs to go into bamboo construction techniques for it to find acceptance and appreciation.

ACKNOWLEDGEMENTS

This paper is dedicated to Late Mr. Vinoo Kaley.

The author thanks Mr. V. Giriraj and Mr. A.G. Rao for their inspiring and continuing support, the INBAR family for providing succor in times of need, and Mr. David Farrelly for turning him into 'a bamboo bum'.

Bamboo for Sustainable Development, pp. 689–701
Eds. A. Kumar, I.V. Ramanuja Rao and Ch. Sastry

Hyperbolic paraboloid shelters and retractable structures

VLADIMIR MAUZIT
Architect, Paris, France

Abstract—Hyperbolic paraboloid, a geometric shape used in several applications, offers a range of possibilities in building construction. This paper discusses the application of hyperbolic paraboloid form to construct low-cost houses by saving material and labor costs. The article is of interest to the bamboo world since bamboo is one of the materials considered for the construction of these hyperbolic paraboloid structures, especially using prefabricated elements that are easily stored, transported and erected. By joining the cells in a honeycomb pattern, each cell can be extended to cover 16 m^2. The hyperbolic paraboloid shells can also be used to develop domes of varying geometry.

INTRODUCTION

Curt Siegel (1955), in his book on modern structures, had discussed hyperbolic paraboloid, mentioning that this form can have numerous variations and that this very special mathematical/geometric form had inspired some architects (Candela). He also stated that the few architects who use it, despite the difficulties of understanding it in the beginning, find in it a richness of expression they did not imagine when they started.

This inspired the author, especially as there was a great need for low-cost houses in Morocco where he was working (1950–1962), to study some hyperbolic paraboloid configurations in terms of low-cost houses.

The work involved slum rehabilitation program for a quasi-governmental company and the housing being made had walls of cement blocks and a flat roof of concrete slab, in Moroccan tradition (Medina). Several 4-level houses were also built, replacing thousands of shanty houses. The hyperbolic paraboloid attempts, however, never got beyond the theory stage, as there was a psychological barrier to it among the staff of the company as well as among the users.

When his return to France in 1962, the author built an office for himself, using the research to make full-scale example of the drawings. The first hyperbolic paraboloid elements had to be constructed by the author himself, as the form was not traditional and the carpenters could not understand its shape or the way to work with it (plywood boards stuck and stapled onto a specially curved frame in the hyperbolic paraboloid form). The author invented a turning guide for a belt saw to make carpenters understand these twisted frames. The result was very good and a clever carpenter and a boy helped the author complete his office. With each cell having 15 m^2, 60 m^2 of area could be obtained with four cells made with 33 hyperbolic paraboloids, all in the same configuration. The floor was made with 15 mm plywood and glass was used all around, except for some blind walls.

This type of construction has been used by several of the author's customers, allowing the opportunity to diversify the construction methods and materials. In the last 25 years, the author has had the occasion to examine various theoretical questions related to hyperbolic paraboloids and low-cost houses. The following discussion centers on some of these issues.

THE TYPE OF LOW-COST DWELLING

There could be a wide range of low-cost dwellings, based on several factors. However, the question of how low the cost could be is answered by three main factors that govern the construction cost: the cost of land (the plot); the cost of materials; and the cost of labor.

The cost of land

This factor, of course, is beyond an architect's ambit; nevertheless, it is perhaps the most important deciding factor in terms of construction cost, particularly in view of the ever-shrinking per capita land availability and a high incidence of land speculation that drive up the land costs the world over. What to do against the land speculation? This question was asked some 30 years ago by the Swiss magazine *Architecture Forms & Functions* (1971, No. 16) to 58 leading thinkers in the fields of architecture, art, medicine and psychology. Nothing concrete came out of the exercise, and the same question could now be asked to contemporary thinkers.

The cost of materials

There are two relevant questions to be asked. One, what sort of shelter saves the most on materials? The answer is: shell-type construction (using metal, wood, bamboo, inflatable forms, composite material, etc.). Two, what technique to choose? The answer is: it depends on the material chosen and the geometry of the shell.

These aspects need some elaboration. Let us choose wood or bamboo, placed as a ruled surface. This ruled surface will be a hyperbolic paraboloid (HP) owing to its rigidity (all HPs are ruled surfaces; that is, the tangent plane at any point of the surface cuts the surface in two lines, called rulings). The plan can be designed on a screen, shaped as a hexagon in view of the need for repetition and the greater variety it provides than a square screen.

The construction type designed by the author uses one-element HPs (5 m × 2.9 m), which can be connected to each other to obtain a closed and covered volume (see Annex). A few other elements are added for partitions, floor, plumbing provision, etc.

The main characteristics of the system can be listed as follows:

- Identical HP-shaped prefabricated parts;
- Serves the purpose of covering and closing, waterproofing, and structure (self-supporting, with no other framework is needed); and
- Can be stacked up, and is light enough for easy transport (by rail, road, sea or air).

Pricing can be easily made, based on the cost of a hexagon and the number of hexagons covered by the building.

In terms of materials, this system promotes:

- Wood, bamboo or reed, which may be used up to 99%;
- Second-quality wood may be used in composite material;
- Shell-shaped architecture, as against flat surfaces that have a certain rigidity from an aesthetic point of view;
- Infinite diversity of forms and their coverings and, at the same time, unity of character whether for a small house, a settlement or larger structures (sports, commercial, industrial complexes); and
- Possibility for industrial-scale production as prefabricated elements.

The cost of labor

Use of prefabricated elements obviously saves labor costs, especially if the production is on an industrial level. As hyperbolic paraboloid structures present this opportunity, the use of these would contribute to lower construction costs.

POSSIBLE EXTENSIONS

This research on possible extensions holds infinite applications for HP. It is enough to change the dimension of an edge or an angle (or both) to create another geometric form and the scale of the building, and thus obtain meeting halls, warehouses, sheds.

If the structure is permanent – that is, if provision for dismantling is not required – then prefabrication is not necessary and there could be some leeway in the precision of the elements.

A working group (in a school of architecture or some professional association) could be encouraged and financed to carry our research and development on the following themes:

- HP membranes (wood, bamboo, reed);

- Ruled membranes (wood, bamboo, reed);
- Composite materials membranes (wood + plaster, wood + resins, wood + resin foam);
- Glass reinforced concrete (GRC) membranes;
- Polystyrene shells in sandwich + GRC;
- Practical approach to tensile lightweight structures (bamboo) and inflatable forms.

The author ran a trial on constructing a bamboo dwelling unit at the Research Institute of Human Settlements (Cileunyi) from July to November 1995. With part help, he was able to perfect a process of building light bamboo HP elements and assemble them into a 16 m^2 dwelling unit, which could be replicated N-times to make large settlements. A larger experiment needs to be carried out, as the next logical step, with several finished units as part of one complex because it is to difficult to visualize the resulting volumes just by looking at photos, plans or even models.

The author is sure that all housing-related aspects – such as water proofing, wind load resistance, earthquake resistance, aesthetic appeal, speed of construction, enlarging the unit for more space, the cost of construction, etc. – would work out positive when materials such as bamboo are used.

The use of bamboo has also been tried out in folding (retractable) systems. Semi-prefabricated HP elements, in their simplified form, can be folded and unfolded by introducing two tensile strings. In the opened position (tensile equilibrium), the structure can take a bamboo mat prepared in the same HP surface form. This method will help facilitate transport and erection for small builders.

Most people might have come across the spectacular use of (at least in photos) of vast bamboo scaffoldings several stories high. With more sophisticated nodes, it would be possible to erect large bamboo structures as an alternative to steel 3D structures.

The calculation of nodes (metal, plastic, ferrocement) and of the bamboo culms required needs to be done by structural engineers. The technology and theory of such structures have already aroused great interest among scientists who are members of spatial structure associations in many countries, as proved by the last spatial structure congress in Surrey University in 1994 which had the participation of several hundred participants from over 40 countries.

The author has constructed several folding models for units that would use bamboo as the main construction material. These folding shapes (flat structures, domes) could take several forms by simple alterations. These folding domes, if covered with industrial waterproof textile, could cover meeting halls, temporary exhibition halls, temporary shops/stalls – in short, any temporary large or medium size covered surface, including emergency shelters.

CONCLUSION

Construction, especially housing construction, needs to be seen as a part of a more global approach for developing shelters aimed to achieve bodily and spiritual health, called bio-construction by some ecologists. As hyperbolic paraboloid structures, which can use natural materials in perfect harmony with the environment, fulfill this criterion, the formation of a working group for research and development in this area would be a fruitful activity. It is suggested that this be implemented in coordination and cooperation with the different national and international associations studying spatial and other similar structures.

ANNEX: VLADIMIR'S BAMBOO HOUSE

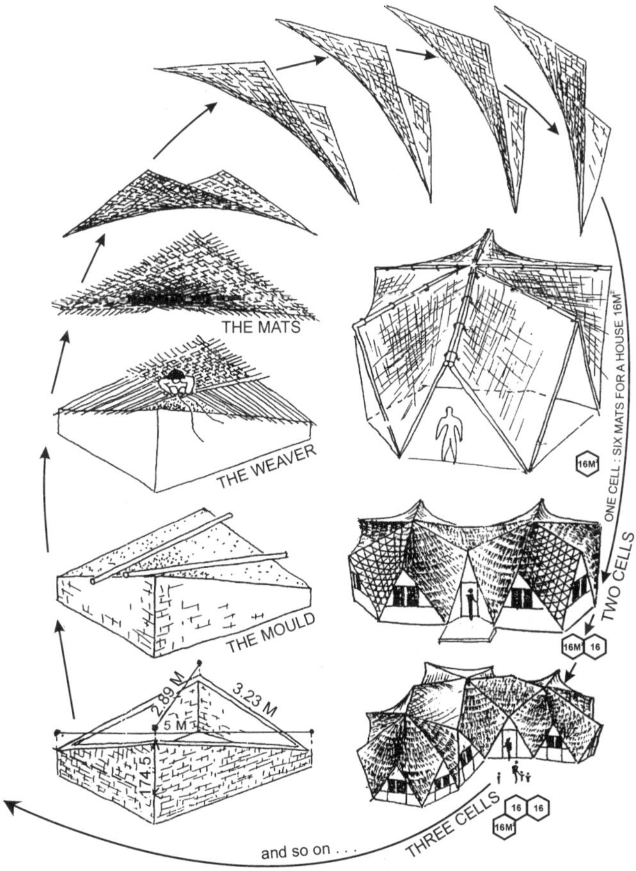

Figure 1. Different stages in the construction of a hyperbolic paraboloid bamboo house

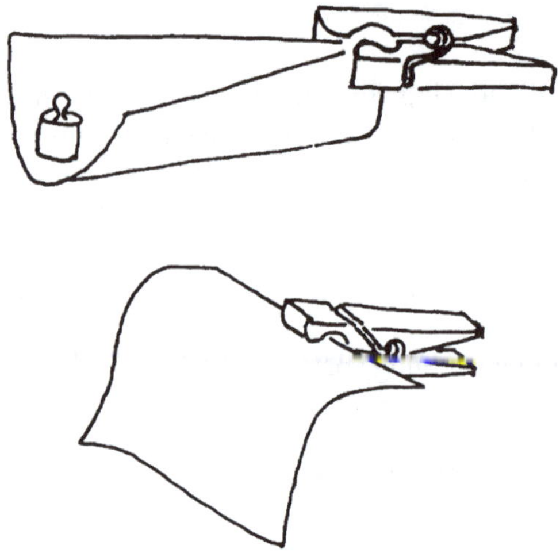

Figure 2. (a) Paper sheet folded as in shell construction; and (b) paper sheet as in flat construction

A saddle-shaped shell for the house

Why this saddle shape? Mainly because with this shape one can make a sturdy construction using few raw materials, and it can easily be done with self-help. The strength of a shell construction can be understood by comparing the two sheets of paper shown in Fig. 2. The paper sheet in Fig. 2a is like shell construction and can withstand loads, while the sheet shown in Fig. 2b is like a flat construction that cannot withstand loads. Although this is a simplistic illustration, it serves to put across the superior strength of shell construction.

Similarly, using bamboo strips, one can weave a flat bamboo mat or a shell-shaped bamboo mat (Fig. 3). The saddle-shaped shell made of bamboo mat (Fig. 3b) is a hyperbolic paraboloid (HP).

Construction of HP shell

While a flat bamboo mat is woven on a flat floor, an HP bamboo mat has to be woven on a surface in HP shape. To obtain the correct HP surface, 4 walls may be erected to make a box as shown in Fig. 4. This box (500 × 289 cm) can be made with a sturdy material like bricks, stones or wood. The upper side is not horizontal – the points B and D are 144.5 cm higher than points A and C.

When the box construction is completed, it may be filled with earth and crushed stones mixed with water, and then packed down by beating with a hammer. As a finish to the last layer, much like the floor of a house, some cement or other binding material may be added. When spreading the final layer, use a ruler (a straight pipe of 3–5 cm diameter and 330 cm length will serve the purpose) to get the correct floor shape. Place the ruler across the AB axis and slide it along the AD and BC axes, making sure that the ruler is always at the same distance

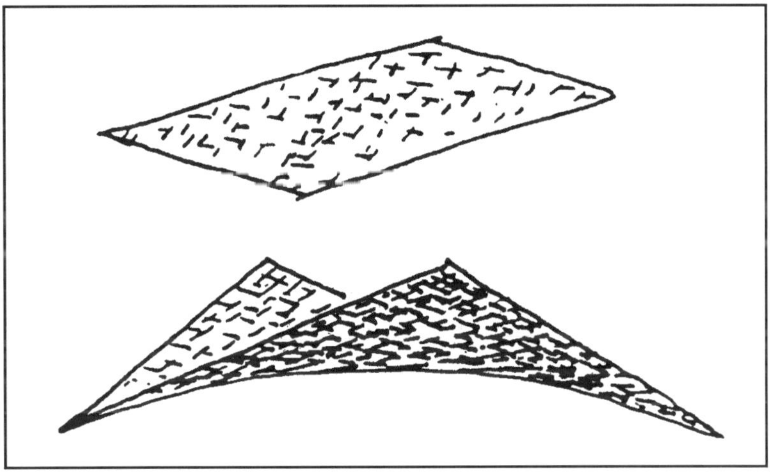

Figure 3. Bamboo mat weaving – (a) flat; and (b) hyperbolic paraboloid

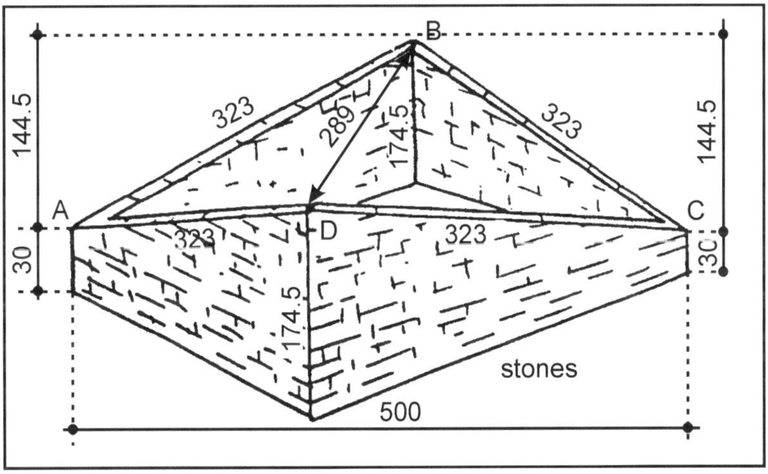

Figure 4. Preparing an HP surface for mat weaving – box construction

from A as from B and is in contact with the finished floor. Thus, AX1 = BX2 and AY1 = BY2 (Fig. 5).

The surface is now ready for bamboo mat weaving (Fig. 6). The length of the bamboo strips must be about 350 cm. As with the ruler, make sure that the ends of the strips are always the same distance from A as from B.

While weaving, one can notice that the strips are tighter towards the center of the mat and looser towards the edges, leaving square holes between the strips. By adding shorter strips (SS) these holes can be filled (Fig. 7). The completed HP

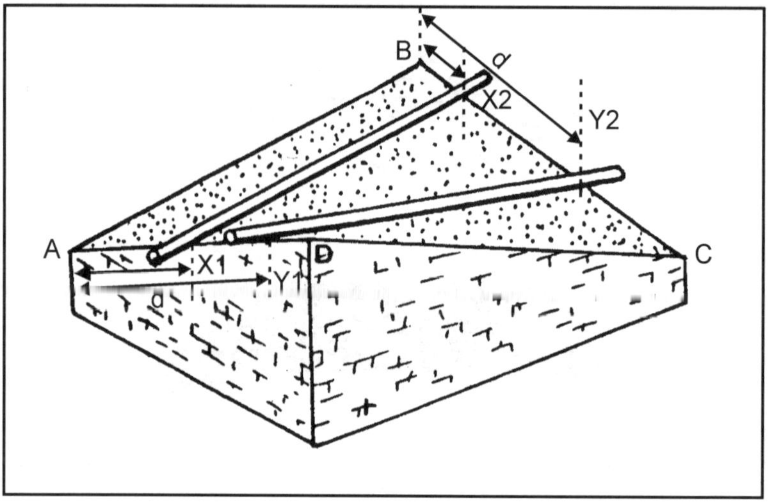

Figure 5. Checking the surface with a ruler

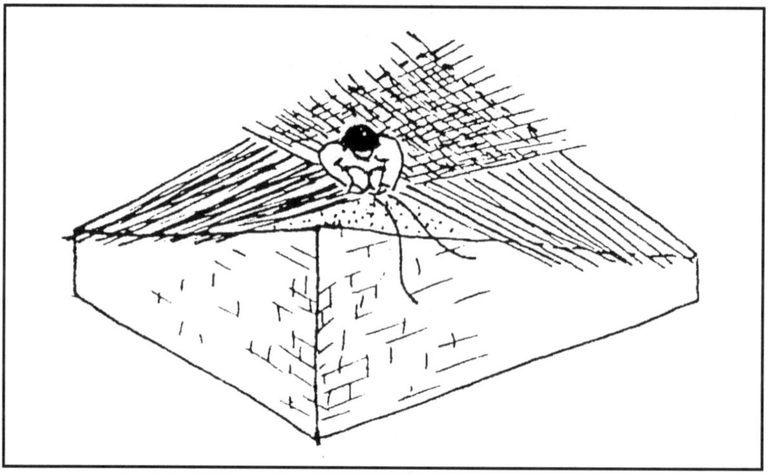

Figure 6. Weaving the bamboo mat on the HP surface

bamboo mat can be reinforced with 4 bamboo poles fixed along the edges (Fig. 7). A useful way to do this is to cut a groove of 1 cm width along the length of the bamboo culm, push the mat as deep as possible into this groove, and fasten the two together using galvanized wire. At the corners where the bamboo poles meet, the ends could by sawed off in half and the poles fastened together using galvanized wire passing through four holes drilled on the poles in alignment (Fig. 7). This completes one HP shell.

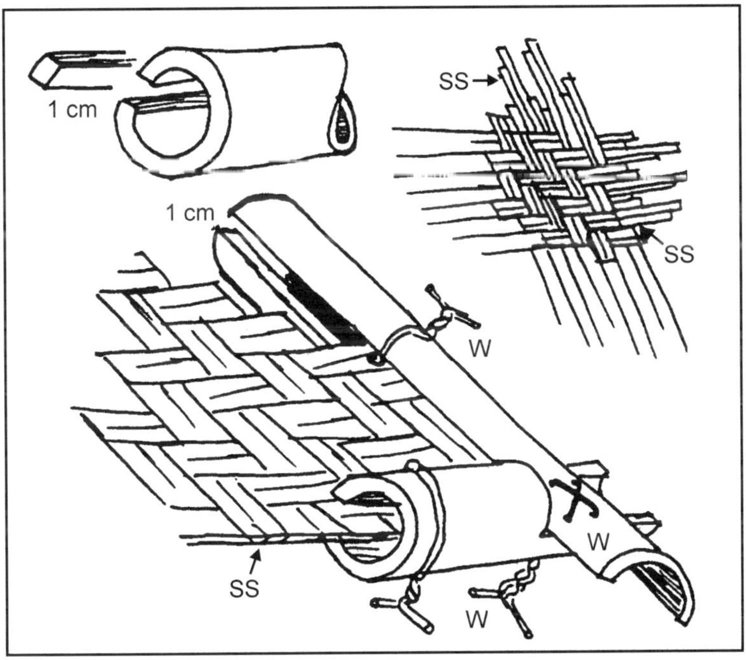

Figure 7. Filling the holes in the mat using short strips, and preparing the ends of the mat

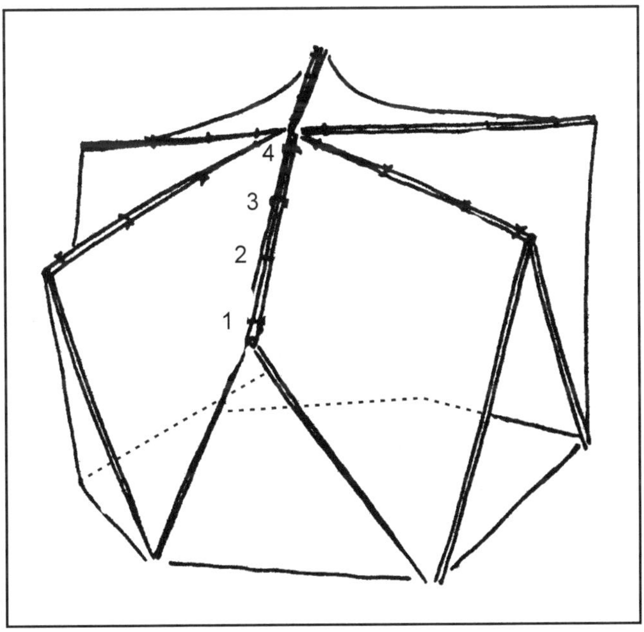

Figure 8. Erection of the HP shells

Erection of the structure

When six such HP shells are completed, a 16 m^2 house can be erected. The six elements have to be fastened together with wires, attached to the elements before erection, with each two adjacent elements requiring four wires (Fig. 8). The 6 feet of the shells must be anchored to the ground, using large iron nails or concrete foundations, or fixed to a laid floor.

This 16 m^2 house will have a height of 430 cm at the center (Fig. 9). The six triangular openings can be built up either as solid walls or as walls with a window or a door. The floor plan is a hexagon so that it is possible to extend this house by joining one or several similar 16 m^2 structures (Fig. 10).

The rain proofing of the HP shells is done before the erection of the structure. There are several ways of doing this:

- Straw covering;
- Mortar made of reinforced earth, cement, lime, coal tar and resin;
- Shingles (leaves, tiles, wood, bamboo).

Treated wooden slats can also be used for the construction in place of bamboo mat, in which case no water proofing will be needed.

Light HP bamboo structure, without mat weaving, can also be made to save labor, time and raw materials. The structure can be made using small-diameter bamboo culms with, for example, 7 culms in one direction and 7 on the other, making a second layer (Fig. 11). A third layer is then laid on the top so that the poles in this layer will cross the first layer to make a triangular grid.

After binding with galvanized wire, the HP elements become a rigid shell (Fig. 12). The edges need to be reinforced and the HP grid finished with a traditional rainproof thatch covering.

A variation in the first basic dimensions employed (AC = 5 m; AD = 3.23 m; BD = 2.89 m) will give an infinite number of HP shells, the only limitation being imagination (Fig. 13). Very large HP surfaces would need structural calculations by engineers, though this is easy because of the properties of HP shells (Fig. 14).

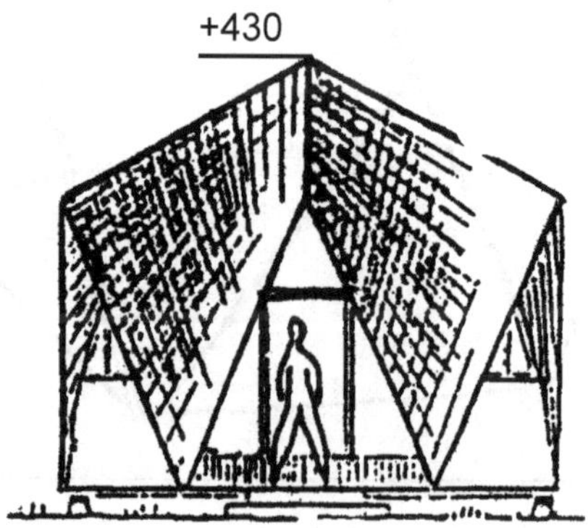

Figure 9. The completed HP structure

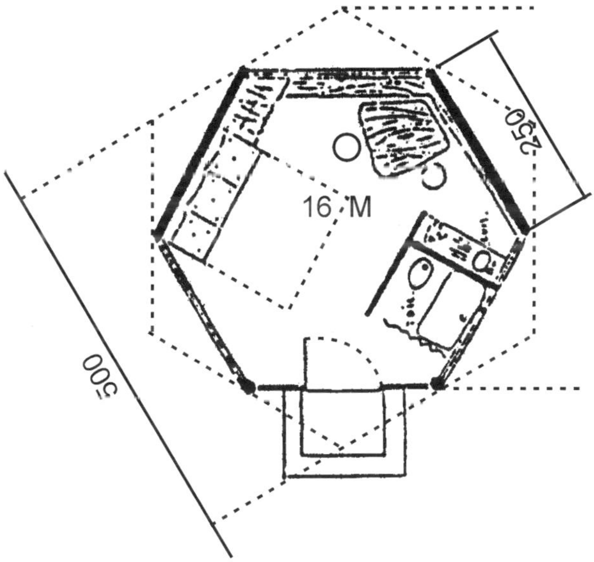

Figure 10. Floor plan of the unit

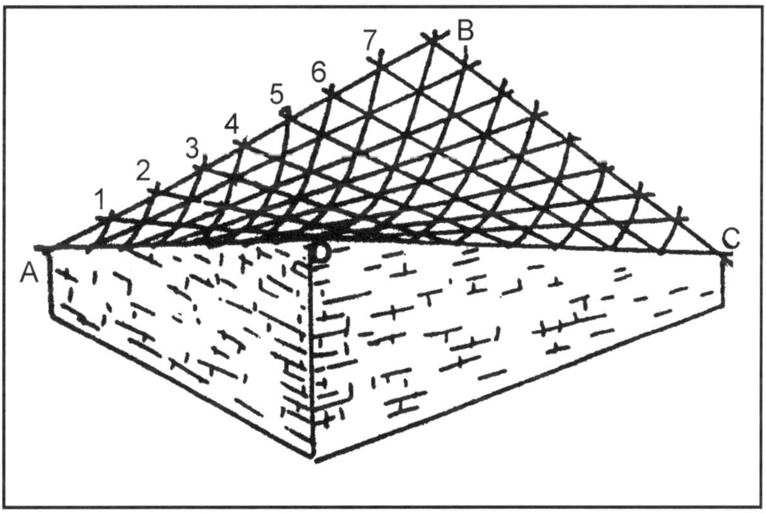

Figure 11. Using bamboo poles in place of bamboo mat

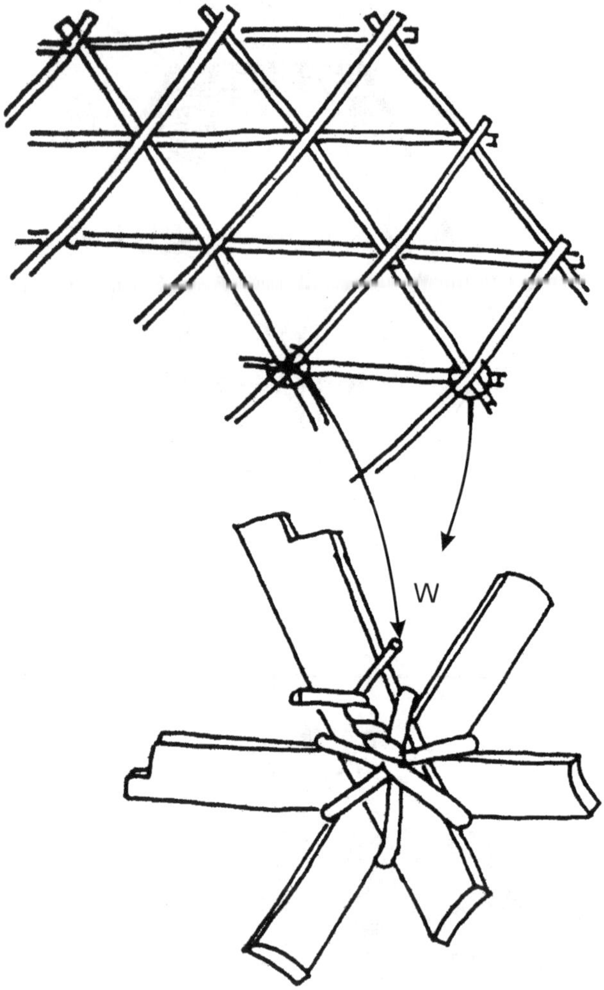

Figure 12. Binding the poles with galvanized wire

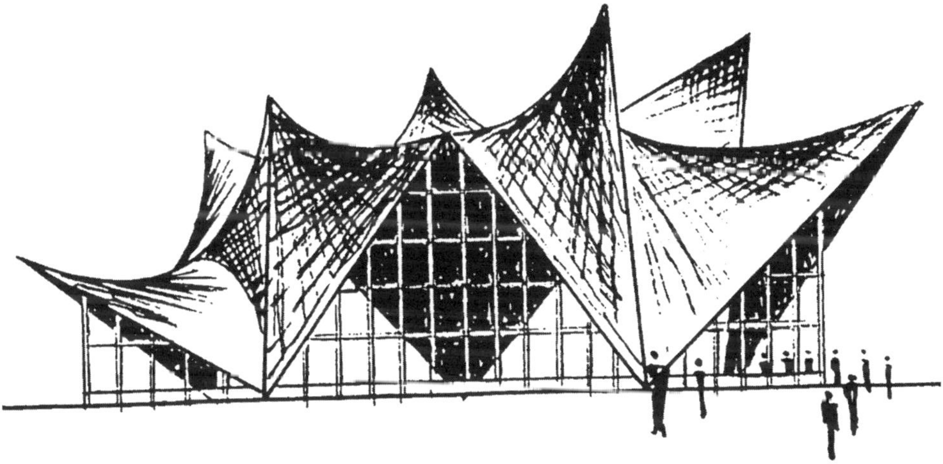

Figure 13. A variation in the basic dimensions gives infinite number of HP shells

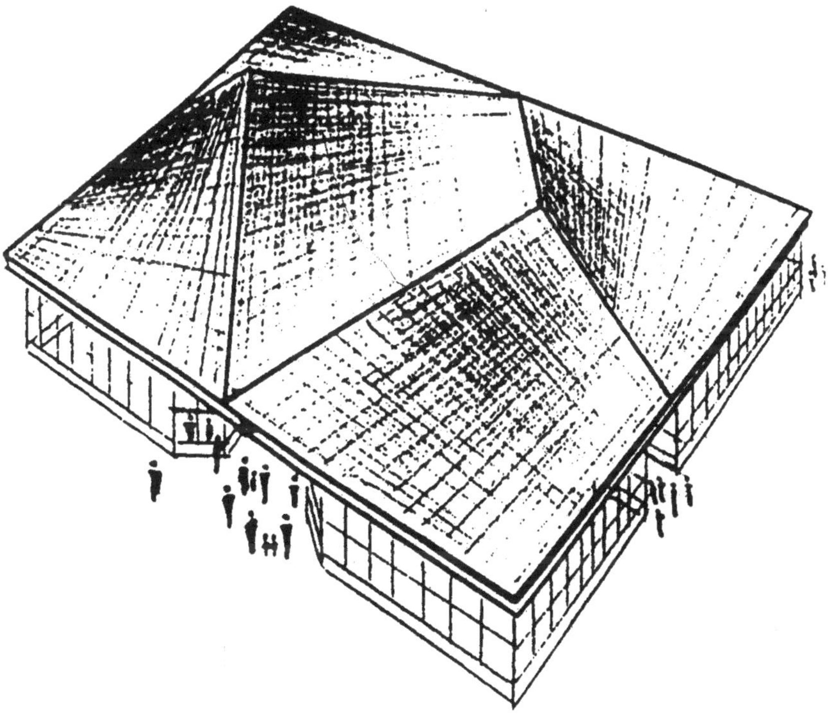

Figure 14. A large HP construction

Bamboo for Sustainable Development, pp. 703–705
Eds. A. Kumar, I.V. Ramanuja Rao and Ch. Sastry

A model to predict strength properties of bamboo

JULES J.A. JANSSEN
Faculty of Building and Architecture, Eindhoven University of Technology, The Netherlands

Abstract—Hundreds of thousands of people live in the rural areas of bamboo-growing countries. It is these people who predominantly use bamboo for construction. Yet, they are far away from the assistance of a laboratory to determine the strength properties of the bamboo being used for building their homes. In finding a remedy to this situation, two issues have to be borne in mind: (1) the current scientific knowledge on the subject has to be applied to find a method that does not require the technical tools used in laboratories; (2) the method has to be translated to a level at which it would be clearly understood by the rural people. Solving these two issues is not an easy proposition; this paper is an attempt to solve the two issues. Very simple tests are proposed to determine some properties, like the E-modulus, followed by some rules that would help deduce other mechanical properties.

INTRODUCTION

In rural areas, people build with bamboo according to local traditions, in the best case, or according to their own intuition, in the worst case. While this practice might lead to self-reliance of the local population, it might also result in unsafe constructions and/or in uneconomical solutions. In the follow-up of the decision taken at the International Bamboo Congress and Workshop at Bali in 1995 to prepare an international standard for bamboo, the question whether it might be appropriate to write also a manual for people in rural areas on how to build safer and more economical construction was considered. This manual was conceived as a bridge between the results of modern research in laboratories and the practices in the field. More often than not, the environment in research laboratories where the solutions emerge is vastly different from the environment in the field where those solutions are to be applied. For instance, while we have developed for bamboo a series of reliable tests using modern tools, the application of these test results are mostly in rural areas that are far from the laboratory facilities. For this purpose, the current scientific knowledge has to be applied and the ground reality thoroughly understood to reach an outcome without the technical tools used in laboratories. This, then, needs to be translated to a level that would be understood in the rural areas. Solving these two problems is difficult.

The document "A model to predict the properties of bamboo in a rural situation" approaches the issue thus: first, the background for each subject is described; this is followed by a trial of one or more lessons on the subject provided in the manual. Before the draft is finalized, it is essential to organize some trials in a rural situation to check the feasibility of this procedure: do the villagers understand the lessons in the manual, and do they consider it as an improvement on their tradition? Their opinion is the only validation and criterion.

THE MANUAL

After much study of reference works and reflection on his own experience, the author has selected three items for the manual.

1. Estimation of the quality of bamboo by determining its mass per volume (density) and by finding the ratio between this mass per volume and the E-modulus.
2. Design of a simple bending test, which could be carried out in a rural situation, to determine the bending stiffness *EI* of a bamboo culm.
3. A simple method of calculations to determine how many bamboo culms would be needed to support a floor, a bridge, etc.

The word "simple" in the above may be based on wishful thinking; but as said earlier, only a field trial can verify this.

A draft text of seven chapters has resulted from the efforts. In brief, these chapters deal with the following aspects.

1. As a beginning, a study was made on the ratio between the mass per volume (density) and the E-modulus. It found that the E-modulus (in N/mm^2) is 24 times the mass per volume (in kg/m^3). The question is whether it is possible to determine the mass per volume in a rural situation. This chapter deals with an idea that the author developed for a procedure suited to rural situation, and the tests carried out to check the reliability of the procedure.
2. This chapter discusses about the quality of bamboo, which, in effect is the same as the E-modulus. Explanations are given on various aspects such as: how to cut a piece of bamboo for testing; how to measure the dimensions; and how to calculate the volume. Once the volume and the weight have been determined, the "quality" of the bamboo can be calculated.
3. A bending test, which can be carried out in a situation outside a laboratory, is described next. It is shown that the bending stiffness *EI* can be determined as a function of the load *F*, the span *L* and the deformation *f*. Methods of calculation to check the assumptions made and the reliability of this mathematical simplification are described.
4. The following chapter explains how to perform a bending test according to the said simplification. It is a four-point bending test on a bamboo with a free span of 50 times the mean outside diameter, and the load is applied till the deformation equals 1.5 times the diameter.

5. Calculations can be limited to deformations, provided the deformation in practice does not exceed 1/300th of the span. In this area one has not to check the strength, which simplifies calculations considerably.
6. The sixth chapter gives some examples on how to make calculations, once the diameter and the wall-thickness of the bamboo are known, and how to interpret the "quality" of bamboo or the result of the bending test.
7. The last chapter is a reflection on the manual itself – on the methodology used, the level of dissemination required, the needs of rural people in terms of bamboo-related literature, etc.

CONCLUSION

A draft manual on the determination of properties of bamboo in a rural situation has been developed. This has to be checked in several trials in rural situations to check whether villagers understand the manual and if yes, whether they consider it helpful or not.

Bamboo for Sustainable Development, pp. 707–710
Eds. A. Kumar, I.V. Ramanuja Rao and Ch. Sastry

International standard on properties of bamboo

JULES J.A. JANSSEN
Faculty of Building and Architecture, Eindhoven University of Technology, The Netherlands

Abstract—One of the recommendations of the International Bamboo Congress and Workshop, held at Bali in 1995, was for INBAR to organize a task force to discuss and finalize a building code for bamboo. A task force constituted by INBAR in 1996 decided to prepare several documents, and one of these was to be on the determination of the mechanical properties of bamboo. The goal was to prepare an International Standard for the determination of mechanical and physical properties of bamboo in laboratory, as standardization of test methods was urgently needed so that tests on bamboo could be performed in similar ways across the world. The target group of this exercise was identified as laboratory staff, and the full name of the standard was to be the "International Standard on Testing of Bamboo for the Determination of Mechanical Properties". The subjects covered by the draft standard include: compression, static bending, shear, tension, moisture content, mass per volume and shrinkage. After discussion and approval the finalized document will be submitted to the International Organization for Standardization (ISO).

INTRODUCTION

The state of the art for the determination of the properties of bamboo is that every laboratory, unfortunately, is performing tests according to its own "standards". Researchers have done and are still doing a good job, even under difficult circumstances, in determining the properties of bamboo. Compared with timber research, however, a major disadvantage in bamboo research is the lack of common standards. This is a problem for the researcher, who has to decide how a particular test is to be performed, and for the users of the research results who find it impossible to compare the results coming from different sources. This, in a way, is a negation of the effort of the researchers and the need of the users; it is also an obstacle for the recognition of bamboo as an engineering material. Although the absence of standards has not hindered the construction of bamboo houses at low-income levels, it has certainly impeded the wider acceptance of bamboo as a reliable construction material.

Arce (1993) expressed this concern as follows: "Engineers and architects prefer to work with the determinacy of a well known system or material, supported by solid knowledge of its properties, backed by the existence of a minimum code of specifications on which they can base their judgement and design decisions". A similar statement has been made by the Indian Standards Institution, in the preface to IS 6874 (ISI 1973) Methods of tests for round bamboos: "In order to ascertain the suitability of bamboos for various structural purposes, their physical properties are required to be tested. Standard methods of tests have been developed for the purpose and the same are described in this standard. The data obtained by these methods may be made use of with confidence where strength factors are involved".

Such reflections have been the motivation for many participants of the previous International Bamboo Workshops to discuss whether it would be desirable to develop International Standards for the determination of properties of bamboo in laboratory. One discussion, in particular, has been a stimulus for the author. At a timber congress, he expressed his concern about the lack of an international standard for bamboo. His colleagues in the timber sector told him not to worry because even in the world of timber, different standards exist in different countries and a single world standard would now be almost impossible. For bamboo, however, it is quite possible to agree on one single international standard, which will be a huge advantage for the recognition of bamboo as a reliable engineering and building material.

For the reasons mentioned above, the previous International Bamboo Congress and Workshop at Bali in 1995 recommended that the International Network for Bamboo and Rattan (INBAR) constitute a task force to discuss and finalize a common standard for bamboo. In the following years, discussions led to the conclusion that it is better to divide this standard into two parts: one for architects and engineers, a "Building Code"; and one for laboratory staff on how to determine properties of bamboo. This paper deals with this latter part of the proposed standard for bamboo.

DEVELOPMENT OF THE STANDARD

The optimism that accompanied the decision at the Bali Congress and Workshop soon turned into disappointment. Activities could not be started owing to lack of funds, and it was almost two years before this problem was solved. In the meantime, a modus operandi was decided upon. It was agreed that the author would write a first draft, and a group of about 15 referees from several bamboo-growing countries would give their comments on this draft. The first draft was ready in February 1998, and distributed to the said referees. Within a few months most of them gave very valuable comments to improve the contents of the draft.

Another round of discussion will be held soon in San José, Costa Rica, to finalize the text of this standard. From that moment onwards this standard would already have some authority within INBAR: no researcher would willingly ignore this document when starting a new research program on bamboo. But for the recognition at the international level, two more steps need to be taken: recognition by National Standard Organizations of bamboo-growing countries, and recognition by the International Organization for Standardization (ISO). Invitations have been sent to National Standard Organizations in 18 countries (Bangladesh, Brazil, China, Colombia, Costa Rica, Ecuador, Ethiopia, India, Indonesia, Japan, Malaysia, The Philippines, Sri Lanka, Tanzania, Thailand, Vietnam, Nepal and Peru), requesting their participation in a discussion on this standard, and many positive reactions have already been received. ISO has been contacted as well, mainly enquiring about the formal procedure to be followed. ISO prescribes that an initiative like this should be submitted to the Technical Committee in charge of the matter, in this case TC 165 on “Timber Structures”.

CONTENTS OF THE STANDARD

The guiding principle for the draft standard has been to follow, as much as possible, the same outline as the existing standards on timber (like some ISO standards, ASTM and Eruocode 5), to facilitate its acceptance. Where differences between timber and bamboo exist, these have been clearly recorded. Much use has been made of the Indian Standard IS 6874 on testing of round bamboos, and of some recent scientific publications on testing of bamboo, like Arce (1993). The contents of the draft standard are given below.

1. General items: foreword, with a general introduction; contents; and scope.

- Terminology, units and symbols, for which the ISO standards are used as far as possible.
- Organization and use of the standard, including the determination of size and weight of specimens, temperature and moisture content, and the rate of loading in tests.
- Sampling and storage of specimens. This is a very important item, as incorrect sampling would render the test series useless. A notable difference is made, depending on whether the sample is for scientific research or for commercial purpose. If it is for scientific study, the selected culms have to be totally defect-free; if it is for commercial purposes, the sample has to be representative of the lot in all aspect, including defects. A good book-keeping of the culms is prescribed, with emphasis on how to divide full culms into test pieces and how to record from which culm each test specimen has been taken.

2. Tests on compression, static bending, shear, tension, moisture content, mass per volume (the modern name for density), and shrinkage. Much guidance has been obtained from the Indian Standard IS 6874, Method of tests for round bamboos. Data from publications on research results and relevant ISO standards have been taken into account as well. Each item has the same structure: scope and field of application, references, principle, apparatus, preparation of test pieces, procedure, calculation and expression of results, and test report.

In the text on compression, much attention has been given to a method to avoid strengthening of the test piece by friction between the bamboo and the steel plates at both ends, as this is a problem while testing bamboo specimens. For bending tests, a minimum free span has been prescribed to avoid failure by transversal forces instead of bending moments. For tests on tension, special attention has been given to a form of the test-piece that will fail by tension but not by lateral compression at the grips.

CONCLUSION

The recommendation of the International Bamboo Congress and Workshop at Bali in 1995 – to constitute a task force to prepare standards for building code and for testing bamboo properties in laboratory – has been fulfilled (see this volume for more information). A draft standard for testing properties of bamboo in laboratory has been written and reviewed by a group of specialists. Contacts have been made with National Standards Organizations, requesting their participation in the final formulation of the standard, and with ISO for determining the formal procedure for issuing the standard prepared as an International Standard.

ACKNOWLEDGEMENTS

Many specialists from several parts of the world have assisted the author in performing this task. He is grateful to all of them, particularly to N.S. Adkoli, R. Gnanaharan, K. Ghavami, M.A. Latif, S. Rajput, and J. Siopongco.

REFERENCES

Arce, O. 1993. Fundamentals of the design of bamboo structures. Ph.D. thesis, Eindhoven University, The Netherlands. 261 pp.

ISI (Indian Standard Institute). 1973. IS 6874, 1973, Method of tests for round bamboos. Indian Standard Institute, New Delhi, India.

Bamboo for Sustainable Development, pp. 711–713
Eds. A. Kumar, I.V. Ramanuja Rao and Ch. Sastry

Production of a manual on testing methods for the use of bamboo in engineering applications

JULES J.A. JANSSEN
Faculty of Building and Architecture, Eindhoven University of Technology, The Netherlands

Abstract—Tests on different properties of bamboo are being conducted in many laboratories in bamboo-growing countries. In the absence of standards that prescribe definite methods, the laboratory staff often has to improvise and innovate to find clever methods or tools. In many cases, such useful information stays inside the originating lab owing to lack of exchange of such knowledge. To remedy this, a laboratory manual is proposed with the purpose of publishing testing methods being practiced in laboratories in different parts of the world. Such a publication will also form a practical "how-to-do" manual, explaining the steps involved in performing tests according to the "International Standard on Testing of Bamboo for the Determination of Mechanical Properties" (see this volume).

INTRODUCTION

Tests on the properties of bamboo are being carried out in many laboratories in bamboo-growing countries, frequently using ingenious ways to surmount problems fostered by the lack of ideal working conditions. People who are accustomed to laboratory facilities in the so-called developed countries will be surprised time and again by the clever solutions derived by the staff working in such developing country laboratories to overcome, for instance, a shortage in measuring or loading equipment. The laboratory staff working in developing countries has an eye for the physical reality of the bamboo material, having had the advantage of growing up with bamboo right from their cradle.

A very important drawback of the working methods in these laboratories, however, is that the methods employed for the same test may differ from one laboratory to the next. This means that the results from similar tests from different laboratories cannot be compared; the results are influenced by the test methods. Some examples of this sad reality are:

- Bending test can be carried out as a three or four-point test, which gives different outcome for strength and stiffness;
- In shear tests, the form of the test piece is critical because many forms introduce bending moments in the shear area, and the preparation of the test piece itself is sometimes complicated; and
- The test speed influences the resulting strength.

DEVELOPMENT OF THE MANUAL

In the International Bamboo Congress and Workshop, held at Bali in 1995, it was decided to prepare a building code for bamboo. In the discussions that followed, a decision was taken to develop a "Building Code" and a standard for the determination of properties. While the author, with the assistance of many referees, was preparing a draft for this standard, it was felt that a standard should be like a legal document, at a rather abstract level. This, however, meant that many informal but valuable pieces of information have to be discarded. Therefore, it was decided that a manual, which follows the said standard but gives practical information, should be developed. This would provide the laboratory staff with enough how-to-do information on different tests, while the formulation in the standard would leave an opening for any future development.

The first draft of the manual written by the author was sent to a group of referees in bamboo-growing countries. Many useful improvements have been proposed by this group, and taken into account in a second draft. In a meeting at the end of October 1998 in San José, the group of referees is expected to finalize the text of this manual. It is proposed that the finalized manual, after editing, be published by the International Network for Bamboo and Rattan (INBAR) as a Technical Report. According to developments in practice, updated editions can be issued.

CONTENTS OF THE MANUAL

The "Laboratory manual on testing of bamboo for determination of mechanical properties", will act in tandem with the "International standard on testing of bamboo for determination of mechanical properties". The contents of the manual, which gives practical solutions and explanations on how to work with the international standard, are given below.

- The proposed standard prescribes, for example, that culms shall be marked clearly before felling; the manual gives a model for a form that can be used in the field, and ideas on how to mark the culms and how to make a sketch of each culm.
- The manual gives some elementary explanation about the statistical background of the prescriptions in the standard: for instance, why the number of test specimens should be not less than such and such.
- In compression test, steel plates are often used to load the bamboo test piece till failure. The standard, however, prescribes the use of a friction-free intermediate layer, and the manual explains why it should be like this and gives a practical solution.

- For the bending test, it explains how to support and load the bamboo culm with wooden saddles in order to create failure by bending stresses and not by forces perpendicular to the bamboo culm.
- Examples of data sheets are provided for several tests, which can be used while reporting test results.
- Some examples of calculations are also given, with a special emphasis on the correct use of units.

In general, much use has been made of data from the Indian Standard IS 6874 (ISI 1973) and the handbook of Sotela (1990). The latter, in particular, contains valuable practical experience emanating from a good laboratory, the Laboratory for Forest Products at the University of Costa Rica.

CONCLUSION

The recommendation of the 1995 International Bamboo Congress and Workshop at Bali has been fulfilled to a level better than that could be foreseen three years ago. After the finalization of the text of the manual, INBAR would be able to publish it as a technical report for bamboo laboratories. From that time onwards, the staff in these laboratories would be able to improve the level of their tests. The beneficiaries would be the millions who use bamboo.

ACKNOWLEDGEMENTS

Many specialists from several parts of the world have assisted the author in performing this task. He is grateful to all of them, particularly to N.S. Adkoli, R. Gnanaharan, K. Ghavami, M.A. Latif, S. Rajput, and J. Siopongco.

REFERENCES

ISI (Indian Standard Institute). 1973. IS 6874, 1973, Method of tests for round bamboos. Indian Standard Institute, New Delhi, India.

Sotela, J. 1990. Manual de ensayos fisico-mecanicos para especimenes de bambu (Manual for physical-mechanical tests on bamboo specimens). Laboratory for Forest products, University of Costa Rica.

Bamboo for Sustainable Development, pp. 715–719
Eds. A. Kumar, I.V. Ramanuja Rao and Ch. Sastry

International building code for bamboo

JULES J.A. JANSSEN
Faculty of Building and Architecture, Eindhoven University of Technology, The Netherlands

Abstract—One of the recommendations in the International Bamboo Congress and Workshop, held at Bali in 1995, requested the International Network for Bamboo and Rattan (INBAR), "to organize a task force to discuss and finalize a building code for bamboo". Consequently a draft was prepared under the title, "An International Model for National Building Codes, Section Bamboo". The target group of this Code comprises architects, builders, designers and engineers. These people are already familiar with national building codes for timber, masonry, etc. For bamboo, however, a similar Code is missing and this is an obstacle for the recognition of bamboo as a building and engineering material. After approval by the next International Bamboo Congress and Workshop, the document will be submitted to the International Organization for Standardization (ISO). Once it has been accepted as an ISO-Standard, it can serve as the model or basis for developing National Building Codes by countries that grow bamboo and use it in construction. The topics included in the Code are: philosophy, design concepts, structural design, beams, columns, joints, trusses, panels, durability, fire protection, grading and quality control.

INTRODUCTION

Why does bamboo need an international building code? Typically, bamboo is a material used in the informal sector in rural areas, for which application a code is not needed. It is also true that in many cases building codes are being used to declare the houses of the poor people sub-standard and demolish them. These objections are valid, but there is much more to bamboo and codes. Firstly, bamboo can be developed as a widely recognized building and engineering material, which maintains its value for the informal sector as it is but adds more value for other groups of the population. Secondly, good building codes or standards do not harm any section of the society; they protect consumers, workers and the environment, and promote economic growth. The issue is not whether bamboo codes should be implemented, but who defines them. Standards can play a role as part of the development solution. We are living in an increasingly regulated world. This urges developing countries to join their efforts to develop standards that meet their specific requirements. The central issue is that standards regarding production and consumption recognize the environment in developing countries.

Looking more specifically into building codes or standards, they should aim at the following goals:

- Be in line with local tradition;
- Recognize that building is an incremental process;
- Take into account the needs of the masses;
- Recognize the needs of different sections of the society, and assist in improving their dwellings;
- Take into account the lack of funds;
- Take into account people's priorities;
- Work with local materials.

A bamboo building code should also improve the position of bamboo as a recognized building and engineering material, with access to the international market. This would promote banks and other financial institutions to support the use of bamboo in construction.

DEVELOPMENT OF THE CODE

Following the Bali Congress and Workshop, quite some thoughts have been given to the Code: What should be the structure of an international model code for bamboo? How would it fit in with existing National Building Codes? What documents and information are available? In the course of this process, several documents have been collected, including: standards for timber, like ISO and ASTM standards and the Eurocode 5; articles about standards for developing countries; and results of scientific tests conducted on bamboo.

The importance of an international model code cannot be underplayed. If the development of national bamboo code is left to national authorities, a wide diversity among different national codes is certain. This point is amply illustrated by the case of timber: standards have been individually developed accepted in the United States, Europe and Australia, and they can never agree on a common world standard any more. An international model for national standards, on the other hand, would strengthen the position of bamboo considerably by providing a platform of universal acceptance and facilitate the work of national authorities. For bamboo, the opportunity for a world standard is open, and if we succeed in bringing it about, bamboo would be half a century ahead of timber.

The differences between commercial and non-commercial sectors are very important in the case of bamboo. The purposes of regulations for the commercial sector are:

- To provide a healthy environment;
- To reduce the frequency of structural and other functional failures with regard to the environment;
- To derive improvements in the quality and economy of construction by setting up standards that provide a reasonable and readily usable basis for contracts between builders and owners, while permitting innovation and development; and

- To protect the interests of neighborhoods, and those of the community as a whole, from irresponsible development of sites that may otherwise take place.

The purposes of regulations for the non-commercial sector are:

- To provide improvements in the standards of hygiene;
- To mitigate suffering and loss in the event of natural disasters, such as earthquakes, winds, storms, floods and fires;
- To meet the needs of the owner or the occupier of the building, rather than to serve the convenience of the regulating authority; and
- To educate owners, occupiers and builders about the means with which the above can be achieved economically.

THE INTERNATIONAL MODEL CODE

A draft of the international model code has been prepared. A summary of its contents, given below, would be helpful in understanding the goal and direction of this effort.

1. The document starts with a general philosophy about the purpose of this model, a description of the general requirements for the design of structures (mostly with regard to safety), a description of the scope, and assumptions and definitions.
2. Design concepts are dealt with next. For the commercial sector, design concepts as stated in the code shall be based on calculations; but for the non-commercial sector, the code is open to the use of good traditions and evaluation reports. This is exceptional for a building standard, but bamboo needs such an approach. The use of good tradition is defined as follows:

"Experience from previous generations, well preserved in a local tradition and carefully transmitted to people living today, by means of a reliable system of oral transfer. This expertise can be considered as an informal, non-codified standard".

The criteria for the reliability of a transfer system for traditions are:

- The content shall be generally known and accepted;
- It shall be considered as old and pure tradition, as general wisdom; and
- The community shall be characterized by an undisturbed social structure, with a well-recognized social pattern.

The limitations of such a system are identified as:

- The content cannot be transferred to other localities, to different botanical species, etc.; and
- After migration, the presence of this tradition will no longer be self-evident.

Evaluation reports referred to in the definition of good traditions are defined as follows:

"Reports based on evaluations made after disasters like earthquakes and hurricanes. If these reports contain descriptions about structures that did survive a quantitatively described disaster, similar structures shall be considered as adequate for similar disasters in the future."

The criteria for the reliability of evaluation reports are:

- The report shall be composed by acknowledged engineers, with adequate experience in the field;
- The report shall be accepted by the international technical community and/or proven by referees; and
- The report shall give full details and complete information, with which one can build similar buildings.

The limitation of such reports is that they cannot be transferred to other localities, to different botanical species. etc.

3. Criteria for structural design, covering limit states, serviceability, material properties, design requirements, allowable stresses, and other items needed in any building code or standard.
4. Schematization – the process of "translating" the physical reality of a building structure into a schematized system of symbols that can be used during the calculation process. Schematization is based on the theory of applied mechanics. Compared with materials like steel and timber, this process is even more important for bamboo because of its properties like elastic behavior, the thin culm walls and the form of most bamboo culms – not perfectly straight, tapered and having nodes at different intervals.
5. Structural elements – beams and columns – are discussed next. Many elements from timber standards are applicable here, but with special emphasis on bamboo's unique properties.
6. The next subject is "joints", the most difficult item in bamboo-based construction. In most examples of bamboo structures, the strength and stiffness of bamboo culms are reduced by the weakness and deformation of the joint. This is a drawback for the safety and economy of any bamboo structure. The model code describes a minimum level of safety, and promotes development of joints that are both safer and more economical. The treatment of trusses that follows is an extrapolation of the discussion on joints.
7. Panels form the next item, and with this the model enters the area of modern industrial applications of bamboo. The production of panels from bamboo is a good option for job creation in new industries in any region where bamboo grows. It can also improve the economy of a bamboo-growing country, as it is an option to obtain foreign exchange through exports to Western Europe and North America. The importance of standard for the international market is, of course, obvious.
8. Durability is the weak spot of bamboo as a building material. Fortunately, in the past years sufficient progress has been made in bamboo preservation, and

based on these results it might be possible to prepare a draft standard that shows the way for a gradual improvement.

9. Grading and quality control are the last items in the model. These subjects are essential for the recognition of any building material, but difficult to describe in a draft standard. These areas still require a wider discussion.

CONCLUSION

The draft is still under preparation, as some of the areas mentioned above require more thoughts and a wider discussion of the different aspects involved. After completion of the first draft, it will go through the following steps: comments by the referees; writing an improved second draft that incorporates the comments of referees; submitting the second draft to the referees and National Standards Organizations of bamboo-growing countries; finalization of the document based on feedback from referees and National Standards Organizations; and submitting the final document to ISO.

Section 4:

Bamboo Design and Utilization

Bamboo for Sustainable Development, pp. 723–737
Eds. A. Kumar, I.V. Ramanuja Rao and Ch. Sastry

Design technology, training and marketing (DTTM) link for survival of bamboo craft in the 21st century

A.G. RAO

Industrial Design Center, Indian Institute of Technology Bombay, India

Abstract—Bamboo craft has been a livelihood for a large number of people in South countries. In recent times, industrialization and general neglect have caused the decline of this craft. Bamboo being an eco-friendly material, the potential of bamboo craft to offer sustainable employment in future is immense. A vision is necessary to see the craft as a trade of the future, not only to check the population flow to the urban centers but also to offer a creative occupation for the future. Linking of design, technology, training and marketing with insights in each one of these facets offers new scope for bamboo craft.

Weave forms resulting from the study of bamboo craft can open up new areas of pedagogy in design. Post-modernism at one end and green design at the other can give a new expression to bamboo craft design. Combining bamboo craft with other materials and craft processes can lead to new experiments by seeing design as an art form. Linking design to marketing and production can ensure effective inputs into design transfer at the grassroots level.

Product-based craft technologies for laminations and finishes with natural dyes can add a new dimension to product look. Linking technology to design and marketing can result in new product strategies for bamboo-crafted products. Training of craftsperson in the use of jigs and fixtures can result in enhanced quality of the crafted product. Visual communication through print and electronic media opens new modes for the training of the craftsperson. Marketing via Internet can open up global markets. The paper examines these possibilities in detail.

INTRODUCTION

Bamboo craft has been a livelihood for large number of people in bamboo-growing developing countries. In India alone, 1.3 million workers are estimated to be dependent on bamboo. Increasing industrialization and neglect of traditional occupations have caused the decline of this craft. A considerable number of persons has abandoned the craft in favor of other more profitable occupations. Unemployment is one of the major problems in the South countries, and environmental damage owing to excessive use of natural resources is another. In this context, bamboo craft has three important implications for the future:

1. Bamboo craft can create eco-friendly employment in the rural areas, thereby checking the disproportional exodus of population to urban centers, affecting the quality of life in general.

2. Bamboo, being one of the fastest growing biomass, can act as a substitute for wood. Reduction in demand for wood could save already threatened forests.
3. Bamboo craft can become a cultural tool to promote healthy attitudes to craft, environment and life in general.

Many South countries have recognized the importance of bamboo as a raw material, but little focus has come to bamboo craft as a trade of the future. Rich craft traditions, with amazing skills, already exist in certain parts of the world like India, particularly in its North-East. Learning from this tradition and repositioning bamboo craft as a viable, healthy, self-reliant trade is essential for its very survival in the 21st century.

Seen from the current situation, however, the future of bamboo craft looks bleak. To fully understand this scene one needs to trace the historical background. Modern design developed from crafts. Bauhaus in Germany, the famous school of design, had laid considerable emphasis on combining crafts and arts. As the industrialization process rose to dominance for a period, crafts took a back seat. Bamboo craft had no place in the pedagogy of design even at the time of Bauhaus, as the material and skills needed were unfamiliar. Potentials of three-dimensional weave forms thus remained dormant to professional design.

Post-modernism has opened up new dimensions in design, with a considerable debate going on in professional design circles. As Peter Dormer observes, "Reexamination of craft aesthetics and methods is one way that diversity and originality may be institutionalized in design". (Thackara 1988).

Excessive profit motivation of industries, which make huge investments in new product introductions, has frustrated designers. New technologies, such as computer-aided design and rapid prototyping, have minimized the creative role of the designers in an industrial set-up. Post-modern movements like "Memphis" have rebelled against the earlier aesthetic hold. Looking at product as an art form has brought new life into design thinking, and bamboo craft design can offer new means to this process.

Green design is a concern of the immediate future. It has started looking at the total process of manufacture with concepts like Life Cycle Analysis (LCA) of products. Each process and source of each material is being looked at from the points of view of sustainability and environmental damage. Concepts such as zero waste in production are gaining ground. From these concerns, craft-based production in rural areas assumes great significance for the future. To make such concepts viable in the current socio-economic structure, a new understanding and linking of design, technology, training and marketing is essential.

DESIGN

Large gaps exist between the knowledge frames of a traditional bamboo craftsperson and a modern industrial designer, and it is imperative to bridge this gap for the survival of the craft. Today's training in industrial design is heavily biased towards using industrial materials and processes related to metals and plastics, with the result that very little is learnt about bamboo or bamboo craft. Weave forms in bamboo have acquired sophistication in using geometry. The very structuring and the process of weaving control the form of bamboo basketry. Various weaves and forms generated in bamboo craft have been studied by Ranjan (1985). Traditional ways of rim treatments, joining of legs and handles, etc. have much to offer to modern design. Weave forms open up a new chapter in the study of form courses in design. On the other side, a traditional craftsperson is little exposed to the ways of contemporary design. Modern concepts of design – such as user-friendliness, ergonomics, stackability, packaging and use of other materials – have to be exposed to craftspersons for the survival of bamboo craft in the competitive markets.

Post-modern movement in design offers a new impetus to bamboo craft. To start with, it has questioned all forms dominated by the constraints of industrial mass production and distribution. The tacit bindings of these visual orders, hitherto taken for granted and justified by the economic notions, are now being questioned. In the post-modern ventures, we find notions of symmetry, homogeneity of material and forms, violated intentionally to reveal new orders and meanings. Movements in literature, such as "de-construction", have started influencing design leading to unrelated visual elements in the products. Messages that come in fragments have come to be accepted as part of communication. Interestingly, today's youth have readily accepted this as a reality. What would be the advantage for bamboo craft in this new world scenario? It can bypass modernism and move towards post-modernism, which is looking at design as an art form (Fig. 1).

So far, bamboo's future has been seen as that of a cheap substitute to wood. An urgent paradigm shift is hence necessary to see bamboo as a rich material. No doubt, proper treatments and new finishes will help in such a shift; but new experimentation in design by combining different materials with bamboo and creating products with new messages is the vital need of the hour. In categories of products for home and personal use, where cultural identities are sought for, new designs using bamboo craft have immense scope.

Eco-friendliness

Green design has become the current concern following the damage caused by our increasingly unsustainable consumption patterns. Bamboo, being eco-friendly, offers a ready material that addresses this concern. Bamboo craft can rely on this intrinsic value of bamboo for finding wider markets in the North, where consumers are becoming increasingly aware of such issues (Whiteley 1993). Design developments using bamboo craft as a venture in green design need to become a concern for the design community all over the world.

Figure 1. Influences on art and craft

Design transfer

The process of knowledge transfer from a professional designer to a craftsperson or a crafts group and vice versa continues to pose unresolved problems. Crafts groups and leaders of non-government organizations (NGOs) need to understand the trends of professional design, no doubt newly interpreted for the use of bamboo craft. The following inputs to craftspersons are crucial for effecting design transfer:

1. Exposure to product planning, and evaluation of designs at the concept level as well as with actual feedback from dealers and users;
2. Exposure to concepts of user-friendliness, convenience of use over a time, product maintenance, etc.;
3. Exposure to process planning and to the use of mechanical devices such as jigs and fixtures to reduce drudgery and increase quality;
4. Exposure to packaging aspects, such as stackability, visibility in a store, product information, etc.;
5. Exposure to finishes in bamboo; and
6. Exposure to other materials, which can combine well with bamboo – ceramics, rattan, cast iron, jute, brass, copper, steel, etc.

There is a need for schemes to link professional design inputs to crafts groups. The development of a tray by Sign Design at Pune, India, provides a good example of how practicing designers can get involved with crafts groups. Sign Design received a brief for the design of a file tray that had a potentially good demand in the large city of Mumbai. It took the help of Mr. Deca, a crafts trainer at the small town of Sangli in Maharashtra, to develop the tray and execute sample orders. A short workshop for a crafts group was conducted at Sangli jointly by the Industrial Design Centre (IDC) faculty and Sign Design. However, when the orders were actually placed, the crafts group was not organized to supply.

Another group near Mumbai, Academy for Developmental Sciences, lost an order for bamboo trays with steel wire frame owing to lack of detailing, which can come only from professional design inputs. It is important to develop suitable schemes to involve professional designers in bamboo craft at different levels. Many designers and design offices will be willing to work in bamboo craft for a concessionary fee.

TECHNOLOGY

So far, technology development in the bamboo sector is akin to that in many other areas: developments are based on increased production and elimination of workers. For instance, an automatic chopstick-making machine needs no craftsperson, as it is very capital-intensive.

Bamboo ply board manufacture is based on bamboo mats made by hand. Though the technology looked highly promising in India, several problems have surfaced. For example, ensuring quality mats at low costs as initially assumed by the industry has become a problem. Even in such ventures, craftspersons not only end up doing uncreative jobs but also become dependent on the success of the industry for their survival. Poor management of the industry and inadequate marketing of the boards, both of which are not in the hands of crafts group, can affect their living. Further, as the industry is small, it faces distribution and post-sale service problems. The bamboo ply, which has a high proportion of resin, is strong but difficult to cut with a handsaw. This affects the end use, as most carpenters who use hand tools refuse to work with bamboo board.

The Chinese experience in bamboo ply, with 200 factories providing employment to 26 000 people, has been positive. But even in China, "bamboo based boards find use only in concrete form work, packaging and as floor for vehicles and houses" (Zhu 1996).

It would be better to develop technologies to make molded bamboo board as an end-product rather than a raw material competing with plywood. This would ensure entry of bamboo ply into other value-added products. Products such as molded chair shells, furniture parts, suitcase shells, shelving systems, trays and containers that use bamboo mat board would have a high market potential.

An interesting craft-based industry exists in a few villages like Cherilam and Koimura near Agartala, the capital of Tripura, a north-eastern Indian state. About 150 families in each village are engaged in basket weaving. They all make identical baskets with star weave. Surprisingly, a self-organized segmentation has taken place in making the baskets. Some people are involved only in making the bamboo strips, while some are making only rims (Figs. 2, 3). Some others weave the baskets taking strips and rims from others. This village industry faces a different problem; it is totally dependent on government organizations for marketing and new designs. Introduction of new designs and technologies linked to design and marketing can be readily absorbed in such village industrial set-up.

Lamination technology for bamboo is available at industrial level in countries like Japan (Lam 1996); but lamination technology suitable for craft can be developed. Simple jigs, fixtures and hand-operated machines can lead to solutions that are more creative. Laminated products developed at IDC have clearly illustrated the potential (Figs. 4, 5).

Appropriate technology for bending bamboo in a controlled fashion is another need. Steam bending or hot air blowers at high temperature can facilitate the manufacture of many products. A mechanized gadget for controlled smoking and appropriate spray booths for spraying lacquer and other finishing materials are the other urgent needs.

Figure 2. An artisan making bamboo strips

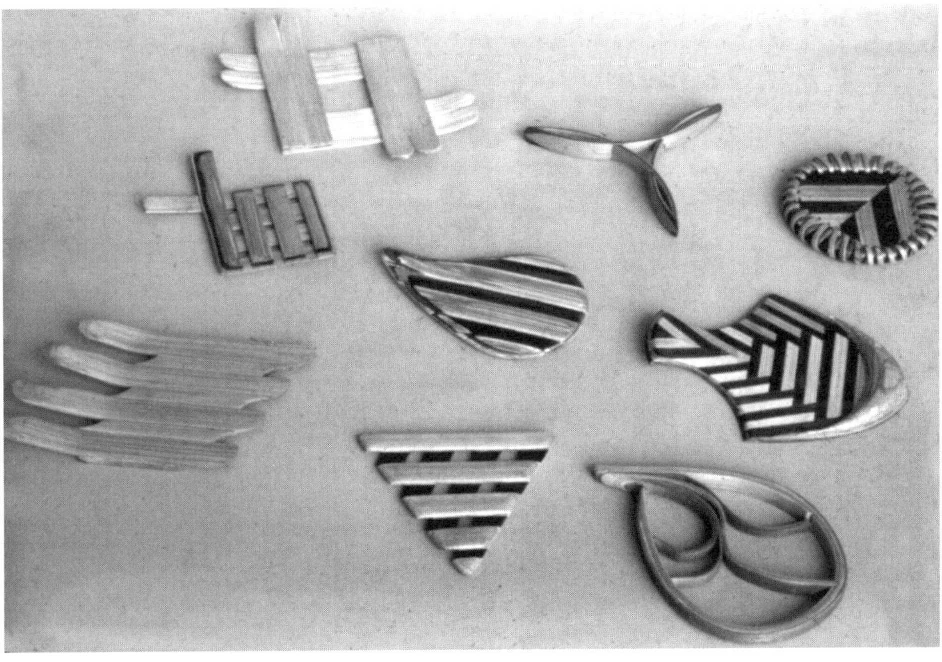

Figure 3. A basket weaver at work

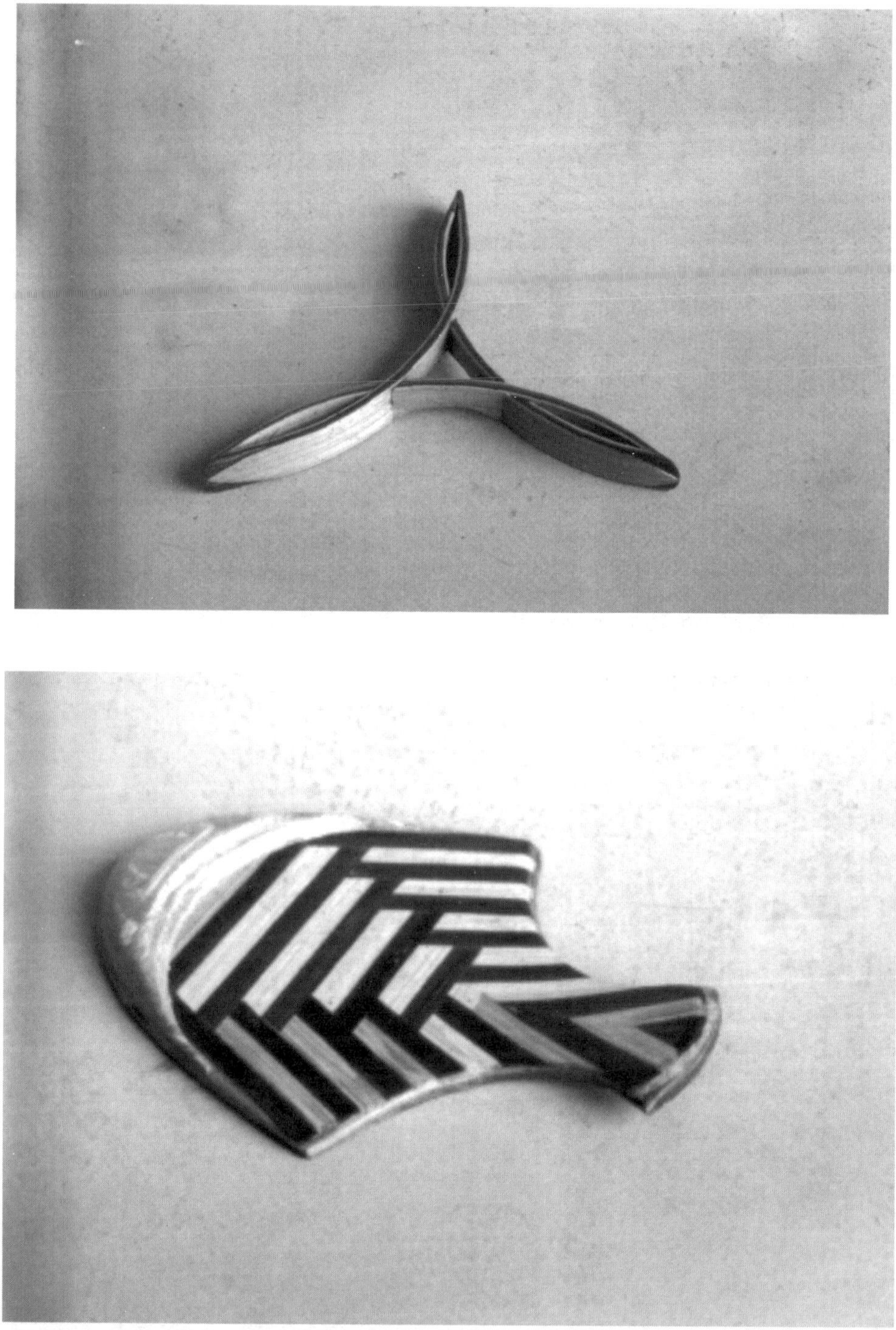

Figure 4. Laminated bamboo products

Figure 5. Laminated bamboo products

Various jigs and fixtures are required to adopt some of the wood working machines for bending and sanding bamboo. Again, it is important to link this effort with design so that a product-oriented technology can be developed for bamboo craft. Joints in bamboo are critical in several product applications. Standard screws, nails, latches and hinges cannot be easily used with bamboo. There is an urgent need to develop joining elements that are easy to fix and have an aesthetic appeal. Here too, it would be important to link this effort with product design and market feedback. In products such as hair clips and earrings, bamboo craftspersons are already using available metal fittings. Specially designed hooks, rims, etc. can bring in a new quality to bamboo-crafted product. Also, elements in other materials need to be developed for bamboo craft. Such efforts need to be connected to market feedback.

TRAINING

Bamboo craft, like many other crafts, is learnt traditionally within families. Certain communities in India, like the Buruds in Maharashtra, have bamboo craft as their sole profession. Although almost every person in tribal communities learns traditional craft, the communities that take up bamboo work are few. The advent of modern education has put further stress on bamboo craft; whenever a craftsperson gets "educated" in the modern sense, he or she leaves the traditional learning.

There was no organized formal learning in bamboo craft in India earlier. Efforts by the government have resulted in institutions such as the Bamboo Cane Development Institute in Agartala, where a six-month rigorous course in bamboo craft is available to 20 persons at a time. Still, bamboo craft has not become a part of technical training. Strangely, craft, which was the predecessor of technology, got segregated from modern technology. Results are that today:

- There are no higher-level courses in crafts; and
- Improved technical inputs do not reach craftspersons at the grassroots level owing to the absence of a suitable mechanism.

The lack of higher-level courses has alienated designers and architects from bamboo craft learning. Surprisingly, even in premier educational institutions such as the Shilpa Sadan at Shantineketan, where 3-year craft training is offered after school education, bamboo craft was removed for lack of demand. An educational backup by an institutional structure gives hope and security in a country like India. Lack of it is a serious problem even for a traditional craft like bamboo craft. Those persons, who are more talented and are able to take initiative, tend to leave bamboo craft in favour of other occupations. Introduction of study of bamboo craft at various levels of education can create faith in the craft and help build careers in it. At school level, bamboo craft can contribute significantly towards developing dexterity of hands and creativity in children. Such inputs at school education are important for all further education, including science and engineering (Rao 1994).

Introduction of diploma courses in bamboo craft and design after school education (10+2) can create a human resource pool that can supply entrepreneurs. Offered as an elective subject at college for students of architecture, design and engineering, bamboo craft can consolidate the knowledge base. This would also facilitate higher-level projects in bamboo craft, leading to Master's and Ph.D. degrees. There is also great demand in urban centers for learning bamboo craft as a hobby or creative enterprise.

Since a large percent of bamboo craftspersons still depends on craft for living, it is imperative to reach the enhanced skills and techniques to them at the village itself. NGOs and field workers have a major role to play in this. Development of teaching material on bamboo craft, updating its level to help make craft products that can be sold in competitive markets, is essential. Research into the design and technology of craft products and visual communication are crucial for this task. Use of videos and multimedia will have additional advantage. Rapid progress in mass media is creating interesting new situations: for example, bamboo craft groups, which cannot get electricity to increase their production, have access to television through community centers.

Emphasis has to be on the quality of educational material generated. Firstly, the training needs to emphasize on attitudes and techniques to update actual making of marketable products. Secondly, the link between design, technology and marketing with training has to be strengthened. This needs the bridging of gaps between craft skills and industrial design skills, marketing during local events and web marketing, craft technology and computer technology.

MARKETING

Bamboo-crafted products are being sold for generations. In a rural setup, people go to the crafts groups and buy or order the products needed. This practice continues till today in India. To cite an example, Navinbhai, a bamboo craftsperson in the river island of Majuli in Assam, was continuously engaged to make trays for marriages. His workload was so high that he could not make new designs, a work that requires more time and effort. He was not willing to hike the price of bamboo trays, though he could very well have done it, as all customers were known to him since his childhood. Probably, these customers lend him assistance when the need arises. Such healthy traditional relations continue to exist in a few pockets. However, in many places, the new generation is not aspiring or willing to take up bamboo craft as a career.

Local fairs

Another major traditional marketing avenue for crafts is *melas,* or fairs held in conjunction with some local event of importance. *Melas* take place all over India at different times, mostly during religious festivals. *Kumbha Mela,* which happens once in 12 years and attracts millions of people, is one well-known example. A wide variety of products is sold in these *melas.* Bamboo craft items are a usual fare in these markets.

Weekly markets

Weekly markets are a common practice in many big villages or towns all over India (Fig. 6). All kinds of goods – from cows and goats to vegetables and baskets to pots and furniture – are sold in these markets (Fig. 7). This is an important avenue for marketing rural produce.

Traditional craft marketing is rooted in one-time selling in a festive environment, a trait that can be observed all over the world. Such marketing offers several advantages for the craftspersons:

1. The festive atmosphere induces people to make purchases; several save money and plan to buy things at these festive gatherings;
2. New ideas are readily accepted;
3. No extra publicity is required as the thousands that come to take part in such festivities provide a captive clientele; and

Figure 6. A scene from a weekly market

Figure 7. Baskets and other produce on sale

4. The events, adjusted to certain seasons, take place at the "right time" for craftspersons.

One significant attempt to recreate such a tradition needs mentioning. At Shantiniketan, Calcutta, the visionary efforts of Rabindranath Tagore, its founder, had provided for craft fairs fifty years back. Even today, one can see at Shantiniketan fairs like *Pouskar Mela* in December and *Ratindra Mela* (Fig. 8) in November taking place every year. University students of arts and crafts, as well as various crafts groups from nearby places, participate in these fairs that create a festive atmosphere. Buyers come from distances as far as 300 km.

Craft exhibitions in cities and urban centers, often subsidized by government, have come up in recent times. The Indian government has large budget allocations for this purpose. Unfortunately, such subsidies have done much damage to the cause of crafts. Firstly, the power has shifted to the hands of government officials who have no stake in the business. Secondly, the government rule that a craftsperson once invited at government expense cannot be invited again discourages good craftspersons. Thirdly, crafts groups often do not factor out the subsidies while pricing their products; this makes it difficult for the groups to come to city centers later and sell on their own. In addition, the subsidies and the consequent underpricing of the craft articles have given the urban customer a notion that bamboo craft items should be inexpensive. A careful analysis f the situation will show that

Figure 8. Scenes from *Ratindra Mela*

it is the urban customer who is getting the benefit of subsidy; not the craftsperson. The craftspersons are, however, happy with these exhibitions because they are able to sell large volumes in a couple of days. While this looks good on a short-tern perspective, it makes the craftsperson complacent in product design and development. Unable to understand the reality, craftspersons often identify their main problem as "not getting a chance to be in the government-sponsored exhibition".

Some attempts have been made to over come this; Dilli Haat is one such experiment. It is a permanent craft market created in Delhi at considerable cost with government support. Crafts groups are given a place for hire for 4–6 weeks at a time. Dilli Haat has made an impact, as many foreign tourists visit Delhi and such a large and well-organized crafts market facilitates buying. The rate of sale, however, is slow as customers have no urgency to buy like in fairs or craft exhibitions. To customers, it is a regular market from where one can buy what is needed when the occasion arises. To most crafts groups, however, it is a one-time selling in a year as they may not be able to hire the place more than once a year.

If bamboo craft has to be a regular occupation, it is important to explore regular markets and new marketing modes. Regular marketing demands accountability for long-term performance. Hence, preservative treatment of bamboo and good joining details become essential. Marketing has to be linked to design, which has to take into account factors such as shelf display and floor space. For example, most products need to be stackable as the floor space in city centers is very expensive. Intermediaries or brokers play an important role in craft marketing. Many NGOs and funding agencies consider brokers as an evil necessity. This notion has cost bamboo craft dearly, as the role of brokers as a risk taker, advertiser and new market seeker has been underplayed. There is a need to promote a healthy system involving intermediaries. In the long run, if bamboo craft has to survive in the coming years, it is better to initiate actions that would make the craft competitive in the regular business channels.

BAMBOO LINK PROJECT

Bamboo Link, a project financially supported by the International Network for Bamboo and Rattan, is being attempted at the Industrial Design Centre, Indian Institute of Technology, Bombay. In the project, bamboo products will be designed and developed with crafts as the focus. A suitable design strategy to fit into different markets will be evolved. Two to three crafts groups will be participate in the project, which will have a series of in-house workshops for them.

Trial orders will be placed on new designs as part of training-cum-earning process. The craft products made will be used for market testing and feedback. Simultaneously, crafts groups will also be exposed to mechanisms of banking, loan procurement and forming cooperatives. All training materials will be designed for ease of communication. The material, with feedback on the project, can be used by other groups later. Thus, the Bamboo Link Project will act as a testing ground for the many ideas elaborated earlier in this paper.

REFERENCES

Lam, Y.H.T. 1996. Bamboo and modern design: application and design innovation in a traditional Asian market in bamboo. *In* Ganapathy, P.M.; Janssen, J.A.; Sastry, C.B. ed., Bamboo, people and the environment, Vol. 3, Engineering and utilization. Proceedings of the Vth International Bamboo Workshop, Ubud, Bali, Indonesia, 19–22 June 1995. International Network for Bamboo and Rattan, New Delhi, India. p. 305.

Ranjan, M.P. 1985. Structure of bamboo baskets. *In* Gass, S.; Drusedau, H.; Hennike, J., ed., IL31 Bamboo. Institute für Leichte Fachentragwerke, Stuttgart, Germany. pp. 356–367.

Rao, A.G. 1994. An approach to bamboo craft. *In* Rao, A.G.; Madhavi Koli, ed., Bamboo craft design. Industrial Design Centre, Indian Institute of Technology, Bombay, India. pp. 101–112.

Thackara, J., ed. 1988. Design after post-modernism. Thames and Hudson, UK. p. 24.

Whiteley, N. 1993. Design for Society. Reaktion Books.

Zhu, H.M. 1996. Bamboo-based boards in China: an introduction. *In* Ganapathy, P.M.; Janssen, J.A.; Sastry, C.B. ed., Bamboo, people and the environment, Vol. 3, Engineering and utilization. Proceedings of the Vth International Bamboo Workshop, Ubud, Bali, Indonesia, 19–22 June 1995. International Network for Bamboo and Rattan, New Delhi, India. pp. 150–151.

Bamboo for Sustainable Development, pp. 739–748
Eds. A. Kumar, I.V. Ramanuja Rao and Ch. Sastry

Teaching architecture with bamboo

HORACIO SALEME, JOSÉ A. MÉNDEZ and SUSANA COMOGLIO
Universidad Nacional de Tucumán, Tucumán, Argentina

Abstract—Bamboo is not only appropriate for buildings but also a fast renewable and extremely cheap resource. Furthermore, it withstands very well the tremors caused by earthquakes. For these reasons, bamboo was chosen as a construction material to be studied in a research project at the Tucumán National University, Tucumán, Argentina. The project seeks to find the best method for comprehension of structural problems and for teaching practical aspects. It has the participation of students, making good use of the young students' creativity. It is a challenge to find economical alternative building materials and identify construction possibilities with renewable materials. Bamboo has been found to be a very good alternative building material, which can be effectively used to construct as well as to teach architecture.

INTRODUCTION

The Province of Tucumán is located in the North-Western Region of Argentina (NOA). Although the smallest province in the country, with an area of 22 524 km^2 (0.6% of the national territory), it is the most densely populated (1.3 million inhabitants). It is a subtropical region, 300 km away from the Tropic of Capricorn. Its climate offers wide variations in temperature between a rainy summer and a dry winter.

Tucumán has a tradition in sugarcane cultivation and sugar manufacture. The nature of this economical activity and its instability has caused deep and recurrent crisis in the province, compelling it to get engaged in a long process of rehabilitation and diversification into citrus, soya bean, tobacco and vegetable crops, and into metal-working and heavy industries. Despite these widely varied activities, the sugar industry remains first in importance.

The geographical diversity of the different regions conditions the kind of dwelling and popular constructions, which generally conform to the landscape. Thus, in the Calchaquí Valleys, houses are made of adobe with roofs of earth and *jarilla* (a sort of cane), while in the East Plains the predominant kind of constructions are huts made with hollow canes forming frames with or without mud plastering.

On several occasions, Tucumán has suffered the effects of earthquakes with their epicenters located within the province. The National Institute for the Prevention of Earthquakes (INPRES) has classified the province as a moderate seismic hazard zone. The expansion of the agropecuary boundaries throughout the present century has brought about a change in river basins, turning them into critical basins, owing to the harmful results of the overflowing of rivers during the rainy seasons. These alterations in the ecology have made the province pay very high costs.

The capital city of Tucumán, San Miguel de Tucumán, is located in the geographical center of the province, at 436 m above sea level. It has a population of 500 000 and it is 6th in order of importance among the most important cities of the country. Tucumán holds a great cultural tradition, and its local National University (founded in 1914) is the most significant in the Northwestern Region.

THE BAMBOO PROJECT

With the work of interdisciplinary teams in the rural areas of the province, the National University seeks a way to correlate scientific and technological knowledge to the life experiences of the inhabitants. This would benefit both sides, helping one grow and profit as a rural community and the other as an academic one. The latter would then readjust its curricula in research and teaching, and base them on the knowledge of true life in its human, geographical, economical and cultural aspects.

Within this frame of reference, the first *Cátedra de Estructuras* of the School of Architecture started to research on constructions made of hollow canes (bamboos), widely known in the East plains. The aim was to improve popular technologies, and thus ensure the continuity of a construction tradition and the optimization of its applications.

Bamboos

A wide variety of bamboos grow in the East Plains and to a lesser extent, at the foot of the Aconquija Mountain. These species are not native of the Americas; the pioneers of the sugar cane industry probably brought them to Argentina. At present, they are widely used in rural dwellings, tobacco driers, stands, furniture and fences. The most widespread species in Tucumán are *Bambusa vulgaris* (bambú amarillo), *Dendrocalamus* sp. (bambú común) and *Bambusa tuldoides* Munro (caña tacuara). Some species have a culm diameter of up to 15 cm (even 20 cm, though rare), reaching up to 25 m in height. The way bamboo rhizomes grow makes them suitable for the rehabilitation and enrichment of eroded soils, helping them recover relatively fast, thanks to the biodiversity generated by bamboo plantations.

Project

When starting the study of different varieties of bamboo that grow in the region and on analysing their physical, mechanical, chemical and technological properties, it becomes obvious that these plants have an enormous potential for usage, not only in rural housing and social construction but also in more important constructions such as wide-spanned structures, bridges, supporting walls and large buildings. Furthermore, they can be used in furniture, boards, and for fuel and food (shoots). Their great flexibility and strength, besides light weight, make them adequate for seismic-resistant constructions. Because of bamboo's light weight, seismic loads loose their relative importance; its great strength allows it to rapidly absorb any type of loading (depending, of course, on adequate structural design and proper construction); and its flexibility provides the ability to dissipate seismic energy. Bamboo houses are very easy to build and repair, require very little changes in rural techniques used for low-cost construction, and provide a quality comparable to that of the houses made with traditional systems.

The main drawbacks of bamboo as a construction material are damages caused by humidity, biological agents (fungi, insects) and fire, similar to wooden buildings. In Tucumán, bamboo constructions are usually very poor, without sanitary services, running water or other facilities. These are put up following very rudimentary techniques and hence, it is only natural that they deteriorate quickly. The improvement of material properties and construction techniques will result in an improvement in the life quality of thousands of inhabitants of rural areas. At the same time, provided it spreads on a regional scale, bamboo could become a powerful impulse to the uplift regional economy through new technologies and industries. It will also help in the rehabilitation of critical eroded basins, the recovery of ecological balance and the enrichment of biodiversity.

OBJECTIVES

The aims of the Bamboo Project are:

- To study local bamboos and determine their structural, technological and chemical properties;
- To improve and optimize existing low-cost construction techniques;
- To develop new structural systems and construction procedures;
- To disseminate among governmental and community organizations the advantages of using bamboo, with the aim of promoting production alternatives that will foster a sustainable development;
- To include university students in the process of technological research, as a way of initiation and of making them aware of the practical realities of the region.

Study of local bamboos

One of the most serious problems that Latin America faces on studying local bamboos is that there are no local botanists or taxonomists with expertise in their identification, classification or nomenclature. Even after identifying the best-known species in Tucumán, there is still a lot to be done. The lack of taxonomists is less serious than the lack of interest of scientists in the subject. In view of this, several Colombian, Peruvian and Ecuadorian professionals were contacted for overcoming this drawback. Despite these problems, several laboratory experiments have been conducted to determine tensile, compression and shearing strengths of local species. Experimenting with bamboos under tensile stress has been difficult as the samples collapse because of shear in the joint. It is not surprising then that the main technological problem with bamboo structures is the jointing of members under tensile stress. This is made up for to some extent by the tensile efficiency and the considerable length of the culms, which allow joints to be reduced. The guiding idea is to change the tensile stresses on the joints into compression on the tympan through an adequate construction technique. The resistance bamboo offers to tensile stress, as tested with culm wall samples in the same as wood is tested, is comparable to that of steel (breakage stress is about 3 200 kg/cm^2).

Improvement of low-cost construction techniques and development of construction and structural alternatives

Construction procedures are very elementary in the region: generally, systems of beams with simple supports. Obviously, there is a great deflection because of the flexibility. In the first experiments, it was enough to stress the beams with high-strength wires (or wire nets) and increase their resistance considerably. After that, from elementary triangulations of bamboo alone or bamboo and wire, new forms and structural typologies were generated to obtain different kinds of roof trusses. These experiments, generally for flat structural types, were first conducted using students and then transferred to rural areas.

Some experiments with spatial truss models were also carried out. These systems, although they are a little complex because of the multiplication of joints, are suitable for the continuity of upper and lower grids. Because it is possible to increase the inner moment of resistance, they can also be split in such a way that the stresses get considerably reduced to achieve structural efficiency.

Traditional jointing techniques almost everywhere are based on binding. New kinds of joints with different materials have been tested. Naturally, the most affected joints are those acted upon by tensile stresses, something that has to be avoided while designing the structure. Once this problem was recognized, joint members of steel, wood and concrete with metallic web were designed. Joints of members under compression are easier to construct. The most satisfactory test results for joints acted upon by tensile stress were obtained when steel plates were used as clamps in the joints. When these were made similar to the connectors of the American system (of the gang-nail type), the behavior of the joint improved noticeably. It is convenient to place the clamps at the joints because that point offers shearing strength as well. Obviously, it is not easy to make them coincide but with an adequate amount of members those that offer more advantages can be selected. The joints with concrete are also quite efficient but depend to a great extent on the bamboo's moisture content, which conditions the absorption of humidity from the concrete by culm walls and the efficiency of adhesion. Although it is necessary to make small slots next to the joints to fill concrete, these small cuts get covered on filling, and the tensile stress gets transferred by a homogeneous thrust on the tympan of the joint, which is very strong.

Models to determine the efficiency of elementary bamboo structures were also constructed. Although the efficiency depends on the loads acting upon the structure, the structural models studied were found to be highly efficient owing to their geometry and the strength of bamboo in relation to the ratio of load supported to the weight of the structure. For example, a *quebracho* (a durable hardwood found in South America) cylinder of 10 cm diameter and 30 cm length breaks under compression with a 46 tons load (average value). A bamboo culm with similar dimensions breaks under 21 tons load. A *quebracho* cylinder of the dimensions mentioned weighs 2.95 kg on average while the bamboo culm weighs only 600 g. The load-to-weight ratios show that *quebracho* breaks under a load 15 590 times its own weight, while bamboo breaks under a load 35 000 times its weight. That is, bamboo is much more efficient. As we are dealing with organic materials, there is a natural dispersion of values, but the efficiency of bamboo remains constant. One should not forget that bamboo culm is hollow while *quebracho* is solid, which means that the same shape of the section is more efficient *per se* though it is influenced by bending of the walls. Such deflection is conditioned, in turn, by wall thickness and the space between joints. From the point of view of enclosed structures, bamboo trusses plastered with mud noticeably improve on using soil-cement plaster or lime plaster reinforced with cement. This technique is closely related to the South American systems like the Peruvian *quincha* or the Colombian *bahareque*.

Propagation of Guadua

The Guadua, native of South America, is the most important bamboo genus in the continent. This genus with about 15 giant species, including *Guadua angustifolia* (tacuaruzú) which has the greatest economical value for the continent, is economically very important owing to the variety of applications and because Guadua species are among the most outstanding bamboo species in the world in terms of strength and durability. For this reason, they are the most appropriate for building construction. Although these species do not grow in Tucumán, they are being cultivated in the province through agreements with the Instituto Nacional de Tecnología Agropecuaria (INTA), as they grow naturally in the provinces of Misiones and Chaco (North-Eastern region of the country).

The teaching of architecture

Jean Piaget, a Swiss epistemologist, researched on the way in which human beings acquire knowledge. He was concerned with the psychogenesis of knowledge, which tries to explain the intellectual mechanisms by which science (in all its different varieties) searches for the real; that is, he studied in what way knowledge is built up. Piaget postulated that the acquisition of knowledge is based on the interaction between the subject with the object of knowledge. The process shows the way in which pre-existent concepts of the subject are modified until new ones are acquired and rebuilt, concepts that took mankind a long time to process. According to Piaget, objective knowledge stands as an accomplishment not as initial input. There is an important difference between the idea of the learning subject as receptor of knowledge coming from outside and the idea of that same subject as producer of knowledge. This distinction is what differentiates "Conductism" from Piaget's conception of the learning-subject.

Piaget himself summarizes his ideas as follows:

> "The main goal of education is to generate new men capable of carrying out new enterprises, not mere repetitors of what other generations have done: creative, inventive men. Second in significance is the goal of shaping critical minds, capable of verifying and selecting what is presented to them before acceptance".

College education demands that the knowledge provided to the students be more than closed packages of information; an education that should stress the conceptual and methodological aspects as tools for the permanent education of the future professional. This is based on the swiftness of knowledge renewal and the huge amount of information one has access to today. In the field of technology, it is necessary to stress on heuristics (the art of creating) more than on formulations, models or recipes that can be outdated before the student graduates. On the other hand, the access of great masses of students to the system has conditioned the learning-teaching processes imposing sometimes a theoretical procedure, which never gets to "realize" the fundamental objective of technology. Most of the times, students get a degree knowing only what others have done without having experienced a technological process themselves. At the most, the learning process ends in a project without its actual realization.

From the beginning, the schools and colleges of architecture in Argentina have stressed the "creative formal" approach to architectural design, disregarding the fact that a work of architecture is such only when, in the words of Nervi, it becomes "a material's living reality".

This contradictory attitude from those who need materials for their work obeys, to a large extent, conductivism in the teaching of structures, inherited from old concepts in the teaching of engineering. The methods for teaching structures have been always controversial. Classical teaching almost exclusively based on "mathematical models" has resulted in failure in the schools of architecture. Direct contact with materials, laboratory experimentation on their physical properties, the building of scale models and the study of structures through didactic models, in which everything from deflection to stress is analysed, have proved much more adequate for the development of "the heuristics of structures". Most students feel a natural rejection towards technological subjects, particularly when they are based exclusively on mathematical models. A modest material like bamboo has proved quite adequate to motivate students to study structural design. Besides, the lack of prejudices on the part of the student makes it feasible to obtain notable achievements and ingenious solutions as a result of applying his/her talent and common sense. The psychogenesis of structural concepts takes nourishment from the experience of the student, who is able to manage the structure of his/her own body as well as a wide repertoire of instrumental, sporting, and other elements of common use. If we add to the above-mentioned the observation of nature, basis of bionics, so much more enriching would be the experience and so much more sound the learning would become. What is still valid as an effective pedagogical system is the old Socratic discussion about midwifery: Socrates thought that teaching is like the art of assisting women in childbirth.

These principles are put into practice in an experimental workshop, whose objective is the generation of structural forms suitable for bamboo to show the close relationship between function and shape, stresses, structural systems and building procedures. The characteristics of bamboo that prompted its selection for the project have been explained earlier in this paper. Another reason is that its low cost enable students to build models on full scale. As the project envisages interaction with local institutions other than the university, students can make contact with the practical aspects of the profession, at the same time learning the fundamental issues and technological research methodology. Information dissemination through the press keeps farmers and artisans informed on the proper use of bamboo in different kinds of productive processes, such as: the propagation of bamboo for the rehabilitation and improvement of farmland; use of bamboo in building construction; and applications of bamboo in other productive activities. Besides, studying bamboo also involves working with other professionals such as botanists, agronomists, chemists, structural engineers and artisans. In other words, it necessarily implies interdisciplinary work, so much valued by Piaget.

The active participation of students in research and extension projects thus constitutes a powerful pedagogical tool for the development of both creative skills and the knowledge of reality to which experience leads. The results have so far been outstanding in terms of acquisition of knowledge, abilities and skills. Students of architecture usually consider technological subjects as mere obstacles when these are just theoretical. When dealing with practical reality and concrete materials, however, they get a grip on these subjects and feel sincerely enthusiastic enough to present their own ideas. From the elementary typologies used by rural workers, the students themselves have generated different structural systems. From the round culm used as a single beam to different flat and spatial trusses (Fig. 1), arches, rigid frames, hyperbolic paraboloids (Fig. 2), simple and complex geodesic domes (Fig. 3), and various kinds of lineal frameworks of single and double bending (Fig. 4). The coverings of these structures are made of ferrocement, or concrete reinforced with fiberglass and synthetic webs, which are still at an experimental stage. Laboratory tests enabled the determination of admissible strengths, the development of various tables for easy calculations, and the design of special types of joints, the main technological problem of bamboo. Various curing and fire retarding treatments as well as use of pesticides were studied.

CONCLUSION

Jorge Sábato, a distinguished Argentine technologist, postulated that in all modern economies production depends on traditional factors: capital, work and raw materials, as well as technological innovation. Such innovation is produced if knowledge is added to production in order to modify a construction process or to create new processes. There are socio-cultural (prejudices, vested interests, fear of the unknown, etc.), economic, financial and scientific factors acting as obstacles to innovation. Therefore, and following Sábato, we have to assume that: "…the development of a Nation will never be achieved if science and technology do not stop from being imported magic to become a habit of its people".

Despite the modesty of the material studied, adverse criticisms and the few resources available, the Bamboo Project of the National University of Tucumán tries to be a means of meeting the main objectives of the University: Teaching, Research and Extension.

Figure 1. Students on a spatial truss

Figure 2. A combined hyperbolic paraboloid structure

Figure 3. A geodesic dome

Figure 4. Students working on a double-bending framework

Bamboo for Sustainable Development, pp. 749–755
Eds. A. Kumar, I.V. Ramanuja Rao and Ch. Sastry

Holistic relationship between humankind and environment through bamboo

VICENTE GUILLERMO NORIEGA MORENO and DANIELA FAVILLA
Forest House Environment Planning Research Center, Campinas, Brazil

Abstract—This paper describes a project in Campinas, Brazil, aimed at the development of a self-sustainable model community system built around bamboo plantation and processing. The system is envisaged as an "open university" for environment planing and sustainable development, offering cultural and technical knowledge to community members to enable them to understand and participate in sound environmental management. The project takes into account the different factors required to make the bamboo-based "Agricultural Community" a self-sustaining one.

INTRODUCTION

The project consist of graded steps of building up environmental awareness, divulging information and personnel involvement in working towards an "Agricultural Community", transforming the present way of life in the direction of a holistic relationship between humankind and the environment.

This project envisages the participation, in particular, of cultural leaders, associations of outlying urban area and living centers in agricultural areas to generate a self-sustainable model of agricultural life, fulfilling the role of a central cell around which the society can organize its space in relation to a reconstructed forest.

Seven different programs have been developed under the project. The common element in all these programs, and the one that interrelates them, is bamboo. Moreover, the some courses proposed make use of bamboo for the reconstruction of the forest (through seedlings), rehabilitation of degraded areas and manufacture of artesanal objects and buildings. The project aims to interact in a constructive way with the environment through bamboo, its features and its applications. It emphasizes the use of the *Guadua* species, ideal for use in construction owing to its physical features and mechanical characteristics. Guadua bamboo use has a history in the area – in bridges, fences, stairs, houses, and equipment and articles of daily use.

Through partnerships with NGOs (Mata Ciliar), Civil Air Patrol, the Agricultural Institute of Campinas and two universities – Puccamp and Unicamp – the project has reached the stage of setting up courses in building with bamboo, growing saplings and developing plantations.

AGRICULTURAL COMMUNITY

New paradigms

The present model for our civilization requires a rethinking in terms of urban development, focusing on its failures with regard to the exhaustion of natural resources, extermination of native forests, and all the consequences that we are experiencing.

This project proposes an Agricultural Community, a new dynamic model for social organization, whereby people will come to relate intimately with nature (Moreno 1993). The Agricultural Community, beyond acquiring knowledge and involving an increasing number of people in environmental aspects, will intervene directly in the reconstruction of the forest. The environmental education is conceived within a set of new ethics that searches the deepening of the environment educational processes, constructing a new rationality that makes possible modalities of relations between the society and the nature. These modalities cover global scientific and technical knowledge and techniques, as well as relationships between diverse social groups and between countries, centered on the respect and the right to life in all its aspects.

In this creative process, human beings as a group is conceived as the one that betters prevalent situations in a dialectic way, surpassing established logics, given the practical possibilities of construction arising from organically structured ideas in relation to the dimensions of space and time. This is an interactive, cooperative, constructive and representative process, wherein the operating citizens interactively grasp its features and moorings, visualizing its logic and its feelings.

In this proposal, art is introduced as the creator of situations where the citizens feel close to the essential elements of the life and being able to work with them in different ways, surpass existing situations and advance the architecture of new spaces. These new spaces form the elements of "life and culture", and the existing relations and interactions with the environment become the rejuvenating force.

The project

The project will be developed simultaneously on different fronts, involving different groups of citizens and presented through seven different and independent though inter-related programs. All programs are centered on the key ideas of "Environmental Planning" and "Art Education". While one set of programs emphasizes art education, another set aims at the rehabilitation of hill slopes with green cover and reestablishing the riverside vegetation.

The area selected for the project consists of Sousas and Joaquim Egídio, two sub-districts of Campinas, in the state of São Paulo, Brazil. The region is characterized by the historical heritage, associated with its old coffee farms, and by its environmental richness, reflected by its forested areas and abundant water bodies that supply the City of Campinas. The region is also the site of Santa Mônica Farm of Mr. Wollfin Schimidt – a pioneer in forest reconstruction – the Capricórnio Observatory, and the natural forest area that the Civil Air Patrol supervises. The region that comprises Sousas and Joaquim Egídio comes under the Municipal Protected Area for the City Hall of Campinas since 1993, and is covered by the Management Plan drawn up in 1996. In view of these factors, the area was considered as the ideal site for setting up an Agricultural Community, which can grow in the future to other regions of the State.

The natural forest area supervised by the Civil Air Patrol presents the features necessary for the project: a healthy environment, abundant water and adjacent deforested areas. The Agricultural Community will be developed in the deforested areas, without upsetting environmental balance, which would also be the starting point for forest reconstruction.

Bamboo: the renewable, accessible resource

The very nature of the project dictates that the materials used need to be readily accessible to the social groups involved; they should also be environmentally suitable and able to support the different facets of the project.

Bamboo, by virtue of its rapid growth cycle, shows potential for reforesting hill slopes and riverbanks. As a basic material for infrastructure construction, it also forms the core of the project (Moreno and Allonso 1997). Owing to the special properties of this graminaceous plant, balanced autochthonous techniques and the new technologies are recommended for its planting, harvesting and use, taking into account the requirements of the project, the localities and specific situations.

Implementation plan

The seven programs of the Agricultural Community Implantation Plan comprise activities that aim to organize and construct a model system (the "Open University"), as well as activities for different localities in the project area where the knowledge centered on the Agricultural Community is necessary. The following pages present these programs in brief.

Program 1: Environmentally sustainable activities with financial returns for the low-income population

This program is organized in the community proper and includes theoretical lessons (general knowledge of environmental sustainability, compatible activities), and practical lessons under the specially designed course on "Construction with Bamboo" (see later in this article). This course is of 3 months duration. It can include activities such as rehabilitation of hill slopes through bamboo reforestation and reconstitution of riverbank vegetation. This course is meant for adult community members; children and adolescents will be organized and enrolled in the Forest House Program, designed for these groups (see Program 2).

Program 2: Theoretical and practical environmental education, with emphasis on environmentally sustainable activities
Under this program, the activities will take place in the Forest House, specifically constructed for the purpose. The activities are designed for adolescents and children. The educational course is of 6 months duration, and will have the participation of groups of students of municipal schools or groups organized by associations of inhabitants. The basic course will, initially, comprise theoretical lessons (environmental education, ecology), followed of practical lessons on "Construction with Bamboo" (see later in this article). The latter will involve the regeneration of the forest by raising seedlings and using them for reforestation. It will subsequently teach techniques for making various musical instruments and domestic utility articles.

If sufficient area is available, the program will also develop greenhouses and reforestation activities with, if needed, the cooperation of the NGO, Mata Ciliar.

Program 3: Mobile Workshop for teaching the "Course on Construction with Bamboo" at different localities and to diverse groups
This program envisages the construction of workshop facilities and fitting them in a van to form a Mobile Workshop. The activities under this program are arranged in a previously set place, such as a school, a municipal gymnasium, a large public area, a less-crowded street square, etc. The place must be of easy access to the groups that will attend the workshop. These groups would register for activities with the Mobile Workshop through associations of inhabitants or other such institutions that represent them.

The course content is the same as in Program 2, with alterations made to suit different situations, designed for adolescent and children. The 4-week course has theoretical lessons on environment and ecology at the beginning of each day. The practical lessons on construction with bamboo (see later in this article) are shortened to activities that can be done in 3 weeks, period of duration of the workshop, and will be aimed at providing all basic training required for craftworkers. The workshop will emphasize aspects – such as proper sizing and handling of bamboo for artisanal use, production of articles for daily use, etc. With adequate notice, the practical lessons can be modified to cover planting of bamboo for reforestation, restoration of riverbanks and rehabilitation of other environmentally degraded areas.

Program 4: Reforestation of degraded areas and areas identified for forest cover, and restoration of riverbank vegetation
Partnerships with NGOs will happen giving technical support. This program, which will work in accordance with the Campinas Management Plan, is meant to absorb and use the great amount of seedlings that will be generated under the "Course on Construction with Bamboo".

Program 5: Volunteers for program facilitation
This program aims to recruit volunteers/monitors for working with the communities in program locations – the Forest House, the Mobile Workshop and reforestation areas.

The universities will be involved in this program to promote extension courses on environment planning, renewable resources, environmental sustainability, reforestation and construction with bamboo. These courses will be proposed by Professors on the universities – PUCCAMP and/or UNICAMP – and accepted by the universities. The Municipal City Hall of Campinas can participate in the process expressing the existing demand for such courses.

Program 6: Construction of a basic cell of the Forest House
Under this program, one OIKOS unit of the Forest House (for details, see Moreno and Allonso 1997) will be constructed in the Municipal Protected Area of Sousas and Joaquin Egídio, Campinas. OIKOS unit is the basic cell that will mark the beginning of the construction of the Agricultural Community model system. Different activities and physical spaces will be later developed from it. Under the current program, the unit will host the "Course on Constructing with Bamboo", and will generate a green house for the species used in the work.

Program 7: Implementation of permaculture
The seventh program will gradually implement the techniques of permaculture, to develop a system that does not squander resources and makes full use of the wastes generated.

COURSE ON CONSTRUCTING WITH BAMBOO

The general objective of this course is to spread awareness of the environmental questions among an increasing number of people and to obtain their involvement in activities – such as reforestation and rehabilitation of degraded areas – aimed at environmental sustainability. The course will disseminate – in an informal and engaging manner – information on the techniques of constructing with bamboo, and on topics such as rehabilitation of degraded land, combating soil erosion, restoring riverbank vegetation, and planting and management of bamboo. Different species of bamboo will be used, though the emphasis will be on *Guadua* sp. and ways and means to achieve its wide dissemination in Brazil will be explored.

The course will culminate with the building of the Forest House. The objectives of the Forest House are to promote the environmental education to a heterogeneous public comprising children, adolescents and adults. Within the spectrum of education and research, the course has the possibility of improving environmental planning methods by refocusing them on environmentally sustainable employment opportunities for the people. The course is thus designed for the professionals in the areas of agriculture, engineering, architecture and urban planning, as well as for community leaders.

All through the course, bamboo will be the link between small-scale operations (such as the production of utility items) and medium-scale operations (such as construction of buildings) with macro environmental planning.

Stages of the course

1. Information and Marketing. Participants must arrange lectures in entities such as social clubs, professional associations and colleges to publicize the course and kindle interest among those who will are likely to attend the course. The target will be construction engineers, architects, interior decorators, contractors, environmental NGOs, and craft designers and workers.
2. Global Environmental Aspects. The course will begin with imparting general information on bamboo, and its characteristics and diverse uses. Bamboo utilization will be presented as a practical model of sustainable use of the environment and the resources. This stage of the program includes raising of bamboo seedlings, replantation of seedlings, plantation management, and the harvest of culms of diverse species that will be used in the course. A greenhouse, which would allow exotic and native fruit species to be grown, will be constructed during this stage, involving the community to work with the environment.
3. First Module. The first stage of imparting empirical knowledge of the material to the students – called First Module here – will be through the production of small musical instruments (wind, string or percussion). In this stage, all the basic techniques of working with the material and techniques for finishing will be taught.
4. Second Module. After developing contact with the material in its natural state (during the production of seedlings, raising plantation and harvesting) and learning the methods of handling the material (during production of musical instruments and utility items for the house, kitchen or garden), the next stage – the second stage of imparting empirical knowledge – is commenced. In this stage, learning of drawing techniques and ergonomics is required, as the objects need to be properly designed, guaranteeing its functionality, before its execution.
5. Third Module. This stage of empirical knowledge transfer focuses on the use of the bamboo in construction, specifically of houses. The output from this stage is the construction of the Forest House.

Each group of students will work, during this stage of the workshop, in one part of the Forest House, working in the construction of that part from beginning to conclusion. Models of the buildings of the Forest House complex are prepared and discussed to reach an agreement on its design among the students. Before beginning the construction, concepts of foundation, walls, pillars, shear forces, roof, plastering and finishing will be discussed in detail.

CONCLUSION

Bamboo can be grown with minimal care and resources. It is easy to handle and work with, and has high strength, flexibility, lightness, shape and beauty. It is also durable if adequately treated. Its use human lives is diverse and numerous, and it is also useful in environmental protection. Because of these, bamboo is used in the programs aimed at establishing a self-sustaining Agricultural Community, which will provide the citizens with a clear understanding of the environment and how to live in harmony with it through the use of renewable resources.

ACKNOWLEDGEMENTS

The author thanks Profa. Allonso Luiza Da Silva, who besides her scientific contribution to the Forest House project, assisted in the development of the Agricultural Community project.

REFERENCES

Moreno, V.G.N. 1993. Constructing freedom – ambient art and architecture education course. *In* Proceedings of the Ist World Congress on Guadua Bamboo, Pereira, Colombia, 8–15 August 1992. (In Spanish).

Moreno, V.G.N.; Allonso, L. 1997. Forest house: projeto de arte-educação arquitetura-ambiente casa da fioresta. Pontifícia Universidade Católica de Campinas, São Paulo, Brazil.

Bamboo for Sustainable Development, pp. 757–761
Eds. A. Kumar, I.V. Ramanuja Rao and Ch. Sastry

Simple mechanical processes and tools for the efficient and environment-friendly manufacture of high-quality Guadua products

GERMÁN RUBIO-LUNA
Consulting Mechanical Engineer[1], *Santafé de Bogotá, Colombia*

Abstract—In Cundinamarca, Colombia, the contrast between the lush forests of Guadua (*Guadua angustifolia*) and the severe economic and technological challenges of the rural human population is striking. Guadua forests are an important ecological asset of the region, but failure to value them as an economic asset threatens their very existence. Consequently, their biodiversity as well as the erosion control and watershed protection they offer are also endangered. This report covers the initial results of a project designed to increase household income by the development and transfer of affordable, appropriate technology for the transformation of Guadua into a variety of high-quality products. By adding economic value to Guadua, incentives for the local conservation and enhancement of the Guadua forests may develop.

INTRODUCTION

The severe economic and technological challenges faced by the Colombian peasants, who live in the forested region where Guadua (*Guadua angustifolia*) grows, can be reduced by developing and transferring appropriate technologies for processing the Guadua wood. Producing household and commercial products from Guadua could alleviate, to some extent, the livelihood problems of the people, while providing incentives for the protection and enhancement of Guadua forests. Concomitantly, a second goal related to the local quality of life could be approached, by reducing deforestation, soil erosion and watershed destruction in the region.

The traditional way of processing Guadua culms (7 inches in diameter with 1.5 inches wall thickness) is primitive and inadequate. The common woodworking techniques and tools used for processing are wasteful and inefficient. The use of machines and tools is also slow and costly. Therefore, a project was undertaken to fulfil three objectives:

1. To develop specific processing techniques and tools in order to set up a completely equipped Guadua workshop to meet technical criteria for efficiency and

[1] Also Member, National Bamboo Committee, Colombia.

product quality. This should allow commercial production using simple low-cost tools, technologies and materials available in any small Colombian city.

2. To design and test (product development) Guadua products in four areas:

   a. Handicrafts and household items;
   b. Furniture;
   c. Rural architecture; and
   d. Structural systems for erosion control.

3. To design and produce educational (training) handbooks and VHS videos for the transfer (at the instructor level) of the above techniques, and the wide dissemination of the art of working with Guadua.

RESULTS

Research under the project took place between February 1992 and March 1998. It included the development or adaptation of specific techniques, equipment and tools to obtain appropriate technologies. Tools and techniques were developed or adapted for the following aspects.

Press
Three usual wood vises and mechanical jigs were adapted, and a frame-type vise was designed to press in three different directions Guadua culms with diameters up to 7 inches.

Cutting
A regular hacksaw was adapted with an 18-teeth pitch blade for effecting straight cuts. For curved cuts, a hacksaw was designed with a wider frame and narrow 18-teeth pitch blades.

Drilling
Standard twist drills up to 3/8 inch in diameter and wood drill bits up to 1 inch in diameter were used for drilling purposes.

Grooves, slots, rabbets and mortises
The required processes were defined to make these in the grain direction and perpendicular to the grain.

Splitting
A manual splitting tool was designed and its use described, and the various uses of the natural flexibility of the splits were defined.

Dowels
A manual splitting tool was designed and the procedure for its use to make dowels was defined.

Form correction after the cutting

An elastic wheel grinder was adapted for an electric grinder of 1 200 to 1 500 RPM.

Joints

Ten permanent joints and 6 removable joints were selected. The use of polyester resins in the joints, manual cutting of threads for the long joining screws, and the use of dowels in the joints were detailed.

Sanding

Use of elastic grinding wheels, wire brush wheels, hand wire brushes and broken glass bottles to sand the cut edges were described.

Finishing

Finishing processes described included the use of different colors and pyroxylin lacquers, preparation and use of waxes, and imparting a gloss using buffing wheels.

Table 1 provides the results at a glance.

Table 1.
Results of the project

Process	Traditional method	Processes, techniques and tools developed
Pressing	Helper's hand Tying by hand	Adapted three common vises; designed a frame-type vise.
Straight cuts	Regular handsaw, machete	A regular 18-teeth pitch hacksaw was adapted.
Curved cuts	Machete	A hacksaw with a wider frame and narrow 18-teeth pitch blade was designed.
Drilling	Red hot rod, machete, hand drill	Adapted twist drills and drill bits.
Grooves, slots, rabbets, mortises	Non-existent	Processes were developed that take into account the grain of the culm.
Splitting	Machete	A manual splitting tool was designed; uses of natural flexibility were defined.
Dowels	Non-existent	A manual splitting tool was designed; use of dowels for joints was developed.
Form correction	Chisels, manual sanding	Use of elastic wheel grinder was developed.
Joints	Tying, nailing, nut-and-bolt (developed by the Colombian Arch. Simon Velez)	16 types of joints were designed or adapted; use of polyester resins was developed; Manual cutting of threads on long screws and use of dowels for joints were developed.
Sanding	Hand sanding	Developed the use of elastic wheel grinder, wire brush wheels, broken glass.
Finishing	Lacquers and varnishes	Developed the use of colours, pyroxylin lacquers, polishing waxes, and buffing wheel for imparting glossy finish.

Table 2.
Guadua products developed by the project

Handicrafts and household products
20 different kinds of containers
8 different kinds of pincers
15 different kinds of silverware or flatware
10 different kinds of garden tools
10 different kinds of flower vases
6 different kinds of flowerpots
2 different kinds of wine bottle holders
40 different kinds of tools for ceramics
2 different kinds of knife grinders
Home furniture
5 different kinds of chairs
2 different kinds of couches
5 different kinds of beds
4 different kinds of tables
2 different kinds of stools
1 sofa bed
1 hammock holder
1 school desk
Structural systems for rural construction
3 basic trusses for:
- house construction
- grain storage areas
- shelters
- booths
- greenhouses, schoolhouses and others
Soil erosion protection structures
Soil ladders
Soil retainers
Structures to slowdown water flow
Land contouring structures
Control of cattle path

Five different handbooks, both in print and video formats (45–60 min. each), were prepared for instructor training and dissemination. The handbooks cover five subjects:

1. Establishment and use of the Guadua workshop;
2. Guadua household items and handicrafts;
3. Guadua home furniture;
4. Guadua structural systems for rural constructions; and
5. Guadua structures to control soil erosion.

Products were developed, tested to meet high technical standards, and produced using the tools and techniques developed by the project. Table 2 lists some of these products.

CONCLUSION

In the field of technological research and development, it is known that there is always something more to improve or to design. The current results provide, in relation to the objectives of the project, a basis for further development. The next logical phase, involving extensive field-testing, fine-tuning and final release of this initiative, will take place subject to the availability of funds.

Bamboo for Sustainable Development, pp. 763–775
Eds. A. Kumar, I.V. Ramanuja Rao and Ch. Sastry

Exponent moderation of shape in plants

JOHAN GIELIS
Geniaal Bvba., Nottebohmstraat 8, Antwerp, Belgium

Abstract—In nature, optimization is very important. A special case of optimization is via shape and form. "Super shapes" provide the missing link between form and growth. They have not been reported before in nature, but can be found throughout. Within the complexity of geometrical design in plants, super shapes provide the simple governing principles. Their straightforward mathematical properties give new insights into the architecture, growth and morphogenesis in plants. Using bamboo as a model plant, this paper attempts to show that super shapes can lead to particularly attractive bio-mechanical properties, such as material minimization, efficient space utilization and maximization of mechanical resistance. Super shape optimization also inherently allows for adaptive growth in changing environmental conditions.

INTRODUCTION

Form of organisms has attracted both biologists and mathematicians since ancient times. In organisms, in general, form is the combined result of hereditary information and the everchanging environment in which they grow and develop. In many cases, the shape or form of an organism seems to be governed by simple mathematical or biophysical laws. Circular and cylindrical forms in particular are often observed in plants (d'Arcy Thompson 1942). Moreover, very complex forms can be thought of as combinations of sine functions and analysed using Fourier analysis.

It should be, however, stressed that a true organism or its building blocks always deviate from this perfect mathematical or "algorithmic" beauty. The mathematical and biophysical laws have to be interpreted as rules that are never exactly met. One cannot find two completely identical leaves; neither do perfect circles exist in nature. With algorithms, one can obtain plant-like figures (Prusinkiewicz and Lindenmayer 1989), but given a certain plant, it is impossible to find the exact algorithm that generated the plant (Van Oystaeyen et al. 1996).

Nevertheless, the search for mathematical rules remains highly interesting, especially when they are simple and straightforward, both mathematically and in regard to functional aspects. Such simple but general rules are always surprising, both by their intrinsic value and beauty as well as by their application, giving insight to why plants or organisms grow as they do. An additional requirement to rules should be that they must allow for changes, or developmental adaptations.

Among the most beautiful of all mathematical forms are "super shapes". The mathematics underlying these shapes is relatively simple and straightforward. Instead of using quadratic equations – such as circles, ellipses or cylinders – the exponents of these equations can easily be modified in a creative or adaptive way. Remarkably, plants and organisms have been doing this for a very long time and it is not surprising that super shapes find wide application in botany, both for modelling and to understand the physical processes underlying morphology, anatomy and morphogenesis in plants. What is surprising is that it took so long to discover it.

Plants make use of these simple principles in various ways, and most often super shapes involve some kind of optimization strategy, such as minimization of material or maximization of space or strength. Instead of obeying 'human-made' mathematical rules of quadratic equations and perfect geometrical objects, plants experiment freely with mathematical equations by moderating the exponents, generating shapes that are no less mathematically perfect: this paper will give ample examples to prove this point. Since super shapes and sub-shapes can be defined in many shapes described by quadratic equations, the general term Exponent Moderation (EM) is used. EM almost inherently implies or allows creative adaptation of form to environmental conditions. It is more general since it refers to any shape, including super and sub-shapes, as well as quadratic shapes. In this respect, one can speak of EM-circles, EM-ellipses, EM parabolas, EM paraboloids of revolution, etc.

This paper concentrates on some mathematical details and examples in plants that have been studied in more detail in recent years. Mathematical focus is limited to EM shapes of conic sections (circles, ellipses and parabolas), extended also into three dimensions. A lot more is to be said about mathematical aspects of EM shapes, but this is beyond the scope of this paper. It will be shown that EM optimization is simple, yet widespread in plants. Most examples will be taken from bamboos, the author's favorite groups of plants (Gielis 1996).

MATHEMATICAL ASPECTS OF SUPER SHAPES

Super and sub-circles and ellipses

Circles and ellipses are described by equation (1), with n = 2, and for the circle a = b. The name 'Lamé-ovals' has been suggested for shapes generated by moderating the exponent in the quadratic equations for circle or ellipse. These shapes were first studied by the French mathematician Gabriel Lamé (1795–1870), who used curvilinear coordinates for several applications in physics (Lamé 1857). In the 1960s and 1970s, Lamé-ovals have been used in architecture and design (Allard 1964; Gridgeman 1970). The generalized equation for Lamé or EM ovals is:

$$|x^n| / a^n + |y^n| / b^n = R^n \qquad \text{(Equation 1)}$$

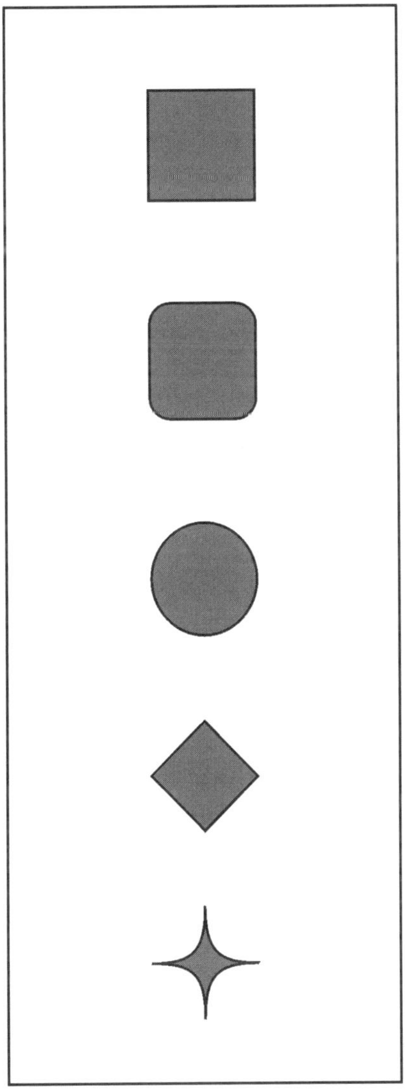

Figure 1. Super circles and sub-circles

In fact, this equation defines a whole set of shapes of which circle and ellipse (the "quadratic members") are only special cases. Circles and ellipses are special cases in the sense that only they have a finite radius of curvature in every point, and a circle is the object that can enclose the largest area for a given perimeter. When the exponents 'n' are varied in Equation 1, other members of this set are generated (Figure 1). When 'n' approaches infinity, circles transform into squares and ellipses into rectangles. For n = 0, circle and ellipse are reduced to a cross, formed by X-axis and Y-axis.

Shapes with $n > 2$ are called super circles, with $n < 2$ sub-circles. The circle with $n = 2$ has exactly four points in common with the circumscribed square ($n\rightarrow\infty$). For all super circles a *flachpunkt*, a region with infinite radius of curvature, is generated. Because of this *flachpunkt,* a zone of stability is created which allows super shapes to stabilize, contrary to circles. Only when $n\rightarrow\infty$, this *flachpunkt* transforms into true line segments.

One could imagine the increase of exponent 'n' as inflating a balloon in a box. But theoretically, just as the circle, all super shapes with exponent $n < \infty$ have only 4 points in common with the circumscribed square. Nevertheless, this zone of the *flachpunkt* provides stability and mathematically an infinite number of super ellipses, each 100 m high and 1 mm wide, can be piled to form a stable structure. Interestingly, EM shapes can be generated in all regular polygons, and the number of *flachpunkten* equals the number of angles in the polygon.

Super and sub-parabolas

Parabola and hyperbola are conic sections. A parabola is a special case of the ellipse, since one of the two focal points is situated at infinity. For the simplest case $y = x^2$, the parabola crosses the origin and is symmetrical about the Y-axis. Super or sub-parabolas (EM parabolas) are obtained by:

$$y = |x^n| \qquad \text{(Equation 2)}$$

While parabolas have only one point in common with the X-axis, namely the origin, for $n > 2$ a *flachpunkt* is created as well, resulting in a parabola, which rests stable on the axis. For $n < 2$ a *flachpunkt* is created on the Y-axis (Fig. 2). On the X-axis, however, a double cusp is formed in this case.

A very important property is that parabolas with exponent 'n' and '1/n' are fully symmetrical about the $n = 1$ parabola, the latter being straight lines at angles of 45° with the axes. The proof is as follows:

1. Let $y = |x^n|$.
2. Area A + Area B = 1 in the segment (1; 1), the simplest possible integration, with A the area enclosed by the parabola and B outside the parabola.
3. $y = |x^n|$ so $x = |y^{1/n}|$.

This full symmetry is not true for circles and ellipses.

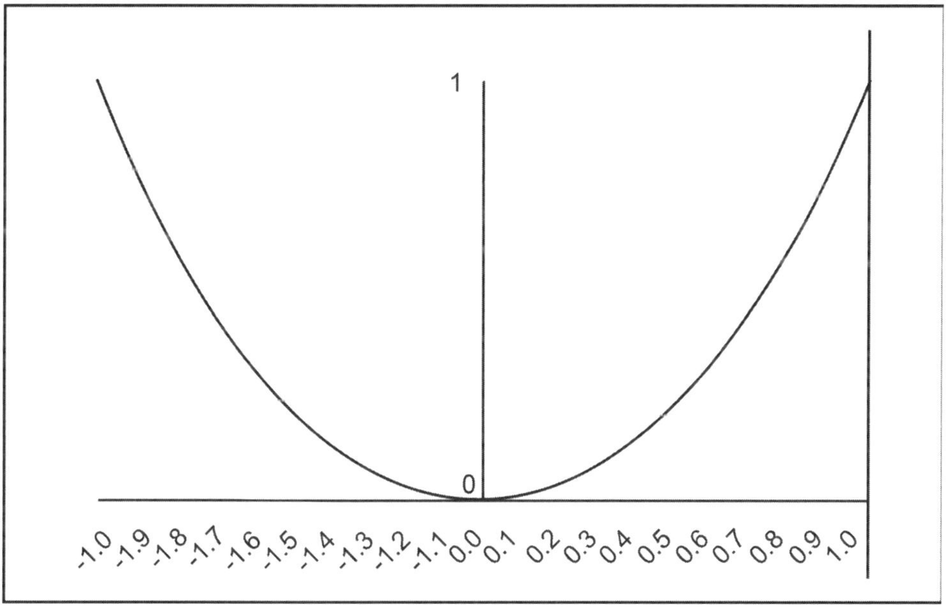

Figure 2a. Parabola ($y = x^2$)

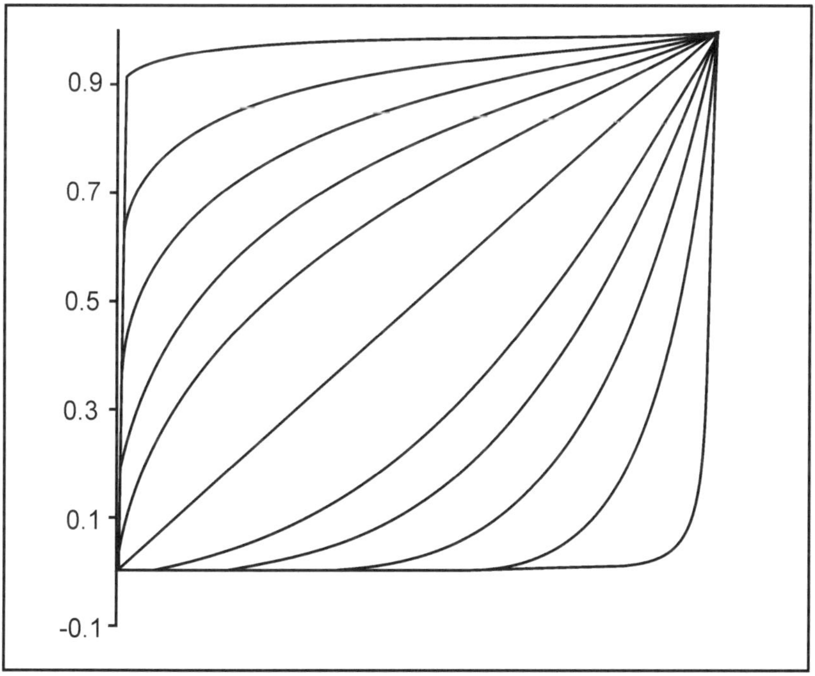

Figure 2b. Super and sub-parabolas in the first quadrant

Gaining from exponent moderation

By moderating the exponents of simple quadratic equations, it is possible to obtain shapes that have some biomechanical advantages, as compared with circles, ellipses or parabolas. In Fig. 3, a set of parabolas is shown with $x,y \in [0; 1]$. The parabola with $n = 2$ divides this square into a part $B = 1/3$ under the curve and a part $A = 2/3$ in the curve, as was shown already by Archimedes. For exponents different from 2, the area B is given by $1 / n+1$, the area A is given by $n/n+1$ (see proof of theorem 1), with $A + B = 1$. But this increase in area is realized by only a moderate increase in length of the arc – e.g. going from $n = 1$ to $n \rightarrow \infty$, the area doubles, but the length of the arc increases only from $\sqrt{2}$ to 2n (about 39%).

Even more impressive is the gain in mechanical resistance. Every shape has a certain resistance against external forces, such as bending and torsion. In plants also, shapes determine this resistance (Niklas 1992). The advantages of EM are shown in Fig. 3. For super circles with $n = 5$ the area increases by about 21% compared to the circle, but the resistance against bending (the moment of inertia I) increases by about 50%. So by moderating the exponent of the super shape, a small increase in area results in a large increase in resistance. Even with an increase of 'n' from 2 to 2.5, 'I' increases by about 16%, while the area increase is about 7% only. These calculations assume that isotropic materials are used, meaning that material characteristics are the same in any point of the area. The effect would thus be more impressive if stronger material would be at the periphery of the shape or object. And this is precisely what bamboo does: in the periphery the fibers are omnipresent, and hardly any parenchyma is found.

Super shapes in three dimensions

Equation 1 can be extended into any number of dimensions. For instance, super ellipsoids can be generated by:

$$|x^n| / a^n + |y^n| / b^n + |z^n| / c^n = R^n \qquad \text{(Equation 3)}$$

Equation 3 is a general equation for EM shapes in three dimensions. But the exponents can be different in the different directions (Equation 4). When $m = n$, EM ellipsoids are generated; but when $n = 2$, the body of revolution of a super shape around the Z-axis is described.

$$|x^n| /a^n + |y^n| / b^n = [(1- (|z^m| / c^m))^{1/m}]^{\,n} \qquad \text{(Equation 4)}$$

Objects in 3-D can also be defined for parabolas. When $n = 2$, a paraboloid of revolution is obtained; but when 'n' approaches infinity, a geometrical pyramid is obtained for $z =$ constant (Allard 1964).

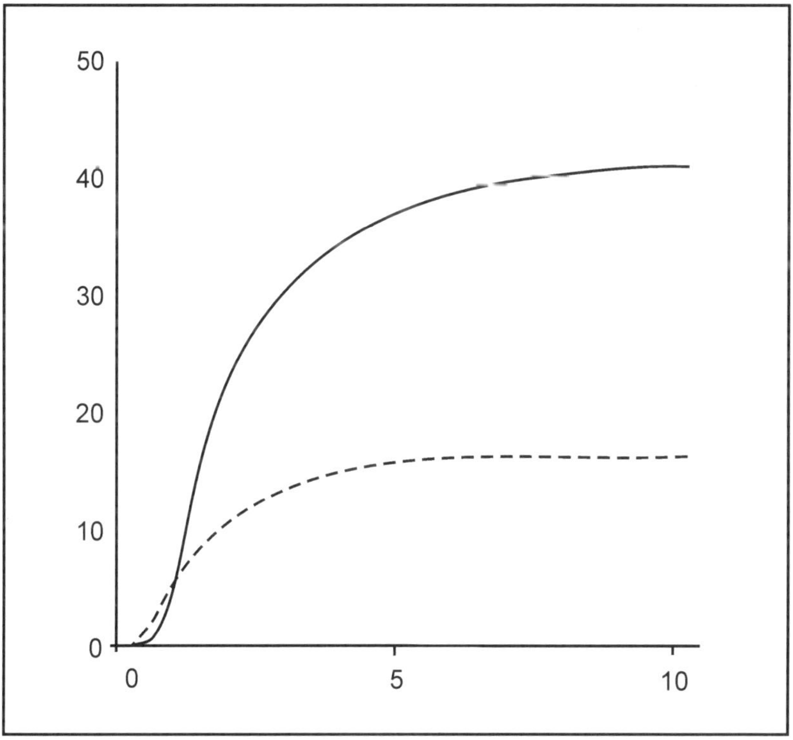

Figure 3. Area (dotted line) and moment of inertia I (solid line) for super circles (radius 2) as function of the exponent n (0.5–10)

SUPER SHAPES IN PLANT MORPHOLOGY: SOME EXAMPLES

Square bamboos

Bamboos are a highly interesting subfamily of the grasses. Their culms are very flexible but very strong and can be as tall as 30 m. In most bamboos, the culm is cylindrical or nearly so. The square bamboo *Chimonobambusa quadrangularis* is the most notable exception (McGowan 1886). In a recent review of the genus (Wen Tahui 1994), the genus *Chimonobambusa* was divided into three sections. About 2/3 of the described species have square stems in the lower 1/3 part of the culm: precisely the part that has to withstand most forces acting on the culm (weight of foliage and stem, wind, snow and rain). The square stems are super-shaped in cross section (Gielis 1996) with exponents in Equation 1 roughly between 2.5 and 4. The biomechanical advantages of such a shape have been discussed earlier. As pointed out, calculations of 'I' are for isotropic materials. In culms of bamboos fibers are concentrated at the periphery, so that the super shape effect is attenuated. Also, a lot of other plants use this simple but highly effective strategy of square stems. Square stems are highly valued and they can be produced artificially (Ueda 1960; Hashimoto 1962; FAO 1978).

In other bamboos, only a part of the culms is flattened. A good example is *Phyllostachys* where the sulcus is flattened throughout the internode, at the side of the branches or buds. Because of the distichous arrangement of leaves and branches in bamboo, the result is a compound increased mechanical resistance, necessary for these giant bamboos with the most northerly distribution.

Hollow internodes and parenchyma cells

The internodal regions of the majority of bamboo culms are hollow. The laminae form very early in the elongation process of the bamboo culm. In longitudinal section, the hollow internodes resemble super ellipses very well, described by Equation 1 with exponents generally ranging about 4–10. The advantage of maximizing the hollow area is obvious: the material needed is much less than for an ellipsoid. At the same time, angles of 90° (a rectangle) and high investment in energy are avoided. Moreover, both vascular bundles, which branch in the nodes, as well as stress can be diverted smoothly around these corners. An almost perfect super ellipse is found in the logo of the VI International Bamboo Workshop and the V International Bamboo Congress. So, even the womb of the human race is super-elliptical!

In three dimensions, the hollow internodes in cylindrical culms are described by Equation 4. For cylindrical culms m = 2 (a body of revolution of a super ellipse-shaped longitudinal section), but for square bamboos, as well as for artificially shaped rectangular culms of bamboo, m > 2 or equal to 'n' (also super-shaped in 3-D). This also counts for long and short parenchyma cells, which resemble vertical rows of super ellipse-shaped cells. In 3-D, the shape of cells can be adequately described by Equation 4.

The elongation of bamboo culms is very rapid. A culm can reach a height of 30 m in about 30 days only in some species. A record increment was measured in *Phyllostachys edulis:* 121 cm in 24 hours! The process of elongation is almost exclusively by cell elongation. Therefore, all parameters have to be optimized, including material use, already at a very early stage. Besides optimizing material use, super shapes reduce the elongation process to a mathematically very simple process: only the length of the axis in vertical z-direction (Equation 4) changes. All other parameters have been fixed by the time the shoot has been formed in the soil: the length of axes in the horizontal planes 'a' and 'b', and the exponents in all three dimensions. The process should indeed be as simple as possible. Bamboo cannot afford too many mistakes during elongation. It has no secondary thickening growth, nor any cambial activity for correcting mistakes (except for the intercalary meristem and the pulvinus to maintain negative geotropic growth).

The hollow internodes of bamboo, with a *flachpunkt* both at the upper and the lower ends, provide also another highly intriguing architectural approach. A bamboo culm is a collection of "super ellipsoids" or building blocks of air (or emptiness). Bamboo culms both contain and traverse the air (Stephen Glassman). Because of the presence of *flachpunkt,* several such super eggs can be piled upon each other, without affecting the mathematical stability in any way. The bamboo wood surrounding these super ellipsoids, made of emptiness, has as function to stabilize this pile in the real world. Just like the function of a cup lies in its emptiness (so that a liquid can be poured into it), the specific shape of hollow culms is (at least partly) to be found in its emptiness. Parenchyma cells, on the other hand, try to maximize their volume. As a result, the intercellular space is minimized.

EM parabolas and leaf shapes

Another straightforward example of EM moderation in plants is encountered in leaf shapes. Nature plays a very simple but ingenious part here. In Fig. 4, some parabolas of the general form are depicted (only the quadrant with $x, y \in [0;1]$). As proved in theorem 1, the parabolas for 'n' and '1/n' are perfectly symmetrical, with the bissetrices of the quadrant ($n = 1$) as the symmetry axis. The similarity of the compound figure of parabolas in Fig. 4 and laminae of bamboo foliage leaves, attached to the sheath of the leaf through a pseudo-petiole, is striking. The perimeter, the mid-vein and all parallel veins can be represented by parabolas, the mid-vein being the axis of symmetry of the two longitudinal halves of the blade (Gielis 1997).

In many variegated leaves of bamboo in particular, the variegated sectors are contained between two veins or parabolas. This is seen in many other variegated grass leaves as well (the difference with bamboo is that the base of the blades in other grasses is implanted around the stem, unlike bamboo with its pseudo-petioles). The presence of such parallel veins in bamboo and other monocots is related to the elongation of the blade, almost exclusively by elongation from the basal (i.e. located proximally) cells. EM shapes also reduce this process into a simple one. Moreover, the parabolas can account for the bilateral symmetry in the blades.

In many dicots, EM leaf blades can be observed as well. In *Plantago,* the parallel veins most probably function to strengthen the leaf. In other dicots with elliptic, ovate or lanceolate leaves, the perimeters are also described by EM parabolas. The main difference with monocots is that in dicots during cell division and meristematic activity are found throughout the blade. This results in an inverted EM parabola, i.e. the origin is located distally. This conforms to the age of these distal parts, which are ontogenetically older than proximal parts of the blade.

Root architecture

In roots, super shapes play a predominant functional role in two ways. First, cells in cross sections of the root try to maximize the available space and try to minimize the intercellular space, just like parenchyma cells in culms. Secondly, super shape optimization is of crucial importance for the quiescent center in roots. The author studied the occurrence of super shapes in roots of corn and *Arabidopsis.*

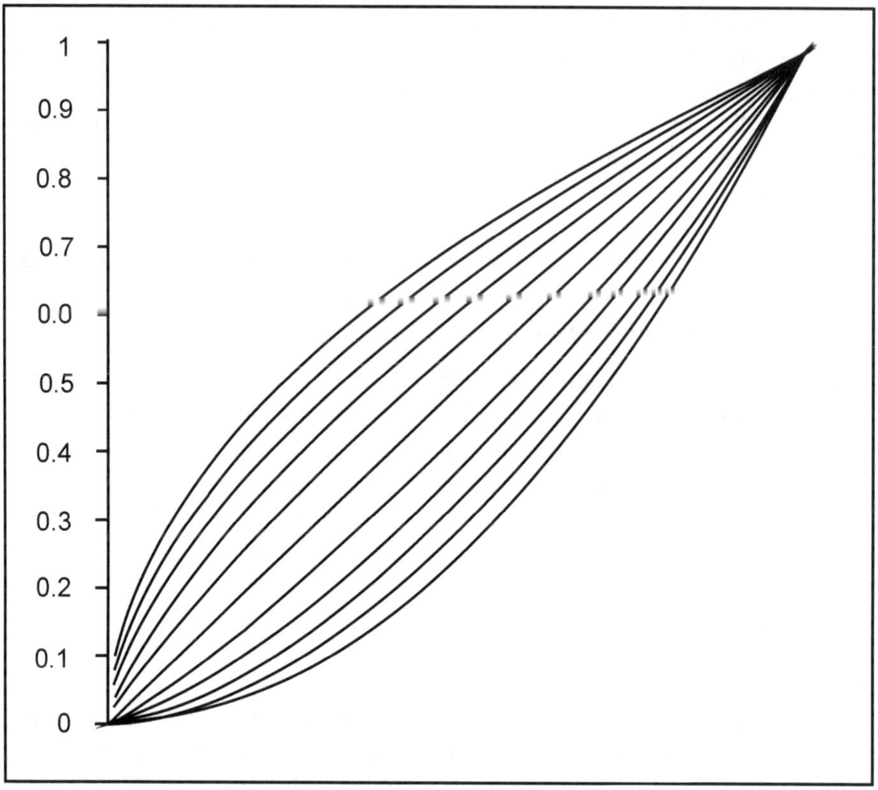

Figure 4. A bamboo leaf with veins as super and sub-parabolas (values for n = 1, 1.2, 1.4, 1.6, 1.8, 2 and 1/n)

Earlier, corn roots were the best-studied roots, featuring in most if not all botanical textbooks. At the molecular genetic level, this position now belongs to *Arabidopsis*, mainly because its genome is small. Its root architecture has been studied in detail and the molecular mechanisms underlying root growth and development are being unveiled.

In both root types, it is seen that a *flachpunkt*-like shape occurs at the junction of the root cap and the root proper. Fig. 5 depicts an ellipse, which models the outer layer of the root tip; an ellipse was chosen because it is a closed shape but similar results can be obtained using super parabolas (Van Oystaeyen et al. 1996; Gielis, 1997). The functional significance of the super shape in roots is three-fold (Beirinckx 1997).

1. Space optimization: Since the QC has to be located in the root tip. At its periphery, cell divisions occur, giving rise to the cells for both root body and root cap. Cell elongation and periclinal division occur later, giving rise to the Kör-

per-Kappe. To accommodate all cell types of the QC, such a super shape is ideal since they are all located in a sac with a flattened bottom.

2. Stress distribution: A tyre of a bicycle is flattened if subjected to a load and the reaction force is distributed evenly over the contact surface. Similarly in roots, by assuming a *flachpunkt* shape, the zone that harbors the QC and its peripheral zones does not come under much stress since any force is distributed over a larger surface, resulting in a low pressure. Finite Element Analysis has proved that for any root-like shape and for two different zones (body and cap) with different material properties (elasticity) a QC is formed where the internal stresses are zero during growth when the soil exerts a reaction force, and super shapes strengthen this effect. If growth and penetration of the root into the soil is more difficult, the QC diminishes in size and cell division is increased. This conforms well to anatomical and physiological studies of root growth.
3. Since cap and the root proper are two distinct entities, changes in the direction of forces exerted on the tip can result in dislocation of the cap. The existence of a *flachpunkt* effectively prevents rotation. Indeed, the stability between the cap and the root is increased by the *flachpunkt.*

CONCLUSION

Mathematics is a very important tool for studying and understanding natural phenomena, especially in morphology, since many believe that "Nature is written in the book of Geometry" (quoted in d'Arcy Thompson 1942). The notion of super and sub-shapes provides a new and challenging approach to modelling and understanding of plant morphology and anatomy. Again, a proof that nature had conceived such a very useful strategy, long before we found it and applied it in physics or architecture. The examples shown here are based on conic sections only. But it can be shown that many other shapes are generated by super shapes, but of different rotational symmetry.

As nature does little in vain (Isaac Newton), the choice of a super strategy is very good in terms of return on investment. In shape optimization, a super circle has a longer perimeter to enclose a given area than a circle; but then, super circles are very good compromises in terms of efficient space utilization and resistance against mechanical forces. Another very important consequence of this super strategy is that variation is allowed: whether the culm of a square bamboo has exponent of 3 or 7 is not important. This principle itself is very important since it allows natural growth in natural conditions, and adaptation to changing conditions, at the same time preserving the necessary robustness.

The discussion in this paper has been restricted to conic sections and their occurrence in plants. Here, the rotational symmetry is limited to 4 (super circles and squares) or less (ellipses and parabolas). However, the super formula can easily be extended to include other functions and different rotational symmetries. Then it would turn out to be a giant formula to develop many different shapes,

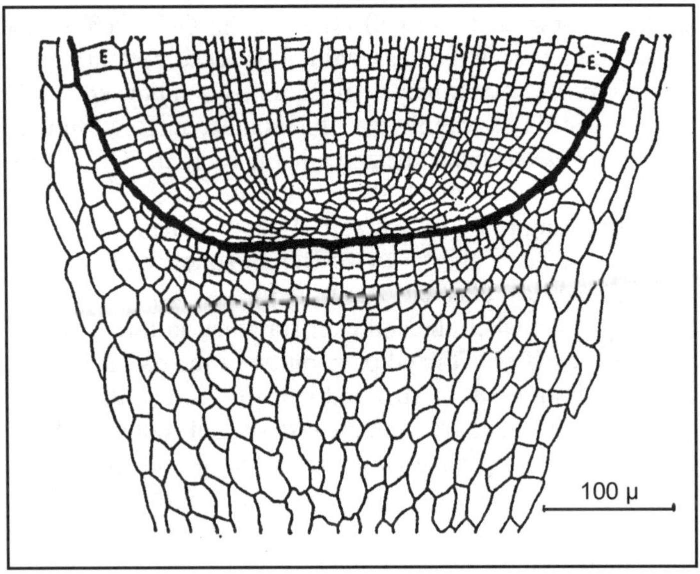

Figure 5a. The *flachpunkt* of corn roots

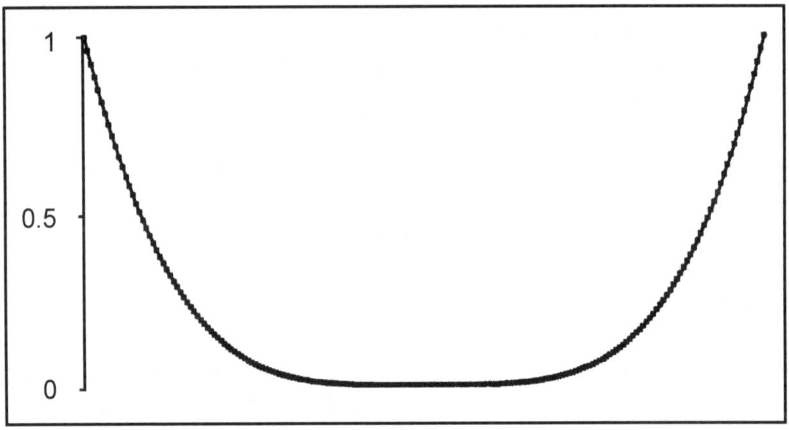

Figure 5b. Super parabola with exponent n = 4

including the shape of eggs, star fishes, flowers, spider webs, shells of molluscs, birds, etc. It would also allow one to explain how phyllotaxis arises in plants. The definition of this formula is probably a major scientific discovery, capable of transforming our views on nature and human culture most profoundly.

ACKNOWLEDGEMENTS

The author thanks: Nancy Van Acker, whose computer program helped in the initial computations on super shapes; Bert Beirinckx, who did marvellous work on FEM analysis of roots; Fred Van Oystaeyen and Tom Gerats for extending constant moral support; and Chris Stapleton for suggesting the term "moderation".

REFERENCES

Allard, J. 1964. Notes on squares and cubes. Mathematics Magazine, 37, 210–214.

d'Arcy Thompson, W. 1942. On growth and form. Cambridge University Press, Cambridge, UK.

Beirinckx, B. 1997. Supervormen, Wortels en Bamboe. MSc thesis, Antwerp, Belgium.

FAO (Food and Agriculture Organization), 1978. Artificial production of square bamboos. Forest News for Asia and the Pacific, 2(3), 28.

Gielis, J. 1996. Supervormen bij bamboes. Newsletter of the Belgian Bamboo Society, 14, 21–26.

Gielis, J. 1997. Supervormen bij bamboes, deel 2. Newsletter of the Belgian Bamboo Society, 16, 18–21.

Gridgeman, N.T. 1970. Lamé-ovals. The Mathematical Gazette, 54, 31–37.

Hashimoto, 1962. On the producing method of quadrangular bamboo culms. Reports of the Fuji Bamboo Garden, 6, 24.

Lamé, G. 1857. Leçons sur les coordonnées curvilignes et leurs divers applications. Paris, France.

McGowan, D.J. 1886. Square Bamboo. Nature, 33, 560.

Niklas, K.J. 1992. Plant biomechanics: an engineering approach to plant form and function. The University of Chicago Press, Chicago, USA.

Prusienckiewicz, P.; Lindenmayer, A. 1989. The algorithmic beauty of plants. Springer Verlag, Germany.

Ueda, K. 1960. Studies on the physiology of bamboo. Bulletin of the Kyoto University Forests, No. 30.

Van Oystaeyen, F.; Gielis, J.; Ceulemans R. 1996. Mathematical aspects of plant modelling. *In* Gielis, J.; Gerats, T. ed., Aspects of morphogenesis of leaves, flowers and somatic embryos. Scripta Botanica Belgica, 13. pp. 7–27.

Wen Tahui, 1994. The taxonomy and cultivation of *Chimonobambusa makino*. The Journal of the American Bamboo Society, No. 1 and 2.

Bamboo for Sustainable Development, pp. 777–782
Eds. A. Kumar, I.V. Ramanuja Rao and Ch. Sastry

Rehabilitation of mined areas through bamboo afforestation: A case study

D.N. TEWARI and KAUSHAL KUMAR
UTTHAN, Center for Sustainable Development and Poverty Alleviation, Allahabad, India

Abstract—Land degradation is a major problem confronting India, as it has affected about 174 million ha of agricultural and non-agricultural lands and forests. Bamboo, with its unique ability to stitch and repair damaged soils, is ideal for use in rehabilitating degraded lands. A bamboo afforestation project was launched to test and transfer appropriate bamboo-based plantation technologies to reclaim degraded land, improve productivity through soil and water conservation, and upgrade the skills for making bamboo and grass crafts and utility items for the socio-economically disadvantaged communities. The project involved two villages located in the upper stream of the confluence of the Ganga and Yamuna Rivers at Allahabad, where rampant mining of topsoil by the local brick industry over long periods had ravaged the land. This paper presents a case study of the project.

INTRODUCTION

With the increasing pressure of ever-growing human and cattle population, land degradation has started owing to deforestation, shifting cultivation, loss and destruction of catchments, soil erosion, water logging, salinization, poor agricultural practices, mining (including surface mining for brick manufacture) and pollution. One of the major problems confronting India is land degradation, which has affected about 174 million ha of agricultural and non-agricultural lands and forests.

Land and water have a symbiotic relationship, and integrated watershed programs with people participation can enhance productivity of biomass, generate employment and income and improve environment to bring about a healthy overall development.

Investments in direct rehabilitation projects are supported by complementary investments targeted to raise the income of villagers and farmers affected by the project and upon whom the long-term sustainability of the improvements depends. Private investments in forest-related ventures can range from conventional Poplar, Eucalyptus and Casuarina plantations through agro-forestry and now bamboo plantations, and their improved utilization for early quick return. Bamboo, with its unique ability to stitch and repair damaged soils, is ideal for use in rehabilitating degraded lands.

The objective of the bamboo afforestation project has been to test and transfer appropriate bamboo-based plantation technologies to reclaim degraded land, improve productivity through soil and water conservation, and upgrade the skills for making bamboo and grass crafts and utility items for the socio-economically disadvantaged communities.

It also aimed to develop community-based institutions to manage and sustain community initiatives on integrated micro-watershed management for sustainability and poverty alleviation. Depending on the availability of funds, the technology can be extended to rehabilitate the complete watershed covering an area of 4 000 ha.

DEGRADED LAND AND IMPOVERISHED PEOPLE

The land at 20 villages situated in the upper stream of the confluence of the rivers Ganga and Yamuna at Allahabad had been ravaged over time from the rampant mining of topsoil by the local brick industry. Over the last 2–3 decades, the poor villagers have either leased their land for excavation of topsoil (up to 3 m in depth) or sold their land outright to the brick industry. Bricks are made from the topsoil and after 6–10 years, a plot of land so leased or purchased is rendered useless for any agricultural production.

About 4 000 ha of degraded land in the above villages is facing the problems of:

1. Soil erosion and ravine formation;
2. Floods and droughts every year;
3. Water scarcity and drop in ground water table;
4. Severe dust storms starting from the area and engulfing Allahabad city;
5. Lack of moisture in the air and acute moisture stress in the soil from February to June;
6. Erratic rains and fluctuating temperatures;
7. Heavy storms originating in such areas because of excessive heating of refractory soil, causing extensive damage and total disruption of normal life in the city for few days;
8. Poor people exploiting the marginal resources available to them through low inputs, low productivity, overgrazing and destruction of all vegetation;
9. Migration of the people to cities in search of casual employment; and
10. Malnutrition, diseases and poverty.

Poor people and poor areas (degraded by surface mining for brick manufacture) have become locked in a vicious cycle of mutual destruction. Only integrated watershed approach can help rehabilitate the degraded land and regenerate its productive capabilities for supporting increasing human and cattle population through adequate biomass production, productive employment generation, people's empowerment, opportunity for economic advancement of all, and social and gender justice.

MICRO-WATERSHED APPROACH FOR DEVELOPMENT

Owing to limited finances, Kotwa-Rahimbad villages at the head of the catchment (4 000 ha degraded area being the catchment of Sasur-Khaderi River) were selected for the project. The basic socio-economic data of the area are given below:

Total area of the micro-watershed	106 ha
Total area mined up to a depth of 3 m	60 ha
Common degraded land available	12 ha
Total population of these villages	2 340
Total households	311
Landless families	125

The strategy adopted for micro-watershed management included:

- Rational land use to prevent further degradation, erosion and silting of riverbed;
- Thrust on construction of strategically located structures along drainage lines to reduce velocity of run-off and impound water in the depressions;
- Use of indigenous technologies, local skills and materials in construction of gully plugs and check-dams;
- Involvement of people (farmers on their land and landless on common land) in planning and execution of various component of watershed development;
- A wide range of components included in the rehabilitation program, including soil and water conservation, plantation of bamboo along field boundaries (bunds), agro-forestry, low-input sustainable agriculture, application of organic manures and bio-fertilizers;
- Afforestation of common land with multi-purpose species providing usufructs, which motivated landless people (the position of landless people was found to be critical, especially for watershed approach, which is essentially a land-based program) to participate in the watershed development program; and
- Priority given to occupational training and skill upgrading in traditional production-to-utilization activities on bamboo and munjh grass (*Saccaram munja*).

ROLE OF BAMBOO IN RECLAIMING DEGRADED LAND

For developing Kotwa-Rahimabad micro-watershed, a research project on "Eco-rehabilitation of degraded farm/common lands through bamboo plantations and improved crafts making" was started by UTTHAN Center for Sustainable Development and Poverty Alleviation, Allahabad. The International Network for Bamboo and rattan (INBAR) financed the project.

Bamboo was preferred because it is:

1. A fast-growing species;
2. Hardy and best for reclaiming marginal and surface-mined soils;
3. Invaluable for rapid control of soil erosion (one plant can bind up to 6 m^3 of soil), conservation of moisture and protection of watershed (a vast underground rhizome network that may cover up to 1 000 m^2 around the plant);
4. Produces 30% its volume in biomass;
5. Ideal for controlling land acidity;
6. Tall, strong and elastic, and ideal for use as windbreakers;
7. A good noise and climate buffer;
8. Converts sulphur dioxide and other toxic gases into oxygen;
9. Multi-utility species, providing food (young shoots and seeds consumed by human beings) fodder, bamboo houses, and about 1 500 traditional uses;
10. Valuable crop having comparatively short production cycles and yielding cost-effective, short-term return;
11. Yields versatile raw material for a wide range of industries, such as pulp and paper, rayon (six times more cellulose than pine trees), bamboo mat board, and cottage/handicraft industry;
12. Lightweight yet strong material (about 5 000 pounds more tensile strength than steel);
13. Easily workable with simple tools;
14. Comes with readymade finish;
15. Forms part of traditional medicine systems;
16. Good for anti-seismic construction;
17. Has ever-increasing local demand and rising export potential;
18. Excellent substitute for wood in many applications;
19. Has the ability to rejuvenate degraded soil; and
20. Lends itself to intercropping.

A germplasm collection of bamboo species, which may be agro-climatically suitable for the project area, was attempted. The species selected were: *Bambusa bambos, B. vulgaris, B. nutans, B. tulda, Dendrocalamus strictus, D. giganteus* and *D. asper.*

An irrigated nursery was developed to raise bamboo seedlings in raised soil beds. When the seedlings had attained a height of about 25 cm, they were transplanted in polythene bags. Planting was done during the rainy season, when seedlings were about one m in height.

Replicable models of different bamboo plantations (strip plantation, mixed plantation, block plantation) were developed for small, medium and large farmers, as well as for common lands. Refined technologies were developed for soil and water conservation, rejuvenation of degraded soil, improving microclimate, agroforestry, etc.

To develop the skills of bamboo artisans, they were specially trained to produce: (1) domestic market requirements of utility and decorative items (such as lampshades, shopping and picnic baskets, wall fixtures and trays); and (2) prototypes of handicraft products for export.

New planting technologies (planting of full grown culms with rhizome, and application of hormone and bio-fertilizers), adopted yielded good results. In the second year, on an average 15 culms per clump were produced. During third year, cutting of 5 culms per clump was possible.

People have started using the bamboo produced for construction, scaffolding, ladders and articles of daily use. Cottage and handicraft industry will develop soon.

SUSTAINABILITY IN WATERSHED DEVELOPMENT

Sustainability of food and fodder production increasingly depends on sound and efficient soil and water conservation and use with respect to rain-fed areas, livestock rearing and agro-forestry. The restoration of vegetal cover with bamboo and grasses, and construction of small check-dams and small storage tanks across all drainage lines have yielded spectacular results.

Better resource management was possible because all farmers, and even the landless people, were stakeholders of the watershed development project. Ground water recharge and availability of electricity facilitated supply of water for drinking, livestock development and irrigation of agro-forestry crops.

Cultivation of bamboo increased the ecological security of the micro-watershed, food security of people through increased production of agricultural crops – moringa (*Moringa pterygosperma*) fruits and bamboo shoots – milk and meat, and livelihood security through employment generation and income from sale of bamboo and grass handicrafts and other products.

Availability of clean water, fuel-wood and fodder reduced the suffering of women, who were travelling long distances to collect these items. Women were specially trained to start home-based production of various household and handicraft items using bamboo. To start with, UTTHAN procured bamboo from outside and made them available. Although at present there is no problem of marketing bamboo products, special arrangements will be required in future.

Traditional community institutions were having weakness and deficiencies to varying degrees, which impaired their suitability for micro-watershed management. Modifications were made to ensure:

1. More focus on vulnerable groups, especially landless and resourceless people;
2. Greater representation and involvement of all groups of community;
3. Primary healthcare of people in an organized manner;
4. Marketing of the produce; and
5. Human resource development for availing new developmental opportunities.

Thus, bamboo has helped in reclaiming the degraded land and micro-watershed. It prevented the loss of topsoil, rejuvenated the degraded soil, conserved moisture, and increased the vegetal cover and food security. Integration of various measures for comprehensive development increased the sustainability and helped alleviate poverty.

Bamboo for Sustainable Development, pp. 783–791
Eds. A. Kumar, I.V. Ramanuja Rao and Ch. Sastry

Use of bamboo in silk industry

N.S. ADKOLI
Executive Director, Bamboo Society of India, Bangalore, India

Abstract—Production and use of silk in India is known from mythological times. Silk fabric is made of yarn spun from the thin and delicate strands derived from cocoon of a moth. Rearing of silk worms, production of yarn and weaving into fabric engage hundreds of thousands of rural families in India. Green leaves of mulberry plant, *Morus alba*, cultivated on selected lands in a few states, form the food for the larvae of silk moth, *Bombyx mori*.

Out of an annual production of 14,500 tons of raw silk in India, Karnataka State has the major share – 8,450 tons.

A circular tray, called *thatte* and made from woven slivers of bamboo, is used for rearing the juvenile larvae for about 21 days. As the larvae increase in size, they are transferred to other trays to reduce congestion. The fat larvae are transferred after 21 days in to a compartmentalized tray, called *chandrika*, where they are fed with more mulberry leaves for another 5–6 days, after which the larvae enter the stage of pupae inside the cocoon spun by them using delicate thin and long silk strands. The cocoons are boiled in water to kill the pupae and separate the long silken strands, which are then spun and reeled into yarn before it is used for weaving into fabric.

The quantity and value of bamboo annually used in Karnataka for the silk industry are given in the paper. The employment generated in the process of manufacture of the bamboo articles used in sericulture is also estimated.

INTRODUCTION

Silk

Production and use of silk in India is known from mythological times. Silk apparels are used during religious functions and festivities by both the rich and the poor. India has been and still remains a leading country in the production and use of silk. Saris, shirts, pants, neckties and other kinds of daily and fancy wear are very popular in India, as well as abroad.

Silk is a delicate fiber, produced by the larvae of an insect named *Bombax mori*, belonging to the family *Lepidoptera*. The egg of the insect hatches into a larva. The larvae feed on tender green leaves of the mulberry plant (*Morus alba*) for about 29 days, during which they molt a few times as they grow in size. The grown-up larvae spin their egg-shaped cocoon in which they transform into pupae, before emerging as the adult moths.

To obtain silk, the cocoons are boiled before the emergence of the moths, to separate and reel the delicate thin continuous strands of silk. The strands are reeled to produce the silk thread used in weaving the silk cloth. Mulberry plants are grown by a large section of farmers for either sale of green leaves or for feeding to larvae for the production and sale of cocoons. Production of cocoons, reeling of yarn, weaving of fabric, etc. together forms a major cottage industry in Karnataka State, India, and provides employment to hundreds of thousands of rural families in the country.

Silk production

India produces about 14 500 tons of silk annually, of which Karnataka State alone accounts for 8 500 tons. Other states that produce silk are Tamilnadu, Andhra Pradesh, West Bengal and Jammu & Kashmir. Kerala, Madhya Pradesh and Orissa also produce silk, but in small quantities. An idea about the role of silk production in rural India can be obtained from Table 1.

Table 1.
Silk industry statistics for Karnataka

	1994–95	1995–96	1996–97	Average/year
No. of villages involved in sericulture	18 583	18 583	18 633	18 608
No. of families involved in sericulture	334 793	333 339	333 339	333 823
Silk production (ton)	8 845	8 169	8 328	8 454
Cocoons production (ton)	77 384	71 209	70 678	72 090
Silkworm rearing centers	1 585	1 585	1 585	1 585
Mulberry cultivation (ha)	167 778	167 422	167 422	167 540
Egg layings (in million)	187.8	195.0	184.2	189.0
No. of cocoon markets	64	63	64	64
No. of government-owned silk farms	95	95	96	95

(Source: Annual Reports of the Department of Sericulture, Karnataka)

There are 1 545 silkworm rearing centers (*chawki*) in Karnataka. Apart from cultivation of mulberry, rearing of silk worms and marketing of cocoons, silk industry also supports other income-generating activities such as spinning, reeling, weaving, and manufacture and sale of garments. The industry is so large and important that the Karnataka State has a separate government department with a minister in charge and a network of offices. A Central Silk Board, with its main office in Bangalore, coordinates production, research, sales promotion and other such activities.

Research stations with a large number of scientists are established for continuous research on different aspects of silk industry, from raising improved leaf fodder to processing and production of superior quality garments for indigenous sale and exports.

BAMBOO IN SILK INDUSTRY

Bamboo strips and slivers constitute ideal material for three main requirements in rearing of silk worms. A specially woven bamboo basket is used for collection of mulberry leaves from the fields and transportation to the worm rearing centers. A circular tray (*thatte*), made of woven bamboo slivers and strengthened with round or split bamboo, is used for rearing the silkworm. A 'mountage' (*chandrika*), which is a large bamboo tray with several concentric compartments, is used for housing adult larvae. The larvae spin their cocoons in the compartments, using the partitioning ribs of the tray for support. All the three woven bamboo articles are in regular and continuous use by the villagers engaged in silk production. Besides these articles, bamboo is used to a smaller extent in reeling, storage and transport of cocoons and silk threads.

Basket for collection of leaves

The basket is wide in the middle and tapered on both ends. It is woven using bamboo slivers 2 cm wide and 3 mm thick, hemmed into 5–6 mm thick strips of bamboo used at intervals, both longitudinally and horizontally. The rim of the opening at the top, the wide middle and the flat bottom are all strengthened with 2–5 cm thick bamboo splits. The basket is 50 cm wide in the middle, 55 cm high and about 20 cm at both ends (Fig. 1). Each basket weighs 500 g. Three such baskets are needed for gathering of leaves from each hectare of mulberry cultivation.

Rearing tray

The rearing tray (*thatte*) is a wide, circular, shallow basket with a flat bottom. It is woven using 3 cm wide and 2 mm thick slivers (between 1.5–2 cms thick splits). The tray is strengthened at the rim and criss-crossed at the bottom, at regular intervals, using 2–3 cm thick bamboo strips. The tray is 100 cm in diameter and the vertical edge is 6 cm high (Fig. 2). In air-dry state, it weighs about 2 kg. The trays are made by expert weavers (*medars*) who earn their livelihood by weaving bamboo trays. The weavers sell their products to traders, who market them to the users. Each tray needs about 40 slivers and strips, each of 4 m in length. Wastage of split bamboo in the process of weaving trays is about 15%.

A female moth lays 350 to 550 eggs, with an average of 450 eggs per laying. These eggs are spread out in the tray for hatching. Hygienic conditions are maintained in the rearing centers to prevent disease conditions and reduce larva mortality. After the eggs hatch into larvae, they are regularly fed with chopped green mulberry leaves for 10–12 days, during which they molt twice. To relieve congestion, the growing larvae are transferred to more trays and fed with leaves for another 9–10 days. About 35 trays are needed for 100 layings, which equal about 45 000 eggs. The trays can be used five times on an average in one year and have a service life of three years.

Figure 1. Basket for mulberry leaf collection

Figure 2. Tray for rearing silkworm

Figure 3. Front view of mountage (cocoon tray)

Cocoon tray (mountage)

The base of a cocoon tray (referred to in the industry as "mountage" and locally as *chandrika*) is made of a strong, woven, rectangular bamboo mat measuring 2 × 1.2 m (Fig. 3). Round and split bamboos are used to strengthen its four sides and the center. Two diagonally tied bamboo rounds at the back of the mat add to the strength. Woven bamboo ribs of 6 cm width, fitted on the upper surface of the mat in a concentric fashion, divide the mat into a series of compartments. The ribs, splits and round bamboos making the frame of the tray are fastened in position by means of coir (Fig. 4). The slivers used for the mat are 3–4 cm wide and 2–3 mm thick. The round bamboo used for the outer frame, diagonals and horizontals are 3–5 cm in diameter (splits of the same width are sometimes employed). The average dry weight of each mountage is 13 kg. The mat is generally woven from big

Figure 4. Use of coir as fastener (rear view of mountage)

bamboo (*Bambusa bambos*) while the frame is made of small bamboo (*Dendrocalamus stictus*).

The larvae, when they are 21 days old, are carefully transferred to the mountage and evenly spread between the concentric ribs. Feeding of leaves is continued for 7–8 days more, during which time the larvae considerably grow in size. The fat larvae, which are 28–29 days old, start spinning their cocoons individually to pupate (Fig. 5). The spinning takes about one or two days after which the cocoons are removed from their anchorage on the ribs and sold by weight to those engaged in the job of separation of silk strands.

These cocoons are then placed in pans containing boiling water to kill the pupae and separate silk filaments. The dead pupae are dried and used in the manu-

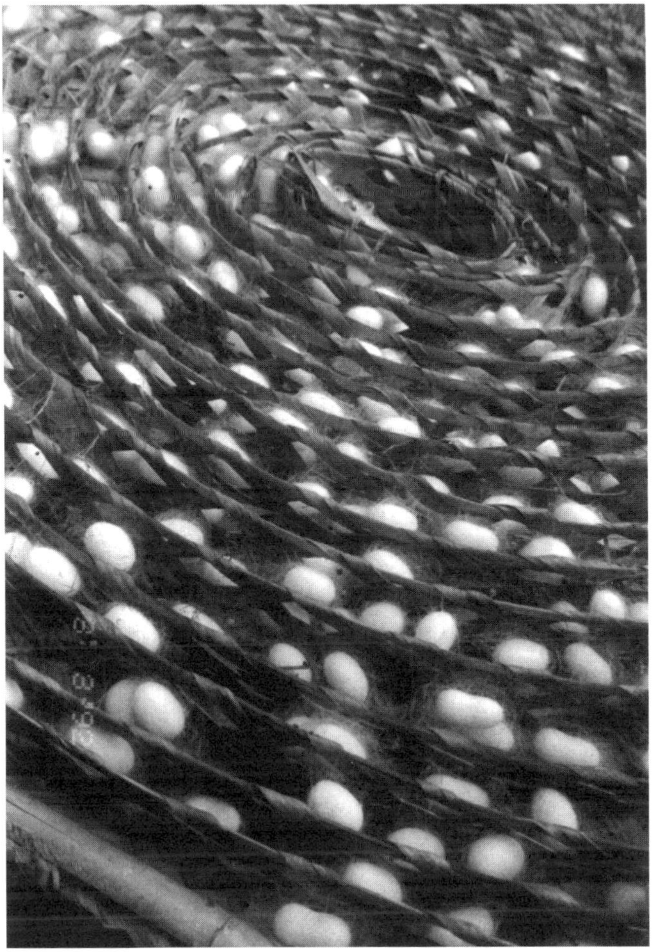

Figure 5. Cocoons inside mountage

facture of poultry feed. Each mountage can be used 6–8 times in a year and has a service life of three years.

BAMBOO REQUIREMENT FOR SILK INDUSTRY

Calculations of the bamboo requirement of the industry based on the three major articles – baskets, rearing trays and mountages – are given below.

Baskets

Area of mulberry cultivation = 167 580 ha
Number of baskets used = 167 540 × 3 = 502 740
Weight of bamboo baskets = 502 740 × 0.5 kg = 251.37 tons

Rearing trays
35 trays × 2.5 kg = 87.5 kg (for 100 layings)
Mountages
25 × 13 kg each = 325 kg (for 100 layings)

Total dry weight of bamboo used annually for the three articles described above, to cover 189 million layings (Table 1), therefore works out to 77 962 tons. Considering the service life of the articles is three years, the dry weight of bamboos of these three articles comes to an annual figure of 26 238 tons. The green weight of round bamboo required can be calculated by adding 15% for wastage and 40% for moisture content. The approximate weight of green culms used in sericulture in Karnataka State is thus 42 000 tons. This figure exceeds 50% of recorded annual production, indicating that a large quantity used by this industry comes from unrecorded sources.

VALUE

The average market prices (at 1997 price level) of the three manufactured articles are used for the calculation of total value of the products given in Table 2.

Table 2.
Annual value of bamboo articles used by the silk industry

Item	Unit price (INR)[1]	Annual use (million units)	Annual value (million INR)
Basket	30	0.17	5.10
Rearing tray	60	22.05	132.30
Cocoon tray	250	15.75	393.75
Total			531.15

It can be seen from the above that the annual value of the three major bamboo articles alone amounted to approximately USD 14.8 million (at mid-1997 exchange rate).

EMPLOYMENT GENERATION

Investigations have that an average weaver can prepare five baskets, three rearing trays or one cocoon tray per day. Based on this, the number of workdays generated in the manufacture of these three items has been estimated to total 23.13 million workdays per year (Table 3).

[1] INR (Indian Rupee) 1 = approx. USD 0.024

Table 3.
Workdays created per year in the manufacture of bamboo articles

Item	Number manufactured (million)	Workdays generated (million)
Basket	0.17	0.03
Rearing tray	22.05	7.35
Mountage	15.75	15.75
Total		23.13

Thus, 23.13 million workdays are created per year in the manufacture of three main bamboo articles needed by the silk industry in Karnataka. The figure will be much higher if the labor involved in the harvest and transport of bamboo is also takcn into account. Almost all of this cmploymcnt is in thc rural scctor and without any distinction of age, sex or even physical ability.

In view of such a high employment potential and value generation in the rural economy, it is necessary to conserve and develop bamboo resources, and promote the use of bamboo by the silk industry in the country.

Bamboo for Sustainable Development, pp. 793–795
Eds. A. Kumar, I.V. Ramanuja Rao and Ch. Sastry

Bamboo: A cellulosic and energy raw material

ANISIO AZZINI[1] and ANTONIO LUIZ DE BARROS SALGADO[2]
Instituto Agronômico de Campinas, São Paulo, Brazil

Abstract—For centuries, bamboo has been an important material in many tropical countries in Asia, South America and Africa: for crafts, energy, building and construction, food, pulp, paper and many others. In Brazil, however, bamboo utilization is not widespread and the natives species are not well known. The most used species are the introduced ones: *Bambusa vulgaris, B. vulgaris* var. *vittata, B. tuldoides, Dendrocalamus giganteus* and *Phyllostachys* sp. The only industrial utilization for bamboo in the country, particularly in the north-east region, is to produce pulp for paper. The objective of this paper is to present some data regarding the combined production of cellulosic fibers and ethanol or cellulosic fibers and starch from the culms of *B. vulgaris* Schrad. The yields of cellulosic fibers and ethanol were determined using shredded chips, which have been treated with diluted sulfuric acid solution. Laminated chips with reduced thickness (0.5 mm) were used for starch extraction.

The average combined results obtained from a ton of dry chips were about 495 kg of cellulosic fibers and 136 l of ethanol or the same quantity of cellulosic fibers and 64 kg of starch.

INTRODUCTION

Bamboo was one of the first raw materials used in the tropical countries to increase human comfort and well-being. In Asian countries, bamboo is called "the poor man's timber" owing to its widespread utilization as a relatively cheap material for hundreds of uses. According Liese (1986), bamboo forests in certain regions of Asia are more important for the rural people than the goods derived from trees. The increasing interest for bamboo at present is because of its easy propagation, vigorous regeneration, fast growth, high productivity and quick maturity, all of which make harvesting possible within a short cycle (2–4 years).

Besides the traditional utilization, bamboo can be used as an important industrial raw material to produce cellulosic fibers and energy. The objective of this paper is to provide data on the combined production of cellulosic fibers and ethanol or cellulosic fibers and starch from the bamboo *Bambusa vulgaris* Schrad.

---

[1] Working under a grant from the National Council of Research and Development (CNPq)
[2] Retired from Instituto Agronômico de Campinas

MATERIALS & METHOD

For the purpose of this study, the species *Bambusa vulgaris* was considered because it is the most important bamboo used in the production of pulp and paper in Brazil. Ten culms between 1 and 5 years old were harvested from the bamboo stand at the Experimental Station of the Instituto Agronômico de Campinas (IAC). Internodes from the basal, middle and top portions were converted into laminated chips with a thickness of about 0.5 mm. The yields of cellulosic fibers and starch were determined in terms of culm age (1 to 5 years) and sampled portion (base, middle and top).

Chip shredding was then carried out in a domestic liquefier. The ratio between the quantities of chips and water was 1:33, and the shredding time was 30 seconds. A fibrous fraction and a liquid phase with granular starch in suspension resulted. The fibrous fraction was converted into cellulosic fibers in an oxidizing solution composed of concentrated acetic acid (50%), hydrogen peroxide (40%) and distilled water (10%). Granular starch was obtained by decantation from the liquid phase (Azzini 1984). Ethanol was obtained from shredded chips by treating them with a diluted solution of sulfuric acid (1.0%) (Azzini et al. 1987). The yields of fibrous fraction, cellulosic fibers, ethanol and starch were calculated over the chip dry matter.

The average yields of fibrous fraction, cellulosic fibers, extracted starch and ethanol in terms of culm age and culm portions are given in Tables 1 and 2.

Table 1.
Average yields of fibrous fraction, cellulosic fibers, extracted starch and ethanol in function of culm age

Determinants	Culm age		F Test	Tukey (5%)	VC (%)
	1 year	5 years			
Fibrous fraction (%)	76.13a	69.35b	11.04**	3.95	6.64
Cellulosic fibers (%)	50.73a	48.89b	19.39**	2.18	5.48
Ethanol (liter/100 kg of chips) (1)	13.08b	14.79a	3.54*	2.85	14.10
Extracted Starch (%)	6.52a	6.16a	0.38n.s.	–	25.32

Notes: (1) = calculated values; ** = highly significant at P<0.01; * = significant at P<0.05; ns = not significant. a, b = if the same alphabet repeats across the columns, the difference is insignificant at P<0.05.

The fibrous fraction obtained after chips shredding changed in function of culm age, from 76.13% (1 year) to 69.35% (5 years), but not in function of culm regions (72.16 to 76.35%). The chips treatment with diluted sulphuric acid solution promoted the starch saccharification without effecting fibrous fraction. Darkening was the only visible alterations noticed after the acid treatment. The average cellulosic fibers yields changed significantly in function of age (48.89 to 50.73%) and culm regions (46.85 to 51.26%).

Table 2.
Average yields of fibrous fraction, cellulosic fibers, extracted starch and ethanol in function of culm regions.

Determinations	Culm portion			F test	Tukey (5%)	VC (%)
	Base	Middle	Top			
Fibrous fraction (%)	76.35a	72.81a	72.16a	2.97ns	–	7.52
Cellulosic fibers (%)	51.26a	49.86a	46.85b	19.39**	1.76	4.44
Ethanol (liter/100 kg of chips) [(1)]	12.77b	13.40ab	14.32a	22.15**	2.23	11.01
Extracted Starch (%)	6.94a	6.72a	5.78a	0.93ns	–	21.89

Notes: ([1]) = calculated values; ** = highly significant at P<0.01; * = significant at P<0.05; ns = not significant. a, b = if the same alphabet repeats across the columns, the difference is insignificant at P<0.05.

The ethanol yield varied in relation to culm age (13.08 to 14.79 liter/100 kg of chips) and culm portions (12.77 to 14.32 liter/100 kg of chips). When compared with the conventional process that produces only cellulosic fibers, the combined production of cellulosic fibers (for paper making) and ethanol is a new possibility for optimizing bamboo utilization.

Another such option is the combined production of cellulosic fiber and starch. As mentioned earlier, in the present study, starch was obtained by decantation of the liquid phase. The starch granules obtained resembled rice starch in size and shape. Its high absolute density (1.531 g/cm^3) indicated a compact granule. According Toledo et al. (1987), bamboo starch may find an industrial use in products where low retrograding is desirable, and/or as a substitute for rice starch when small granules are desired.

CONCLUSION

The combined production of cellulosic fibers and ethanol or cellulosic fibers and starch from bamboo chips was found to be technically feasible. The average production from a ton of dry chips was about 495 kg of cellulosic fibers and 136 liter of ethanol or the same quantity of cellulosic fibers and 64 kg of starch.

REFERENCES

Azzini, A. 1984. Amido a partir de bambu. Bragantia, Campinas, 43(1), 45–50.

Azzini, A.; Arruda, M.C.Q.; Ciaramello, D.; Salgado, A.L.B.; Tomazello Filho, M. 1987. Produção conjunta de fibras celulósicas e etanol a partir do bambu. Bragantia, Campinas, 46(1), 17–25.

Liese, W. 1986. Characterization and utilization of bamboo. *In* Proceedings of the XVIII IUFRO World Congress, Ljubjana, Yugoslavia, September 1986. pp. 11–16.

Toledo, M.C.F.; Azzini, A.; Reyes, F.G.R. 1987. Isolation and characterization of starch from bamboo culm. Starke, Weinheim, 39, 158–160.

Bamboo for Sustainable Development, pp. 797–806
Eds. A. Kumar, I.V. Ramanuja Rao and Ch. Sastry

Potential of producing biofuels from bamboo

N. EL BASSAM[1], D. MEIER[2], CH. GERDES[2] and A.-M. KORTE[1]

1 Federal Agricultural Research Center, Institute of Crop Science, Braunschweig, Germany
2 Federal Research Center for Forestry and Forest Products, Institute for Wood Chemistry and Chemical Technology of Wood, Hamburg, Germany

Abstract—Bamboo was introduced into Europe about 200 years ago, mainly for horticultural uses. Since then, a wide variety of bamboos has been developed for European conditions. Bamboo's ability to tolerate temperatures as low as -25°C is of great importance in northern European countries. Owing to its fast growth rate and excellent chemical and physical characteristics, this plant species holds the potential for being developed as a new alternative crop for the production of raw materials for the industry and for energy.

In 1990, the Federal Agricultural Research Center in Braunschweig started collection and evaluation of several bamboo genotypes for studying their performance under the dominating climatic conditions. In field trials, 17 bamboo genotypes were planted and included in the evaluation procedure.

The bamboo produced approximately 7 t DM/ha/year. The stem height of about 3 m and the shoot number of about 200 shows a potential for high biomass production. The energy content of bamboo stems (17.1 MJ/kg DM) and the cellulose content (over 40%) emphasize the suitability of bamboo as a raw material for energy, pulp and paper production. Bamboo material, stems (90%) mixed with some leaves (10%), was flash-pyrolysed at 487°C in a continuously operating fluidized-bed reactor. About 70% of the feed material was converted into a pyrolysis liquid (bio-oil), which can be used as a fuel or for chemical feedstock.

Selection of adapted bamboo genotypes and further breeding, as well as the development of bamboo crop management from the time of planting to harvest (including harvesting systems), are factors that can improve the yield of bamboo which is the basic requirement for the introduction of bamboo as an energy source. Positive effects on environment, climate and soil erosion will be the logical consequence.

INTRODUCTION

It is estimated that more than two billion people worldwide currently lack access to modern energy resources. Although we know that energy is absolutely essential for development, relatively little attention has been devoted to this matter at national and international levels.

The magnitude of energy consumed per capita has become one of the indicators of 'modernization' and progress of a country. Thus, energy issues and policies have been strongly concerned with increasing the supply of energy. The strategic and environmental consequences of energy consumption patterns have been neglected for a long time. The world continues to seek energy to satisfy its needs without considering the social, environmental, economic and security impacts of energy use.

Many of the crises on our planet arise from the desire to secure supplies of raw materials, particularly energy sources, at low prices. The pressure will become even greater as fossil energy feedstock and uranium are depleted. Although some of these resources might last a little longer that predicted, especially if additional reserves are discovered, the main problem of scarcity will remain, and this represents the greatest challenge to humanity (El Bassam 1998).

It is now clear that current approaches to energy are unsustainable and non-renewable. Of all renewable energy sources, generally the largest contribution, especially in the short and medium range, is expected to come from biomass. Fuels derived from energy crops are not only potentially renewable, but are also sufficiently similar in origin to the fossil fuels (which also began as biomass) to provide direct substitution. They can be converted into a wide variety of energy carriers using existing and novel conversion technologies, and thus have the potential to be significantly new sources of energy into the 21st century.

Energy plant species are understood to mean those annual and perennial species that can be cultivated to produce solid, liquid or gaseous energy feedstock. Renewable energy is of growing importance in satisfying environmental concerns of fossil fuel usage. Lignocellulosic biomass is one of the main renewable energy resources available and the only one that can be converted into liquid, solid and gaseous fuels. Biomass can be used in several ways to provide energy by:

- Combustion (straw, briquettes, pellets, bales, etc.) to provide heat for heating systems and electricity via steam production;
- Gasification to provide a fuel for heating systems and engines to produce electricity; and
- Flash pyrolysis to provide a liquid fuel that can substitute for fuel oil to produce not only heat and power, but also chemical feedstock (Fig. 1).

Over the last several years, substantial research and development work has been carried out to improve the flash pyrolysis process for biomass (summaries can be found in Bridgwater 1997; Bridgwater and Boocock 1997; Kaltschmitt and Bridgwater 1997). There is only limited information available about the pyrolysis of annual or perennial plant material. Therefore, bamboo was selected to study its behavior in flash pyrolysis.

Currently, more than 130 genera of woody bamboos and 25 grass bamboos have been named worldwide, though some other estimates place these number as lower. These bamboos are made up of over 1 300 species of the tribe Bambusoideae, which are distributed among tropical, subtropical and mild temperate zones covering an area of more than 25 million hectares (Fig. 2).

The high calorific values (4–6 kcal/kg) and the relatively low ash contents of bamboo biomass offer the best requirements to produce renewable energy feedstock (Chen et al. 1987).

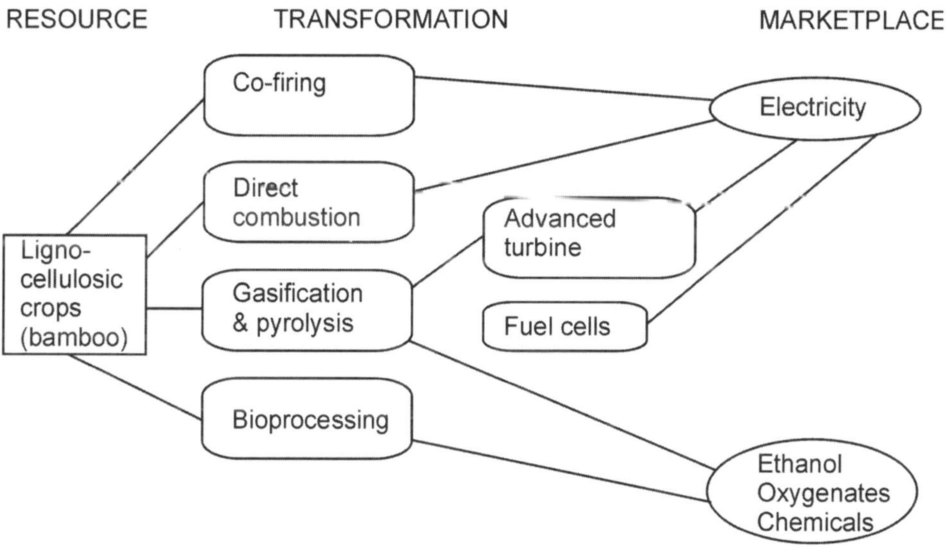

Figure 1. Possibilities of lignocellulosic Crops to different modern biofuels (Overend 1995)

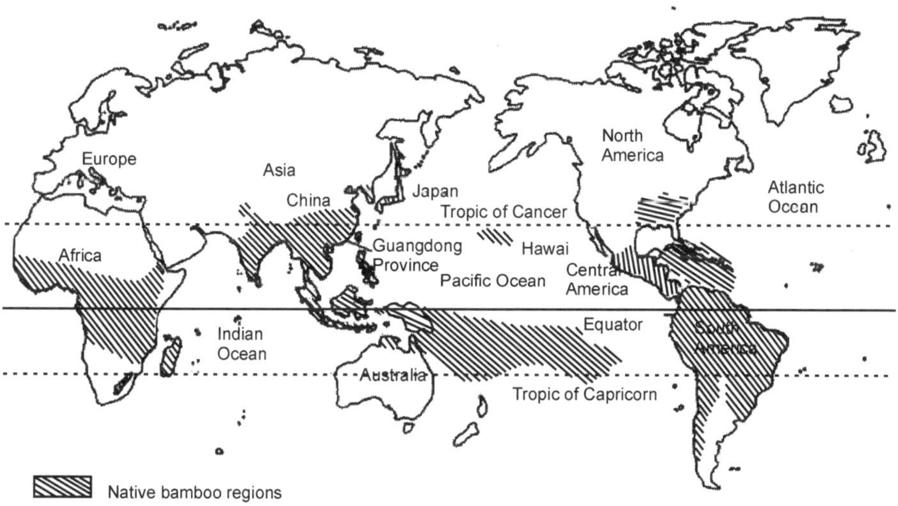

Figure 2. Native bamboo regions (El Bassam and Jakob 1996)

Bamboo has been used for producing carbon in different parts of the world. It is worthwhile in this context to mention that the first electric bulb by Edison had a bamboo carbon filament (Liese 1985). Bamboo has the potential to be an energy cash crop in a sustainable agricultural farming system. Fast-growing and early-maturing bamboos are the most promising and suitable species for producing raw material for biofuel production.

Bamboo was introduced into Europe about 200 years ago, mainly for horticultural uses. Since then, a wide variety of bamboos has been developed for European conditions. In 1990, the Federal Agricultural Research Center (FAL) started bamboo collection and evaluation activities to study the performance of the species under the domain climatic conditions and to investigate the possibilities of using and converting bamboo into biofuel.

MATERIAL AND METHODS

The evaluation of 17 bamboo genotypes has been carried out in field trials. Measurement of plant height, counting of number of tillers and chemical analysis of the plant material were conducted to establish the most important parameters for bamboo as energy feedstock – energy contents, proportion of cellulose and lignin, and nitrogen, potassium and phosphorus percentages. Bamboo material (90% stems) was flash-pyrolysed at 487°C in a continuously operating fluidized-bed reactor. Table 1 shows the characteristics of the bamboo material. The whole plant material was separated into leaves and stems and only the stem fraction was subjected to flash pyrolysis.

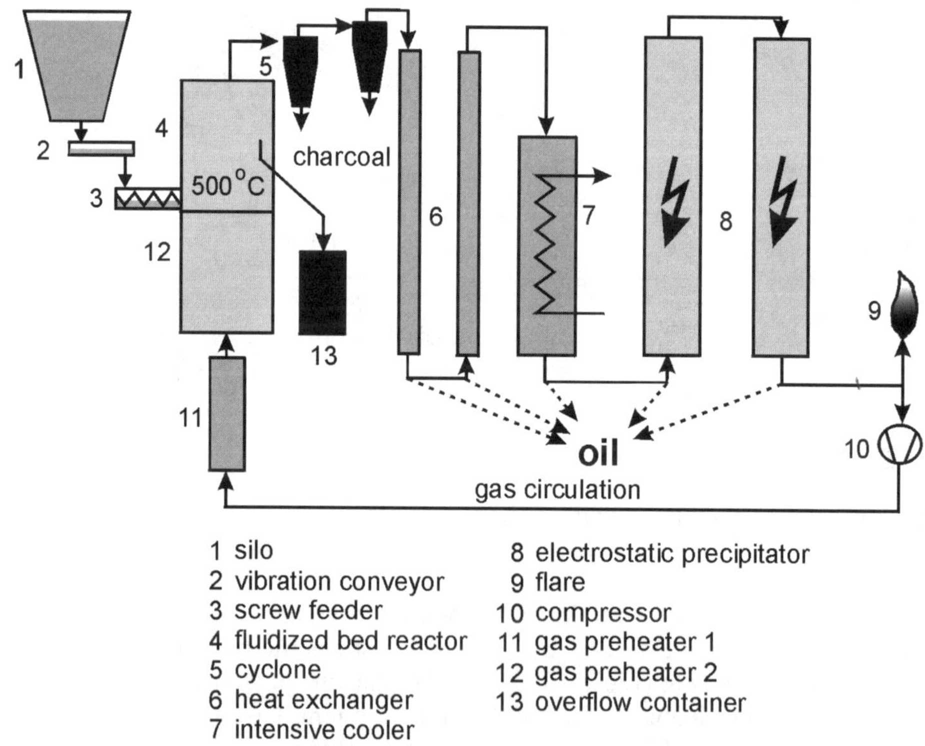

Figure 3. Scheme of the flash-pyrolysis plant for biomass

Table 1.
Specifications of the feedstock

Feedstock	Bamboo stems
Particle size	2–4 [mm]
Glow residue (850°C@4h)	≈ 1.03 [wt.%]
Moisture content	7.1 [wt.%]

A flow scheme of the flash-pyrolysis plant is given in Fig. 2, and the main experimental parameters of the thermal decomposition of bamboo are presented in Table 2.

Table 2.
Experimental (TP29) conditions

Parameter	Value
Temperature	488°C
Vapor residence time	1.36 s
Particle size	
- feed	3 mm
- sand bed	0.55 mm
Run time	1.5 h
Gas flow	
- cold	5.23 m^3/h
- hot	12.93 m^3/h
Throughput	1 961 g/h

RESULTS AND DISCUSSION

Field trials established 1990 with 17 genotypes of bamboo have been monitored for their morphological characteristics. Table 3 indicates some tillers or stems were more than 3 m long and the number of the tillers was up to 286.

Bamboo produced approximately 7 tons of dry matter (biomass) per hectare per year; thus, its yield potential is lower than that of conventional crops. Both stem height and the number of shoots are important for high biomass production. The average dry matter content was 83%. The proportion of stems was 70–80%, the side tillers 8–15% and the leaves 12–15% of the total biomass. Analysis of the stems and the leaves (Table 4) indicated that leaves have much higher nitrogen, phosphorus and potassium content than the stems.

The energy content of bamboo stem (17.1 MJ/kg DM) and the high cellulose content (over 40%) emphasize the suitability of bamboo as a raw material for energy, pulp and paper production (Table 5).

Table 3.
Some morphological data of the 17 genotypes after 5 years

Bamboo genotype	Plant height (cm)	Longest tiller (cm)	No. of tillers
Fargesia murielae	100	130	144
Fargesia nitida 'eisenach'	140	189	87
Fargesia nitida 'nymphenburg'	170	223	256
Phyllostachys aureosulcata 'spectabilis'	128	163	19
Phyllostachys humilis	115	185	10
Phyllostachys nigra	175	280	143
Phyllostachys praecox	70	135	6
Phyllostachys viridiglaucens	85	118	81
Phyllostachys vivax	104	117	22
Phyllostachys 'zwijnenburg' (?)	178	217	286
Pseudosasa japonica	40	73	60
Sasa disticha (syn. Pleiblastus distichus)	167	227	127
Sasa keguma	215	325	79
Sasa kurilensis	30	30	9
Sasa palmata	48	60	6
Sasa pumila (syn. Pleiblastus pumilus)	205	273	130
Semiarundinaria kagamiana	53	68	18
Average	119	165	87

Table 4.
Percentage of N, P, K and Mg in the stems and the leaves of bamboo

	N	P	K	Mg
Leaves	1.54	0.13	0.76	0.073
Stems	0.47	0.08	0.54	0.036

Table 5.
Laboratory analysis of bamboo lignin, cellulose and energy contents

	Stems	Leaves
Lignin (%)	10.7	8.7
Cellulose (%)	40.9	22.0
Energy (MJ/kg DM)	17.1	–

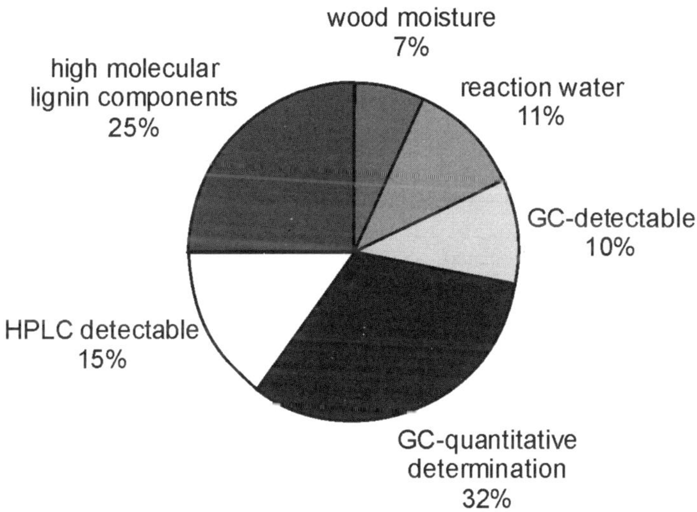

Figure 4. Analytical accessibility of the bio-oil from bamboo

Pyrolysis

Flash pyrolysis of bamboo material gave three product fractions – a liquid (or bio-oil), a gas and a charcoal fraction – in the following composition (wt.%): [1]

Oil	: 57%
Gas	: 21%
Charcoal	: 22%

Bio-oils are complex mixtures of degradation products of cellulose, hemicellulose and lignin having a wide range of molecular weight distribution and polarities. Various analytical methods are necessary to give a chemical description of a bio-oil. A typical scheme of analytically accessible fractions is presented in Fig. 4:

The volatile compounds of bamboo bio-oil were separated by gas chromatography, quantified with a FID detector and identified by a mass-selective detector. The main compounds are presented in Table 6.

As mentioned above, bio-oils can be used to substitute liquid fuels. In contrast to fossil fuels, bio-oils have unique characteristics owing to their completely different composition. As can be seen in Table 6, all compounds have at least one oxygen functional group making the oils water miscible to a certain extent.

Table 7 gives an overview of the main differences between bio-oil and mineral oil. Despite their relatively low heating value, successful trials have been made to burn bio-oil in burners, diesel engines and gas turbines (Kaltschmitt and Bridge-

[1] Oil percentage is corrected by the wood moisture 7.37 wt.%, water content of oil 18.69 wt.%, reaction water 11.32 wt.%.

water 1997). Parallel to the activities in using bio-oil for energy, investigations on its potential as a chemical feedstock are also being carried out (Radlein and Piskorz 1997).

Table 6.
Volatile compounds in pyrolysis-oil (corrected by wood moisture)

Volatile compounds	Wt.% (based on dry oil)
Carbohydrates (main components)	
Acetic acid	24.18
Hydroxypropanone	9.95
Hydroxyacetaldehyde	5.76
Levoglucosan	3.49
2-Furaldehyde, 2-Furfural	2.49
(5H)-Furan-2-one	0.96
2-Hydroxy-1-methyl-1-Cycolpentene-3-one	0.70
Dihydro-methyl-furanone	0.69
Alpha-angelicalactone	0.51
Gamma-butyrolactone	0.43
2-Furfuryl alcohol	0.33
4-Hydroxy-5,6-dihydro-(2H)-Pyran-2-one	0.30
3-Methyl-2-Cyclopentene-1-one Öl	0.14
2,5-Dimethoxy-tetrahydrofuran (cis) Öl	0.10
Σ	50.04
Decomposition products of lignin: Guajacyl components	
Isoeugenol (trans)	1.07
Guaiacol	0.71
4-Methyl guaiacol	0.51
Vanillin	0.33
Isoeugenol (cis)	0.26
Guaiacyl acetone	0.13
Coniferylaldehyde	0.13
Acetoguaiacone	0.12
4-Propyl guaiacol	0.07
4-Ethyl guaiacol	0.07
Eugenol	0.06
Σ	3.46
Syringyl components	
Syringol	0.65
Acetosyringone	0.36
4-Vinyl syringol	0.35
4-Methyl syringol	0.31
4-Propenyl syringol (trans)	0.31
Syringaldehyde	0.24
4-Allyl- and 4-propyl syringol	0.19
4-Propenyl syringol (cis)	0.16
4-Ethyl syringol	0.10
Syringyl acetone	0.08
Sinapaldehyde	0.04
Sinapyl alcohol isomer	0.02
Propiosyringone	0.01
Σ	2.83

Other lignin components	
3-Methoxy catechol	3.87
1,2 Ethanediol	1.04
Phenol	0.66
4-Hydroxy benzaldehyde	0.26
Meta-cresol	0.22
4-Vinyl phenol	0.17
2,4- and 2,5-Dimethyl phenol	0.14
4-Methylanisol	0.12
Resorcin	0.12
Ortho-cresol	0.11
3- and 4-Ethyl phenol	0.10
3-Methyl catechol	0.08
Acetophenone	0.05
Benzyl alcohol	0.03
2,6-Dimethyl phenol	0.02
Σ	6.99
Σ, Total	63.33

Table 7.
Typical properties of bio-oil and mineral oil

	Bio-oil	Mineral oil
pH	2.6	–
Water content (wt.%)	20.0	0.03
Viscosity at 50°C (cSt)	30.0	6.0
Density at 15°C (g/cm^3)	1.24	0.89
Heating capacity (H_u, MJ/kg)	17.0	40.0
Ashes (wt.%)	0.03	0.01

The value of bamboo as a fuel source is connected to its dry matter content. Although bamboo has a very wide range of uses, its utilization as firewood has been limited. Bamboo's use for a variety of processed products signifies that there is hardly any scope for using it as an energy and fuel source. To a small extent, bamboo is burned for heat production; however, this does not approach the scale of using the plant as a substantial energy or fuel producer. Nevertheless, it is a possible candidate for bioenergy production because of its high biomass yield and its high heating value – about 4 600 cal/g as compared with 4700–4900 cal/g for wood (Chen et al. 1995). Besides briquetting, charcoal and production of powder for use as biofuel, gasification is a solution that needs to be examined for the future because of reduced exhaust gases (such as NO_X) and less adverse influence on the environment.

Selection of adapted bamboo genotypes and further breeding, as well as the development of bamboo crop management from the time of planting to the harvest (including harvesting systems), are factors that can improve the yield of bamboo, which is the basic requirement for the introduction of bamboo as an energy source.

REFERENCES

Bridgwater, A.V. 1997. Biomass fast pyrolysis and applications in Europe. *In* Overend, R.P.; Chornet, E., ed., Making a business from biomass in energy, environment, chemicals, fibers and materials. Proceedings of the Third Biomass Conference of the Americas. Vol. 1, 2. Elsevier Science, Oxford, New York. pp. 797–809.

Bridgwater, A.V.; Boocock, D.G. B., ed. 1997. Developments in thermochemical biomass conversion. Vol. 1, 2. Blackie Academic & Professional, London, UK.

Chen, Y.; Qin, W.; Li, X.; Gong, J.; Nimanna. 1987. The chemical composition of ten bamboo species. *In* Rao, A.N.; Dhanarajan, G.; Sastry, C.B. ed., Recent Research on Bamboo. Proceedings of the International Bamboo Workshop, Hangzhou, China, 6–14 October 1985. Chinese Academy of Forestry, Beijing, China; International Development Research Centre, Ottawa, Canada. pp. 110–113.

El Bassam, N. 1998. Energy plant species. James & James (Science Publishers) Ltd., London, UK.

El Bassam, N.; Jakob, K. 1996. Bambus – eine neue Rohstoffquelle – Erstevaluierung-Landbauforschung Völkenrode. Heft, 2, 76–83.

Kaltschmitt, M.; Bridgwater, A.V., ed. 1997. Proceedings of the International Conference on Gasification and Pyrolysis of Biomass, Stuttgart, Germany, 9–11 April 1997. CPL Press, Newbury, UK.

Liese, W. 1985. Bamboos - biology, silvics, properties, utilization. Deutsche Gelleschaft für Technische Zusammenarbeit (GTZ), Nr. 180, TZ Verlagsges., Roßdorf 1, 132 pp.

Radlein, D.; Piskorz, J. 1997. Production of chemicals from bio-oil. *In* Kaltschmitt, M.; Bridgwater, A.V., ed., Proceedings of the International Conference on Gasification and Pyrolysis of Biomass, Stuttgart, Germany, 9–11 April 1997. CPL Press, Newbury, UK. pp. 471–480.

Bamboo for Sustainable Development, pp. 807–814
Eds. A. Kumar, I.V. Ramanuja Rao and Ch. Sastry

Bamboo: Alternative food and fiber crop for the Pacific Northwest

CORAL A. MILES, PHILIP NORTHCRAFT and WILLIAM MATTHEWS
Washington State University Cooperative Extension Center for Sustaining Agriculture and Natural Resources, Northcraft Farm, Washington, USA

Abstract—This study investigates the potential of bamboo as an alternative food and fiber crop for the Pacific Northwest region of North America. To determine which varieties are best suited to the region, on-farm research plots of 26 bamboo species were established in Tenino, southwest Washington in October 1997. Composted dairy manure was applied to plants as a source of nutrients and organic matter. Measurements will be taken, over the next 6 years, of plant height, number of new culms and spatial distribution of culms, and harvest time and yields will be determined. After less than one year in the field, there has been a 1.2–8.6 fold increase in the number of culms per plant and a 1.04–1.88 fold increase in plant height. At this time, the most vigorous species are *Phyllostachys nigra*, *P. nigra* '*henon*', *P. aurea* and *Pseudosasa japonica*. Projected outcomes of this project include variety recommendations for shoot and pole production, recommendations for in-field nursery bed production, and guidelines for sustainable cultural practices including soil fertility management for small-scale commercial bamboo production in the region.

INTRODUCTION

Timber has long been the major natural resource product of the Pacific Northwest region of North America. Pressures resulting from development and urbanization of productive timberlands, as well as agricultural lands, are affecting the ability of timber companies and farmers to implement traditional production practices (Lee et al. 1997). As a result, landowners are seeking alternative food and fiber crops, which are compatible with a rapidly urbanizing environment and are well suited to small-scale sustainable farming systems. Bamboo may be a viable option owing to its long-term productivity, regenerative growth habit and the high market value of bamboo shoots and poles.

Because of increased land values in the region, landowners require crop alternatives suitable for small averages with high economic returns. The current market value of fresh organic bamboo shoots is USD 0.90–1.40 per kilogram, and small-diameter (3–5 cm) poles sell locally for USD 1.00 per meter, on average. There are no established bamboo farms in the region, though managed small groves are yielding 2–5 tons of shoots per year (Lewis 1993). The potential minimum annual income from shoots alone is USD 3 240 per hectare.

Bamboo was introduced to the Pacific Northwest region by Asian immigrants more than 100 years ago (Cooper 1994). Today, over 190 varieties of bamboo are grown in the coastal regions of Washington and Oregon (Jaquith et al. 1995). Bamboo has been planted, however, as a landscape material. Variety selections have been for ornamental rather than production purposes, and plant nutrition needs and site selection have not been investigated (USARS 1961; Liese 1987; Shanmughavel and Francis 1996). Only over the last few years is bamboo being viewed in the region as a source of food, fiber and farm income.

To determine which bamboo varieties are best suited for shoot and pole production in the region, on-farm research plots of 26 bamboo species have been established in Tenino, southwest Washington. Plant growth, harvest time and yields will be measured.

The region is also home to many dairies; thus, dairy manure is a readily available nutrient resource. Bamboo is a shallow-rooted, heavy nutrient feeder, and may respond well to topical applications of manure (Muller 1996; Shanmughavel and Francis 1997). Composted dairy manure will be applied to bamboo plots both as a source of plant nutrients and as organic matter. This component of the project will help identify how applications of dairy manure affect bamboo growth. If successful, this will provide dairy farmers with a new outlet for surplus manure.

Projected outcomes of this project include: variety recommendations for shoot and pole production; recommendations for in-field nursery or propagation bed establishment and production; and guidelines for sustainable cultural practices and soil fertility management for small-scale commercial bamboo production in the region. The project site will be made available for bamboo field days and special educational events.

MATERIALS AND METHODOLOGY

In October 1997, 26 bamboo species were established in on-farm research plots at the trial site in Tenino, Washington (Table 1). The species included in this experiment were identified by members of the Pacific Northwest Bamboo Society (PNBS) as having the greatest potential for food and fiber production for this region. All plants were donated by PNBS members. Most plots have two plants of each species, but some have one or three plants. Totally 89 plants were planted in 49 plots, each plant spaced 0.9 m apart and each plot at a distance of 4.5 m from the adjacent. The total experimental area measures 22 × 66 m, or 0.15 ha. Alleys are maintained between plots by mechanical cultivation for weed control and to prevent spread of bamboo into neighboring plots. Weed control within plots is manual. Since the beginning, plots and alleys have been cultivated two times, in May and July 1998.

Table 1.
Bamboo species planted, and number of plots and plants at the on-farm research sites in Tenino, Washington, USA

No.	Genus and species	No. of plots	Total no. of plants
	Fargesia		
1	*F. diacocephala*	1	3
2	*F. nitida*	2	3
	Phyllostachys		
3	*P. atrovaginata*	3	6
4	*P. aurea*	3	6
5	*P. aureosulcata*	1	1
6	*P. aureosulcata alata*	1	2
7	*P. bambusoides*	1	2
8	*P. bambusoides* 'castillonis'	1	1
9	*P. bissetii*	3	6
10	*P. dulcis*	4	4
11	*P. flexuosa*	1	2
12	*P. heteroclada* (water bamboo)	1	2
13	*P. heteroclada* 'solid stem'	1	2
14	*P. heteroclada* 'straight stem'	1	2
15	*P. heterocycla pubescens* (moso)	2	4
16	*P. humilis*	3	6
17	*P. meyeri*	1	3
18	*P. nigra*	2	3
19	*P. nigra* 'bory'	2	3
20	*P. nigra* 'henon'	4	9
21	*P. nigra* 'megurochiku'	3	4
22	*P. platyglossa*	1	2
23	*P. rubromarginata*	3	5
24	*P. vivax*	2	4
	Pseudosasa		
25	*P. japonica*	1	2
	Semiarundinaria		
26	*S. fastuosa viridis*	1	2

Owing to the high cost of plant material (USD 10–50 per plant), only half the species are present in more than one trial plot. Therefore, this is not a replicated field trial. A long-term goal is to utilize plants from these plots to establish a replicated field trial. In the short term, the results are expected to give an estimate of species performance, and help set general guidelines for nursery or propagation bed establishment and production. This information will be helpful to farmers who do not have the capital to invest in plant material for a full-scale planting.

Currently, prospective farmers are being encouraged to establish and maintain a few plants in the field, and divide and transplant these after several years. This is an affordable though protracted method of producing bamboo offsets to increase plant numbers, especially as the production potentials are still unclear.

Manure

Composted dairy manure was applied directly to bamboo plants twice, once at planting in October 1997 and again in May 1998. At both applications, each plant received 16.5 kg of wet manure, which was equivalent to 10 144 kg per hectare. Nitrogen application rate was equivalent to 39 kg total nitrogen per hectare in October and 45 kg total nitrogen per hectare in May, or a total nitrogen application of 84 kg per hectare (Table 2).

Table 2.
Percentage total nitrogen and available nitrogen of composted dairy manure on a wet weight basis in October 1997 and May 1998, in Tenino, Washington, USA

	October 1997	May 1998
Total nitrogen	0.39%	0.44%
Available nitrogen (ammonia + nitrate)	0.02%	0.02%

Soil and temperature

Soil type is Spanaway gravelly sandy loam, about 152 cm deep, well drained, with moderately rapid permeability and low available water capacity. No supplemental drainage was installed at the site. Summer and winter temperatures throughout the region are generally mild and many perennial crops are grown without irrigation. Range of mean temperatures are: January–March, -0.2° to 12.2°C; April–June, 2.3° to 21.6°C; July–September, 7.1° to 25.1°C, and October–December, 0.0° to 15.8°C.

Moisture

Plants receive no supplemental moisture. Annual rainfall at the trial site is about 1 143 mm. In general, rainfall occurs every month of the year throughout the region. Highest rainfall is from October through March – 160 mm per month on average – while July has the lowest rainfall – 20 mm on average. Immediately bordering the plots on one side is a low spot in the field, which accumulates standing water during the winter months. However, this standing water does not extend into the plots. Soil throughout the trial is 80–100% saturated for approximately 4 months, December through March. It appears that the high soil moisture is adequate for sub-irrigation but not high enough to cause anaerobic soil conditions or disease. Many agricultural fields throughout the region exhibit these conditions. This site was selected to investigate the suitability of bamboo for these soil conditions.

Results

In December 1997 and July 1998, total number of culms and maximum height of each plant were measured. In July 1998, plant diameter at the soil surface was also measured (Table 3).

Table 3.
Mean number of bamboo culms, maximum plant height (m) and plant diameter (cm) in December 1997 and July 1998 at the on-farm research sites in Tenino, Washington, USA

No.	Species	No. of culms		Max. plant height		Plant diameter
		Dec 97	Jul 98	Dec 97	Jul 98	Jul 98
1	*Fargesia diacocephala*	n/a	14.7	0.5	0.6	7.0
2	*F. nitida*	4.5	17.0	2.3	2.3	42.8
3	*Phyllostachys atrovaginata*	2.8	7.0	1.0	1.3	16.7
4	*P. aurea*	2.8	15.8	1.6	1.9	23.7
5	*P. aureosulcata*	5.0	10.0	2.4	2.5	52.0
6	*P. aureosulcata alata*	3.0	9.0	2.5	2.9	19.5
7	*P. bambusoides*	2.5	9.5	0.8	0.9	11.5
8	*P. bambusoides* 'castillonis'	1.0	1.0	0.5	0.5	2.0
9	*P. bissetii*	3.7	14.3	1.0	1.2	20.2
10	*P. dulcis*	2.3	7.3	1.0	1.6	18.8
11	*P. flexuosa*	5.0	9.5	1.6	1.7	25.0
12	*P. heteroclada* (water bamboo)	3.0	6.5	1.3	1.5	16.0
13	*P. heteroclada* 'solid stem'	2.5	8.5	0.9	0.9	24.0
14	*P. heteroclada* 'straight stem'	2.0	2.0	1.8	2.2	16.5
15	*P. heterocycla pubescens* (moso)	3.0	11.8	1.6	1.6	30.8
16	*P. humilis*	3.2	10.0	1.7	2.2	31.8
17	*P. meyeri*	5.3	9.0	2.9	3.2	25.0
18	*P. nigra*	1.3	10.3	0.8	1.5	10.0
19	*P. nigra* 'bory'	3.3	5.5	1.2	1.4	12.8
20	*P. nigra* 'henon'	3.3	7.9	1.1	1.9	21.1
21	*P. nigra* 'megurochiku'	3.0	8.5	1.8	2.1	23.5
22	*P. platyglossa*	2.5	3.0	0.5	0.5	4.5
23	*P. rubromarginata*	4.8	10.7	2.1	2.2	21.7
24	*P. vivax*	1.5	2.5	1.1	1.6	11.0
25	*Pseudosasa japonica*	5.0	34.0	1.0	1.3	33.5
26	*Semiarundinaria fastuosa viridis*	2.5	9.0	1.4	1.5	16.5

Plant diameter here refers to the widest distance of separation at soil level between culms of the same mother plant. As a growth pattern indicator, it will be used to determine the spatial distribution over time of each species in this trial.

Spatial distribution over time is important when estimating how long it will take for each species to fill an area, and will help determine optimum planting densities for each species in field situations.

After less than one year in the field, the number of culms per plant has increased 1.2–8.6 fold, and mean increase in culm number is 3.1 fold. The very large increase in culm number is due entirely to the production of short and bushy culms, which are not likely to grow beyond 0.5 m tall. Plants that showed a mean to moderate increase in culm number have produced single, straight culms, which will be suitable for pole harvest though culm diameters are generally small, less than 2 cm (data not shown).

Plant height has increased 1.04–1.88 fold, and mean increase in plant height is 1.21 fold. All increases in plant height can be attributed to the growth of culms that emerged in 1998. Old culms, which were transplanted into the field, have not increased in height and in some cases, have died and broken off the clump.

In this trial, *Phyllostachys bambusoides* 'castillonis' has exhibited no new growth; *P. heteroclada* 'straight stem' did not have a net increase in culm number but did increase in plant height; and both *P. pubescens* and *Fargesia nitida* increased number of culms, but this did not result in greater plant height. At this time, the most vigorous species are *Phyllostachys nigra, P. nigra* 'henon', *P. aurea* and *Pseudosasa japonica.*

CONCLUSIONS

Overall, all bamboo species in this experiment are growing well. The results, though preliminary, are very promising. At this time, there are three keys to the success of in-field bamboo plant establishment: site selection, weed control and application of manure.

Site selection

Field planting of bamboo was carried out without supplemental irrigation or drainage. Bamboo growth in the plots does not appear limited by lack of irrigation in the summer or saturated soils in the winter. Many farmers do not have the ability to irrigate because of restricted water right permits for agricultural use. Therefore, for bamboo to be a viable crop alternative in the region, it is necessary that it grow without irrigation. At the same time, saturated soils during winter months can limit crop choices. Thus, a viable crop must be able to adapt to the variable moisture conditions. A field site, which may be saturated during winter months but does not have standing water for more than one week at a time, with good natural drainage, appears to be a viable choice for bamboo.

Weed control

The trial site has been kept weed-free by means of mechanical cultivation between the plots and hand cultivation in the plots. Sowing of a cover crop mixture of grass and legume is planned in the alleys in the fall for weed control. Hand cultivating will be continued in the plots as needed. As the amount of inter-plant competition was limited, bamboo plants have been able to optimally utilize soil moisture and nutrient resources. After one more year, it is expected that the plants will be able to shade out many of the weeds in the plots.

Manure

All bamboo species in this experiment appear to be responding well to applications of composted dairy manure. However, the amount of manure and total nitrogen that have been applied to plants may not be optimal for plant growth (Shanmughavel and Francis 1996). It is likely that bamboo is capable of utilizing more nitrogen and phosphorus than the quantities applied. In early September, when the fall rains begin, 16.5 kg of wet manure will be applied to each plant. Total application rate in the first year will then be equivalent to 122 kg total nitrogen per hectare. In 1999, an equivalent rate of 200 kg total nitrogen per hectare will be applied, over three equal applications – in April, June and September.

ACKNOWLEDGEMENTS

This project is supported by the Pacific Northwest Bamboo Society with donations of plant material, the United States Department of Agriculture with a Grower/Rancher Sustainable Agriculture Research and Education grant, and a travel grant from Washington State University Center for Sustaining Agriculture and Natural Resources.

REFERENCES

Cooper, G. 1994. American bamboo industry – past, present and future: comparison/contrast to China. *In* Proceedings of the Pacific Northwest Bamboo Agro-forestry Workshop. pp. 1–15.

Jaquith, N.; Flemmons, K.; Remmick, M. 1995. The bamboo garden: a nursery catalog. Portland, Oregon, USA. 33 pp.

Lee, A.; Bai, X.; Bangi, A. 1997. Flexural properties of bamboo-reinforced southern pine OSB beams. Forest Products Journal, 47(6), 74–78.

Lewis, D. 1993. Bamboo on the farm. Bamboo Gardens of Washington (publishers), Washington, USA. 48 pp.

Liese, W. 1987. Research on bamboo. Wood Science and Technology, 21(3), 189–209.

Muller, L. 1996. Can a private *Gigantochloa* timber plantation be profitable? American Bamboo Society Newsletter, 17(2), 4–9.

Shanmughavel, P.; Francis, K. 1996. Biomass and nutrient cycling in bamboo (*Bambusa bambos*) plantations of tropical areas. Biology and Fertility of Soils, 23, 431–434.

Shanmughavel, P; Francis, K. 1997. Balance and turnover of nutrients in a bamboo plantation (*Bambusa bambos*) of different ages. Biology and Fertility of Soils, 25, 69–74.

USARS (United States Agricultural Research Service). 1961. Growing ornamental bamboo. Home and Garden Bulletin No. 76, Crops Research Division, USARS, USA. 8 pp.

Bamboo for Sustainable Development, pp. 815–824
Eds. A. Kumar, I.V. Ramanuja Rao and Ch. Sastry

Bamboo for sewage treatment: A sustainable strategy

DAVID DEL PORTO
Sustainable Strategies, Concord, Massachusetts 01742–2968, the United States

Abstract—The contamination of soil and water by human and animal wastes is becoming a major problem in many parts of the world, causing algal blooms, choking off of aquatic life, death of reefs, pollution of drinking water and disease outbreaks. Sustainable Strategies, a US-based ecological engineering firm, has developed specially engineered systems that tackle this problem, using bamboo as a key tool. In the engineered WasteWater Gardens and Bamboo Forests, pretreated wastewater is pumped into an evapo-transpiration system strategically planted with bamboo. The bamboo and its microbially active rhizosphere transform pollutants, including nitrogen (which causes unwanted plant and pathogen growth), into valuable biomass. They also evaporate and transpire the remaining water.

INTRODUCTION

The contamination of soil, groundwater and waterways by human sewage and animal wastes – mostly pig and chicken manure – is becoming a crippling and even life-threatening problem in some parts of the United States and around the world. The results include algae blooms and eutrophication (the choking off of aquatic life) in lagoons, dying reefs, contaminated drinking water wells and outbreaks of gastro-intestinal disease and cholera. Sustainable Strategies, an ecological engineering firm in Massachusetts, the United States, developed a solution: specially engineered systems that employ bamboo as a key component. These are called WasteWater Gardens and Bamboo Forests.

Conventional sewage treatment options, such as wastewater treatment plants and septic systems, often threaten the quality of ground and surface waters by essentially placing harmful pathogens and nutrients from wastes directly into or over the water table. Although in certain soils and under the right conditions some treatment can take place, nitrogen and other pollutants still generally remain in a form that presents potential health and environmental hazards.

In the engineered Bamboo Forest system, pretreated wastewater is pumped into a specially engineered evapo-transpiration system strategically planted with bamboo. The bamboo and its microbially active rhizosphere transform pollutants, including nitrogen (which causes unwanted plant and pathogen growth), into valuable biomass. They also evaporate and transpire the remaining water. Bamboo is chosen as the primary operative in these systems, as it is quite phreatophytic, which means it voraciously consumes nitrogen to produce protein, making it an ideal nutrient sink.

The growing demand for this material means that this treatment system results in a marketable product. This has the potential to offset the construction and maintenance costs of the system. In this way, the farmers who use it may profit from their wastes while protecting public and environmental health.

DESIGN ISSUES

Bamboo can take anywhere from six months to three years to reach full working size (depending on species and maturity of bamboo stock planted), so the system will not be working at full capacity at first. In the beginning, unused effluent will flow to a storage pond or a holding tank in closed-loop, zero-discharge systems. The need for overflow backup diminishes as the plants mature and the active microbial population in the rhizosphere develops. The system works best when it's sheltered from rain and extreme cold by citing it close to buildings, or, in some instances, in greenhouses. Riparian and coastal bamboo forests and buffer strips will improve lake and river water and provide an economical and sustainable strategy for the management of agricultural manures.

SYSTEM DESCRIPTION

AMRET (Aerobic Microbial Respiration and Evapo-transpiration)
WasteWater Garden™ irrigation system for wastewater.

Key Factors

Key factors making the wastewater garden more effective than a typical irrigation system:

1. Sand layer evaporation.
2. Aerobic environment.
3. Higher temperatures encouraged/maintained.
4. Specially chosen high-water-use plants.
5. Controlled water characteristics.

Wastewater from all gray-water and/or blackwater sources at a site drains or pumps into ecologically engineered gardens where aerobic microbes digest the organic molecules and transform them into heat, carbon dioxide, water vapor, and oxidized, water soluble compounds. This wastewater is pre-filtered, as the lint from modern clothing is primarily non-biodegradable fabric such as polyester and nylon which will not decompose biologically, thus causing the clogging of the holes in the distribution pipes and void spaces in the sand and gravel of the Garden bed. It is also pre-treated, which is necessary for blackwater to alter the wastewater characteristics to optimally feed the microbes and plants in the bed, hence increase the system efficiency, as well as to prevent clogging of the distribution pipes with solids.

Evergreen bamboo and/or broad-leafed vascular plants utilize these oxidized nutrients, minerals and water for their growth and transpire the water as vapor.

AMRET systems are sized to contain, treat, and recycle all effluent generated. Plant species are selected for their all-weather hardiness, rapid growth rates and high water requirements.

Biomass is harvested, thereby removing the nutrients in the wastewater through pruning and natural leaf drop.

The benefits of AMRET systems

Total Aerobic Microbial Respiration and Evapo-transpiration (AMRET) ensures that none of the wastewater can be infiltrated into the ground. A watertight liner is employed to seal the bottom and sides of the garden bed so no liquid can escape the system. However, if adequate pervious soil is available to safely allow for groundwater recharge, then it is not necessary for the system have a waterproof liner.

In practice, the landscaped WasteWater Gardens are shallow, flat-bottomed lined trenches or beds, typically filled with six inches of crushed stone and covered with 12 inches of sharp sand. Perforated pipes on the stone and beneath the sand distribute the filtered wastewater along the length of the trenches. The trenches should be no more than 18 to 24 inches deep to maximize the uptake of water and nutrients by plant roots and to allow proper oxygenation. Air is introduced by venting the distribution pipes with riser vents or "breezers" (these may be augmented by blowers to increase aeration, heat the beds and remove CO_2 and water vapor). Aeration is critical to support aerobic bacteria and cilliata, which transform any remaining pollutant-nutrients in the wastewater at Exceptional speeds to compounds that the plants can utilize. If a smaller area and/or higher evapo-transpiration rates are desired, the WasteWater Garden can be enclosed and/or heated.

Evaporation occurs as water moves upward through the sand by capillary action and is evaporated at the surface. Evaporation also occurs within the bed as aerobic bacteria release heat at they metabolize the nutrients in the effluent, and this heat drives water into vapor, which is forced to the surface.

Transpiration occurs as water absorbed by plant roots is drawn up the stalk and stems of plants into the leaves, where it is released as pure water vapor.

AMRET of wastewater depends on the energy available for the change of state from a liquid to a gas. The resultant gasses must be mobilized out of the bed, so that a partial pressure increase does not inhibit the transfer of the water vapor and CO_2 gas (a by-product of microbial respiration). This is enhanced by the void spaces and surface area of the stone and sand granules in the upper part of the bed and the surface area of the leaves of the plants in the WasteWater Garden.

In wastewater AMRET, the microbial energy from respiration of bacteria, protozoan and other cilliata increases the rate of evaporation as they metabolize the nutrients in the effluent.

Increasing temperature of the bed by heating (from the surface and/or from beneath the trenches) has a super charging effect on the AMRET rate:

1. First, by increasing the available energy, greater AMRET rates will occur. One pound of Carbonaceous Biological Oxygen Demand (CBOD) produces 8,000 to 10,000 Btu as the heat of respiration when digested by aerobic microbes. About 9,000 Btu will evaporate one gallon of effluent over time (rate determined by temperature).
2. Second, increasing temperatures doubles or triples the release of energy from the respiration of aerobic microbes. As the resident microbes are poikilothermal, their metabolic rate varies with the temperature of their surroundings. Therefore, the warmer they are, the faster they consume oxygen and digest the molecules of organic matter. This is known as "the Q10 temperature coefficient". It holds that for every 10°C (18°F) rise in temperature, within certain limits, the biochemical, respiration and transpiration rate of reaction is doubled or, in some instances, tripled. This principle allows the area required for AMRET to be reduced as the temperature is increased.
3. Third, increased temperature increases the rate of transport of water from the roots of plants to the surfaces of the leaves where it is released as a vapor, primarily from the stomata (and to a small extent through the cuticles and lenticles). Transpiration results in an upward movement of water and dissolved mineral salts flowing through the plants' xylem cells to the leaves.

Surface area of the leaves, i.e. the number and size of the leaves per plant and the species of plant, are significant contributing factors to the elevated AMRET rates. Key to a high-rate AMRET system is the selection of plant species that have high transpiration rates, such as broad-leafed wetland or rainforest plants, dense-leafed evergreens, or high-growth-rate plants, such as bamboo and many grasses and reeds. These factors combine in an insulated and heated AMRET trench to increase the normal rate of evapo-transpiration. Typical annual average outdoor AMRET rates in the northeastern United States are from 0.05 to 0.2 gallons per square foot of surface area of WasteWater Garden. However, this rate can be significantly increased (0.25–5.0 gal/sft /day) to tropical AMRET rates by altering the environment, such as by enclosing the AMRET beds in a greenhouse or translucent cover of sufficient size to allow for the growth and periodic harvest of the biomass and/or heating of the AMRET beds. Heat and water recovery equipment may then be employed to increase the economic efficiency of the whole system.

AEROBIC BACTERIA VS. ANAEROBIC BACTERIA

Aerobic microbial organisms require an oxygenated environment. This requires the effluent treatment system to be aerated, providing the soil-contained organisms access to atmospheric oxygen, which is essential to their metabolic process.

Why use aerobic bacteria, which are environmentally specific? These microorganisms provide essential benefits to the garden system as a whole. Anaerobic bacteria do not.

The by-products of the aerobic metabolic processes are oxygenated compounds, plant food, which are readily absorbed by plant roots in the system for their nutritive value. Thus, there is no toxic build-up. Anaerobic metabolism results in non-oxygenated compounds, except phosphates and water, such as methane, ammonia and hydrogen sulfide. These by-products are not plant food, and instead serve as a system contaminant.

The use of aerobic bacteria ensures natural population control and no clogging of voids occurs in the system. Owing to the aerobic environment, larger microorganisms (protozoa) such as ciliata and amoebae are able to exist. Protozoa (50–100 microns in size) prey upon the smaller aerobic microbial organisms (1–5 microns in size) of the system devouring daily about one million bacteria, thus, naturally controlling their ever-increasing numbers. This predator-prey relationship prevents the clogging of the system with excessive bacteria, which would lead to inhibition of the capillary rising of nutrients and water to the plant root systems. These protozoa cannot exist in an anaerobic environment.

PERFORMANCE DATA

The water and nutrient uptake rate is a function of a complex of variables that interact and coevolve over time. They include, but are not limited to, the following:

- Maturation of the rhizosphere, stem and leaf system (which includes the bioplex in the planting bed);
- Ambient temperature and moisture content of planting bed and air;
- Genus, species and variety of bamboo selected;
- Light levels and phototropic issues;
- Dissolved oxygen sag curve, influent characteristics and dosing rates; and
- General health and maintenance of the system.

Owing to the complexity of these and other factors, their combined effect is represented in gallons of ET (evapo-transpiration) per square foot of planted area per day. In the cold climates of Massachusetts, it would be better to shelter the planting system to eliminate rain and minimize the impacts of dormancy from the very cold winters. The number 0.25 gal/sft of planted area/day is taken as the conversion value to rough out the preliminary design.

PLANTS USED

The following are the plants that are best for the Washwater Garden:

A. For a summer garden or tropical greenhouse then the following are best:

1. *Phragmites australis* or *P. communis* (Common reed, Disturbance weed)
2. *Deconia verticillatus* (Swamp loosestrife)
3. *Hibiscus rosa-sinensis* (tropical), *H. syriacus* (Rose of Sharon), *H. moscheutos* (Rose mallow), *H. cannabinus* (Kenaf)
4. The large order Zingiberales are the most thirsty and beautiful plants that can be found for tropical or greenhouse applications. The family members include: Musaceae (Banana), Strelitziaceae (Bird-of-Paradise), Lowiaceae, Heliconiaceae, Zingiberaceae (Ginger), Cannaceae (Canna lilly) and Marantaceae (Prayer plant). There are so many to choose from, but we like:
   Zanteceschia aethiopica (Canna or Arum lilly)
   Heliconia (*standeyi Macbride* or *psittacorum*)
   Strezlitzia reginae (Bird-of-Paradise)
   Musa paradisiaca (common Banana) or *Musa coccinea* (Flowering banana)
   Alpina zerumbet (Shell ginger) or *Alpina purpurata 'rosea'* (Rose ginger)
5. *Typha* sp. such as *glauca, angustifolia, latifolia, minimus* (Bull-rush, Cat-tail)
6. *Monstera deliciosa* (Monstera)
7. *Hosta fortunei* (Plantain lilly)
8. Aroids such as *Alocasia macrorrhiza* (Elephant ear) or *Colocasia esculenta* (Taro)
9. *Melaleuca quinquenervia* (Cajeput tree) or *M. nesophylla* (Western tea myrtle)
10. *Cyperus papyrus* (Egyptian paper plant)
11. *Eichornia crassipes* or *E. azurea* (Water hyacinth)
12. *Rhizophora* sp. (Mangrove)
13. Bamboo (genus *Arundinaria, Bambusa, Phyllostachys, Sasa* or *Shibataea*)
14. *Tamarix ramosissima* (Salt cedar)

B. If this is for a winter Washwater Garden then the following are best:

1. *Phyllostachys aureosulcata* (Yellow groove bamboo)
2. *Phyllostachys bissetti*
3. *Phyllostachys nuda* (Snow Bamboo)
4. *Fargesia murielae* (Umbrella Bamboo)
5. *Fargesia nitida* (Fountain Bamboo) (subspecies include Ancepts, de Belder, Eisenbach, Ems River, McClure, and Nymphenburg)
6. *Arundinaria* or *Pleioblastus pygmaea* (Dwarf bamboo) (this can be used for ground cover together with the tall bamboo)
7. *Thuja occidentalis* (Emerald or American dark Arborvitae)
8. *Ilex verticilatus* (Swamp holly) or the hybrid "Blue or Meserve"
9. *Klamia latifolia* (Mountain laurel)
10. *Taxus canadensis* (American yew)

11. *Tsuga canadensis* (American hemlock)
12. *Salix lanata* (All willows are very thirsty, but deciduous)
13. *Tamarix ramosissima* (salt cedar-very thirsty halophyte)

C. Grasses: such as Phalaris (canary grass), Lagarus (Hare's tail), Briza (quaking grass), Pancium (red switch grass), Zizania (wild rice), *Erianthus saccharoides* (spreading witch grass) are excellent seasonal or warm-weather plants.

D. Flowers: such as Nasturium, Nicotiana, Cosmos, Snap dragons, Salvia, Swan river daisy, Passiflora and Blue splendor will add color to you summer garden.

E. Vines: such as Plumeria, Ipomoea, and Dioscorea will add fragrance and drama for the summer garden.

We will mix in seasonal with winter-hardy plants and any of our clients favorite plants for beauty, but these are the "living machinery" that will be hardy to the point of being invasive if you don't contain them, such as bamboo. This is important because washwater is, after all, an effluent that contains human disease organisms and the potential for complex compounds that will require rugged plants to deal with them.

Owners may need to use fertilizer for nitrogen, phosphorous, potassium and micronutrients, and maintain the proper soil pH (acid, neutral or alkaline) for the specific plants' requirements. As sodium and boron are toxic to most plants (except saltwater species and halophytes or salt-loving), the fertilizers must not contain these chemicals.

SOME WASTEWATER MANAGEMENT PROJECTS

1. From pig manure to bamboo forests: turning animal waste problem into a cash crop opportunity

Animal waste that flows to conventional effluent lagoons, outclass and discharge pipes, often ends up-be it partially treated or untreated-into environmentally sensitive "receptors", such as ground or surface waters. Outbreaks of new emerging pathogens, such as *Pfiesteria*, are linked to the effects of swine and chicken manure pollution in warm water estuaries and rivers, killing fish and causing disease in fishermen who come in contact with *Pfiesteria*-contaminated water.

In North Carolina, unplanned growth of swine feedlots has devastated the state's rivers and estuaries. Legislation has been enacted to clean up North Carolina's waterways by using tougher pollution restrictions and a $1 billion plan to modernize water and sewage treatment. According to a North Carolina newspaper, the new legislation will spread the responsibility for clean water among hog farmers, local governments, golf course owners and others whose activities have polluted the Neuse River, Pamlico Sound and other waterways. Other legislation has been proposed emphasizing land-use planning, better sewage treatment and criminal penalties for persistent polluters. The problem is so severe that the Governor placed a two-year moratorium on new swine facilities.

Sustainable strategy

Sustainable Strategies was invited to discuss solutions at a North Carolina gathering of hog farmers, bamboo growers, bamboo materials manufacturers (flooring, plywood, particle board), bamboo shoot marketers, university researchers, USDA agents, environmental groups and state and federal water pollution control officials.

Sustainable Strategies used the ecological paradigm to safely utilize the polluting components of animal manures to ultimately grow plants that have economic value. Planning, engineering and design with the ecological paradigm as our template is the work of Sustainable Strategies.

In the "Bamboo Forest" plan, pretreated wastewater drains into an appropriately engineered forest of phreatophytes-plants known for fast growth and high water usage rates. These plants and their microbially active rhizosphere (roots) will transform pollutants, including the nitrogen (which causes unwanted plant and pathogen growth) into valuable biomass and evaporate and transpire (plant respiration) the remaining water. Bamboo is a workhorse plant in this regard, and the growing demand for this material means that this treatment system results in a marketable product, which offsets the construction and maintenance costs of the system. The organic loop is thus closed and the farmers profit from their wastes. Most importantly, the Bamboo Forest does this while protecting public and environmental health. An "eco-nomic" solution! The Bamboo Forest design is modelled after the Wastewater Garden™, a zero-discharge, ecologically engineered, aerobic microbial evapo-transpiration wastewater utilization system developed by Sustainable Strategies. When soils are not suited to absorbing wastewater, the Bamboo Forest will be zero-discharge system with an impermeable liner and a recycle tank with overflow alarm (to signal the need for pumping).

A pilot program in 1998-1999 will show which bamboo species are the most effective and best serve as an agro-forestry crop. We'll also work with different irrigation and planting techniques.

Multifarious bamboo

Bamboo, the tallest member of the grass family of plants, could be called a "wonder plant". It is quite phreatophytic, which means it voraciously consumes nitrogen to produce protein, making it an ideal nutrient sink.

Bamboo is well known in the Americas, as well as in tropical Asia, for its use as building and construction materials due to its uncommon strength, hardiness and rapid growth. It even has food value in the form of its nutritious shoots. Bamboo leaves are also an excellent animal fodder crop, as they have an 18–22% protein content. In India, bamboo fiber provides the basis for paper and corrugated media. It is also used for fuel: the rapid growth and high energy content of bamboo also make it a leading source of biomass-derived fuels in developing countries.

Bamboo can take up to three years to reach full working size (depending on species and maturity of bamboo stock planted). So the system will not be working at full capacity at first. In the beginning, unused effluent will flow to a storage pond or a holding tank in closed-loop, zero-discharge systems. The need for overflow backup diminishes as the plants mature and the active microbial population in the rhizosphere develops. The system works best when it is sheltered from rain and extreme cold by constructing it close to buildings or (in some instances) in greenhouses. Riparian and coastal bamboo forests and buffer strips will improve lake and river water and provide an economical and sustainable strategy for the management of agricultural manures. Once again, nature provides when we listen to her!

2. Small-scale piggery waste system grows pig feed in Micronesian islands

For the Pacific island state of Pohnpei, Sustainable Strategies has developed a way to transform pig waste into pig feed.

On Pohnpei, many families keep 1–10 pigs. Pig manure has been contaminating soil and waterways, resulting in diseases, such as Leptospirosis. To address this problem, Sustainable Strategies has designed and will demonstrate a small-scale piggery pollution prevention system. This low-cost zero-discharge treatment system will use both Wastewater Garden™ and composting technology to convert the nutrient-rich pig waste into valuable plants. This ecologically integrated system first composts the manure and spilled feed. Any liquid (urine, water) that drains from the composter is evaporated and is transpired by plants in a small Wastewater Garden.

Among the plants grown in the Wastewater Garden will be kangkon and fast-growing water hyacinth, banana and papaya, which can be used for high-protein, low-cost pig feed. Construction- and furniture-grade bamboo will be grown to use up the remaining water and nutrients.

Store-bought pig feed is expensive in the islands, and the production of feed from waste will serve as an incentive to use of this technology. Commercial feed currently sells on-island for about $0.20 to 0.25 per pound; an average pig is fed five pounds a day. Given this expense (more than $350 per pig per year), the cost of constructing a Wastewater Garden would be returned over a relatively short period of time, as plants grown in the garden replace a substantial amount of commercial feed, closing the organic loop.

3. Washwater Garden comes to the rescue of vehicle wash facilities

In 1995, when the town of Harwich, Massachusetts, needed to build another fire station, town officials found out that they had a big, expensive wastewater problem. They discovered that the water left over from the daily washing down of the fire department vehicles is categorized by the federal government as hazardous waste. Federal law requires that this water be drained to a holding tank, then pumped and treated by a firm specializing in toxic waste management. This would cost the town $ 1.25 per gallon!

By Harwich's calculations, their vehicle wash-water disposal costs would exceed $125 000 a year. Clearly, an alternative solution was needed.

The Fire Department, with the approval of the Massachusetts Department of Environmental Protection, brought in Sustainable Strategies. A Washwater Garden was designed for Harwich, to treat the washwater in a safe, specially engineered controlled environment. It was installed in 1995 by a landscape company and planted by community groups. The Washwater Garden has been treating the wastewater successfully for since 1995.

4. Washwater Garden uses Phyllostachys aurea and local fill plants to produce a permanent solution to groundwater contamination

Jim and Marilyn MacDonald, remodeling contractors, thought they had lost thousands of dollars on the four-bedroom home they had renovated as an investment property. When they prepared to sell the home, the septic system was deemed failed under Title 5, the newly tightened on-site wastewater disposal regulations of Massachusetts. The leaching field was right on bedrock, and wastewater drained into wetlands. There was no alternative site for the leaching bed. At first, it seemed that the only option was to install a holding tank. Installation cost would be only about $2 000, but the pumping fee at 16 cents a gallon would cost $19 272 a year, or $385 440 over a 20-year mortgage. That made the home virtually non-saleable.

The MacDonalds had heard that Title 5, in addition to raising standards for waste treatment systems, had also approved several alternatives to septic and holding tanks. The couple contacted Sustainable Strategies, seeking a legal and effective solution to their problem.

Within four months, the couple had a cost-effective solution. The home, now sold, has an ultra-low-flush toilet that flows to a Vera Carousel Composter in the basement. The home's gray water drains into a Washwater Garden – an aerobic, trench-based planting system – where the water evaporates and is absorbed by plants such as bamboo, specially chosen for their high transpiration rates. The home's new owners will still have to service the composter once a year; but composting will reduce the material in it to 10% of its original volume. Once a year, the homeowners can either empty the composter themselves (the resulting humus has the look and odor of garden soil) or hire a hauler to empty it for a minimal fee.

Bamboo for Sustainable Development, pp. 825–830
Eds. A. Kumar, I.V. Ramanuja Rao and Ch. Sastry

Control of beach erosion using bamboo

J. DRAKE
T-Square (Pvt) Ltd., Sri Lanka

D.L. JAYANETTI and P.R. FOLLETT
TRADA Technology, Buckinghamshire, UK

Abstract—Coral mining along the South-west Coast of Sri Lanka has resulted in severe beach erosion along the tourist-oriented beaches and threatens destruction of roads and rail routes. Existing erosion control methods rely on the placement of boulders or gabions or the manipulation of vegetation growth limits. These methods are costly and can be unsightly, as they protrude above the beach surface at all times. A popular holiday resort, located on a 5-km sand spit, required a different erosion control solution that was convenient to implement, easy to maintain, cost-effective and remained hidden.

A system devised to address these needs consisted of a bamboo frame, boulder-filled gabion and pre-cast concrete slabs. The bamboo frame played a major role, that of an anchor to the system, which could be completed at one-third the cost of a conventional boulder system.

INTRODUCTION

Beach erosion in Sri Lanka is a major problem. Coral mining along the South-West Coast of Sri Lanka has caused severe beach erosion along the tourist-oriented beaches and threatens destruction of roads and rail routes. This erosion occurs seasonally as the South-West monsoon winds blow strongly on shore. The beaches rebuild, only partially in some cases, as the winds drop and switch to the North-East direction during that monsoon.

Existing erosion control methods rely on:

- Massive boulders placed in large numbers against embankments or, for beaches, at vegetation growth limits set under and above the natural beach surface level; or
- Less substantial structures that use rock-filled gabions backed with manmade fiber filters able to retain the sand particles, laid to the pitch of the beach.

Both methods are costly, and unsightly as they protrude above the beach surface at all times.

The site in question – The Robinson Club resort in Bentota – is on a 5-km long sand spit extending from the river mouth for 1 km of beach frontage. Access is only by boat across the river. During the North-East monsoon period, the gently sloping beach extends some 75–100 m from the lowest line of vegetation. This is the major attraction of the resort. During the 1996 monsoon, the sea eroded away 10 m of the 30-m protective indigenous plant strip, and invaded the gardens at some points. The protection provided by this plant strip is vital to the success of the resort. For among other things, it hides the existence of a large resort from those on the natural beach.

The erosion solution required to be one that:

- Has pre-prepared components that are easily transportable across the river and across the grounds of the fully operational resort;
- Is visible only during extreme erosion times;
- Can be laid within the periods between erosions;
- Can be sited 15 m away from the lowest line of present plant growth so that some lost planting can be reclaimed;
- Can be maintained easily and repaired periodically by a competent maintenance team;
- Makes optimum use of local materials and skills; and
- Is cost-effective compared with the other methods in use.

A professional team was put together to devise a system consisting of specialists in harbors, structures, irrigation and hydro-mechanical civil engineering, and an instigating architect. The proposed system was approved by the Sri Lanka Coast Conservation Department. Funds were provided for a demonstration project along a 90-m stretch of beach.

FLEXIBLE EROSION CONTROL SYSTEM

The system devised (Figs. 1 and 2) consists of:

- An inclined bamboo frame anchored back to deep-sunk, rock-filled gabions;
- Sea-facing pre-cast reinforced concrete slabs (type A) consisting of voids, flats and protuberant flat-topped pyramids (a) lying on, and loosely tied to the bamboo frame and (b) loosely tied together with plastic rope passing through toothed edge links, and resting on a boulder-filled gabion foundation toe;
- Behind the bamboo frame, rubberized coir matting stitched together to minimize the passage of sand particles from back to front;
- An apron of pierced pre-cast reinforced concrete slabs (type b), rope tied and located at the top of the bamboo frame, extending up-beach to prevent back mining by over-breaking waves.

Preformed sandwiches, 4.5 m long and 1.2 m wide are laid side by side in a excavated cut at 30° to the horizontal. The sandwiches comprise interwoven grids (200 mm cells) of 50-mm wide bamboo strips laid on either side of 35 mm thick latex/coir matting, which is dense enough to stop sand particles from passing

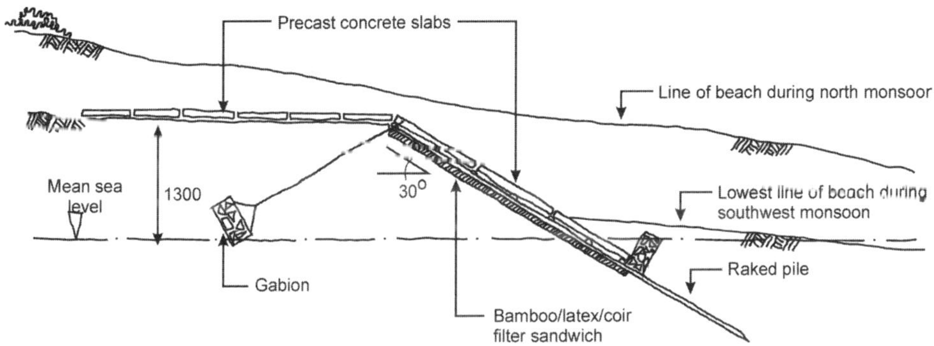

Figure 1. General sectional arrangement of bamboo beach protection system

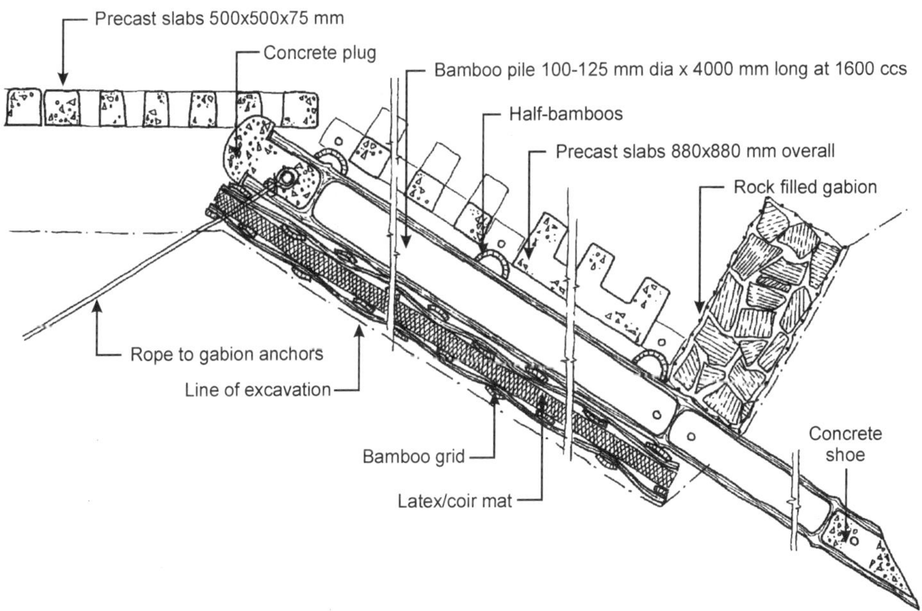

Figure 2. Detail of bamboo beach erosion protection system (rope ties omitted - not to scale)

through. The matting edges are half-cut, lapped and stitched together firmly. The whole assembly is sewn together with plastic rope (Fig. 3).

Bamboo piles, 4 m long and 100–125 mm in diameter, are driven into the sand at 1.6 m apart on the upper face of the sandwiches using a pre-sunk steel guide. The piles have a pointed concrete tip, held in place by a dowelling peg. At the top of the pile is a removable shock-absorbing, hardwood plug. Halved bamboo culms (75–100 mm diameter) made into a grid (300 mm cells) forms a support for the pre-cast slabs (Fig. 4). This mat is tied with a plastic rope (10 mm diameter) to

Figure 3. Preformed sandwiches of rubberized coir and bamboo strip mats

Figure 4. Lay-up on slope, Type A slabs (with bamboo grid, piles and mats)

dowelling pegs passing through the piles either side of the node points. The gabion foundation toe is positioned at right angles to the piles at the base of the mat with its top edge 600 mm below mean sea level. The gabion is tied to the locating dowels in the piles and to the mat with plastic rope.

Three type A pre-cast slabs, each measuring 880 × 880 mm (including toothed links) × 75 mm thick, are laid one above the other on the mat with protuberances uppermost and loosely tied together and to adjacent slabs with plastic rope. The resulting inclined plane – 2.4 m long with voids, flats and protuberances mounted on a flexible frame – is intended to reduce the force of the broken wave. At the head of the inclined plane, six rows of simple pre-cast slabs 500 × 500 mm consisting of alternating voids and flats are laid at the natural pitch of the beach (Fig. 5). Below these pre-casts, two rows of latex/coir matting sandwiches are laid, overlapping the top end of the inclined sandwiches. When the wave breaks, these receive the water deposited beyond the head and pass it through to the back of the frame.

Figure 5. Lay-up at head of slope, Type B slabs (with bamboo grid, piles and mats)

MATERIALS

Bamboo

Mature bamboo (*Bambusa vulgaris*) is freely available along or near freshwater courses in Sri Lanka. It is strong, highly flexible, lightweight, easily transported in long lengths and can be accurately split lengthways. Sri Lankans have an intimate knowledge of the material. It is, however, susceptible to both insect and fungal attack if not properly treated.

Treatment of bamboo can be effectively carried out using the modified Boucherie process. In this process, the sap is pushed out of the entire length of the bamboo pole and replaced with a water-soluble preservative such as a solution of boric acid, copper sulphate and sodium dichromate. It is important that bamboo is treated within 12 hours of harvesting, in the wet season. The ends should be off back immediately before treatment commences. When all the sap has been replaced, culms should be allowed to dry in the shade with good air circulation all round for two weeks before use.

Rubberized coir

Coir, a readily available by-product of the coconut industry, has been impregnated locally for many years with homegrown latex for mattresses and other articles.

Pre-cast concrete

There is a sophisticated pre-cast concrete industry in Sri Lanka using imported steel. A refinement might be the use of bamboo in place of steel. In this regard, and lacking funds for reinforced concrete structures, the Robinson Club has sewage treatment ponds constructed from bamboo reinforced puddled-clay, which have now been in use for 15 years without problems.

Gabions

The gabions used for the demonstration were imported at short notice, and were galvanized with only a limited choice of sizes available. Locally manufactured, plasticized gabions can be expected to be available once the system finds increasing applications in canal and other embankment work.

Plastic rope and string
Rope and string were specified as made of UV Protected Polypropylene.

DIFFICULTIES ENCOUNTERED

Funding difficulties imposed severe time restraints not matching seasonal changes. The North-East monsoon season was well advanced when a combined contract could be let for the provision of material and works. This left only four months before the extreme limit of erosion was anticipated. Setting up of the Boucherie process by a specialist chemical treatment company took time and in the event, not all bamboo could be treated in this way. A large proportion of the matting bamboo was pre-split and subjected to pressure treatment in a sealed tank. Timing complications also arose from the imposed timber transport licensing process, which includes limitations on the transporting of bamboo.

By the time sufficient supplies were assembled to commence construction, the beach erosion process had pushed the sea water table some 200 mm above MSL and excavation took place in heavy, wet sand with continuous pumping. The 30° slope could not be maintained and flattened to 20° in some places.

It was found necessary to carry out modifications to the pre-cast slabs. With the original profile, the break-up of the wave was not complete and the replacement slabs lacked curing time. Difficulties were encountered in threading rope through positioned pre-cast slabs even when a threading needle was devised out of bamboo sliver.

Finally, the maximum extent of erosion was limited in the 1997 season. The system was, therefore, not fully tested and had to be back-filled to await the hopefully more active 1998 season. However, the observations were encouraging enough for the design team to take out a patent for the system.

The cost per meter run was found to be one-third of that for the boulder system.

Bamboo for Sustainable Development, pp. 831–840
Eds. A. Kumar, I.V. Ramanuja Rao and Ch. Sastry

A bamboo wheelchair for the developing world

J.R. O'BRIEN, R. DEVLIEG, D. AI-DAKHEEL and A.K. COUSINS
Department of Mechanical and Manufacturing Engineering, Seattle University, Seattle, USA

Abstract—There is a great need for inexpensive orthotic and prosthetic appliances in the developing world, where an affordable artificial leg or wheelchair can make a tremendous improvement in the life of a disabled person. The use of appropriate technology – devices built by local artisans using readily available materials – seems to be the most sensitive way to fill this need. This paper describes one such device – a wheelchair with a frame built almost entirely out of bamboo using simple hand fabrication methods. Bamboo was chosen as the wheelchair material because of its excellent structural properties and easy availability; bamboo is also a familiar material to craft workers throughout the Third World, and they should be very comfortable in working with and repairing it. The development and testing of several bamboo wheelchair prototypes carried out by the authors are presented here.

INTRODUCTION

The work described here was sponsored by the Vietnam Veterans of America Foundation (VVAF), which currently operates a wheelchair and prosthetics factory in Cambodia. Owing to the large number of unexploded land mines in the agricultural areas of Cambodia (a legacy of the Vietnam War), farmers and their families are still losing limbs on a regular basis, keeping up the demand for wheelchairs. A bamboo wheelchair could potentially replace the more expensive steel wheelchairs being distributed by VVAF at present. Steel wheelchairs require not only the steel itself but also welding equipment and expertise, which is not universally available. In remote jungle area, it would be too hazardous to fly in the oxygen necessary for welding, which is a further incentive to develop a wheelchair based on local materials and fabrication methods.

This paper reports on the results of the bamboo wheelchair development efforts undertaken by the authors over the last two years. Several prototype wheelchairs have been built, and the last two versions have been subjected to a rigorous series of tests to eliminate weaknesses in their design. The wheelchairs have shown excellent durability and are comparable to expensive commercial wheelchairs in their ride and handling characteristics. At this point, they are ready for testing in the field.

BAMBOO AS A WHEELCHAIR MATERIAL

Modern wheelchairs in the developed world tend to be made of aluminium or steel, generally costing the user over US$ 1 000, a prohibitive price for the average citizen of a developing nation. Historically, early wheelchairs were made of solid hardwood, but they were relatively heavy and difficult to propel. Recently, Banerji and Banerji (1984) have reported on their use of cane (rattan) and bamboo to fabricate inexpensive splints, artificial limbs and a wheelchair. However, their wheelchair seems to have been built of the more pliable and easily worked cane rather than bamboo, and they report the wheelchair to have been of "limited use". The authors chose to use the Chinese bamboo *Arundinaria amabilis* for its superior stiffness and strength.

Bamboo has long been used for construction purposes in the tropics (von Dunkelberg 1985; Janssen 1991). While its strength and stiffness have values about one-tenth those of steel, its density is also about one-tenth that of steel (Janssen 1981, 1991). Therefore its "strength-to-weight" and "stiffness-to-weight" ratios (crucial engineering parameters) are actually very close to those of steel or aluminium. The natural tubular form of bamboo has evolved for good engineering reasons – a hollow tube is the most efficient distribution of material in building a structure to withstand transverse loads. Presumably, this shape evolved to withstand wind loads, but it serves our present purpose of supporting a wheelchair user's weight admirably as well.

There is at least a 100-year history of using bamboo in bicycles whose frames are similar in many important respects to wheelchair frames. During the 1890s, several bamboo bicycle frames were in commercial production in Europe and the United States; bamboo bicycles are still manufactured in China, and in recent years some bicycle frame builders have constructed their own bamboo bicycles.[1] In his classic work on bicycle design, originally published in 1896, Sharp (1979) shows a bamboo bicycle and mentions that in many ways, bamboo is a more ideal material than steel for bicycle frame construction. This is due to the fact that the stresses in the members of a well-constructed frame are so low that ideally only a very thin-walled steel tube would be required to build the frame. However, such a thin steel tube would be susceptible to local denting and buckling. On the other hand, a thicker-walled bamboo frame of the same weight would be much more resistant to denting and buckling failures, and would be strong enough to support the stresses encountered in practice.

[1] Craig Calfee of Carbonframes Inc., Santa Cruz, California, in personal communication in 1997.

TECHNICAL ISSUES

Splitting and cracking

While bamboo has many virtues as an engineering material, it does have certain weaknesses that need to be dealt with in order to produce a durable and viable wheelchair. One issue relates to its durability – resistance to insect damage, rotting and splitting on drying. The first two issues have been dealt with extensively elsewhere (such as Cusack 1998a, 1998b), and there are well-known pre-treatment methods to protect bamboo against insect damage.

The splitting induced by climatic changes and gradual drying of the bamboo is more problematic. Naturally, it is desirable to avoid any splitting in a wheelchair, since this may put the wheelchair rider at risk. The splitting problem can also be particularly difficult when making joints (see below) since many methods of joining bamboo (especially those involving drilling through the joints) tend to initiate longitudinal splits in the bamboo. Splitting from joining methods can be avoided by careful fabrication process; however, splitting caused by gradual drying must be dealt with in some other fashion. When working with small diameter (1.5 inches or less) Chinese bamboo, it was found that bamboo would invariably develop major longitudinal splits when left to dry naturally in a room.

This type of splitting was completely avoided by finishing the bamboo members with several coats of marine varnish. This seems to be an effective method of retarding the drying of the bamboo, although there are no data beyond one year as to the long-term effectiveness of this treatment. This problem is expected to be very much less in the uniformly humid climate of regions such as the Southeast Asia. Finally, splitting is not necessarily disastrous, a partially split member can be mended effectively by lashing it with rope soaked in epoxy, applying a splint made of a split piece of larger-diameter bamboo, or simply using a rope or wire lashing to bind the split section together.

Joining methods

Developing appropriate and effective methods of joining bamboo members is the essence of the problem in designing bamboo wheelchairs. Many cultures have developed ways of joining bamboo over the centuries; von Dunkelberg (1985) and Janssen (1981) give extensive descriptions of traditional joining methods. However, most of these joining methods have been developed with building structures in mind. In the wheelchair application, bamboo of much smaller diameter must be used to keep weight low, and the dynamic nature of the loads a wheelchair is subjected to on a daily basis require a different type of construction.

In the end, it was decided to use hardwood dowels inserted into the ends of the bamboo members as a joining method (Fig. 1). The ends of the bamboo were first slit with a saw, and a hardwood dowel with a diameter matching the inner diameter of the bamboo was inserted a few inches into the bamboo. In the first prototype, automotive hose-clamps were used to fit the bamboo tightly around the

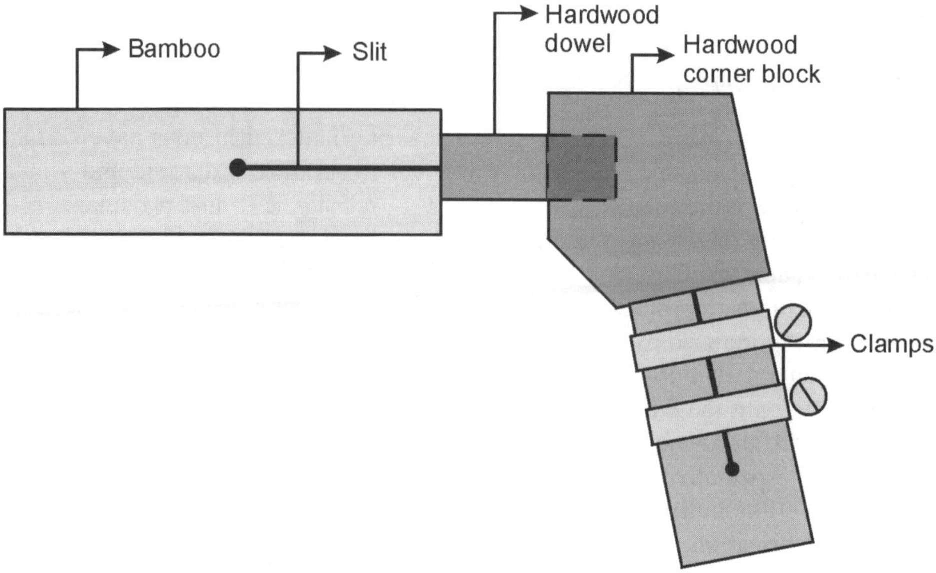

Figure 1. Joint using slit bamboo, hardwood dowels, corner block and hose-clamps

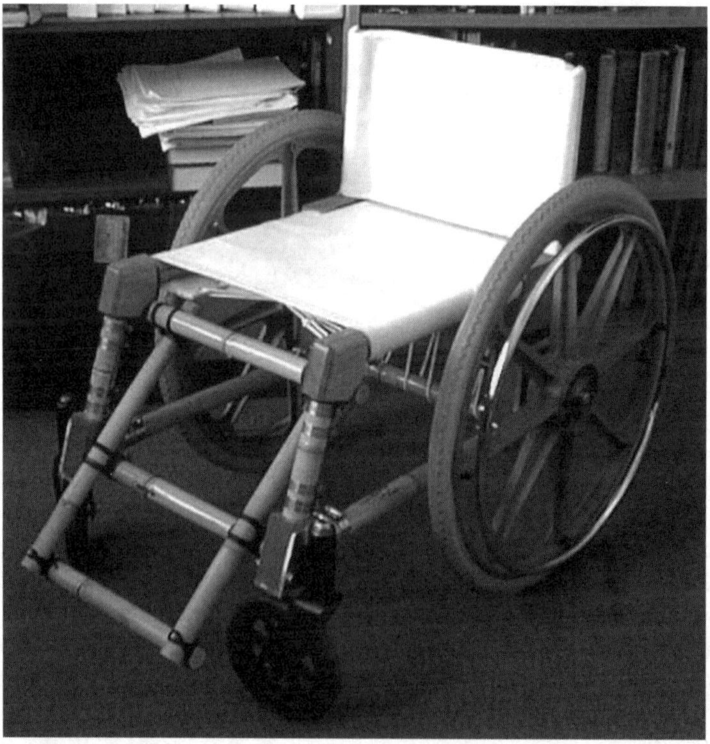

Figure 2. The initial prototype, using hose-clamps to bind joints together

dowel. The other end of the dowel was glued into hardwood corner blocks (Fig. 2). This joining method provides a rigid frame, while still allowing the chair to be easily disassembled and repaired.

Hose-clamps have the great advantage of allowing rapid disassembly and rebuilding of the frame. However, part of the ultimate objective has been to eliminate as much as possible any hardware not available at the village level. As automotive hose-clamps may be too difficult or expensive to obtain in a village, the current prototype of the chair substitutes them with twine lashing around the joints (Fig. 3). The lashing is tightened using bamboo wedges driven underneath the lashing after it has been tied (a traditional technique) (Fig. 4). This has helped eliminate much of the cost of the metal hardware. A push bar (Fig. 5) is attached to the back of the seat using wooden dowels and twine lashing.

It was important for us to know how much force was required to destroy these joints by pulling a dowel longitudinally out of the end of the bamboo clamped around it. Laboratory tests indicated that the joint using hose-clamps could withstand a longitudinal force of more than 250 kg (550 lb) without any slipping of the joint; this is far above any forces expected in an actual wheelchair. The joint using twine lashing and bamboo wedges did remarkably well in the same test; it began to slip only after a 170 kg (375 lb) load was applied to it, which again is quite adequate to withstand typical forces encountered in a wheelchair.

Alternative joining methods, which would allow the elimination of the internal dowels and some of the wooden corner blocks in the frame, were explored. This approach (Fig. 6) also involves redesigning the frame as a truss structure. Trusses are known to be more structurally efficient than designs involving bending of the members, as all loads on the members act axially in the former and the joints between members are relatively free to rotate like hinges. In the experimental wheelchair side frame (Fig. 7), the bamboo members are fitted together at their ends and lashed together using twine and pins fabricated by cutting the heads off nails.

Finally, an alternative joining method, lashing the members together with jute or hemp soaked in marine epoxy (without using any internal dowels), was tested but not adopted in the end. This approach, suggested by a bicycle frame builder[1], provides a very rigid and lightweight joint, but its drawback is that it is impossible to replace damaged frame members because of the permanent nature of the rope/epoxy-reinforced joint. However, this approach probably deserves further exploration.

[1] Craig Calfee of Carbonframes Inc., Santa Cruz, California, in personal communication in 1997.

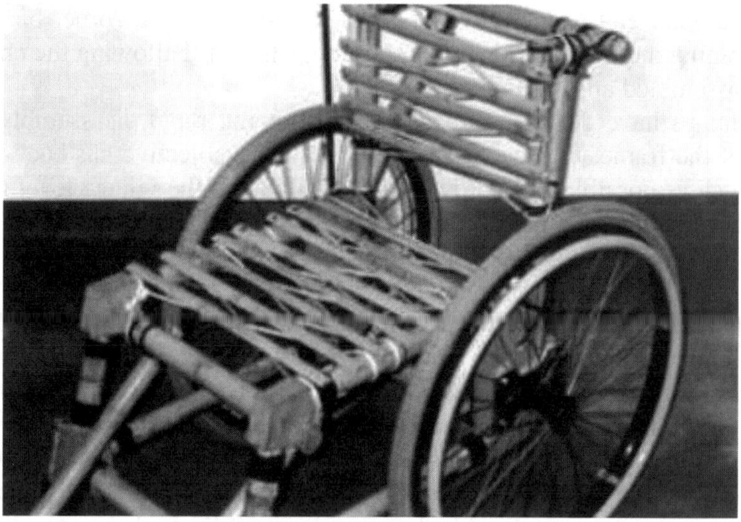

Figure 3. Current prototype, with lashing joints and push bar

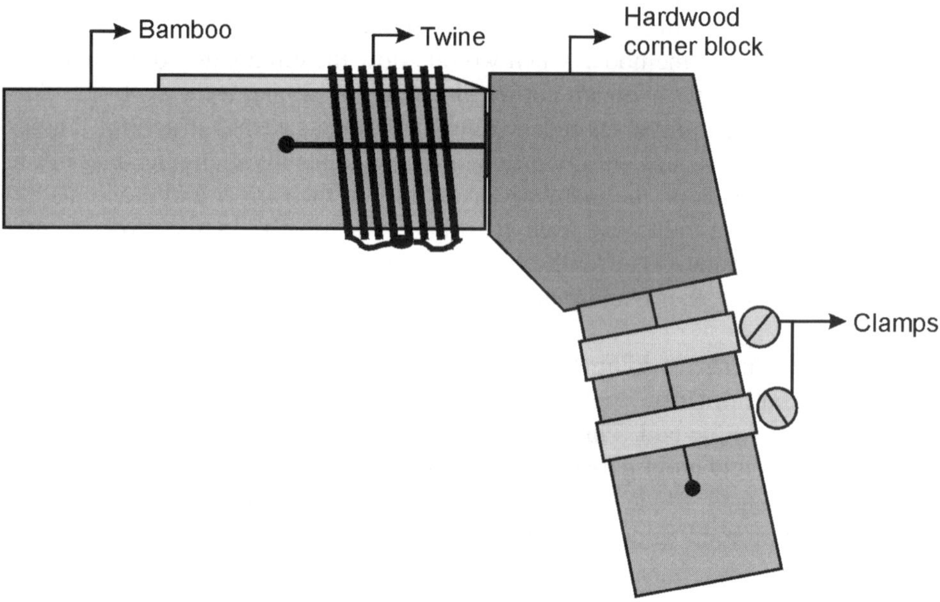

Figure 4. Hose-clamps replaced by twine lashing with bamboo tightening wedges (upper joint)

Figure 5. Rear push bar; lashed and pinned construction

Figure 6. Experimental wheelchair side frame, lashed using twine and nails as pins

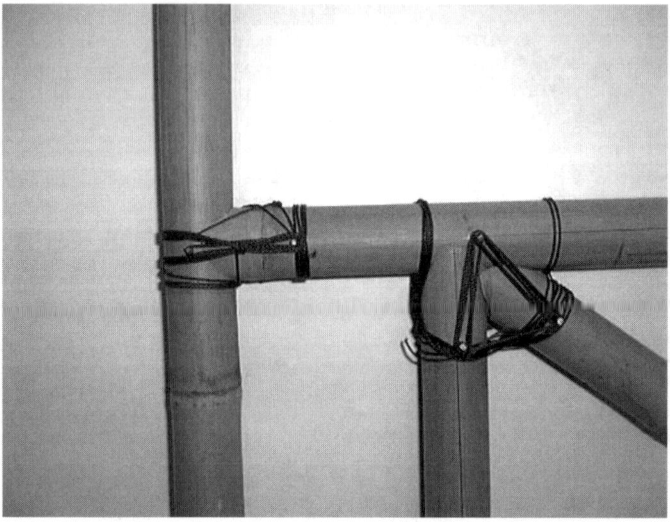

Figure 7. Detail of lashing, experimental side frame

DURABILITY TESTS

The typical wheelchair is used all day long, every day, and subjected to continuous stress in negotiating sidewalks, curbs, broken pavement, unpaved paths, etc. Wheelchairs are sometimes even used as "battering rams" to force open doors that are not designed for wheelchair accessibility. Under these conditions, any wheelchair, even an expensive commercial steel or aluminium model, will eventually break down and require repair. The bamboo wheelchairs will also fail with use, but one of their advantages is that they could easily be repaired by local craftspeople.

It was necessary to ensure that the bamboo wheelchairs could withstand rigorous use conditions to which commercial metal wheelchairs are subjected. For this, several test procedures were designed that closely imitated the actual use conditions. The first series of tests was based on recommendations developed at the Wheeled Mobility Center at San Francisco State University under the leadership of Ralf Hotchkiss (Ballard et al. 1997). In these tests, the wheelchair was loaded with 91 kg (200 lb) of sandbags distributed to simulate a human rider. The tests included dropping the loaded wheelchair of a 5-inch (12.7 cm) curb and smashing it into a concrete barrier at a speed of 7 mph (11.3 km/h). After minor modifications, the bamboo wheelchairs passed these tests with no signs of damage or joint slippage.

The second major test series involved long-term durability and fatigue resistance of the wheelchair frame. A set of testing standards (the ANSI/RESNA Wheelchair Testing Standards) has been developed in the US for this situation. A wheelchair treadmill tester (Fig. 8) was built in our laboratory for this purpose according to the ANSI/RESNA specifications. The treadmill was operated by an electric motor, and each wheel of the wheelchair rested on a roller with a bump of specified size (12 mm high × 36 mm long) at a random location. For this test, the wheelchair was loaded with 68 kg (150 lb) of sand to simulate a human rider's weight, and the machine was operated at a speed equivalent to 1.4 mph (2.25 km/h) travel speed for four hours. At the end of this time, the wheelchair showed absolutely no signs of damage or incipient failure. This test is remarkably violent – on the initial test the entire machine began to "walk" around the laboratory and had to be restrained, and ultimately the gears of the testing machine were stripped owing to the violence of the bumping.

CONCLUSION

The prototype bamboo wheelchair has performed quite well thorough laboratory testing, comparing favorably with commercial aluminium wheelchairs. It provides a comfortable and reliable ride, and handles as well as a good commercial wheelchair. It weight is about 10% more (3–4 lb) than that of a similar aluminium wheelchair, but its cost is very much less (roughly US$75, exclusive of the wheels). The frame of the wheelchair is constructed entirely of bamboo, a small amount of hardwood for joints, and two pieces of 2 × 4 timber. The frame construction was done using only hand-powered tools. The wheel axles and axle bushings are made of inexpensive steel bolts, and several pieces of aluminium are used to form mounting points for the rear axles and the front caster wheels. The seat is constructed of two pieces of canvas. The prototype is ready in all respects for actual field-testing.

That the fabrication process is labor-intensive, particularly the process of turning down the hardwood dowels to match the inner diameter of particular bamboo pieces, should not be a drawback to the use of this design. Indeed, in much of the Third World there is an abundance of inexpensive skilled labor. Local craft workers will not hesitate to modify, improve upon and repair the design as necessary since bamboo is such a familiar material in much of the Third World. For all these reasons, the authors believe that the chair should be of use in alleviating some of the problems of the disabled population of the developing world.

ACKNOWLEDGEMENTS

The authors are grateful for the support of Bob Eaton of the Vietnam Veterans of America Foundation, who motivated and funded this project. Craig Sorensen, Jeff Gould and Joe Tynan, all Seattle University students, built the wheelchair endurance-testing machine. Care Medical Equipment and Mission Medical, both of Seattle, provided parts for the project.

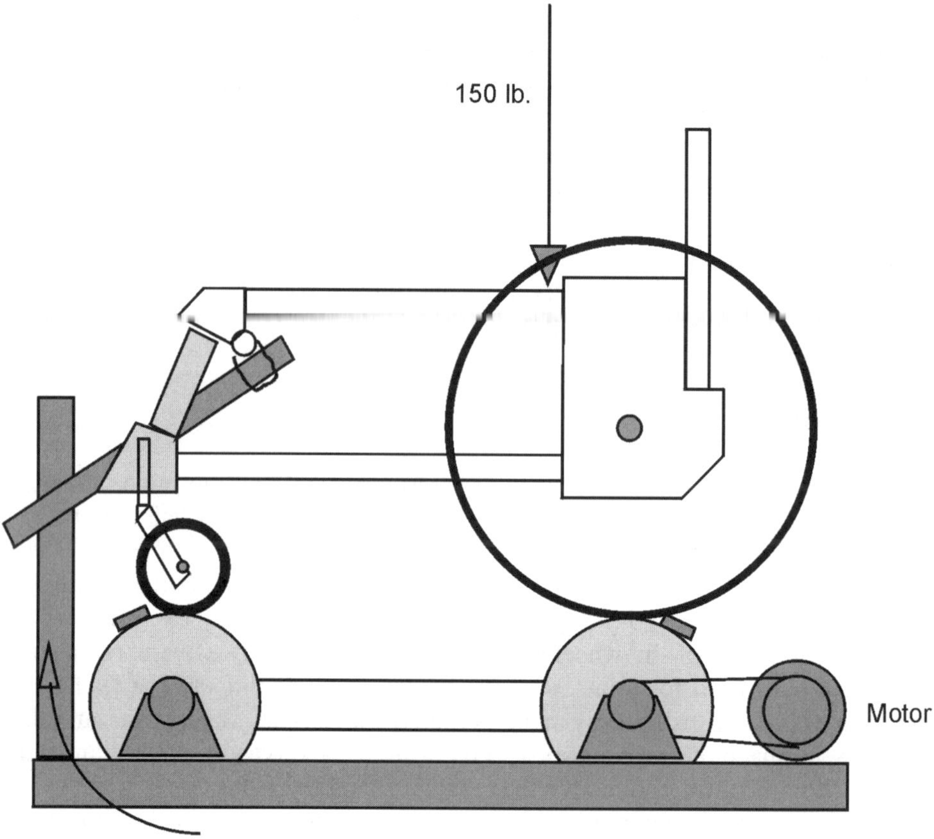

Figure 8. ANSI/RESNA treadmill fatigue tester, with bumps on wheels

REFERENCES

Ballard, R.; Yap, D.; Concepcion, A. 1997. Strength testing of manual wheelchairs. Report No. 697, San Francisco State University, School of Engineering, San Francisco, California, USA.

Banerji, B.; Banerji, J.B. 1984. A preliminary report on the use of cane and bamboo as basic construction material for orthotic and prosthetic appliances. Prosthetics and Orthotics International, 8, 91–96.

Cusack, V. 1998a. First plant your bamboo – then build your house! Permaculture International Journal, 65, 18–21.

Cusack, V. 1998b. Bamboo rediscovered. Earth Garden Press, Trentham, Victoria, Australia.

von Dunkelberg, K. 1985. Bambus (Bamboo). *In* Gass, S.; Drusedau, H.; Hennike, J., ed., IL31 Bamboo. Institute für Leichte Fachentragwerke, Stuttgart, Germany.

Janssen, J.J.A. 1981. Bamboo in building structures. Ph.D. Thesis, Eindhoven University of Technology, Eindhoven, the Netherlands. 235 pp.

Janssen, J.J.A. 1991. Mechanical properties of bamboo. Kluwer Academic Publishers, The Netherlands. 134 pp.

Sharp, A. 1979. Bicycles and tricycles: an elementary treatise on their design and construction. MIT Press, Cambridge, MA, USA.

Bamboo for Sustainable Development, pp. 841–846
Eds. A. Kumar, I.V. Ramanuja Rao and Ch. Sastry

Bamboo drippers for sub-surface irrigation systems

EDUARDO LUIZ DE OLIVEIRA

Faculdade de Engenharia e Tecnologia, Departamento de Engenharia Civil, UNESP Campus de Bauru, São Paulo, Brazil

Abstract—Irrigation problems form one of the factors that render agriculture non-remunerative in many areas. The development of a low-cost and effective irrigation system is hence necessary to ensure that small and medium-scale farmers remain competitive in agricultural production.

The present research on the use of bamboo drippers for sub-surface irrigation was completed in two phases: laboratory research and development; and on-field experimentation and testing. The laboratory work involved the hydraulic study of bamboo culm segments used as sub-surface dripping irrigators, mainly to establish the variations in the flow rate (rate of dripping) with variations in pressure applied on the bamboo segments.

In the field experiments, comparative studies were performed between surface dripping system using commercial (Queen-Gil) drip tubes, and sub-surface dripping system using bamboo dripping tubes first and polypropylene black hose with bamboo drippers later, for the production of cucumber and kidney beans, respectively.

The results obtained were excellent, considering the low uniformity of the sub-surface irrigation system with bamboo drippers and that the way the water was applied in this system (sub-surface) might have offset the effect of lack of uniformity in the distribution of water in the ground.

INTRODUCTION

The development of agriculture, as well as its strengthening in terms return on investments, depends much on productivity improvement. Irrigation systems play a decisive role in reaching high productivity.

The present study focuses on the use of bamboo culms as drippers, coupled with bamboo and black recyclable plastic tubes (polypropylene hoses), for low-cost sub-surface irrigation systems. This method aims at supplying small and medium-scale farmers with conditions to overcome, at comparatively lower costs, technical difficulties in the usage of irrigation systems (Oliveira 1997). This would make it possible for them to not only to enjoy the obvious benefits of a good irrigation system, but also to earn money for the much-needed investments to upgrade their farms.

BAMBOO IRRIGATION SYSTEM

Initially, 10 bamboo stems from among 80 were tested at random and selected. These stems were taken from the first 2/3rd of the secondary branches of *Bambusa vulgaris* culms, by sawing near the nodes in order to make the two ends naturally closed by the nodal diaphragms. Later, to allow water into the dripper, a 1 mm hole was made at one end of each stem. At the other end, 2 cm from the edge of each stem, four radial 3 mm holes were drilled at 90° angle (perpendicular to the length of the stem) to let the water flow out. In some stems the water inlets were made to 1.5 mm diameter.

In a second version, the opening for water inlet was made 3–5 mm from the node, keeping the diaphragm of the node intact. The water outlets (4 radial, 3-mm holes) were made the same way as described previously.

The tests were carried out for 7 different pressures – 1.0, 1.5, 2.0, 2.5, 3.0, 3.5 and 4.0 meters of water column (mwc). To determine the flow rate of the drippers for these pressure levels, the volume of water was measured at 60-second intervals. All stems were tested in the same way.

The data obtained indicated the need for a modification to the 1 mm top holes (inlet), as these were found to be totally soaked after 24 hours under water. This modification was made to all second version stems (see above), using a 1-mm drill bit. Tests were carried out once more using the modified stems, and the flow rate for the above levels of pressure determined. The hydraulic functioning of the bamboo stems was found to be steadier, but the flow rate, as expected, was found to have increased.

The thickness of the nodal diaphragm did not allow for the holes to be precisely made at the top of stems. Drilling the hole before the node did not make any appreciable difference in this aspect, possibly because of the way they were made and also the small drill diameter.

The stems that showed a better hydraulic performance for the studied purpose were those that had 1-mm inlet holes made on the stem and not on the diaphragm, particularly when these were not yet soaked with water. Reducing (irregular) flow rate indicated the diminishing size of holes owing to the expansion of bamboo on soaking.

Four prototype drippers with similar features were made, and tested at the pressure levels used in the experiments with bamboo stems.

While the laboratory tests were being carried out, a greenhouse (5 m wide × 21 m long) was built for field research. Four longitudinal rows of crops – cucumber (*Cucumis sativus*) and kidney beans (*Phaseolus vulgaris*) – were cultivated in the greenhouse and each row was divided into 4 portions of 4.5 m each.

Commercially available drip tubes (Queen-Gil) were used for drip irrigation for comparing with the bamboo-based sub-surface irrigation system.

Bambusa vulgaris culms were chosen and treated as indicated by Oliveira (1992), but without curing under plastic cover because the experiment was only for a short duration. Bamboo drippers made from secondary branches of the said species were fixed onto the bamboo culm tubes as described below.

The edges of the stems (with a 1 mm hole drilled) were cut level and straight with a knife, taking care to keep them as round as possible (Fig. 1). A small length (about 1 cm) of flexible surgical tube (latex tube) was used for the improved fastening of the bamboo stem to the bamboo tube, and for better sealing and flexibility of the joint. The stems were introduced into the bamboo tubes with the help of lubricant oil applied around the latex tube (Fig. 2). A latex tube piece was used also to cap the free end of the stem coupled to the tube (Fig. 1, on the right).

Some drippers were coupled also to plastic tubes, in the same way they were coupled to the bamboo tubes. The end of the stems was not cut to same thickness, as in bamboo tube coupling. Instead, the node geometry, with the prominent nodal rings, was used to work as an anchor in relation to the walls of the tube (Fig. 3).

Cucumber plant (hybrid Hokushin variety) was chosen for productivity tests. Eight cucumber vines were planted per portion (see above), making a total of 128 vines. Half of them were irrigated using the bamboo sub-surface drip irrigation system and the other half by the surface dripping system using Queen-Gil drippers.

At the end of the first experiment with cucumber plants, the bamboo tubes were replaced with 2 cm black plastic tubes (recyclable polypropylene) in order to implement the second experiment.

Kidney bean plants were chosen for the productivity test in this phase. They were planted in 10 holes per portion, making a total of 160 plants. Half of them received irrigation through the bamboo sub-surface drip irrigation system, while the other half were irrigated by the surface dripping system with Queen-Gil drip tubes.

After the experiments, the performance of the two irrigation systems used was studied based on the behavior of the two plant species.

Statistical studies applied to random block experiments revealed that in both experiments the plants which were submitted to the sub-surface drip irrigation system showed better productivity than the ones received water through the surface drip irrigation system (Queen-Gil drip tubes).

The average yield of the cucumber crop in sub-surface irrigation was 4 181.56 g, whereas in surface dripping it was only 2 957.34 g. The total (adjusted) production was 456 889.80 g – 267 619.98 g (59%) produced in the portions irrigated by bamboo sub-surface drip irrigation system and 189 269.82 g (41%) produced by the surface drip irrigation system with Queen-Gil drip tubes.

The average yields for the kidney beans were 1 448.26 g/vine in bamboo sub-surface drip irrigation and 1 326.89 g/vine in surface drip irrigation. The total (adjusted) production until 20 December 1996 was 221 8249.15 g – 115 672.76 g (52%) produced in the portions irrigated by bamboo sub-surface drip irrigation and 106 151.39 g (48%) produced in the portion serviced by the surface drip irrigation system with Queen-Gil drip tubes.

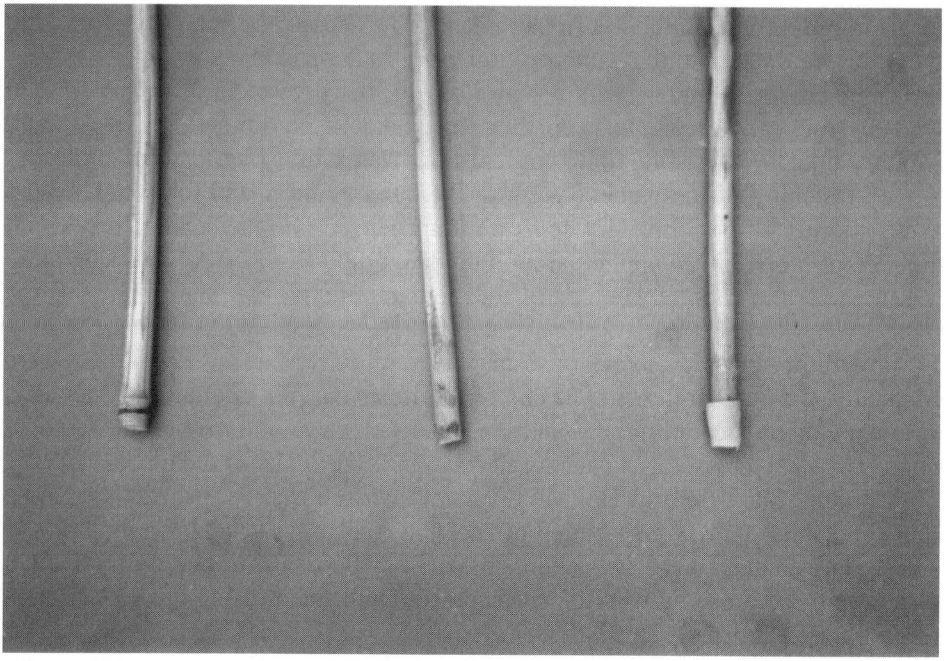

Figure 1. The edges of the stems cut to equal thickness and the latex tube capping the end

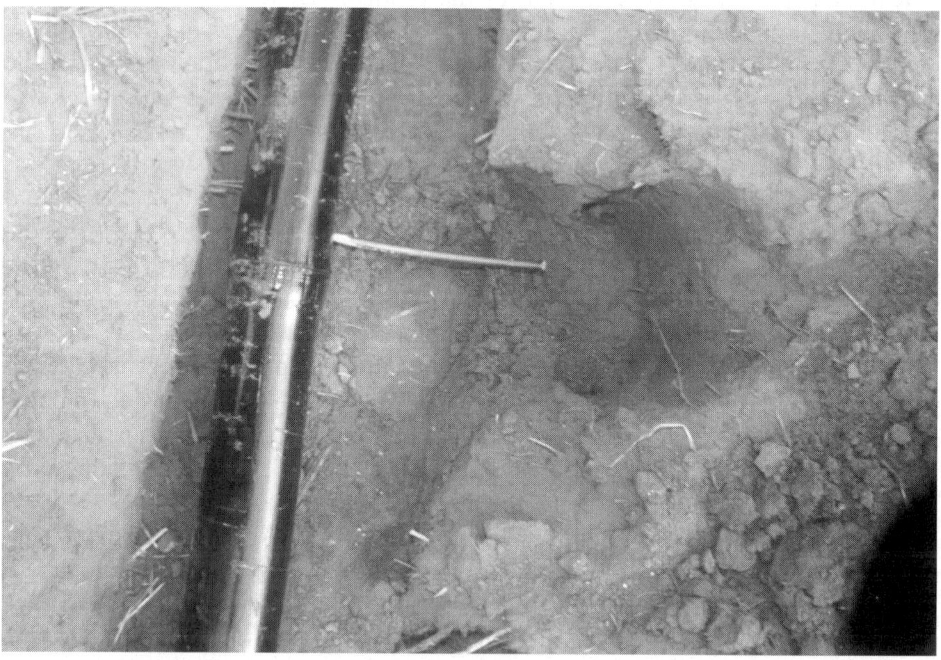

Figure 2. A dripper (prepared bamboo stem) coupled to a bamboo tube

Figure 3. Dripper coupled to plastic tube

DISCUSSION

The research carried out in the laboratory indicated that bamboo drippers had great flow rate variation among them for the same pressure level (25–100%). The hole pattern, internal and external diameters and stem length did not have any bearing on the flow rate variation. The hydraulic performance related to the 1 mm hole at the top of the stems (on the diaphragm) may have varied owing to a combination of factors such as the node thickness in the place the hole was made and the structural composition of the node (higher or lower density). The relative roughness (ε/d) can be higher or lower, thus affecting the loss of head in the hole because the Reynolds number there at a flow rate of 4 liters per hour will be about 1 414.

The sub-surface irrigation process is necessary for the bamboo drippers and tubes as well as for the plastic tubes (polyethylene) because their lifespan in the open air is shorter and a layer of earth would offer better protection.

CONCLUSION

The results obtained, the adopted methodology and the experimental conditions in which this research was carried out allow for the following conclusions:

- The use of bamboo stems for sub-surface irrigation is viable;
- The low pressure level needed for operating the system represents a great advantage in its application;
- The variation in the flow rate for the bamboo drippers did not negatively affect plant growth or productivity when compared with the results obtained for the commercial sub-surface drip irrigation system;
- The drippers are easily made and placed in bamboo and plastic tubes and no sophisticated technique is needed; and
- The sealing system with the latex tube proved to be safe since there was no leakage, and it provided certain flexibility between tube and dripper, thus eliminating the risk of developing fissures in the tube when buried.

REFERENCES

Oliveira, E.L. 1992. Utilização do bambu como condutor de água para fins de irrigação – uma solução de baixo custo. Dissertação (Mestrado em Agronomia), Faculdade de Ciências Agronômicas, Universidade Estadual Paulista, Brasil. 104 pp.

Oliveira, E.L. 1997. Gotejadores de bambu para utilização em sistemas de irrigação – uma solução para pequenos produtores. Tese (Doutorado em Agronomia), Faculdade de Ciências Agronômicas, Universidade Estadual Paulista, Brasil. 113 pp.

FURTHER READING

Alam, M. 1991. Leaky tubing for sub-surface irrigation. Journal of the American Society of Agricultural Engineering, Vol. 91, 2158.

Ayars, J.E., et al. 1995. Impact of bed location on the operation of sub-surface drip irrigation systems. *In* Proceedings of the International Microirrigation Congress 5, Orlando, USA, 1995. American Society of Agricultural Engineering, Michigan, USA. pp.141–146.

Hidalgo, L.O. 1974. El bambu: su cultivo y aplicaciones en fabricación de papel, construcción, arquitectura, ingeneria e artesania. Estudos Técnicos Colombianos, Bogota, Colombia. 318 pp.

Bamboo for Sustainable Development, pp. 847–858
Eds. A. Kumar, I.V. Ramanuja Rao and Ch. Sastry

Bamboo as a water conductor: Hydraulic characteristics

MARCO ANTONIO DOS REIS PEREIRA
Departamento de Engenharia Mecânica, UNESP Campus de Bauru, São Paulo, Brazil

Abstract—The giant bamboo species (*Dendrocalamus giganteus*) seems to be suitable to be used as a water conductor, concerning its good physical and mechanical characteristics. In this research, the species was tested as a water conductor when submitted to a pressurized flow, verifying some hydraulic characteristics – such as the f and C roughness coefficients, rupture pressure and rubber joint pressure behavior. The results showed a value of C = 63 and an expression of $f = 0.3459\ Re^{-0.117}$ for a manual tool and a value of C = 101 and an expression of $f = 0.1107\ Re^{-0.097}$ for an electric tool in terms of the diaphragm removal. A minimum value of 0.51 MPa was observed for rupture pressure, and a maximum value of 0.22 MPa can be recommended for the rubber joints.

INTRODUCTION

Bamboo is a tropical plant of fast growth and great productivity; it is self-renewing and has physical and mechanical characteristics that make possible its use in thousands of things, including water conductor for minor irrigation purposes (Pereira 1992). This material is easy to work with using simple and cheap tools. One can say that if a farm has a bamboo clump, it has a small factory of bamboo tubes – tubes that can serve for its irrigation purpose. The author used giant bamboo (*Dendrocalamus giganteus*) tubes as main and lateral water conductor lines for a small sprinkler irrigation system. The bamboo tubes were joined with rubber, buried in the soil and put to work at a pressure of about 15 mwc (meters of water column) or 0.15 MPa.

According to some authors (Farrelly 1984; Liese 1985; Hsiung 1988), bamboo has been used for centuries in people's daily life in many tropical countries, principally because of its straightness, lightness, strength, fiber content, flexibility and workability – characteristics that are ideal for diverse technical purposes that are already known or remain to be discovered.

Among the thousands of bamboo species known, the dimensions and characteristics of the giant ones make them suitable for use in irrigation purposes. In this study, *Dendrocalamus giganteus* species was employed because it is common in Brazil. Despite its availability, the species is hardly used because of lack of research and knowledge on the bamboo in general, and about its characteristics and uses in particular.

Although it is known that adequate irrigation improves crop yield, it is not much practiced in Brazil because of the high cost of traditional irrigation systems, the low purchasing power of farmers and/or lack of information.

There is not much information related to bamboo's hydraulics characteristics, its use as water conductor, head-loss and roughness coefficients. The information, where it exists, is related to bamboo in general and the values given always vary with the variation in the methods used to obtain those values. The aim of this work is to contribute towards the development of bamboo as a water conductor for irrigation purposes, through the determination of the roughness coefficients f (Darcy-Weisbach formulae) and C (Hazem-Williams equation), as well as to present some data about the rupture pressure of bamboo tubes and the pressure exerted on the bamboo-rubber joints used in the system.

MATERIALS AND METHODOLOGY

Bamboo sections

Whole bamboo culms were divided into three 4-m-long sections: base, middle and top. The selected sections had an internal diameter of about 10 cm (4 inches), which is the average diameter of the species used.

Diaphragm removal

The culm sections were removed of their nodal diaphragm using two tools specially made for the purpose. For one set of culm sections, a manual impact tool that effects a coarse removal was used, while for the other set an electric rotary cutting tool that gives a finer and closer cut was employed.

The manual impact tool, based on a similar tool presented in a prior work (Pereira 1992), comprises an iron tubular piece – 63 mm (2½ inches) diameter and 300 mm (12 inches) length – with both ends sharp enough to cut through the diaphragm, welded onto a 2-m iron pipe with a diameter of 32 mm (1¼ inch). The iron pipe serves as the handle for the tool (Fig. 1). The tool is operated by push-and-pull motion.

The electric rotary cutting tool consists of a rotating body, with three cutting heads symmetrically placed in a cylindrical aluminum core. The cutting width of

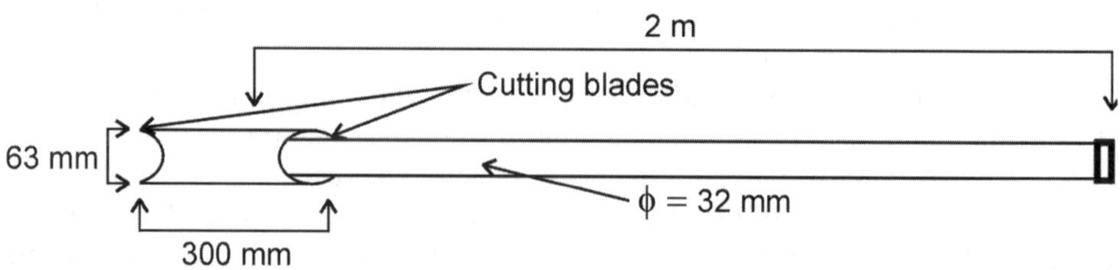

Figure 1. The manual impact tool for removing diaphragm

Figure 2. Electric rotating tool

the tool can be adjusted by a spring mechanism to work with materials of different diameters. The tool is linked to a steel tube that plays the role of a torque transmitter axis. A 3-hp electric motor powers the tool (Fig. 2).

Determination of diameter

The internal diameters of the bamboo tubes were determined by measuring the inner volume (V), as bamboo tubes have a cylindrical shape with length (L) and diameter (D). The inner volume was determined by filling the bamboo with water, and the diameter obtained was termed the average diameter (Dm) of the bamboo tube (Equation 1).

$$Dm = \sqrt{\frac{4V}{\partial L}} \qquad (1)$$

where:

Dm = Average diameter [m]
V = Inner volume [m^3]
L = Length [m].

Determination of discharge, velocity and static pressure

For the measurement of discharge Q (m^3/s), an orifice-plate flow meter was constructed with areas relation of A_1 (conductor area)/A_2 (orifice area) equal to 0.6. The measurer had a previous calibration as shown in Equation 2.

$$Q = 0{,}01668\ H^{0{,}51922} \qquad (2)$$

where :

Q = discharge (m^3/s)

H = manometer level difference (m)

The average velocity (Vm) through the bamboo tubes was determined using the discharge (Q), average diameter (Dm) and continuity equation (Equation 3).

$$\mathrm{Vm} = \frac{4\mathrm{Q}}{\partial \mathrm{Dm}^2} \tag{3}$$

where:

Vm = Average velocity (m/s)
Q = Discharge (m^3/s)
Dm = Average diameter (m)

Four piezometric rings (h_1, h_2, h_3 and h_4 in Fig. 3) were installed over the PVC tubes to measure the static pressure.

Head-loss in bamboo tubes

Test bench

To measure the head-loss and to verify the measurement system, a test bench was constructed (Fig. 3). It consists of a water tank with a constant 2 000 liters level and connected to a motor pump. A PVC tube (10 mm diameter, 20 m length) carries the water from and returns it back to the tank, maintaining a closed system. The piezometric rings and the orifice plate flow meter were installed in the PVC line. The 4-m-long bamboo tube was placed in the test section. The test bench was previously calibrated placing a 4-meter-long PVC tube in that section and reproducing the classic literature results for PVC (Blasius equation) with error less than 5%.

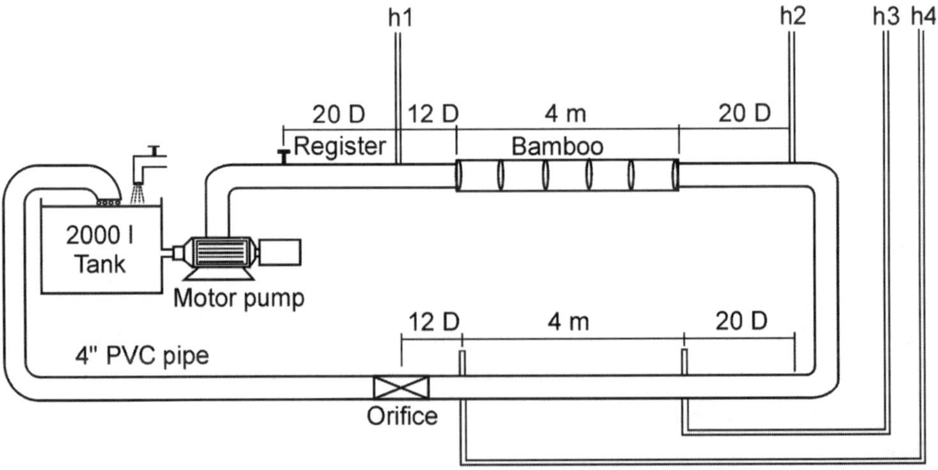

Figure 3. Test bench

Formulas used

Among many formulas for head-loss calculation, the Darcy-Weisbach formula and Hazem-Williams equation are more used in irrigation projects. These were applied in this case also.

The Darcy-Weisbach formula is as follows:

$$hf_{bamboo} = \frac{fLV^2}{Dmg} \tag{4}$$

where:

hf_{bamboo} = Bamboo head-loss (m)
L = Bamboo length (m)
f = Roughness coefficient
g = 9,81 (m/s)
Dm = Bamboo average diameter (m)

The Hazem-Williams equation is represented as follows:

$$hf_{bamboo} = \frac{10{,}646 \times L \times Q^{1{,}85}}{C^{1{,}85} \times Dm^{4{,}87}} \tag{5}$$

where:

hf_{bamboo} = Bamboo head-loss (m)
L = Bamboo length (m)
Q = Discharge (m^3/s)
Dm = Bamboo average diameter (m)
C = Roughness coefficient

As for a determined flow L and Dm are constant, the C coefficient can be obtained as a linear regression of $hf_{bamboo} \times Q^{1.85}$.

Experimental procedure

The head-loss was determined for both sides of 4-m-long bamboo tubes. The side with the larger diameter was named A and the smaller B. Each bamboo tube was tested first in A→B direction and then in the B→A direction. In the final results, A→B and B→A overlapped (Figs. 5–8). The procedure was carried out separately for both types of bamboo tubes – the one for which the manual impact tool was used to remove the diaphragm and the one for which electric rotary cutting tool was employed.

Rupture pressure and pressure on rubber joint

An apparatus was constructed to verify the resistance of the walls of the diaphragm at the last node and of the rubber joint. It consists of a metal plug placed at the end of the bamboo tubes and linked to a manual pump. Through this pump the internal pressure was raised and the values monitored by a manometer (Fig. 4).

Figure 4. Apparatus measuring the pressure on the bamboo tube and rubber joint

RESULTS AND DISCUSSION

Head loss and roughness coefficients

Figs. 5–8 give examples (combined) of the J × Q and f × Re curves obtained for two sets of the bamboo tubes tested (diaphragm removal by two different tools). They also present the C coefficient and equations for f as a function of Reynolds number.

Table 1 shows the values obtained for the C coefficient respectively to the tool used. The final expressions obtained for the f coefficient are given in Figs. 9 and 10.

Table 2 provides some calculated f coefficients for bamboo and PVC (Blasius equation) within the Reynolds' number band normally used in irrigation.

Some authors (Herrera 1974; Gubler 1979) reported the C coefficient for bamboo in general as 43–75 in the case of coarse diaphragm removal and 70–90 in the case of fine diaphragm removal. For *Dendrocalamus giganteus* species, Almeida Neto (1992) reported the values as 48 and 89, respectively. The authors obtained the values 63 and 101, respectively, for the same species.

Fig. 11 compares the f × Re curves obtained for bamboo tubes (manual tool and electric tool) and PVC tubes.

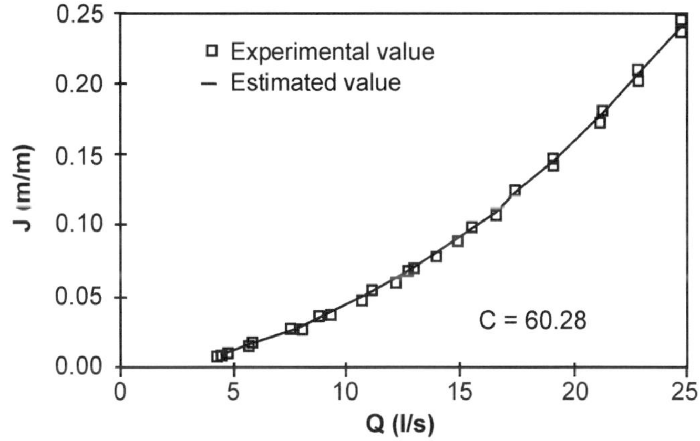

Figure 5. J × Q curve and C coefficient (A–B, B–A combined) (Bamboo data: L = 4.02 m; Dm = 106.83 mm; nodes = 9; diaphragm removal - manual impact tool)

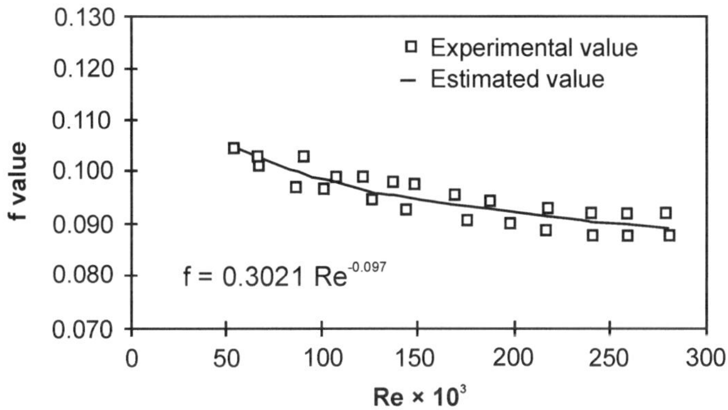

Figure 6. f × Re curve and equation f = f (Re) [A–B, B–A combined] (Bamboo data: L = 4.00 m; Dm = 106.83 mm; nodes = 9; diaphragm removal - manual impact tool)

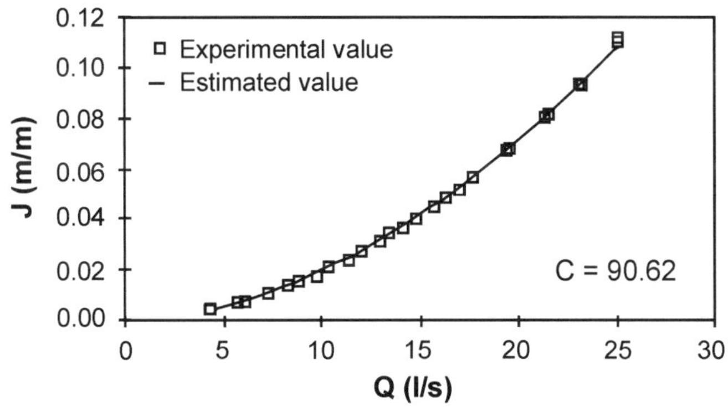

Figure 7. J × Q curve and C coefficient (A–B, B–A combined) (Bamboo data: L = 4.00 m; Dm = 113.80 mm; nodes = 9; diaphragm removal – electric rotary cutting tool)

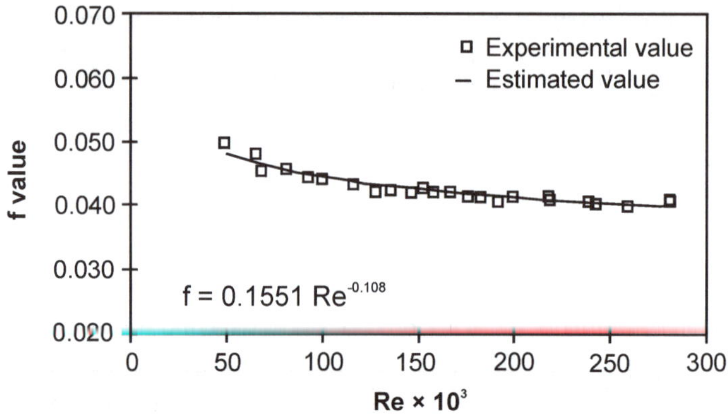

Figure 8. f × Re curve and equation f = f (Re) [A–B, B–A combined] (Bamboo data: L = 3.94 m; Dm = 107.98 mm; nodes = 9; diaphragm removal – electric rotary cutting tool)

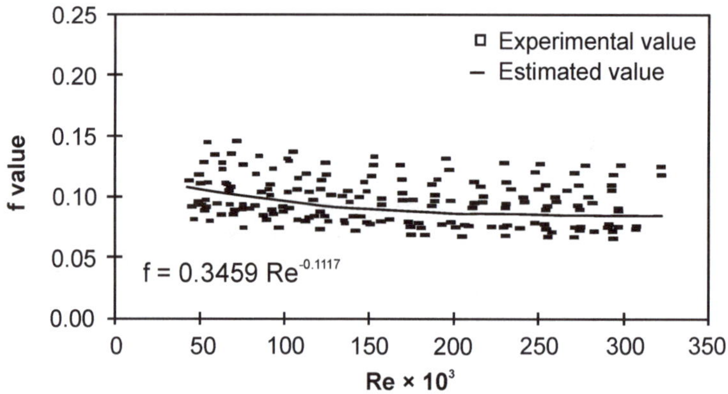

Figure 9. Final expression f = f (Re) [manual impact tool]

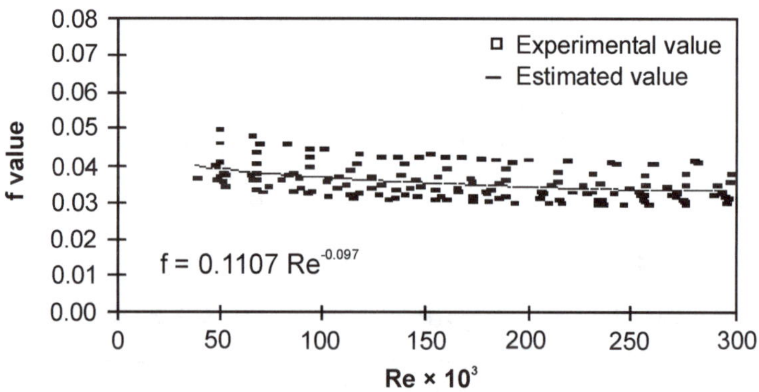

Figure 10. Final expression f = f (Re) [electric rotary cutting tool]

Table 1.
C coefficient and average C value obtained bamboos

Sample A–B & BA combined	Length L (m)	Number of nodes	C coefficient	
			Manual tool	Electric tool
Bamboo 1	4.00	11	51.23	–
Bamboo 2	3.94	9	55.15	106.48
Bamboo 3	–	9	60.28	90.62
Bamboo 4	4.02	7	68.35	106.92
Bamboo 5	4.00	10	65.82	–
Bamboo 6	4.00	10	61.08	95.91
Bamboo 7	4.00	8	65.66	–
Bamboo 8	3.97	7	71.00	102.95
Bamboo 9	4.00	8	65.75	101.31
Average value	–	–	63.00	101.00

Table 2.
Calculated f values for bamboo and PVC tubes

Re ($\times 10^3$)	PVC ($f = 0.3164 \times Re^{-0.025}$)	Bamboo (electric tool) ($f = 0.1107 \times Re^{-0.097}$)	Bamboo (manual tool) ($f = 0.3459 \times Re^{-0.117}$)
50	0.0212	0.0388	0.0975
100	0.0178	0.0362	0.0899
150	0.0161	0.0348	0.0858
200	0.0149	0.0339	0.0829
250	0.0141	0.0331	0.0808
300	0.0135	0.0326	0.0791

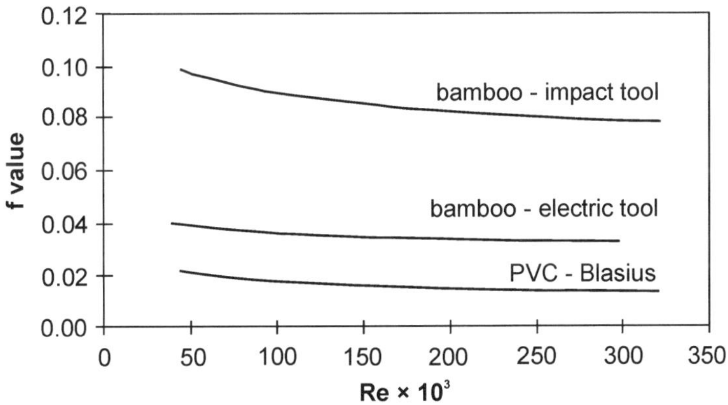

Figure 11. f × Re curves for bamboo and PVC (Blasius equation)

Rupture pressure in bamboo tubes

Table 3 shows the values obtained for the rupture pressure (p) and normal tension (σ) for bamboo tubes.

Bamboo rubber joint

Table 4 shows the pressure values that were operating when leakage occurred in bamboo-rubber joints.

Table 3.
Rupture pressure (p) and normal tension (σ) for bamboo tubes

Culm part	Age (y)	Node number	Length (m)	Diameter (m)	Wall thickness (m)	Rupture pressure MPa)	Rupture point	Normal tension (MPa)
Sample 01								
Bottom	1.5 – 2.5	12	3.75	0.094	0.019	0.81	Wall	2.01
Sample 02								
Bottom	>=3.5	6	2.16	0.105	0.016	0.96	End node	3.16
Sample 03								
Middle	>=3.5	9	3.92	0.097	0.013	0.66	Wall	2.46
Sample 04								
Upper	>=3.5	5	2	0.086	0.010	0.56	Wall	2.30
Sample 05								
Bottom	1.5–2.5	9	3.7	0.102	0.018	0.51	Wall	1.42
Sample 06								
Bottom	1.5–2.5	9	3	0.109	0.016	0.61	Wall	2.09
Sample 07								
Upper	>=3.5	7	3.4	0.097	0.010	0.51	Wall	2.51
Sample 08								
Bottom	1.5–2.5	10	3.3	0.120	0.022	0.51	Wall	1.38
Sample 09								
Middle	1.5–2.5	8	3.5	0.117	0.015	0.51	End node	2.00
Sample 10								
Upper	1.5–2.5	7	3.1	0.111	0.013	0.51	Wall	2.09
Sample 11								
Bottom	1.5–2.5	11	3.7	0.106	0.020	0.66	Wall	1.75
Sample 12								
Middle	1.5–2.5	7	2.9	0.101	0.014	0.66	Wall	2.42
Sample 13								
Upper	1.5–2.5	4	1.7	0.091	0.011	0.61	Wall	2.59
Sample 14								
Bottom	>=3.5	8	3.12	0.115	0.020	0.13	Wall	3.78
Sample 15								
Middle	>=3.5	6	2.81	0.116	0.017	0.81	Wall	2.79
Sample 16								
Upper	>=3.5	7	3.2	0.097	0.012	0.76	Wall	3.01

Table 4.
Leakage pressure in bamboo-rubber joint

Sample part	Length (m)	Average diameter (m)	Wall thickness (m)	No. of rubber strip	Leakage pressure (MPa)
Bottom	3.2	0.109	0.018	4	0.16
Bottom	3.2	0.109	0.018	4	0.18
Bottom	3.2	0.109	0.018	6	0.25
Bottom	3.2	0.109	0.018	6	0.25
Bottom	3.2	0.109	0.018	6	0.26
Bottom	3.2	0.109	0.018	6	0.26
Bottom	2.7	0.113	0.016	8	0.22
Bottom	2.7	0.113	0.016	8	0.25
Bottom	2.7	0.113	0.016	8	0.28
Bottom	2.7	0.113	0.016	8	0.26

CONCLUSION

As far the head loss is concerned, the bamboo tubes displayed a hydraulic behavior clearly inferior to that of PVC tube. Bamboo tubes on which the electric rotary cutting tool was used for diaphragm removal showed better hydraulic characteristics than those on which the manual impact tool was used. Despite these results, it was clear that it is possible to use bamboo tubes as water conductor.

The rupture of the bamboo tubes occurred mostly in the culm wall instead rather that at the end node diaphragm. This indicates the viability of using the end node diaphragm as a plug. The rupture pressure values obtained were adequate for minor irrigation purposes.

The pressure values obtained for the rubber joints were inferior to the rupture pressures, indicating that the rubber joint has to be improved. However, these values were adequate for minor irrigation.

ACKNOWLEDGEMENTS

The author is grateful to the Fundação de Amparo à Pesquisa do Estado de São Paulo (FAPESP), which supported the presentation of this work.

REFERENCES

Almeida Neto, J.A. de. 1992. Contribuição ao estudo do bambu : o comportamento hidráulico de tubos de bambu gigante (*Dendrocalamus giganteus*). Master dissertation, Feagri/Unicamp., Campinas, São Paulo, Brazil. 126 pp.

Farrelly, D. 1984. The book of bamboo. Sierra Club Books, San Francisco, USA. 332 pp.

Gubler, D. 1979. Hydraulic properties of bamboo pipes. Tanzania Hydraulics Laboratory, University of Dar es Salaam, Tanzania.

Herrera, E.O. 1974. Feasibility study on the use of bamboo in pressurized water systems. Part of the master dissertation, University of Massachussetts, Massachussetts, USA. 24 pp.

Hsiung, W.Y. 1988. Prospects for bamboo development in the world. Paper presented at IBC 88, Prafrance.

Liese, W. 1985. Bamboos - biology, silvics, properties, utilization. Deutsche Gelleschaft für Technische Zusammenarbeit (GTZ), Nr. 180, TZ Verlagsges., Roßdorf 1, 132 pp.

Pereira, M. A. dos R. 1992. Viabilidade da utilização do bambu para fins de irrigação - aspectos técnicos. Master dissertation, FCA/UNESP, Botucatu, São Paulo, Brazil. 103 pp.

Bamboo for Sustainable Development, pp. 859–863
Eds. A. Kumar, I.V. Ramanuja Rao and Ch. Sastry

Wine from the bamboo species *Oxytenanthera braunii*

T.N. LIPANGILE
Timber Products and Engineering Service Company Ltd., Iringa, Tanzania

Abstract—*Oxytenanthera braunii* bamboo species that yields a drinking wine grows in Tanzania and is not found elsewhere in the world. Tanzanians have been growing this bamboo species since many years for tapping wine from these bamboos. The daily yield of wine per culm is about 1 liter, and the wine can be harvested for a prolonged period by starting during the rain season and continuing to some time after the passing of the rainy season. Bamboo wine is marketed in Tanzania in same manner as other brewed alcoholic drinks such as beer. It is consumed in almost all rural and urban areas in the country. The wine has an alcoholic content almost the same, or even higher, than standard bottled beer.

This paper briefly outlines methods of establishing plantations of *O. braunii* species in the rural areas, the procedure for tapping wine from the culms, method of collecting the harvested wine from the plantations, and its distribution and sale. It also discusses the social impact of this product on the daily life of the people. The financial gains of the rural people who are cultivating bamboo wine plantations are so high that the economic position of these people has been steadily improving.

INTRODUCTION

The wine bamboo *Oxytenanthera braunii* grows abundantly in the southern areas of Tanzania. It is a clump-type bamboo (Pachymorph), a typical clump consisting of about 15–30 culms that grow to a height of about 3–4 m. Each culm measures 5–7.5 cm in diameter from the basal up to the middle position. For many years, the natives of these areas have used this bamboo as a source of drinking wine. Today, about 5 million Tanzanians are estimated to be consuming bamboo wine.

THE CLIMATE

O. braunii has been found growing well at altitudes of 1 640 m to 1 920 m above mean sea level. The ideal rainfall per annum for obtaining the best wine yield ranges between 740 mm to 1 440, depending on the soil and its water requirement. Suitable temperatures for *O. braunii* is 10°–25°C.

PLANTATION ESTABLISHMENT

New wine bamboo plantations are established using rhizome offsets. The best season for offset planting is during the rainy season.

Some soil samples were collected at the sites where *O. braunii* was providing a good wine yield, and these were then subjected to laboratory testing to determine the soil type. Table 1 presents the results of the laboratory analysis.

Table 1.
Analysis report of soil samples from *O. braunii* plantations

Soil and Texture	Base saturation (%)	CEC meq/100g	Nutrients	pH	Management characteristics
Top soil dark brown	250	6	C = 1% K = 0.2 mg P = 8 ppm N = 218 mg/l	5.5	Low carbon and nitrogen, moderate potassium
Sub soil Brown sandy clay					Low nutrition retention capacity, good workability. Possible phosphate fixation. Without N, P and K application, productivity in the long run will be very low despite periods of fallow. High leaching.

PLANTATION MANAGEMENT

New shoots arise from the rhizomes every year. In the first and second year after planting a new grove, shoots are allowed to grow to full height. In the third year, some of these shoots are retained to expand the clump, while others are tapped for wine when they reach a height of about 30–50 cm. Tapping of wine is carried out regularly from the third year.

All bamboo branches, leaves and grasses found near the clumps are placed around the shoots as fertilizer. Shoots are tapped daily up to about one year and abandoned as they will no longer produce wine.

The cost of establishing wine bamboo plantations in Tanzania works out to about US$ 30 per ha. The activity is usually carried out by a individual rural families through manual labor.

HARVEST OF WINE

When shoots reach a height of about 30–50 cm, they are cut at the apex using a sharp knife or a razor blade. The cutting has to be made daily, and about 5 mm will thus be cut each day for 7 days. On the 8th day, a bamboo jar is tied with a string at the cut end to collect the bamboo wine that starts to trickle down. Bamboo leaves are wrapped around the part where the wine jar meets the shoot, mainly to protect the wine from any possible contamination.

The yield of wine from one single bamboo shoot is about 0.5 liter for larger shoots and 0.25 liter for smaller shoots during morning hours. The same amount of wine can be harvested during evening hours. Thus the maximum yield is about 1 liter per day per single shoot.

Although cutting of the shoot top is done also in the afternoon, wine collection is done only towards the evening. The wine collected daily during morning and evening hours is then stored at home.

CHARACTERISTICS OF BAMBOO WINE

The best brew is obtained after the wine is stored for at least about 12–14 hours. Storing for longer periods will cause the wine to ferment more, increasing its alcohol content. However, such wine is preferred by most men, as well-fermented wine is less sweet and more intoxicating. Women and children prefer the freshly tapped wine for the opposite reasons.

A recent chemical analysis conducted on bamboo wine has proved that it possesses a high alcoholic content, and that it can effectively compete against the standard beer brewed and bottled in Tanzania. Table 2 presents the results of the chemical analysis.

Table 2.
Chemical analysis report of bamboo wine

Total Solids	Total acidity (acetic acid)	Protein content	Total sugars	Fat-ether extract	Alcohol (ethanol)	Coliform, count	pH
2.29%	72% M/V	1.531%	3%	Nil	7.01% V/V	< 3/ml	5.0

The alcoholic content of standard bottled beer available in Tanzania is 5% V/V. Bamboo wine has an alcoholic content ranging from 4.9% to 7.1%, depending on the fermentation period of the wine.

Temporary diarrhea has been reported in some people who are consuming freshly tapped bamboo wine for the first time. This problem is very temporary and disappears soon without any treatment.

COMMERCIAL AND SOCIAL ASPECTS

Until recently, bamboo wine drinking was mostly a part of the rural social activities. In recent years, however, the cost of bottled beer has increased very sharply in both rural and urban areas of Tanzania. Consequently, bamboo wine has secured a healthy market demand in these areas, competing directly with bottled beer. In certain areas, large pubs that can accommodate up to 150 wine customers have come up. These and other pubs sell bamboo wine in standard tin containers, bottles and plastic jars.

The sale price of bamboo wine falls into four groups:

1. Price in areas where bamboo wine is produced – US$ 0.05 per liter;
2. Price in villages located far away from the production area – US$ 0.15 per liter;
3. Price in township areas – US$ 0.40 per liter; and
4. Price of the wine in 500 ml tin containers – US$ 0.30 per tin container.

The cost variations are the result of transportation and canning costs.

The means of transportation of the wine from production centers to marketing centers are:

1. Within the village, women carry by head 20-liter plastic containers;
2. From one village to another, bicycles are employed for transportation;
3. From village and nearby townships, pick-up trucks of 1.0–1.5 tons capacity are used; and
4. Trucks of up to 8 tons capacity are employed to transport bamboo wine collected from production centers to villages and townships situated at distances of 300 km and above. Plastic tanks, each with a capacity of up to 450 000 liters, are loaded on these trucks.

CONCLUSION AND RECOMMENDATIONS

Bamboo wine technology has received public acceptance since many years in Tanzania, especially in areas where the bamboo species *Oxytenanthera braunii* grows. As the wine is also competitive in terms of price, particularly when compared with bottled beer, its export to other countries appears quite feasible. Tanzania can also offer propagules and the technology to other countries where this technology might find application.

ACKNOWLEDGEMENTS

The author wishes to extend special thanks to Dr. Walter Liese of Germany for his personal support and keen interest in guiding the preparation of this paper for presentation during the International Bamboo Congress and Workshop.

FURTHER READING

Lipangile, T.N. 1992. Bamboo wine from *Oxytenanthera braunii* species. Newsletter of the American Bamboo Society, Vol. 13, October 1992.
Lipangile, T.N. 1997. Tanzania report on *Oxytenanthera braunii*, the wine bamboo.

Bamboo for Sustainable Development, pp. 865–868
Eds. A. Kumar, I.V. Ramanuja Rao and Ch. Sastry

Bamboo processing in forestry and industry sectors

RAFAEL BEJARANO
Division of Industrialization, FUNBAMBU, San Sebastián, Costa Rica

Abstract—This paper describes steps involved in the harvest, treatment and industrial processing of bamboo for different applications and uses, as practiced in FUNBAMBU, Costa Rica. The description includes some basic equipment and parameters for quality, dimensions and sizes, and other measurements. The potential uses and manufacture of bamboo slats, panels, plybamboo, paper pulp, bamboo for food, rolled or I-beams, etc. are covered. The paper also examines applications of sawdust, shavings and other processing wastes in the production of soil fertilizer, floor parquet, toothpicks and skewers.

INTRODUCTION

Bamboo industrialization process in FUNBAMBU starts with the coordination of harvesting with the Forestry Department, which holds the right to harvest bamboo. Bamboo poles of specified lengths and diameters – suitable for the intended end-use – are selectively cut. After felling, the bamboo culms are immediately transported to the preservation plant of the bamboo slats factory.

BAMBOO PROCESSING

Preservation of logs

Preservation of the harvested bamboo culms is initiated within 24 hours of harvest, to prevent lack of penetration of preservative owing to the closure of conducting vessels in the culms. Harvested culms are cut to logs of basal, middle or top portions for preservation, which is done by the Boucheri System (see Liese, Gutiérrez and González, this volume). FUNBAMBU has the capacity to preserve bamboo logs of any diameter and length. The equipment employed for this is fully portable: it can be easily transported on a pick-up truck and installed in any place that has provision for the water needed. Electricity can be supplied by a small generator.

The equipment is fitted with a pump for supplying the preservative under pressure, a pressure tank and a filter. It can carry out the preservation of about 120 bamboo logs at a time, depending on the number of nozzles installed.

An important step, if not the most important, that follows is quality control, effected piece by piece. Once preservation treatment and quality test are completed, the logs are stacked and air-dried for 2–3 months. When completely dry, they are ready for use.

Bamboo slats factory

Bamboo culms selected and harvested for slat-making enter the factory and are screened for length and diameter. Depending on the latter, the culm is sawed to obtain bamboo slats of 1, 1–1/2 and 2 inches. If slats to be manufactured need to be of first quality, these are sawed to a few extra millimeters because to compensate for the shrinking that occurs during the drying process. Once the pieces are dry, they are gauged to obtain the desired uniform size. Second-quality slats do not need to be gauged because they will be used to manufacture wall panels that would later be covered with mortar.

Slats are preserved through immersion in a preservative of higher concentration than the one used in Boucheri process. Once the slats are preserved, they are dried in kilns or are open air-dried. Kiln drying is much faster and particularly suitable during wet climate.

Panels factory

The raw material for making panels comes from the bamboo plantation or the slat factory. The manufacturing is fully standardized to make panels of size 10 × 240 cm. Raw material is pre-cut according to the required size of the panel type planned to be built. The main customer of this factory is the Construction Division of FUNBAMBU. At a smaller scale, some private customers are also catered to. Each house model designed and developed at FUNBAMBU require a specific number of prefabricated panels, which have been assigned codes for easy identification. Hence, when a request for panels for a housing project is received, it is easy to assess, with the aid of a computer, the number of panels required. Consequently, batch production of panels can be carried out, ensuring better efficiency and greater productivity. State-of-the-art tools are used to achieve steady output, quality production and high safety conditions for the employees. Most tools – such as nail guns and clampers – are pneumatic.

With 24 employees, the factory could build in one week the panels needed to make 100 houses. It has also been the policy to design and develop workshops that can be easily set up and run: for example, a workshop that can be set up in two days and can produce in a week panels required for 60 houses. All that is needed is a covered area and electricity. Once finished, the panels are transported to the housing projects on 40-feet flatbed trucks, which can carry up to eight sets of panels for the same number of houses.

What has been achieved in this factory can be termed as a technological breakthrough. A significant amount of wood was initially used for the structural part of panels. Today, the panels are 100% bamboo, and there are several techniques to provide different finishes to the houses being built.

Research and development are routine at FUNBAMBU. Some of the products at diiferent stages of study and development are presented below.

BAMBOO PRODUCTS

Plybamboo

A pre-feasibility study has been concluded by the Technological Institute of Costa Rica, specifically the CIVCO (Center for Housing and Construction Research). This study mainly addressed technical development aspects and the results have been very encouraging.

Bamboo-based paper

Tests have been made on bamboo-based paper, and cottage-level production exists in hand-made and shop-crafted paper for tourism, arts and crafts markets. Although theoretically bamboo-based production is a possibility of great interest, the large production scale and huge investments required put such projects beyond countries like Costa Rica.

Bamboo as food

The technologies involved in and the potential of this industry have been studied, as it is a major revenue generator in many Eastern countries, with the intention of attracting potential investors. FUNBAMBU received support from the Technical Mission of Taiwan-China for this study. A demonstration plot has been set up for the purpose of the study.

Bamboo rolled or I-beams

The bamboo slats factory has an estimated capacity of 1.2 million slats per year. Assuming that 30% of the slats produced are of first quality and that this entire output is for the production of rolled beams or I beams, some estimations can be made. The I-beam can be assumed as having a cross-section of 50 × 100 mm (although actually beams are made in different sizes, according to the need). Considering that one 50 × 100 mm beam requires about 12 slats, then it is possible to produce 1 200 000 × 0.30 ÷ 12 = 30 000 rolled beams. The remaining slats could be used for the production of panels for houses.

According to studies, this plant can be easily expanded to produce 150 000 rolled beams. This requires a plantation of 70 ha of Guadua bamboo solely for this purpose. The second-quality slats remaining after the appropriation of first-quality slats for beams are enough to make panels for 1 700 houses per year.

The study also examines other important aspects, such as the use of waste materials resulting from the production process. Thus, for example:

- Sawdust, together with the diaphragms removed from the nodes of culms that are processed in the slat producing machines can be used as fuel in the boiler furnace of the bamboo-drying kiln;
- Shavings and other leftovers from the smooth planing and edge planing operations, together with the ashes from the furnace, can be mixed with urea and used as fertilizer for the bamboo plantations or for other crops; and
- The cuttings – small end-pieces of bamboo slats, leftovers from the longitudinal sizing operation, etc. – can be used to make floor parquets.

Toothpicks and meat skewers

The manufacture of toothpicks and meat skewers has been studied. An agreement has been entered into with a private company that makes these products. Under the agreement, FUNBAMBU is responsible only for the supply of the raw material.

Thus, it can be seen that an integrated approach is possible in the utilization of bamboo as a raw material for the manufacture of several products in the forestry and industry sectors.

Bamboo for Sustainable Development, pp. 869–880
Eds. A. Kumar, I.V. Ramanuja Rao and Ch. Sastry

Bamboo furniture construction using Guadua

BRIAN ERICKSON

Designer, Furniture Workshop, Fundación del Bambú (FUNBAMBU), Costa Rica

Abstract—Guadua bamboo strips are lightweight, strong and linear, permitting the development of flat or curved surfaces in varied furniture designs, some of them knock-down, through the use of rope, cable or rivets. Extensive experimentation with furniture prototypes at FUNBAMBU, using standardized Guadua strips in combination with mechanical fasteners, has yielded interesting results, which are discussed in this paper.

INTRODUCTION

The Fundación del Bambú (FUNBAMBU) of Costa Rica, builder of low-cost housing for the rural poor, is founded mainly for the cultivation and exploitation of Guadua, an American native bamboo genus that includes some giant species. Strips of Guadua bamboo was initially viewed as a more economical means of exploiting the large culm for furniture construction. One Guadua culm can yield up to 144 meters of strip material, thus greatly extending its usefulness. Other notable benefits of using strips are: reduced weight, greater comfort and simpler jointing.

After an experimental phase, during which four different bamboo saws were built, the FUNBAMBU now produces Guadua strips on a regular basis at its plantation in Guápiles, Costa Rica. It is used for prefabricated housing panels or furniture, or is sold for trim.

Extensive work with furniture prototypes has been carried out, and now a small furniture production unit has been set up in Guápiles. This paper details some of the methods used in the making of Guadua furniture with the aim that these methods might be used and/or enhanced by others working in the same field.

MATERIALS AND METHODS

Guadua

A tropical bamboo, Guadua is well suited to the hot, humid climate prevailing in much of Costa Rica. It is a strong-walled, tall and straight bamboo with a fairly open clump formation, characteristics that make it excellent for use as a construction material.

Two species have been planted on a large scale in the plantations of FUNBAMBU: *Guadua angustifolia* and *G. chacoensis*[1]. The former is smaller in diameter (8–12 cm.) with few lower branches, while the latter is larger (12–16 cm.) and taller with many long, thorny branches at the base. Because of the larger diameter, *G. chacoensis* is the species that is sawn longitudinally to produce bamboo strip; *G. angustifolia* is best used in its round form.

Guadua strips

The transformation of round bamboo into long narrow strips is a process that has been described in a previous paper, along with a description of the bamboo saw (Erickson 1992).

The freshly harvested bamboo is strip-sawn green and immediately treated by immersion in a boron solution. The sawn surfaces facilitate the absorption of the solution and lower the cost of treatment considerably[2] (González and Gutiérrez 1996).

The rough-sawn, treated, uncalibrated bamboo strip is stapled to the frames of prefabricated housing panels, later to be covered by stucco. The bamboo strip to be used in furniture and trim is treated, dried thoroughly and then calibrated to widths of 1, 1½ and 2 inches. It is this last step, the standardization of the material, which permits its utilization in planar series. This opens the way to relatively low-tech solutions for large flat or curved surfaces in beds, benches, tables, chairs and so on.

Some examples

A simple example of strip repetition is the bamboo swing (Fig. 1). A 5/16-inch hole has been drilled at both ends of 36 bamboo strips, each measuring 1×42 inches. A drill press facilitates making the holes with precision. The holes are centered on the width of the strips and are equidistant from the sides so that they all line up evenly on both sides. An 8-mm rope is passed through these holes to link the strips. A second rope coils twice around the first rope between each bamboo strip, passing under each strip on to the next, on both ends of the strips to evenly space them out. This makes a series of narrow flat surfaces (the hammock-style seat and back) that hang in a looped curve. As the sitting surface of this swing is the sawn edge of the bamboo strip, the edges need to be sanded smooth for comfort and good finish.

---

[1] Tentative identifications by Dr. Lyon Clark.

[2] Any round bamboo used in this furniture has been treated with the same solution but by the modified Boucherie method.

Figure 1. Bamboo swing

The chair shown in Fig. 2 is made entirely of calibrated strips. The legs are made of glued-up strips (planed flat on the inner surfaces) with a flat bamboo or a wood spacer 5/16 inch thick between them (Fig. 3). Properly positioned gaps or mortises are left for the insertion of tenoned wide-boards. The same mortise-and-tenon jointing is used in combination with wide-boards in the construction of other types of chairs and tables. Stout round bamboo can also be used for the legs instead of glued strips.

The wide-boards are made up of two 2-inch strips or three 1½-inch strips glued and clamped together on the edges to make a sufficiently broad board that allows for tenons and tenon shoulders (Fig. 3). These are nailed or pinned and glued into place for a secure and stable joint in the base of the chair.

The seat of the chair is a series of 1-inch strips nailed into place on thick transversals pinned into the lateral wide-boards. The back of the chair is a series of close-fitting parallel 1-inch strips joined by a length of 1/16-inch galvanized braided cable (used in bicycles and motorbikes) running (not visible) through a linear series of widthwise holes near the top of each strip. These boards descend through a slot behind the chair seat into another slot at the level of the chair runners, where they are fixed in place by nails using a pneumatic nailgun. The back of the chair thus constructed is free of any framework above the chair base and is free to bend. The strips, joined by cable, bend to the contour and position of the sitter making this a responsive and comfortable chair.

Figure 2. Bamboo strip chair

The same kind of back construction can be seen amplified in the knockdown sofa shown in Fig. 4. The back is a curtain formed by a row of 1½-inch strips unified by two strands of cables. It is simply pinned in place on to a long horizontal beam (formed by three culms riveted together), which serves as a kind of fulcrum. The lower end of the curtain extends down into a slot formed by four more horizontal culms near the floor, forcing the upper portion of the back to flex and return like a spring in response to the sitter's weight and position.

Figure 3. Mortise-and-tenon jointing

The use of mechanical fasteners to join bamboo was increased greatly by the introduction of a very simple rivet system. A length of 1/4-inch smooth steel construction rod fitted with two 3/16-inch washers becomes a rivet with a few glancing hammer blows to the rod ends. This fastening method was used in the knockdown sofa to form the seat, a series of bamboo strips standing on edge and riveted at three points (Fig. 5). The seat is dropped in and held in place by the

Figure 4. Knockdown sofa being assembled

structure of the arm modules that are themselves riveted. When installed, the seat presses against the sofa back, trapping it between the seat and the back fulcrum (Fig. 5).

The triangle chair (Fig. 6) is made almost entirely from bamboo strips. It uses four rivets in its construction, running front to back through the round chair arms above, and front to back through the lower corners of the triangles in the base below. The rivet ends are not visible as they are hidden under glued-on strips or caps.

Figure 5. Sofa seat being made of bamboo strips

Figure 6. Triangle chair

Just as in the sofa discussed earlier, the flat supporting surface of a bed of any size can be made by riveting together many 1½-inch strips, through their thickness, with wound rope spacers. Single and full-size beds are made with only one section, while queen and king sizes are two sections bolted side to side. The two sections thus mounted have two rod inserts between them so that they flex in unison (Fig. 7). A glued-up 2-inch bamboo beam is fastened with hook bolts to the bottom side of the bed surface at the foot end (Fig. 8). Two large round culm sections are then bolted on to the beam using threaded rod and tee-nuts (Fig. 9). A plywood disc, fitted tight against the node, provides the clamping surface necessary to hold the round culm against the bottom surface of the bed.

The headboard is self-standing, made from a combination of strips and round culms riveted together. Usually the structure has a pair of bamboo beams integrated to provide the rigid support, on to which the bed surface sections can be bolted (Fig. 10).

Guadua strips mounted on parallel rivets can be used in structures calling for curved surfaces such as an armchair, like the fence chair in Fig. 11. Smaller Guadua poles can also be connected by rivets and rope to form stiff but flexible screens (Fig. 12).

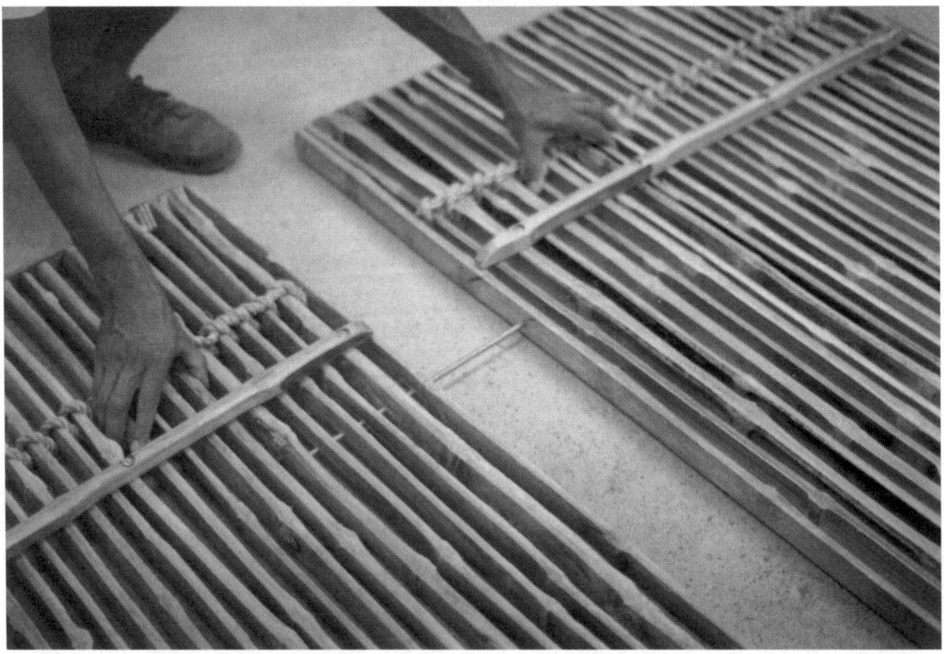

Figure 7. Two sections of bed surface being bolted side to side

CONCLUSION

The use of rivets, cable, hooks and bolts has proved to be a viable way to construct furniture using Guadua strips assembled in series. Such assembly can be in a wide range of dimensions, and either flat or curved. The same system of rivets serves as the basis of the joinery of parts perpendicular to the series (such as legs) by means of hook bolts and tee-nuts, allowing the shipping of furniture in knocked-down condition.

Although this furniture is structurally sound, comfortable and attractive, its construction is dependent on power tools (saws, drills, sanders, etc.) and on the availability of steel rod, cable, etc. As such bamboo furniture is not very cheap to make, it is necessary to achieve a certain economy of scale in production for it to be a viable enterprise. It offers an excellent alternative to those who can afford it and who want to avoid buying wood or plastic. So far, however, it has not been economical enough to appeal to lower-income customers. What remains to be done is to simplify or refine these methods in order to lower the costs enough to make the furniture appealing to a broader public.

ACKNOWLEDGEMENTS

The author is grateful to the workshop crew members for their assistance: Héctor, Omar, Willi, Carlos and Óscar. Thanks are also due to Joaquín Solís and Ing. Rafael Bejarano for their continued support through the offices of FUNBAMBU in San José.

Figure 8. Bamboo beam fastened with hook bolts under the bed surface

Figure 9. Bamboo culm sections being fastened onto the beam

Figure 10.

Figure 11.

Figure 12.

REFERENCES

Erickson, B. 1992. Experimental Guadua furniture in the context of Costa Rica. *In* Zhu, S.; Li, W.; Zhang, X.; Wang, Z. ed., Bamboo and its use. Proceedings of the International Symposium on Industrial Use of Bamboo, Beijing, China, 7–11 December 1992. International Tropical Timber Organization; Chinese Academy of Forestry, Beijing, China. pp. 180–191.

Gonzales, G.; Gutierrez, J. 1996. Bamboo preservation at the Costa Rican National Bamboo Project. *In* Ganapathy, P.M.; Janssen, J.A.; Sastry, C.B. ed., Bamboo, people and the environment, Vol. 3, Engineering and utilization. Proceedings of the Vth International Bamboo Workshop, Ubud, Bali, Indonesia, 19–22 June 1995. International Network for Bamboo and Rattan, New Delhi, India. pp. 121–129.

Bamboo for Sustainable Development, pp. 881–886
Eds. A. Kumar, I.V. Ramanuja Rao and Ch. Sastry

Japanese bamboo crafts: A sample of bamboo's history

NANCY MOORE BESS
Artist, Nancy Moore Bess Co., New York, USA

Abstract—Throughout the world, bamboo is considered the most economically efficient plant material for manufactured and crafted goods. As such, it makes a significant contribution to the cultural history of Asia. Nowhere have bamboo crafts proliferated as diversely as in Japan, where they remain widely practiced to this day. However, the importance of bamboo to the Japanese people goes far beyond the use of bamboo as craft material. The historical place of bamboo in Japanese culture is reflected in design, art and literature. Bamboo is an integral part of the rituals that enhance the spirit – festivals, tea ceremony and religious ceremonies – and its symbolism as an emblem of purity, straightforwardness and flexibility plays a key role. Even some favorite Japanese sports – kendo, archery and fishing – have bamboo elements. Bamboo is part of daily life in Japan and enjoys a rich history there. That bamboo still maintains an active role in Japan enriches this history and any discussion on its continued presence and development.

INTRODUCTION

Within the international bamboo community, many issues and concerns must be addressed. Each region of the world has its own needs, economic considerations and resources, and each has its own bamboo history and mandate for the future.

Why look to the seemingly ancient history of another culture, why look outside our own culture? We examine beyond our own boundaries for enrichment, for solutions and to place ourselves in a larger context. As part of an international community, we have resources and the possibility of information exchange outside our national borders. However, this exchange can only take place, and take place with true understanding, if we understand each other and know each other's history.

As a professional craftsperson, the author sees the rich history of bamboo use in Japan as a resource for information to build an equally rich future of bamboo use – in Japan and elsewhere. To establish an image of Japan, scholars, authors and researchers look to old books, artwork (scrolls, screens, prints), fabric design and collections (toys, household items, sports equipment, etc.). They analyse legends and myths, examine festivals and explore rituals. This is a similar attempt, to establish the ongoing role of bamboo in Japan.

BAMBOO CRAFTS IN JAPAN

For the benefit of those who have not travelled yet to Japan, it is better to begin with an illustration of bamboo's physical presence throughout the country. Seen from the train window, along the roadside, bamboo edges the property line, fills the hillsides and beautifies garden and historical sites. Bamboo is almost everywhere – a constant visual reminder of its natural beauty and continued importance.

Because bamboo is so plentiful, and because it is tall, straight and lightweight, the simplest of bamboo crafts take advantage of these characteristics. Bamboo is used to dry the rice and persimmons, fly the carp banners and flags, and hang out the laundry. Bamboo becomes the handle, the broom and the rake very easily. Smaller hand tools are also made of bamboo. Split bamboo becomes a brush, bamboo sheaths cover the batten in printmaking, smoothed bamboo makes the creases in everything from *tabi* footwear to the paper covers on umbrellas. These tools and crafts, as well as many others, will continue to be made and used, but they rarely warrant mention in books that emphasize the collectible. Nevertheless, these simple pieces help illustrate the wide range of bamboo crafts and uses.

Bamboo crafts historically have been part of everyday life in Japan and remain visible from the most public examples (architecture, garden elements) to the more intimate items affecting one's spiritual enrichment. In Japan, unlike many other Asian cultures, bamboo is not the principal building material. Traditionally, constructions were primarily of wood and paper. Still, bamboo has been present, if invisible, as the grid structure within "plaster" walls, a technique still used today. More visible are the oft-photographed ceiling supports under the thatched roof of *minka* farmhouses. Such usage is now limited to reconstructions or the building of "traditional-type" structures, such as tea-ceremony buildings, historical sites, etc. These buildings may also incorporate bamboo roof tiles, gutters and downspouts, not otherwise seen in contemporary architecture.

Interior elements (even in modern apartment buildings and small suburban houses) may include a small, *tokonoma* alcove with bamboo support pole or shelf. In order to evoke a traditional feeling in an otherwise modern, stark space, bamboo may be used in a transom design or a special window covering. Bamboo mats (*sudare*) filter and soften sunlight. Those that grace religious buildings may be made by hand and bound in brocade, but even those made by machine require a skilled worker. Bamboo and paper lanterns in Japan are used frequently in festivals and are especially visible during the August celebration of *O-Bon*. In everyday home life, a single lantern or two may be used within the home, especially at the entrance or in a non-Western room. The famous *Akari* lanterns designed by Isamu Noguchi are still made in Gifu by the same company are still distributed worldwide. Bamboo and rush mats make seating more comfortable in summer heat, and a small balcony or verandah area may be constructed to include a section of whole bamboo culms, thus, uniting the exterior garden area with the home itself – a long-standing tradition in Japan. Seemingly difficult to achieve in a high-rise apartment building, small "garden" spaces using bamboo can create the harmony achieved in earlier times by careful blending of buildings and garden into a single whole.

Bamboo fencing – sometimes used alone, sometimes in concert with plaster and stone walls, sometimes used with wooden structures and other times with landscape plantings – offers a barrier and a property divider of considerable beauty. Fencing can be used to effectively camouflage an undesirable view or to emphasize a borrowed one! The variety of styles in recent years has been enhanced by recombining materials and styles in new ways. These examples enrich the living environment. In some cases, bamboo fences and gates are now supplemented or replaced by those constructed of wire and plastic, but the fencing industry remains a viable one. A business concentrating on bamboo fences might also be responsible for bamboo installations in department stores and restaurants. Smaller versions of fencing are made for use in tea ceremony, food display and presentations. Larger fencing units must be constructed on site, but smaller units ("sleeve fences") can be constructed indoors all year around. So also can the smaller bamboo garden elements like the *shishi-odoshi* deer scarer used to introduce sound in the garden, as well as ladles, brooms and rakes. Each of these crafts, as well as the baskets used for picking up leaves and moving rocks and rubbish within the garden, requires the hand of a skilled craft worker.

The Japanese are known for their sensitivity to seasonal changes, reflected in everything from food to clothing to hobbies. In food presentation, this seasonal aspect is illustrated by use of bamboo in food packaging and presentation. The appearance of bamboo shoots (*takenoko*) in the market marks the beginning of spring. *Takenoko gohan*, a rice dish, is served at this time of the year and appears everywhere from the finest restaurants to the corner "take out" convenience stores. In early summer, the department stores and advertisements begin displays of new bamboo food containers. Cold, delicious treats are sold in green, fresh bamboo tubes displayed on ice. In fall, persimmons are dried and sold on bamboo. Displays of them hang in department store windows. Bamboo remains year round a popular material for food packaging – sometimes it is a unit of the culm used to hold pickles or fish paste, sometimes it is the sheath (*takekawa*) folded to hold rice cakes or woven into a mat to hold a steamed fish, sometimes it is the split bamboo woven into a mat to hold dried foods, or woven into a basket to display sweets.

Bamboo continues its presence behind the scenes in the Japanese kitchen, be it at home or in a restaurant. The bamboo *zaru* is a simple woven basket tray used in most kitchens in Japan, both modern and traditional. It is used for everything that involves rinsing with water, from vegetable to noodles. *Zaru* appears in ads displaying sweets, pickles, family specialties and more. With a *Sasa* bamboo leaf tucked in the corner, its presence implies freshness and handmade quality. Bamboo mats drain blocks of tofu, bamboo tongs lift *senbei* crackers from the grill, bamboo mats held in bamboo frames serve *soba* noodles and bamboo baskets drain vegetables. Bamboo graters, chopsticks, *O-bento* baskets and *zaru* fill the bamboo market around the New Year, alongside bamboo backpacks and baskets used in the garden. There is a skilled craft worker behind each of these crafts, from simple to complex.

Bamboo crafts are also part of one's personal daily life. It would be easy to think of the well-documented fans and umbrellas made of paper and bamboo as nostalgic remnants of an earlier period of time and no longer relevant. However, one summer day in Japan offers convincing proof that this is not the case. Every hot day sees both men and women using convenient folding fans (*sensu*) as they wait for air-conditioned buses and trains. At summer festivals people tuck colorful flat fans (*uchiwa*) into their belts and sashes to keep them handy or use them to greet a friend, fire up the embers on the grill, or cool the *sushi* rice. The umbrellas for daily use are more likely to be made of fabric, but the beautifully lacquered festival umbrellas are a feast to the eye, and their larger versions that grace tea gardens and restaurants create a cool and inviting place to rest. Umbrellas continue to be made by hand, with those from Gifu being among the most famous. Different stages of the umbrella are completed at different sites, rather than one craftsperson completing the process from start to finish. Some of these people have been working for over sixty years in the business. Customers return to their favorite shops, where they can rely upon the skill of the craftsperson to produce an umbrella that even smells right!

Summer weather brings sales of bamboo cricket cages as well. Traditionally the sound was a reminder of seasonal change and the beginning of summer. It was appropriate for women to keep a caged chirping insect in the bedroom. Cricket cages are still sold each summer, and the same techniques needed to make the round bamboo elements used in these cages are also used to construct baskets, vases and picture frames. This is a very labor-intensive process requiring multiple stages of construction. One region of Japan, Shizuoka, is famous for its products utilizing this technique.

Japan is known to the West as a culture of dedicated workers; less is known about Japan's sense of play. However, think about fishing and archery and kite-flying, and another image of Japan emerges. Bamboo, of course, plays a part in these activities of leisure time. Sports and kite-flying are not limited to childhood years. The kite-maker is often the kite-flyer as well. Kite competitions in Japan are full of color, energy and dedication. To Westerners, archery and kendo events appear more traditional and more formal, reflecting the self-discipline involved. The equipment is more specialized and costly, the training most severe. Without the bamboo kendo stick, the bamboo fishing pole, the bamboo bows and arrows, and the bamboo-and-paper kites, all these activities become less appealing and less Japanese in our eyes. It is true that some of these crafts are replaced by plastic versions, but the authentic bamboo examples are still crucial to the activity and are always used at important events.

Bamboo appears as an element in many festivals and religious rituals in Japan. Sometimes bamboo is used simply because it is lightweight and strong. On those occasions, it becomes the pole that carries the lanterns during the parade and the ladder the firefighters use to display their talents during New Year's celebrations; bamboo forms provide the structure for the elaborate decorations held high overhead during *Tanabata* (a frequently photographed summer festival) and bamboo rakes hold the auspicious ornaments designed to bring good fortune during the *Tori-no-ichi* celebrations. On other ceremonious occasions, bamboo is chosen because it represents flexibility combined with strength and its green color symbolizes a new beginning. The *kadomatsu* is the perfect example of such use. Before each New Year, businesses place two of these "gate pines" outside their door. Three pieces of bamboo are combined with pine and other auspicious symbols to provide the spirits with a happy place to dwell during this most important holiday. In another use, bamboo is pulled into a curved ornament called *mikikuchi*, used as a stopper in the two sake bottles placed on the home-shrine shelf. This ornament is designed to catch the spirits and transport them into the sake for the holiday. Green bamboo ladles are placed at temple sites so that everyone can purify hands and mouth with water before entering.

Bamboo is often grouped with plum blossoms and pine in designs. This grouping, called *Sho-Chiku-Bai*, is often referred to as the "Three Friends of Winter" and is used in design motifs year round. At the New Year, this grouping has special significance and is considered good luck. Pine and bamboo remain growing and green even in snow, and the plum is the first to bloom, sometimes even when there is still snow. Bamboo appears alone as a design motif in fabrics, printed papers and advertisements. Bamboo designs are painted on ceramics, etched in metal and inlayed in lacquer. New Year postcards show bamboo in snow, while those sent in summer are illustrated with bamboo fans, *geta* (traditional foot wear) and insect cages! The use of bamboo motifs ranges from paper wrapping on chopsticks to intricately woven kimonos.

As the traditional craftspersons become older, and as fewer young people are willing to dedicate years of their lives to prepare for this older life style, there is mounting concern that the traditional bamboo crafts will not survive. As the crafts are increasingly made of plastic and other manmade materials, there is apprehension that bamboo may no longer be necessary. As cheaper and cheaper bamboo products are imported into Japan, there is a fear that these products will replace Japan's own. However, while sharing the concern, there could be a more positive outlook, as there is evidence that steps are being taken to address these problems. Regional governments, through vocational schools and business subsidies, are supporting and expanding interest in pursuing bamboo work as a professional option. No longer relying only on family members to maintain the business in the next generations, businesses are hiring and training the next generation of workers and craft workers. University and government agencies are sending bamboo scholars abroad to share their knowledge and learn new technical information. Research groups are developing new techniques, which will result in easier and more efficient curing procedures, a wider range of laminated products, etc. Younger participants in the craft fields are experimenting with new designs and combining new materials with bamboo, all of which will result in new products that appeal to a wider market.

At the same time, there will always be a role for the Japanese traditional bamboo crafts. Flowers will continue to be arranged in bamboo baskets, food will continue to be displayed and sold in bamboo, bamboo-and-paper kites will fly at *Hamamatsucho*, and bamboo-and-paper fans will continue to cool in summer.

There is something to be learned here from the history of bamboo crafts in Japan. There is certainly technical information to be shared with and skills to be taught to those in other cultures. There is also a sense of "progression" to be learned an attitude about the positive influence change can exert and about adapting to new demands while maintaining traditions as well.

ACKNOWLEDGEMENTS

The research was supported by La Bambouseraie d'Anduze, and by The American Bamboo Society and several of its regional chapters (Northeast, Southeast, Northern California, and Texas). Special thanks to the Japanese craftspersons, bamboo business people and training school staff members who contributed with such enthusiasm to this project.

FURTHER READING

Anonymous. 1987. Japanese tradition in color and form architecture. Graphic-sha Publishing Company Ltd., Tokyo, Japan.

Anonymous. 1996. Japan crafts sourcebook: a guide to today's traditional handmade objects. Kodansha International Ltd., Tokyo, Japan.

Austin, R.; Dana, L.; Ueda, K. 1970. Bamboo. John Weatherhill Inc., New York, USA.

Cort, L.A.; Nakamura, K. 1994. A basketmaker in rural Japan. John Weatherhill Inc., New York, USA; Arthur M.Sackler Gallery, Smithsonian Institution, Washington, D.C., USA.

Farrelly, D. 1984. The book of bamboo. The Sierra Club, San Fransico, California, USA.

Moes, R. 1995. Mingei: Japanese folk art from the Montgomery Collection. Art Services International, Alexandria, VA, USA.

Bamboo for Sustainable Development, pp. 887–904
Eds. A. Kumar, I.V. Ramanuja Rao and Ch. Sastry

Application of bamboo in sculptural forms

LUIS FERNANDO GONZÁLEZ BRENES
Plastic Arts Faculty, Liceo León Cortés Castro, Grecia, Alajuela, Costa Rica

Abstract—Bamboo is a material that can be used in the creation of sculptures. This material has some special characteristics, such as hardness, strength and flexibility. Its cylindrical and hollow form also offers possibilities. These qualities of the material allow the application of different technical processes – such as sawing, rasping, carving and heating – for sculpturing. Bamboo can be integrated with other materials such as wood, stone and metal, among others. Sculptures can also be created using available handmade objects, which have plastic and space qualities besides their utilitarian function, along with bamboo. Bamboo is a valuable material particularly when handmade resources constitute an important part in the creation of sculptures.

INTRODUCTION

Bamboo is a strong, flexible, natural material that can be easily cultivated. It is also a low-cost, lightweight and easily transported organic material that admirably represents the "green" concept. It is a woody plant, which grows rapidly to guarantee a continuous supply of culms. For this reason, it is considered a wood substitute, particularly because it is getting increasingly difficult to obtain wood as time passes.

This material offers many sculptural possibilities because it allows the application of several technical processes and its cylindrical form, with its characteristic nodes, offers a very strong structure. Bamboo has a range of applications in sculpturing since the whole culm, culm segments or split bamboo strips can be used. One can carve, saw or bend it, among other procedures. It also allows integration with other materials.

This study on the application of bamboo in sculptural forms was presented as a Graduation Project at the University of Costa Rica, and the following objectives were achieved:

1. Description of the techniques used to work with the bamboo; and
2. Application of the technical processes employed with bamboo in the construction of sculptural objects.

To achieve these objectives, it was necessary to carry out different experiments that are listed below:

1. Creation of volume with bent bamboo strips where the air space is part of the voluminous mass;
2. Application of carving technique on the material;
3. Confirmation of the ability of the bamboo to resist weight and tension in sculptural forms;
4. Construction of mobile structures that use space depending on the way they move;
5. Integration of bamboo with wood, iron and stone;
6. Use of weaving in sculptural structures;
7. Use of coping technique on culms; and
8. Use of crafts in three-dimensional objects.

TECHNIQUES USED

Bamboo is a material that can be worked using several techniques, as its characteristics allow the use of different tools to make a variety of objects. These techniques are discussed below.

Sawing

The sawing process involves cutting the harvested bamboo culms into sections of the appropriate size. For cutting, one can use a handsaw or pruning saw, which are tools that are used manually, or a circular saw or a band-saw, both of which are electrically operated. A small motor-saw is another tool that can be used for the purpose.

Rasping

Bamboo culms have a smooth protective cuticle on the outer surface. For some work, this outer skin may have to be eliminated by rasping. This procedure is needed because when the cuticle dries, it may produce dark stains that spoil the uniform and permanent yellow-ocher tone of the culm, or develops cracks that mar the beauty of the culm. The work is executed using a special tool called rasper, which is a curved metal blade with bevelled underside and one handle at each end.

Production of bamboo strips

Making bamboo strips is a laborious process. First, the chosen culm is cut to suitable length. Then it is halved lengthwise, and each piece is further split to obtain strips of the needed width. The inner soft layer and parts of the diaphragm are removed, and the part with the outer skin (the hardest part of bamboo) intact is used. To obtain strips of even width, they are introduced one by one between two small knives that are fixed onto the workbench. To obtain the same thickness, one by one the strips are passed through a press in which a blade shaves off the extra material. Some strips are made wide and thick.

The thin bamboo strips are used to make woven objects such as: baskets, rings, handbags, lamps, jewelry, hangers, plates, furniture, etc. There are many types of weaves used with bamboo strips in the production of articles: square, net-knitted, hexagon, crowns, turtle, sea wave, rope, octagon, double hexagon, hemp, double hemp, flower, etc. As will be discussed later in this paper, sculptural structures can be built with bamboo strips using techniques other than weaving. For instance, they can be bent by heat, stuck with a glue to any other material, and jointed together using wicker, rivets, screws, etc.

Heating and bending

These procedures are closely related to each other, since to bend the material it must be softened by heating. Bamboo strips can be heated with a heating iron. When the iron is hot, the bamboo strip is placed on the iron, just for a few seconds so that the strip does not burn. The heat makes the strip pliable enough to bend in any desired curvature. The strip can then be dipped in water to cool so that it becomes hard in its new form. The operation may be repeated as necessary. This technique is used mainly with thin strips that are more flexible, as they can be bent in different places and in different directions.

Another way of heating the bamboo consists of using a blowtorch or gas burner. This is used with wide and thick strips, to achieve more open and wider curves, as it is easier to manage such pieces with the blowtorch than with the heating iron.

When heating a strip with blowtorch, one should work quickly so to not burn the piece. It is advisable to lay the strip flat on the ground while using the torch. When the strip softens under heat, it is kept pressed against cylindrical surfaces of different diameters (metallic tubes, plastics or wooden concave surfaces) to obtain the curve required.

Carving

This technique is similar to the one used commonly with wood, and consists of removing some mass from the plane surface with a gouge (or chisel) and a mallet to create a figure in intaglio or a relief.

The carving of bamboo is a procedure that requires precise and delicate work since the culm wall has limited thickness. Each stroke needs to be delivered with a force just enough to remove the material but not break through the wall. Bamboo carving is a work that needs to be done in several steps. First, it is necessary to trace the contour of the figure on the surface by means of an angular cut. This is followed by one or more steps that involve chiseling away material at a uniform depth, following the contour traced. Modeling of the figure based on the drawing is the last step in the process.

The work with the gouges on bamboo is different from carving in wood because bamboo tends to splinter more easily and its fibers are straight along the culm length. For this reason, the chiseling direction needs to be controlled. Moreover, bamboo is structured like a hollow cylinder, and carving is done on a convex surface, the thickness of which depends on the species. If the bamboo is thick-walled, the carving can be deep; otherwise, it should be superficial. The bamboo piece should be held firmly but in a way that would allow it to be easily turned for work around the surface. Some of the carving requirements can be accomplished using a straight grinder.

Coping

Another process used is the coping of the culms. To carry out this procedure, geometric figures are marked on the culm piece. Small holes are then drilled on each figure, on a part to be removed, as the point of cutting (coping) for inserting the jigsaw (or coping saw). Jigsaw is used to carry out certain cuts where there are curves, circles and other irregular contours.

When compared with coping plane surfaces, working with bamboo is different because of the convex surface. It is necessary to work with the jigsaw maintained at an angle of 90° with regard to the contour of the culm. The bamboo piece might have to be turned as the machine is cutting.

SOME BAMBOO-BASED SCULPTURES

Round strips of bamboo integrated with wood

The objective of this sculptural proposal was to integrate bamboo with wood. This integration was achieved using a wooden piece, and round bamboo strips with a diameter of 5 mm and different lengths.

One piece of wood was placed vertically on another piece that served as the horizontal support or base (Fig. 1). The vertical piece was maintained perpendicular with regard to the support, forming a 90° angle with the base, with the lowest bamboo strip holding it leaning on to the base in perfect balance.

The bamboo strips were placed in the board by pressing them into holes of the same diameter. The strips were made with a knife, and shaped to uniform diameter by pulling through a hole made on a steel sheet. The drill bit used to perforate the sheet was of the same size as the one used to make the holes in the wood, so that the fit was precise. The strips, aligned in an arc, were placed parallel to each other and passing through the vertical wood piece so that their ends or tips projected a curved line that can be seen on profile.

The bottom-most bamboo strip was the longest, and its longer (with respect to the vertical wood piece) end leaned on the base, balancing the vertical structure without bending under its weight. Since the horizontal and vertical wood pieces were not fastened together, the strip could change position when the structure moved.

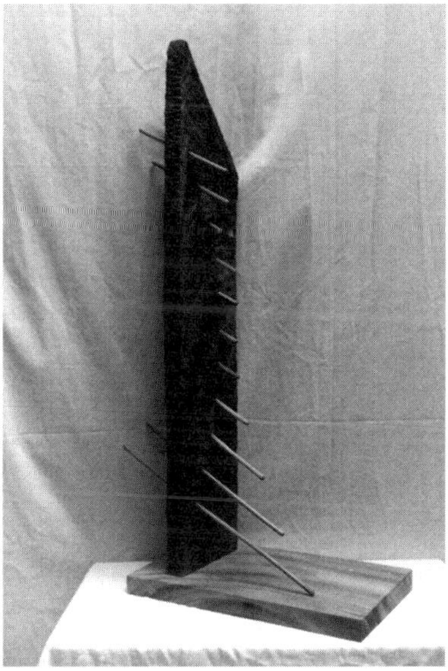

Figure 1. Round strips of bamboo integrated with wood

In the finished piece, the strips and the upright wooden piece together create a structure that is located in the space in vertical form, with the empty spaces between the slanting strips suggesting parallel bands projecting towards the sides. When the structure is placed towards one end of the base, the space on the other side is freed and all the ends of the aligned strips then constitute projecting lines that create open forms.

Stone on vertical bamboo strips

The purpose of this object was to demonstrate the capacity of the bamboo strips to resist weight. The resistance of the material was proved using three vertical strips of bamboo that held a stone weighing approximately 2 kg.

One end of three bamboo strips was pressed into a heavy cylindrical and hollow iron piece that functioned as the base. The strips, with the inner part facing inwards, were then tied together with wicker at three equidistant points to form a single structure. For the bottom and middle joints, small holes were drilled on the strips to pass the wicker through. At the top, angular notches made on the sides of the strips held the wicker.

The top end of the structure was heated to bend each strip outwards, separating them in such a way that each strip formed roughly 120° with the other two. The ends of the strips were sharpened so that they could get into some crevice of the stone placed on top and thus provide a better grip on it.

The strips must have the same thickness, width and length so that the tension and force is the same on all; this is important because otherwise the stone will not

Figure 2. Stone on bamboo strips

remain stable. The stone should be chosen carefully so that the load it places on the strips is appropriate for their load bearing capacity; the idea is to achieve a balance between the weight of the stone and the resistance of the material, challenging gravity. The stone can be sculpted or chosen just for its shape. The stone shown in Fig. 2 was sculpted to have triangular shape and a concave lower part, with three small holes drilled to insert the ends of the bamboo strips.

In this object an organic material like bamboo – which is soft, fragile and flexible – is integrated with an inorganic material like stone – which is inert, heavy and very hard. The object juxtaposes organic and inorganic, softness and hardness, flexibility and inflexibility, and lightness and heaviness, establishing a balanced relationship between these antagonisms.

This sculpture demonstrates how bamboo strips can stand a lot of weight. Thick strips produce more tension than the thin ones since they are more resistant. When placed vertically and joined as a single structure, they are able to support weight. The weight must be well distributed at different supporting points to achieve a good balance. The base is also important, and must be heavy enough to hold the whole structure in stable condition.

The object possesses little volume since it is a thin structure of about 1 m height. The top end has a larger volume because of the stone. Space surrounds the thin structure of the object freely, but finds on the top a solid, heavy and balanced mass.

Figure 3. Carved wood on top of a bamboo culm

Carved wood and bamboo culm

This sculptural object unites a carved wood piece with a bamboo culm, integrating the two materials with a metal piece that serves as a circular base.

The preparation of the culm involved rasping, sanding, lacquering, and painting the inside to prevent fungal attack. The wood used was guachipelín. A figure shaped like a flame was designed on the wood piece. Some parts were made concave and these were textured with cuts made with a gouge. These parts were tinted dark to produce a contrast with the flat surfaces that were retained in the natural color of the wood. The tinting was made after carving and in the end, the carved piece was waxed to give it a smooth finish.

The carved wood piece was then pressed into the prepared bamboo culm, after making necessary corrections on the diameter of the wood to get a close fit. A thick metal ring was placed between the bamboo culm and the wood, to act as a bridge between the two materials. The whole structure was conceived as a column (Fig. 3).

The object is rigid, and its volume is determined by the thickness of the cane and the wood. The structure is compact and light. Its vertical disposition and its small volume allow the space to freely surround the structure.

Relief carving on bamboo

The purpose of this sculpture was to apply the technique of relief carving on a bamboo culm piece, and to set it as a column on a metal base (Fig. 4).

Figure 4. Relief-carved bamboo column

The selected bamboo culm piece was rasped, and the circumference and height of the piece were taken. A design was drawn on paper to suit the area defined by the circumference and height. Using gouges and an electric grinder, this design was copied onto the culm surface as seven simplified drawings. The gouges were used to define the contours (the marking) and to model (carve interspaces) the different figures. The bottom part was worked on with a semi-plane gouge and textured with a straight grinder to obtain a rough texture. It was not sanded to preserve intact the print of the tool; only the figures in relief were sanded to a flat texture.

In specialty of this object is that carving was applied on bamboo, a material with peculiar characteristics such as its cylindrical form, nodes, hollowness and straight fibers. The tools were always used in the direction of the fiber. The texturing of the bottom was a slow work, which often blunted the carving blade.

The metal base was a cylinder, 27 cm high and 12 cm in diameter. The assembly of the column on the base was made using a rod welded to the base, so that the bamboo column can rotate on this axis. The base was painted black and was filled with cement to increase its weight so that it would hold the bamboo column.

The column is hollow, and the top end cut horizontally at the internode to reveal its internal hollow space. The volume corresponds to the thickness of the bamboo culm, with the relief carved on the surface creating volume and texture on the plane. The free space around the bamboo allows movement to appreciate the wholeness of the relief, which is the most relevant plastic resource in this object.

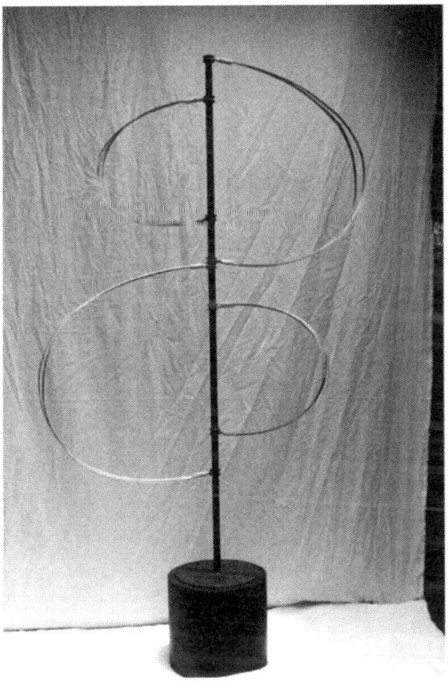

Figure 5. Bamboo strip arches on metal

Bamboo arches

This is a voluminous structure in which bamboo strips were used in the form of arches that produce tension and force. It consists of a vertical metallic tube on which four arches, made with bamboo strips, are fixed. The tube was welded onto a circular metal piece and the whole structure was mounted on a cylindrical wooden base (Fig. 5).

The idea behind this object was to make the bamboo arches rotate around the tube, which works as the axis, in such a way that they create varying forms in space. The two small and two large arches are arranged in such a way that each small one is inside a large one. The arches were formed when the bamboo strips were bent and fastened to the metal tube, producing certain tension in maintaining the semi-circular form. Each end of the arches was fixed onto the metal tube using a small piece of bamboo strip, bent around the tube and the ends fastened to the arches with wicker, and a pair of washers.

The arches can rotate around the axis; a person can hence move the arches to place them in different positions to form diverse figures. The arches create spaces with their circular forms, which can be placed opposite to or inside one another.

In this sculpture, the flexibility of bamboo is highlighted, as the bamboo strips were bent to form a semicircle without heating them. Another important aspect is that the strips have been connected to the metallic axis using bamboo rings. This procedure effected the integration of bamboo and metal.

Figure 6. Bamboo mobile atop a bamboo column with woven end

Bamboo column with woven top and mobile

This object was designed with two purposes: to create volume with bamboo strips and to use woven bamboo in a sculptural object.

The sculpture consists of two parts: a bamboo strip mobile and a bamboo column with a woven top. The first of them consists of 17 bamboo strips connected to each other and bent in irregular forms. These strips were tied in the center with wicker, and were then bent one by one using heat to obtain and curved, circular and oval forms. This construction, measuring about 50 cm in length, 13 cm in width and 30 cm in height, hangs from a thread like a mobile.

The other part of this sculpture functions as a base, and consists of a piece of bamboo culm 1.09 m in height and 35 cm in circumference. Its top 33 cm is woven like a cone to end with a circumference of 15 cm (Fig. 6).

To prepare this woven piece, several longitudinal cuts were made on the culm piece from the top end to 33 cm downwards so that slivers that remained attached to the culm were formed. Then the soft inner layers of the culm were progressively removed until only 1-mm thick strips of the outer layer remained. Other strips of the same thickness were prepared separately and inserted to form the horizontal pieces of the weave. After making a simple weave of approximately 14 cm height, alternating strips were cut to reduce the circumference and shape the woven part into a rough cone.

The completed structure is placed vertically on a flat surface. The mobile of strips is suspended on a thread from the ceiling or from any other similar structure to a height just about 15 cm above the bamboo.

The bamboo culm is a cylindrical object with a conical end that remains immobile, as a point of reference to the bent bamboo strips that form the mobile, the strips of which constitute a three-dimensional structure where the space is part of the volume. The mobile is a very fragile, flexible and light object that moves easily in the wind, rotating indefinitely on its own axis.

Group of five carved-through bamboos

The construction of these five pieces was aimed at using the technique of carving through to create opened spaces in the bamboo culms. This sculpture is constituted by a group of five bamboo culm pieces of different heights. Each piece has two internodes, of which the top internode was the portion carved.

The figures made were different on each bamboo. They were drawn first on paper, then traced on the internode, and later carved through. The culms can be placed in different positions, exploring the several space possibilities available.

The pieces can be set in an ascending order because they possess different heights (Fig. 7). Placed this way, a progressive rhythm where each cane is higher than the previous one can be achieved. They may be placed in long-short sequence to create an irregular order. The distance between each piece can also be altered to get different space relationships. Also, one or two bamboo pieces can be placed horizontally to highlight the vertical sense of the sculpture.

Several bases or supports – such as sand, cement or metal – can be used. Different effects are achieved when the pieces are placed on different surfaces. For instance, when placed on the ground, the organic value of the bamboo is accentuated. If the group is placed on cement or sand, there is a contrast between the inorganic and the organic, the inert and the live. If placed on metal or other materials, there would be other effects.

This bamboo sculpture proves that the material can be carved through easily, and delicate and light figures can be created. The carved-through figures create diverse hollow spaces that bring the interior into focus.

Classic carving on bamboo

The purpose of this object is to apply different carving techniques on a bamboo piece.

A drawing was made of scrolls and leaves, which are ornamental elements used traditionally by carvers. First, a full-size (appropriate to the height and circumference of the bamboo piece) drawing was made on paper. This was then traced onto the surface of a piece of Guadua bamboo, 28 cm in height and 15 cm in circumference. The design was then cut using relief carving and carving through techniques. The wall thickness of this wood-like bamboo species makes it possible to carve high reliefs (Fig. 8).

Ornamental carving is usually carried out on seats, beds, cabinets, cupboards and other furniture items, as well as on articles like doors and pillars. Carving through the culm achieves a greater effect in the volume and fineness of the work.

Figure 7. A group of five carved-through bamboo pieces

Figure 8. A classic carving on Guadua bamboo

The decoration covers the full circumference of the bamboo in an endless rhythm. The holes of the carved-through spaces make the bamboo piece to look light and delicate.

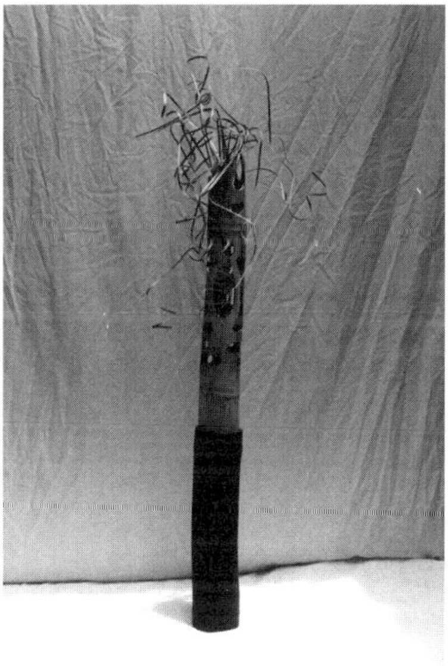

Figure 9. Carved-through bamboo with cascading strips

Carved-through bamboo with strip inserts

The purpose of this sculptural project was to create volume with bamboo strips, which have been inserted at the center of a carved-through culm, so that parts of them can be seen through the carved out spaces.

A geometric design was used for carving. The figures were first drawn on a culm with two nodes and then carved. After this, the carved incisions were sanded smooth. Several bamboo strips were inserted through holes drilled at the center of the diaphragms of the two nodes. These strips were tied together and glued to the bottom end of the culm. At the top, these strips were bent in different angles using heat to obtain a voluminous mass of twisted strips.

The object is set as a vertical structure of carved-through figures, with several bamboo strips at the top extending in different directions. Some long strips cascade down the sides of the bamboo piece under their own weight (Fig. 9).

The carved-through figures become open spaces that allow one to see through the culm the strips bunched together and running vertically. The hollow spaces are important in the object, following the words of Henry Moore in *Sculpture in the XX Century:* "A hole can have as much formal importance as a solid mass" (quoted by Andrey 1981). The strips used also create surfaces in three dimensions.

The incised spaces on the culm gives it a more delicate and lighter aspect. The figures establish a solid-hollow relationship and let one perceive the inner space of the culm.

Figure 10. Bamboo-and-stone sculpture

Bamboo culm with inserted stone

The objective of this sculpture was to integrate stone with bamboo, and this was achieved by inserting a sculpted rock slantingly on a vertical piece of bamboo.

A bamboo culm piece was prepared with one node and one internode. The piece was cut close to the node to form the bottom part, and the internode was cut diagonally at the top. The stone chosen was 30 cm long, 20 cm wide and 8 cm thick. It was sculpted a little to give it a flowing shape, and the bottom surface was chiselled flat. A hole large enough to accommodate the culm was made on the stone. After inserting the stone on the culm at an angle that matched the diagonally cut top of the culm, its position was fixed with glue. This whole assembly was then placed on a black iron base into which the bamboo was pressed down (Fig. 10).

In this structure, two materials that possess different characteristics are integrated: bamboo, which is soft, delicate, long, cylindrical, hollow, light, organic and has a smooth texture and yellow-ocher tone; and stone, which is hard, tough, stocky, solid, heavy, inorganic and has a dark gray tone and a rough texture. Together, these two materials form a voluminous structure in which the slanting placement of the stone breaks the vertical rigidity of the bamboo culm.

Bamboo strips on a wooden base

This object aimed to create a three-dimensional and voluminous structure using wide and thick bamboo strips held on to a plane surface.

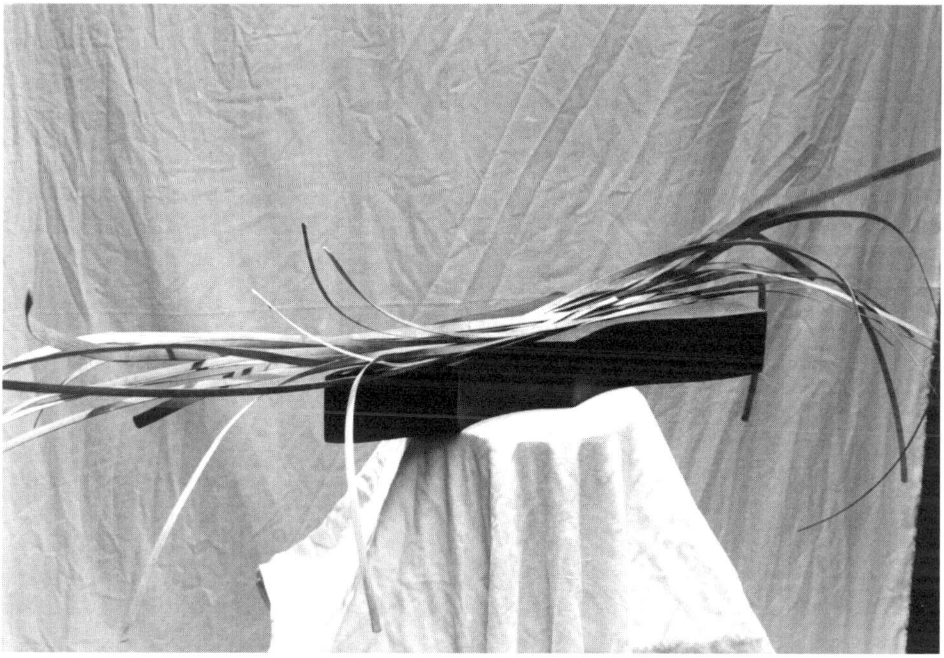

Figure 11. Bamboo strips on wooden base

It was made using 14 bamboo strips of approximately 1.5 m length and 8–15 mm width. The strips were bent by means of heat to form different shapes, mainly curves and semicircles, so that their tips pointed to either sides, up or down, embracing the space despite the force of gravity. The midpoints of the strips were fixed on to the horizontal wood base using screws and glue. (Fig. 11).

The completed structure is approximately 150 cm long, 40 cm high (in its highest point) and 50 cm wide. It mainly characterizes fragility. It has a lot of movement, and both flexibility and low weight of the material are highlighted. The structure has a lengthened and semi-curved shape that suggests the continuation of a circle upwards.

Mobile with bamboo objects

The purpose of this project was to use handmade bamboo articles in the construction of a sculptural mobile.

Two baskets, one large and one small, woven with bamboo strips were used in the production of this sculpture. These objects were made with very thin bamboo strips, using a weave known as "crown". A bamboo strip – 150 cm long, 76 cm wide and 2 mm thick – was placed horizontally and the large basket was tied with wicker to the left of the midpoint of the strip. The small basket was tied in the

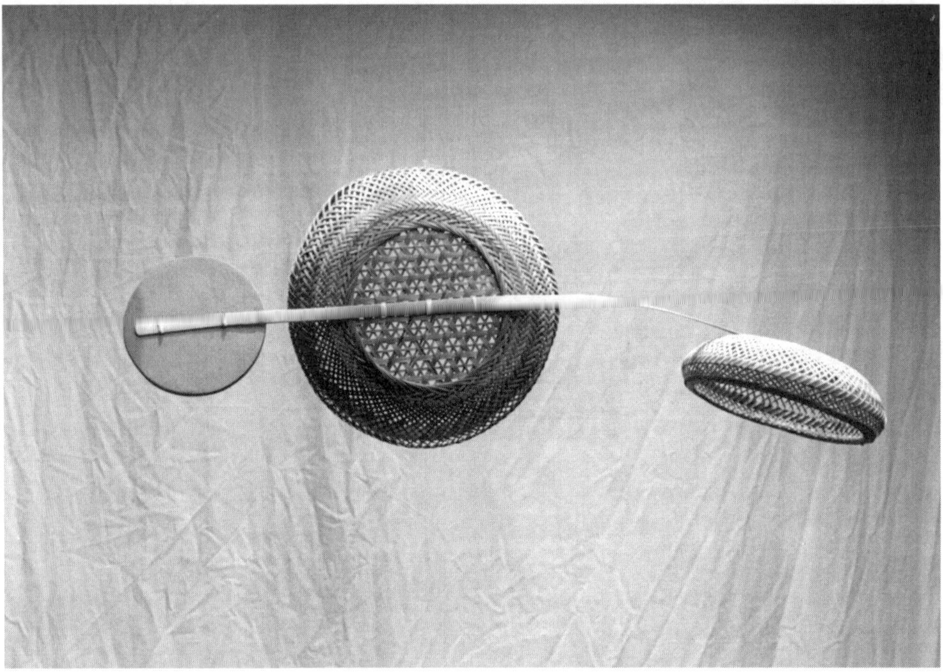

Figure 12. Mobile made with bamboo articles

same way to the longer end of the strip, and a wooden disc – with a size and weight that balance the baskets – was tied to the shorter end.

The assembly, which is hung from a thin but strong thread, can rotate on the impulse caused by a person or the wind.

The baskets used were the ones made for household use but as part of this sculptural object, they became voluminous structures on a horizontal axis, possessing plastic-space qualities above their utility. "For this reason, every object that is created, every product, points necessarily to some use and it is necessary to make considerable efforts to contemplate it only in its objective existence and its aesthetic meaning, independently of all its practical functions". (Andrey, 1981).

The wooden disc is nearer the large basket, thus balancing the structure. It is a horizontal structure with three aligned circular objects with the largest object located nearer to the center and the other ones to the sides. All three objects were placed vertically but, after some time, the small basket turned and ended up horizontally facing downwards, as the strip that holds them did not bear the weight and bent towards the end (Fig. 12).

The thread that holds the structure is tied to the large basket and it is hung from the ceiling or any other high point to a point about 1 m above the floor. When the structure moves, therefore, the smaller objects rotate around the large basket, which can rotate only on itself.

The sculpture is a light and delicate one with a certain amount of transparency. The wooden circle was painted violet to contrast with the yellow-ocher color of the bamboo. The volume is constituted by two fragile objects (made of bamboo) and another more solid wooden one. All three elements are circular and move in space in a revolving movement around a central point.

CONCLUSION

Different sculpturing techniques were used during this study to ensure that the bamboo objects made are varied. The potential of bamboo to build three-dimensional structures is demonstrated in each one of them. When using bamboo strips, especially when heating and bending them, voluminous masses emerge in space. They constitute delicate and light structures. By means of carving through one can make holes in the culm, creating negative shapes that are as important as the figures of solid material. These holes enable one to penetrate the inner space of the culm and lighten the material.

Bamboo's fineness and flexibility allow sculptural alternatives that are not obtained with other materials. Its exceptional qualities facilitate the use of varied techniques such as cutting, relief carving, through carving, polishing, weaving, bending, twisting, painting, rasping, sanding, etc. Carving is a very useful technique for working on bamboo. It can create different volumes, depending on the thickness of the culm, and varied textures on the surface.

From the plastic point of view, bamboo cannot be substituted by wood, stone or metal as each one has different characteristics; for example, wood and the stone are compact and solid while bamboo is cylindrical and hollow. Bamboo, however, possesses enough useful properties to offer many plastic alternatives to an artist.

In this investigation, bamboo was integrated with wood, metal and stone, traditional materials used in sculpture. Bamboo has an organic relationship with wood since both are of vegetal origin. With stone, however, bamboo has the antagonistic relationship between the inert and the live. In the case of metal, there is a contrast between the natural and the artificial. Nevertheless, the possibilities of combining bamboo with other materials are enormous. Its characteristics allow integration with plastic, resin, glass, paper, acrylic and many other materials. In his Technical Manifesto of the Futurist Sculpture, published in 1912, Umberto Boccioni says that sculptors should work with twenty different materials: glass, wood, cardboard, iron, cement, leather, fabric, mirrors, electric light, etc. (Plancton 1974).

The sculptural object can be made with varied materials that allow new possibilities in the creation of volume and control of space. From the compact blocks, we jumped to the construction of delicate structures in which the space is important as a real form in connection with the solid material.

By carving through bamboo culms, it is possible to penetrate space, creating positive and negative forms. With bamboo strips, the material becomes light and fragile. Voluminous structures can be built with these, where the strips can be twisted by means of heat to embrace the space in all directions.

Thick strips produce tension and force, and they demonstrate the resistance of bamboo. By means of weaving, it is possible to build very resistant but light structures that can be moved easily by the wind. Carving allows the creation not only of volume on a plane surface but also of different kinds of texture.

To fasten bamboo to another material, it can be riveted, screwed, nailed, pasted or tied. It can also be assembled by pressure with certain ease.

The straight and rigid culm is a cylindrical and hollow column, whose volume depends on its size and circumference. Many culms put together create diverse space relationships, using which structures of great size can be built. Bamboo can be cut into segments to create modular units, with which three-dimensional structures can be built.

Bamboo is a valuable material where manual techniques constitute an important part in the creation of sculptures.

REFERENCES

Andrey, C. 1981. La Escultura en el Siglo XX. Stripcelona, España, Editorial Blume, 122 pp.

Plancton, 1974. La Escultura Moderna. La Magia del Arte, Enciclopedia Temática Plancton, Editorial Plancton, Stripcelona España, 1° Edición, 193 pp.

Bamboo for Sustainable Development, pp. 905–916
Eds. A. Kumar, I.V. Ramanuja Rao and Ch. Sastry

Portable research laboratory for learning with living models

ANA BRANCO[1]

Laboratory and Workshop for Training and Development of Prototypes (LOTDP), Art Department, PUC, Rio de Janeiro, Brazil

Abstract—The study deals with the construction and usage processes of a portable environment used for learning activities in design with natural elements. The result of the study is a portable laboratory for research on learning. It comprises a shelter structured out of bamboo poles and a nylon sheet, and instruments and tools for designing and constructing using natural materials collected from the environment where the laboratory is set up. The structure provides a shelter for research activities related to observation, handling and transformation of "living models" with the aim of obtaining information from the non-verbal codes present in Nature. The laboratory indicates, through its shape and materials, a group of activities and attitudes oriented towards the application of specific knowledge of the environment, where the project is temporarily located.

INTRODUCTION

The Laboratory and Workshop of Training and Development of Prototypes (LOTDP), of the Art Department at Pontifícia Universidade Católica of Rio de Janeiro, conducts studies in the field of "living design", which justifies the use of materials that are readily available in Nature, such as bamboo, clay, vegetables and seeds. Such materials are simultaneously elements of investigation and materials for design and construction. Shelters, equipment and tools have been designed and constructed using natural materials – such as bamboo – and information generated by their use. These objects together form the portable laboratory for research on learning.

MATERIALS AND METHODOLOGY

For developing the furniture described in this paper, special attention was paid to various aspects such as function, usage, comfort and autonomy. It is important to note that this work was possible because of the low cost of the project, precluding

[1] In collaboration with the scholarship students and volunteers of LOTDP, and under the coordination of Prof. José Luiz Mendes Ripper.

the need for external funding, and the initiatives of all those who voluntarily participated in the study.

The project was based on the theses of Maturana (1997) and Emerson (1981) in the areas of autopoiesis and intuitive archaeology, respectively. It also drew up on the long history of research on "Living Design", related to the study and construction of structures designed to influence the learning activities that occur in them.

The previous experiments point to a dilution of the traditional teaching environments, as can be noticed in the objects constructed with a minimum of obstacles between the interior and the exterior, with the purpose of gathering knowledge that comes from outside the academic space.

The portable laboratory for research on learning was constructed with a self-tensioned bamboo structure, which promotes freedom and allows closeness with the environment. These qualities are reflected in the movement of the air, of the people and of the knowledge as revealed in the use of the laboratory.

In the manner in which they were designed and constructed, the objects (furniture, designs, etc.) symbolize the presence of a group in activity in the environment. Besides defining the space where the action took place, they also emphasize the result of collective design. Their use promotes the awareness of singularities, for example, through the division of the tasks of assembly, disassembly and transportation.

The organic nature and the lightness of the materials used provide an example of the attitude necessary for reading the non-verbal codes present in Nature. These materials also promote the opportunity to reestablish the direct connection between man and the earth, by highlighting the environment in which they are set.

Design of the tent

The portable tent (Fig. 1) is designed to accommodate groups of up to 25 people, outdoors on earth soil, using an area of approximately 25 m^2. Its main characteristics are lightness, ease of handling and transportability. It is constituted of four curved bundles of bamboo (Fig. 2), fixed to the ground at the base and converging at the center. The stability of the structure is provided by eight ropes, each of 2.5 m, used as ties, while protection is provided by a nylon cover stretched underneath the bamboo structure (Fig. 2).

The bamboo bundles are constructed using two strips of *Dendrocalamus giganteus*, 4.80 m long and 0.05 m wide, juxtaposed and fixed to one another using five braces (Fig. 3). In between the strips, where the braces are used, metal clips are placed on the inside to prevent the deformation of the structure and to limit the independent flexing movement of the strips.

Figure 1. The portable bamboo-and-nylon tent

Figure 2. Stretching of the nylon sheet under curved bamboo bundle using ropes

Figure 3. A brace used for joining the two bamboo strips

Figure 4. Prefabricated PVC cross joint and cylinders

The top joint is made of four hollow PVC cylinders – 40 mm diameter and 280 mm long –attached and glued to a prefabricated cross joint (Fig. 4). The joint is internally reinforced using a wooden peg. It is placed at the center of the nylon sheet, which measures 4 × 4 m, and the ends of the bamboo bundles are inserted into the cylinders (Fig. 5).

The structure is anchored and kept in tension using four pairs of 5 mm polypropylene ropes tied to 8 pegs driven in the ground. The nylon sheet has eyelets in the corners through which the ropes pass. In each of the diagonals, 8 flexible Velcro straps are fixed to restrict the movement of the support bundles and to keep them and the joint in place (Fig. 6). The nylon sheet has four pockets, made parallel to the sides and meeting in the middle, to accommodate two bamboo strips that cross each other at the top. These strips help define the geometrical shape of the tent.

Design of furniture

Seats for the students, who attend design workshops held inside the tent, were designed with the same concern as in the case of the tent. The chairs share with the ground the task of supporting the weight of the user. The chair is structured out of three culm pieces of the bamboo *Phyllostachys aurea* and a canvas seat. Each culm piece is 40 mm in diameter 800 mm long. The pieces are perforated on a node halfway through their length, and are held together by a 4-mm polypropylene rope (Fig. 7).

The arrangement of the culm pieces as a tripod defines two triangular plans: one in relation to the ground and the other in relation to the seating body. The seat is made of cotton canvas in a roughly triangular shape, reinforced internally along the edges with structured nylon fabric (Fig. 8a). The seat cover holds in place the top ends of the bamboo pieces, which are prevented from sliding by a string that runs through the hem of the cover and determines the opening limit of the balanced segments (Fig. 8b).

The furniture is completed with a table, although, because of the values underlined in the project, it is meant to support only the weight of the drawings and compositions the students make. Its shape, material and structure clearly communicate that it is not possible for the students to transfer their body weight on to the structure, and thus reminds them of the need for lighter body postures at work (9a).

The table has four *Phyllostachys aurea* culm pieces, each 1 100 m long and 40 mm in diameter. A 4-mm polypropylene rope passes through a hole at the node located halfway the length of each piece, and ties them together at the middle as a collapsible tetrapod. The limit of opening of the tetrapod is set by four ropes fixed to the top ends of the bamboo pieces (Fig. 9b). The tabletop is made with a mat of bamboo strips, each measuring 30 mm wide and 800 mm long, held together by a 3-mm polypropylene rope. Fixed to the ropes are four Velcro straps for fixing two other bamboo strips of the same size, placed parallel to each other and at right angles to the others, which provide stability to the tabletop (9c).

Figure 5. Placement of cross joint and bamboo bundles

Figure 6. Fixing the bamboo bundles with Velcro straps

Figure 7. (a) Three bamboo culm segments joined by a rope; (b) detail of the joint

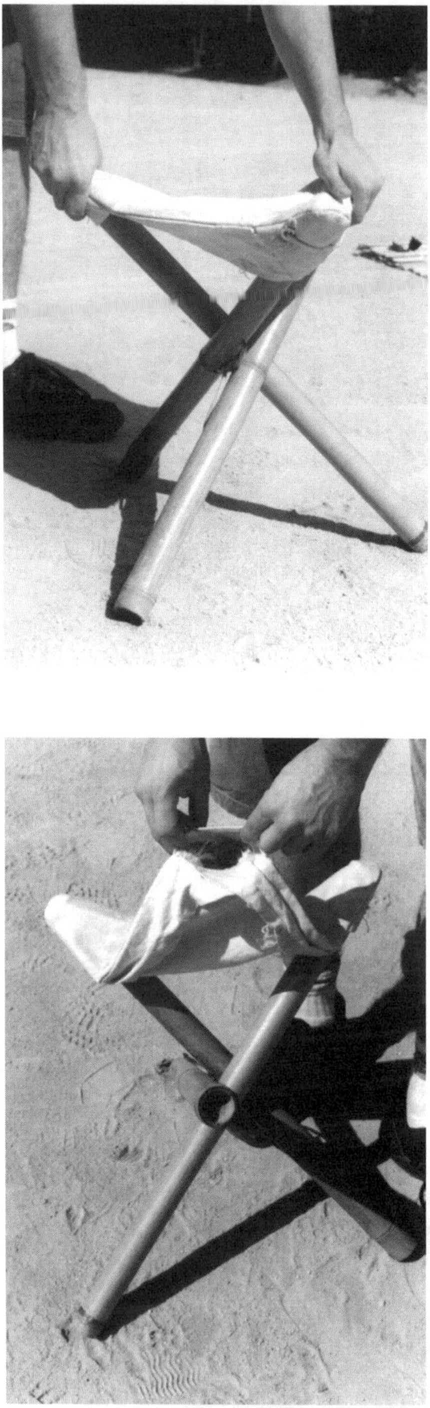

Figure 8. (a) The chair has three culm pieces that work as a tripod; (b) a triangular canvas piece serves as the seat and holds in place the top ends of the tripod

Figure 9a. Bamboo table

Figure 9b. The tetrapod structure and the ropes that control the opening

Figure 9c. Two parallel bamboo strips that provide stability

Usage of the objects

As the utility of the laboratory was in its portability, its design placed emphasis on transportation, assembly and disassembly, installation and maintenance aspects.

Transportation: The whole set – 25 chairs, 4 tables, 25 instrument cases, nylon cover and the structural joints – is sorted by the instructor and packed into eight bags. These objects, together with the bamboo bundles, are transported to the site of use (Fig. 10). The laboratory was designed to favor its installation in natural soil, in the open air. It takes an area of 25 m^2 and the choice of the site is made according to natural available shelters such as trees or hillsides.

Installation: Installation is easy and can be handled by the participants themselves, with minimum supervision. The learning – related to the reading of non-verbal codes present in the environment – uses techniques that stimulate remembrance through direct physical contact with natural elements (such as vegetables, clay, sand). Through intuitive archaeology (Emerson 1981), the materials provide an understanding of the universe as a dynamic system. The participants, after they come back from the field research of the environment, are encouraged to use chairs arranged in a circle inside the tent, supporting the study boards on their legs. The "living designs" they create are transported to the table only for collective design, wherein the individual designs will form part of a bigger design that will be displayed for discussion.

Figure 10. Materials for the laboratory in knocked-down state

CONCLUSIONS

The laboratory has been in use since April 1997 for research in different parts of the state of Rio de Janeiro. As the users are encouraged to assemble the laboratory themselves, a sense of involvement is being kept alive. The information gathered by the user in the process of construction has a bearing on the result, establishing a direct relation between production and usage and facilitating the absorption of practical information by conceptual intelligence. Thus, the handling of materials, the movement of the body involved in the construction, the limits of that body, all take an active part in defining the form of the object, as the use defines the changes in the laboratory.

The process is a path of learning wherein the maintenance is not oriented towards the conservation of the form, but to the continuation of the design movement that brings together action and reflection. For "it is in the attentive action of making that we design our process of self-construction".

REFERENCES

Emerson, J.N. 1981. Intuitive archeology: a psychic approach. Journal of Transpersonal Anthropology, Vol. V, No. 1.

Maturana, H.R. 1997. De máquinas e seres vivos. Autopoiese a Organização do Vivo. Artes Médicas, P.A., Brazil.

FURTHER READING

Hidalgo, O. 1981. Manual de construccion con bamboo. Estudios Técnicos Colombianos Ltda., Bogota, Colombia.

Maturana, H.R. 1995. Árvore do conhecimento. Psy II, S.P., Brazil.

Otto, F. 1971. Tension structures. Conrad Roland Praeger Publishers, Inc., New York, USA.

Bamboo for Sustainable Development, pp. 917–922
Eds. A. Kumar, I.V. Ramanuja Rao and Ch. Sastry

Bamboo and aeronautics: The past and future

MICHEL ABADIE
The World of Bamboo, Paris, France

Abstract—History tends to forget that man reached the skies with the help of bamboo. To develop the art of flying, aviation pioneers used bamboo extensively till 1908. Ninety years later, the Flyboo Project to make an ultra light aircraft entirely of bamboo was launched to promote the plant as material for the future. This paper traces the history of usage of bamboo in aviation, up to the making of Flyboo aircraft.

INTRODUCTION

The legend of Icarus and his father Daedalus notwithstanding, bamboo certainly led man to take off for the first time: probably in Northern China under the Qi Dynasty (550–577 AD). The first one to experiment with flight was Emperor Gao Yang, who practiced the Chinese religious tradition of "liberation of all living beings" that demands to set free all captured fish or birds. In his case, however, the principle gave birth to the questionable practice of giving human captives the order to fly down to the ground from a 30-meter high tower with the help of bamboo mats. Trial and error led to an improvement of the process and in 559 the Emperor flew Prince Wei tied to an owl-shaped kite over a distance of 3 km. The experiments are recorded in the "Mirror of history to assist in government" compiled after the annals of that time.

However, if we are to believe the historian Needham, it was the Taoist wise man Ge Hong (288–343 AD) who first flew on a kite "using clouds like stairs". Was it made of bamboo like the ones made today throughout Asia? Ge Hong perhaps benefited from earlier experiments, such as the ones alluded to by Marco Polo: "One sees only fools or drunkards clinging to flying kites for those of sane mind wouldn't expose themselves to such peril". The same Ge Hong describes a helicopter, which seems to have been a popular toy in China at that time, under the name "bamboo dragonfly". It consisted of a rotor around the axis of which a string was wound. Pulling of the string set the toy in motion and made it take off.

Sir George Cayley, one of the fathers of modern aeronautics, was introduced to this "dragonfly" by two French men named Launoy and Bienvenu. He studied the toy and left behind plans to construct a model helicopter capable of climbing up to 30 m height. Even if it is true that the principles of the rotor and the aircraft were taken from a Chinese toy, Leonardo da Vinci would not have to turn in his grave: that wonderful visionary led the western world into technical studies of flying objects.

Let us leave those precursors and their dreams to get back to the use of bamboo by the pioneers of modern aviation. In the early stages, in the absence of powerful engines at their disposal, they turned to balloons and dirigibles. In France, where the siege of Paris in 1870 had demonstrated the strategic importance of aerial observation, research on balloons developed. As early as 1883, Gaston and Albert Tissandier built and flew a balloon propelled by an electric motor with a bamboo nacelle, but the two pilots failed to get back to their starting point. In 1884, "France", the first dirigible to make a full round-trip flight, was built near Paris by Renard and Krebs. It was made entirely of bamboo and had a nacelle 30 m long (it is exhibited in the workshop of the French Air Museum in Meudon). A few years later, the German dirigible "Deutschland" was built by Dr. Woelfert. In November 1899, the Brazilian Alberto Santos-Dumont's flight around the Eiffel Tower caught people's imagination. His dirigible was made of bamboo! On May 1902, his compatriot Augusto Severo however didn't share the same luck. Severo's dirigible "La Pax", constructed with a 30-m long bamboo structure, crashed on Du Maine Avenue (Paris), killing its crew of two.

The adventure of the airplane followed. One early success was that of Clement Ader. He took off aboard "Eole", a 295-kg replica of a bat, with a 20 hp engine and a bamboo propeller. On 9 October 1890, his assistant saw the airplane fly about 50 meters at the height of a few centimeters above the ground. However, because Ader wanted to keep his experiment for the "eyes only" of the French War Ministry, no official commission witnessed it and the man and his deed failed to be entered in the history of flight.

It is not known whether Lilienthal, Chanute and Herring used bamboo in their experiments with flight. Otto Lilienthal, a German engineer, flew on a glider about 2,500 times from 1890 till 1896, when he died during a flight. He used *Arundo donax* cane for his glider, according to some sources. Octave Chanute, an American born in France, was one of the aviation pioneers. Like Herring, he constructed several flying machines during 1895–1901. It is not known whether he was using bamboo or a similar material. In the United States, around 1898, Samuel Langley was thinking to build a flying machine that is capable of carrying one man, for use during a war. In 1899, a prototype was built with "artificial bamboo", hollowed out wooden pieces held together with glue.

History was to play tricks on the efforts of a pioneer: the German-born American Gustave Whitehead could well have flown a propelled aircraft as early as 1901 around Bridgeport, Fairfield and Stanford, Connecticut (USA). The high-wing monoplane was described by Stanley Yale Beach in the 8 June 1901 edition of the *Scientific American:*

> "A novel flying machine has just been completed by Mr. Gustave A. Whitehead of Bridgeport, Connecticut and is now ready for the prelimi-

nary trials. Several experiments have been made, but as yet no free flights have been attempted. The machine is built after the model of a bird or bat. The front wheels are connected to a 10-horsepower engine to get up speed on the ground and the rear wheels are mounted like casters so that they can be steered by the aeronaut. On either side of the body are large aeroplanes, covered with silk and concave on the underside, which give the machine the appearance of a bird in flight. The ribs are bamboo poles and are braced with steel wires. The wings are so arranged that they can be folded up. The 10-foot rudder, which corresponds to the tail of a bird, can also be folded up and can be moved up and down, as to steer the machine on its horizontal course. A mast and bowsprit serve to hold all the part in their proper relation. In front of the wings and across the body is a double compound engine of 20 horsepower, which drives a pair of propellers in opposite directions, the idea being to run the machine on the ground by means of the lower motor until it has necessary speed to rise from the ground. Then the upper engine actuates the propellers so as to cause the machine to progress through the air to make it rise on its aeroplanes. The wings are immovable and resemble the outstretched wings of a soaring bird. The steering will be done by running one propeller faster than the other..."

On a full page of the 19 August 1901 issue of *Bridgeport Sunday Herald,* the sports editor Dick Howell made an original sketch of the plane of Whitehead flying with power for a distance of "one half mile" on 14 August 1901. Howell never used photos; he was an art purist who rendered other featured reports with sketches. Stanley Y. Beach wrote on 19 September 1903 in the *Scientific American:* "...By running with the machine against the wind after the motor had been started, the airplane was made to skim along above the ground at heights of from 3 to 16 feet for a distance, without the operator touching, of about 350 yards. It was possible to have traveled a much longer distance, without the operator touching terra firma, but for the operator's desire not to get too far above it. Although motor was not developing its full power, owing to the speed not exceeding 1 000 rpm, it developed sufficient to move the machine against the wind...". This was before what history records as the first powered flight, made on 17 December 1903 at Kitty Hawk by Orville and Wilbur Wright. The two brothers produced photos of the flight only in 1908 in "an understandable effort to first obtain a patent for their design". In *Flight Journal,* October 1998, Major William J. O'Dwyer, a USAF Reserve, wrote in a brilliant article: "The definition of 'flight' is being applied to history in a subjective manner, and that must cease, if for no other reason than it confuses the issue. A machine that rises off level ground under its own power with no catapulting devices and stays there is 'flying'. Examining the records with that definition, it becomes obvious that Whitehead 'flew' prior to December 17, 1903". And the bamboo flew with Whitehead.

The replica of Whitehead plane has been made in USA and in Germany to prove that the plane was "flyable". The first attempt since 1901 to study the aerodynamics of Whitehead's design took place at Sikorsky Memorial Airport, Stratford, Connecticut, in 1986. The German reproduction made its successful demonstration for the media on 18 February 1998, at Bavaria's famed Manchings Luftwaffe military air base. Distances flown ranged up to 500 m in relatively still air.

In France, in 1906, the Brazilian Alberto-Santos Dumont constructed an aircraft called 'Santos-Dumont 14 Bis'. It was based on six cells of Horgrane's kite. The fuselage was made of bamboo that came from the dirigible 'Santos-Dumont 14'. The engine at the rear was an Antoinette with a power of 24 hp and a weight of 40 kg. The wingspan was 12 m and the length 10 m. The total weight was 160 kg for a wing surface area of 80 m^2. With this plane, Santos-Dumont flew for 21 seconds and 220 m in November 1906, at Bagatelle Park close to the Champs Elysées (Paris), under the official attendance of a jury from the Aeroclub of France.

His main challenger was the French Captain Ferdinand Ferber. He did not succeed mainly because of lack of money to buy a powerful engine. But his Model 9, which was completely made of bamboo, flew in 1908 at Issy-les-Moulineaux in the suburbs of Paris. The model is in display at the Aeronautic and Space Museum in Le Bourget. Ferber accounted for the use of bamboo to make airplane by saying, "as different from wood, when it breaks it never does make a splinter that can hurt the pilot..."

In the same period, M.M. Etrich and Wels in Oberalstad (Austria) constructed a model with strips of bamboo tied and glued together.

The most revolutionary airplane using bamboo was certainly the "Demoiselle" of Santos-Dumont, the first plane constructed in series with nearly a hundred models. This was built in 1907 with the code name of n°19. It was using bamboo for the main part of the fuselage. The wing had a span of 5.10 m and a surface area of 10 m^2, and the aircraft had a length of 8 m. The first model weighed only 56 kg and was fitted with a 22 hp engine. During the period of production, the "Demoiselle" was tested with different types of engines. As one of the goals of Santos-Dumont was to produce an affordable airplane, he refused to have any patents on the plane and let the constructors compete with each other to produce cheaper planes, at about 5 000 French francs at that time.

The use of bamboo in airplanes ceased when the development of aviation moved out of the hands of pioneers, till an experimental bamboo airplane was made in the Philippines in 1947.

The experimentation led by engineer Antonio J. de Léon of the Institute of Science and Technology, Manila (Philippines), had the purpose of using bamboo mats panel to cover a classical airplane structure. The bamboo mat material used to make the experimental "XL-14 Maya" was called Wobex, for Woven Bamboo Experimental, and was made from *Bambusa espinosa* Roxb and *Bambusa vulgaris* Schrad. The maximum size of Wobex panels used was 25 x 25 cm. The aircraft used for the experiment was the well-known Piper Cub. A second experimentation was made with another plane, the 'Tagak', using bamboo panels to cover the fuselage, the wings, the controls and the nose. Tests showed that the resistance fatigue and flexile strength of bamboo to be much superior to wood material.

The Flyboo project (for FLYing bamBOO), started 90 years after the "Demoiselle", aimed at an ultra-light aircraft, entirely made of bamboo. The idea was proposed by The World of Bamboo, a Paris-based organization that promotes bamboo as material for the future. Flyboo, that uses state-of-the-art aeronautical designs and an innovative shape, has four butt-jointed wings of 7.70 m span. The fuselage is built with two bamboo poles supported by two laminated bamboo rings of 2.20 m. The 64-hp engine and the propeller are located inside the back ring. The overall length is 5.50 m and the weight 230 kg.

The pilot and the passenger are protected by the first ring. The very original shape of the wings highlights the harmony between the material and the design of the plane. The use of these natural and resilient materials makes the manufacturing quite inexpensive. Flyboo is easy to dismantle and rebuild. Very light, the whole plane can be carried aboard a commercial aircraft. Flyboo uses bamboo in different forms: natural, woven and laminated. The project has already received the Henry Ford European Award for Nature Protection in 1996. On 26 September 1998, Flyboo was displayed for the first time to the public at Champs Elysées (Paris), during the show of the Centenary of Aeroclub of France.

Centuries after its first use in an aircraft, bamboo continues to help humankind realize its dream to fly; and Flyboo is going to lead a worldwide campaign with the message: "Let forests breathe; use bamboo".

ACKNOWLEDGEMENTS

Thanks to: Delphine Herscovici for her patience and help during this long journey through the history of flight and the development of Flyboo; Jean Paul Charbuillet and his team for the timely help during the construction of Flyboo; Christian Rausch for the daily consultations; Benoit Treilhou for the studies of scale models; M. Marty and Noël from Supaero; M. Robin from Estaca and Louis Huchez, Pierre Jung; Professor Lam Cong Dinh (Vietnam) for the furniture; Qian Qiang (China) for net-based information and the good bamboo material; M. Ishii (Japan) for bamboo stress treatment; Douglas Lewis and Bamboo Harwoods; Henri Abadie, Michel Puech, Laurent Ferrari, Jean Dupin, Bernard Pinoteau, Ikoumi; and all the others who supported this work and the creation of Flyboo.

FURTHER READING

Abadie, M. 1997. Le bambou volant. Bambou, EBS-France, France.
Chambe, R. 1958. L'histoire de l'aviation. Flammarion, France. pp. 39–49.
Hidalgo Lopez, O. [n.d.]. Bambu. pp. 244–251.

de Masfrand, A. March 1906. L'Aerophile. Paris, France. pp. 66.
Nicolaou, S. 1997. Santos-Dumont. Etai/Musée de l'air. France.
O'Dwyer, W.J. 1998. The "who flew first" debate. Flight Journal, USA: 48–55.
Temple, R.K.G. 1987. Quand la Chine nous précédait. Bordas, France. pp. 173–192.

Bamboo for Sustainable Development, pp. 923–930
Eds. A. Kumar, I.V. Ramanuja Rao and Ch. Sastry

Some bamboo uses in Indonesia

ANANG SUMARNA
Galeri 16, Indonesian Bamboo Development Foundation, Bandung, Indonesia

Abstract—West Java is one of the 27 provinces of Indonesia and is also known as "Parahyangam". It lies in the tropical region and is blessed with a great number of flora and fauna, including an abundant bamboo resource.

West Javanese people, also known as Sundanese since a long time ago, have been utilizing the flexible and durable bamboo in every almost aspect of their lives. Today, there are about 165 bamboo articles – craft items, furniture and household appliances – that the Sundanese use. It is expected that the utilization of bamboo would increase in the near future, with the development of more articles in more varieties, especially in the fields of art and culture including paintings and modern sculptures.

INTRODUCTION

Bamboo in Indonesian life

Bamboo is a plant that has had, and still has, many utilities and functions in the lives of Indonesians and much of other Asians since ancient time. It is so inalienably linked to their lives that it is considered a symbol of good fortune and several myths have evolved around it. There is some evidence that prove bamboo's role in certain historical events of Indonesia, as material for both weapon and defense fortifications, during the last colonial period. Bamboo is today utilized as a construction material for buildings, bridges, hedges, furniture, fishing and agricultural implements, musical instruments, cooking ware and even some ornaments. It still serves out its prime role in many rites and rituals of age-old traditions, as does in the more modern industrial pursuits of paper pulp making and prefab housing panels. The Sundanese thus continue to maintain their intimate and traditional connection with bamboo, while looking to advance the material's use in the more contemporary facets of today's society.

The following sections present a brief account of some of the uses of bamboo in Indonesia, especially among the Sundanese.

Musical instruments
When the use of bamboo in musical instruments began in Indonesia is not exactly known. However, some of the bamboo musical instruments had been developed long before the Buddhist influence. This is supported by the Borobudur relief,

Figure 1. A bamboo flute (made of *Schizostachyum brachycladum*) being played

which contains pictures of *calung, angklung, gambang* and some other bamboo musical instruments. Some assumptions have that the development of bamboo musical instruments is closely related to the Malayo-Polynesian ancestors hundreds of years ago.

There are about 20 kinds of bamboo musical instruments in West Java, including: *calung, angklung, celempung, karinding, gambang awi, rengkong, karinding, goong, guyon, bolon, hatong, bangsing, suling, eleot, taleot (surilit)* and *kolecer*. Usually, the bamboo species employed in the making of these instruments is tamiang (Sund.). An exception would be the flute (*suling*) that usually accompanies the *kecapi suling*, a Sundanese traditional string and flute harmony performed in extreme tones, which is made from *Schizostachyum brachycladum* (Fig. 1).

Another flute, shorter than the flute for *kecapi suling*, with four holes can produce very deep tones and is prominently used in the *gamelam degung*, the classical Sundanese music. The *angklung* music consists of 9 pieces of instruments accompanied by two *dog-dog* and one flute called *angklung bandung*. A variation of this is in Sumedang, where *angklung* consists of 12 instruments with four pieces of *dog-dog*.

Calung is also a form of traditional Sundanese music, which consists of a number of hollow bamboo poles of varying but carefully selected lengths are used to produce different tones by tapping the bamboo with a stick. The performance of

Figure 2. *Calung gamelan* orchestra giving a performance

this music usually involves five people who play the instruments, interspersed with jokes and songs, to entertain audience for several hours.

There are three kinds of *calung*.

1. *Calung gambang* (also called *calung renteng*) is a xylophone-type instrument comprising hollow bamboo tubes lined up on a rope. One end of the rope is tied to the player's waist and the other to the tree. The tubes are then tapped with a bamboo stick.
2. *Calung gamelan* is an orchestra consisting of several instruments like *calung gambang, calung saron, calung slentem, suling, goong (gong)* and *kendang* (Fig. 2). In Sumedang it is also called a *salentrong*. Bamboo species such as *Gigantochloa atroviolacea* are used for this.
3. *Calung jinjing* is a modernized *calung*, which is more practically designed than the other two. It can therefore be easily held by the players and allows some dynamic and energetic movements. This musical performance is actually accompanied by songs and jokes. This music consists of five instruments: *melodian, pangiring, jenglong, calung gede* and *goong*.

One of the most famous and talked about Indonesian traditional bamboo instrument is *angklung*, an instrument consisting of three or four separate hollowed bamboo pipes. The pipes are closed at the bottom end, but open at the upper end. Thinner bamboo sticks frame the pipes. Single sticks passes through all the pipes loosely. *Angklung* produces sounds when shaken. *Gigantochloa atroviolacea* species is used to make this instrument.

Bamboo is also used for *kokoprak*, an instrument employed to scare away birds in rice fields. The instrument, usually tied to a small pillar, produces sounds when wind blows on it.

Ritual implements

In Indonesia, bamboo has a role in most rituals such as wedding ceremony, seventh month of pregnancy, birth of a baby and circumcision. The bamboo musical instrument *angklung* is played in the several ritual occasions. Some of the ritual roles of bamboo are given below.

1. At the *ngareureueus para* ceremony conducted to honor Dewi Sri, the Hindu goddess of rice, when the planting of rice seedlings in the fields is about to begin. After sacrificial offering of food sacrifice preparations and burning of incense sticks, the eldest shaman makes five holes on a piece of sharp-ended bamboo pole, which is then filled with some rice seeds by several women. This ceremony accompanied by *angklung* performance with dancing and songs.
2. *Helaran*, the ceremony to escort boys who are going to be circumcised, is always accompanied by *angklung* performance. *Ngadu angklung*, a grand ceremony celebrated at the end of a year, involves physical fighting contests preceded by a performance of the musical instrument.
3. In Bali, there is a specific bamboo called "Bali bamboo", which is used to make the bier that carries the dead body during the *ngaben* ceremony. Balinese use some other bamboo species – such as *Bambusa vulgaris, Gigantochloa atter, G. apus* and *Schizostachyum brachycladum* – during cremation ceremonies.
4. People in Buton consider that "bambu betung" (*Dendrocalamus asper*) as embodying the spirit of their ancestors and they honor the plant. There is also a popular myth that a goddess was born from the young shoots of that bamboo.
5. *Tedak sinten* is the ceremony of ritually making a seven-month old baby to tread on the ground. The articles used for the ceremony include a bamboo cage and a five-stepped ladder, which is leaned against a chair. The ceremony begins with the baby being made to step on the top rung of the ladder, and continue down each rung of the ladder till the baby's feet come to touch the ground.

In Indonesia – as in the Philippines, China and Japan – there are some rituals when the baby's umbilical cord must be cut using a bamboo knife. Different types of bamboo knife are used for girls and boys; the knife used on a girl is made from male bamboo while the one used for a boy comes from a female bamboo.

Handicrafts and household implements

There are three main bamboo handicrafts works: plaiting, engraving and carving. Bamboo plaiting is an expertise of the Southeast Asian people. Plaited bamboo is usually used for making containers, lunch boxes, pets' cages, fans and hats. The plaiting varies from loose to dense, from thick to thin or from big to small.

Figure 3. A bamboo sculpture that uses the bottom part of the stem and the roots

Engraving is employed to decorate bamboo items. It is done on dyed or singed parts of bamboo, and follows traditional patterns. Engraved bamboo boxes are usually used for placing jewelry, beads and other valuable items.

Bamboo stem offers a unique potential for carving. The stems and rhizomes of bamboo are very good materials for carving. As engraving, carving is commonly employed on bamboo containers used to keep valuable goods. Some bamboo carvings, usually relief carving around bamboo stem, tell popular tales and legends. The rhizome of bamboo is a very special material that allows the creation of artistic and aesthetics sculptures (Fig. 3).

Recently, bamboo processing has improved with the availability of advanced cutting, plaiting and straightening machines. Nailing of bamboo is now easily done using bamboo nails, and never metal ones that will cause breakage in bamboo. In utilizing bamboo for handicrafts, a very important factor is its durability. How can bamboo materials be preserved against attacks by insect pests? Some chemicals (most of which are toxic) can be used for preservation, and the growth of mold can be prevented by covering bamboo with a coat of paint or varnish. Another, and more common among villagers, method of preservation is to soak the bamboo in a lake, river or pond for a specified period. Most villages in Indonesia prefer salty water for soaking, as this is supposed to enhance durability. Another simple way is to boil bamboo in hot water at 40°C for some time. Three spoons of caustic soda per 15 liters of water or 5% boric acid solution is added to the water before boiling. This process is known locally as *direbus*.

Another process, *dunun*, involves heating the upper portion of the bamboo pole over a fire to eliminate the sap. A simpler and possibly less effective process, *dijemur*, involves drying bamboo poles in direct sunlight for one month or till they lose the characteristic smell.

The household implements made from bamboo include a wide range of vases, containers, brooms, egg mixer, rice steamers, rice bowls, baskets, strainers and so on. Bamboo shoots form part of Indonesian diet, and bamboo tube is used in the cooking of some dishes. *Lemang*, a dish of rice mixed with spices, is one such (Fig. 4). This kind of food is a traditional Sumateran fare, especially during religious ceremonies. The bamboo species used as the cooking "vessel" is either *Gigantochloa atter* or *G. apus*. Bamboo is widely used to manufacture elegant and very functional furniture (Fig. 5).

Sailing and fishing equipment and weaponry

The Javanese are thought to have started using bamboo rafts for river navigation centuries ago, even reaching Australia. This type of raft still plies along Indonesian rivers. Harvested bamboo poles are transported by tying them together in raft fashion. Everyday many such rafts reach Jakarta, where the bamboo poles will be made into everything from furniture to scaffolding. The Mollucan traders used to employ bamboo canoes to send their spices to busy trading ports until the early 19th century.

In the calm rivers, lakes and coasts of Indonesia, some traditional fisherman use the trap *bubu* – a bottle-like plaited bamboo basket with the open end narrowing and folding inwards – to catch fish. The design of *bubu* makes escape impossible for the fish that enters the trap. The fish caught is put on *korang*, a bamboo basket similar to *bubu* but with a rounded shape and a flat bottom.

Sumpitan or blowgun is a Southeast Asian bamboo weapon used extensively for hunting by tribal people, such as the Dyaks, Kubus and Hmong. It is made using the species *Bambusa wrayi*, which has very long (over 7 ft) and straight internodes. The projectile commonly used is a small dart cut from bamboo or rattan with a poison applied to its tip. Once used as a weapon in tribal wars, its use is now confined to hunting birds and small animals.

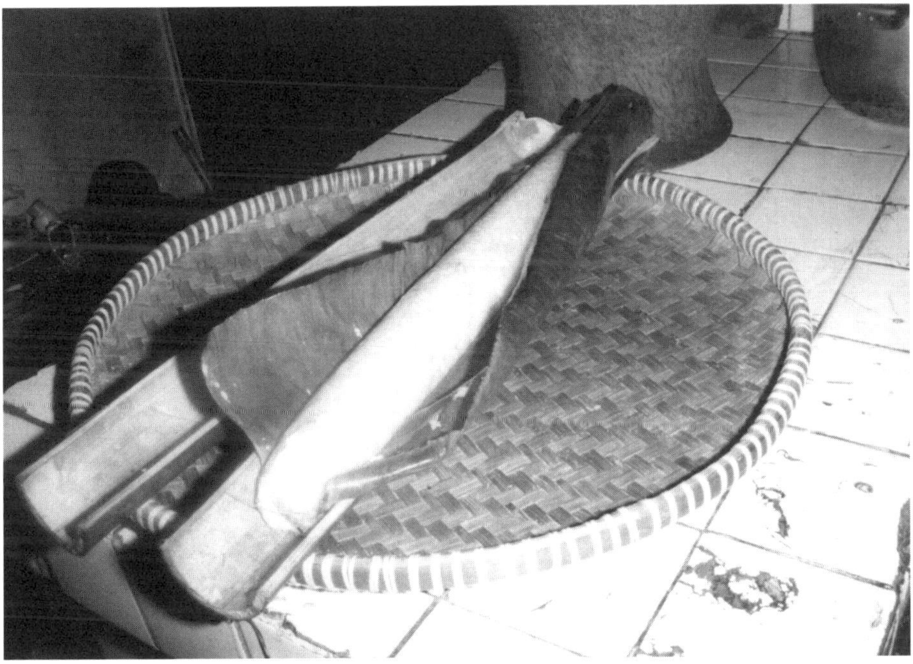

Figure 4. *Lemang*, a delicious rice dish cooked in bamboo

Figure 5. Bamboo furniture

Figure 6. A bamboo bridge

Bridges and construction

There are many bridges made by bamboo in Indonesia (Fig. 6). These bridges are usually supported by either bamboo pillars or bamboo ropes. Bamboo rope bridges are known as *jembatan gantung*, meaning hanging bridges.

The Food and Agriculture Organization (FAO) and the Directorate of Forestry report that almost 70% of houses in Indonesia use bamboo constructions, and in Java, it is about 95%.

Bamboo is utilized for covered walls, such as in Sukabumi, while in Bali and Toraja people use bamboo for roofs. According to the native people in Toraja, their bamboo houses survive for up to 100 years without any major rework.

FURTHER READING

Anonymous. 1997. The National Geographic Magazine, May 1997, 103–104.

Farrelly, D. 1984. The book of bamboo. Sierra Club Books, San Francisco, USA. 332 pp.

Pipe, J.M. 1992. Bamboo and rattan: traditional uses and beliefs. Oxford University Press, Singapore.

Sumarna, A.; Widjaja, E.A. 1987. Bamboo. Angkasa Publishing, Bandung, Indonesia, 203 pp.

Sumantera, W.I. 1996. Bamboo in Balinese rituals. *In* Belcher, B.; Karki, M.; Williams, T., ed., Bamboo, people and environment, Vol. 4, Socio-economics and Culture. Proceedings of the Vth International Bamboo Workshop, Ubud, Bali, Indonesia, 19–22 June 1995. International Network for Bamboo and Rattan, New Delhi, India. pp. 97–100.

Widjaja, E.A. 1994. Endemic bamboo from Sumatra. *In* Bamboo in Asia and the Pacific. Proceedings of the 4th International Bamboo Workshop, Chiangmai, Thailand, 27–30 November 1991. International Development Research Centre, Ottawa, Canada; Forestry Research Support Programme for Asia and the Pacific, Bangkok, Thailand. pp. 5–6.